Mechanical Systems, Classical Models

MATHEMATICAL AND ANALYTICAL TECHNIQUES WITH APPLICATIONS TO ENGINEERING

Series Editor
Alan Jeffrey

The importance of mathematics in the study of problems arising from the real world, and the increasing success with which it has been used to model situations ranging from the purely deterministic to the stochastic, in all areas of today's Physical Sciences and Engineering, is well established. The progress in applicable mathematics has been brought about by the extension and development of many important analytical approaches and techniques, in areas both old and new, frequently aided by the use of computers without which the solution of realistic problems in modern Physical Sciences and Engineering would otherwise have been impossible. The purpose of the series is to make available authoritative, up to date, and self-contained accounts of some of the most important and useful of these analytical approaches and techniques. Each volume in the series will provide a detailed introduction to a specific subject area of current importance, and then will go beyond this by reviewing recent contributions, thereby serving as a valuable reference source.

Series Titles:
MECHANICAL SYSTEMS, CLASSICAL MODELS, Volume I: Particle Mechanics
  P.P. Teodorescu, ISBN 978-1-4020-5441-9
INVERSE PROBLEMS IN ELECTRIC CIRCUITS AND ELECTROMAGNETICS
  N.V. Korovkin, ISBN 0-387-33524-7
THEORY OF STOCHASTIC DIFFERENTIAL EQUATIONS WITH JUMPS AND APPLICATIONS
  Rong Situ, ISBN 0-387-25083-2
METHODS FOR CONSTRUCTING EXACT SOLUTIONS OF PARTIAL DIFFERENTIAL EQUATIONS
  S.V. Meleshko, ISBN 0-387-25060-3
INVERSE PROBLEMS
  Alexander G. Ramm, ISBN 0-387-23195-1
SINGULAR PERTURBATION THEORY
  Robin S. Johnson, ISBN 0-387-23200-1

Mechanical Systems, Classical Models

Volume I: Particle Mechanics

by

Petre P. Teodorescu
*Faculty of Mathematics,
University of Bucharest,
Romania*

 Springer

A C.I.P. Catalogue record for this book is available from the Library of Congress.

ISBN-13 978-90-481-7369-3
ISBN-10 1-4020-5442-4 (e-book)
ISBN-13 978-1-4020-5442-6 (e-book)

Published by Springer,
P.O. Box 17, 3300 AA Dordrecht, The Netherlands.

www.springer.com

Printed on acid-free paper

Translated into English, revised and extended by Petre P. Teodorescu

All Rights Reserved

© 2010 EDITURA TEHNICĂ
This translation of "Mechanical Systems, Classical Models" (original title: Sisteme mecanice. Modele clasice, Published by: EDITURA TEHNICĂ, Bucuresti, Romania, 1984-2002), First Edition, is published by arrangement with EDITURA TEHNICĂ, Bucharest, ROMANIA.

© 2010 Springer
No part of this work may be reproduced, stored in a retrieval system, or transmitted in any form or by any means, electronic, mechanical, photocopying, microfilming, recording or otherwise, without written permission from the Publisher, with the exception of any material supplied specifically for the purpose of being entered and executed on a computer system, for exclusive use by the purchaser of the work.

To
Oncuța,
Delia, Edu
and Ana,
my dear family

CONTENTS

PREFACE			xi
1.	NEWTONIAN MODEL OF MECHANICS		1
	1. Mechanics, science of nature. Mechanical systems		1
		1.1 Basic notions	1
		1.2 Mathematical model of mechanics	45
	2. Dimensional analysis. Units. Homogeneity. Similitude		57
		2.1 Physical quantities. Units	57
		2.2 Homogeneity	62
		2.3 Similitude	70
2.	MECHANICS OF THE SYSTEMS OF FORCES		75
	1. Introductory notions		75
		1.1 Decomposition of forces. Bases	75
		1.2 Products of vectors	79
	2. Systems of forces		89
		2.1 Moments	89
		2.2 Reduction of systems of forces	95
3.	MASS GEOMETRY. DISPLACEMENTS. CONSTRAINTS		115
	1. Mass geometry		115
		1.1 Centres of mass	115
		1.2 Moments of inertia	125
	2. Displacements. Constraints		162
		2.1 Displacements	162
		2.2 Constraints	166
4.	STATICS		201
	1. Statics of discrete mechanical systems		201
		1.1 Statics of the particle	201
		1.2 Statics of discrete systems of particles	224
	2. Statics of solids		229
		2.1 Statics of rigid solids	229
		2.2 Statics of threads	275
5.	KINEMATICS		287
	1. Kinematics of the particle		287
		1.1 Trajectory and velocity of the particle	287
		1.2 Acceleration of the particle	293
		1.3 Particular cases of motion of a particle	300

	2.	Kinematics of the rigid solid	305
		2.1 Kinematical formulae in the motion of a rigid solid	305
		2.2 Particular cases of motion of the rigid solid	310
		2.3 General motion of the rigid solid	316
	3.	Relative motion. Kinematics of mechanical systems	330
		3.1 Relative motion of a particle	330
		3.2 Relative motion of the rigid solid	333
		3.3 Kinematics of systems of rigid solids	340
6.	DYNAMICS OF THE PARTICLE WITH RESPECT TO AN INERTIAL FRAME OF REFERENCE		353
	1.	Introductory notions. General theorems	353
		1.1 Introductory notions	353
		1.2 General theorems	365
	2.	Dynamics of the particle subjected to constraints	384
		2.1 General considerations	384
		2.2 Motion of the particle with one or two degrees of freedom	389
7.	PROBLEMS OF DYNAMICS OF THE PARTICLE		401
	1.	Motion of the particle in a gravitational field	401
		1.1 Rectilinear and plane motion	401
		1.2 Motion of a heavy particle	406
		1.3 Pendulary motion	422
	2.	Other problems of dynamics of the particle	440
		2.1 Tautochronous motions. Motions on a brachistochrone and on a geodesic curve	441
		2.2 Other applications	452
		2.3 Stability of equilibrium of a particle	457
8.	DYNAMICS OF THE PARTICLE IN A FIELD OF ELASTIC FORCES		469
	1.	The motion of a particle acted upon by a central force	469
		1.1 General results	469
		1.2 Other problems	476
	2.	Motion of a particle subjected to the action of an elastic force	480
		2.1 Mechanical systems with two degrees of freedom	481
		2.2 Mechanical systems with a single degree of freedom	494
9.	NEWTONIAN THEORY OF UNIVERSAL ATTRACTION		543
	1.	Newtonian model of universal attraction	543
		1.1 Principle of universal attraction	543
		1.2 Theory of Newtonian potential	549
	2.	Motion due to the action of Newtonian forces of attraction	555
		2.1 Motion of celestial bodies	555
		2.2 Problem of artificial satellites of the Earth and of interplanetary vehicles	577
		2.3 Applications to the theory of motion at the atomic level	587
10.	OTHER CONSIDERATIONS ON PARTICLE DYNAMICS		595
	1.	Motion with discontinuity	595
		1.1 Particle dynamics	595
		1.2 General theorems	606

2.	Motion of a particle with respect to a non-inertial frame of reference		615
	2.1	Relative motion. Relative equilibrium	615
	2.2	Elements of terrestrial mechanics	629
3.	Dynamics of the particle of variable mass		659
	3.1	Mathematical model of the motion. General theorems	659
	3.2	Motion of a particle of variable mass in a gravitational field	674
	3.3	Mathematical pendulum. Motion of a particle of variable mass in a field of central forces	682
	3.4	Applications of Meshcherskiĭ's generalized equation	689

APPENDIX 693
1. Elements of vector calculus 693
   1.1 Vector analysis 693
   1.2 Exterior differential calculus 705
2. Notions of field theory 709
   2.1 Conservative vectors. Gradient 710
   2.2 Differential operators of first and second order 716
   2.3 Integral formulae 723
3. Elements of theory of distributions 731
   3.1 Composition of distributions 731
   3.2 Integral transforms in distributions 733
   3.3 Applications to the study of differential equations. Basic solutions 736

REFERENCES 745
SUBJECT INDEX 765
NAME INDEX 771

PREFACE

All phenomena in nature are characterized by motion; this is an essential property of matter, having infinitely many aspects. Motion can be mechanical, physical, chemical or biological, leading to various sciences of nature, mechanics being one of them. Mechanics deals with the objective laws of mechanical motion of bodies, the simplest form of motion.

In the study of a science of nature mathematics plays an important rôle. Mechanics is the first science of nature which was expressed in terms of mathematics by considering various mathematical models, associated to phenomena of the surrounding nature. Thus, its development was influenced by the use of a strong mathematical tool; on the other hand, we must observe that mechanics also influenced the introduction and the development of many mathematical notions.

In this respect, the guideline of the present book is precisely the mathematical model of mechanics. The classical models which we refer to are in fact models based on the Newtonian model of mechanics, on the five basic principles, i.e.: the inertia, the forces action, the action and reaction, the parallelogram and the initial conditions principle, respectively. Other models, e.g. the model of attraction forces between the particles of a discrete mechanical system are part of the considered Newtonian model. Kepler's laws brilliantly verify this model in case of velocities much smaller than the light velocity in vacuum. The non-classical models are relativistic and quantic.

Mechanics has as object of study mechanical systems; this notion is emphasized throughout the book, no matter the systems we are working with are discrete or continuous. We put into evidence the difference between these models, as well as the specificity of the corresponding studies; the generality of the proofs and of the corresponding computations yields a common form of the obtained mechanical results for both discrete and continuous systems. On the other hand, the discrete or continuous mechanical systems can be non-deformable (e.g., rigid solids) or deformable (deformable particle systems or deformable continuous media); for instance, the condition of equilibrium and motion, expressed by means of the "torsor", are necessary and sufficient in case of non-deformable and only necessary in case of deformable systems.

A special accent is put on the solving methodology as well as on the mathematical tool used; vectors, tensors and notions of the field theory. Continuous and discontinuous phenomena, various mechanical magnitudes are presented in a unitary form by means of the theory of distributions. Some appendices give the book an autonomy with respect to other works, special previous mathematical knowledge being not necessary.

Passing by non-significant details, one presents some applications connected to important phenomena of nature, and this also gives one the possibility to solve problems of interest from the technical, engineering point of view. In this form, the book becomes – we dare say – a unique outline of the literature in the field; the author wishes to present the most important aspects connected with the study of mechanical systems, mechanics being regarded as a science of nature, as well as its links to other sciences of nature. Implications in technical sciences are not neglected.

Starting from the particle (the simplest problem) and finishing with the study of dynamical systems (including bifurcation, catastrophes and chaos), the book covers a wide number of problems (classical or new ones), as one can see from its contents. It is divided in three volumes, i.e.: I. Particle mechanics. II. Mechanics of discrete and continuous systems. III. Analytical mechanics.

The book uses the known literature, as well as the original results of the author and his more than fifty years experience as a Professor of Mechanics at the University of Bucharest. It is devoted to a large circle of readers: mathematicians (especially those involved in applied mathematics), physicists (particularly those interested in mechanics and its connections), chemists, biologists, astronomers, engineers of various specialities (civil, mechanical engineers etc., who are scientific researchers or designers), students in various domains etc.

7 January 2006   P.P. Teodorescu

Chapter 1

NEWTONIAN MODEL OF MECHANICS

The mathematical models of mechanics, *the first basic science of nature*, are shortly presented in this chapter; stress is put on the notion of *mechanical system*. Considerations concerning the units are made and some results concerning the theory of homogeneity and similitude are given. Using these introductory notions, one can pass to a mathematical study of the mechanical systems.

1. Mechanics, science of nature. Mechanical systems

All the phenomena of nature are characterized by *motion*; this is an *essential property* of matter, having infinity of aspects. For instance, the motion can be *mechanical, physical, chemical* and *biological*; there correspond various *sciences of nature*. Among them, *mechanics* studies the objective laws of the *mechanical motion* of material bodies, the simplest form of motion of matter; we will deal with such a study in what follows.

The scientific study of matter has put in evidence its existence in the form of *physical systems*. The simplest physical system is the *substance*, which is met in nature in the form of *material body*; we will consider these bodies in their real form, as well as in their idealized form of *discrete systems of particles* (atoms, molecules, usual bodies, planets etc.) or of *continuous systems* (solids, fluids), under the denomination of *mechanical systems*.

Mathematics plays a very important rôle in the study of a science of nature. *Mechanics* has been *the first science of nature* which was strongly mathematicized, by considering various mathematical models, which have been proved to correspond to the surrounding reality. To mechanical systems there correspond thus mathematical models of organized matter, having certain properties of structure, form etc. The development of mechanics has been thus much influenced by the use of a strong mathematical tool; as a matter of fact, we observe that mechanics also played an important rôle in the introduction and development of many mathematical notions.

1.1 Basic notions

Introducing the notions of system and mathematical model, we consider the main systems and models of such kind used in mechanics; we put thus in evidence the hypotheses that are introduced and the basic notions which appear. The object of study of mechanics is thus emphasized. We must point out that the notions which will be introduced are *basic notions*; these ones cannot be defined with the aid of other notions, and they must be described by their various properties, obtaining thus the considered mathematical models.

1.1.1 Notion of system. Notion of mathematical model

In the study of a science of nature, more and more the connection between the modern methods of the theory of systems and the mathematical modelling of this science is put into evidence. One must first of all lighten the notion of system, which may be different from that suggested by the usual language; on the other hand, this notion must be used always when one has to do with complicated phenomena, the search of which is proved to be difficult. At the same time, one must take into consideration the possibility of coupling characteristic phenomena from various sciences of nature (mechanics, physics, chemistry, biology etc.).

A *system* is defined by a *set of elements* (component parts), which influence (condition) each other (which *interact*), on which external influences (actions, denominated *input*) take place, and the actions of which towards the exterior (towards other systems) are denominated *output*. One obtains thus the schema

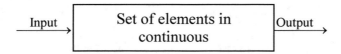

Among the most important properties of a system we must mention its *change* (*transformation, motion*), as well as *the possibility to be* influenced (*conducted*) by a convenient choice of the inputs (forces). In classical mechanics, the connection between the influences (actions) exerted on the system and the changes of the latter one is called *dynamics of systems*. It is understood that these considerations refer first to mechanical systems; but, in general, taking into account the connection between various systems, dynamics of systems is an *interdisciplinary science*.

To study these systems, following problems are put: i) *the construction of the mathematical model* corresponding to a given system; ii) *the study of the model* thus created (putting into evidence the properties of the system); sometimes, because of the difficulties of the considered problems, one cannot give a complete solution to the problem of motion and one emphasizes only some of its properties or one can give only approximate solutions (e.g., by linearizing non-linear phenomena); iii) *the choice of the main* (dominant) *inputs*, which govern the considered phenomena; iv) *the simulation of the behaviour of the* considered *system*; one can thus put in evidence the deficiencies of the chosen model as well as of the considered inputs, improving then the model and the inputs. We will deal particularly with the first two problems in our study, accidentally with the third one and less with the fourth one.

Usually, by *model* we understand an object or a device artificially created by men, which is similar in a certain measure with another one, the latter being an object of search or of practical interest. The scientific notion of model constitutes a possibility of knowledge of the surrounding reality, consisting in the representation of the phenomenon to be studied with the aid of a system artificially created. Hence, the most general property of a model is its capacity to reflect, to reproduce things and phenomena of the objective world, their necessary order and their structure.

From the very beginning, the models can be divided in two great classes: *technical* (material) *models* and *ideal* (imaginary) *models*; this classification is made considering the construction of the models as well as the possibilities by which the objects to be studied may be reproduced.

The technical models are created by man, but they do exist objectively, independent of his conscience, being materialized in metal, wood, electromagnetic fields a.s.o. Their destination is to reproduce for a cognitive goal the object to be studied, to put in evidence its structure or certain of its properties. The model can maintain or not the physical nature of the object which is studied or the geometrical similitude to this one. If the similitude is maintained, but the model differs by its physical nature, we have to do with *analogic systems*. E.g., electrical models may reproduce processes analogous to those encountered in mechanics, qualitatively different, but described by similar equations. These models, as others of the same kind, take part in the class of *mathematical models*.

One can construct such models, for instance, to study the torsion of a cylindrical straight bar of arbitrary simply or multiply connected cross section. If the bar is isotropic, homogeneous and linear elastic (subjected to infinitesimal deformations), then the phenomenon is governed by a Poisson type equation in B. de Saint-Venant's theory. L. Prandtl showed that the same partial differential equation is met in case of a membrane which rests on a given contour and is subjected to an interior constant pressure; if this contour is similar to the frontier of the plane domain corresponding to the cross section of the straight bar, then we obtain a correspondence of the boundary conditions, hence the classical *membrane analogy* (or of the *soap film*). One uses also other analogies for the same problem, i.e.: electrical modelling, modelling by optical interference, hydrodynamical modelling a.s.o.

Another type of technical models used in mechanics corresponds to the *intuitive notion of model*. Various elements of construction are performed partially or in totality at a reduced scale, obtaining thus results concerning the maximal stresses and strains which appear. These models can be built of the same material as the objects to be studied or of other materials, so that quite difficult problems of similitude must be solved.

Generally, *the ideal models* are not materialized and – sometimes – they neither can be. From the viewpoint of their form, they can be of two types.

The models of first order are built by using intuitive elements, which have a certain similitude with the corresponding elements of the real modelled phenomenon; we observe that this similitude must not be limited only to space relations, but can be extended also to other aspects of the model and of the object (for instance, the character of the motion). The intuitiveness of these models is put into evidence first of all by the fact that the models themselves, formed by elements sensorial perceptible (plates, levers, tubes, fluids, vortices etc.), are intuitive, and – on the other hand – by the fact that they are intuitive images of the objects themselves. Sometimes, these models are fixed in the form of *schemata*.

The models of second order are systems of signs, their elements being special signs; logical relations between them form – at the same time – a system, being expressed also by special signs. In this case, there is no similitude between the elements of the model and the elements of the corresponding objects. These models do not have intuitiveness in the sense of a spatial or physical analogy; they have not, by their physical nature, nothing in common with the nature of the modelled objects. The models of second order reflect the reality on the basis of their isomorphism with this reality; we suppose a one-to-one correspondence between each element and each relation of the model. These models reproduce the objects under study in a simplified form, constituting thus – as all models – a certain idealization of the reality.

The types of ideal models mentioned above can be considered as *limit cases*. In fact, there exist ideal models that have common features to both types of models which have been described; they contain systems of notions and axioms which characterize quantitatively and qualitatively the phenomena of nature, for instance representing *mathematical models*. Such models are extremely important and their systematic use has led to the great development of mechanics of deformable bodies in the last time.

The basic dialectical contradiction of the model (the model serves to the knowledge of the object just because it is not identical with the latter one) is useful, for instance to put into evidence the properties of continuous deformable media. In fact, a model contains more information about the object if it is closer to it. But physical reality is very complicated; the solving of the contradiction is realized by the use of a sequence of models, more and more complete, each one having its contribution to the knowledge of the real continuous deformable media. We try to put into evidence just this process of continuous improvement of the models in general mechanics as well as in mechanics of continuous deformable media, process that constitutes *the main tenor* of the development of *mechanical systems*.

In general, after adopting a model, it is absolutely necessary to compare the results obtained by theoretical reasoning to physical reality. If these results are not satisfactory (sometimes it happens to be between some limits, which can be sufficiently close), then it is necessary to make corrections or to improve the chosen model. In fact, on this way, mechanics, the theory of mechanical systems developed itself, the word "model" being more and more used by researchers dealing with this science of nature.

1.1.2 Vectorial modelling of mechanical quantities. Vector space

A great part of the quantities which appear in mechanics are mathematically modelled as vector quantities; in fact, the concept of vector appeared together with the development of mechanics. We will thus introduce the notion of vector space, a space of mechanical quantities, by considering three types of vectors useful in the study of mechanical systems, i.e.: free vectors, bound vectors and sliding vectors.

The free vector is a mathematical entity characterized by *direction* and *modulus* (magnitude). The direction includes all straight lines *parallel* one to each other, as well as a *sign*; in case of opposite signs, the directions are *opposite*. It is denoted by **V** (sometimes $\vec{V}$ or $\overline{V}$ or $\underline{V}$); the modulus of this vector will be $|\mathbf{V}| = V$, $V \geq 0$ (it is denoted also by $\|\mathbf{V}\|$, and it is called *the norm* of the vector).

We fix a right-handed frame of reference, formed by an origin O and by three orthogonal axes $Ox_i, i = 1, 2, 3$, in the Euclidean three-dimensional space E_3. We mention that by a *right-handed frame of reference* we understand that one for which an observer situated along the axis Ox_i, in its positive sense, sees the superposition of the axis Ox_j onto the axis Ox_k, after a rotation of angle $\pi/2$ in the positive sense (from right to left, the right-hand rule); we admit that

$$(i, j, k) = (1, 2, 3), \tag{1.1.1}$$

the indices i, j, k taking the distinct values $1, 2, 3$ in the given order or after a cyclic permutation of them. Projecting the free vector **V** on the axes (we use the definition

Newtonian model of mechanics

that will be given later, in this subsection), we obtain its canonical co-ordinates $V_i, i = 1, 2, 3$ (Fig.1.1,a); these co-ordinates have a certain sign, as the direction of the

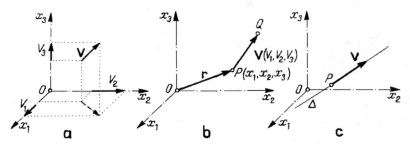

Figure 1.1. Free (a), bound (b) and sliding (c) vectors.

projections coincides or not with the direction of the co-ordinate axes (which are directed axes), and constitute an ordered triplet of numbers, which entirely characterize the free vector. We can write

$$V = \left(\sum_{i=1}^{3} V_i^2 \right)^{1/2} = \sqrt{V_i V_i} = \sqrt{V_1^2 + V_2^2 + V_3^2} , \qquad (1.1.2)$$

where we use *Einstein's convention of summation*, in conformity with which the double existence of an index (called *dummy index*) in a monomial indicates the summation with respect to this index. An index, which appears only once, will be a *free index*; in the frame of the adopted convention, an index cannot appear three times.

The *bound vector* is a mathematical entity, characterized by *direction*, *modulus* (magnitude) and *point of application* (*origin*). In the frame of reference $Ox_1x_2x_3$, the point of application P (the origin of the vector) is given by the *position vector* $\mathbf{r}(x_1, x_2, x_3)$ (which is also a bound vector, the point of application of which is the origin O); the point Q represents the *extremity* of the vector (Fig.1.1,b). A bound vector $\mathbf{V} = \overrightarrow{PQ}$ is characterized by the bound vector $\mathbf{r}$ and the free vector $\mathbf{V}$, hence by two ordered triplets of numbers (x_i and $V_i, i = 1, 2, 3$). Thus, $\overrightarrow{PQ}$ is a *directed* (*oriented*) *segment*; we notice that $\overrightarrow{QP}$ is the *opposite directed segment*.

The *sliding vector* is a mathematical entity, characterized by *direction*, *modulus* (magnitude) and *support* (the vector lies on an axis Δ, without mentioning the point of application, but having the possibility to slide along this axis). From the mathematical point of view, a sliding vector will be characterized also by two ordered triplets of numbers (x_i and $V_i, i = 1, 2, 3$), between which there exists a certain relation, hence by five independent numbers (a sixth number is necessary to can have the position of the point of application P on the axis Δ, with respect to an origin chosen on it, to obtain a bound vector) (Fig.1.1,c).

We say that between two vectors $\mathbf{V}$ and $\mathbf{W}$ of the same type there exists a relation of *equality* if they are characterized by the same elements (direction, modulus, point of application, support) or the same canonical co-ordinates. For instance, for the free vectors $\mathbf{V}(V_i)$ and $\mathbf{W}(W_i)$ we can write

$$\mathbf{V} = \mathbf{W}, \quad V_i = W_i, \quad i = 1,2,3; \tag{1.1.3}$$

for the sliding vectors we must have also the same support, while for the bound vectors the same point of application. We mention following properties ($\mathbf{V}$, $\mathbf{W}$ and $\mathbf{U}$ are vectors):

i) $\mathbf{V} = \mathbf{V}$ (*reflexivity*);
ii) $\mathbf{V} = \mathbf{W} \Leftrightarrow \mathbf{W} = \mathbf{V}$ (*symmetry*);
iii) $\mathbf{V} = \mathbf{W}, \mathbf{W} = \mathbf{U} \Rightarrow \mathbf{V} = \mathbf{U}$ (*transitivity*)

valid for each type of vector.

If we bound two equal free vectors to two distinct points of application, then we obtain two bound vectors, which are *equipollent*.

Let be vectors $\mathbf{V}_1$ and $\mathbf{V}_2$; the sum

$$\mathbf{V} = \mathbf{V}_1 + \mathbf{V}_2 \tag{1.1.4}$$

is a vector for which the direction and modulus are given by the diagonal of the parallelogram constructed by means of these vectors, admitting that they are applied at

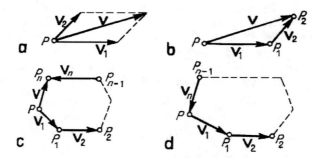

Figure 1.2. Addition of two vectors (a, b). Addition of n vectors: non-vanishing resultant (c); vanishing resultant (d).

the same point (Fig.1.2,a). The above given definition corresponds to free vectors, to bound vectors (if they have the same point of application), as well as for sliding vectors (if their supports are concurrent). The sum can be obtained also by closing a triangle, the first two sides of which are $\mathbf{V}_1$ and $\mathbf{V}_2$ (Fig.1.2,b); this observation can be used to construct the sum of n vectors (Fig.1.2,c)

$$\mathbf{V} = \mathbf{V}_1 + \mathbf{V}_2 + ... + \mathbf{V}_n. \tag{1.1.5}$$

We can write also *Chasles' relation*

$$\overrightarrow{PP_n} = \overrightarrow{PP_1} + \overrightarrow{P_1P_2} + ... + \overrightarrow{P_{n-1}P_n}. \tag{1.1.5'}$$

Newtonian model of mechanics 7

The resultant **V** closes the vectors' polygon (a skew polygon in E_3); if this polygonal line is closing, then the resultant vanishes (Fig.1.2,d). We mention following properties:

i) $\mathbf{V}_1 + \mathbf{V}_2 = \mathbf{V}_2 + \mathbf{V}_1$ (*commutativity*);
ii) $\mathbf{V}_1 + (\mathbf{V}_2 + \mathbf{V}_3) = (\mathbf{V}_1 + \mathbf{V}_2) + \mathbf{V}_3$ (*associativity*).

The inequalities

$$||\mathbf{V}_1| - |\mathbf{V}_2|| \leq |\mathbf{V}_1 + \mathbf{V}_2| \leq |\mathbf{V}_1| + |\mathbf{V}_2|, \quad |\mathbf{V}_1| \geq |\mathbf{V}_2| \tag{1.1.6}$$

result from Fig.1.2,b.

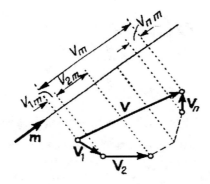

Figure 1.3. Projection of a vector on a directed axis.

Let V_{im} be the projection of vector $\mathbf{V}_i$, $i = 1, 2, ..., n$, on an directed axis, of sliding vector **m**, while V_m is the projection of vector **V** on the same axis (by *projection of a bound vector* $\overrightarrow{PQ}$ on a directed axis we understand the length of the directed segment $\overrightarrow{P'Q'}$ of this axis, limited by planes normal to the given axis, passing through points P and Q; this length is a positive number if $\sphericalangle(\overrightarrow{PQ}, \mathbf{m}) < \pi/2$ and negative in the contrary case). We observe that (Fig.1.3)

$$V_m = V_{1m} + V_{2m} + ... + V_{nm}; \tag{1.1.7}$$

if the vectors' polygon is closing, then the sum of the projections on an axis vanishes. Thus, for the sum of two vectors one obtains the canonical co-ordinates

$$V_i = V_{1i} + V_{2i}, \quad i = 1, 2, 3. \tag{1.1.8}$$

We say that vector **V** is equal to *zero*

$$\mathbf{V} = \mathbf{0} \tag{1.1.9}$$

if and only if

$$V_i = 0, \ i = 1,2,3, \ \text{or} \ V = 0; \tag{1.1.9'}$$

in this case, the properties

$$V + 0 = 0 + V = V \tag{1.1.9''}$$

are verified.

We observe that *the law of internal composition* of vectors (addition of vectors, introduced above) is essential to define them; we must complete the definition given in this subsection, because there exist quantities which can be represented by directed segments, but which do not verify such a law of composition, so that they are not vectors in the sense considered above (for instance, the finite rotation of a rigid solid about an axis).

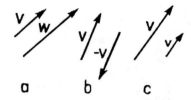

Figure 1.4. The product of a scalar by a vector.

The scalar is a mathematical entity, characterized by *sign* and *modulus* (magnitude); hence, it is a number.

The product of a scalar λ by a vector V is a vector (Fig.1.4,a)

$$W = \lambda V, \tag{1.1.10}$$

which has the same direction as V if $\lambda > 0$ or an opposite direction if $\lambda < 0$, and the modulus

$$W = |\lambda| V. \tag{1.1.10'}$$

In the particular case $\lambda = -1$, we obtain the vector $-V$, which has the same modulus as V, but an opposite direction (Fig.1.4,b); this vector verifies relation

$$V + (-V) = 0. \tag{1.1.11}$$

As a consequence, the subtraction of two vectors is also an addition

$$V_1 - V_2 = V_1 + (-V_2). \tag{1.1.12}$$

We mention following properties:

i) $\lambda_1(\lambda_2 V) = \lambda_2(\lambda_1 V) = \lambda_1 \lambda_2 V$ (*associativity*);
ii) $\lambda_1 V + \lambda_2 V = (\lambda_1 + \lambda_2) V$ (*distributivity with respect to addition of scalars*);
iii) $\lambda V_1 + \lambda V_2 = \lambda (V_1 + V_2)$ (*distributivity with respect to addition of vectors*).

Newtonian model of mechanics 9

The product (1.1.10) vanishes ($\lambda \mathbf{V} = \mathbf{0}$) if and only if $\lambda = 0$ or $\mathbf{V} = \mathbf{0}$.

We observe that two vectors $\mathbf{V}$ and $\mathbf{W}$ have the same (or an opposite) direction (they are *collinear*, even if they have not the same support) if and only if they verify a relation of the form (1.1.10) or of the form

$$\lambda \mathbf{V} + \mu \mathbf{W} = \mathbf{0}, \quad \lambda, \mu \neq 0. \tag{1.1.13}$$

Let be a vector $\mathbf{V}$. We denote by $\mathbf{v}$ a vector which has the same direction as $\mathbf{V}$, but of modulus equal to unity (Fig.1.4,c); this vector is called the *unit vector* (*versor*) of $\mathbf{V}$ and is denoted by

$$\mathbf{v} = \operatorname{vers} \mathbf{V}. \tag{1.1.14}$$

We may write

$$\mathbf{V} = V \operatorname{vers} \mathbf{V} = V\mathbf{v} = |\mathbf{V}|\mathbf{v}. \tag{1.1.14'}$$

The unit vectors of the co-ordinate axes Ox_j are denoted by $\mathbf{i}_j$, $j = 1,2,3$.

Let be the set of free vectors $\{\mathbf{V}\}$, for which an operation of *internal composition* (addition of vectors) is defined, so that the sum of two elements of this set is an element of the same set; this operation is associative and admits an inverse.

Vector $\mathbf{0}$ is the *neutral element*, which verifies relations (1.1.9''), while vector $-\mathbf{V}$ is the *inverse* of $\mathbf{V}$, verifying relation (1.1.11); these vectors belong to set $\{\mathbf{V}\}$. Hence, the set of free vectors $\{\mathbf{V}\}$ forms an *Abelian* (*commutative*) *group* (the operation of internal composition of vectors is commutative).

Because in the set $\{\mathbf{V}\}$ was defined also the operation of multiplication of one of its elements by a scalar (*external composition*), this set constitutes a *three-dimensional vector space* (*the linear space* L_3). Analogously, one can introduce also the *n-dimensional vector space* L_n.

1.1.3 Space. Time

Space and *time* are *basic notions*, which appear in a mathematical model of mechanics. To study the mechanical motion, representations of space and time are necessary; so, in classical mechanics, the physical space is the three-dimensional Euclidean space E_3, while time is assimilated also to a Euclidean space, but to a unidimensional one E_1. It was thus admitted that material bodies have three dimensions (e.g., length, width and height); but if the bodies to be studied are reduced to two dimensions or to only one dimension, so that they may be modelled in such a form, then we use a two-dimensional Euclidean space E_2 or a unidimensional one E_1, respectively. The geometric models for space and time used in classical mechanics reflect thus properties of real space and time, as objective forms of existence of matter.

In a classical model of mechanics, the *space* is considered *infinite, continuous, homogeneous, isotropic* and *absolute*, being independent of matter. Material bodies are immersed into space, where they fill a certain position with respect to other objects of the material world; they have also a certain *extent*. One can thus say that the space represents a *support of matter*. If, after a series of mechanical phenomena (for instance, after a succession of motions), material bodies maintain their properties of extent and relative situation with respect to other material objects, one considers that they have the same position. *The physical space* is formed by the totality of these positions, having thus an *absolute character*. The properties of *homogeneity* and *isotropy* put into evidence the fact that local properties of this space are independent of the position and of the direction, respectively.

An observer who perceives mechanical phenomena of the material world from a "laboratory" can put in evidence a "relation of order" for them; so appears the notion of *individual physical time*. By synchronization of time scales of two observers one passes from the individual physical time to the *universal physical time*; in this case, the transmission of signals has an *instantaneous character*. Thus, in a classical model of mechanics, the time has an *absolute character*, being the same for all observers who study a certain mechanical phenomenon. Any material process has certain *duration*. We say that an elementary material process, which has no duration (is of null duration), is an *event*. If two events take place at the same time, then they are *simultaneous*; in the contrary case, they are *successive*, one of them taking place before the other one. Thus, the set of events is *ordered*, and the time, characterized by duration, simultaneity and succession, is unidimensional. Admitting, in a classical model of mechanics, that time is independent of matter, it results that – as in the case of the space – it has no structure; hence, the time "flows" uniformly, being *homogeneous*. We mention also that the time is *infinite*, *continuous* and *irreversible*.

The modelling of the space, as well as the modelling of the time, represents, in classical mechanics, an approximation of the material reality; they are valid only in case of small distances and small velocities with respect to the velocity of propagation of light in vacuum.

1.1.4 Motion. Rest. Frames of reference

By *mechanical motion* we understand the change of position of a material body $\mathcal{B}_1$ with respect to another material body $\mathcal{B}_2$. In this case, at least a distance from a point of the body $\mathcal{B}_1$ to a point of the body $\mathcal{B}_2$ varies in time; we say that the corresponding points undergo certain *displacements*. This mechanical phenomenon has a *relative character*; indeed, if the body $\mathcal{B}_1$ moves with respect to the body $\mathcal{B}_2$, then the latter one moves also with respect to the first one.

If we refer the motion (in what follows we leave out the adjective "mechanical", because we have to do only with such motions) of a body $\mathcal{B}$ (we omit also the adjective "material" in subsequent formulations) to another body $\mathcal{R}$, the latter one will be called *frame (system) of reference*; in what follows, the frame of reference $\mathcal{R}$ will be considered to be *rigid* (the distances between any two points of it are constant in time), being made up at least of three non-collinear points. If the distances of all the points of the body $\mathcal{B}$ to the frame $\mathcal{R}$ (to three non-collinear points of it) are constant in time, then we say that this body is at *rest* with respect to this frame.

We cannot speak about *absolute motion* (only in the sense that matter is in continuous motion) because an *absolute frame of reference* does not exist, but only about *relative motion*; as well, we cannot speak about *absolute rest*, but only about *relative rest*. Analogously, we cannot speak about *absolute space* or about *absolute time*, but only about *relative space* (referring to a spatial frame), and about *relative time* (referring to a temporal frame).

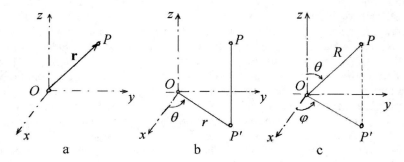

Figure 1.5. Orthogonal Cartesian (a), cylindrical (b) and spherical (c) co-ordinates.

The position of a point P of a body will be given by the position vector $\mathbf{r}$ with respect to the *pole* (*origin*) O of a given frame of reference. We will consider only frames the bases of which are *positive*, using various systems of co-ordinates. In general, we use *right-handed orthonormed frames* (see Chap. 2, Subsec. 1.1.2); for instance, in case of a system of *orthogonal Cartesian co-ordinates*, the position of the point P will be given by the co-ordinates x_i, $i = 1,2,3$, or by *the abscissa* x, *the ordinate* y and *the applicate* z (Fig.1.5,a), which are – at the same time – the components of the position vector. In *cylindrical co-ordinates*, we use *the polar radius* r, *the polar angle* θ and *the applicate* z (Fig.1.5,b), while in *spherical co-ordinates* we introduce the *polar radius* R, *the colatitude* θ and *the azimuth* (*longitude*) φ (Fig.1.5,c) (the notations do not correspond in the two systems of co-ordinates, but we like better to use well known notations); in general, we introduce arbitrary *curvilinear co-ordinates* q_i, $i = 1,2,3$.

The distance between two points $P(x_1, x_2, x_3)$ and $Q(x_1, x_2, x_3)$ will be given by

$$\left|\mathbf{r}_P - \mathbf{r}_Q\right| = \sqrt{(x_i - y_i)(x_i - y_i)} = \sqrt{(x_1 - y_1)^2 + (x_2 - y_2)^2 + (x_3 - y_3)^2},$$
(1.1.15)

satisfying all the properties of the norm.

The time will be characterized by a *variable* t, *monotonous increasing*, being independent of any physical phenomenon and frame (observer). This variable can take values on all the time axis ($t \in (-\infty, \infty)$) or we can admit the existence of an *initial moment* t_0, from which starts the study of the considered phenomenon ($t \in [t_0, \infty)$); we can consider also that $t \in (-\infty, t_1]$ or that $t \in [t_0, t_1]$. We can admit, without loosing the generality, that $t_0 = 0$ (we choose the initial moment as origin for the time axis).

Long time, the scientists searched an absolute frame of reference in Universe, admitting that – with respect to such a frame – the mathematical model of mechanics takes its simplest form. So, in *the geocentric frame of reference* (*Ptolemy's frame*) one admits that the origin is placed at the centre of mass of the Earth; usually, the equatorial plane is taken as principal plane, one of the axes in this plane being straighten in the direction of *the vernal equinoctial point* (at the intersection of the equatorial plane with the ecliptic one, which contains the trajectory described by the Earth), while the third axis is normal to this plane (hence, it is the rotation axis of the Earth). Later, one considered *the heliocentric frame of reference* (*Copernicus' frame*), in which the origin is in the centre of mass of the Sun. In such a frame, the plane of the ellipse described by the Earth, hence the ecliptic plane, is taken as principal plane (*ecliptic heliocentric frame of reference*), or a plane parallel to the equatorial plane of the Earth is taken as such a plane (*equatorial heliocentric frame of reference*), the third axis being normal to the respective principal plane; in both cases, one of the axes contained in the principal plane is along the direction of the vernal point (at the intersection of the ecliptic plane with a plane parallel to the equatorial plane of the Earth, passing through the centre of mass of the Sun), hence it is parallel to the axis considered in case of the geocentric frame. Obviously, this last frame represents a progress with respect to the previous one, leading to simpler properties of motion of planets. But the Sun is only one of more than two hundred milliards of stars of our Galaxy, having – at this moment – a relatively peripheral position in it; all these stars, including the Sun, move with respect to the centre of mass of the Galaxy. We are thus led to a frame with the origin at this centre of mass (a *galactocentric frame of reference*); the principal plane is the galactic median plane, one of the axes contained in this plane being along its intersection with a plane parallel to the equatorial plane of the Earth. In all these cases one uses the so-called "fixed stars" (very far stars, the position of which are approximately fixed with respect to observations which can be made on the Earth); knowing the positions of these stars (their co-ordinates), one can use any of the frames mentioned above. A catalogue of approximately 1500 such stars, called *basic stars*, has been elaborated. Theoretically, only four stars are sufficient to identify a frame; practically, one considers a greater number of stars, because – in any case – their positions vary in time, so that their determination can be erroneous.

The laws of classical models of mechanics are sufficiently well verified in a galactocentric frame of reference. But we observe that the motion of the heliocentric frame with respect to the galactocentric one can be considered – with a good approximation – to be a uniform and rectilinear translation, even for a relatively long interval of time; one can thus use – in many cases – the heliocentric frame, obtaining very good results, for instance in the study of motion of objects launched in the cosmic space. For usual motions on the surface of the Earth, the geocentric frame of reference leads also to very good results; the influence of the rotation and revolution motions of the Earth can be introduced subsequently, every time it is necessary.

The frames of reference with respect to which the basic laws of mechanics can be verified are called *inertial frames of reference*. If, in a classical model of mechanics, its laws are verified in a certain frame of reference, then they are verified with respect to any other frame in rectilinear and uniform motion with respect to the latter one; thus, one obtains a *class* of inertial frames of reference. An "absolute space" in the sense of Newton cannot be identified on the basis of a mechanical experiment.

The galactocentric frame of reference is an inertial one; as we have seen above, for different cases of mechanical motion one can admit that the heliocentric frame of reference or even the geocentric frame of reference are inertial frames. On this way, we

Newtonian model of mechanics 13

can construct also other frames of reference (considering a larger region of the universe, at a greater scale), which are also inertial; but this is not necessary from the point of view of classical models of mechanics.

1.1.5 Trajectory. Velocity. Acceleration

In what follows, let us consider the motion of a body with respect to a frame of reference $\mathscr{R}$ which, by convention, is admitted to be "fixed". The motion of a point P of the body with respect to this frame of reference is defined if, for any value of the time

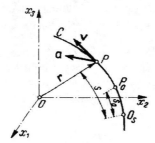

Figure 1.6. Trajectory, velocity and acceleration of a particle P.

t in E_1, we obtain its position in E_3; the point P describes thus, in its motion, a *trajectory C* (the geometric locus of positions occupied by point P) (Fig.1.6), of *vector equation*

$$\mathbf{r} = \mathbf{r}(t) \qquad (1.1.16)$$

or of *parametric equations*

$$x_i = x_i(t), \quad i = 1, 2, 3. \qquad (1.1.16')$$

In cylindrical co-ordinates, these equations are written in the form

$$r = r(t), \quad \theta = \theta(t), \quad z = z(t) \qquad (1.1.16'')$$

and in spherical co-ordinates we have

$$r = r(t), \quad \theta = \theta(t), \quad \varphi = \varphi(t); \qquad (1.1.16''')$$

In arbitrary curvilinear co-ordinates, they are given by

$$q_i = q_i(t), \quad i = 1, 2, 3. \qquad (1.1.16^{\text{iv}})$$

These equations define the *law of motion*. Eliminating the time t between equations (1.1.16'), one obtains the *Cartesian equations* (two equations) of the trajectory. The knowledge of the trajectory in one of the forms mentioned above does not mean that we know how the point P is moving along this trajectory; to state accurately this motion we must define the velocity and the acceleration of the point.

The mechanical nature of the phenomenon of motion imposes certain conditions to the vector function (1.1.16). So, it must be *continuous* (the trajectory is continuous, the point cannot occupy simultaneously several positions in space) and *bounded in modulus* for $t \in [t_0, t_1]$, where we have put in evidence the definition interval; if the condition of boundedness is not fulfilled, then the trajectory has points at infinity. This function must be also *differentiable* (obtaining thus the velocity and the acceleration), exception making – eventually – a finite number of moments. If the vector function admits everywhere derivatives of second order, then the *motion* is *continuous*, otherwise having to do with a *discontinuous motion* (in fact, the denomination refers to the velocity, not to the motion). In the last case, one can introduce *regular distributions* to characterize the velocity, as well as regular distributions or *singular* (if it is necessary) *distributions* to characterize the acceleration.

If the derivatives of first order of functions (1.1.16') are continuous at any moment t in the definition interval, then the trajectory is a *rectifiable curve*; we can introduce – in this case – the curvilinear co-ordinate s, measured from an arbitrary point O_s as origin (Fig.1.6). Taking the direction of motion as a positive one to measure arcs s, we can establish the correspondence

$$s = s(t), \qquad (1.1.17)$$

which will be called *the horary equation of motion*; the graphic representation of this function in a frame of reference Ots will be *the graph of the motion*. From

$$d\mathbf{r} \cdot d\mathbf{r} = (d\mathbf{r})^2 = |d\mathbf{r}|^2 = (ds)^2 = dx_i dx_i = (dx_1)^2 + (dx_2)^2 + (dx_3)^2$$
$$(1.1.18)$$

we deduce

$$s(t; s_0) = s_0 + \int_{t_0}^{t} \sqrt{\dot{x}_i(\tau)\dot{x}_i(\tau)} d\tau, \qquad (1.1.18')$$

s_0 being the length of the trajectory till point P_0, to which corresponds the initial moment t_0. In case of a finite number of discontinuity moments of the derivative of first order of function (1.1.16), the curve is *piecewise rectifiable*.

Let us introduce also the fixed frame of reference $\mathcal{R}'$, rigidly connected with the frame of reference $\mathcal{R}$; the origin O' of the new frame will be specified by the position vector $\mathbf{r}_0 = \overrightarrow{OO'}$ with respect to the first one. The point P will be thus given by the new position vector

$$\mathbf{r}'(t) = \mathbf{r}(t) - \mathbf{r}_0. \qquad (1.1.19)$$

Differentiating successively with respect to time, vector $\mathbf{r}_0$ will be eliminated, and one obtains

$$\dot{\mathbf{r}}' = \dot{\mathbf{r}}, \quad \ddot{\mathbf{r}}' = \ddot{\mathbf{r}}, \quad \dddot{\mathbf{r}}' = \dddot{\mathbf{r}} \ldots \qquad (1.1.20)$$

Newtonian model of mechanics

Hence, one can affirm that *all derivatives of the position vector with respect to time are invariant with respect to a change of a fixed frame of reference*. We mention the first two invariants

$$\mathbf{v} = \dot{\mathbf{r}}, \quad \mathbf{a} = \ddot{\mathbf{r}}, \qquad (1.1.20')$$

which represent the *velocity* and the *acceleration* of the point P, respectively, and play an important rôle in mathematical modelling of classical mechanics.

1.1.6 Mass. Momentum

The mass is a *quantity of state* (quantity which depends on the actual state of the body but not on how one got to this state) which appears in all phenomena of motion, being an *important property of matter*; it does exist objectively and is independent of the place where it is measured, as well as the electric charge in an electromagnetic field. Newton introduced the concept of mass as a *measure of the amount of matter*; this concept has a statical character. But we observe that the mass is present in various phenomena of motion too, and can be put in connection especially with the inertia and the universal attraction of bodies.

Inertia is a general tendency of matter to conserve its state of rest or of rectilinear and uniform motion; from this point of view, the mass represents a *measure of the inertia of the body* in motion. Let us consider a system formed by two homogeneous spheres S_1 and S_2, of negligible dimensions, which become in collision at point O (Fig.1.7,a); they have the velocities $\mathbf{v}'_1$ and $\mathbf{v}'_2$ before collision, and $\mathbf{v}''_1$ and $\mathbf{v}''_2$ after collision. One observes that the variations of velocity $\mathbf{v}''_1 - \mathbf{v}'_1$ and $\mathbf{v}''_2 - \mathbf{v}'_2$ are vectors on the same

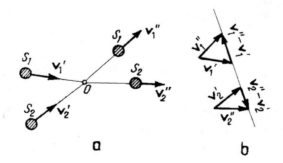

Figure 1.7. Collision of two small homogeneous spheres (a).
Variations of velocity (b).

support but of opposite sign (Fig.1.7,b). Ș. Țițeica has made such an experiment, using two homogeneous small spheres fixed at the same fixed point by two inextensible threads of the same length; one obtains thus two simple (mathematical) pendulums. The isochronism of small oscillations (they take place in the same interval of time) ensures the property of tautochronism (one arrives at the same position in the same time interval, independently of the initial position), so that the spheres, starting from different points, become in collision in a point corresponding to the vertical of the fixed point;

the velocities before and after collision will be proportional to the respective amplitudes, so that we can measure them. We may thus write

$$\mathbf{v}_1'' - \mathbf{v}_1' = -\mu_{12}(\mathbf{v}_2'' - \mathbf{v}_2'), \qquad (1.1.21)$$

where μ_{12} is a positive scalar; repeating the experiment with the same spheres but other velocities $\mathbf{v}_1'$ and $\mathbf{v}_2'$, one observes that this scalar is constant, depending only on the chosen spheres (it is a characteristic of their interaction, this fact being put in evidence by the two indices). Let us consider now three homogeneous spheres S_1, S_2 and S_3, of negligible dimensions; repeating the experiment with these spheres taken two by two, we find a relation of the form

$$\mu_{12} = \mu_{13}\mu_{32}. \qquad (1.1.21')$$

Supposing now that $\mu_{12} = \mu_{12}(X_1, X_2)$, where X_1 and X_2 are quantities which characterize the spheres S_1 and S_2, respectively, it follows from (1.1.21') that $\mu_{22} = 1$ if the spheres S_2 and S_3 are identical; if the spheres S_1 and S_2 are identical and we take into account the previous result, then we can write also $\mu_{13} = 1/\mu_{31}$. Relation (1.1.21') becomes

$$\mu_{12} = \frac{\mu_{32}}{\mu_{31}} = \frac{\mu_{32}(X_3, X_2)}{\mu_{31}(X_3, X_1)}$$

and is valid for any sphere; there results that μ_{12} is the ratio of two scalars, each one depending only on quantities characterizing the spheres S_1 and S_2, respectively. We can thus write

$$\mu_{12} = \frac{m_2}{m_1}, \quad m_1, m_2 > 0. \qquad (1.1.22)$$

In this case, relation (1.1.21) shows that, for the same scalar m_2 and for the same variation of velocity $|\mathbf{v}_2'' - \mathbf{v}_2'|$ (in modulus), if the scalar m_1 is greater, then the variation $|\mathbf{v}_1'' - \mathbf{v}_1'|$ is smaller; the velocity of the sphere S_1 is subject to a smaller variation, its inertia being greater. The inertia of the sphere S_1 increases thus at the same time as the scalar m_1; the latter one can be considered as a measure of the inertia of the respective sphere (analogously, m_2 measures the inertia of the sphere S_2). So, one can attach to a homogeneous sphere S of negligible dimensions an *inertial mass* $m_i > 0$.

In what concerns the universal attraction of bodies, let us consider their falling on the surface of the Earth (the action of the terrestrial gravitational field). We can measure the attraction, by the Earth, of two homogeneous spheres S_1 and S_2, of negligible dimensions (their weight), using a dynamometer (hence, measuring some lengths); the two gravity forces $\mathbf{G}_1$ and $\mathbf{G}_2$, which are thus put in evidence, have the same direction, so that we can write

$$\mathbf{G}_1 = \nu_{12}\mathbf{G}_2, \quad \nu_{12} > 0. \tag{1.1.23}$$

Considering three homogeneous spheres S_1, S_2 and S_3, of negligible dimensions, we can establish a relation of the form

$$\nu_{12} = \nu_{13}\nu_{32}; \tag{1.1.23'}$$

by a reasoning similar to that above, we attach to the sphere S a *gravitational mass* $m_g > 0$. This scalar does not depend on the position on the surface of the Earth in which the experiment takes place.

Very fine experiments made by L. Eötvös in 1890 and taken again by Zeeman in 1917, led to *proportionality* between the inertial mass and the gravitational one (a body is as much inertial as it is heavy). Taking conveniently the units, we may have a proportionality coefficient equal to unity, hence the relation

$$m_i = m_g = m > 0, \tag{1.1.24}$$

which emphasizes the equality between the inertial mass and the gravitational one; in what follows, it will no more be necessary to characterize the respective property of the mass using the adjectives "inertial" and "gravitational". This equality has a quantitative (numerical) aspect in classical mechanics; in relativistic mechanics (general relativity), its qualitative aspects will also appear.

We mention an experiment made by W. Pohl, which puts in evidence the aspects of inertial and gravitational mass, respectively. A homogeneous sphere is linked in two

Figure 1.8. W. Pohl's experiment on inertial and gravitational mass.

diametral opposite points A and A' by two identical threads AB and $A'B'$; the end B is connected at a fixed point, while at the end B' acts a force $\mathbf{F}$ (Fig.1.8). If the force $\mathbf{F}$ acts suddenly (from the very beginning with its entire intensity), then the thread $A'B'$ breaks (because of the inertial aspect of the mass), while if this force has a static action (the intensity of the force growth in a short but finite time), then the thread AB breaks (the influence of the force $\mathbf{F}$ is added to the weight of the sphere, and the gravitational aspect of the mass is emphasized).

The Newtonian character of mass, i.e. to be a *measure* of the quantity of matter of a body, may be experimentally verified; indeed, the mass of several bodies is equal to the sum of the masses of the bodies, as it can be seen if we put in evidence the gravitational aspect of mass (the weights of bodies, measured with the aid of a dynamometer, in a certain place of the Earth, can be added). This is a *property of additivity*.

Thus, in a mathematical modelling, we associate to each body $\mathcal{B}$ a quantity of state $m(\mathcal{B})$ called *mass*, having the following properties:

i) the mass of the body is a positive scalar ($m(\mathcal{B}) > 0$);

ii) for each division (disjoint parts) $\mathcal{B}_i$, $i = 1,2,...,n$, of the body we can write (the property of additivity)

$$m(\mathcal{B}) = \sum_{i=1}^{n} m(\mathcal{B}_i). \tag{1.1.25}$$

In a classical model of mechanics we admit also that:

iii) the mass of the body remains constant during the motion ($\dot{m} = 0$); the property is modified in case of a body of variable mass.

Space, time and mass are independent one of each other in a classical model of mechanics.

Newton has introduced the notion of *momentum* (which he called *quantity of motion*, denomination of a statical character; it will not be used in what follows) $m\mathbf{v}$, representing the product of the mass by the velocity of a point of the body $\mathcal{B}$, as a measure of the mechanical motion; we observe that, in a motion of translation, the property of additivity takes place for any division $\mathcal{B}_i$ of the body. Taking into account (1.1.22) in relation (1.1.21), one obtains

$$m_1 \mathbf{v}_1' + m_2 \mathbf{v}_2' = m_1 \mathbf{v}_1'' + m_2 \mathbf{v}_2''; \tag{1.1.26}$$

that means that the momentum of the two homogeneous spheres of negligible dimensions remains constant during the phenomenon of collision. It is thus justified to consider the momentum to be a characteristic of mechanical motion, a *measure* of it. One can state also that relation (1.1.26) puts in evidence the capacity of a mechanical motion to be transformed into another mechanical motion.

1.1.7 Mathematical modelling of discontinuous phenomena. Distributions. Stieltjes integral

To study discontinuous phenomena together with continuous ones it is necessary to introduce the notion of distribution and Stieltjes integral in distributions; in fact, this mathematical modelling is the only one which allows a correct representation of the respective phenomenon. In this order of ideas, we will give some basic results and formulae to be used in what follows.

We call *functional* a mapping of a vector space X (with respect to Γ) into Γ. If Γ is the corpus of real numbers $\mathbb{R}$, then we say that we have to do with a *real functional*. We call *distribution* a continuous linear functional defined on a topological vector space X (*fundamental space*).

By definition, the fundamental space K is constituted of the functions of real variables $\varphi(x)$ ($x \equiv (x_1, x_2, ..., x_n)$ represents a point in $\mathbb{R}^n$), indefinite differentiable (of class C^∞) and vanishing together with all their derivatives in the exterior of certain bounded domains; these domains, together with their boundaries, determine the supports of these functions, called *fundamental functions* (by *support* of a function $\varphi(x)$ we mean the smallest closed set which contains the set of points x for which $\varphi(x) \neq 0$). By an extension of the space K we reach another class of functions, which determine the fundamental space S. The functions belonging to this class have also the property to be indefinite differentiable; for $|x| \to \infty$ these functions tend to zero together with their derivatives of any order, more rapidly that any power of $1/|x|$. We introduce also the space K^m, which includes the functions with compact support, having continuous derivatives up to and including the *m*th order (of class C^m). The distributions defined on spaces K, S and K^m are called *of infinite order*, *temperate* and *of finite order* $p \leq m$, respectively.

Let $f(x)$ be a function defined on the real axis $\mathbb{R}$; we say that this function is *absolutely integrable* in a closed interval $[a,b]$ of $\mathbb{R}$, if the integral

$$\int_a^b |f(x)| \, \mathrm{d}x < \infty \tag{1.1.27}$$

exists.

If the function $f(x)$ is absolutely integrable in any finite interval of $\mathbb{R}$, then we say that $f(x)$ is a *locally integrable function*. We mention that an absolutely integrable function is also *integrable*, i.e. the integral

$$\int_a^b f(x) \, \mathrm{d}x \tag{1.1.27'}$$

exists.

Locally integrable functions generate an important class of distributions; we assume thus that to any fundamental function $\varphi(x) \in K$ there corresponds a real number

$$(f, \varphi) = \int_{\mathbb{R}} f(x)\varphi(x) \, \mathrm{d}x = \int_a^b f(x)\varphi(x) \, \mathrm{d}x, \tag{1.1.28}$$

where $f(x)$ is a locally integrable function and $[a,b]$ is the support of $\varphi(x)$. It is easy to see that the functional thus defined is linear and continuous. The functional defined on space K by means of the locally integrable function $f(x)$ represents a *distribution* on this space, which will be denoted also by $f(x)$, like the generating function. Such distributions are called *regular distributions* (*distributions of function type*). Analogously, one can define *temperate regular distributions* on space S.

The distributions, which are not of function type, are called *singular distributions*. If to any function $\varphi(x) \in K$ we attach its value in the origin (the value $\varphi(0)$), then we see that the respective functional is linear and continuous, hence it is a distribution which is not regular; this is *the Dirac distribution*, which will be denoted by the symbol $\delta(x)$. Symbolically, we write

$$(\delta(x), \varphi(x)) = \varphi(0). \tag{1.1.29}$$

We can define the Dirac distribution also on the space K^0; it will be called, in this case, *the Dirac measure*.

If the fundamental functions $\varphi \in K$ are defined in $\mathbb{R}^n$, then we have

$$(\delta(x_1, x_2, \ldots, x_n), \varphi(x_1, x_2, \ldots, x_n)) = \varphi(0, 0, \ldots, 0). \tag{1.1.29'}$$

The relation

$$(f(x - x_0), \varphi(x)) = (f(x), \varphi(x + x_0)), \quad x, x_0 \in \mathbb{R}^n, \tag{1.1.30}$$

defines a *translated distribution* $f(x - x_0)$. In particular, for the Dirac distribution $\delta(x - x_0)$ we may write

$$(\delta(x - x_0), \varphi(x)) = \varphi(x_0), \quad x, x_0 \in \mathbb{R}^n. \tag{1.1.30'}$$

For distributions subjected to a *homothetic transformation* with respect to the independent variable, we shall use – by definition – the formula

$$(f(\alpha x), \varphi(x)) = |\alpha|^{-n} \left(f(x), \varphi\left(\frac{1}{\alpha} x\right) \right), \quad x \in \mathbb{R}^n. \tag{1.1.31}$$

For $\alpha = -1$ we obtain the property of *symmetry* and we may write

$$(f(-x), \varphi(x)) = (f(x), \varphi(-x)), \quad x \in \mathbb{R}^n. \tag{1.1.32}$$

In particular, we observe that $\delta(-x) = \delta(x)$, hence the Dirac distribution is *even* with respect to the independent variable $x \in \mathbb{R}^n$.

The equality of two distributions $f(x)$ and $g(x)$ is defined by the relation

$$(f, \varphi) = (g, \varphi), \quad \forall \varphi \in K; \tag{1.1.33}$$

hence, we may write

$$f = g. \tag{1.1.33'}$$

A distribution $f(x)$ is *equal to zero* ($f = 0$) if, for any fundamental function $\varphi(x)$, we have $(f(x), \varphi(x)) = 0$. If the distributions f and g are generated by continuous functions $f(x)$ and $g(x)$, then the equality (1.1.33') occurs in the usual sense, i.e. punctual, because – in this case – the distributions f and g coincide everywhere with the functions f and g. If the functions f and g are locally integrable and coincide almost everywhere, then the distributions generated by them will be equal and relation (1.1.33') holds.

We define *the Heaviside function* (*the unit function*) on the real axis in the form (Fig.1.9)

$$\theta(x) = \begin{cases} 0, & x < 0, \\ 1, & x \geq 0; \end{cases} \qquad (1.1.34)$$

the distribution generated by it will be called *the Heaviside distribution*. If $f(x)$ is a function of variable x, defined on the real axis, then we will call its *positive part* the function f_+ defined by the relation (Fig.1.10,a)

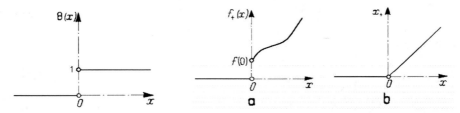

Figure 1.9. The Heaviside function.   *Figure 1.10.* The positive part of function $f(x)$ (a) and of function $f(x) = x$ (b).

$$f_+(x) = f(x)\theta(x) = \begin{cases} 0, & x < 0, \\ f(x), & x \geq 0. \end{cases} \qquad (1.1.35)$$

In particular, for the function $f(x) = x$ we can introduce the positive part (Fig.1.10,b)

$$x_+ = x\theta(x) = \begin{cases} 0, & x < 0, \\ x, & x \geq 0; \end{cases} \qquad (1.1.36)$$

obviously, to these functions correspond certain distributions of function type. We introduce also the distribution generated by the function $1(x) = 1$, $\forall x \in \mathbb{R}$ (Fig.1.11).

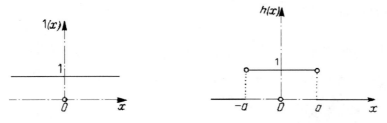

Figure 1.11. The function $1(x) = 1$.   *Figure 1.12.* The characteristic function.

The characteristic function corresponding to the interval $[-a, a]$ is defined in the form (Fig.1.12)

$$h(x) = \begin{cases} 0, & |x| > a, \\ 1, & |x| \leq a, \end{cases} \quad a > 0, \tag{1.1.37}$$

being thus led to a distribution of function type; this distribution may be expressed also by the Heaviside distribution

$$h(x) = \theta(x + a) - \theta(x - a) = \theta(a - |x|), \quad x \in \mathbb{R}. \tag{1.1.37'}$$

If $\psi(x)$ is a function of class C^∞, then we can write the equality

$$\psi(x)\delta(x) = \psi(0)\delta(x); \tag{1.1.38}$$

in particular, for $\psi(x) = x^n$, we obtain

$$x^n \delta(x) = 0, \tag{1.1.39}$$

so that the product of a distribution by a function indefinite differentiable may be equal to zero, even if no one of its factors vanishes. Analogously,

$$\psi(x)\delta(x - a) = \psi(a)\delta(x - a). \tag{1.1.38'}$$

We mention also the decomposition formula of a distribution

$$\delta(x^2 - a^2) = \frac{1}{2x}[\delta(x - a) - \delta(x + a)], \quad a > 0; \tag{1.1.40}$$

using formula (1.1.38'), we may write

$$\delta(x + a) + \delta(x - a) = 2a\delta(x^2 - a^2), \quad a > 0. \tag{1.1.40'}$$

We also have

$$\delta(x - a) + \delta(x - b) = |a - b|\delta[(x - a)(x - b)]. \tag{1.1.41}$$

In $\mathbb{R}^2$ we obtain

$$\delta(x_1^2 - a_1^2, x_2^2 - a_2^2) = \frac{1}{4a_1 a_2}[\delta(x_1 - a_1, x_2 - a_2) + \delta(x_1 + a_1, x_2 - a_2)$$
$$+ \delta(x_1 - a_1, x_2 + a_2) + \delta(x_1 + a_1, x_2 + a_2)], \quad a, b > 0. \tag{1.1.42}$$

We mention also formula (in $\mathbb{R}^3$)

$$2x_1 \delta\left[x_1^2 - (x_1^0)^2, x_2 - x_2^0, x_3 - x_3^0\right] = \delta(x_1 - x_1^0, x_2 - x_2^0, x_3 - x_3^0)$$
$$- \delta(x_1 + x_1^0, x_2 - x_2^0, x_3 - x_3^0). \tag{1.1.43}$$

Newtonian model of mechanics

We call *support of the distribution* f ($\operatorname{supp} f$) the complement of the union of open sets on which this distribution vanishes; therefore, the support of a distribution is a closed set. If the support of a distribution is contained in a set A, then we say that the distribution is *concentrated on the set* A. Thus, we may say that the Dirac distribution $\delta(x)$ is *concentrated at a point* (the origin). Analogously, one can define distributions concentrated on curves or surfaces, in general distributions concentrated on a manifold of a space $\mathbb{R}^n$.

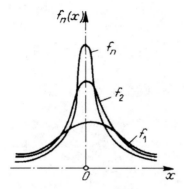

Figure 1.13. A δ representative sequence.

We say that a sequence of functions $f_n(x_1, x_2, ..., x_m)$ is a δ *representative sequence* if, in the sense of the topology of K' (K' is the *topological dual* of K, containing the distributions defined on this fundamental space) we have (in Fig.1.13 is given a δ representative sequence for $m = 1$)

$$\lim_{n \to \infty} f_n(x_1, x_2, ..., x_m) = \delta(x_1, x_2, ..., x_m); \tag{1.1.44}$$

obviously, this condition is equivalent to

$$\lim_{n \to \infty} (f_n(x_1, x_2, ..., x_m), \varphi(x_1, x_2, ..., x_m)) = \varphi(0, 0, ..., 0). \tag{1.1.44'}$$

From the very beginning, we mention that *the distributions admit derivatives of any order*, which is a great advantage with respect to usual functions. Let $f(x)$ be a function of class C^1 and $\varphi(x)$ a fundamental function belonging to the fundamental space K; considering the corresponding distribution of function type, we obtain *the rule of differentiation* in the form

$$(f', \varphi) = -(f, \varphi'). \tag{1.1.45}$$

In particular, we have

$$\theta'(x) = \delta(x). \tag{1.1.46}$$

In case of a distribution of several variables $f(x_1, x_2, ..., x_n)$, we can write

$$\left(\frac{\partial}{\partial x_i} f(x_1, x_2, ..., x_n), \varphi(x_1, x_2, ..., x_n) \right)$$
$$= -\left(f(x_1, x_2, ..., x_n), \frac{\partial}{\partial x_i} \varphi(x_1, x_2, ..., x_n) \right), \quad i = 1, 2, ..., n; \quad (1.1.47)$$

one obtains also the property

$$\frac{\partial^2 f}{\partial x_i \partial x_j} = \frac{\partial^2 f}{\partial x_j \partial x_i}, \quad i, j = 1, 2, ..., n, \quad (1.1.48)$$

which shows that, in case of distributions, the derivatives do not depend on the order of differentiation.

Let $f(x)$ be a function of class C^1 everywhere, excepting at the point x_0, where the function has a discontinuity of the first species, the corresponding *jump* being given by

$$s_0 = f(x_0 + 0) - f(x_0 - 0). \quad (1.1.49)$$

We denote by $f'(x)$ *the derivative of the distribution* $f(x)$ *(in the sense of the theory of distributions)* and by $\tilde{f}'(x)$ *the distribution corresponding to the derivative of the function, which generated the distribution, in the usual sense,* wherever this derivative exists; we obtain the relation

$$f'(x) = \tilde{f}'(x) + s_0 \delta(x - x_0). \quad (1.1.50)$$

It is worth to note that if the function $f(x)$ is continuous at the point x_0, then the jump s_0 vanishes, and formula (1.1.50) becomes

$$f'(x) = \tilde{f}'(x); \quad (1.1.50')$$

i.e., the derivative in the sense of the theory of distributions coincides with the derivative in the usual sense.

If the function $f(x)$ is of class C^1 everywhere excepting the points x_i, $i = 1, 2, ..., n$, where it has discontinuities of the first species and if we denote by s_i the jump of the function at the point x_i, then, by a similar procedure, we obtain a more general formula, namely

$$f'(x) = \tilde{f}'(x) + \sum_{i=1}^{n} s_i \delta(x - x_i). \quad (1.1.51)$$

A last property, which is worth to be revealed, is the following: *If the derivative of a distribution is equal to zero, then the distribution is a constant.*

Let $f(x)$ be a continuous function on $[a,b]$, and $g(x)$ a function with an integrable derivative on the same interval; one can show that, in this case, *the Stieltjes integral* $(S)\int_a^b f(x)\mathrm{d}g(x)$ exists, its computation being reduced to the computation of a *Riemann integral*

$$(S) \int_a^b f(x)\mathrm{d}g(x) = (R) \int_a^b f(x)g'(x)\mathrm{d}x . \qquad (1.1.52)$$

It is important to mention that, in general, the symbol $\mathrm{d}g(x)$ which appears in the Stieltjes integral does not represent *the differential* of the function $g(x)$; the fact that the function $g(x)$ has an integrable derivative in the conditions mentioned above allows us to admit that $\mathrm{d}g(x)$ is the differential of $g(x)$ in the formula (1.1.52), hence one can pass easily from the Stieltjes integral to the Riemann one. If $f(x)$ and $g(x)$ are distributions and the Stieltjes integral does exist, then the symbol $\mathrm{d}g(x)$ can be also considered as a *differential in the sense of the theory of distributions*; in this case, if $f(x)$ is a continuous function on $[a,b]$, excepting a finite number of points $a = c_0 < c_1 < c_2 < ... < c_m = b$, where there are discontinuities of the first species, then there exists the Stieltjes integral and the relation

$$(S) \int_a^b f(x)\mathrm{d}g(x) = (R) \int_a^b f(x)g'(x)\mathrm{d}x + f(a)[g(a+0) - g(a)]$$
$$+ \sum_{k=1}^{m-1} f(c_k)[g(c_k+0) - g(c_k-0)] + f(b)[g(b) - g(b-0)] \qquad (1.1.53)$$

holds.

Analogously, if $f(x_1, x_2)$ is a continuous function in a two-dimensional interval Δ and if the function $g(x_1, x_2)$ is a non-decreasing function with respect to each variable and has the property

$$\Delta_2 g(x_1, x_2) \equiv g(x_1 + h_1, x_2 + h_2) - g(x_1 + h_1, x_2) - g(x_1, x_2 + h_2)$$
$$+ g(x_1, x_2) \geq 0, \quad \forall h_1, h_2 > 0, \qquad (1.1.54)$$

then *the double Stieltjes integral* $\iint_\Delta f(x_1, x_2) \mathrm{D}_2 g(x_1, x_2)$ exists. If the function $g(x_1, x_2)$ admits second order continuous partial derivatives, then the computation of the double Stieltjes integral reduces to the computation of a *double Riemann integral*

$$(S) \iint_\Delta f(x_1, x_2) \mathrm{D}_2 g(x_1, x_2) = (R) \iint_\Delta f(x_1, x_2)\, g''_{x_1 x_2}(x_1, x_2)\mathrm{d}x_1 \mathrm{d}x_2 ; \qquad (1.1.55)$$

this result remains valid if, instead of the continuity of the mixed derivative $g''_{x_1 x_2}(x_1, x_2)$, its integrability only is asked. On the basis of this theorem, one can pass from the double Stieltjes integral to the double Riemann integral directly if, instead of the symbol $D_2 g(x_1, x_2)$ one considers the expression $g''_{x_1 x_2}(x_1, x_2) dx_1 dx_2$. In fact, the symbol $D_2 g(x_1, x_2)$ represents a differential operator specific to the double Stieltjes integral, which is nothing else than *the two-dimensional differential* of the function $g(x_1, x_2)$; we mention also that $g''_{x_1 x_2}(x_1, x_2)$ represents the *two-dimensional derivative* of this function. We can write

$$D_2 g(x_1, x_2) = g''_{x_1 x_2}(x_1, x_2) dx_1 dx_2. \tag{1.1.56}$$

If the function $f(x_1, x_2)$ defined on Δ is continuous, then we have

$$(S) \iint_\Delta f(x_1, x_2) D_2 \theta(x_1 - x_1^0, x_2 - x_2^0) = f(x_1^0, x_2^0). \tag{1.1.57}$$

Let be the function $g(x_1, x_2)$ of the form

$$g(x_1, x_2) = \tilde{g}(x_1, x_2) + \sum_{k=1}^{n} g_k \theta\left(x_1 - x_1^{(k)}, x_2 - x_2^{(k)}\right), \tag{1.1.58}$$

where g_k are constants and $\tilde{g}(x_1, x_2)$ is *the continuous part of the function* $g(x_1, x_2)$ on Δ; taking into account the relations (1.1.55), (1.1.57) as well as the property of additivity of the Stieltjes integral, we get

$$(S) \iint_\Delta f(x_1, x_2) D_2 g(x_1, x_2) = (R) \iint_\Delta f(x_1, x_2) \tilde{g}''_{x_1 x_2}(x_1, x_2) dx_1 dx_2$$
$$+ \sum_{k=1}^{n} g_k f\left(x_1^{(k)}, x_2^{(k)}\right). \tag{1.1.59}$$

We may write also the relation

$$D_2 g(x_1, x_2) = D_2 \tilde{g}(x_1, x_2) + \sum_{k=1}^{n} g_k D_2 \theta\left(x_1 - x_1^{(k)}, x_2 - x_2^{(k)}\right) \tag{1.1.60}$$

or the relation

$$D_2 g(x_1, x_2) = \tilde{D}_2 g(x_1, x_2) + \sum_{k=1}^{n} (\Delta g)_k \, \delta\left(x_1 - x_1^{(k)}, x_2 - x_2^{(k)}\right) dx_1 dx_2, \tag{1.1.61}$$

where $D_2 g(x_1, x_2)$ and $\tilde{D}_2 g(x_1, x_2)$ represent *the two-dimensional differentials in the sense of the theory of distributions* and *in the usual sense*, respectively, while

$$(\Delta g)_k = g\left(x_1^{(k)} + 0, x_2^{(k)} + 0\right) - g\left(x_1^{(k)} + 0, x_2^{(k)} - 0\right)$$
$$- g\left(x_1^{(k)} - 0, x_2^{(k)} + 0\right) + g\left(x_1^{(k)} - 0, x_2^{(k)} - 0\right) \tag{1.1.61'}$$

is *the two-dimensional jump* of the function $g(x_1,x_2)$ at the point of discontinuity $\left(x_1^{(k)}, x_2^{(k)}\right)$, $k = 1,2,\ldots,n$.

Let $g(x_1,x_2,x_3)$ be a non-decreasing function with respect to each variable, defined on a three-dimensional interval Δ, and which verifies the inequality

$$\Delta_3 g(x_1,x_2,x_3) \equiv g(x_1+h_1, x_2+h_2, x_3+h_3) - g(x_1, x_2+h_2, x_3+h_3)$$
$$-g(x_1+h_1, x_2, x_3+h_3) - g(x_1+h_1, x_2+h_2, x_3) + g(x_1+h_1, x_2, x_3)$$
$$+g(x_1, x_2+h_2, x_3) + g(x_1, x_2, x_3+h_3) - g(x_1,x_2,x_3) \geq 0, \quad \forall h_1, h_2, h_3 > 0; \tag{1.1.62}$$

if $f(x_1,x_2,x_3)$ is a continuous function on Δ, then *the triple Stieltjes integral* $\iiint_\Delta f(x_1,x_2,x_3) D_3 g(x_1,x_2,x_3)$ does exist. If the function $g(x_1,x_2,x_3)$ admits continuous partial derivatives of third order, then the triple Stieltjes integral is reduced to a *triple Riemann integral*

$$(S) \iiint_\Delta f(x_1,x_2,x_3) D_3 g(x_1,x_2,x_3)$$
$$= (R) \iiint_\Delta f(x_1,x_2,x_3) g'''_{x_1 x_2 x_3}(x_1,x_2,x_3) dx_1 dx_2 dx_3. \tag{1.1.63}$$

Hence, in the above-mentioned conditions, one may pass directly from the Stieltjes integral to the Riemann one, admitting the relation

$$D_3 g(x_1,x_2,x_3) = g'''_{x_1 x_2 x_3}(x_1,x_2,x_3) dx_1 dx_2 dx_3; \tag{1.1.64}$$

$D_3 g(x_1,x_2,x_3)$ represents here *the three-dimensional differential* of the function $g(x_1,x_2,x_3)$, while $g'''_{x_1 x_2 x_3}(x_1,x_2,x_3)$ is *the three-dimensional derivative* of this function. As in the case of the one- or two-dimensional Stieltjes integral, if $f(x_1,x_2,x_3)$ and $g(x_1,x_2,x_3)$ are distributions for which the product $f(x_1,x_2,x_3)$ $\cdot D_3 g(x_1,x_2,x_3)$ makes sense, then this interpretation is always possible. Introducing the Heaviside function, one can show that

$$(S) \iiint_\Delta f(x_1,x_2,x_3) D_3 \theta(x_1 - x_1^0, x_2 - x_2^0, x_3 - x_3^0) = f(x_1^0, x_2^0, x_3^0), \tag{1.1.65}$$

where $f(x_1,x_2,x_3)$ is a continuous function on Δ. Let be now the function

$$g(x_1,x_2,x_3) = \tilde{g}(x_1,x_2,x_3) + \sum_{k=1}^n g_k \theta\left(x_1 - x_1^{(k)}, x_2 - x_2^{(k)}, x_3 - x_3^{(k)}\right), \tag{1.1.66}$$

where g_k are constants, while $\tilde{g}(x_1,x_2,x_3)$ is *the continuous part* of this function; one may show that

$$(S) \iiint_\Delta f(x_1,x_2,x_3) D_3 g(x_1,x_2,x_3)$$
$$= (R) \iiint_\Delta f(x_1,x_2,x_3) g'''_{x_1 x_2 x_3}(x_1,x_2,x_3) dx_1 dx_2 dx_3 + \sum_{k=1}^n g_k f\left(x_1^{(k)}, x_2^{(k)}, x_3^{(k)}\right).$$
(1.1.67)

It results

$$D_3 g(x_1,x_2,x_3) = D_3 \tilde{g}(x_1,x_2,x_3) + \sum_{k=1}^n g_k D_3 \theta\left(x_1 - x_1^{(k)}, x_2 - x_2^{(k)}, x_3 - x_3^{(k)}\right);$$
(1.1.68)

we can also write

$$D_3 g(x_1,x_2,x_3) = \widetilde{D}_3 g(x_1,x_2,x_3)$$
$$+ \sum_{k=1}^n (\Delta g)_k \, \delta\left(x_1 - x_1^{(k)}, x_2 - x_2^{(k)}, x_3 - x_3^{(k)}\right) dx_1 dx_2 dx_3 ,$$
(1.1.69)

where we have put in evidence *the three-dimensional differentials in the sense of the theory of distributions* and *in the usual sense*, respectively, as well as *the three-dimensional jumps* $(\Delta g)_k$ of the function $g(x_1, x_2, x_3)$ at the points of discontinuity $\left(x_1^{(k)}, x_2^{(k)}, x_3^{(k)}\right)$, $k = 1, 2, ..., n$.

1.1.8 Mechanical systems

The mathematical models of bodies have frequently ideal characteristics; one may thus consider *material points* (without dimension), *material lines* (one-dimensional manifolds), or *material surfaces* (two-dimensional manifolds) in the space E_3. For example, a body whose dimensions are negligible with respect to the dimensions which appear in the considered problem (a planet with respect to the dimensions of the solar system, an electron with respect to the dimensions of the atom, or the homogeneous spheres considered in Subsec. 1.1.6) and for which the motion of rotation is immaterial admits as mathematical model a material point; the mechanical materialization of the abstract concept of material point is called also *particle*, denomination which we will use pre-eminently above all in what follows.

Hence, a *particle* (*material point*) is a geometric point to which we attach a finite mass $m > 0$. The mass of this point is indivisible, as well as its support.

Frequently, a thread can be considered to be a *material line* (a one-dimensional continuous medium); a membrane will also be modelled by a *material surface* (a two-dimensional continuous medium). The material line and the material surface represent a geometric line and a geometric surface, respectively, to which we attach a positive mass that may depend on one variable or on two variables, respectively.

In general, a *continuous medium* (a *continuous material*) will be a mathematical model of a body, formed by a domain (one-, two- or three-dimensional) to which we attach a positive mass (which can depend on one, two or three variables); all the quantities in connection with such a medium will be represented by continuous functions. The particle is a limit case of a continuous medium (the case in which the domain is reduced to a geometric point). In a computation, it is customary to refer to a point of a continuous medium, but this one is not a particle and has not a finite mass.

By a *mechanical system* $\mathscr{S}$ we mean a set of bodies, modelled as continuous media or as particles; hence, a mechanical system is formed, in the most general case, by a finite number of continuous media and a finite number of particles. Usually, we will consider *discrete mechanical systems* of particles (formed by a finite number of particles) or *continuous mechanical systems*. The main mechanical results which are obtained for a mechanical system have the same form either the system is discrete or continuous; hence, the demonstrations which we will give, as well as the computations we will make will refer to any mechanical systems. Various particular results will be put in evidence if they have a certain importance. In general, a mechanical system is in mechanical interaction with other mechanical systems; otherwise, we have to do with an *isolated mechanical system*.

Let Ω be the *geometric support* of a mechanical system $\mathscr{S}$; in case of a discrete system of particles, this support will be formed by a finite number of geometric points, while in the case of a continuous mechanical system this support will be a domain in which the body is immersed. Taking into account the property of additivity (1.1.25) and the definition of the integral, we may express the total mass (usually we will say only mass) of the mechanical system in the form

$$M = \int_\Omega \mathrm{d}m \, ; \tag{1.1.70}$$

the integral is a Stieltjes integral and the mass $m = m(\mathbf{r})$ is a distribution.

Differentiating in the sense of the theory of distributions, *the unit mass* (the mass of a unite volume, *the density*) is given by relation

$$\mathrm{d}m = \mu \mathrm{d}V \, , \tag{1.1.71}$$

where the element of volume to which the corresponding mass is related can be an

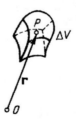

Figure 1.14. Element of volume.

element of line or of area; as well, $\mathrm{d}m$ may be the *three-*, *two-* or *one-dimensional differential* considered in the previous subsection. In general (Fig.1.14)

$$\mu = \mu(P;t) = \mu(\mathbf{r};t) \, , \tag{1.1.71'}$$

where the variation in time of the density is also put in evidence. Returning to formula (1.1.51), we may write

$$\mu(\mathbf{r}) = \tilde{\mu}(\mathbf{r}) + \sum_{i=1}^{n} m_i \delta(\mathbf{r} - \mathbf{r}_i), \qquad (1.1.71'')$$

where the sign "tilde" corresponds to *the differentiation in the usual sense*, $\tilde{\mu}$ is the density corresponding to a continuous system, m_i are the masses of n particles of position vectors $\mathbf{r}_i$, $i = 1, 2, ..., n$, while δ is *the Dirac distribution*.

The density of a particle of mass m and position vector $\mathbf{r}_0$ is thus given by

$$\mu(\mathbf{r}) = m\delta(\mathbf{r} - \mathbf{r}_0); \qquad (1.1.72)$$

in the case of a discrete mechanical system we have

$$\mu(\mathbf{r}) = \sum_{i=1}^{n} m_i \delta(\mathbf{r} - \mathbf{r}_i), \qquad (1.1.72')$$

while for a continuous mechanical system we obtain

$$\tilde{\mu}(\mathbf{r}) = \frac{\mathrm{d}m}{\mathrm{d}V}. \qquad (1.1.72'')$$

The function $\mu(\mathbf{r}) \geq 0$ is continuous at least partially; in general, if no confusion is possible, then we omit the sign tilde. Obviously, we can introduce also the notion of *mean density*

$$\mu_{\mathrm{mean}}(\mathbf{r}) = \frac{\Delta m}{\Delta V}, \qquad (1.1.72''')$$

where Δm is the mass of a finite volume element ΔV (Fig.1.14); there results

$$\mu(\mathbf{r}) = \lim_{\Delta V \to 0} \mu_{\mathrm{mean}}(\mathbf{r}). \qquad (1.1.72^{\mathrm{iv}})$$

The mass is given by

$$M = \int_{\Omega} \mu \, \mathrm{d}V \qquad (1.1.73)$$

for an arbitrary mechanical system, the integral being a *Stieltjes integral*; in case of a *discrete mechanical system*, we have

$$M = \sum_{i=1}^{n} m_i, \qquad (1.1.73')$$

while for a *continuous mechanical system* we may write

$$M = \int_{V} \mu(\mathbf{r}) \, \mathrm{d}V, \qquad (1.1.73'')$$

the integral being a *Riemann integral* ($\mu = \tilde{\mu}$).

In this last case, the density varies from one point to another one, and the continuous mechanical system is *non-homogeneous*; in the case in which the density μ is the same for all the points of the continuous medium, this one is *homogeneous* (besides, in this case all its mechanical and physical properties are the same in any points) and we may write

$$M = \mu V. \qquad (1.1.73''')$$

We remark that, in general, the results obtained for a discrete mechanical system may be enounced also for a continuous mechanical system by replacing the sign $\sum$ by the sign $\int$. Thus, two types of mathematical models of mechanics have been put in evidence, i.e. *the discrete model* and *the continuous model*. We mention that one can construct also other more sophisticated mathematical models, for example *quasi-continuous models* (continuous models which take in consideration the discrete structure of matter); the above general considerations remain valid also in this case. One can study also mathematical models which start from the discrete character of material bodies; with the aid of a sufficient great number of particles, they introduce statistical methods of computation, using *aleatory variables*.

1.1.9 The principle of mass conservation

The geometric support of a continuous mechanical system can vary in time ($V = V(t)$). The formula (1.1.73'') is of the form

$$M = \int_{V(t)} \mu(x_1, x_2, x_3; t) \, dV, \qquad (1.1.74)$$

where $x_j = x_j(t)$, $j = 1,2,3$, and $dV = dx_1 dx_2 dx_3$. By a change of variables of the form

$$x_i = x_i\left(x_1^0, x_2^0, x_3^0; t\right), \quad i = 1,2,3, \qquad (1.1.75)$$

the Jacobian J of the transformation being different from zero

$$J \equiv \det\left[\frac{\partial(x_1, x_2, x_3)}{\partial(x_1^0, x_2^0, x_3^0)}\right] \equiv \det\left[\frac{\partial x_i}{\partial x_j^0}\right] \neq 0, \qquad (1.1.75')$$

it follows that

$$M = \int_{V_0} \mu J \, dV_0, \qquad (1.1.74')$$

where $x_j^0 = x_j(t_0)$, $j = 1,2,3$, $V_0 = V(t_0)$, $dV_0 = dx_1^0 dx_2^0 dx_3^0$ correspond to the initial moment t_0. If we write the formula (1.1.74) for $t = t_0$, and if we take into account (1.1.74'), admitting that the function under the integral sign is continuous and choosing the domain V_0 arbitrarily, then the principle of mass conservation leads to

$$\mu J = \mu_0, \tag{1.1.76}$$

where $\mu = \mu(t)$ and $\mu_0 = \mu(t_0)$. This is *d'Alembert's condition of mass continuity*. One observes from (1.1.76) that μ is a *relative scalar*, i.e. a *scalar density*, the denomination given to such a quantity being thus justified.

If at any moment we have a relation of the form $J = 1$ or $\mu = \mu_0$, then *the motion is incompressible* (or *the body is incompressible*).

One may also put the condition of mass conservation (invariance), equating to zero the derivative of the mass (1.1.74') with respect to time; one obtains

$$\frac{dM}{dt} = \int_{V_0} \frac{d}{dt}(\mu J) dV_0 = \int_{V_0} \left(\frac{d\mu}{dt} J + \mu \frac{dJ}{dt} \right) dV_0 .$$

We notice that we may write (see Chap. 2, Subsec. 1.2.3)

$$\frac{dJ}{dt} = \frac{d}{dt} \det\left[\frac{\partial x_i}{\partial x_j^0}\right] = \frac{\partial}{\partial t} \det\left[\frac{\partial x_i}{\partial x_j^0}\right]$$

$$= \frac{1}{6} \epsilon_{ijk} \epsilon_{lmn} \frac{\partial}{\partial t}\left(\frac{\partial x_i}{\partial x_l^0} \frac{\partial x_j}{\partial x_m^0} \frac{\partial x_k}{\partial x_n^0}\right)$$

$$= \frac{1}{6} \epsilon_{ijk} \epsilon_{lmn} \left(\frac{\partial v_i}{\partial x_l^0} \frac{\partial x_j}{\partial x_m^0} \frac{\partial x_k}{\partial x_n^0} + \frac{\partial x_i}{\partial x_l^0} \frac{\partial v_j}{\partial x_m^0} \frac{\partial x_k}{\partial x_n^0} + \frac{\partial x_i}{\partial x_l^0} \frac{\partial x_j}{\partial x_m^0} \frac{\partial v_k}{\partial x_n^0}\right),$$

where we used the formula (2.1.36"), corresponding to the development of a determinant; if we interchange indices i and j and indices l and m with each other, respectively, in the second sum, and indices i and k, and indices l and m with each other, respectively, in the third sum, and if we take into consideration the properties of Ricci's tensor, then we obtain

$$\frac{dJ}{dt} = \frac{1}{2} \epsilon_{ijk} \epsilon_{lmn} \frac{\partial v_i}{\partial x_l^0} \frac{\partial x_j}{\partial x_m^0} \frac{\partial x_k}{\partial x_n^0} = \frac{1}{2} \epsilon_{ijk} \epsilon_{lmn} \frac{\partial v_i}{\partial x_p} \frac{\partial x_p}{\partial x_l^0} \frac{\partial x_j}{\partial x_m^0} \frac{\partial x_k}{\partial x_n^0},$$

where we expressed the components of the velocity of a point of the mechanical system as functions of the co-ordinates x_j, $j = 1,2,3$. Taking now into account the relation (2.1.37'), it follows that

$$\frac{dJ}{dt} = \frac{\partial v_i}{\partial x_p} J \delta_{ip} = J \frac{\partial v_i}{\partial x_i},$$

where we have introduced the velocity components of a point of the mechanical system

Newtonian model of mechanics

$$v_j = \frac{\mathrm{d}x_j}{\mathrm{d}t} = \frac{\partial x_j}{\partial t} = \dot{x}_j, \quad j = 1, 2, 3; \tag{1.1.77}$$

we obtain thus *Euler's formula* (here and in what follows, see App., §2)

$$\frac{\mathrm{d}J}{\mathrm{d}t} = J \operatorname{div} \mathbf{v}, \tag{1.1.78}$$

which is verified in any point of a continuous mechanical system and by any particle in motion; this relation may be written also in the form

$$\frac{\mathrm{d}}{\mathrm{d}t} \ln J = \operatorname{div} \mathbf{v}. \tag{1.1.78'}$$

Returning to the derivative of the mass M with respect to time, which we equate to zero, we can write

$$\frac{\mathrm{d}M}{\mathrm{d}t} = \int_{V_0} \left(\frac{\mathrm{d}\mu}{\mathrm{d}t} + \mu \operatorname{div} \mathbf{v} \right) J \mathrm{d}V_0 = 0,$$

where the volume V_0 is arbitrary. If the function under the integral sign is continuous, then we may write *the condition of mass continuity in Euler's form*

$$\frac{\mathrm{d}\mu}{\mathrm{d}t} + \mu \operatorname{div} \mathbf{v} = 0. \tag{1.1.79}$$

Observing that

$$\frac{\mathrm{d}\mu}{\mathrm{d}t} = \frac{\partial \mu}{\partial t} + \operatorname{grad} \mu \cdot \frac{\mathrm{d}\mathbf{r}}{\mathrm{d}t} = \frac{\partial \mu}{\partial t} + \mathbf{v} \cdot \operatorname{grad} \mu,$$

$$\operatorname{div}(\mu \mathbf{v}) = \mu \operatorname{div} \mathbf{v} + \mathbf{v} \cdot \operatorname{grad} \mu,$$

we obtain the condition (1.1.79) also in the form

$$\frac{\partial \mu}{\partial t} + \operatorname{div}(\mu \mathbf{v}) = 0. \tag{1.1.79'}$$

If the density is constant at any time, we obtain

$$\operatorname{div} \mathbf{v} = 0, \tag{1.1.78''}$$

hence, the field of velocities is *solenoidal* (the velocities form a *field of curls*). The relation (1.1.79) can be expressed also in the form

$$\frac{\mathrm{d}}{\mathrm{d}t} \ln \mu + \operatorname{div} \mathbf{v} = 0. \tag{1.1.79''}$$

The relations (1.1.78') and (1.1.79") lead to

$$\frac{\mathrm{d}}{\mathrm{d}t}\ln(\mu J) = 0;$$

hence, the product μJ is constant at any moment, being equal to μ_0 (corresponding to the initial state). We return thus to the condition of mass continuity given by d'Alembert. The forms in which the principle of mass conservation may be expressed are thus obtained.

The transformation relation (1.1.75) puts in evidence the position vector $\mathbf{r}_0(x_1^0, x_2^0, x_3^0)$, corresponding to *the initial position* $t = t_0$, and the position vector $\mathbf{r}(x_1, x_2, x_3)$, corresponding to *the actual position* (an arbitrary moment t); this relation may be written in the vector form

$$\mathbf{r} = \mathbf{r}(\mathbf{r}_0; t). \qquad (1.1.75")$$

If the continuous mechanical system $\mathscr{S}$ is immersed into the domain D of the volume V in the actual state, then we admit that one may pass continuously from D_0 to D, by means of the geometric transformation (1.1.75); we admit also that the inverses of these functions

$$x_i^0 = x_i^0(x_1, x_2, x_3), \quad i = 1, 2, 3, \qquad (1.1.75''')$$

are univocally determined (in particular, $x_i^0 = x_i(x_1^0, x_2^0, x_3^0; t_0)$). The functions (1.1.75''') are univocally determined, at least in a neighbourhood of the considered point, if and only if the functions (1.1.75) are of class $C^1(D)$ and the Jacobian (1.1.75') does not vanish in this domain. This hypothesis is known as the *continuity axiom*, which expresses the *indestructibility* of matter. A domain to which corresponds a finite positive volume cannot be transformed in a domain of zero or infinite volume; this implies also the *impenetrability* of matter. The motion (1.1.75) or (1.1.75''') transforms each domain in a domain, each surface in a surface and each curve in a curve. We admit that the domain D_0 is limited by a sufficiently regular surface S_0, which verifies, for instance, conditions of the Lyapunov type. The equations (1.1.75) lead to the surface S, which is the frontier of the domain D and the image of S_0 by this transformation; this surface also verifies conditions of Lyapunov's type.

We admit that the variables x_i^0, $i = 1, 2, 3$, corresponding to the domain D_0 (in the initial state), are independent variables, and call them *Lagrange co-ordinates* (or *material co-ordinates*); this denomination was given because, in the phenomenon to study, a point of the continuous mechanical system is followed in its motion. They are *referential co-ordinates*; indeed, they give the position of the point with respect to the initial state, considered as a frame of reference. In such a formulation, the co-ordinates of the domain D (the material continuum in the actual state) are the unknown functions of the problem, which must be determined. As well, we may consider variables x_i, $i = 1, 2, 3$, as independent variables, calling them *Euler co-ordinates* (or *spatial co-ordinates*); one has given this denomination because, at a point in space, there are observed points of the continuous mechanical system which pass through this spatial position in their motion. The unknown functions can be, for instance, the components of the velocity at this point.

1.1.10 Classification of bodies

We have seen at Subsec. 1.1.8 that bodies can be modelled as mechanical systems (discrete or continuous). In the case in which the mutual distances of all pairs of points of a mechanical system remain invariant in time, we say that we have to do with a *non-deformable mechanical system* (in particular, a *rigid solid*, which is a non-deformable continuous medium); otherwise, we have to do with a *deformable mechanical system* (in particular, a *deformable continuous medium*).

The deformable continuous media may be *deformable solids* (which change little their form under external loads, for instance under the action of forces) or *fluids* (which change very much their form, taking the form of the surrounding bodies, under the action of external loads). The fluids may be *liquids* (which change little their volume) or *gases* (which change very much their volume, being expansive). The boundaries of these classical types of continuous deformable media cannot be perfectly settled (as one could expect from the definitions given above), because there exist media which have simultaneously, for instance, properties of a solid body as well as properties of a liquid one (in the same physical conditions).

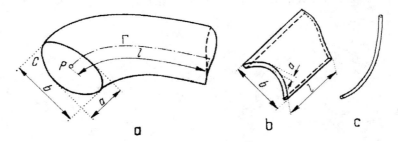

Figure 1.15. Curved bar (a). Thin walled bar (b). Thread (c).

In what concerns the deformable solids, one can put in evidence the properties of *elasticity* (if the causes which have led to the deformation of the solid cease their action, then this one returns to its initial form; the deformation is reversible), *plasticity* (if, after moving off the causes which have led to the phenomenon of deformation, one obtains a permanent deformation) and *viscosity* (variation of the deformation as a function of time); we mention thus elastic solids, elastoplastic solids, viscoelastic solids, elastoviscoplastic solids a.s.o. The fluids can have also properties of elasticity and viscosity, justifying thus the affirmation made above concerning bodies which can have simultaneously properties corresponding to the two states of aggregation.

We mention also that a fourth state of aggregation of matter has been put in evidence: *the plasma*. This is a state of the matter at high temperature; it is piecewise electroconductive and in average neutral. Matter is thus a mixture of particles (some of them have a positive or negative electric charge, and other ones are neutral), which are directed along the force lines of a magnetic field. The plasma is thus a deformable and electroconductive continuous (or discrete) medium (for instance, a gas piecewise or totally ionized).

If the body does not change its volume during the motion, then we say that it is incompressible; this is an approximation of the physical reality. In general, the bodies are compressible. We make now a classification (a modelling) of the deformable solids according to their dimensions.

Thus, the bodies which have a dimension (the length) much more longer than the other two dimensions (corresponding to the cross section) are called *bars*. We will give a constructive definition of this notion. Let be a curve Γ (*the axis of the bar*) of finite length l. In the plane normal to Γ at the point $P \in \Gamma$, let us consider a closed curve C, bounding a plane domain D (*the cross section* of the bar); we suppose that the centre of gravity of the domain D is at the point P. If the point P describes the curve Γ, then the curve C, which may be deformable, generates a surface, the boundary of a three-dimensional domain, support of a body called *bar* (Fig.1.15,a). Corresponding to its axis, a bar may be *straight* or *curved*; the last one can be a *three-dimensional curved bar* or a *plane curved bar*. The two (mean) dimensions a and b of the cross-section are considered of the same order of magnitude, with the condition $a, b \ll l$. If all three dimensions are of different order of magnitude ($a \ll b \ll l$), we have to do with *thin walled bars* (Fig.1.15,b). Finally, if the two dimensions of the cross-section are completely negligible with respect to the length of the bar, so that this one be perfectly flexible and torsionable (without opposition to bending or torsion), then we obtain a *thread* (Fig.1.15,c).

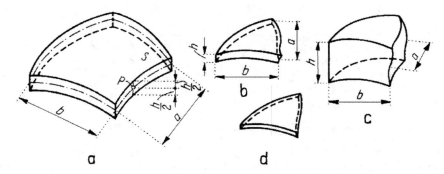

Figure 1.16. Plate (a). Thin plate (b). Plate of mean thickness (or thick plate) (c). Membrane (d).

The body, a dimension (the thickness) of which is much smaller than the other two dimensions (corresponding to the middle surface) is called a *plate*. Let be given an arbitrary surface S (*the middle surface* of the plate) of finite area. Along the normal to S at a point $P \in S$, let us consider a segment of length h (*the thickness* of the plate), the middle of which is on S, at the same point P. If the point P describes the surface S, then the extremities of this segment, which may be of variable length, generate two surfaces, boundary of a three-dimensional domain, support of a body called *plate* (Fig.1.16,a). Corresponding to the form of the middle surface, the plates may be *plane* (denominated simply *plates*) or *curved* (denominated *shells*). The ratio between the (mean) dimensions a and b of the middle surface and the thickness h of the plate leads to: *thin plates* (for $h \ll a,b$) (Fig.1.16,b), *plates of mean thickness* and *thick*

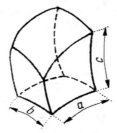

Figure 1.17. Block.

plates (for which $h < a,b$) (Fig.1.16,c). We observe that one cannot make a very good delimitation between these categories of plates; they depend on the possibilities of computation of the states of stress and strain, and can be different from case to case. Finally, if the thickness h is completely negligible with respect to the other two dimensions, so that the plate is perfect flexible, then we obtain a *membrane* (Fig.1.16,d).

The bodies for which all three dimensions a, b and c are of the same order of magnitude are called *blocks* (Fig.1.17).

Generally, the real bodies have as supports finite three-dimensional domains. But we will consider also infinite domains, the study of which present a special interest; indeed, such domains are ideal models for real cases which one may often encounter in practice or may be used as intermediary steps in solving problems for other domains.

1.1.11 Force. Field. Classification and representation of forces. Internal forces

Another element appearing in the mathematical modelling of a mechanical phenomenon is the cause which yields the mechanical motion, being mentioned in previous subsections. From a mechanical point of view, bodies are acting one on the other and often it is difficult to put in evidence the physical nature of such an action; in general, it received the name of *force*. This denomination, resulting from the action of the human being on the external world, gets a special meaning in mechanics; it is a measure for the transmission of motion. The origin of forces is not always of mechanical nature, but the goal of mechanics is not to determine their nature; mechanics admits that forces are given.

Taking into account the property of inertia of a material point, its natural state is the state of rest or uniform rectilinear motion; the force is the cause which is changing this state. It appears thus the connection between the force and the variation of the velocity (or of the momentum). We have seen in Subsec. 1.1.6 that the momentum is a measure of the mechanical motion, remaining constant if the mechanical system is not acted upon by another mechanical system (e.g., the case represented by relation (1.1.26)); if the mechanical system is subjected to such an action, then appears a variation of the momentum, hence a variation of the velocity (an acceleration). Thus, the force **F** applied to a point of the mechanical system is a vector quantity (as well as the variation of the momentum to which it corresponds). From the experiments, it follows that – in the most general case – the force may be modelled by a bound vector of the form

$$\mathbf{F} = \mathbf{F}(\mathbf{r}; \mathbf{v}; t), \qquad (1.1.80)$$

where $\mathbf{r}$ is the position vector of the point of application (belonging to the mechanical system), while $\mathbf{v}$ is its velocity; in components, we have

$$F_i = F_i(x_1, x_2, x_3, v_1, v_2, v_3, ; t), \quad i = 1,2,3. \qquad (1.1.80')$$

The force corresponds to the interaction of bodies and is emphasized in various manners; thus, we distinguish between *contact actions* and *actions at distance*. The action of a homogeneous sphere, which is in collision with another homogeneous sphere (case considered in Subsec. 1.1.6), is a contact action. The force by which the Earth acts upon material bodies, in such a way that they fall on it, represents an action at distance.

Such an action at distance is called a *field*. In general, a field is a domain D of the space, which constitutes a zone of influence of a mechanical system $\mathscr{S}$ or of a certain mechanical phenomenon; in the presence of $\mathscr{S}$, the domain D acquires a special state, so that the evolution of any other mechanical system, which has properties permitting to be influenced by D and passing through D, is changed. For instance, by its property of mass or of electrical charge, a body generates a gravitational or an electric field, respectively; any other body, which passes through such a field, is influenced in its motion. In a continuous model of mechanics, the action of such a field upon a mechanical system is characterized by a force; it is supposed that this action is instantaneously propagated (otherwise, one must introduce a factor of delay in the mathematical modelling of the phenomenon). So, we do not introduce in computation the real action, which took place with a time lag. In a non-classical model, the hypothesis of propagation step by step is used; indeed, the experience shows that no action is propagated instantaneously. In fact, the hypothesis of the instantaneous propagation corresponds to the hypothesis of the universal (absolute) time.

Modelling the force by a bound vector (1.1.80) (it is characterized by direction, intensity (modulus) and point of application, properties corresponding to an intrinsic definition of such a mathematical entity), this one will enjoy all the properties of bound vectors (for instance, the summation of vectors). So, the action of two forces applied at a point of the mechanical system $\mathscr{S}$ can be replaced by the action of only one force applied at the same point, along the diagonal of the parallelogram formed by the mentioned forces; the reciprocal of this affirmation is, obviously, true: the action of a force at a point can be replaced by the action, at the very same point, of two forces representing the sides of the parallelogram, the diagonal of which is the given force.

Attaching to each point (of position vector $\mathbf{r}$) of a domain D a vector $\mathbf{F}$, at any moment, we obtain a *field of forces* of the form (1.1.80). If $\mathbf{F}$ does not depend explicitly on t $(\partial \mathbf{F}/\partial t = \dot{\mathbf{F}} = \mathbf{0})$, then the field of forces is called *stationary*. In fact, for any field and any mechanical phenomenon, which have this property, the same denomination is used.

Because the forces are modelled by bound vectors, we may consider also *systems of forces* (discrete systems) in action upon a mechanical system (discrete or continuous), which will be modelled by systems of bound vectors; obviously, all the results obtained for these systems of vectors may be used for the corresponding systems of forces. Let $\mathscr{F} \equiv \{\mathbf{F}_i, i = 1,2,...,n\}$ be such a system of forces; it corresponds rigorously to a

Newtonian model of mechanics 39

discrete mechanical system. In the case of a continuous mechanical system, the support of which is the domain D, the system $\mathscr{F}$ represents only an approximate modelling of the physical reality. If the subdomain $\mathscr{D} \subset D$, to which the load is transmitted, has negligible dimensions with respect to the dimensions of the domain D, its localization can be punctual; in this case, the load is called *concentrated* (or *punctual*). If at least one of the dimensions of the subdomain $\mathscr{D}$ cannot be neglected, the load is called *distributed* (or *continuous*); one considers that it is transmitted continuously to the respective subdomain. We may thus define *linear, superficial* or *volume loads*, the intensity of which is proportional to the elements of line, area or volume, respectively. A distributed load constitutes a vector field, the geometric support of which is the subdomain to which the load is transmitted.

The forces acting upon a mechanical system may be *external forces*, which can represent the action of other systems on the given one (eventually the action of systems of non-mechanical nature, e.g., temperature fields, electromagnetic fields, fields of radiation etc.), or *internal forces*, which represent the action of certain points of the given system upon other points of the same system (these forces can also be of non - mechanical nature). Let be the points P_i and P_j of the given system; the force $\mathbf{F}_{ij}$ will represent the action of the point P_j upon the point P_i, having the support $P_i P_j$ ($\mathbf{F}_{ij} = \lambda \overrightarrow{P_i P_j}$, λ scalar). We introduce also the internal force $\mathbf{F}_{ji}$, defined analogously; these forces appear always as a pair and are linked axiomatically by the relation (Fig.1.18)

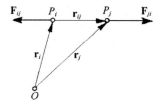

Figure 1.18. Internal forces $\mathbf{F}_{ij} = -\mathbf{F}_{ji}$.

$$\mathbf{F}_{ij} + \mathbf{F}_{ji} = \mathbf{0}. \qquad (1.1.81)$$

1.1.12 Conservative forces

Upon a mechanical system (discrete or continuous) can act a field of forces, which may be conservative or non-conservative. *The conservative forces* (which derive from a *potential*) are modelled by conservative vectors, expressed in the form

$$\mathbf{F} = \operatorname{grad} U = \nabla U = U_{,j} \mathbf{i}_j, \quad F_j = U_{,j}, \qquad (1.1.82)$$

where $U = U(\mathbf{r}) = U(x_1, x_2, x_3)$ is *the force function* (*potential function* or *potential*). One can have $U = U(\mathbf{r};t) = U(x_1, x_2, x_3;t)$ in the representation (1.1.82); the function is called *quasi-potential*, and the forces are *quasi-conservative* in this case.

A field of conservative forces is *stationary*, while a field of quasi-conservative forces is *non-stationary*. We observe that

$$\text{curl } \mathbf{F} = \mathbf{0}, \qquad (1.1.82')$$

so that a field of conservative or quasi-conservative forces is *irrotational*. As an example of conservative field of forces we may consider the *terrestrial gravitational field* of the form

$$\mathbf{G} = m\mathbf{g} = \text{grad}(-mgx_3 + \text{const}), \qquad (1.1.83)$$

where m and $\mathbf{G}$ are the mass and the weight of a particle, respectively, while $\mathbf{g}$ is the *gravity acceleration*; the axis Ox_3 is along the gravity acceleration, but in the opposite direction of it. Galileo proved experimentally that, in the same place on the Earth, all the bodies have the same acceleration $\mathbf{g}$ by a free falling in vacuum; but this acceleration depends on the place in which this measurement is made (the latitude), taking into account the fact that this place can be closer (at the poles) or further away (at the equator) from the centre of the Earth (the vector $\mathbf{g}$ is considered approximately directed towards this centre, but – in reality – it is directed along the vertical of the place; the greatest deviation is given by an angle of approximate six minutes with respect to this vertical at the 45° parallel). The magnitude (modulus) of this acceleration is $g \cong 9,781 m/s^2$ at the equator and $g \cong 9,831 m/s^2$ at the poles, at the sea table; experimentally, at the University of Bucharest it was obtained $g \cong 9,806 m/s^2$, corresponding approximately to the 45° parallel.

The formula (1.1.72^{iv}) allows us to introduce *the mean unit weight* and *the unit weight*, respectively, in the form

$$\gamma_{\text{mean}} = \mu_{\text{mean}} g, \quad \gamma = \lim_{\Delta V \to 0} \gamma_{\text{mean}} = \mu g, \qquad (1.1.83')$$

and we may use all the considerations of Subsec. 1.1.8.

Another important example is given by the field of *Newtonian attraction forces*; Newton admitted that two particles P_i and P_j of masses m_i and m_j, respectively, are acted upon by an attraction force

$$\mathbf{F} = \mathbf{F}_{ji} = -\mathbf{F}_{ij} = -f \frac{m_i m_j}{r_{ij}^2} \text{ vers } \mathbf{r}_{ij} = -f \frac{m_i m_j}{r_{ij}^3} \mathbf{r}_{ij}$$

$$= \text{grad}\left(f \frac{m_i m_j}{r_{ij}} + \text{const} \right), \qquad (1.1.84)$$

where $\mathbf{r}_{ij} = \overrightarrow{P_i P_j}$; $\mathbf{F}_{ij}$ and $\mathbf{F}_{ji}$ are internal forces, which verify the relation (1.1.81). Observing that $\mathbf{r}_{ij} = \mathbf{r}_j - \mathbf{r}_i$, it is obvious that $\mathbf{F}_{ji} = \mathbf{F}_{ji}(\mathbf{r}_j)$ and $\mathbf{F}_{ij} = \mathbf{F}_{ij}(\mathbf{r}_i)$ (the second position vector $\mathbf{r}_i$ or $\mathbf{r}_j$, respectively, playing the rôle of a parameter) in the

Newtonian model of mechanics 41

calculus of the gradient in (1.1.84); if $P_i \equiv O$ (the corresponding particle having the mass M), and $\mathbf{r}_{ij} = \mathbf{r}_j = \mathbf{r}$ (the particle $P_j \equiv P$ being of mass m), we may write

$$\mathbf{F} = -f\frac{mM}{r^2}\text{vers }\mathbf{r} = -f\frac{mM}{r^3}\mathbf{r} = \text{grad}\left(f\frac{mM}{r} + \text{const}\right), \qquad (1.1.84')$$

and the force $\mathbf{F}$ is applied at the point P. The coefficient $f > 0$ is a *universal constant* (sometimes denoted by G), which can be obtained experimentally in a particular case; such a determination has been made for the first time by Henry Cavendish in 1798, using the balance of torsion of Coulomb. In the CGS system, $f \cong 6.673 \cdot 10^{-8}$ cm/g·s². This is a very small value; the Newtonian attraction force is not practically perceptible for bodies on the Earth surface (their masses are relative small); this force has an intensity, which must be taken into consideration in case of bodies with great masses (the case of cosmic bodies). As well, this force has an intensity which cannot be neglected in case of a gravitational field, if one of the masses is great; thus, the terrestrial gravitational field can be considered as been generated by the Newtonian attraction forces, between bodies on the Earth surface and the Earth itself. Formulae (1.1.83) and (1.1.84') lead to

$$\mathbf{G} = -f\frac{mM}{(R+r)^2}\text{vers }\mathbf{r} = -mg\text{ vers }\mathbf{r},$$

where $\mathbf{G}$ is the weight of a material body of mass m, at the distance r from the Earth surface (along the vertical of the place, neglecting the deviation mentioned above), while R is the distance from the Earth surface to the centre (we admit that the Earth is approximately spherical, its radius being R) of the Earth, the mass of which is M; we may thus write

$$fM = gR^2\left(1 + \frac{r}{R}\right)^2 = gR^2\left[1 + 2\frac{r}{R} + \left(\frac{r}{R}\right)^2\right].$$

But $r \ll R$, so that we may neglect the ratio r/R with respect to unity; finally, we obtain

$$fM \cong gR^2. \qquad (1.1.85)$$

The forces which arise between two *electric charges* q_i and q_j (attraction if they are of contrary sign and repulsion in case of the same sign) have the same structure as the forces (1.1.84) (*Coulomb's law*)

$$\mathbf{F} = \mathbf{F}_{ji} = -\mathbf{F}_{ij} = k\frac{q_i q_j}{r_{ij}^2}\text{vers }\mathbf{r}_{ij} = k\frac{q_i q_j}{r_{ij}^3}\mathbf{r}_{ij} = \text{grad}\left(-k\frac{q_i q_j}{r_{ij}} + \text{const}\right),$$

$$(1.1.84'')$$

where $k = \varkappa/4\pi\varepsilon_0$, ε_0 being *the electric permittivity of vacuum* and $\varkappa$ *the rationalization factor* ($\varkappa = 1$ in rationalized units, in the SI system (see Subsec. 2.1.3), while $\varkappa = 4\pi$ in non-rationalized units).

In general, any force (we use formula (A.2.16))

$$\mathbf{F} = F(r)\text{vers }\mathbf{r} = \text{grad}\int_{r_0}^{r} F(\rho)\mathrm{d}\rho, \qquad (1.1.86)$$

exerted by a pole O upon a particle P and the intensity of which depends only on the distance r is conservative; such forces are encountered, for instance, in the kinetic theory of gases. The forces of the form (1.1.86) are called *central forces*, because their supports pass through the pole O (in fact, it is a particular case of central forces; an arbitrary central force, for which $F = F(r,\theta)$, in polar co-ordinates, because in such a case the trajectory of the particle is a plane curve, is – in general – no more a conservative force). The force is attractive or repulsive as we have $F \lessgtr 0$; the Newtonian attraction force (1.1.84') represents a particular case. Another particular case is obtained for $\mathbf{F}(\mathbf{r}) = -k\mathbf{r}$, $k > 0$. One obtains thus an *attractive elastic force* of the form

$$\mathbf{F} = -k\mathbf{r} = \text{grad}\left(-\frac{1}{2}kr^2 + \text{const}\right), \quad k > 0; \qquad (1.1.87)$$

if we take the sign plus ($\mathbf{F} = k\mathbf{r}$, $k > 0$), then we have a *repulsive elastic force*. Such forces appear in case of a linear elastic spring and are used for the modelling of a linear elastic continuous medium; as well, they can lead to linear oscillations, which allow the construction of the simplest models of atoms.

The constants in the formulae (1.1.83), (1.1.84) – (1.1.84"), (1.1.87) and the constant r_0 in the formula (1.1.86) are determined by a given value of the potential U (which may – eventually – be equal to zero) at a given point.

Besides the forces given by (1.1.82), we may consider also quasi-conservative or conservative forces, the components of which are expressed in the form

$$F_j = U_{,j} - \frac{\mathrm{d}}{\mathrm{d}t}\left(\frac{\partial U}{\partial \dot{x}_j}\right), \quad j = 1,2,3; \qquad (1.1.88)$$

in this case, the function $U = U(x_1, x_2, x_3, \dot{x}_1, \dot{x}_2, \dot{x}_3; t)$ is called a *generalized quasi-potential* or a *generalized potential*, as it depends or not explicitly on time. Such forces appear in the motion of the electron in an electromagnetic field (*Lorentz's forces*). Supposing that the force $\mathbf{F}$ does not depend explicitly on the acceleration $\ddot{\mathbf{r}}$, as it will be seen in Subsec. 1.2.1, it follows that the generalized quasi-potential U must have a linear dependence with respect to the components of the velocity

$$U = U_j\dot{x}_j + U_0, \quad U_j = U_j(x_1, x_2, x_3; t), \quad j = 1,2,3,$$
$$U_0 = U_0(x_1, x_2, x_3; t). \qquad (1.1.88')$$

Newtonian model of mechanics 43

Unlike the generalized potential (quasi-potential), a potential (quasi-potential) is called also a *simple potential* or a *simple quasi-potential*. A system of forces (corresponding – for instance – to a field of forces) which admits a simple potential (quasi-potential) or a generalized potential (quasi-potential) is called also a *natural system*. The forces which are not conservative are called *non-conservative forces*.

1.1.13 Elementary particle interactions

If a mechanical system $\mathscr{S}$ is subjected only to the action of internal forces (e.g., the solar system, considered as independent of the other systems of the Galaxy, or a system of elementary particles), then the respective system is closed (neither inputs, nor outputs take place). The physical phenomena which occur in nature can be, in general, described by *interactions* between particles (we consider, in what follows, *the elementary particles*, which are basic constituents of the matter).

Corresponding to the experimental data, till now only four basic (elementary) interactions (which, in fact, are forces) have been encountered, i.e.: the gravitational interaction, the weak interaction, the electromagnetic interaction and the strong (nuclear) interaction, which determine all the forces known in the Universe. An interaction is characterized by its *intensity* (expressed by a number) with respect to one of them taken as unit, and determined by the properties of the particles which they transport.

The gravitational interaction maintains the Earth as a whole, as well as the Sun and the planets in the solar system, and connects the stars in the frame of the Galaxy; the intensity of this interaction is $GM^2/\hbar c \cong 10^{-39}$, where $G = f$ is the universal gravitational constant, M is the mass of the nuclear particle, c is the velocity of light in vacuum, and $\hbar = h/2\pi$, h being *Planck's constant*. This interaction is the weakest of all interactions; it acts between all particles and is exerted at any distances (small, great or very great, theoretically infinite), being susceptible to be propagated by a quantum of vanishing mass, called *graviton*. The corresponding forces are the only *attraction forces* between particles at great distances.

The weak interaction is an interaction of contact (at a small distance of approximate 10^{-15} cm, smaller than the nuclear dimensions which are of approximate 10^{-13} cm), its intensity being 10^{-9} (its intensity in an elementary disintegration volume is of approximate 10^9 times smaller that the intensity of the strong interaction). *The vector bosons* are carrying agents of those interactions, which are exerted between *leptons* and *quarks*, the corresponding forces between identical particles being *repulsive*; thus, the weak interaction cannot form stable states. In the absence of a conservation law of *barions* (heavy particles), the weak interaction could lead to the decay as electrons and neutrons of matter in the Universe, in an interval of time less than 10^{-3} s.

The electromagnetic interaction is an interaction at distance between all particles with electric charge or/and magnetic moment. It ensures the existence of atoms, molecules and crystalline systems as stable systems. The intensity of this interaction is given by *the constant of fine structure* $\alpha = e^2/4\pi\varepsilon_0\hbar c$, where e is *the elementary charge* and ε_0 is *the dielectric constant* of vacuum; we have $\alpha \cong 7.297351 \cdot 10^{-3}$ $\cong 1/137.030602 \cong 1/137$, its intensity being 10^{-2}. *The photons*, which can be coupled with any charged particle, are carrying agents of those interactions, the corresponding forces between identical particles being *repulsive*; as the gravity forces, these ones have a large radius of action and may be studied in the frame of quantum electrodynamics.

The strong (nuclear) interaction leads to the building of all elements, by linking the nuclear particles, and is acting at a small distance upon quarks, the carrying agent being *the gluons*. It corresponds a *repulsive force*, the intensity of which is – conventionally – taken equal to unity, with respect to the intensity of the other forces (the interactions have been presented in order of growth of their intensities).

The idea of *unifying* the four basic interactions, so that to use a single set of equations to predict all their characteristics, is old and, at present, it is not known whether such a theory can be developed. However, the most successful attempt in this direction is *the electroweak theory* (*the Weinberg-Salam model*) proposed during the late 1960s by Steven Weinberg, Abdus Salam, Samuel Glashow (Nobel prize, 1979). The carrying agents are *the vector bosons*, with electric charge $W^\pm$, and *the neutral vector boson* Z^0. In the frame of this model, the masses $m_W = (82 \pm 2.4)$ GeV and $m_Z = 75$ GeV (Gigaelectronvolts) have been theoretically calculated. In December 1982, the mass of the vector boson has been determined experimentally, its value being $m_W = (85 \pm 5)$ GeV, in excellent agreement with the theoretical prediction of the Weinberg-Salam model. The measures have been effected at CERN (Switzerland) and represent the greatest discovery of this important European centre of research in the area of physics. In June 1983, the existence of *the third vector boson* Z^0 has been confirmed experimentally. Recent experimental values of the masses of vector bosons are $m_W = (80.423 \pm 0.0039)$ GeV and $m_Z = (91.1876 \pm 0.0021)$ GeV.

The electroweak theory is *a unified theory* of electromagnetic and weak interactions, based on the SU(2)×U(1) symmetry. It regards the weak force and the electromagnetic force as different manifestations of a new fundamental force (electroweak), similarly to electricity and magnetism that appear as different aspects of the electromagnetic force. We mention that the electroweak theory is – in essence – *a gauge theory*.

We may infer that there are only *three fundamental interactions* in Nature: *gravitational, electroweak* and *strong (nuclear)*. Further, one tries to unify these three interactions in the frame of a *unitary theory*, reducing – eventually in a first stage – their number to two; in this order of ideas, theories which try to link the electroweak interaction to the strong one (e.g., *Giorgi's theory*) have been developed. As well, *the supergravitation* describes gravitational phenomena in the frame of a quantum theory of field; the problem is very difficult also because the gravitational interaction, which plays the most important rôle in the study of mechanical systems, is the only interaction which leads only to forces of attraction (described, e.g., by the Newtonian model). Certain researchers assume the existence of *a fifth type* of basic interaction too.

Now, the theory of electroweak interactions is a component of *the Standard Model*, that includes also the theory of strong interactions; but it seems that the current models of strong interactions need to be revised.

However, the recent discovery of *exotic quark systems* suggests that the Nature is much more complicated; thus, in July 2003, nuclear physicists in Japan, Russia and the USA have discovered *the pentaquark*, and in November 2003, a new subatomic particle has been discovered, while Daniel Gross, David Politzer and Frank Wilczek have been awarded Nobel prize, 2004.

In conclusion, stress is put to set up new models for unifying the three remaining basic interactions.

1.2 Mathematical model of mechanics

By means of the fundamental notions introduced above, one can formulate – in a new form – Newton's principles, which constitute the basis of the mathematical model of mechanics. The object of study of this science of nature may be thus emphasized. A short history is also given.

1.2.1 Newton's principles

Continuing the results obtained by his predecessors, especially those of Galileo Galilei (who intuits the principle of inertia and the principle of the initial conditions), Sir Isaac Newton enounced, in his famous work "Philosophiae Naturalis Principia Mathematica" (the first fascicle appeared in London at 5th of July 1686), the three laws which form the basis of classical models of mechanics.

Lex I. *Corpus omne perseverare in statu suo quiscendi vel movendi uniformiter in directum, nisi quantenus illud a viribus impressis cogitur statum suum mutare* (Any body preserves its state of rest or of uniform rectilinear motion if it is not constrained by induced forces to change its state) (after the last enunciation of Newton, in the third edition (1726) of his treatise).

Lex II. *Mutationem motus proportionalem esse vi motrici impresse et fieri secundum lineam rectam, qua vis illa imprimitur* (The variation of motion (of the quantity of motion) is proportional to the induced moving force and is directed along the straight line, which is the support of this induced force).

Lex III. *Actioni contrariam semper et aequalem esse reactionem, sive corporum duorum actiones in se mutuo semper esse aequales et in partes contrarias dirigi* (The reaction is always opposite and equal to the action or the reciprocal actions of two bodies are always equal and directed in contrary directions).

In *Newton's laws* (principles) (called sometimes the *Galileo-Newton principles*, because of the important contribution of Galileo to the building of the classical model of mechanics) by body we understand a particle (material point). As well, we admit the existence of an inertial frame of reference with respect to which these principles are verified and the motion is described. In these conditions, we may enounce the principles of classical mechanics in the following new form:

i) **Principle of inertia.** *A particle, which is not subjected to the action of any force, is in a uniform and rectilinear motion or is in rest.*

The motion or the rest are considered with respect to the inertial frame of reference, whose existence, independent of the particle, was postulated. Practically, we cannot take out a body of the influence of other bodies in the Universe. But we can reduce this influence, observing that it is diminished if one moves away the other bodies from that on which we focus our attention or that one can annul an action by another one; other actions (for instance, friction, resistance of the medium etc.) may be sensibly reduced by certain technical means. We may thus imagine some experiments (reducing friction, experiments in vacuum etc.) leading closer to the ideal conditions of this principle; this one is thus verified with a good approximation. It follows that there can exist a motion (an inertial motion) even in the absence of a given force; we observe that the acceleration of a particle, which is not in interaction with other particles, vanishes. Hence, not the motion (not the velocity), but the variation of the motion (variation of the velocity, that is of the momentum) must be put in connection with the force, as we have seen in Subsec. 1.1.11.

ii) **Principle of action of forces.** *Between a force acting upon a particle and the acceleration induced by it takes place a relation of the form*

$$m\mathbf{a} = m\ddot{\mathbf{r}} = \mathbf{F}, \qquad (1.1.89)$$

where m is the mass of the particle.

This constitutes *the basic law of mechanics*, discovered by Newton, observing various phenomena of mechanical nature: the falling of bodies (phenomenon of gravitational nature), the oscillation of the pendulum, simple devices (studied previously by Galileo and Huygens), motion of the Moon around the Earth etc. Tullio Levi-Civita admits that $d\mathbf{v}$ is along the force $\mathbf{F}$, $d\mathbf{v}$ being proportional to $\mathbf{F}$ and to dt and in inverse proportion to the weight $G = |\mathbf{G}|$ of the particle; hence

$$d\mathbf{v} = \frac{k}{G}\mathbf{F}dt.$$

We suppose that k is a constant, which does not depend on the force $\mathbf{F}$, but only on the considered particle (affirmation corresponding also to *the principle of equivalence* enounced by Albert Einstein). In case of a free falling in vacuum, we see that

$$\mathbf{g} = \frac{k}{G}\mathbf{G},$$

where $\mathbf{G}$ is the gravity force; it follows that $k = g$ ($\mathbf{g}$ is the gravity acceleration), hence

$$\frac{d\mathbf{v}}{dt} = \mathbf{a} = \frac{g}{G}\mathbf{F}.$$

Observing that the ratio

$$\frac{g}{G} = m > 0$$

is constant anywhere on the Earth, no matter where G and g are measured, m being – in fact – the mass of the particle, we obtain the law (1.1.89). In case of gravity forces (taking into account the equality between the gravitational and the inertial mass), we may write this law in the form

$$m\mathbf{g} = \mathbf{G}. \qquad (1.1.89')$$

The mass of the bodies on the Earth is determined with the aid of the balance; for cosmic bodies we use the law (1.1.89) and *the principle of universal attraction*, which leads to the Newtonian attraction forces (1.1.84). It is interesting to notice that, in fact,

Newtonian model of mechanics 47

Newton enounced the second law in the form (he intended by $m\mathbf{v}$ the *quantity of motion*)

$$\frac{\mathrm{d}}{\mathrm{d}t}(m\mathbf{v}) = \mathbf{F}, \qquad (1.1.89'')$$

relation equivalent to (1.1.89) if $m = \text{const}$; this conception is the more valuable as a relation of the form (1.1.89'') is maintained also in relativistic mechanics. If $\mathbf{F} = \mathbf{0}$, then the relation (1.1.89) leads to $\mathbf{a} = \mathbf{0}$, hence to a uniform and rectilinear motion with respect to an inertial frame of reference; that does not mean that the principle i) is a particular case of the principle ii). Indeed, the principle of inertia can lead to the right solution if the solution of the differential equation (1.1.89) is not unique.

We will consider an example given by Jean Chazy in 1941, that is a particle of mass m, acted upon by a force $F(x) = 6mx^{1/3}$ along the Ox-axis; the differential equation of motion is

$$\ddot{x} = 6x^{1/3}, \qquad (1.1.90)$$

with three possible solutions

$$x = 0, \quad x = -t^3, \quad x = t^3, \qquad (1.1.90')$$

which verify the initial conditions

$$x = 0, \quad \dot{x} = 0 \text{ for } t = 0. \qquad (1.1.90'')$$

Hence, the problem has an infinity of continuous solutions of class C^2 for $t \in [0,T]$, $T > 0$, i.e.

$$x(t) = \begin{cases} 0 & \text{for } 0 \le t \le t_1, \\ \pm(t - t_1)^3 & \text{for } t > t_1, \end{cases} \qquad (1.1.91)$$

where $0 \le t_1 \le T$.

Using the principle of inertia, Victor Vâlcovici solves, in 1951, the problem. Indeed, this principle leads to the solution $x = 0$, which verifies the conditions (1.1.90''); for any t, the particle remains in the initial position of rest. For $t_1 \le T$, any other solution (1.1.91) is eliminated by the first principle of mechanics, so that $x = 0$ for $t \in [0,T]$; hence, this principle must be considered as an independent one.

iii) **Principle of action and reaction.** *To any action of a particle towards another particle, there corresponds a reaction of the second particle towards the first one, having the same support and magnitude, but opposite direction.*

If we denote by $\mathbf{F}_{12}$ *the action* of a particle P_2 upon a particle P_1 and by $\mathbf{F}_{21}$ *the reaction* of the particle P_1 upon the particle P_2 (an action can be a reaction or inversely, the denominations being relative), then the relation (these actions are, in fact, forces)

$$\mathbf{F}_{12} + \mathbf{F}_{21} = \mathbf{0}, \quad \mathbf{F}_{21} = \pm F_{21} \text{ vers } \overrightarrow{P_1 P_2}, \quad F_{12} = F_{21} > 0, \tag{1.1.92}$$

where we take the sign + or − as the forces are repulsive or attractive, holds; we notice that these forces are of the nature of internal forces and enjoy the properties put in evidence in Subsec. 1.1.11. The action and the reaction are applied to two different particles, so that we cannot say that they are in equilibrium. Let us consider, for instance, an agent exerting a force $\mathbf{F}$ on a particle P, inducing a certain motion of it; writing the equation (1.1.89) in the form $\mathbf{F} - m\mathbf{a} = \mathbf{0}$ and taking into account the principle of action and reaction (formula 1.1.92)), we can introduce a reaction $-m\mathbf{a}$ applied on the considered agent. The force $-m\mathbf{a}$ is called *force of inertia* and is due to the inertia of the particle. This principle may be extended passing from a particle acting upon another particle to a mechanical system acting upon another mechanical system; obviously, in this case, the action as well as the reaction are systems of forces.

iv) **Principle of the parallelogram of forces.** *If a particle is acted upon separately by the forces $\mathbf{F}_1$ and $\mathbf{F}_2$, which induce the accelerations $\mathbf{a}_1$ and $\mathbf{a}_2$, respectively, and then it is acted upon simultaneously by the two forces, then the latter ones can be summed vectorially* (using the parallelogram rule, $\mathbf{F} = \mathbf{F}_1 + \mathbf{F}_2$), *the acceleration induced being obtained analogously* ($\mathbf{a} = \mathbf{a}_1 + \mathbf{a}_2$).

Indeed, starting from

$$\mathbf{F}_1 = m\mathbf{a}_1, \quad \mathbf{F}_2 = m\mathbf{a}_2, \tag{1.1.93}$$

we may write

$$\mathbf{F} = \mathbf{F}_1 + \mathbf{F}_2 = m(\mathbf{a}_1 + \mathbf{a}_2) = m\mathbf{a}, \tag{1.1.93'}$$

the independence of the action of forces being thus put into evidence; this principle may be easily verified experimentally and, together with the principle iii), allows − for instance − to consider simultaneously the discrete mechanical systems in the Universe. The principle iv) was presented by Newton as a corollary to the second law. We must notice also that this principle is applied admitting that the particle is in the same initial state (position, velocity, physical or chemical state etc.) if it is acted upon either by the force $\mathbf{F}_1$ or by the force $\mathbf{F}_2$. This principle allows us to affirm that the force and the acceleration are vector quantities.

v) **Principle of initial conditions.** *The initial state* (initial position and velocity) *of a particle determines entirely its motion at an arbitrary given moment* (if one uses the first four principles).

In fact, it is not necessary to know the initial state (at time $t = t_0$); it is sufficient to know the state of a particle at a given moment. This observation is due to Galileo, who showed that the initial position is not sufficient to determine the motion of the particle, but it is necessary to know also its initial velocity; indeed, we can write the development

Newtonian model of mechanics 49

$$x_i(t) = x_i(t_0) + (t-t_0)\dot{x}_i(t_0) + \frac{1}{2}(t-t_0)^2 \ddot{x}_i(\bar{t}), \quad \bar{t} = t_0 + \varepsilon(t-t_0),$$
$$\varepsilon \in (0,1), \quad i = 1,2,3, \tag{1.1.94}$$

for the components of the position vector **r**. Hence, we see that – the initial state being given – it is necessary to calculate also the acceleration of the particle for the determination of its motion. This principle may be enounced also in another form, equivalent to that given above.

v') *If two particles are isolated from the rest of the Universe, then the forces in interaction at a given moment are determined in direction* (including support) *and modulus if at this very moment the positions and the relative velocities are known.*

In this form, the principle of initial conditions shows that the given forces **F** depend only on time, position vector and velocity, hence

$$\mathbf{F} = \mathbf{F}(\mathbf{r},\dot{\mathbf{r}};t). \tag{1.1.95}$$

In this context, we may admit that the basic law (1.1.89) of mechanics constitutes *the representation in a normal form* (in which we express the acceleration $\ddot{\mathbf{r}}$, that is the derivative of highest order as a function of the position **r**, the velocity $\dot{\mathbf{r}}$ and time t) of a law of the form

$$\mathrm{f}(\mathbf{r},\dot{\mathbf{r}},\ddot{\mathbf{r}};t) = \mathbf{0}; \tag{1.1.95'}$$

hence, in the Newtonian model of mechanics do not intervene *accelerations of higher order*, as it is experimentally confirmed by all mechanical phenomena thus modelled.

In conclusion, from a philosophical point of view, one can state that the classical model of mechanics is a *deterministic* one.

1.2.2 Mathematical models of mechanics

Once the basic notions of mechanics (space, time, mass, motion, force) introduced, *the classical* (*Newtonian*) *model* of this science of nature is born, adopting the principles discussed in the previous subsection. This model was verified in practice in case of bodies moving with *velocities v relatively small* (negligible with respect to the velocity c of light in vacuum).

But the Newtonian model of classical mechanics was only a step in the process of knowledge; indeed, it was proved (by the famous experiments of Albert Abraham Michelson in 1881, as well as together with Edward William Morley in 1887 and of E. W. Morley and Dayton Clarence Miller in 1904, who showed the inexistence of the ether) that, at greater velocities (comparable with the velocity of light in vacuum), the principles of Newton must be changed or completed. To describe more exactly the properties of the real space, *non-Euclidean geometries* are introduced. As well, the Newtonian concept of the universal time must be replaced by the relativistic representation, which takes into account *the individual physical time*, the fact that time depends on the motion itself. The mass is also no more constant, depending on the velocity and, implicitly, on time. Thus, following the papers of Albert Einstein of 1905 and 1916, appear *the Special* and *the General Relativity*, respectively, which constitute new steps in the process of knowledge of the motion of mechanical systems.

We notice that these new models (which tried to maintain the principles of classical mechanics and to adapt them, (for instance, the principle of inertia) have been verified by many important experiments (*the crucial experiments* in General Relativity). But there exist still some contradictions (for instance, *the horologes' paradox* in Special Relativity), so that it arises the necessity of some improved models. It is also possible that the actual theory of relativity do correspond to velocities comparable to the velocity of light in vacuum, but not to velocities close to this *superior limit velocity*; it is possible that at such velocities more complex models be necessary. There have been conceived also models in which the velocities are always greater than the velocity of light in vacuum, which is thus an *inferior limit velocity*. We notice that for velocities v negligible with respect to the velocity c of light in vacuum ($v \ll c$), the results of relativistic mechanics tend to the classical ones; there exists thus a matching of the two mathematical models of mechanics, which is necessary for the viability of the improved model (the relativistic model).

We mention also that there have been created mathematical models, leading to *quantum mechanics* (non-relativistic and relativistic models) to describe other mechanical phenomena, connected to bodies with "microscopic" character (elementary particles a.s.o.), in opposition to those connected to bodies with "macroscopic" character, with which we have to do usually. As a matter of fact, relativistic and quantum mechanics are the two basic non-classical models of mechanics. Other mathematical models led to *undulatory mechanics, statistical mechanics, invariantive mechanics, nanomechanics* a.s.o.

As we have seen above, the usefulness of the notion of mathematical model has been put into evidence, as well as the great variety of such models, constructed to can study mechanical phenomena, which – in general – are enough complex. We notice also the so-called *generalized models* of deformable continuous media.

1.2.3 Division of classical mechanics

Synthetizing the above exposure, we may state the object of study of mechanics.

Mechanics deals with the study of the equivalence of systems of forces and with the study of motion of mechanical systems subjected to the action of these forces.

Methodically and conventionally, and taking into account its historical development too, we may admit that classical mechanics has three great divisions.

Statics deals with the study of equivalence of systems of forces (in particular, the equivalence to zero) acting upon a mechanical system. Statics studies a particular case of motion, that case in which a mechanical system, subjected to the action of certain forces, remains in rest (or has a uniform and rectilinear motion) with respect to a given frame of reference (considered as an inertial one).

Kinematics considers the motion of mechanical systems without taking into account the forces which act upon them; hence, the object of study of kinematics is the *geometry of motion*. Sometimes, kinematics is called also the *geometry of mechanics* (*phoronomy*).

Dynamics deals with the motion of mechanical systems subjected to the action of given forces; in fact, this is the object of study of mechanics.

These divisions of classical mechanics allow a systematization of our study; in the frame of those divisions, the problems of *mechanical vibrations*, of *stability of equilibrium* and of *motion*, of *optimal trajectories* etc. did develop very quickly. Chronologically, first appeared statics (in antiquity), then dynamics (at the same time as the Italian Renaissance); kinematics arises only in the XIXth century.

From a methodical point of view, it is more convenient to consider separately *classical models* (Newtonian) and *non-classical models* (relativistic, quantum (the last one can be non-relativistic or relativistic), undulatory, statistical, invariantive etc.) *of mechanics*. For *deformable continuous media* (in opposition to what – usually – is called *general mechanics*, containing classical models of mechanics) it is necessary a separate study; the mathematical models of such mechanical systems must be completed either they are classical models or non-classical ones. In this case too one can put into evidence statics, kinematics and dynamics.

Actually, the searching in the frame of classical and non-classical models of mechanics or in the domain of mechanics of deformable media is developing very much, in multiple directions, comprising new methods of computation, analytical and numerical ones. One of the tendencies of today is represented by the extension and the generalization of the models of mechanics. We mention thus the search in the direction of *axiomatization of classical mechanics*. A particular impetus takes the theory of continuous deformable mechanical systems, either by solving boundary value problems or by *coupling* the corresponding problems to other non-mechanical phenomena (a thermal field, an electromagnetic field etc.).

1.2.4 Short history

From the earliest time A.C., man has been put in connection with mechanical phenomena; empirical knowledge allowed him to set up tools and mechanical devices, as it can be seen also from the drawings and the basso-rilievos of Assyria and ancient Egypt.

The beginnings of mechanics in the ancient eve date from VI-Vth centuries A.C., together with the oldest notings. Mo Tsy, in ancient China, gives in his writings some notions of mechanics, concerning time, motion and force. By Anaximander (610-546 A.C.), Anaximenes (585-525 A.C.), Heraclit of Efes (540-470 A.C.), as well as by Pythagoras' school (571-477 A.C.) we meet referrings about matter in motion. Thales at Miletus (640-546 A.C.) took much interest in astronomy and Herodotus (484-425 A.C.) tells us that he has even succeeded in predicting an eclipse. Anaxagoras of Clazomenae (499-427 A.C.) admits that there are some forces, which do not allow celestial bodies to fall one upon the other. The arguments about motion (Achilles and the tortoise, the arrow, the stadium, the dichotomy), which are the paradoxes of Zeno of Elea (490-430 A.C.), contribute to the stimulation of studies about space, time, problems of continuity and discontinuity a.s.o. Œnopides of Chios (b. 465 A.C.) was one of the leading astronomers of his time; he is thought to have learned the science of the stars and the obliquity of the ecliptic from the priests and temple astronomers of Egypt. He is said to have invented the cycle of 59 years for the return of the coincidence of the solar and lunar years, giving the length of the solar year as 365 days and somewhat less than 9 hours. Parmenides of Elea (b. 460 A.C.) taught at Athens in the middle of the 5th century A.C., and among his theories on the Universe was the one that the Earth is a sphere; Meton, Phaeinus and Euctemon dealt with mathematical astronomy in the same century. Archytas of Tarentum (430-360 A.C.) discovers the pulley, the screw as well as some mechanical tools to set up curves; he applied mathematics in any noteworthy way to mechanics. The first written works about mechanics are assigned to him. It is said that Eudoxus of Cnidus (408-355 A.C.), at one time a pupil of Plato (430-349 A.C.), introduced the study of spherics (mathematical astronomy) into Greece and made known the length of the year as he had found it given in Egypt, from where he brought also the theory of motion of the planets; he found that the diameter of the Sun is nine times that of the Earth. Aristotle of Stageira (384-322 A.C.) wrote a voluminous work on mechanical problems; he puts in evidence the relative character of motion and intuits the principle of virtual displacements in the study of equilibrium of levers. To him we owe the first known definition of continuity: "A thing is continuous when of any two successive parts the limits at which they touch are one and the same and are, as the word implies, held together". Democritus of Abdera (460-370 A.C.) is considered to be the founder of the atomic theory of the ancient philosophy, which asserted: "the original characteristics of matter are functions of quantity instead of quality, the primal elements being particles homogeneous in quality, but heterogeneous in form". Other atomists have been Epicur (341-270 A.C.) and Carus Lucretius (99-55 A.C.); they gave incipient results concerning the falling of bodies and the structure of matter.

The greatest mathematical centre of ancient times was neither Crotona nor Athens, but Alexandria; here it was, on the site of the ancient town of Rhacotis, in the Nile Delta, that Alexander the Great (356-323 A.C.) founded a city worthy to bear his name. Under Ptolemy's Soter (Ptolemy the Preserver) benevolent reign (323-283 A.C.), Alexandria became the centre not only of the world's commerce but also of its literary and scientific activity; here was established the greatest of the world's ancient libraries and its first international university. Of all the great names connected with Alexandria, that of Euclid (320-270 A.C.) is the best known; he was the most successful textbook writer that the world has ever known, over one thousand editions of his geometry having in print since 1482. Euclid's greatest work is known as the "Elements", and represents the basis of the geometry bearing his name. We notice the interesting contributions of Euclid to statics; he wrote also "Phoenomena", dealing with the celestial sphere and containing twenty-five geometric propositions. Eratosthenes of Cyrene (274-194 A.C.) dealt with the measurement of the diameter of the Earth and found 7850 miles, that is 50 miles less than the polar diameter, as we know it; he considered also the problem of distances in the solar system, i.e., the distances from the Earth to the Moon and to the Sun, respectively. Archimedes (287-212 A.C.) was born and died at Syracuse in Sicily; he may be considered as to be the founder of statics. Archimedes discovers various simple devices and defines the centre of gravity. Archimedes is the most important researcher of ancient mechanics; his work "On the equilibrium of plane surfaces and on centres of gravity" is comparable to Euclid's "Elements". After Ctesibios (approx. 200 or 180 A.C.), Hypsicles of Alexandria (approx. 180 A.C.), Hypparchus of Rhodes (190-120 A.C.), Filon of Byzantium (second half of IInd century A.C.) and Poseidonius of Rhodes (135-84 A.C.), the most important name is that of Heron of Alexandria (first century A.C.); he invented the pneumatic device, commonly known as *Heron's fountain*, a simple form of the steam engine, and wrote on pneumatics and mechanisms. Of the Romans, no one is more prominent than Marcus Vitruvius Pollio, commonly known as Vitruvius (50 A.C.-20 A.D.), who published the textbook "De architectura".

At the beginning of Christian era (from now on, all data are A.D.), Ptolemy (Claudius Ptolemaeus) (70-147) did for astronomy what Euclid did for plane geometry, Apollonius Pergaeus (Pamphylia) (approx. 225 A.C.) for conics and Nicomachus of Gerasa (approx. 100 A.C.) for arithmetic; his greatest work, commonly known as the "Almagest" (with the Arabic "al" (the)) contains much information about the history of ancient astronomy as well as his concept on the geometric stellar system. Ma Kim (IIIrd century) created in China mechanisms with geared wheels. Pappus of Alexandria (approx. beginning of IVth century) gives a synthesis of known results in mathematics, astronomy and mechanics in his great work "Mathematical Collections"; we find here the famous theorems concerning centres of gravity of bodies of revolution (taken over by the Swiss Paul (original first name Habakuk) Guldin (1577-1643)). To Proclus of Byzantium (412-485) (surnamed the Successor, because he was looked upon as the successor of Plato in the field of philosophy) are due many comments on mechanics. Joannes Philiponus of Alexandria (known also as Joannes Grammaticus) (VIth century) develops the notion of impetus (because of which the motion continues without any external action). The Arabic science is that which takes again and amplifies these results till the XIIth century; we mention thus Mohammed ibn Mohammed ibn Tarkhan ibn Auzlag, Abu Nasr al Farrabi (of Farab, in Turkestan) (870-951), who wrote "About the eternal motion of the celestial sphere", Al-Hosein ibn Abdallah ibn al-Hosein (or Hasan) ibn Ali Abu Ali al-Sheich al-Rais (known in Christian Europe as Avicenna) (980-1037), born in Safar, not far from Bokhara (Uzbekistan), philosopher and physician, which considered the motion of matter as fundamental, Mohammed ibn Ahmed ibn Mohammed ibn Roshd Abu Velid (commonly called Averroes in the Middle Age) (1126-1198) of Cordova, to whom we due interesting comments on Aristotle, and Nured-din al-Betruji Abu Abdallah (called Alpetragius by the Christians) (XIIth century) simplified the cosmic system of Ptolemy and applies the "impetus" to the motion of celestial bodies. Rabbi Moses ben Maimum (called also Maimonides) (1135-1204), a Jewish writer native of Cordova, physician to the sultan, is an astronomer of prominence, and to him is due a Jewish calendar.

Beginning with the XIIIth century, appear the great European schools, where are translated the most important works of the Antiquity and of the Middle Age; the ideas of the time are thus developed. The scholasticism accommodates many of these ideas to the needs of the Catholic Church. A critical interpretation of Aristotle is made, among others, by Siger of Brabant (1235-1281), Thomas d'Aquino (1226-1274) and Richard of Middletown (beginning of the XIVth century). The nominalist current reflects certain liberties of thinking and position with respect to the church, initiating methods of research based on experiments; we mention thus, Pierre Abélard (Petrus Abaelardus) (1079-1142), Roger Bacon (1214-1294), the most prominent scholar in England, a man of erudition and of prophetic vision, and – especially – Jordan of Namur (Jordanus de Saxonia, Jordanus Nemorarius) (XIIIth century), who studies the motion of heavy bodies along a curve and leads a true school of mechanics. Beginning with the end of the XIIIth century till the XVth century appear the antischolastics; among them we mention Jean Buridan (1300-1360), Nicolas Oresme (1323-1382), Thomas Bradwardine (1290-1349), representative of the school of mechanics of Oxford, Filippo Bruneleschi (1377-1446), Florentine sculptor and architect, Biagio Pelacani (Blasius of Parma) (d. 1416) and the German cardinal Nikolaus Krebs (Nicolaus Cusanus) (1401-1464).

The XVth century represents the beginning of the Renaissance; research is made on experimental basis and statics becomes a complete discipline. Niklas Koppernigk

(Nicholaus Copernicus of Thorn on the Vistula) (1473-1543) published his heliocentric concept on the world "De revolutionibus orbium coelestium" (Nürnberg, 1543) (The orbits of the planets are circles, the Sun being in the centre); other works are published by Leone Battista Alberti (1404-1472) and by Giorgio Valla (1447-1500). The most prominent figure of this epoch is Leonardo da Vinci (1452-1519), who said that "mechanics is the paradise of mathematical knowledge, because – through the agency of it – one bears the fruits of mathematics"; he studies the laws of the free falling and of friction, introduces the notion of moment and applies the principle of virtual displacements. To the progress of mechanics in this period contributed also Jules César Scaliger (1484-1558), who adopts the denomination of *motion* for impetus, Girolamo Cardan (1501-1576), Nicolo Fontana (Tartaglia) (1499- or 1501-1557), Frederico Commandin di Urbino (1509-1575) and Giovanni Battista Benedetti (1530-1590). The XVIth and XVIIth centuries bring a new raising of mechanics; the technologies in various branches of production bear a particular development and bring into life the premises of the great industry in the future. The Academies of Sciences do appear (Accademia dei Lincei in Rome (1590), Accademia del Cimento in Florence (1651), Royal Society in London (1622-1663), Académie des Sciences in Paris (1666) etc.), contributing to the promotion of theoretical and experimental research. The mathematical model of classical mechanics is completed in the XVIIth century. Simon Stevin (1548-1620), Flemish mathematician and physicist, solves the problem of the inclined plane and enounces the rule of composition of forces, Giordano Bruno (1548-1600) brings arguments against Aristotle, while Luca Valerio (1552-1618) determines a great number of centres of gravity. Galileo Galilei (1564-1642) is one of the founders of the modern dynamics. To him are due the principle of inertia and the principle of initial conditions, the formulation of which represents a revolutionary step in the development of mechanics, and thus is put an end to Aristotelic conceptions; he states the so-called "golden rule of mechanics", that is: how much is won in force is lost in velocity, and studies the motion of a projectile in vacuum. His most important ideas are contained in "Dialogo di Galileo Galilei delle due massimi sistemi del mondo, il Tolemaico e il Copernicano" (1632) and in "Discorsi e dimonstrazioni mathematiche intorno a due nuove scienze attenanti alla meccanica e i movimenti locali" (Leyda, 1638). Francis Bacon (1561-1626) considers the motion as a property of the matter and criticizes the scholastic conceptions in "Novum organum" (1620). Evangelista Torricelli (1608-1647), disciple of Galileo, develops the theory of motion of heavy bodies and of the stability of equilibrium. Johannes Kepler (1571-1630) starts from observations of Tycho Brahe (1546-1601) on Mars and in "De revolutionibus orbium coelestium" enounces his three famous laws, which replace the Copernican motion of planets, supposed to be circular and uniform, by a motion on an ellipse. Other contributions to the construction of the mathematical model of mechanics are due to Tomaso Campanella (1568-1639), to Paul (Habakuk) Guldin (1577-1643), who founds again the theorems of Pappus, to Marin Mersenne (1588-1648), who introduces Galileo's work in Paris, to René Descartes (1596-1650), who studies the collision of bodies and enounces the theorem of conservation of momentum, and is the first to put into evidence the infinitesimal character of virtual displacements, to Pierre Fermat (1601-1665), to whom belongs the "theorem of minimum time", to Gilles Personne de Roberval (1602-1675), who dealt with levers and balances and research in statics and kinematics, to Giovanni Alfonso Borelli (1608-1679), a monk who laid the bases of biomechanics, to John Wallis (1616-1703), who dealt with the collision of inelastic bodies, to Jacques Rohault (1620-1675), to Pierre Varignon (1654-1722), who develops the theory of moments, to Christian Huygens (1629-1695), who studies the physical pendulum, discovers the isochronism of the cycloidal pendulum and invents the clock with pendulum, to Cristoph Wren (1652-1723), to Nicole Malbranche (1638-1715) and to Gottfried Wilhelm Leibniz (1646-1716), who considers the "vis viva" to be the basic mechanical characteristic of motion.

But the scientist who attached his name to the first mathematical model of mechanics, the model of classical mechanics, has been Sir Isaac Newton (1642-1727), illustrious English mathematician, physicist and astronomer. He defined the notion of mass, generalized the notion of force and formulated mechanics' laws in "Philosophiae Naturalis Principia Mathematica" (London, 1686-87), published under pressure from his good friend Edmund Halley (1656-1742), the astronomer, fighting for priority with Robert Hooke (1635-1703); Newton discovered the law of universal attraction, verifying thus the laws of motion of planets around the Sun, given by J. Kepler. But Newton sketched also an ample plane for the development of the science thus created; today, classical mechanics is developing in the large frame of the three fundamental laws as a theoretical and experimental science, which must permanently introduce new concepts and confront theoretical constructions to reality.

The XVIIIth century represents the beginning of the industrial revolution, leading to an intensive development of classical mechanics; the first superior technical schools are founded, for instance "l'École Polytechnique" of Paris (1794), which – due to the activity of Gaspard Monge (1746-1818) and Louis Poinsot (1777-1859) – played an important rôle in the development of mechanics. Jacques (Jacob) Bernoulli (1654-1705) dealt with dynamics of mechanical systems subjected to constraints and gave the solution to the problem of the oscillation centre; some corrections are brought by Guillaume François de l'Hospital (1661-1704). Jean (Johannes) Bernoulli (1667-1748) gives a correct formulation to the principle of virtual displacements, while his son Daniel Bernoulli (1700-1782) studies the conservation of the "vis viva" principle. Jacob Hermann (1678-1733) studies the dynamics of constrained systems, Colin Mclaurin (1698-1746) is the first to enounce, in 1742, the second law of Newton in the form (1.1.89), while Pierre Louis Moreau de Maupertuis (1698-1759) enounces in 1744 the principle of least action. A special contribution is due to Leonhard Euler (1707-1783), one of the most prominent scientists in mathematics and mechanics of the century, who introduced the notions of material point, moment of inertia and moment of "amount of motion" (moment of momentum); he studies the motion of the rigid solid, gives the differential equations of motion of the rigid with a fixed point and integrates them in a particular case. He also enounces, in 1748, the second law of Newton in the form (1.1.89). Mikhail Vasilievich Lomonosov (1711-1765) extends the "principle of conservation of matter" to the "conservation of motion". Samuel Koenig (1712-1757) polemizes with Maupertuis about the priority on the principle of least action; Euler and François Marie Arouet (Voltaire) (1694-1778) are also participants to this dispute. Alexis Claude Clairaut (1713-1765) deals with the relative motion and with the motion of the Moon. Jean le Rond d'Alembert (1717-1783) states in his "Traité de dynamique" (Paris, 1743) a general method to study the discrete mechanical systems (d'Alembert's principle). Paul d'Arcy (1725-1779) enounces the theorem of conservation of the kinetic moment for a conservative system (1747), Charles-Augustin de Coulomb (1736-1806) does important research on sliding and rolling friction and Pierre Simon de Laplace (1749-1827) publishes his famous "Traité de mécanique celeste" (Paris, 1799-1825), in five volumes, where important cosmogonic hypotheses are made. "La mécanique analytique" (Paris, 1788) by Joseph-Louis Lagrange (1736-1813) had a great influence for the subsequent directions of development of mechanics; in this volume, which contains the famous equations bearing Lagrange's name, are studied discrete mechanical systems, without figures, using only methods of mathematical analysis. He deals with the study of small motions of a mechanical system around a stable position of equilibrium; we notice also his results on "ideal constraints" and the introduction of the so-called Lagrange's multipliers. Fundamental contributions to the development of analytical methods in mechanics are brought by Wiliam Rowan Hamilton (1805-1865), who finds the canonical equations of analytical mechanics bearing his name and enounces the most important variational principle of mechanics and by E.J. Routh (1831-1907), the equations of which are a generalization of Lagrange's and Hamilton's ones. Carl Gustav Jacob Jacobi (1809-1882), Siméon-Denis Poisson (1781-1840) and Joseph Liouville (1809-1882) have valuable contributions concerning the integration of the canonical system; Peter Gustav Lejeune-Dirichlet (1805-1859), Jules Henry Poincaré (1854-1912), Adolf Hurwitz (1859-1919) and Aleksandr Mikhailovich Lyapunov (1857-1918) deal with the problem of stability of motion. Gustav Gaspard Coriolis (1792-1843) discovers the complementary acceleration, used by Jean Bernard Léon Foucault (1819-1868), who puts in evidence the motion of rotation of the Earth, by means of the pendulum bearing his name. George Atwood (1746-1807) verifies the laws of the bodies' falling, using the device to which was given his name, and Lazare Nicolas Carnot (1753-1823) gave the first analytical formulation of the principle of virtual displacements and enounced the theorem of dissipation of energy in case of collision. Carl Friedrich Gauss (1777-1855) elaborates a method to calculate the elliptical orbits of planets. Jacques Philippe Marie Binet (1786-1856) deals with the action of central forces and Mikhail Vasilievich Ostrogradskiĭ (1801-1862) enounces a generalization of Hamilton's principle, independently of him, and develops the mechanics of systems with unilateral constraints. André Marie Ampère (1775-1836) studies many problems of kinematics, introducing this denomination. Jean Baptiste Joseph Fourier (1768-1830) tries to give a general demonstration to the principle of virtual work. Jean Victor Poncelet (1788-1867) connects the notion of work to that of energy. Rodrigues B. Olinde (1794-1851) deals with the principle of least action, Jacob Steiner (1796-1863) studies moments of inertia, while Nicolas Léonard Sadi Carnot (1796-1832) calculates the mechanical equivalent of heat. Eugen Dühring (1833-1921), Felix Klein (1849-1925) and Ernst Mach (1839-1916) contribute to a criticism of classical mechanics, and Vladimir Ilich Ulyanov (Lenin) (1870-1924) is against them. Ivan Vsevolod Meshcherskiĭ (1895-1935), Konstantin Eduardovich Tsiolkovskiĭ (1857-1935) and Tullio Levi-Civita (1873-1941) elaborate basic works concerning the mechanics of bodies of variable mass. To William-John Macquorn Rankine (1820-1872) is due the exact expression of kinetic energy, eliminating the notion of "vis viva". Hermann Ludwig Ferdinand von Helmholtz (1821-1894) generalizes the principle of least action to non-mechanical phenomena. Sonya Krukowsky (Sophia Kovalevsky) (1850-1891) deals with the motion of a rigid solid with a fixed point. Valuable contributions are due to James Clerk Maxwell (1831-1879), William Thomson (lord Kelvin) (1842-1907), Eötvös Loránd (1848-1919), Isaac Todhunter (1820-1884), Karl Pearson (1857-1936), Paul Émile Appel (1855-1930), Heinrich Rudolph Hertz (1857-1894), Paul Painlevé (1863-1933), Sergei Alekseevich Chaplygin (1869-1942) and Georg Hamel (1877-1955). Gheorghe Vrănceanu (1900-1979) constructs *non-holonomic geometries*, suitable to the study of non-holonomic mechanical systems.

Max Karl Ernst Ludwig Planck (1858-1947) gets up in 1900 to the conclusion that the change of energy between radiation and substance (the radiation problem of the black body) is discontinuous, in quanta, and *quantum theory* is thus born. Erwin Schrödinger (1887-1961), Paul Adrien Maurice Dirac (1902-1984) and Werner Karl Heisenberg (1901-1976) worked (in 1925-26) to the bases of *quantum mechanics*. Starting from the experiments of Albert Abraham Michelson (1852-1931) and Edward Williams Morley (1838-1923), and from the interpretations of Hendrick Anton Lorentz (1853-1928) and of others, Albert Einstein (1879-1955) puts, by his paper "Zur Elektrodynamik der bewegter Körper" (Annalen der Physik, 1905), the bases of the *Special Relativity*, and, by the paper "Die Grundlagen der allgemeinen Relativitätstheorie" (Annalen der Physik, 1916), the bases of the *General Relativity*; Hermann Minkovski (1864-1909), H. Poincaré and M. Planck contribute to the development of these theories. Josiah Dixon Willard Gibbs (1839-1903) and Ludwig Eduard Boltzmann (1844-1906) put, by their studies, the bases of *statistical mechanics*. To Louis Victor Pierre Raymond prince de Broglie (1892-1987) is due the birth of *undulatory mechanics*. The name of Octav Onicescu (1892-1983) states for the *invariantive mechanics*.

Nowadays, research in the frame of classical and non-classical models of mechanics or in the frame of continuous deformable media is very much developed, in many directions. One of the actual tendencies is represented by the extension and generalization of mechanical models. Efforts have been made by Clifford Ambrose Truesdell (1919-2000), Walter Noll (b. 1925) and Bernard Coleman for the *axiomatization of classical mechanics*. In the so-called frame of general mechanics rise studies in vibration, stability, dynamical systems and optimal control. A particular development is noticed in the theory of continuous mechanical systems, by solving many boundary value problems, as well as by coupling the respective mechanical phenomena with other non-mechanical ones (corresponding to a thermic or electromagnetic field etc.). Due to this enormous development, it is very difficult to review, even summarily, the most important researchers dealing with these problems. So that we preferred to mention only scientists who became classic and to mention only research till the first decades of the XXth century.

2. Dimensional analysis. Units. Homogeneity. Similitude

Dimensional analysis deals with the study of relations which describe mechanical, physical, chemical phenomena a.s.o.; this study is based on the property of dimensional homogeneity, which must be verified by all relations obtained on a rational or empiric way. To do this, we will put into evidence the physical quantities which appear in the study of mechanical systems, as well as the corresponding units. We consider also the properties of homogeneity and similitude and the possibility of using them.

2.1 Physical quantities. Units

In what follows, we introduce the notion of physical quantity (in particular, the notion of mechanical quantity), as well as the basic and derived quantities and units. We may thus define the systems of units.

2.1.1 Physical quantities

Considering a set of physical objects of the same nature from the point of view of a certain property, we reach the notion of *physical quantity*. A process of abstraction defines this notion by: i) the statement of an *equivalence relation*, which allows to distribute the respective objects in *equivalence classes*; ii) the statement of a *relation of order* between these equivalence classes, to can appreciate if the objects belonging to one class are "greater" or "smaller" than the objects belonging to another class, from the point of view of the considered property; iii) the statement of a *criterion of comparison* by which one can estimate "how many times" an object of a class is greater or smaller than an object belonging to another class, from the point of view of the same property. A physical quantity M, which appears in the study of a mechanical system (the velocity of a particle, the intensity of a force etc.), is a mechanical quantity.

The direction of a force, considered as a physical object, does not represent a property to allow the introduction of the notion of physical quantity; indeed, one can obtain classes of parallel forces, but one cannot state a relation of order. But the property of intensity of a force permits the introduction of a physical quantity; the corresponding property can be put into evidence with the aid of a dynamometer.

By the comparison criterion, to any class of equivalence is attached a real number. The function f, which states the correspondence between the classes of equivalence and the set $\mathbb{R}$ of real numbers, must be strictly increasing, while the ratio between two quantities of the same nature must remain the same; it follows that the function f must be linear, hence of the form

$$f(x) = kx + f(0), \quad k > 0. \qquad (1.2.1)$$

One can assign arbitrary numbers (corresponding to the arbitrary constants k and $f(0)$) only to two equivalence classes; for the other classes, the numbers, which must be attached, result from (1.2.1). These arbitrary constants may correspond to a conventional unit equal to *zero* and to a conventional intensity equal to *unity*. In physics (in particular, in mechanics) the zero convention is made for a same class of equivalence, which is naturally imposed. For instance, if upon a dynamometer does not act any force, then its deformation is equal to zero; we can say that a force of null intensity is acting. Conventionally, one considers also the class to which the number one is assigned; a physical object of this class is called a *unit* U of the respective physical quantity. The real number attached to a class of equivalence is an abstract (non-dimensional) number, called the *numerical value* n *of the physical quantity* M considered; this value, obtained by an operation of measurement (on an experimental way, with a certain degree of approximation), will be given by *the basic equation of measurement*

$$M = nU. \qquad (1.2.2)$$

2.1.2 Quantities and basic units. Quantities and derived units

The unit U, corresponding to a certain physical quantity, must be really reproduced with the best precision possible (depending on the technical possibilities at a certain moment), in the form of a *gauge*. For certain quantities (length, mass, time) one can

Newtonian model of mechanics 59

easily construct units (metre, kilogram, second); these quantities can be measured *directly*. For other quantities (velocity, acceleration etc.) one cannot easily realize units, a direct measurement being difficult to do. In the last case, a quantity M is measured *indirectly*, measuring directly quantities of a different nature with numerical values $n_1, n_2, \ldots$; a relation of the form

$$M = f(n_1, n_2, \ldots) \tag{1.2.3}$$

takes place. We must judiciously choose the units, so that coefficients depending on these units do not appear.

The numerical values n_1, n_2 of two physical quantities M_1, M_2 of the same nature can be added only if they proceed from measurements with the same units U; in this case

$$M = nU = M_1 + M_2 = n_1 U + n_2 U, \quad n = n_1 + n_2. \tag{1.2.4}$$

Also

$$\frac{M_1}{M_2} = \frac{n_1}{n_2}. \tag{1.2.4'}$$

If we measure the same physical quantity M with different units U_1 and U_2, then the corresponding numerical values n_1 and n_2 are linked by the relation

$$M = n_1 U_1 = n_2 U_2, \quad \frac{n_1}{n_2} = \frac{U_2}{U_1}. \tag{1.2.5}$$

Certain relations take place between the physical quantities which appear in nature. We may thus choose a relatively restrained number of independent physical quantities, so that the other quantities be expressed by certain relations depending on the first ones; thus, using the above relations, one can express the units of all physical quantities with the aid of a restrained number of units. The physical quantities thus chosen are called *basic (primitive) quantities*, the corresponding units being *basic (primitive) units*. All the other quantities and units are *derived quantities* or *units*, respectively.

Usually, one takes as basic quantities the physical ones which naturally appear, that is quantities corresponding to space, mass and time, for instance, in case of mechanical systems; such quantities are called also *primary quantities*, the other ones being *secondary quantities*. But one can choose as basic quantities also other quantities than the primary ones. For instance, the force can be also a basic quantity, useful in case of mechanical systems.

We notice that a relation describing a physical phenomenon in which appear quantities the units of which are fixed (basic and derived units) contains a proportionality coefficient, and any other relation deriving from it will contain the same coefficient.

2.1.3 Systems of units

The basic and derived units form a *system of units*. Two systems of units may differ by the chosen basic quantities and by the units corresponding to these quantities. We mention thus *the physical systems* in which the basic quantities are *the length* (of unit L), *the mass* (of unit M) and *the time* (of unit T), and *the technical systems* to which correspond the length, *the force* (of unit F) and the time. Unlike the physical systems of units, independent of the point in which we are on the surface of the Earth (independent of the presence of a gravitational field), the technical systems of units depend on the latitude of the place (depend on the presence of a gravitational field).

Among the physical systems we mention *the CGS system*, in which the units are *the centimetre*, cm, *the gram*, g and *the second*, s, and *the international system*, SI, which uses *the metre*, m, *the kilogram*, kg and the second; in the latter physical system one introduces also, as basic quantities: *the intensity of the electric current* (ampere, A), *the thermodynamical temperature* (kelvin, K), *the quantity of substance* (mol, mol) and *the light intensity* (candle, cd). The SI system was adopted at the XIth International Conference of Measures and Weights (Paris, 1960). The most used technical system is the MKfS system, where one introduces the metre, *the kilogram force*, kgf and the second. We notice that in the SI system the unit of force is the newton, N (the necessary force to induce to a mass of 1 kg an acceleration of 1 m/s^2 in vacuum; 1 kgf $\cong$ 10 N).

One uses *the decimal system* for space and mass, while for time remains *the* classical *sexagesimal system*. Theoretically, the metre is defined as $1/40\ 000\ 000$ of the length of the Paris meridian. Practically, the metre is equal to 1 650 763.73 wave lengths of the radiation which corresponds to the transition of the atom of krypton 86 between the energy levels $2p_{10}$ and $5d_5$ in vacuum (Paris, 1960); the centimetre is the hundredth part of the metre defined above. Before 1960, the metre was defined as the length, at a temperature of 0°C, of the international prototype in irradiate platinum, sanctioned by the General Conference of Measures and Weights in 1889, preserved at the International Bureau of Measures and Weights, at the pavilion of Breteuil (Sèvres, France). The kilogram represents the mass of the international prototype in irradiate platinum sanctioned at the same time and preserved in the same place as the metric prototype; the gram is the thousandth part of the kilogram defined above. The mass of the mentioned prototype represents theoretically the mass of a decimetre cube of distilled water at 4°C, at a pressure of an atmosphere; the weight of this prototype at 45° boreal latitude, at the sea level, represents a kilogram force. Till 1960, the second was defined as the 86400th part of the mean solar day, considered constant and defined with respect to the tropical year (the mean interval between two consecutive passages of the Sun at the spring mean equinox), admitted to be also constant. Observing that the latter one is not constant (it is expressed by a polynomial formula with respect to time), after 1960 the tropical year, which has 365.242 198 79 solar days or 31 556 925.974 7 seconds, corresponding to 1900, has been taken into consideration; the second thus defined is the ephemerides' second. In the international system SI the second is defined with the aid of atomic horologes; thus, a second is an interval of time equal to 9 192 631 770 oscillation periods of radiations emitted by the transition between two hyperfine energy levels (F=4, M_F=0; F=3, M_F=0) of the basic state ($^2S_{1/2}$) of the atom of caesium 133, in the absence of perturbations of a magnetic field. The standard hour (implicitly the second) is conserved by the International Bureau of the Hour in Paris.

The units of geometrical and mechanical quantities encountered in the physical systems CGS and SI and in the technical system MKfS are given in Table 1.1. As well, in Table 1.2 are given the decimal multiples and submultiples of these units, exception the time, which has other units (second, minute, hour, day, year a.s.o). We notice that in the SI the unit for pressure is the pascal ($1 \text{ Pa} = 1 \text{ N/m}^2$); as well, dyn, rad, Hz, erg, J, W, are used for dyne, radian, hertz, erg, joule and watt, respectively.

Table 1.1

Quantity	Symbol	Dimensions in the system		Units in the system		
		LMT	LFT	CGS	SI	MKfS
Length	l	L	L	cm	m	m
Mass	m	M	$L^{-1}FT^2$	g	kg	$kgf \cdot s^2/m$
Time	t	T	T	s	s	s
Force	F	LMT^{-2}	F	dyn	N	kgf
Area	A	L^2	L^2	cm^2	m^2	m^2
Volume	V	L^3	L^3	cm^3	m^3	m^3
Plane angle	$\alpha, \beta, \ldots, \varphi$	1	1	rad	rad	rad
Period	T	T	T	s	s	s
Frequency	f	T^{-1}	T^{-1}	Hz	Hz	Hz
Pulsation (circular frequency)	ω, p	T^{-1}	T^{-1}	s^{-1}	s^{-1}	s^{-1}
Angular velocity	ω	T^{-1}	T^{-1}	rad/s	rad/s	rad/s
Angular acceleration	ε	T^{-2}	T^{-2}	rad/s^2	rad/s^2	rad/s^2
Velocity	v	LT^{-1}	LT^{-1}	cm/s	m/s	m/s
Acceleration	a	LT^{-2}	LT^{-2}	cm/s^2	m/s^2	m/s^2
Unit mass (density)	μ	$L^{-3}M$	$L^{-1}FT^2$	g/cm^3	kg/m^3	$kgf \cdot s^2/m^4$
Unit weight	γ	$L^{-2}MT^{-2}$	$L^{-3}F$	dyn/cm^3	N/m^3	kgf/m^3
Moment of inertia of the mass	I	L^2M	LFT^2	$g \cdot cm^2$	$kg \cdot m^2$	$m \cdot kgf \cdot s^2$
Pressure	p	$L^{-1}MT^{-2}$	$L^{-2}F$	dyn/cm^2	N/m^2	kgf/m^2
Percussion (impulse of the force)	$P, \int F dt$	LMT^{-1}	FT	$dyn \cdot s$	$N \cdot s$	$kgf \cdot s$
Moment of a force (couple)	M	L^2MT^{-2}	LF	$dyn \cdot cm$	$N \cdot m$	$kgf \cdot m$
Impulse of the moment of a force	$\int M dt$	L^2MT^{-1}	LFT	$dyn \cdot cm \cdot s$	$N \cdot m \cdot s$	$kgf \cdot m \cdot s$
Momentum (linear momentum)	H	LMT^{-1}	FT	$g \cdot cm/s$	$kg \cdot m/s$	$kgf \cdot s$
Moment of momentum (angular momentum)	K	L^2MT^{-1}	LFT	$g \cdot cm^2/s$	$kg \cdot m^2/s$	$kgf \cdot m \cdot s$
Work	L	L^2MT^{-2}	LF	erg	J	$kgf \cdot m$
Energy (mechanical)	E, T, V	L^2MT^{-2}	LF	erg	J	$kgf \cdot m$
Power (mechanical)	P	L^2MT^{-3}	LFT^{-1}	erg/s	W	$kgf \cdot m/s$

The force is a derived quantity in case of a physical system of units; it is determined by the basic law of mechanics (1.1.89), as well as by the law of universal (Newtonian) attraction (1.1.84). The universal constant f has a value which depends *univocally* on the arbitrary magnitude of the basic units; if we wish that the relation (1.1.84) be written so as to have $f = 1$, then the mass must no more be considered as a basic quantity, being defined by means of length and acceleration. Hence, the unit of mass becomes a derived unit, and can be expressed with the aid of units of length and time, the latter ones remaining arbitrary, thus, the number of basic units is reduced with a unity (from three to two). To reduce the number of basic units with a unity more, one must take into consideration a relation between length and time, containing a universal constant (for instance the velocity of light in vacuum); if in the considered system of units the second remains the basic unit for time, then one obtains approximately 300 Mm as unit for length. The unit of length becomes thus a derived unit. If we take, for instance, Planck's constant h equal to unity, then we obtain a system of units without any basic unit. One may imagine also other systems of units having this property.

Table 1.2

Submultiples			Multiples		
Prefix	Symbol	Multiplicative factor	Prefix	Symbol	Multiplicative factor
deci	d	10^{-1}	deca	da	10^{1}
centi	c	10^{-2}	hecto	h	10^{2}
mili	m	10^{-3}	kilo	k	10^{3}
micro	μ	10^{-6}	mega	M	10^{6}
nano	n	10^{-9}	giga	G	10^{9}
pico	p	10^{-12}	tera	T	10^{12}
femto	f	10^{-15}	peta	P	10^{15}
atto	a	10^{-18}	exa	E	10^{18}

2.2 Homogeneity

In what follows, after some general considerations and the introduction of dimensional equations, we enounce the basic theorem (the theorem Π) of dimensional analysis; some important applications of the property of homogeneity will be also given.

2.2.1 General considerations

The laws of nature establish certain relations between various physical quantities, i.e., between the numerical values $n_1, n_2, \ldots$ respectively; such a relation can be written in the implicit form

$$f(n_1, n_2, \ldots) = 0 \tag{1.2.6}$$

or in the explicit form (emphasizing a numerical value)

$$n = \varphi(n_1, n_2, \ldots). \tag{1.2.6'}$$

Obviously, instead to write such relations between numerical values, we may write

Newtonian model of mechanics

analogous relations between the corresponding physical magnitudes. The above relations represent an objective truth. We may thus affirm that the laws of physical phenomena are invariant to any change of units; in other words, if $n_1', n_2', \ldots$ are new numerical values of the same magnitudes, then we can write

$$f(n_1', n_2', \ldots) = 0.$$

This condition is fulfilled if an only if

$$f(n_1', n_2', \ldots) = Cf(n_1, n_2, \ldots),$$

where C is a constant which depends only on the ratios between the new and the old basic quantities; but the above condition is verified only if relations of the form (1.2.6) or (1.2.6') are homogeneous. Thus, the property of *homogeneity* appears to be an essential property in the dimensional analysis.

By a variation of the basic units takes place also a variation of the derived physical unit; but one must express the variation of the numerical value of the derived physical quantity as a function of the variation of the units used. A relation which expresses the derived unit as a function of the basic units is called by Maxwell *the dimensional equation* of the considered physical quantity. Birkhoff, Charcosset and others show that the derived unit is expressed as a function of basic units by a *relation of monomial type*, fact considered also by Maxwell. Indeed, let be an arbitrary derived dimensional quantity; let us suppose also that y is a geometric quantity depending only on the lengths $x_1, x_2, \ldots, x_n$ ($y = f(x_1, x_2, \ldots, x_n)$). Let us denote by y' the quantity corresponding to the arguments $x_1', x_2', \ldots, x_n'$. The numerical values of y and y', respectively, depend on the unit of the considered length. Let us make now the unit λ times smaller; in this case, the ratio

$$\frac{y'}{y} = \frac{f(x_1', x_2', \ldots, x_n')}{f(x_1, x_2, \ldots, x_n)} = \frac{f(\lambda x_1', \lambda x_2', \ldots, \lambda x_n')}{f(\lambda x_1, \lambda x_2, \ldots, \lambda x_n)}$$

must not depend on the chosen scale for the basic units. It follows

$$\frac{f(\lambda x_1, \lambda x_2, \ldots, \lambda x_n)}{f(x_1, x_2, \ldots, x_n)} = \frac{f(\lambda x_1', \lambda x_2', \ldots, \lambda x_n')}{f(x_1', x_2', \ldots, x_n')}$$

or

$$\frac{y(\lambda)}{y(1)} = \frac{y'(\lambda)}{y'(1)} = \varphi(\lambda);$$

hence, the ratio of numerical values of a derived geometric quantity, measured by distinct units of length, depends only on their ratio. Noting that

$$\frac{y(\lambda_1)}{y(1)} = \varphi(\lambda_1), \qquad \frac{y(\lambda_2)}{y(1)} = \varphi(\lambda_2),$$

we obtain

$$\frac{\varphi(\lambda_1)}{\varphi(\lambda_2)} = \varphi\left(\frac{\lambda_1}{\lambda_2}\right),$$

because for $x_1' = \lambda_2 x_1$, $x_2' = \lambda_2 x_2,\ldots,$ $x_n' = \lambda_2 x_n$ we have

$$\frac{y(\lambda_1)}{y(\lambda_2)} = \frac{y'\left(\frac{\lambda_1}{\lambda_2}\right)}{y'(1)} = \varphi\left(\frac{\lambda_1}{\lambda_2}\right);$$

differentiating with respect to λ_1 and taking $\lambda_1 = \lambda_2 = \lambda$, we get

$$\frac{1}{\varphi(\lambda)}\frac{\mathrm{d}\varphi(\lambda)}{\mathrm{d}\lambda} = \frac{1}{\lambda}\frac{\mathrm{d}\varphi(\overline{\lambda})}{\mathrm{d}\overline{\lambda}}\bigg|_{\overline{\lambda}=1} = \frac{m}{\lambda},$$

where m is an integer or a rational number. It follows

$$\varphi(\lambda) = \lambda^m, \qquad (1.2.7)$$

where a multiplicative constant is taken equal to unity, because $\varphi(1) = 1$. One obtains thus the result enounced above.

2.2.2 Dimensional equations

If to a λ times variation of a basic unit does correspond a λ^m times variation of the derived unit, then we say that the derived unit has *the dimension* m with respect to the basic unit which caused the considered variation; thus, in a physical system, the dimension $[U]$ of the derived unit U is given by the dimensional equation

$$[U] = L^\alpha M^\beta T^\gamma, \qquad (1.2.8)$$

where α, β and γ are the dimensions corresponding to the basic units L, M and T. We notice that, in such a way, one cannot determine the physical nature of a physical quantity, but only the form in which the latter one depends on the basic units. Hence, the dimensional equation of a physical quantity A in the LMT system will be of the form

$$[A] = L^\alpha M^\beta T^\gamma; \qquad (1.2.9)$$

such an equation is established on the basis of the equation of definition of the considered quantity and of the relation connecting the given quantity to the chosen basic quantities. If the numerical value of the quantity A is equal to the product or to the quotient of two quantities B and C ($A = BC$ or $A = B/C$) of dimensional equations

$$[B] = L^{\alpha_1} M^{\beta_1} T^{\gamma_1}, \quad [C] = L^{\alpha_2} M^{\beta_2} T^{\gamma_2},$$

then one obtains

$$\alpha = \alpha_1 \pm \alpha_2, \quad \beta = \beta_1 \pm \beta_2, \quad \gamma = \gamma_1 \pm \gamma_2;$$

if the numerical value of the quantity A is a power n of the numerical value of the quantity B, then

$$\alpha = n\alpha_1, \quad \beta = n\beta_1, \quad \gamma = n\gamma_1.$$

In general, any physical quantity, which appears in mechanics, can be expressed by derived units, given by relations of the form (1.2.8). In some practical problems one has to do also with *the quantity of heat, the temperature* etc.; as unit of the quantity of heat one uses the calorie, while the temperature is measured in Celsius (eventually Réaumur or Fahrenheit) degrees.

The measurement of various physical quantities must be made with the aid of the most convenient units. There arises the problem of changing the units in the frame of a chosen system of units; starting from the dimension $[U]$ of the unit of a derived quantity, given by (1.2.8), we pass to the dimension

$$[U'] = (\lambda L)^{\alpha} (\mu M)^{\beta} (\tau T)^{\gamma},$$

where λ, μ, τ are numbers (non-dimensional), of the same derived quantity, which will be thus given by

$$[U'] = \lambda^{\alpha} \mu^{\beta} \tau^{\gamma} [U]. \tag{1.2.8'}$$

For example, 1 km/h = $(10^3)^1 (3600)^{-1}$ m/s = $(1/3.6)$ m/s.

A more general problem is that of changing the system of units; thus, a quantity A is expressed in the system LMT in the form (1.2.9), and in a new system XYZ, given by

$$[X] = L^{\alpha_1} M^{\beta_1} T^{\gamma_1}, \quad [Y] = L^{\alpha_2} M^{\beta_2} T^{\gamma_2}, \quad [Z] = L^{\alpha_3} M^{\beta_3} T^{\gamma_3},$$

in the form $[A] = X^a Y^b Z^c$. Observing that, in both systems of units, the dimension of the quantity A is the same ($L^{\alpha} M^{\beta} T^{\gamma} = X^a Y^b Z^c$), and taking into account the properties mentioned above for the dimensional equations, we obtain the general equations of transformation of the units in the form

$$\begin{aligned} \alpha &= \alpha_1 a + \alpha_2 b + \alpha_3 c, \\ \beta &= \beta_1 a + \beta_2 b + \beta_3 c, \\ \gamma &= \gamma_1 a + \gamma_2 b + \gamma_3 c. \end{aligned} \tag{1.2.10}$$

The new system of units is correctly chosen if the determinant

$$\Delta \equiv \begin{vmatrix} \alpha_1 & \alpha_2 & \alpha_3 \\ \beta_1 & \beta_2 & \beta_3 \\ \gamma_1 & \gamma_2 & \gamma_3 \end{vmatrix} \neq 0 ; \qquad (1.2.11)$$

for instance, if we pass from the system LMT to the system LFT for which $[F] = LMT^{-2}$, then we have ($\Delta = 1$)

$$a = \alpha - \beta, \quad b = \beta, \quad c = 2\beta + \gamma. \qquad (1.2.12)$$

Let be a relation of the form

$$f(A_1, A_2, \ldots, A_n) = 0 \qquad (1.2.13)$$

between n physical quantities $A_i = L^{\alpha_i} M^{\beta_i} T^{\gamma_i}$, $i = 1, 2, \ldots, n$; the matrix

$$\begin{bmatrix} \alpha_1 & \alpha_2 & \ldots & \alpha_n \\ \beta_1 & \beta_2 & \ldots & \beta_n \\ \gamma_1 & \gamma_2 & \ldots & \gamma_n \end{bmatrix} \qquad (1.2.14)$$

is called *the dimensional matrix* of the considered relation. For instance, in case of the above change of units, the dimensional matrix is

$$\begin{bmatrix} -\alpha & 0 & 0 & \alpha_1 & \alpha_2 & \alpha_3 \\ 0 & -\beta & 0 & \beta_1 & \beta_2 & \beta_3 \\ 0 & 0 & -\gamma & \gamma_1 & \gamma_2 & \gamma_3 \end{bmatrix}. \qquad (1.2.15)$$

This matrix allows the study of the structure of the functional connections which can take place between various physical quantities.

2.2.3 The Π theorem

The physical laws, obtained on theoretical or experimental way, represent certain functional relations between various quantities, which characterize the phenomenon to study; the system of units used has an influence on the numerical values of these quantities, but it is not connected to the corresponding phenomenon. It follows that the mentioned functional relations must have a certain particular structure. Let be a dimensional quantity A of the form

$$A = f(A_1, A_2, \ldots, A_k, A_{k+1}, \ldots, A_n), \qquad (1.2.16)$$

where $A_1, A_2, \ldots, A_n$ are variable or constant independent dimensional quantities. We suppose that the first k quantities ($k \leq n$) are *dimensional independent* (the dimensional equation corresponding to such a quantity cannot be expressed as a

Newtonian model of mechanics

monomial with the aid of the dimensions of the other quantities); the index k is less or at the most equal to the number of basic units. In this case, the dimensional equations of the quantities $A, A_{k+1}, A_{k+2},..., A_n$ are written with the aid of the dimensions of the quantities $A_1, A_2,..., A_k$, in the form

$$[A] = [A_1]^{m_1} [A_2]^{m_2} ...[A_k]^{m_k}, \quad [A_{k+1}] = [A_1]^{p_1} [A_2]^{p_2} ...[A_k]^{p_k},...$$
$$[A_n] = [A_1]^{q_1} [A_2]^{q_2} ...[A_k]^{q_k}.$$

Let us vary the units of the first k quantities $\alpha_1, \alpha_2,..., \alpha_k$ times, respectively; the new numerical values, in the new units, will be

$$A_1' = \alpha_1 A_1, \quad A_2' = \alpha_2 A_2,..., A_k' = \alpha_k A_k, \quad A' = \alpha_1^{m_1} \alpha_2^{m_2} ...\alpha_k^{m_k} A,$$
$$A_{k+1}' = \alpha_1^{p_1} \alpha_2^{p_2} ...\alpha_k^{p_k} A_{k+1},..., A_n' = \alpha_1^{q_1} \alpha_2^{q_2} ...\alpha_k^{q_k} A_n.$$

It follows that

$$A' = \alpha_1^{m_1} \alpha_2^{m_2} ...\alpha_k^{m_k} f(a_1, a_2,..., a_n)$$
$$= f(\alpha_1 a_1,..., \alpha_k a_k, \alpha_1^{p_1} \alpha_2^{p_2} ...\alpha_k^{p_k} a_{k+1},..., \alpha_1^{q_1} \alpha_2^{q_2} ...\alpha_k^{q_k} a_n).$$

The function f is thus homogeneous with respect to the arbitrary scales $\alpha_1, \alpha_2,..., \alpha_k$; to reduce the number of the arguments in the function f, we choose $\alpha_1 = 1/a_1$, $\alpha_2 = 1/a_2,..., \alpha_k = 1/a_k$, supposing that the first k quantities do not vanish or do not tend to infinity (the results may be applied also in these cases if the function f is continuous for these values of the arguments). Noting that, by hypothesis, relation (1.2.16) does not depend on the chosen units, one chooses a system of units with respect to which the first k physical quantities considered have constant numerical values, equal to unity. The values $\Pi, \Pi_1, \Pi_2,..., \Pi_{n-k}$ of the quantities $A, A_{k+1}, A_{k+2},..., A_n$ in the new units, will be given by

$$\Pi = \frac{a}{a_1^{m_1} a_2^{m_2} ...a_k^{m_k}}, \quad \Pi_1 = \frac{a_{k+1}}{a_1^{p_1} a_2^{p_2} ...a_k^{p_k}},...., \quad \Pi_{n-k} = \frac{a_n}{a_1^{q_1} a_2^{q_2} ...a_k^{q_k}}, \quad (1.2.17)$$

where $a, a_1, a_2,..., a_n$ are the numerical values of the corresponding quantities in the old units. We notice that the numerical values $\Pi, \Pi_1, \Pi_2,..., \Pi_{n-k}$ do not depend on the initial chosen system of units; hence, they can be considered to be non-dimensional quantities. In the new system of units, the relation (1.2.16) becomes

$$\Pi = f(1,1,...,1,\Pi_1,...,\Pi_{n-k}). \quad (1.2.18)$$

Using the results given by Vaschy, Buckingham, Federman, Ryabouchinsky, Ehrenfest, Afanasjeva, Bridgman etc., one may state

Theorem 1.2.1 (theorem Π). *A relation of the form* (1.2.16) *between $n+1$ dimensional physical quantities $A, A_1, A_2,..., A_n$, which is independent of the chosen units, may be*

expressed as a relation of the form (1.2.18) *between* $n+1-k$ *quantities* $\Pi, \Pi_1, \Pi_2, ..., \Pi_{n-k}$, *which are non-dimensional combinations of the* $n+1$ *dimensional quantities.*

This theorem, often denominated as the *Buckingham theorem*, is known as *the theorem of powers' product* (explaining thus the denomination of the theorem), and is the basic theorem of the dimensional analysis. We notice that the number $n+1-k$ of the non-dimensional products, which appear in relation (1.2.18), is equal to the difference between the number $n+1$ of the physical quantities and the rank k of the dimensional matrix of the considered relation. We can replace the system of non-dimensional parameters $\Pi_1, \Pi_2, ..., \Pi_{n-k}$ by another system of non-dimensional parameters $\Pi'_1, \Pi'_2, ..., \Pi'_{n-k}$, depending on the first $n-k$ parameters, the form of function f changing in relation (1.2.18); but one cannot form more than $n-k$ combinations of independent powers.

It is thus seen that any physical relation between dimensional quantities can be formulated as a relation between non-dimensional quantities; the necessity to apply the dimensional analysis to problems of mechanical systems is put into evidence. If the number of parameters characterizing the quantity A is small, then the functional dependence is much more simple. If $k = n$, that is if all the quantities $A_1, A_2, ..., A_n$ are independent from a dimensional point of view, then

$$A = CA_1^{m_1} A_2^{m_2} ... A_n^{m_n}, \qquad (1.2.19)$$

where C is an non-dimensional constant, which is determined on a theoretical or experimental way.

We know that one can choose the number of basic units in an arbitrary manner, but their increasing number entails the introduction of supplementary physical constants, while the number $n+1-k$ of the non-dimensional parameters remains constant; one must judge, from case to case, if the enrichment of the number of information brought by the increase of the number of units is useful or not.

2.2.4 Applications

The property of homogeneity, which must be verified by a physical law, leads to a great number of applications. Thus, a non-homogeneous relation between physical quantities cannot be correct; for instance, a relation of the form $s = at$, where s is a length, a is an acceleration while t is the time, cannot take place. On the other hand, the relation $A = 2\pi R^2$, where A is the area of a circle of radius R is homogeneous, but it is not correct from the point of view of the multiplicative numerical coefficient; indeed, the condition of homogeneity is only a *necessary condition* for the correctness of a relation.

The property of homogeneity allows also to determine the nature of some physical quantities. Let be, for instance, a relation of the form

$$s = \frac{1}{2}at^2 + v_0 t + s_0, \qquad (1.2.20)$$

where s is a length, while t is the time; noting that we must have $[at^2/2] = [v_0 t] = [s_0] = L$, it follows that a is an acceleration, v_0 is a velocity, and s_0 is a length.

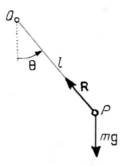

Figure 1.19. Mathematical pendulum.

As well, with the aid of the property of homogeneity and of the Π theorem, we can establish some formulae, abstraction making of certain multiplicative constants. Let be, for instance, the case of *the mathematical pendulum* (the motion on a circle, in a vertical plane, of a heavy particle of mass m, at the end of a thread of length l, fixed at the fixed point O (Fig.1.19)); we denote by θ the angle indicating the position of the particle at the moment t and by **R** the constraint force in the thread, the modulus of which we suppose to be proportional to the modulus of the weight $m\mathbf{g}$ of the particle. We can choose as characteristic parameters the quantities t, l, m, g, θ_0, the last one corresponding to the initial position of the particle. Hence, we write

$$\theta = \theta(t, l, m, g, \theta_0), \quad R = mgf(t, l, m, g, \theta_0), \tag{1.2.21}$$

where θ and f are non-dimensional functions, which do not depend on the units; these functions can be determined starting from the equations of the problem and from the corresponding initial conditions.

Observing that t, l, m are independent physical quantities from the dimensional point of view, we can apply the Π theorem, in which intervene $5 - 3 = 2$ non-dimensional products, for instance $\Pi_1 = \theta_0$ and $\Pi_2 = t\sqrt{g/l}$; it follows that

$$\theta = \theta\left(\theta_0, t\sqrt{\frac{g}{l}}\right), \quad R = mgf\left(\theta_0, t\sqrt{\frac{g}{l}}\right), \tag{1.2.21'}$$

where $\Pi = \theta$, and $\Pi = R/mg$, respectively. The formulae (1.2.21') show that the position of the particle does not depend on the mass m, while the *tension in the thread* (the constraint force) is directly proportional to m.

If T is a characteristic interval of time, for instance *the oscillation period*, we may write

$$T = \sqrt{\frac{l}{g}} \varphi(l, m, g, \theta_0), \tag{1.2.22}$$

where φ is an non-dimensional function; applying once again the Π theorem (l, m, g are independent dimensional quantities), we obtain

$$T = \sqrt{\frac{l}{g}} \varphi(\theta_0). \qquad (1.2.22')$$

The function $\varphi(\theta_0)$ can be determined only theoretically (by solving the corresponding mechanical problem) or experimentally. We notice that one can obtain the formula (1.2.22') also starting from the first formula (1.2.21'). By considerations of symmetry, one has $\varphi(\theta_0) = \varphi(-\theta_0)$, the function φ being an even one; we may thus admit a development of the form (valid for large oscillations)

$$\varphi(\theta_0) = c_1 + c_2 \theta_0^2 + c_3 \theta_0^4 + \dots.$$

In case of small oscillations, there results

$$T = c_1 \sqrt{\frac{l}{g}}, \qquad (1.2.23)$$

the period T being thus determined, abstraction making of a multiplicative constant; the study of this problem leads to $c_1 = 2\pi$. Admitting initially that T depends on l, g, m, but does not depend on θ_0, we can write $T = c_1 l^\alpha g^\beta m^\gamma$; considerations of dimensional homogeneity lead to the same formula (1.2.23).

2.3 Similitude

In what follows we make some general considerations and introduce the models of Froude, Cauchy, Reynolds, Weber etc.; we use thus the geometric, the kinematic and the dynamic similitude.

2.3.1 General considerations

In many problems of the theory of mechanical systems, the physical phenomenon is very complex and depends on many parameters, a study of theoretical nature being extremely difficult; in such cases, it is useful to make an experimental study on *technical models*, which reproduce at a reduced scale the considered construction or element of construction. To do this, we must observe that, generally, any physical phenomenon is expressed by a system of functional equations, which establish relations between various physical quantities which appear; the development of the respective physical phenomenon does not depend on the system of units used.

The models, which will be labelled by index m, will be, at a reduced scale, constructions similar to the *real* ones (*original, prototypes*), which will be labelled by index r. At the basis of a study by modelling, considerations of *similitude* must be taken into consideration. The study of physical phenomena on models is made in a laboratory and is simpler and more economic; but the results thus obtained must be suitable to the real constructions.

In general, two bodies (the real one and its model) are geometrically similar if their corresponding lengths are in the same ratio (have the same scale). Analogously to the geometrical similitude, one can introduce a physical one; we say that two physical phenomena are similar if it is possible to obtain the characteristics of one from the characteristics of the other one, on the basis of the respective scales. If A is an arbitrary physical quantity, we may write

$$A_r = k_A A_m, \qquad (1.2.24)$$

where k_A is a coefficient of similitude corresponding to this quantity. An ideal model must have a perfect similitude (for instance, a *general mechanical similitude*) that is the constant k_A must not depend on the particular physical quantity A; the laws governing the model are in this case identical to those governing the real object. Practically, such a model cannot be realized, because one cannot reduce in the same ratio quantities as lengths, areas, volumes, velocities, accelerations, forces, densities, coefficients of friction, unit weights etc.; one is thus obliged to use an *incomplete mechanical similitude*. Taking into account the importance of a quantity or of another one in the study of a mechanical phenomenon, we may consider a geometric, static, kinematic, dynamic, thermic similitude etc.; thus, various *similitude criteria* are put into evidence. Newton noticed (1686) that the values of the similitude criteria, homologous to two similar physical processes, are equal. Federman shows in 1911 that any physical process can be described with the aid of a functional relation between the respective similitude criteria.

As V.L. Kirpichev noticed in 1874, two physical processes are similar if and only if they are qualitatively similar and their homologous similitude criteria have equal values.

We will denote by λ, μ, τ, χ the similitude coefficients (*the* respective *scales*) for lengths, masses, time and forces, respectively; hence

$$\lambda = \frac{L_r}{L_m}, \quad \mu = \frac{M_r}{M_m}, \quad \tau = \frac{T_r}{T_m}, \quad \chi = \frac{F_r}{F_m}. \qquad (1.2.25)$$

If for a physical quantity of a special interest we put the condition to have the same values for the model and for the real object, then there result certain relations between these coefficients. We put thus in evidence various *laws of similitude* (*modelling laws*), corresponding to various models.

2.3.2 Geometric, static and kinematic similitude

In *the geometric similitude* appears only the space and the only basic quantity is the length. Thus, the scale for lengths will be $l_r / l_m = \lambda$, the scale for areas $A_r / A_m = \lambda^2$ and the scale for volumes $V_r / V_m = \lambda^3$.

In the case of *the static similitude*, besides lengths one introduces also the forces; in this case, the independent scales λ and χ are introduced.

The kinematic similitude introduces the length (scale λ) and the time (scale τ); these scales are also independent. One obtains thus: for velocities $v_r / v_m = \lambda \tau^{-1}$, for accelerations $a_r / a_m = \lambda \tau^{-2}$, for angular velocities $\omega_r / \omega_m = \tau^{-1}$, for angular accelerations $\varepsilon_r / \varepsilon_m = \tau^{-2}$ etc.

2.3.3 Dynamic similitude. Newton's similitude law

The dynamic similitude contains the static similitude, as well as the kinematic one; the ratios between the homologous masses must be also equal, hence the scale μ of the masses intervenes. But only three scales are basic, for instance λ, μ and τ; taking into account the basic law (1.1.89) and the accelerations scale, the relation

$$\chi = \lambda \mu \tau^{-2}, \qquad (1.2.26)$$

called *Bertrand's characteristic equation*, must take place. In this case, the similitude sets also for the forces. One obtains corresponding scales for various derived quantities with a dynamical character. Thus, the scale of the unit masses will be $\lambda^{-3}\mu$, while the scale of the unit weights will have the form $\lambda^{-2}\mu\tau^{-2}$; analogously, one obtains the scale of the work $\lambda^2 \mu \tau^{-2}$, the scale of the powers $\lambda^2 \mu \tau^{-3}$ etc.

To can establish a dynamic similitude between two mechanical systems, one of them being the real one and the other one the model, it is necessary and sufficient that the forces acting upon the model do have the same direction as those on the prototype, the ratio of their moduli being constant and given by the relation (1.2.26); this relation is called also *Newton's law of similitude*.

For instance, let be *the central motion* of a system of particles, the modulus of each central attractive force being proportional to the mass of the particle and to the distance at the *n*th power with respect to a fixed pole; the ratio χ of the attraction forces corresponding to the real mechanical system and to its model, respectively, is given by

$$\chi = \lambda^n \mu. \qquad (1.2.27)$$

Taking into account the relation (1.2.26), we find the condition relation

$$\tau^2 = \lambda^{1-n}; \qquad (1.2.28)$$

noting that the particles describe orbits around the attraction centre, we may affirm that τ is equal to the ratio of the revolution times of these ones. In the particular case of Newtonian attraction forces (in inverse proportion to the square of distances $n = -2$), the relation (1.2.28) becomes

$$\tau^2 = \lambda^3; \qquad (1.2.28')$$

hence, the squares of revolution times are proportional to the cubes of the lengths, corresponding to the third law of Kepler.

Newtonian model of mechanics 73

2.3.4 Particular models

Newton's similitude law asks that all the forces which are acting upon the real mechanical system and upon the model, respectively, be in the ratio χ given by relation (1.2.26). Practically, not all the forces can be reduced in the same ratio; hence, from case to case, one considers only the similitude of those forces which are prevailing in the phenomenon to study. Thus, we may use various similitude laws (particular models), which are – in general – of the form $\tau = \varphi(\lambda)$.

If we put the condition to obtain the same acceleration for the model and for the real object (for example, if *the gravity forces* are predominant), then we may write $a_r / a_m = \lambda \tau^{-2} = 1$, hence

$$\tau = \sqrt{\lambda}; \qquad (1.2.29)$$

we obtain thus *Froude's mechanical model*. For $\chi = \mu = M_r / M_m = \mu_r V_r / \mu_m V_m = (\mu_r / \mu_m)\lambda^3$, in the case in which the model is of the same material as the real object, we may write

$$\chi = \mu = \lambda^3. \qquad (1.2.30)$$

Putting the condition to obtain the same stresses (equal to the ratio between the forces and the areas of the surfaces upon which they are applied) on the model and on the real object, we may write

$$\frac{\sigma_r}{\sigma_m} = \frac{L_r M_r T_r^{-2} / L_r^2}{L_m M_m T_m^{-2} / L_m^2} = \lambda^{-1} \mu \tau^{-2} = 1;$$

if we use the same material for the model as for the real object, hence if relation (1.2.30) takes place, then we obtain the law

$$\tau = \lambda, \qquad (1.2.31)$$

corresponding to *Cauchy's elastic model*.

If the *internal friction forces* have a dominant action (in case of viscous fluids intervene the *kinematic viscosity coefficient* ν), then we impose the condition

$$\frac{\nu_r}{\nu_m} = \frac{L_r^2 T_r^{-1}}{L_m^2 T_m^{-1}} = \lambda^2 \tau^{-1} = 1;$$

obtaining thus *Reynold's hydraulic model* for which the relation

$$\tau = \lambda^2 \qquad (1.2.32)$$

takes place. If the model and the real object are of the same material, the relation (1.2.30) remains still valid.

The condition that the *superficial tension* (equal to the ratio between the force and the length) measured on the model be equal to that on the real object

$$\frac{L_r M_r T_r^{-2} / L_r}{L_m M_m T_m^{-2} / L_m} = \mu \tau^{-2} = 1$$

leads to

$$\mu = \tau^2 ; \tag{1.2.33}$$

for liquids (the same real object and model) the relation (1.2.30) takes place, so that

$$\tau = \lambda\sqrt{\lambda} , \tag{1.2.34}$$

corresponding to *Weber's model*.

Chapter 2

MECHANICS OF THE SYSTEMS OF FORCES

The study of the systems of forces is of a special interest in mechanics because, applying them upon a body, this one maintains its state of rest or motion or is changing this state; moreover, the equivalence of the systems of forces represents one of the objects of study of mechanics, as it was shown in Chap. 1, Subsec. 1.2.3. Taking into account the vectorial modelling of forces, their algebra is – in fact – a *vector algebra*.

1. Introductory notions

In what follows we will deal with some applications of the principle of the parallelogram of forces and of the product of a scalar by a force; one observes thus that the forces are elements of a vector space. As well, we introduce some important products of forces, modelled as vectors.

1.1 Decomposition of forces. Bases

The elementary operations which can be effected with forces are the basis for the study of equivalence of the systems of forces; the decomposition of forces is realized with the aid of such operations and allows the introduction of the notion of basis of a system of forces. We notice that all the results which will be given in this section remain valid for any system of vectors.

1.1.1 Decomposition of forces. Linear dependence

Let us consider the decomposition of a force $\mathbf{F}$ into a sum of n forces $\mathbf{F}_i$, $i = 1, 2, \ldots, n$, of given directions. If $n = 3$ and if we admit that the three directions are non-coplanar, then we obtain a unique decomposition, observing that the force $\mathbf{F}$ is the diagonal of the parallelepipedon formed with the forces $\mathbf{F}_1$, $\mathbf{F}_2$ and $\mathbf{F}_3$ (in general, an oblique parallelepipedon) (Fig.2.1,a); we thus write

$$\mathbf{F} = \mathbf{F}_1 + \mathbf{F}_2 + \mathbf{F}_3 . \tag{2.1.1}$$

If $n > 3$, then we may always choose the forces $\mathbf{F}_4$, $\mathbf{F}_5, \ldots, \mathbf{F}_n$ arbitrarily, and the decomposition is no more unique; as well, the decomposition is not unique neither for $n = 3$ if the three directions initially chosen are coplanar with the force $\mathbf{F}$. If $n = 2$ and the chosen directions are coplanar with the force $\mathbf{F}$, but not collinear, then we obtain a unique decomposition in the plane (the force $\mathbf{F}$ is the diagonal of the parallelogram formed with $\mathbf{F}_1$ and $\mathbf{F}_2$) (Fig.2.1,b); if the two directions are collinear with $\mathbf{F}$, then the decomposition is no more unique.

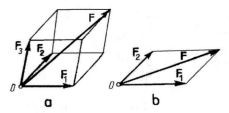

Figure 2.1. Decomposition of forces. Three-dimensional (a) and plane (b) case.

Let be n non-zero forces $\mathbf{F}_1$, $\mathbf{F}_2$,..., $\mathbf{F}_n$. We say that these forces are *linear independent* if the relation

$$\lambda_1 \mathbf{F}_1 + \lambda_2 \mathbf{F}_2 + ... + \lambda_n \mathbf{F}_n = \mathbf{0} \qquad (2.1.2)$$

can take place only for zero values of the scalars λ_1, λ_2,..., λ_n. From the relation (2.1.2), it follows

$$\lambda_1 F_{1i} + \lambda_2 F_{2i} + ... + \lambda_n F_{ni} = 0, \quad i = 1, 2, 3. \qquad (2.1.2')$$

This linear system in λ_j, $j = 1, 2, ..., n$, admits only vanishing solutions if and only if the rank of the matrix of the coefficients of the system is equal to n; because this rank is less or equal to three, it follows that *in the three-dimensional space there exist at most three linear independent forces.*

We say that the forces $\mathbf{F}_1$, $\mathbf{F}_2$,..., $\mathbf{F}_n$ are *linear dependent* if the relation (2.1.2) holds, but not all the scalars λ_1, λ_2,..., λ_n vanish. Thus, for $n = 3$ the condition of linear dependence of the forces $\mathbf{F}_1$, $\mathbf{F}_2$, $\mathbf{F}_3$ is written in the form

$$\lambda_1 \mathbf{F}_1 + \lambda_2 \mathbf{F}_2 + \lambda_3 \mathbf{F}_3 = \mathbf{0}; \qquad (2.1.3)$$

assuming, for instance, that $\lambda_1 \neq 0$, we may express the force $\mathbf{F}_1$ as follows

$$\mathbf{F}_1 = \lambda \mathbf{F}_2 + \mu \mathbf{F}_3, \qquad (2.1.3')$$

where λ, μ are also scalars. It is easy to see that the force $\mathbf{F}_1$ is contained in the plane formed by the forces $\mathbf{F}_2$ and $\mathbf{F}_3$, admitting that these ones are applied in the very same point. Hence, the necessary and sufficient condition for the three forces to be *coplanar* is that they verify the relation (2.1.3), the scalars λ_1, λ_2, λ_3 non-vanishing simultaneously, or the relation (2.1.3'); returning to the system (2.1.2'), the condition is fulfilled if

Mechanics of the systems of forces

$$\begin{vmatrix} F_{11} & F_{21} & F_{31} \\ F_{12} & F_{22} & F_{32} \\ F_{13} & F_{23} & F_{33} \end{vmatrix} = 0. \qquad (2.1.3'')$$

We notice also that the condition (2.1.3") is equivalent to

$$\det[F_{ij}] = \begin{vmatrix} F_{11} & F_{12} & F_{13} \\ F_{21} & F_{22} & F_{23} \\ F_{31} & F_{32} & F_{33} \end{vmatrix} = 0. \qquad (2.1.3''')$$

In the case $n = 2$ (case considered in Chap. 1, Subsec. 1.1.2), we obtain the condition of linear dependence

$$\lambda_1 \mathbf{F}_1 + \lambda_2 \mathbf{F}_2 = \mathbf{0}; \qquad (2.1.4)$$

this is the necessary and sufficient condition (if the scalars λ_1, λ_2 do not vanish simultaneously) for the forces $\mathbf{F}_1$ and $\mathbf{F}_2$ to be *collinear*. This condition can be written in the form

$$\mathbf{F}_1 = \lambda \mathbf{F}_2, \qquad (2.1.4')$$

where λ is also a scalar; returning to the system (2.1.2'), the condition becomes

$$\frac{F_{11}}{F_{21}} = \frac{F_{12}}{F_{22}} = \frac{F_{13}}{F_{23}}. \qquad (2.1.4'')$$

1.1.2 Basis. Canonical representation of forces

An ordered triplet of linear independent forces $\mathbf{F}_1$, $\mathbf{F}_2$, $\mathbf{F}_3$ constitutes a *basis*. In this case, we can express an arbitrary force $\mathbf{F}$ in the form

$$\mathbf{F} = \lambda_1 \mathbf{F}_1 + \lambda_2 \mathbf{F}_2 + \lambda_3 \mathbf{F}_3, \qquad (2.1.5)$$

where λ_1, λ_2, λ_3 are scalars, called the numerical components of the force $\mathbf{F}$ in the given basis. Such a representation can be obtained for an arbitrary vector $\mathbf{V}$, with the aid of some vectors $\mathbf{V}_i$, $i = 1, 2, 3$, which determine a basis. If the basis' vectors are orthogonal one to each other, then the basis is called *orthogonal*. If the basis' vectors are unit vectors, then the basis is called *normed*. An orthogonal and normed basis is called *orthonormed*.

Taking into account the fact that many other mechanical quantities are modelled with the aid of vectors, we will consider, in what follows, the vectors as *abstract mathematical entities*. Let thus be *right-handed* (or *positive*) orthonormed bases (if the vectors are in the order $\mathbf{V}_1$, $\mathbf{V}_2$, $\mathbf{V}_3$, then an observer, situated – for instance – along the vector $\mathbf{V}_1$, sees the superposition of the vector $\mathbf{V}_2$ onto the vector $\mathbf{V}_3$, after a rotation of $\pi/2$ *in the positive direction*, from right to left). More general, a triad of arbitrary vectors $\mathbf{V}_1$, $\mathbf{V}_2$, $\mathbf{V}_3$ forms a positive basis, in the given order, if an observer, situated along the vector $\mathbf{V}_1$, sees the superposition of the vector $\mathbf{V}_2$ onto the vector $\mathbf{V}_3$ by a rotation less than π, in the positive direction; the property must be maintained for a circular permutation of the three vectors.

For a system of Cartesian co-ordinates Ox_j, the orthonormed basis' vectors will be formed by the unit vectors $\mathbf{i}_j$, $j = 1,2,3$, of *canonical co-ordinates* $(1,0,0)$, $(0,1,0)$, $(0,0,1)$, respectively, located at O; in this case, a vector of canonical co-ordinates V_j, $j = 1,2,3$, can be written in the form

$$\mathbf{V} = (V_1, V_2, V_3) = V_1(1,0,0) + V_2(0,1,0) + V_3(0,0,1) = V_j \mathbf{i}_j; \qquad (2.1.6)$$

this is *the canonical representation of the vector* and is unique. In particular, we obtain *the canonical representation of a force*

$$\mathbf{F} = F_j \mathbf{i}_j. \qquad (2.1.6')$$

We notice that the canonical co-ordinates V_1, V_2, V_3 are just the *numerical (scalar) components* of the vector, $V_1\mathbf{i}_1$, $V_2\mathbf{i}_2$, $V_3\mathbf{i}_3$ being its *vector components*; because $\mathbf{i}_1$, $\mathbf{i}_2$, $\mathbf{i}_3$ are fixed given unit vectors, it is sufficient to use numerical components, which – for the sake of simplicity – will be called *components*.

If the vector $\mathbf{V}$ and the unit vectors $\mathbf{i}_j$ are *directed segments*, then the numerical components are numbers; but we mention that, from a dimensional point of view, these components are not always numbers, their dimensions depending on the physical dimension of the vector and of the unit vectors of the chosen basis.

In the case of a bound vector $\mathbf{V} = \overrightarrow{PQ}$, we may write

$$\mathbf{V} = \mathbf{r}_Q - \mathbf{r}_P = Q - P = \left(x_j^{(Q)} - x_j^{(P)} \right) \mathbf{i}_j, \qquad (2.1.7)$$

where we have put into evidence the co-ordinates of the origin and of the extremity of

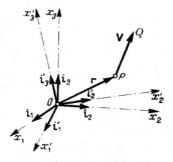

Figure 2.2. A bound vector $\mathbf{V}$ in two right-handed systems of co-ordinate axes.

Mechanics of the systems of forces 79

the vector; this is also the representation of a directed segment. We observe that we may note a position vector also by the point indicated by it.

Let be a positive orthonormed basis of unit vectors $\mathbf{i}_j$, hence a *right-handed system of orthogonal co-ordinate axes* $Ox_j, j = 1,2,3$; let also be a second positive orthonormed basis of unit vectors $\mathbf{i}'_k$, to which corresponds a second right-handed system of orthogonal Cartesian axes Ox'_k, $k = 1,2,3$ (Fig.2.2). Obviously, we have

$$\mathbf{i}'_k = \alpha_{kj}\mathbf{i}_j, \quad k = 1,2,3, \qquad (2.1.8)$$

where we have introduced the cosines

$$\alpha_{kj} = \cos(\mathbf{i}'_k, \mathbf{i}_j), \quad j,k = 1,2,3; \qquad (2.1.9)$$

analogously, we may write

$$\mathbf{i}_j = \alpha_{kj}\mathbf{i}'_k, \quad j = 1,2,3. \qquad (2.1.8')$$

Let be *the position vector* $\mathbf{r}$ of a point P; we obtain, in the two systems of co-ordinates,

$$\mathbf{r} = x_j\mathbf{i}_j = x'_k\mathbf{i}'_k, \qquad (2.1.10)$$

where x_1, x_2, x_3, and x'_1, x'_2, x'_3, respectively, are the co-ordinates of the point in each of the two systems. Taking into account (2.1.8) and (2.1.8'), respectively, we find the relations allowing to pass from a system of co-ordinates to another one

$$x'_k = \alpha_{kj}x_j, \quad x_j = \alpha_{kj}x'_k, \quad j,k = 1,2,3; \qquad (2.1.11)$$

these linear transformations are orthogonal.

1.2 Products of vectors

The conditions of orthogonality or of collinearity of two vectors can be expressed with the aid of their scalar or vector products, respectively; the condition of coplanarity of three vectors introduces their scalar triple product. One may conceive also a vector triple product.

1.2.1 Scalar product of two vectors

The *scalar product* (*internal product*, *dot product*) of the (free, bound or sliding) vectors $\mathbf{V}_1$ and $\mathbf{V}_2$ is a scalar (which does no more belong to the vector space) defined by the relation

$$\mathbf{V}_1 \cdot \mathbf{V}_2 = V_1V_2\cos(\mathbf{V}_1, \mathbf{V}_2). \qquad (2.1.12)$$

In particular, we obtain *the projection* of a vector $\mathbf{V}$ on a directed axis Δ of unit vector $\mathbf{u}$ in the form (Fig.2.3)

$$\mathbf{V}_\Delta = \mathbf{V} \cdot \mathbf{u} = V\cos(\mathbf{V},\mathbf{u}) = \mathrm{pr}_\Delta \mathbf{V}\,; \qquad (2.1.13)$$

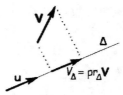

Figure 2.3. Projection of vector **V** on the directed axis Δ.

obviously, the projection (which is a scalar) has a sign, as its direction coincides (sign +) with the direction of **u** or not (sign −). One can thus write the scalar product also in the form

$$\mathbf{V}_1 \cdot \mathbf{V}_2 = V_1 \mathrm{pr}_{\mathbf{V}_1}\mathbf{V}_2 = V_2 \mathrm{pr}_{\mathbf{V}_2}\mathbf{V}_1\,. \qquad (2.1.12')$$

We mention following properties:
i) $\mathbf{V}_1 \cdot \mathbf{V}_2 = \mathbf{V}_2 \cdot \mathbf{V}_1$ (*commutativity*);
ii) $\mathbf{V}_1 \cdot (\mathbf{V}_2 + \mathbf{V}_3) = \mathbf{V}_1 \cdot \mathbf{V}_2 + \mathbf{V}_1 \cdot \mathbf{V}_3$ (*distributivity with respect to the addition of vectors*; consequence of the properties expressed by the relations (1.1.7) and (2.1.13)).

The unit vectors of the co-ordinate axes verify the relations

$$\mathbf{i}_j \cdot \mathbf{i}_k = \delta_{jk}\,, \quad j,k = 1,2,3\,, \qquad (2.1.14)$$

where we have introduced *Kronecker's symbol*

$$\delta_{jk} = \begin{cases} 1 & \text{for } j = k, \\ 0 & \text{for } j \neq k. \end{cases} \qquad (2.1.15)$$

If we express the vectors in the canonical form

$$\mathbf{V}_1 = V_{1j}\mathbf{i}_j\,, \quad \mathbf{V}_2 = V_{2k}\mathbf{i}_k\,, \qquad (2.1.16)$$

then it follows

$$\mathbf{V}_1 \cdot \mathbf{V}_2 = V_{1j}V_{2j}\,, \qquad (2.1.17)$$

where we took account (2.1.14). In particular, if $\sphericalangle(\mathbf{V}_1,\mathbf{V}_2) = 0$ or $\sphericalangle(\mathbf{V}_1,\mathbf{V}_2) = \pi$, then the vectors are collinear and we may write

$$\mathbf{V}_1 \cdot \mathbf{V}_2 = \pm V_1 V_2\,, \qquad (2.1.18)$$

as the vectors have or not the same direction; the square of the modulus (1.1.2) of a vector is given by

Mechanics of the systems of forces

$$\mathbf{V} \cdot \mathbf{V} = \mathbf{V}^2 = V^2 = V_i V_i. \tag{2.1.19}$$

If $\sphericalangle(\mathbf{V}_1, \mathbf{V}_2) = \pi/2$, then there results

$$\mathbf{V}_1 \cdot \mathbf{V}_2 = V_{1j} V_{2j} = 0, \tag{2.1.20}$$

obtaining thus *the necessary and sufficient condition of orthogonality* of the vectors $\mathbf{V}_1$ and $\mathbf{V}_2$ (supposing that $\mathbf{V}_1, \mathbf{V}_2 \neq \mathbf{0}$); obviously, the relation (2.1.20) is verified also if one of the factors vanishes. It is a case in which one has *divisors of zero*.

The square of the relation (1.1.4) leads to *the modulus of the sum of two vectors* in the form

$$V = \sqrt{V_1^2 + V_2^2 + 2V_1 V_2 \cos(\mathbf{V}_1, \mathbf{V}_2)}. \tag{2.1.21}$$

The angle (less than π) *of two directed axes* of vectors $\mathbf{V}_1$ and $\mathbf{V}_2$ is given by

$$\cos(\mathbf{V}_1, \mathbf{V}_2) = \frac{\mathbf{V}_1 \cdot \mathbf{V}_2}{V_1 V_2} = \frac{V_{1i} V_{2i}}{\sqrt{V_{1j} V_{2j}} \sqrt{V_{1k} V_{2k}}}. \tag{2.1.22}$$

We can write *the component of a vector* along an axis in the form

$$V_j = \mathbf{V} \cdot \mathbf{i}_j, \quad j = 1, 2, 3, \tag{2.1.23}$$

so that *the canonical representation* (2.1.6) becomes

$$\mathbf{V} = (\mathbf{V} \cdot \mathbf{i}_j) \mathbf{i}_j. \tag{2.1.6''}$$

As well, the cosines given by (2.1.9) may be expressed by the relations

$$\alpha_{kj} = \mathbf{i}'_k \cdot \mathbf{i}_j, \quad j, k = 1, 2, 3; \tag{2.1.9'}$$

in this case, the scalar products of the relations (2.1.8), (2.1.8') by $\mathbf{i}'_l$ and $\mathbf{i}_l$, $l = 1, 2, 3$, respectively, lead to the relations

$$\alpha_{ij} \alpha_{ik} = \delta_{jk}, \quad \alpha_{ji} \alpha_{ki} = \delta_{jk}, \quad j, k = 1, 2, 3, \tag{2.1.24}$$

verified by these cosines.

1.2.2 Vector product of two vectors

The *vector product* (*external product*, *cross product*) of the (free, bound or sliding) vectors $\mathbf{V}_1$ and $\mathbf{V}_2$ is a vector

$$\mathbf{V} = \mathbf{V}_1 \times \mathbf{V}_2, \tag{2.1.25}$$

the direction of which is normal to the plane formed by the given vectors (supposing that they are applied at the same point), such that the vectors $\mathbf{V}_1, \mathbf{V}_2, \mathbf{V}$ (in this order) form a positive basis, and the modulus of which is given by

$$V = V_1 V_2 \sin(\mathbf{V}_1, \mathbf{V}_2); \qquad (2.1.25')$$

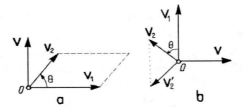

Figure 2.4. Vector product $\mathbf{V}$ of two vectors (a). Vector $\mathbf{V}_2'$ of orthogonal projection (b).

hence, the modulus of the vector product is equal to *the area of the parallelogram defined by the two vectors* (Fig.2.4,a). Thus, we introduce an *oriented plane*, which puts into evidence a *positive direction* for the angle $\sphericalangle(\mathbf{V}_1, \mathbf{V}_2) < \pi$ (so that an observer situated along the vector $\mathbf{V}$ sees the superposition of $\mathbf{V}_1$ onto $\mathbf{V}_2$ by a rotation from right to left, the right-hand rule). *The oriented plane element* (bounded by a closed simple curve), corresponding to the previous definition, is called a *bivector* and is represented by the vector $\mathbf{V}$. In contradistinction to the vector associated to an oriented segment (with the aid of which we have built up the vector space), called also *polar vector*, this vector will be called *axial vector*; the denomination is justified, because it corresponds to a rotation in the oriented plane, about an axis normal to this plane. We notice that one cannot sum a vector with a bivector, but one can set up an analogous *algebra of bivectors*.

Let $\mathbf{V}_2'$ be *the orthogonal projection vector* of vector $\mathbf{V}_2$ on a plane normal to the vector $\mathbf{V}_1$ (Fig.2.4,b); it is easy to see that

$$\mathbf{V} = \mathbf{V}_1 \times \mathbf{V}_2 = \mathbf{V}_1 \times \mathbf{V}_2', \qquad (2.1.26)$$

because the two vector products lead to axial vectors having the same direction and modulus. We mention the following properties:
i) $\mathbf{V}_1 \times \mathbf{V}_2 = -\mathbf{V}_2 \times \mathbf{V}_1$ (*anticommutativity*);
ii) $\mathbf{V}_1 \times (\mathbf{V}_2 + \mathbf{V}_3) = \mathbf{V}_1 \times \mathbf{V}_2 + \mathbf{V}_1 \times \mathbf{V}_3$ (*distributivity with respect to the addition of vectors*; consequence of the property expressed by the relations (2.1.26)).

If we express the vectors in the canonical form (2.1.16), then there results, symbolically,

$$\mathbf{V}_1 \times \mathbf{V}_2 = \begin{vmatrix} \mathbf{i}_1 & \mathbf{i}_2 & \mathbf{i}_3 \\ V_{11} & V_{12} & V_{13} \\ V_{21} & V_{22} & V_{23} \end{vmatrix}, \qquad (2.1.27)$$

Mechanics of the systems of forces

because

$$\mathbf{i}_j \times \mathbf{i}_k = \in_{jkl} \mathbf{i}_l, \quad j,k = 1,2,3, \tag{2.1.28}$$

$\in_{jkl}$ being *Ricci's permutation symbol*

$$\in_{jkl} = \begin{cases} 1 & \text{for } (j,k,l) = (1,2,3), \\ -1 & \text{for } (j,k,l) = (2,1,3), \\ 0 & \text{for } j = k \text{ or } k = l \text{ or } l = j, \end{cases} \tag{2.1.29}$$

where we used the notation (1.1.1). For instance, the component of the vector product (2.1.25) along the axis Ox_l is given by

$$V_l = \in_{jkl} V_{1j} V_{2k}, \tag{2.1.30}$$

so that

$$\mathbf{V}_1 \times \mathbf{V}_2 = \in_{jkl} V_{1j} V_{2k} \mathbf{i}_l; \tag{2.1.30'}$$

this is, in fact, the development of the determinant (2.1.27).

In particular, if $\sphericalangle(\mathbf{V}_1, \mathbf{V}_2) = \pi/2$, then the two vectors are orthogonal and we have

$$|\mathbf{V}_1 \times \mathbf{V}_2| = V_1 V_2. \tag{2.1.31}$$

If $\sphericalangle(\mathbf{V}_1, \mathbf{V}_2) = 0$ or $\sphericalangle(\mathbf{V}_1, \mathbf{V}_2) = \pi$, then the vectors are collinear and we can write

$$\mathbf{V}_1 \times \mathbf{V}_2 = \mathbf{0}; \tag{2.1.32}$$

this relation represents *the necessary and sufficient condition of collinearity* of the vectors $\mathbf{V}_1$ and $\mathbf{V}_2$ (supposing that $\mathbf{V}_1, \mathbf{V}_2 \neq \mathbf{0}$) and is equivalent to a condition of the form (2.1.4") (we use the formula (2.1.27)). Obviously, the relation (2.1.32) is verified also if one of the factors vanishes. It is another case in which one has *divisors of zero*.

By using the formulae (2.1.12) and (2.1.25'), we deduce *Lagrange's identity*

$$(\mathbf{V}_1 \cdot \mathbf{V}_2)^2 = V_1^2 V_2^2 - |\mathbf{V}_1 \times \mathbf{V}_2|^2; \tag{2.1.33}$$

from this relation one can obtain also *Cauchy's inequality*

$$(\mathbf{V}_1 \cdot \mathbf{V}_2)^2 \leq V_1^2 V_2^2. \tag{2.1.34}$$

1.2.3 Scalar triple product of three vectors

The scalar triple product (*mixed product*) of three (free, bound or sliding) vectors $\mathbf{V}_1$, $\mathbf{V}_2$ and $\mathbf{V}_3$ is the scalar defined by the right or by the left member of the relation

$$(\mathbf{V}_1 \times \mathbf{V}_2) \cdot \mathbf{V}_3 = \mathbf{V}_1 \cdot (\mathbf{V}_2 \times \mathbf{V}_3), \tag{2.1.35}$$

which is easy verified if one takes into account the expressions (2.1.17) and (2.1.27) of the scalar and vector products, respectively. The notation of this product in the form

$$(\mathbf{V}_1, \mathbf{V}_2, \mathbf{V}_3) = \det[V_{ij}] = \epsilon_{ijk} V_{1i} V_{2j} V_{3k} = \epsilon_{ijk} V_{i1} V_{j2} V_{k3}, \tag{2.1.36}$$

where we took also (2.1.30') into account, is justified because it is immaterial what member of the definition relation we use. We have thus obtained the development of a determinant of third order too. We may also write

$$\epsilon_{lmn} \det[V_{pq}] = \epsilon_{ijk} V_{li} V_{mj} V_{nk} = \epsilon_{ijk} V_{il} V_{jm} V_{kn}, \quad l,m,n = 1,2,3, \tag{2.1.37}$$

taking into account (2.1.29). Indeed, if two of the indices l, m, n are equal, for instance $l = m$, the product of the quantity $V_{li} V_{mj}$, symmetric with respect to the indices i and j, by Ricci's symbol ϵ_{ijk}, skew-symmetric with respect to these indices, vanishes, as well as ϵ_{mmn}; if all the indices l, m, n are different, for instance $l = 1$, $m = 2$, $n = 3$, then we find again (2.1.36). Analogously, we can prove the relation

$$\delta_{hi} \det[V_{pq}] = \frac{1}{2} \epsilon_{ijk} \epsilon_{lmn} V_{hl} V_{jm} V_{kn}, \quad h,i = 1,2,3. \tag{2.1.37'}$$

We notice also that the scalar triple product of the vectors $\mathbf{U}$, $\mathbf{V}$, $\mathbf{W}$ may be expressed in the form

$$(\mathbf{U}, \mathbf{V}, \mathbf{W}) = \epsilon_{ijk} U_i V_j W_k. \tag{2.1.38}$$

The relations (2.1.35), (2.1.36) show that the scalar product of an axial vector by a polar one has a meaning, because it leads to a scalar; indeed, one can thus introduce the notion of a mixed product of three polar vectors. We mention the following properties:
i) $(\mathbf{V}_1, \mathbf{V}_2, \mathbf{V}_3 + \mathbf{V}_4) = (\mathbf{V}_1, \mathbf{V}_2, \mathbf{V}_3) + (\mathbf{V}_1, \mathbf{V}_2, \mathbf{V}_4)$ (*distributivity with respect to the addition of vectors*);
ii) with three vectors $\mathbf{V}_1$, $\mathbf{V}_2$, $\mathbf{V}_3$ one can form $3! = 6$ mixed products, which verify the relations

$$(\mathbf{V}_1, \mathbf{V}_2, \mathbf{V}_3) = (\mathbf{V}_2, \mathbf{V}_3, \mathbf{V}_1) = (\mathbf{V}_3, \mathbf{V}_1, \mathbf{V}_2)$$
$$= -(\mathbf{V}_2, \mathbf{V}_1, \mathbf{V}_3) = -(\mathbf{V}_1, \mathbf{V}_3, \mathbf{V}_2) = -(\mathbf{V}_3, \mathbf{V}_2, \mathbf{V}_1), \tag{2.1.39}$$

obtaining thus *only two distinct mixed products*, which are of opposite sign.
If we denote $\mathbf{W} = \mathbf{V}_1 \times \mathbf{V}_2$, then we may write

$$(\mathbf{V}_1, \mathbf{V}_2, \mathbf{V}_3) = (\mathbf{V}_1 \times \mathbf{V}_2) \cdot \mathbf{V}_3 = W V_3 \cos(\mathbf{W}, \mathbf{V}_3) = Wh; \tag{2.1.36'}$$

hence, the scalar triple product represents *the volume of the parallelepipedon* formed by the vectors $\mathbf{V}_1$, $\mathbf{V}_2$ and $\mathbf{V}_3$, because W is the area of the parallelogram determined by

$\mathbf{V}_1$ and $\mathbf{V}_2$, while $h = V_3 \cos(\mathbf{W}, \mathbf{V}_3)$ is the height of the parallelepipedon (Fig.2.5). If $\sphericalangle(\mathbf{W}, \mathbf{V}_3) < \pi/2$, then the scalar (2.1.36') is positive, while the relation

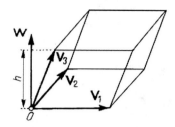

Figure 2.5. Triple scalar product trihedron.

$$(\mathbf{V}_1, \mathbf{V}_2, \mathbf{V}_3) \gtreqless 0 \qquad (2.1.40)$$

represents the condition that the vectors $\mathbf{V}_1$, $\mathbf{V}_2$, $\mathbf{V}_3$, in the given order, form a *positive* (*right-handed*) or *negative* (*left-handed*) *basis*, respectively. If

$$(\mathbf{V}_1, \mathbf{V}_2, \mathbf{V}_3) = 0, \qquad (2.1.41)$$

then the corresponding volume vanishes; hence, this is a *necessary and sufficient condition of coplanarity* of vectors $\mathbf{V}_1$, $\mathbf{V}_2$ and $\mathbf{V}_3$ (supposing that $\mathbf{V}_1, \mathbf{V}_2, \mathbf{V}_3 \neq \mathbf{0}$), and is equivalent to a condition of the form (2.1.3'''). Obviously, the relation (2.1.41) is also verified if one of the factors is equal to zero; if two of the vectors $\mathbf{V}_1$, $\mathbf{V}_2$, $\mathbf{V}_3$ are collinear (in particular, equal), then the mixed product vanishes too (a very important case in practice).

Starting from the formula (2.1.36) and applying the rule of multiplication of two determinants, we obtain

$$(\mathbf{V}_1, \mathbf{V}_2, \mathbf{V}_3)(\mathbf{W}_1, \mathbf{W}_2, \mathbf{W}_3) = \det[\mathbf{V}_i \cdot \mathbf{W}_j]; \qquad (2.1.42)$$

in particular, one has

$$(\mathbf{V}_1, \mathbf{V}_2, \mathbf{V}_3)^2 = \det[\mathbf{V}_i \cdot \mathbf{V}_j] = V_1^2 V_2^2 V_3^2 \big[1 - \cos^2(\mathbf{V}_2, \mathbf{V}_3) - \cos^2(\mathbf{V}_3, \mathbf{V}_1) \\ - \cos^2(\mathbf{V}_1, \mathbf{V}_2) + 2\cos(\mathbf{V}_2, \mathbf{V}_3)\cos(\mathbf{V}_3, \mathbf{V}_1)\cos(\mathbf{V}_1, \mathbf{V}_2)\big], \qquad (2.1.42')$$

the determinant in the right member being *Gramm's determinant*.

Taking into account the geometric signification of the scalar triple product, we may write

$$\in_{jkl} = (\mathbf{i}_j, \mathbf{i}_k, \mathbf{i}_l); \qquad (2.1.43)$$

the formulae (2.1.39) show that there are six components, but only two of them are distinct

$$\in_{jkl} = \in_{klj} = \in_{ljk} = -\in_{kjl} = -\in_{jlk} = -\in_{lkj}, \quad j,k,l = 1,2,3. \tag{2.1.44}$$

The permutation symbol is thus skew-symmetric in all pair of indices (hence, it is *totally skew-symmetric*).

Two permutation symbols lead to the product

$$\in_{ijk} \in_{lmn} = \begin{vmatrix} \delta_{il} & \delta_{im} & \delta_{in} \\ \delta_{jl} & \delta_{jm} & \delta_{jn} \\ \delta_{kl} & \delta_{km} & \delta_{kn} \end{vmatrix}, \quad i,j,k,l,m,n = 1,2,3, \tag{2.1.45}$$

where we took into account the formulae (2.1.42) and (2.1.14). For $k = n$, we obtain

$$\in_{ijk} \in_{lmk} = \begin{vmatrix} \delta_{il} & \delta_{im} \\ \delta_{jl} & \delta_{jm} \end{vmatrix} = \delta_{il}\delta_{jm} - \delta_{im}\delta_{jl}, \quad i,j,l,m = 1,2,3; \tag{2.1.46}$$

if we have also $j = m$, then it results

$$\in_{ijk} \in_{ljk} = 2\delta_{il}, \quad i,l = 1,2,3, \tag{2.1.46'}$$

and if $i = l$ also holds, then we get

$$\in_{ijk} \in_{ijk} = 6. \tag{2.1.46''}$$

Multiplying the relation (2.1.37) by $\in_{lmn}$ and with the aid of the relation (2.1.46"), we obtain

$$\det[V_{pq}] = \frac{1}{6} \in_{ijk} \in_{lmn} V_{il}V_{jm}V_{kn}. \tag{2.1.36''}$$

1.2.4 Vector triple product of three vectors

Let be three (free, bound or sliding) vectors $\mathbf{V}_1$, $\mathbf{V}_2$ and $\mathbf{V}_3$, with which one can form six distinct *vector triple products*, equal – in modulus – two by two. Choosing one of these triple vector products

$$\mathbf{V} = \mathbf{V}_1 \times (\mathbf{V}_2 \times \mathbf{V}_3), \tag{2.1.47}$$

we note $\mathbf{W} = \mathbf{V}_2 \times \mathbf{V}_3$; this is a vector normal to the plane formed by the vectors $\mathbf{V}_2$ and $\mathbf{V}_3$. On the other hand, the vector product $\mathbf{V} = \mathbf{V}_1 \times \mathbf{W}$ is normal to $\mathbf{W}$, belonging to the plane normal to $\mathbf{W}$, hence contained in the plane formed by $\mathbf{V}_2$ and $\mathbf{V}_3$; using the condition of coplanarity of the vectors $\mathbf{V}$, $\mathbf{V}_2$ and $\mathbf{V}_3$, we may write

$$\mathbf{V} = \lambda\mathbf{V}_2 + \mu\mathbf{V}_3, \tag{2.1.48}$$

where λ and μ are – for the moment – indeterminate scalars. This representation of the vector triple product by means of two polar vectors allows us to affirm that the vector product of a polar vector by an axial one has a meaning. The relation (2.1.48) holds for any vectors $\mathbf{V}_1$, $\mathbf{V}_2$ and $\mathbf{V}_3$. We perform a scalar multiplication of both members by $\mathbf{V}_1$; the first member becomes a mixed product with two equal factors, hence vanishing. We may thus write

$$\lambda(\mathbf{V}_1 \cdot \mathbf{V}_2) + \mu(\mathbf{V}_1 \cdot \mathbf{V}_3) = 0.$$

Supposing that $\mathbf{V}_1 \cdot \mathbf{V}_2 \neq 0$ and $\mathbf{V}_1 \cdot \mathbf{V}_3 \neq 0$, it follows that λ and μ must be of the form

$$\lambda = \nu(\mathbf{V}_1 \cdot \mathbf{V}_3), \qquad \mu = -\nu(\mathbf{V}_1 \cdot \mathbf{V}_2),$$

where ν is a still undetermined scalar; replacing in (2.1.48) and projecting on the axis Ox_i, we can write

$$\epsilon_{ijk} V_{1j} \epsilon_{klm} V_{2l} V_{3m} = \nu\left[(V_{1j}V_{3j})V_{2i} - (V_{1j}V_{2j})V_{3i}\right].$$

If we take into account a formula of the form (2.1.46), then we obtain $\nu = 1$; thus, *the basic formula of the vector triple product* is

$$\mathbf{V}_1 \times (\mathbf{V}_2 \times \mathbf{V}_3) = (\mathbf{V}_1 \cdot \mathbf{V}_3)\mathbf{V}_2 - (\mathbf{V}_1 \cdot \mathbf{V}_2)\mathbf{V}_3. \tag{2.1.49}$$

If $\mathbf{V}_1 \cdot \mathbf{V}_2 = \mathbf{V}_1 \cdot \mathbf{V}_3 = 0$, this triple vector product vanishes, but the formula (2.1.49) still holds; if only one of these scalar products is equal to zero, one obtains also an identity.

It follows that *the vector product is not associative*; hence

$$(\mathbf{V}_1 \times \mathbf{V}_2) \times \mathbf{V}_3 \neq \mathbf{V}_1 \times (\mathbf{V}_2 \times \mathbf{V}_3), \tag{2.1.50}$$

in general. But one can verify the relation

$$\mathbf{V}_1 \times (\mathbf{V}_2 \times \mathbf{V}_3) + \mathbf{V}_2 \times (\mathbf{V}_3 \times \mathbf{V}_1) + \mathbf{V}_3 \times (\mathbf{V}_1 \times \mathbf{V}_2) = \mathbf{0}. \tag{2.1.50'}$$

Using the formula (2.1.49), one may easily prove the relations

$$(\mathbf{V}_1 \times \mathbf{V}_2) \cdot (\mathbf{V}_3 \times \mathbf{V}_4) = (\mathbf{V}_1 \cdot \mathbf{V}_3)(\mathbf{V}_2 \cdot \mathbf{V}_4) - (\mathbf{V}_1 \cdot \mathbf{V}_4)(\mathbf{V}_2 \cdot \mathbf{V}_3), \tag{2.1.51}$$

$$(\mathbf{V}_1 \times \mathbf{V}_2) \times (\mathbf{V}_3 \times \mathbf{V}_4) = (\mathbf{V}_3, \mathbf{V}_4, \mathbf{V}_1)\mathbf{V}_2 - (\mathbf{V}_2, \mathbf{V}_3, \mathbf{V}_4)\mathbf{V}_1$$
$$= (\mathbf{V}_4, \mathbf{V}_1, \mathbf{V}_2)\mathbf{V}_3 - (\mathbf{V}_1, \mathbf{V}_2, \mathbf{V}_3)\mathbf{V}_4; \tag{2.1.51'}$$

the last relation leads to

$$(\mathbf{V}_1, \mathbf{V}_2, \mathbf{V}_3)\mathbf{V} = (\mathbf{V}, \mathbf{V}_2, \mathbf{V}_3)\mathbf{V}_1 + (\mathbf{V}, \mathbf{V}_3, \mathbf{V}_1)\mathbf{V}_2 + (\mathbf{V}, \mathbf{V}_1, \mathbf{V}_2)\mathbf{V}_3. \tag{2.1.52}$$

In other notations, we may write

$$V = \frac{\epsilon_{ijk}}{2(\mathbf{e}_1,\mathbf{e}_2,\mathbf{e}_3)}(\mathbf{V},\mathbf{e}_i,\mathbf{e}_j)\mathbf{e}_k = V_k\mathbf{e}_k,$$
(2.1.52')

obtaining thus the representation of the vector $\mathbf{V}$ in a positive basis $\{\mathbf{e}_k,\ k = 1,2,3,\ (\mathbf{e}_1,\mathbf{e}_2,\mathbf{e}_3) > 0\}$; hence, the components of the vector $\mathbf{V}$ in this basis will be V_k (in fact, *contravariant components*, but in what follows we will not use the notions of covariance and contravariance), corresponding to *Cramer's formulae* in the theory of linear algebraic equations. The representation (2.1.52') constitutes a generalization of the canonical representation (2.1.6'').

1.2.5 Applications to the study of certain equations

We will consider now two equations, often used in the vector computation. Let thus be *the scalar equation*

$$\mathbf{a} \cdot \mathbf{x} = m,$$
(2.1.53)

where $\mathbf{a}$ is a given vector, $\mathbf{x}$ is a unknown vector, while m is a given scalar. The solution of *the homogeneous equation* $\mathbf{a} \cdot \mathbf{x} = 0$ is a vector contained in a plane normal to the vector $\mathbf{a}$; hence, the general solution of this equation is of the form $\mathbf{x} = \mathbf{p} \times \mathbf{a}$, where $\mathbf{p}$ is an arbitrary vector. Adding a particular solution of *the non-homogeneous equation*, we obtain the general solution of the equation (2.1.53) in the form

$$\mathbf{x} = \mathbf{p} \times \mathbf{a} + \frac{m\mathbf{a}}{a^2}.$$
(2.1.53')

Let be also *the vector equation*

$$\mathbf{a} \times \mathbf{x} = \mathbf{b}, \quad \mathbf{a} \cdot \mathbf{b} = 0,$$
(2.1.54)

where $\mathbf{a}$ and $\mathbf{b}$ are given vectors, while $\mathbf{x}$ is a unknown vector; we notice that the data of the problem cannot be arbitrary, because the vectors $\mathbf{a}$ and $\mathbf{b}$ must be orthogonal (otherwise, the vector $\mathbf{b}$ cannot be the vector product of the vectors $\mathbf{a}$ and $\mathbf{x}$, and the equation has not solution). The solution of *the homogeneous equation* $\mathbf{a} \times \mathbf{x} = \mathbf{0}$ is a vector collinear with the vector $\mathbf{a}$, hence of the form $\mathbf{x} = \lambda\mathbf{a}$, where λ is an arbitrary scalar. Introducing a particular solution of *the non-homogeneous equation* (which is verified taking into account the canonical formula of the vector triple product and the condition $\mathbf{a} \cdot \mathbf{b} = 0$), one can write the general solution of the equation (2.1.54) in the form

$$\mathbf{x} = \lambda\mathbf{a} + \frac{\mathbf{b} \times \mathbf{a}}{a^2}.$$
(2.1.54')

2. Systems of forces

A system of forces is a set of forces which can be modelled by bound or sliding vectors. The fundamental problem is that of replacing a system of forces by another one of a simpler form, equivalent – from the point of view of its mechanical action – to the first system; to do this, we use the moment of a force (in general, of a vector) and the torsor operator. In what follows, let us deal with discrete systems of vectors, very important in the study of discrete mechanical systems.

2.1 Moments

The notion of moment plays an important rôle in the theory of systems of forces and, in general, in the theory of systems of vectors. We introduce thus the moments with respect to a pole or to an axis, for quantities represented by bound vectors, as well as for quantities represented by sliding vectors; these moments must be considered as being the result of the application of certain operators on the given vectors. For the sake of generality, we will obtain these results for arbitrary vectors.

2.1.1 Moment of a vector with respect to a pole

Let be a pole O and a bound vector $\mathbf{V}$, applied at a point P of position vector $\mathbf{r}$ (Fig.2.6,a). *The moment of the vector $\mathbf{V}$ with respect to the pole O* is, by definition, the vector product (considered as a bound axial vector applied at O)

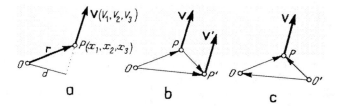

Figure 2.6. Moment of a vector with respect to a pole (a). Variation of the point of application (b) or of the pole (c).

$$\mathbf{M}_O \equiv \mathbf{M}_O(\mathbf{V}) = \overrightarrow{OP} \times \mathbf{V} = \mathbf{r} \times \mathbf{V} ; \qquad (2.2.1)$$

the components of this vector are

$$M_{O,i} = \epsilon_{ijk} x_j V_k, \quad i = 1,2,3. \qquad (2.2.1')$$

The moment $\mathbf{M}_O$ vanishes if $\mathbf{V} = \mathbf{0}$ (banal case) or if $\mathbf{r} = \mathbf{0}$ (the vector is applied at O); as well, it is equal to zero if the vectors $\mathbf{V}$ and $\mathbf{r}$ are collinear, hence if they have the same support. We may thus affirm that the moment of a bound vector with respect to a pole vanishes if and only if the pole is on the vector's support. If the distance of the pole O to the support of the vector $\mathbf{V}$ is d (*the lever arm* of the moment, Fig.2.6,a), then we can write the modulus of the moment in the form

$$|\mathbf{M}_O(\mathbf{V})| = Vd. \qquad (2.2.2)$$

If a vector $\mathbf{V}'$, equipollent to $\mathbf{V}$ (hence $\mathbf{V}' = \mathbf{V}$ as free vectors), is applied at the point P', then we may write (Fig.2.6,b)

$$\mathbf{M}_O(\mathbf{V}') = \overrightarrow{OP'} \times \mathbf{V}' = \left(\overrightarrow{OP} + \overrightarrow{PP'}\right) \times \mathbf{V}' = \overrightarrow{OP} \times \mathbf{V}' + \overrightarrow{PP'} \times \mathbf{V}'$$
$$= \overrightarrow{OP} \times \mathbf{V} + \overrightarrow{PP'} \times \mathbf{V},$$

hence

$$\mathbf{M}_O(\mathbf{V}') = \mathbf{M}_O(\mathbf{V}) + \overrightarrow{PP'} \times \mathbf{V}; \qquad (2.2.3)$$

the variation of the moment of a bound vector with respect to a given pole, by a change of the point of application of the vector is thus emphasized. The moment $\mathbf{M}_O$ remains invariant by a change of the point of application of the vector $\mathbf{V}$ if $\overrightarrow{PP'} \times \mathbf{V} = \mathbf{0}$; this condition holds if $\mathbf{V} = \mathbf{0}$ or $\overrightarrow{PP'} = \mathbf{0}$ (trivial cases) or if P' belongs to the support of the vector $\mathbf{V}$. It results that *the* above given *definition* of the moment of a bound vector with respect to a pole *remains valid also in the case of a sliding vector*, because we can thus take an arbitrary point of application of the latter vector on its support.

Let be a pole O', another one than the pole O. We can write (Fig.2.6,c)

$$\mathbf{M}_{O'}(\mathbf{V}) = \overrightarrow{O'P} \times \mathbf{V} = \left(\overrightarrow{O'O} + \overrightarrow{OP}\right) \times \mathbf{V} = \overrightarrow{O'O} \times \mathbf{V} + \overrightarrow{OP} \times \mathbf{V},$$

hence

$$\mathbf{M}_{O'}(\mathbf{V}) = \mathbf{M}_O(\mathbf{V}) + \overrightarrow{O'O} \times \mathbf{V}; \qquad (2.2.4)$$

this relation shows the variation of the moment of a bound vector by a change of the pole with respect to which it is taken. The moment $\mathbf{M}(\mathbf{V})$ remains invariant if we have $\overrightarrow{O'O} \times \mathbf{V} = \mathbf{0}$; this condition is fulfilled if $\mathbf{V} = \mathbf{0}$ or $\overrightarrow{O'O} = \mathbf{0}$ (trivial cases) or if the vectors $\overrightarrow{O'O}$ and $\mathbf{V}$ are collinear. Hence, the moment $\mathbf{M}(\mathbf{V})$ remains invariant if the pole with respect to which it is calculated is moving along an axis parallel to the support of the vector $\mathbf{V}$.

Let $\mathbf{V}$ be the resultant of n bound vectors V_i, $i = 1, 2, \ldots, n$, applied at the same point P of position vector $\mathbf{r}$ with respect to the pole O. Let us perform a vector product at the left of relation (1.1.5) by $\mathbf{r}$ and take into consideration the distributivity of the vector product with respect to the vector summation; if we denote

$$\mathbf{M}_O(\mathbf{V}_i) = \mathbf{r} \times \mathbf{V}_i, \quad i = 1, 2, \ldots, n, \quad \mathbf{M}_O(\mathbf{V}) = \mathbf{r} \times \mathbf{V}, \qquad (2.2.5)$$

then we may write (Fig.2.7,a)

$$\mathbf{M}_O(\mathbf{V}_1) + \mathbf{M}_O(\mathbf{V}_2) + \ldots + \mathbf{M}_O(\mathbf{V}_n) = \mathbf{M}_O(\mathbf{V}) \qquad (2.2.5')$$

and state

Theorem 2.2.1 (*Varignon*). *The sum of the moments of* n *bound vectors, having the same point of application, with respect to a pole, is equal to the moment of their resultant with respect to the same pole.*

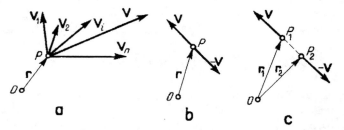

Figure 2.7. Theorem of Varignon (a). Case of two vectors **V** and $-\mathbf{V}$, applied at the same point (b) or at two distinct points on the same support (c).

In the case of two bound vectors **V** and $-\mathbf{V}$, which are applied at the same point P and verify the relation (1.1.11), we can write (Fig.2.7,b)

$$\mathbf{M}_O(\mathbf{V}) + \mathbf{M}_O(-\mathbf{V}) = \mathbf{0};\qquad(2.2.6)$$

the result holds also in the case of two sliding vectors having the same support, as well in the case in which the points of application P_1 and P_2 of two bound vectors are distinct, but belong to their common support (Fig.2.7,c).

2.1.2 Moment of a vector with respect to an axis

Let be an oriented axis Δ, of unit vector **u**, and a bound vector **V**, applied at a point P of position vector **r** with respect to a pole O arbitrary chosen on the axis. *The moment of the vector* **V** *with respect to the axis* Δ is, by definition, the scalar equal to the projection on the axis of the moment of the vector **V** with respect to the point O, hence (Fig.2.8,a)

$$M_\Delta \equiv M_\Delta(\mathbf{V}) = \mathrm{pr}_\Delta \mathbf{M}_O(\mathbf{V}) = \mathbf{u}\cdot(\mathbf{r}\times\mathbf{V}) = (\mathbf{u},\mathbf{r},\mathbf{V});\qquad(2.2.7)$$

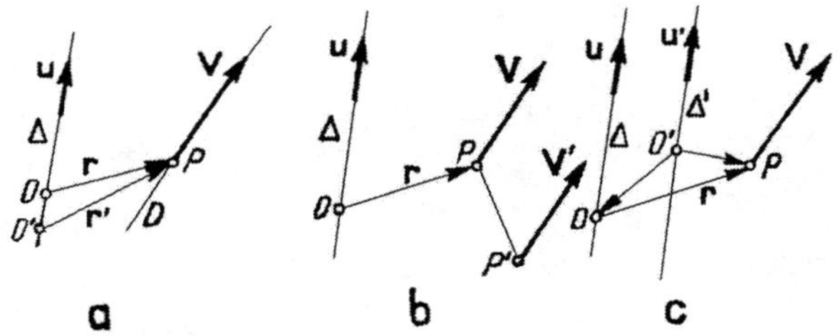

Figure 2.8. Moment of a vector with respect to a directed axis (a). Variation of the point of application (b) or of the axis (c).

this definition is correct only if M_Δ does not depend on the choice of the pole O on the axis Δ. Let be another pole O' on this axis; a scalar product of the relation (2.2.4) by **u** leads to $\mathbf{u} \cdot \mathbf{M}_{O'}(\mathbf{V}) = \mathbf{u} \cdot \mathbf{M}_O(\mathbf{V})$, because $\left(\mathbf{u}, \overrightarrow{O'O}, \mathbf{V}\right) = 0$, hence the definition is correct. Because the definition given to the moment of a vector with respect to a pole holds also for a sliding vector, we can state that *the definition* given for the moment of a vector with respect to an axis *remains valid in the case of a sliding vector* too. In general, the scalar product of the relation (2.2.3) by the unit vector **u** leads to

$$M_\Delta(\mathbf{V'}) = M_\Delta(\mathbf{V}) + \left(\mathbf{u}, \overrightarrow{PP'}, \mathbf{V}\right), \qquad (2.2.8)$$

that is to the variation of the moment of a bound vector with respect to an axis by a change of its point of application (Fig.2.8,b). It results that the moment M_Δ remains invariant if the point of application of the vector **V** is moving along an axis parallel to the axis Δ (in this case $\left(\mathbf{u}, \overrightarrow{PP'}, \mathbf{V}\right) = 0$).

As well, taking an axis Δ' of unit vector $\mathbf{u'} = \mathbf{u}$ (equality as free vectors), a scalar product of the relation (2.2.4) by this unit vector leads to (Fig.2.8,c)

$$M_{\Delta'}(\mathbf{V}) = M_\Delta(\mathbf{V}) + \left(\mathbf{u}, \overrightarrow{OO'}, \mathbf{V}\right); \qquad (2.2.9)$$

hence, the moment $M_\Delta(\mathbf{V})$ remains invariant if we can choose two poles O and O' so that the vectors $\overrightarrow{OO'}$ and **V** be collinear, hence if the axis Δ with respect to which this moment is calculated is moving parallel to itself, in a plane parallel to the vector **V**.

Taking into account (2.2.1'), we notice that the moments of a vector with respect to the co-ordinate axes Ox_i, $i = 1, 2, 3$, are given by

$$M_{Ox_i} = M_{O,i}, \quad i = 1, 2, 3. \qquad (2.2.10)$$

Let u_i, $i = 1, 2, 3$, be the components of the unit vector **u**; taking into account the relation of definition (2.2.7), we may write the moment of the vector $\mathbf{V}(V_i)$ with respect to the axis Δ passing through the point O in the form

$$M_\Delta(\mathbf{V}) = \begin{vmatrix} u_1 & u_2 & u_3 \\ x_1 & x_2 & x_3 \\ V_1 & V_2 & V_3 \end{vmatrix} = \epsilon_{ijk}\, u_i x_j V_k. \qquad (2.2.7')$$

We will consider a plane Π normal to the axis Δ at the point O' of it and let be *the projection vector* $\overrightarrow{P'Q'} = \mathbf{V'}$ of the vector $\overrightarrow{PQ} = \mathbf{V}$ on this plane (Fig.2.9); taking into account the properties of the mixed product, it follows that

$$M_\Delta(\mathbf{V}) = \left(\mathbf{u}, \overrightarrow{OP}, \mathbf{V}\right) = \left(\mathbf{u}, \overrightarrow{OO'} + \overrightarrow{O'P'} + \overrightarrow{P'P}, \overrightarrow{PP'} + \mathbf{V'} + \overrightarrow{Q'Q}\right)$$
$$= \left(\mathbf{u}, \overrightarrow{O'P'}, \mathbf{V'}\right),$$

the eight non-written mixed products (obtained taking into account the property of distributivity of the mixed product with respect to the addition of vectors) vanishing. Because **u** and $\mathbf{M}_{O'}(\mathbf{V}')$ are parallel, we have

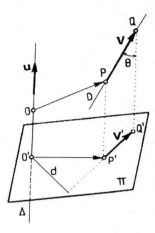

Figure 2.9. Moment of a vector of support D with respect to a directed axis Δ.

$$M_\Delta(\mathbf{V}) = M_\Delta(\mathbf{V}') = \pm |M_{O'}(\mathbf{V}')|, \tag{2.2.11}$$

taking the sign + or − as the rotation indicated by the vector $\mathbf{V}'$ (hence the vector $\mathbf{V}$) about the axis Δ (oriented by the unit vector $\mathbf{u}$) is positive or negative. Hence, the modulus of the moment of a vector $\mathbf{V}$ with respect to the axis Δ is equal to the modulus of the moment of the projection vector of $\mathbf{V}$ on a plane normal to the axis, with respect to the trace of the axis on the plane. If D is the support of the vector $\mathbf{V}$, we denote by $\theta = \sphericalangle(D, \Delta)$ the least angle between the two axes. Let d be the distance from the point O' to the support of the vector $\mathbf{V}'$; we notice that d is just *the length of the common normal* to the axes D and Δ (the least distance between the points of the two axes). Observing that $V' = V\sin\theta$, taking into account the expression (2.2.2) of the modulus of the moment of a vector with respect to a pole and using the formula (2.2.11), we may write

$$M_\Delta(\mathbf{V}) = \pm Vd\sin\theta; \tag{2.2.11'}$$

one takes the sign + or −, using the criterion enounced above. Hence, the moment of a vector with respect to an axis vanishes if $V = 0$ (trivial case), if $d = 0$ (the axes D and Δ are concurrent), or if $\theta = 0$ (the axes D and Δ are parallel); hence, the moment of a vector with respect to an axis vanishes if and only if the support of the vector and the axis are coplanar.

2.1.3 The torsor of a sliding vector

Introducing the moment of a sliding vector with respect to a pole, we may give a new representation for such a vector. Let thus be a sliding vector $\mathbf{V}$, of components V_i,

$i = 1,2,3$, with respect to an orthonormed frame of reference, and let be $\mathbf{M}_O$ the moment of this vector with respect to the pole O (Fig.2.10); we may write the obvious relation

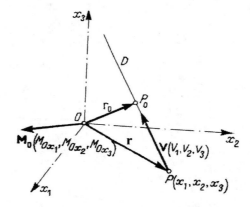

Figure 2.10. Torsor of a sliding vector with respect to a pole O.

$$\mathbf{V} \cdot \mathbf{M}_O = V_i M_{Ox_i} = 0. \tag{2.2.12}$$

Starting from the expression (2.2.1) of the moment $\mathbf{M}_O$, we can write the vector equation

$$\mathbf{V} \times \mathbf{r} = -\mathbf{M}_O, \tag{2.2.13}$$

where $\mathbf{r}$ is an unknown vector; if we take into account the relation (2.2.12), we may affirm that this equation, which must determine the support of the sliding vector $\mathbf{V}$, has a solution. Using the formula (2.1.54'), the general solution of the equation (2.2.13) can be written in the form

$$\mathbf{r} = \lambda \mathbf{V} + \frac{\mathbf{V} \times \mathbf{M}_O}{V^2}, \tag{2.2.13'}$$

obtaining thus the equation of the axis D (the support of the vector $\mathbf{V}$). We may write

$$\mathbf{r}_0 = \frac{\mathbf{V} \times \mathbf{M}_O}{V^2}, \quad r_0 = \frac{|\mathbf{M}_O|}{V} \tag{2.2.13''}$$

for $\lambda = 0$; it follows that $P_0(\mathbf{r}_0)$ is the projection of O on the axis D, because $\mathbf{r}_0 \cdot \mathbf{V} = 0$. Starting from (2.2.13'), we can write the equations of the axis D also in the form

$$\frac{1}{V_1}\left(V^2 x_1 - V_2 M_{Ox_3} + V_3 M_{Ox_2}\right) = \frac{1}{V_2}\left(V^2 x_2 - V_3 M_{Ox_1} + V_1 M_{Ox_3}\right)$$
$$= \frac{1}{V_3}\left(V^2 x_3 - V_1 M_{Ox_2} + V_2 M_{Ox_1}\right). \tag{2.2.13'''}$$

Hence, a sliding vector is characterized by the vectors $\mathbf{V}$ and $\mathbf{M}_O$ which verify the relation (2.2.12); such a vector is thus given by two ordered triplets of numbers $\left(V_1, V_2, V_3, M_{Ox_1}, M_{Ox_2}, M_{Ox_3}\right)$, which verify the relation (2.2.12), hence by five independent numbers. The six numbers mentioned above are called *the co-ordinates of the sliding vector* or *Plücker's co-ordinates*.

Since $\mathbf{M}_O$ is the result of the application of an operator on the vector $\mathbf{V}$, the couple of vectors $\{\mathbf{R}, \mathbf{M}_O\}$ represents the result of the application of another operator on the same vector $\mathbf{V}$; these vectors form *the torsor* (*wrench*) *of the vector* $\mathbf{V}$ *at the point* (*pole*) O,

$$\tau_O(\mathbf{V}) \equiv \{\mathbf{R}, \mathbf{M}_O\}, \qquad (2.2.14)$$

characterizing entirely the sliding vector. The quantity $\mathbf{R} \cdot \mathbf{M}_O$ is called *torsor's scalar* and it vanishes in case of a single sliding vector. Taking into account (2.2.4), we may write

$$\tau_{O'}(\mathbf{V}) = \tau_O(\mathbf{V}) + \left\{\mathbf{0}, \overrightarrow{O'O} \times \mathbf{R}\right\}; \qquad (2.2.15)$$

this relation shows the variation of the torsor of a sliding vector by a change of the pole with respect to which it is calculated. If the torsor of a sliding vector vanishes at a point, then it vanishes at any other point; therefore, the necessary and sufficient condition for a sliding vector to be zero is the vanishing of its torsor at a point.

We also introduce a scalar, called *virial*, by the relation of definition (Fig.2.10)

$$\mathcal{V}_O \equiv \mathcal{V}_O(\mathbf{V}) = \mathbf{r} \cdot \mathbf{V} = x_j V_j; \qquad (2.2.16)$$

in this case, if we add the number $\mathcal{V}_O$ to the co-ordinates of a sliding vector, then we obtain the point of application on the support D. We have thus a new possibility to represent a bound vector (by six independent numbers).

2.2 Reduction of systems of forces

The forces are modelled with the aid of bound or sliding vectors as they are applied upon a deformable or non-deformable mechanical system, respectively; a study of the equivalence of systems of forces (in particular, the equivalence to zero), these ones being modelled correspondingly, is made. We consider also the systems of free vectors, because of their importance.

2.2.1 Systems of free vectors

Let $\{\mathbf{V}\} \equiv \{\mathbf{V}_i, i = 1, 2, ..., n\}$ and $\{\mathbf{V}'\} \equiv \{\mathbf{V}'_j, j = 1, 2, ..., m\}$ be two systems of free vectors. By definition, we say that the two systems are *equivalent* if they have *the same resultant*

$$\mathbf{R} = \sum_{i=1}^{n} \mathbf{V}_i = \sum_{j=1}^{m} \mathbf{V}'_j; \qquad (2.2.17)$$

in this case, we write

$$\{\mathbf{V}\} \sim \{\mathbf{V}'\}. \qquad (2.2.18)$$

The operations of passing from a system of vectors to another system, equivalent to the first one, are operations of vector addition (composition and decomposition of vectors); these operations are called *elementary operations of equivalence*. We consider also the system of free vectors $\{\mathbf{V}''\} \equiv \{\mathbf{V}''_k, k = 1, 2, ..., p\}$. We mention following properties:

i) $\{\mathbf{V}\} \sim \{\mathbf{V}\}$ (*reflexivity*);
ii) $\{\mathbf{V}\} \sim \{\mathbf{V}'\} \Leftrightarrow \{\mathbf{V}'\} \sim \{\mathbf{V}\}$ (*symmetry*);
iii) $\{\mathbf{V}\} \sim \{\mathbf{V}'\}, \{\mathbf{V}'\} \sim \{\mathbf{V}''\} \Rightarrow \{\mathbf{V}\} \sim \{\mathbf{V}''\}$ (*transitivity*).

Taking into account these properties, we may affirm that the set of elementary operations of equivalence forms a *group*.

The simplest system of free vectors equivalent to a given system of free vectors is *the resultant* of the latter one. In particular, the resultant of a system of free vectors can be equal to zero. By definition, we say that a system of free vectors is *equivalent to zero* and we may write

$$\{\mathbf{V}\} \sim \{\mathbf{0}\} \qquad (2.2.19)$$

if its resultant vanishes

$$\mathbf{R} = \mathbf{0}. \qquad (2.2.19')$$

A system of free vectors equivalent to zero can be *eliminated* from computation by elementary operations of equivalence.

2.2.2 Mathematical modelling of systems of forces

The forces acting upon the mechanical systems have been represented by bound vectors, and their points of application are the very same points of the system (the points at which are the particles of a discrete system or the points of a continuous system). We are thus led to the study of a system of forces *modelled by bound vectors*.

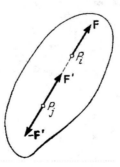

Figure 2.11. Mathematical modelling of a force acting upon a rigid solid.

In the case of a rigid solid, let P_i be a point of it at which acts a force $\mathbf{F}$ (Fig.2.11); we suppose that at the point P_j of this solid, on the support of $\mathbf{F}$, is acting a system of two forces $\{\mathbf{F}',-\mathbf{F}'\} \sim \{\mathbf{0}\}$ (system equivalent to zero), considered as bound vectors and for which holds the relation $\mathbf{F}' = \mathbf{F}$, as free vectors. We may write the following relations of equivalence of the systems of forces

$$\{\mathbf{F}\} \sim \{\mathbf{F}\} + \{\mathbf{F}',-\mathbf{F}'\} \sim \{\mathbf{F},-\mathbf{F}'\} + \{\mathbf{F}'\} \sim \{\mathbf{F}'\},$$

taking into account the relation $\{\mathbf{F},-\mathbf{F}'\} \sim \{\mathbf{0}\}$, because – from a mechanical point of view – the forces are applied at points which are at an invariable distance between them (the modelling of the solid as a rigid); by way of consequence, the effect of the force $\mathbf{F}$ can be replaced by the effect of the force $\mathbf{F}'$. In the case of a rigid solid, the forces will be thus *modelled by sliding vectors*. Hence, a force acting upon a rigid solid can be applied at any point of it, if this point is on the support of the force; this result may be applied also to a non-deformable discrete system if two or several particles of it are on the support of the force. Hence, the necessity to study a system of forces modelled by sliding vectors is put into evidence.

2.2.3 Systems of forces modelled by bound vectors

Let $\{\mathbf{F}\} \equiv \{\mathbf{F}_{ij}, i = 1,2,...,n, j = 1,2,...,n_i\}$ and $\{\mathbf{F}'\} \equiv \{\mathbf{F}'_{ik}, i = 1,2,...,n, k = 1,2,...,n'_i\}$ be two systems of forces modelled by bound vectors; the first index corresponds to the point P_i at which are applied the forces $\mathbf{F}_{ij}$ and $\mathbf{F}'_{ik}$, while the second index individualises the force in the respective system. By definition, we say that the two systems of forces modelled by bound vectors are *equivalent* if they have *the same resultant at each point of application*

$$\mathbf{R}_i = \sum_{j=1}^{n_i} \mathbf{F}_{ij} = \sum_{k=1}^{n'_i} \mathbf{F}'_{ik}, \quad i = 1,2,...,n, \tag{2.2.20}$$

and this is written in the form

$$\{\mathbf{F}\} \sim \{\mathbf{F}'\}. \tag{2.2.21}$$

The operations of passing from a system of forces to another system of forces, equivalent to the first one, are operations of vector addition at each point P_i of the system; these operations are *elementary operations of equivalence* in the case of systems of forces modelled by bound vectors. Let be also the system of forces modelled by bound vectors $\{\mathbf{F}''\} \equiv \{\mathbf{F}''_{il}, i = 1,2,...,n, l = 1,2,...,n''_i\}$. The properties mentioned in Subsec. 2.2.1 still hold; also in this case, the set of elementary operations of equivalence forms a *group*.

The simplest system of forces modelled by bound vectors is formed by the resultants $\mathbf{R}_i$ applied at the points $P_i, i = 1,2,...,n$. If

$$\mathbf{R}_i = \mathbf{0}, \quad i = 1,2,...,n, \tag{2.2.22}$$

then we say, by definition, that the system of forces modelled by bound vectors is *equivalent to zero*, and we may write

$$\{\mathbf{F}\} \sim \{\mathbf{0}\}. \tag{2.2.23}$$

These results hold also for an arbitrary system of bound vectors. In general, a system of bound vectors equivalent to zero can be *eliminated* from computation by elementary operations of equivalence.

2.2.4 Systems of forces modelled by sliding vectors

Let $\{\mathbf{F}\} \equiv \{\mathbf{F}_i, i = 1, 2, ..., n\}$ be a system of forces modelled by sliding vectors. Besides the operations of vector addition (including composition and decomposition of vectors), we introduce also the operations of sliding along the support, obtaining thus *the enlarged set of elementary operations of equivalence*, which forms a *group* too.

Let us consider three non-collinear points O_1, O_2, O_3, so that the plane Π determined by them do not contain the supports of the forces $\mathbf{F}_i$. We choose the point of application P_i of the force $\mathbf{F}_i$ on its support (eventually, we perform a sliding along this support), so that $P_i \notin \Pi$; in this case, the force $\mathbf{F}_i$ can be decomposed univocally along P_iO_1, P_iO_2, P_iO_3 (Fig.2.12,a) (if the support of the force $\mathbf{F}_i$ is contained in the plane Π, then the decomposition remains possible, but it is no more unique). Thus, the system of forces $\{\mathbf{F}\}$, modelled by sliding vectors, may be replaced, after sliding along the supports P_iO_1, P_iO_2, P_iO_3, by three subsystems of forces of the same type, applied at the points O_1, O_2, O_3; summing the forces at these points, we obtain a system of three forces modelled by sliding vectors $\{\overline{\mathbf{F}}\} \equiv \{\overline{\mathbf{F}}_1, \overline{\mathbf{F}}_2, \overline{\mathbf{F}}_3\}$, equivalent to the given system of forces $\{\mathbf{F}\}$. Because of the arbitrariness in the choice of the points O_1, O_2, O_3 and P_i, there exists an infinity of such systems of three forces, which have the above mentioned property. Let Π_2 and Π_3 be the planes determined by the point O_1 and the forces $\overline{\mathbf{F}}_2$ and $\overline{\mathbf{F}}_3$, respectively; the intersection of these planes is a straight line O_1O' (the point O' is arbitrary on this line) (Fig.2.12,b). We decompose the forces $\overline{\mathbf{F}}_2$ and $\overline{\mathbf{F}}_3$, along O_2O_1 and O_2O', and along O_3O_1 and O_3O', in the planes Π_2 and Π_3, respectively; by sliding, these components will be applied at the points O_1 and O', where we are summing them, together with $\overline{\mathbf{F}}_1$. We obtain thus a system of two forces modelled by sliding vectors $\{\overline{\overline{\mathbf{F}}}\} \equiv \{\overline{\overline{\mathbf{F}}}_1, \overline{\mathbf{F}}'\}$, equivalent to the system $\{\overline{\mathbf{F}}\}$, as well as to the system $\{\mathbf{F}\}$; the point O' is arbitrarily chosen, so that there is an infinity of such systems of two forces, modelled by sliding vectors.

Let $\{\mathbf{F}'\} \equiv \{\mathbf{F}'_j, j = 1, 2, ..., m\}$ be also a system of forces modelled by sliding vectors. We say, by definition, that two systems of forces modelled by sliding vectors are *equivalent* if, by operations belonging to the enlarged set of elementary equivalence operations, they can be reduced to *the same system of three* (or *two*) *forces modelled by sliding vectors*, and we may write a relation of the form (2.2.21). We introduce also the system of forces $\{\mathbf{F}''\} \equiv \{\mathbf{F}''_k, k = 1, 2, ..., p\}$, modelled by sliding vectors; then the three properties mentioned at Subsec. 2.2.1 hold.

Mechanics of the systems of forces 99

The simplest system of forces modelled by sliding vectors equivalent to a given one is formed by two forces modelled by sliding vectors. In particular, this system of two vectors modelled by sliding vectors *is equivalent to zero* if the two forces have the same support, the same modulus and opposite directions; in this case, we say that the system $\{\mathbf{F}\}$ of forces modelled by sliding vectors is equivalent to zero, and we write it in the form (2.2.23). These results hold for any system of sliding vectors. In general, a system of sliding vectors equivalent to zero can be *eliminated* from computation by operations belonging to the enlarged group of elementary operations of equivalence.

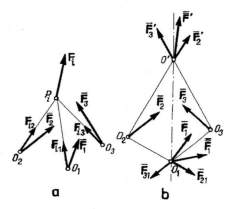

Figure 2.12. Systems of forces modelled by sliding vectors. Equivalent systems of three (a) or two (b) sliding vectors.

2.2.5 Torsor of a system of vectors. The minimal form of the torsor

Because the "torsor" operator appears in connection with any systems of vectors, not only in connection with systems of forces, we will study this notion for arbitrary systems of vectors. Let thus $\{\mathbf{V}\} \equiv \{\mathbf{V}_i, i = 1, 2, ..., n\}$ be a system of n bound vectors, applied at the points P_i of position vectors $\mathbf{r}_i$, or of n sliding vectors on supports passing through these points. We introduce also *the resultant* $\mathbf{R}$ of the system of vectors, in the form of a free vector given by (in this relation the vectors $\mathbf{V}_i$ are considered as free vectors)

$$\mathbf{R} = \sum_{i=1}^{n} \mathbf{V}_i, \qquad (2.2.24)$$

as well as *the resultant moment* of the system of vectors, in the form of a bound vector, applied at the point O and given by

$$\mathbf{M}_O = \sum_{i=1}^{n} \mathbf{r}_i \times \mathbf{V}_i. \qquad (2.2.24')$$

We notice that the pair of vectors $\{\mathbf{R}, \mathbf{M}_O\}$ is the result of the application of an operator τ_O on a system $\{\mathbf{V}\}$ of vectors; this pair of vectors is called, by definition, *the torsor* (wrench) *of the system of vectors* $\{\mathbf{V}\}$ at the pole (point) O

$$\tau_O\{\mathbf{V}\} \equiv \{\mathbf{R}, \mathbf{M}_O\}. \tag{2.2.24''}$$

Because the moment of a bound vector with respect to a pole holds also for a sliding vector, the definition given above is valid as well for a system of sliding vectors. The components of the resultant and of the resultant moment can be written in the form

$$R_j = \sum_{i=1}^{n} V_{ij}, \quad M_{Ox_j} = \epsilon_{jkl} \sum_{i=1}^{n} x_k^{(i)} V_{il}, \quad j = 1,2,3. \tag{2.2.24'''}$$

Let $\{\mathbf{V}\} + \{\mathbf{V}'\} = \{\mathbf{V}_i, \mathbf{V}'_j, i = 1,2,...,n, j = 1,2,...,m\}$ be the sum of two systems of vectors, where $\{\mathbf{V}'\} \equiv \{\mathbf{V}'_j, j = 1,2,...,m\}$ is another system of bound or sliding vectors. As well, let be a scalar λ and a system of vectors $\{\mathbf{V}\}$; the product of λ by $\{\mathbf{V}\}$ is of the form $\lambda\{\mathbf{V}\} \equiv \{\lambda\mathbf{V}\} \equiv \{\lambda\mathbf{V}_i, i = 1,2,...,n\}$. The two operations mentioned above have all the properties which are put in evidence in Chap. 1, Subsec. 1.1.2 concerning the addition of vectors or the product of a scalar by a vector. In particular, we can take $\lambda = -1$, which leads to the operation $\{\mathbf{V}\} - \{\mathbf{V}'\}$. The definition of the torsor of a system of vectors yields the basic properties:

i) $\tau_O(\{\mathbf{V}\} + \{\mathbf{V}'\}) = \tau_O\{\mathbf{V}\} + \tau_O\{\mathbf{V}'\}$;
ii) $\tau_O(\lambda\{\mathbf{V}\}) = \lambda\tau_O\{\mathbf{V}\}$.

Hence, the torsor is a *linear operator*. We notice also that the torsor is *invariant* with respect to the group of elementary operations of equivalence or with respect to the enlarged group of these operations, as we have to do with systems of bound or sliding vectors, respectively.

If a system of vectors $\{\mathbf{V}\}$ is equivalent to zero, then its torsor at an arbitrary pole O is equal to zero

$$\tau_O\{\mathbf{V}\} = \mathbf{0}, \tag{2.2.25}$$

which is equivalent to

$$\mathbf{R} = \mathbf{0}, \quad \mathbf{M}_O = \mathbf{0}. \tag{2.2.25'}$$

Indeed, supposing that we have to do with a system of bound vectors, the resultant at each point of the system must vanish, the affirmation being thus justified. A system of sliding vectors is reduced to a system of two sliding vectors, which verify a relation of the form (1.1.11), the vectors having the same support; we are thus led to a relation of the form (2.2.6) too, hence the affirmation is justified also in this case.

Let be now a system of sliding vectors for which the torsor vanishes. We can reduce this system to a system of two sliding vectors $\{\mathbf{U}\} \equiv \{\mathbf{U}_1, \mathbf{U}_2\}$; but the torsor is an invariant, so that it vanishes also for this system of sliding vectors. Because the resultant of the system $\{\mathbf{U}\}$ is equal to zero, we can state that the vectors $\mathbf{U}_1$ and $\mathbf{U}_2$ have the same modulus, but opposite directions; their supports can be parallel ($\mathbf{U}_1 + \mathbf{U}_2 = \mathbf{0}$, as free vectors). The resultant moment of the system $\{U\}$ also vanishes, so that the two supports must coincide. Indeed the condition $\mathbf{r}_1 \times \mathbf{U}_1 + \mathbf{r}_2 \times \mathbf{U}_2 = \mathbf{0}$ leads to the condition $(\mathbf{r}_2 - \mathbf{r}_1) \times \mathbf{U}_2 = \overrightarrow{P_1P_2} \times \mathbf{U}_2 = \mathbf{0}$, the above affirmation being thus justified (Fig.2.13); by sliding along the common support, it is seen that the system $\{U\}$ is equivalent to zero. Hence, *a system of sliding vectors is equivalent to zero*, and we write this in the form (2.2.19), *if and only if its torsor vanishes*; in the case of a *system of bound vectors*, the condition (2.2.25) is only *necessary*. Taking into account (2.2.24'''), this condition leads to six equations of projection on the three axes of co-ordinates.

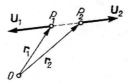

Figure 2.13. Torsor of a system of two sliding vectors equivalent to zero.

Let $\{V\}$ and $\{V'\}$ be two equivalent systems of vectors, so that one can write the relation (2.2.18); it results $\{V\} - \{V'\} \sim \{0\}$, and further $\tau_O(\{V\} - \{V'\}) = \mathbf{0}$. Noting that the torsor is a *linear operator*, we state that *two systems of sliding vectors are equivalent*, and we write the relation (2.2.18), *if and only if their torsors with respect to the same pole are equal*

$$\tau_O\{V\} = \tau_O\{V'\}; \qquad (2.2.26)$$

in the case of *systems of bound vectors*, the condition (2.2.26) is only *necessary*.

The resultant $\mathbf{R}$ of the system of vectors $\{V\}$ is invariant by a change of pole O. In what concerns the resultant moment, by passing to a pole O', we obtain

$$\mathbf{M}_{O'} = \sum_{i=1}^{n} \overrightarrow{O'P_i} \times \mathbf{V}_i = \sum_{i=1}^{n}\left(\overrightarrow{O'O} + \overrightarrow{OP_i}\right) \times \mathbf{V}_i = \overrightarrow{O'O} \times \sum_{i=1}^{n} \mathbf{V}_i + \sum_{i=1}^{n} \overrightarrow{OP_i} \times \mathbf{V}_i,$$

so that

$$\mathbf{M}_{O'} = \mathbf{M}_O + \overrightarrow{O'O} \times \mathbf{R} \qquad (2.2.27)$$

or

$$\tau_{O'}\{V\} = \tau_O\{V\} + \left\{\mathbf{0}, \overrightarrow{O'O} \times \mathbf{R}\right\}, \qquad (2.2.27')$$

a relation of the same form as the relation (2.2.15); we obtain thus the variation of the resultant moment, hence also of the torsor of a system of vectors, by a change of the pole with respect to which this torsor is calculated. The torsor of a system of vectors is invariant by a change of pole O if this pole moves along an axis parallel to the resultant $\mathbf{R}$ or if this one vanishes. Hence, if the torsor of a system of vectors equivalent to zero vanishes at a point ($\tau_O\{\mathbf{V}\} = \mathbf{0}$), then it is equal to zero at any other point ($\tau_{O'}\{\mathbf{V}\} = \mathbf{0}$); otherwise, the equivalence to zero of a system of sliding vectors would depend on the pole with respect to which the torsor is calculated.

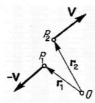

Figure 2.14. Couple of two sliding vectors.

A system of sliding vectors for which $\mathbf{R} = \mathbf{0}$ and $\mathbf{M}_O \neq \mathbf{0}$ is reducible to a system formed by two sliding vectors having the same modulus, parallel supports and opposite directions (Fig.2.14). Such a system is called a *couple*, and its torsor – as we have seen – is invariant by a change of pole.

A scalar product of the relation (2.2.27) by $\mathbf{R}$ leads to

$$\mathbf{R} \cdot \mathbf{M}_{O'} = \mathbf{R} \cdot \mathbf{M}_O, \qquad (2.2.28)$$

observing that we obtain also a mixed product equal to zero; this quantity is called *the torsor's scalar* and is invariant by a change of pole. Let be

$$\mathbf{M}_O = \mathbf{M}_O^{\|} + \mathbf{M}_O^{\perp}, \qquad (2.2.29)$$

where we have decomposed the moment $\mathbf{M}_O$ into two components: a component parallel and a component normal to the resultant $\mathbf{R}$, respectively (Fig.2.15); in this case, the relation (2.2.28) is equivalent to

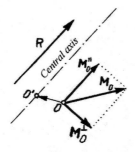

Figure 2.15. Central axis of a system of sliding vectors.

$$\mathbf{R} \cdot \mathbf{M}_{O'}^{\|} = \mathbf{R} \cdot \mathbf{M}_O^{\|}, \qquad (2.2.28')$$

because $\mathbf{R} \cdot \mathbf{M}_O^{\perp} = 0$. Hence, *the component of the resultant moment along the resultant $\mathbf{R}$ is* also *an invariant* by a change of pole

$$\mathbf{M}_{O'}^{\|} = \mathbf{M}_O^{\|}. \qquad (2.2.28'')$$

Mechanics of the systems of forces

In this case, the relation (2.2.27) leads to

$$\mathbf{M}_{O'}^{\perp} = \mathbf{M}_O^{\perp} + \overrightarrow{O'O} \times \mathbf{R}. \qquad (2.2.27'')$$

We put the problem to find a pole with respect to which the torsor of a system of vectors has the simplest form possible (the minimal form). Because $\mathbf{R}$ and $\mathbf{M}_O^{\parallel}$ are invariants, the only quantity varying in modulus, together with the pole O, is $\mathbf{M}_O^{\perp}$; we will try to find a pole with respect to which this component of the resultant moment vanishes. If the pole O' has this property, then the relation (2.2.27'') allows us to write the condition

$$\mathbf{M}_O^{\perp} + \overrightarrow{O'O} \times \mathbf{R} = \mathbf{0}.$$

We obtain thus a vector equation of the form (2.1.54); using the solution (2.1.54'), we may write

$$\overrightarrow{OO'} = \lambda \mathbf{R} + \frac{\mathbf{R} \times \mathbf{M}_O^{\perp}}{R^2},$$

where λ is an arbitrary scalar; we notice that this solution may be written also in the form

$$\overrightarrow{OO'} = \lambda \mathbf{R} + \frac{\mathbf{R} \times \mathbf{M}_O}{R^2}, \qquad (2.2.30)$$

because $\mathbf{R} \times \mathbf{M}_O^{\parallel} = \mathbf{0}$. We obtain thus an axis parallel to the resultant $\mathbf{R}$; the resultant moments with respect to its points have only the invariant component $\mathbf{M}_O^{\parallel}$. We say, in this case, that *the torsor takes its minimal form*; the respective axis is called *the central axis of the system of vectors* (Fig.2.15). For $\lambda = 0$, we obtain $\overrightarrow{OO'} \perp \mathbf{R}$, with

$$\left|\overrightarrow{OO'}\right| = \frac{|\mathbf{M}_O^{\perp}|}{R}. \qquad (2.2.30')$$

A vector product of the relation (2.2.30) at left by $\mathbf{R}$ allows us to eliminate λ, and we obtain

$$\mathbf{R} \times \overrightarrow{OO'} = \mathbf{R} \times \frac{\mathbf{R} \times \mathbf{M}_O}{R^2}. \qquad (2.2.31)$$

The formula of the triple vector product leads to

$$\frac{\mathbf{R} \cdot \mathbf{M}_O}{R^2}\mathbf{R} = \mathbf{M}_O - \overrightarrow{OO'} \times \mathbf{R}. \qquad (2.2.31')$$

Taking the pole O as origin of the co-ordinate axes and denoting the co-ordinates of the point O' by $x_i, i = 1,2,3$, we can write

$$\frac{1}{R_1}\left(M_{Ox_1} - x_2R_3 + x_3R_2\right) = \frac{1}{R_2}\left(M_{Ox_2} - x_3R_1 + x_1R_3\right)$$
$$= \frac{1}{R_3}\left(M_{Ox_3} - x_1R_2 + x_2R_1\right) = \frac{1}{R^2}\left(R_1M_{Ox_1} + R_2M_{Ox_2} + R_3M_{Ox_3}\right); \quad (2.2.31'')$$

these equations represent two planes, the intersection of which is the central axis.

As we have seen, the torsor operator characterizes a system of sliding vectors; thus, taking into account the form of the torsor, we can distinguish several cases of reduction of such a system of vectors to simpler systems, i.e.:

i) If $\mathbf{R} = \mathbf{0}$ and $\mathbf{M}_O = \mathbf{0}$, then we obtain a system of sliding vectors equivalent to zero.

ii) If $\mathbf{R} \neq \mathbf{0}$ and $\mathbf{M}_O = \mathbf{0}$, then the system of sliding vectors is equivalent to a resultant having as support the central axis; indeed, because $\mathbf{M}_O^\perp = \mathbf{0}$, the pole belongs to this axis, so that the affirmation is justified.

iii) If $\mathbf{R} = \mathbf{0}$ and $\mathbf{M}_O \neq \mathbf{0}$, then the system of sliding vectors is equivalent to a couple.

iv) If $\mathbf{R} \neq \mathbf{0}$ and $\mathbf{M}_O \neq \mathbf{0}$, then we distinguish two subcases:

iv') If $\mathbf{R} \cdot \mathbf{M}_O = 0$, then the system of sliding vectors is equivalent to a resultant, the support of which is the central axis; indeed, in this case $\mathbf{M}_O^{\|} = \mathbf{0}$. It is a difference between the cases ii) and iv'), because – in the latter case – the pole O does not belong obligatory to the central axis.

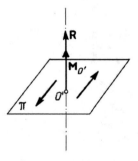

Figure 2.16. Dynam of a system of sliding vectors for which the torsor's scalar does not vanish.

iv'') If the torsor's scalar $\mathbf{R} \cdot \mathbf{M}_O \neq 0$, that is in the general case, then the system of sliding vectors is equivalent to a *dynam* (or a *screw*); the support of the resultant $\mathbf{R}$ is the central axis, while the resultant moment $\mathbf{M}_O$ leads to a couple of moment $\mathbf{M}_{O'}$ (the point O' on the central axis), acting in a plane Π normal to this resultant (Fig.2.16).

2.2.6 Systems of coplanar forces

In the case of a system of coplanar forces $\{\mathbf{F}\} \equiv \{\mathbf{F}_i, i = 1,2,...,n\}$, modelled by sliding vectors, the resultant is – obviously – contained in the considered plane Π

Mechanics of the systems of forces 105

(Fig.2.17). The moment $\mathbf{M}_O$ of the system of forces with respect to a pole $O \in \Pi$ will be a vector normal to the resultant $\mathbf{R}$; hence, $\mathbf{R} \cdot \mathbf{M}_O = 0$ and we are in the case iv'). If $\mathbf{R} \neq \mathbf{0}$, then the system of coplanar forces modelled by sliding vectors is reduced to a resultant along the central axis, contained in the considered plane. If the resultant vanishes, then the system of forces reduces to a couple if $\mathbf{M}_O \neq \mathbf{0}$ (with respect to an arbitrary point in the plane) or is equal to zero if $\mathbf{M}_O = \mathbf{0}$.

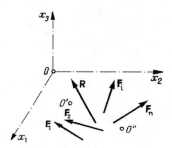

Figure 2.17. Systems of coplanar forces.

Supposing that this system of forces acts in the plane Ox_1x_2, the conditions of equivalence to zero are written in the form

$$R_1 = R_2 = 0, \quad M_{Ox_3} = 0. \tag{2.2.32}$$

In particular, a system of three non-parallel coplanar forces modelled by sliding vectors, the resultant of which vanishes ($R_1 = R_2 = 0$), is *equivalent to zero* if and only if *the three forces are concurrent* (we use the condition $M_{Ox_3} = 0$). But choosing two poles O' and O'' in the plane and writing the conditions

$$\mathbf{M}_{O'} = \mathbf{M}_O + \overrightarrow{O'O} \times \mathbf{R} = \mathbf{0}, \quad \mathbf{M}_{O''} = \mathbf{M}_O + \overrightarrow{O''O} \times \mathbf{R} = \mathbf{0},$$

we state that a system of coplanar forces modelled by sliding vectors is equivalent to zero if and only if the conditions

$$\mathbf{M}_O = \mathbf{0}, \quad \mathbf{M}_{O'} = \mathbf{0}, \quad \mathbf{M}_{O''} = \mathbf{0} \tag{2.2.33}$$

are fulfilled, the poles O, O' and O'' being non-collinear; these conditions lead to three projection equations on the axis Ox_3, which can be useful in applications. We mention that, for the equivalence to zero of a *system of coplanar forces modelled by bound vectors*, these conditions are only *necessary*.

2.2.7 Systems of parallel forces

Let be a direction of unit vector $\mathbf{u}$ and a system of parallel forces $\{\mathbf{F}\} \equiv \{\mathbf{F}_i, i = 1, 2, ..., n\}$, modelled by sliding vectors and given by

$$\mathbf{F}_i = F_i \mathbf{u}; \qquad (2.2.34)$$

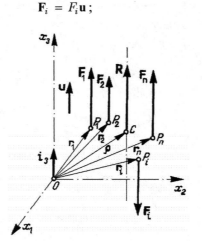

Figure 2.18. Systems of parallel forces.

these forces are applied at the points P_i, of position vectors $\mathbf{r}_i, i = 1,2,\ldots,n$ (Fig.2.18). We notice that F_i represent the components of these forces along the given direction, having vers $\mathbf{F}_i = \pm \mathbf{u}$. The resultant of this system of forces will be

$$\mathbf{R} = \sum_{i=1}^{n} \mathbf{F}_i = F\mathbf{u}, \qquad (2.2.35)$$

where we have used the notation

$$F = \sum_{i=1}^{n} F_i . \qquad (2.2.35')$$

The resultant moment with respect to the pole O is written in the form

$$\mathbf{M}_O = \sum_{i=1}^{n} \mathbf{r}_i \times \mathbf{F}_i = \boldsymbol{\rho} \times \mathbf{R},$$

where we took into consideration the resultant (2.2.35), defining the vector

$$\boldsymbol{\rho} = \frac{1}{F} \sum_{i=1}^{n} F_i \mathbf{r}_i \qquad (2.2.36)$$

and admitting that $F \neq 0$. The position vector $\boldsymbol{\rho}$ determines a point C with respect to which the resultant moment vanishes ($\mathbf{M}_C = \mathbf{M}_O + \overrightarrow{CO} \times \mathbf{R} = \mathbf{0}$); hence, the point C belongs to the central axis, which is determined by the unit vector $\mathbf{u}$. The system of parallel forces modelled by sliding vectors is reduced, in this case, to a resultant along the central axis. We have $\mathbf{R} \cdot \mathbf{M}_O = 0$ and are in the case iv').

If $F = 0$, then $\mathbf{R} = \mathbf{0}$, and the system of forces modelled by sliding vectors is reduced to a couple if $\mathbf{M}_O \neq \mathbf{0}$ or is equivalent to zero if $\mathbf{M}_O = \mathbf{0}$. Supposing that the forces of the system $\{\mathbf{F}\}$ are parallel to the axis Ox_3, i.e., $\mathbf{u} = \mathbf{i}_3$ (Fig.2.18), *the conditions of equivalence to zero* are written in the form

$$R_3 = 0, \quad M_{Ox_1} = M_{Ox_2} = 0. \tag{2.2.37}$$

The above conditions are only *necessary* in the case of *parallel forces modelled by bound vectors*.

The point C is called *the centre of the system of parallel forces modelled by sliding vectors*, through it passing the central axis of the respective system. One obtains the same centre C for the system $\lambda\{\mathbf{F}\}$, λ scalar. Let $\{\mathbf{F}'\} \equiv \{\mathbf{F}'_i, i = 1, 2, ..., n\}$ be another system of parallel forces modelled by sliding vectors, the supports of which pass through the same points, respectively, but have another direction, given by the unit vector $\mathbf{u}'$ ($\mathbf{F}'_i = F'_i \mathbf{u}'$), and which have the same components as the system of forces $\{\mathbf{F}\}$ (that is $F'_i = F_i, i = 1, 2, ..., n$); in fact, it is a rotation of the same angle of all the supports. One obtains the same centre C of the system of parallel forces, the central axis having – obviously – the direction given by the new unit vector $\mathbf{u}'$.

2.2.8 Other considerations concerning systems of forces

Let $\mathscr{F} \equiv \{\mathbf{F}_i, \mathbf{r}_i, i = 1, 2, ..., n\}$ be a system of forces modelled by bound vectors, where we put into evidence also the position vectors of the points of application; the torsor of this system with respect to the pole O is

$$\tau_O\{\mathscr{F}\} = \{\mathbf{R}, \mathbf{M}_O\}, \tag{2.2.38}$$

where the resultant and the resultant moment are given by

$$\mathbf{R} = \sum_{i=1}^{n} \mathbf{F}_i, \quad \mathbf{M}_O = \sum_{i=1}^{n} \mathbf{r}_i \times \mathbf{F}_i. \tag{2.2.38'}$$

We notice that, excepting the so-called forces, a mechanical system is acted upon by a couple of forces (or a moment) too; we use also the generic denomination of *charge* (*load*).

The considered system of forces $\mathscr{F}$ corresponds rigorously to a discrete mechanical system; we have seen in Chap. 1, Subsec. 1.1.11 that, in the case of a continuous mechanical system, which has as support a domain D, the load can be punctual, linear, superficial or volumic, corresponding to the dimensions of the subdomain $\mathscr{D} \subset D$ to which it is transmitted.

The torsor (2.3.38), corresponding to a distributed load on a line L, a surface S or a volume V, respectively will be given by

$$\mathbf{R} = \int_L \mathbf{p}\, ds, \quad \mathbf{R} = \int_S \mathbf{p}\, dS, \quad \mathbf{R} = \int_V \mathbf{f}\, dV, \tag{2.2.39}$$

$$\mathbf{M}_O = \int_L \mathbf{r} \times \mathbf{p}\, ds, \quad \mathbf{M}_O = \int_S \mathbf{r} \times \mathbf{p}\, dS, \quad \mathbf{M}_O = \int_V \mathbf{r} \times \mathbf{f}\, dV, \tag{2.2.39'}$$

where **p** and **f** are continuous functions, representing loads on a unit of line, of area or of volume, respectively; these unit loads are vector quantities. Obviously, we may use also the notation of *mean unit load*

$$\mathbf{p}_{\text{mean}} = \frac{\Delta \mathbf{R}}{\Delta s}, \quad \mathbf{p}_{\text{mean}} = \frac{\Delta \mathbf{R}}{\Delta S}, \quad \mathbf{f}_{\text{mean}} = \frac{\Delta \mathbf{R}}{\Delta V}, \tag{2.2.40}$$

where $\Delta \mathbf{R}$ is the resultant corresponding to a finite element of line Δs, of area ΔS or of volume ΔV, respectively; there results

$$\mathbf{p} = \lim_{\Delta s \to 0} \mathbf{p}_{\text{mean}}, \quad \mathbf{p} = \lim_{\Delta S \to 0} \mathbf{p}_{\text{mean}}, \quad \mathbf{f} = \lim_{\Delta V \to 0} \mathbf{f}_{\text{mean}}. \tag{2.2.40'}$$

We mention that, in general,

$$\mathbf{p} = \mathbf{p}(\mathbf{r};t), \quad \mathbf{f} = \mathbf{f}(\mathbf{r};t). \tag{2.2.39''}$$

In the case of the action of distributed couples **m**, we can write the resultant moment in one of the forms

$$\mathbf{M}_O = \int_L \mathbf{m}\, ds, \quad \mathbf{M}_O = \int_S \mathbf{m}\, dS, \quad \mathbf{M}_O = \int_V \mathbf{m}\, dV. \tag{2.2.39'''}$$

The linear and superficial loads represent, in general, actions of contact (for instance, the action of a fluid upon a solid, on the contact surface between these material bodies). The volume loads correspond to an action at distance of a field of forces on the mass of the continuous mechanical system (forces the intensity of which is proportional to the gravitational or to the inertial mass); we thus have to do with *massic loads*, for which

$$\mathbf{R} = \int_V \frac{\mathbf{f}}{\mu}\, dm, \quad \mathbf{M}_O = \int_V \mathbf{r} \times \frac{\mathbf{f}}{\mu}\, dm, \tag{2.2.39^{iv}}$$

corresponding to a measure induced by the mass and where μ is the density.

As it was shown by W. Kecs and P.P. Teodorescu, to represent concentrated loads, it is convenient to use the methods of the theory of distributions, considered in Chap. 1, Subsec. 1.1.7. Let thus be a field of parallel forces (distributed loads)

$$\mathbf{Q}_\varepsilon(\mathbf{r}) = \mathbf{F} f_\varepsilon(\mathbf{r}), \tag{2.2.41}$$

defined on the sphere of volume V_ε, of centre at the origin and of radius $|\mathbf{r}| = \varepsilon$, f_ε being a δ representative sequence, while **F** is a constant vector. Passing to limit in the sense of the theory of distributions, we obtain

$$\mathbf{Q}(\mathbf{r}) = \lim_{\varepsilon \to +0} \mathbf{Q}_\varepsilon(\mathbf{r}) = \mathbf{F} \lim_{\varepsilon \to +0} f_\varepsilon(\mathbf{r}),$$

so that

$$\mathbf{Q}(\mathbf{r}) = \mathbf{F}\delta(\mathbf{r}), \tag{2.2.42}$$

where the field $\mathbf{Q}(\mathbf{r})$ is a volume density of the concentrated force $\mathbf{F}$. To put in evidence the correctness of this representation, we will calculate the torsor of the field at the pole O, obtaining thus the resultant

$$\mathbf{R}_\varepsilon = \int_{V_\varepsilon} \mathbf{Q}_\varepsilon \mathrm{d}V = \int_{V_\varepsilon} \mathbf{F} f_\varepsilon \mathrm{d}V = \mathbf{F} \int_{V_\varepsilon} f_\varepsilon \mathrm{d}V ;$$

hence, passing to limit ($\varepsilon \to +0$), we have

$$\mathbf{R} = \mathbf{F}. \qquad (2.2.41')$$

The resultant moment is written in the form

$$\mathbf{M}_\varepsilon = \int_{V_\varepsilon} \mathbf{r} \times \mathbf{Q}_\varepsilon \mathrm{d}V = \frac{3}{4\pi\varepsilon^3} \int_{V_\varepsilon} \mathbf{r} \times \mathbf{F} \mathrm{d}V = -\frac{3}{4\pi\varepsilon^3} \mathbf{F} \times \int_{V_\varepsilon} \mathbf{r} \mathrm{d}V = \mathbf{0},$$

where we took into consideration

$$\int_{V_\varepsilon} \mathbf{r} \mathrm{d}V = \mathbf{0}, \qquad (2.2.41'')$$

because the integral corresponds to the static moment of the homogeneous sphere with respect to its centre; obviously, $\mathbf{M}_O = \mathbf{0}$, the representation (2.2.42) being thus justified from a mechanical point of view. We used above a δ representative sequence given by

$$f_\varepsilon(\mathbf{r}) = \begin{cases} \dfrac{3}{4\pi\varepsilon^3}, & r < \varepsilon, \\ 0, & r > \varepsilon, \end{cases} \quad r = |\mathbf{r}| = \sqrt{x_i x_i},$$

the function having as support the sphere of volume V_ε; we notice that one can use any other δ representative sequence, which depends only on ε, for $r \le \varepsilon$. As well, we may consider any other deformable domain of volume V_ε and an arbitrary field of forces $\mathbf{Q}_\varepsilon(\mathbf{r})$, the components of which along the three axes of co-ordinates being three fields of parallel forces.

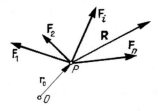

Figure 2.19. System of concentrated forces applied at the same point P.

Let $\mathscr{F}$ be a discrete system of forces applied at a point $P(\mathbf{r}_0)$ (concurrent forces) (Fig.2.19); the equivalent fields are given by

$$\mathbf{Q}_i(\mathbf{r}) = \mathbf{F}_i \delta(\mathbf{r} - \mathbf{r}_0), \quad i = 1, 2, \dots, n. \tag{2.2.43}$$

By means of the resultant (2.2.38'), applied at the same point P, we may write the volume density in the form

$$\mathbf{Q}(\mathbf{r}) = \sum_{i=1}^{n} \mathbf{Q}_i(\mathbf{r}) = \mathbf{R}\delta(\mathbf{r} - \mathbf{r}_0); \tag{2.2.43'}$$

the properties of the concentrated forces applied at the same point are thus preserved in this representation.

In the case of concentrated forces represented by bound vectors, at different points of application, one cannot speak – in general – about their composition. However, in some cases, we can give a representation in distributions of a system formed by two such concentrated forces. So, in the case of a system of $n > 2$ concentrated forces, applied at different points, one can try their composition if they are parallel; otherwise, one can decompose each force along three directions, and one obtains three systems of n parallel forces (components of the forces initially given).

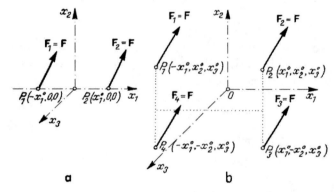

Figure 2.20. System of two (a) or four (b) parallel concentrated forces of the same modulus and direction.

Let be, for instance, two parallel forces of the same direction and the same intensity ($\mathbf{F}_1 = \mathbf{F}_2 = \mathbf{F}$, as free vectors), applied at the points $P_1(-x_1^0, 0, 0)$ and $P_2(x_1^0, 0, 0)$, $x_1^0 > 0$, respectively (Fig.2.20,a); there correspond the equivalent vector fields

$$\mathbf{Q}_1(x_1, x_2, x_3) = \mathbf{F}\delta(x_1 + x_1^0, x_2, x_3),$$
$$\mathbf{Q}_2(x_1, x_2, x_3) = \mathbf{F}\delta(x_1 - x_1^0, x_2, x_3).$$

The volume density of this system of two forces is given by

$$\mathbf{Q}(x_1, x_2, x_3) = 2x_1^0 \mathbf{F}\delta\left(x_1^2 - (x_1^0)^2, x_2, x_3\right), \quad x_1^0 > 0, \tag{2.2.44}$$

where we took into account the relation (1.1.40'); this relation can be used as a rule for the composition of the two equipollent forces (equal as free vectors), applied on the

Mechanics of the systems of forces 111

Ox_1-axis. We have taken $P_1 P_2$ as axis Ox_1, but this is not essential for the problem; e.g., for the points of application $P_1\left(x_1', x_2^0, x_3^0\right)$ and $P_2\left(x_1'', x_2^0, x_3^0\right)$ we obtain

$$\mathbf{Q}(x_1, x_2, x_3) = |x_1' - x_1''|\mathbf{F}\delta\big((x_1 - x_1')(x_1 - x_1''), x_2 - x_2^0, x_3 - x_3^0\big), \quad (2.2.44')$$

corresponding to the relation (1.1.41).

Let be also a system of four parallel forces having the same direction and the same intensity ($\mathbf{F}_1 = \mathbf{F}_2 = \mathbf{F}_3 = \mathbf{F}_4 = \mathbf{F}$, as free vectors), applied at the points $P_1\left(-x_1^0, x_2^0, x_3^0\right)$, $P_2\left(x_1^0, x_2^0, x_3^0\right)$, $P_3\left(x_1^0, -x_2^0, x_3^0\right)$, $P_4\left(-x_1^0, -x_2^0, x_3^0\right)$, $x_1^0, x_2^0 > 0$, respectively (Fig.2.20,b). The equivalent vector field is given by

$$\mathbf{Q}(x_1, x_2, x_3) = \mathbf{F}\big[\delta\left(x_1 + x_1^0, x_2 - x_2^0, x_3 - x_3^0\right)$$
$$+\delta\left(x_1 - x_1^0, x_2 - x_2^0, x_3 - x_3^0\right) + \delta\left(x_1 - x_1^0, x_2 + x_2^0, x_3 - x_3^0\right)$$
$$+\delta\left(x_1 + x_1^0, x_2 + x_2^0, x_3 - x_3^0\right)\big];$$

using the relation (1.1.42), we obtain the composition formula

$$\mathbf{Q}(x_1, x_2, x_3) = 4x_1^0 x_2^0 \mathbf{F}\delta\left(x_1^2 - (x_1^0)^2, x_2^2 - (x_2^0)^2, x_3 - x_3^0\right), \quad x_1^0, x_2^0 > 0. \quad (2.2.45)$$

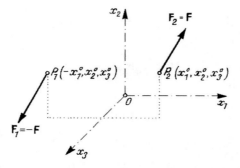

Figure 2.21. System of two parallel forces of the same modulus and opposite directions.

Let us consider also two parallel forces of opposite directions, but of the same intensity ($\mathbf{F}_2 = -\mathbf{F}_1 = \mathbf{F}$, as free vectors), applied at the points $P_1\left(-x_1^0, x_2^0, x_3^0\right)$ and $P_2\left(x_1^0, x_2^0, x_3^0\right)$, $x_1^0 > 0$, respectively (Fig.2.21). The equivalent vector field is

$$\mathbf{Q}(x_1, x_2, x_3) = \mathbf{F}\big[-\delta\left(x_1 + x_1^0, x_2 - x_2^0, x_3 - x_3^0\right) + \delta\left(x_1 - x_1^0, x_2 - x_2^0, x_3 - x_3^0\right)\big];$$

using the relation (1.1.43), we may write

$$\mathbf{Q}(x_1, x_2, x_3) = 2x_1 \mathbf{F}\delta\left(x_1^2 - (x_1^0)^2, x_2 - x_2^0, x_3 - x_3^0\right), \quad (2.2.46)$$

obtaining thus the searched composition formula.

In the case of deformable continuous media there appear various concentrated loads (mechanical quantities which have a punctual support), which play an important rôle, for instance: directed concentrated moments and dipoles of concentrated forces of various orders, centres of rotation, centres of plane or spatial dilatation etc.; all these loads can be expressed by means of distributions, starting from the representation (2.2.42), corresponding to a concentrated force.

One can show that a *linear load* of the form

$$\overline{\mathbf{F}}(u) = \begin{cases} \mathbf{F}(u), & u \in [a,b], \\ 0, & u \notin [a,b], \end{cases} \tag{2.2.47}$$

the support of which is a curve of parametric equations ($\mathbf{r} = \mathbf{r}(u)$)

$$x_i = f_i(u), \quad f_i \in C^\infty, \quad i = 1,2,3, \quad \lim_{u \to \pm\infty} f_i(u) f_i(u) = \infty, \tag{2.2.47'}$$

leads to a volume density

$$\mathbf{Q}(x_1, x_2, x_3) = \frac{\partial^3}{\partial x_1 \partial x_2 \partial x_3} \int_{-\infty}^{\infty} \theta_u(x_1, x_2, x_3) \overline{\mathbf{F}}(u) \sqrt{f_i'(u) f_i'(u)} \, du, \tag{2.2.47''}$$

where $f_i'(u) = \mathrm{d} f_i(u)/\mathrm{d} u$, while

$$\begin{aligned}\theta_u(x_1, x_2, x_3) &= \theta(x_1 - f_1(u))\theta(x_2 - f_2(u))\theta(x_3 - f_3(u)) \\ &= \begin{cases} 1, & x_i \geq f_i(u), \; i = 1,2,3, \; u \in [a,b], \\ 0, & x_1 < f_1(u) \text{ or } x_2 < f_2(u) \text{ or } x_3 < f_3(u) \text{ or } u \notin [a,b], \end{cases}\end{aligned} \tag{2.2.47'''}$$

θ being Heaviside's function.

As well, in the case of a *superficial load*, given by the relation

$$\overline{\mathbf{F}}(u,v) = \begin{cases} \mathbf{F}(u,v), & (u,v) \in D, \\ 0, & (u,v) \notin D, \end{cases} \tag{2.2.48}$$

where D is the definition domain of the parameters u and v, and the support of which is the surface of parametric equations ($\mathbf{r} = \mathbf{r}(u,v)$)

$$x_i = f_i(u,v), \quad f_i \in C^\infty, \quad i = 1,2,3, \quad \lim_{u^2+v^2 \to \infty} f_i(u,v) f_i(u,v) = \infty, \tag{2.2.48'}$$

we are led to the volume density

$$\mathbf{Q}(x_1,x_2,x_3) = \frac{\partial^3}{\partial x_1 \partial x_2 \partial x_3} \int_{-\infty}^{\infty}\int_{-\infty}^{\infty} \theta_{uv}(x_1,x_2,x_3) \\ \times \overline{\mathbf{F}}(u,v) \sqrt{E(u,v)G(u,v) - F^2(u,v)} \, du \, dv, \tag{2.2.48''}$$

where

$$\theta_{uv}(x_1, x_2, x_3) = \theta(x_1 - f_1(u,v))\theta(x_2 - f_2(u,v))\theta(x_3 - f_3(u,v)), \quad (2.2.48''')$$

θ being Heaviside's function, and where we have used the differential parameters

$$\begin{cases} E(u,v) = \left(\dfrac{\partial \mathbf{r}}{\partial u}\right)^2 = \dfrac{\partial f_i}{\partial u}\dfrac{\partial f_i}{\partial u}, \\ F(u,v) = \left(\dfrac{\partial \mathbf{r}}{\partial u}\right)\cdot\left(\dfrac{\partial \mathbf{r}}{\partial v}\right) = \dfrac{\partial f_i}{\partial u}\dfrac{\partial f_i}{\partial v}, \\ G(u,v) = \left(\dfrac{\partial \mathbf{r}}{\partial v}\right)^2 = \dfrac{\partial f_i}{\partial v}\dfrac{\partial f_i}{\partial v}, \end{cases} \quad (2.2.48^{\text{iv}})$$

corresponding to the first basic form of the surface (4.1.15), (4.1.15').

The functions $\overline{\mathbf{F}}(u)$ and $\overline{\mathbf{F}}(u,v)$ considered above are piecewise continuous and lead to regular distributions.

The methods of the theory of distributions allow us to represent continuous loads, as well as discontinuous ones, from a spatial point of view; we obtain also the representation of concentrated loads. One can consider continuous and discontinuous phenomena from a temporal point of view too; thus, the forces appearing in a phenomenon which takes place in a very short interval of time (for instance, the collision of two spheres) can be represented by Dirac's distribution with respect to time.

Returning to a unitary representation in distributions (corresponding to the formula (1.1.73)), we may write, in general,

$$\mathbf{R} = \int_\Omega \mathbf{f}\, \mathrm{d}V, \quad \mathbf{M}_O = \int_\Omega \mathbf{r}\times \mathbf{f}\, \mathrm{d}V, \quad (2.2.49)$$

where Ω is the geometric support of the mechanical system $\mathscr{S}$, while the integrals are Stieltjes integrals. In the case of a discrete mechanical system, we find again the formulae (2.2.38), (2.2.38'), while for a continuous one we obtain formulae (2.2.39), (2.2.39'), and the integrals become Riemann integrals.

In the case of a deformable mechanical system (the forces are modelled by *bound vectors*), the torsor (2.2.38), (2.2.38') leads to *necessary conditions* which can occur in various problems (relations of equivalence, equivalence to zero etc.); in the case of a non-deformable mechanical system (the forces are modelled by *sliding vectors*), the torsor leads to conditions which are also *sufficient* for the respective problems.

We have seen in Chap. 1, Subsec. 1.1.11 that the forces acting upon a mechanical system can be *external forces* or *internal forces*, the latter ones being always pairs (they are applied at the points P_i and P_j), being linked axiomatically by the relation (1.1.81). Taking into account the results of Subsec. 2.1.1, we may affirm that the moment with respect to an arbitrary pole of such a pair of internal forces vanishes, the mechanical system being deformable or even non-deformable; hence, the torsor of these forces with respect to the pole O (Fig.1.18) is equal to zero

$$\tau_O\{\mathbf{F}_{ij}, \mathbf{F}_{ji}\} = \mathbf{0}. \quad (2.2.50)$$

We can make an analogous affirmation for the whole system of internal forces acting upon the given mechanical system. If the relation (2.2.50) holds, then we have

$$\mathbf{M}_O\{\mathbf{F}_{ij}, \mathbf{F}_{ji}\} = \mathbf{r}_i \times \mathbf{F}_{ij} + \mathbf{r}_j \times \mathbf{F}_{ji} = (\mathbf{r}_j - \mathbf{r}_i) \times \mathbf{F}_{ji} = \overrightarrow{P_i P_j} \times \mathbf{F}_{ji} = \mathbf{0};$$

hence, $\mathbf{F}_{ij} = \lambda \overrightarrow{P_i P_j} = \lambda \mathbf{r}_{ij}$, λ scalar, and the considered forces are a pair of internal forces.

A pair of internal forces constitutes a *finite dipole of forces*, which can be represented, in distributions, by a formula of the form (2.2.46).

By decomposing a mechanical system in two subsystems, some forces which – at the beginning – have been internal forces, may become external ones; hence, the classification of the forces in internal and external ones is *conventional*.

The forces acting upon mechanical systems are modelled with the aid of vectors; these ones will thus play an important rôle to determine the *class* of equivalent systems of forces, which lead to the same effects of mechanical order.

Chapter 3

MASS GEOMETRY. DISPLACEMENTS. CONSTRAINTS

It is necessary to introduce some notions which play an important rôle in a static and dynamic study of mechanical systems; we will thus consider problems of mass geometry, as well as problems concerning displacements and constraints, expressed by means of the latter ones.

1. Mass geometry

Mass plays an important rôle in the dynamics of mechanical systems; we are thus led to the study of moments of first order (static moments) or of second order (moments of inertia).

1.1 Centres of mass

We introduce, in what follows, the notions of centre of mass, centre of gravity and static moment (moment of first order); we give also some properties useful for the practical computation.

1.1.1 Centres of mass. Centres of gravity

Let $\mathscr{S}$ be a mechanical system of geometric support Ω. We call *centre of mass* of this system a point C, which may not belong to it, defined by the position vector

$$\boldsymbol{\rho} = \frac{1}{M} \int_{\Omega} \mathbf{r} \, dm, \qquad (3.1.1)$$

the integral being a Stieltjes one, and the mass $m = m(\mathbf{r})$ a distribution. Introducing the density (1.1.71), (1.1.71"), we can write

$$\boldsymbol{\rho} = \frac{1}{M} \sum_{i=1}^{n} m_i \mathbf{r}_i, \qquad (3.1.2)$$

in the case of a discrete mechanical system of n particles P_i, of position vectors $\mathbf{r}_i$ and masses m_i, $i = 1, 2, \ldots, n$, or

$$\boldsymbol{\rho} = \frac{1}{M} \int_V \mathbf{r}\mu(\mathbf{r})\,\mathrm{d}V, \qquad (3.1.3)$$

in the case of a continuous mechanical system of density $\mu(\mathbf{r})$. If the continuum is homogeneous, it results

$$\boldsymbol{\rho} = \frac{1}{V} \int_V \mathbf{r}\,\mathrm{d}V. \qquad (3.1.4)$$

In components, we may write

$$\rho_j = \frac{1}{M} \int_\Omega x_j \,\mathrm{d}m, \quad j = 1,2,3, \qquad (3.1.1')$$

as well as

$$\rho_j = \frac{1}{M} \sum_{i=1}^n m_i x_j^{(i)}, \quad j = 1,2,3, \qquad (3.1.2')$$

or

$$\rho_j = \frac{1}{M} \int_V x_j \mu(\mathbf{r})\,\mathrm{d}V, \quad j = 1,2,3, \qquad (3.1.3')$$

and

$$\rho_j = \frac{1}{V} \int_V x_j \,\mathrm{d}V, \quad j = 1,2,3. \qquad (3.1.4')$$

In the case of a two-dimensional mechanical system, we replace the volume integral by a surface one, using a superficial density; if the mechanical system is plane, then the integral is a double one. As well, in the case of a one-dimensional mechanical system, we replace the volume integral by a curvilinear one, introducing a linear density.

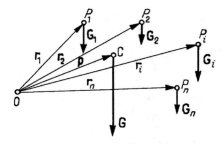

Figure 3.1. Centre of gravity of a heavy discrete mechanical system.

Let us admit that the mechanical system $\mathscr{S}$, supposed to be non-deformable at a given moment, is subjected to the action of a *uniform gravitational field* (for instance, *the terrestrial gravitational field*). To fix the ideas, we consider a discrete mechanical system, the particles of masses m_i being acted upon by the gravity forces

$$\mathbf{G}_i = m_i \mathbf{g}, \quad i = 1, 2, \ldots, n, \tag{3.1.5}$$

which form a field of parallel forces of the same direction. The system of particles being non-deformable at a certain moment, the forces form a system of sliding vectors; the central axis will pass through the point C of position vector (Fig.3.1)

$$\boldsymbol{\rho} = \frac{1}{Mg} \sum_{i=1}^{n} m_i g \mathbf{r}_i,$$

given by the formula (2.2.36). We rotate the whole discrete mechanical system by a given angle; the forces $\mathbf{G}_i$ will not change their supports and their direction. This is equivalent to the supposition that the system did not rotate, but the parallel forces did rotate with the same angle; with the aid of a property put in evidence in Chap. 2, Subsec. 2.2.7, it follows that the central axis passes through the same point C. This point will be called *the centre of gravity* of the mechanical system at a given moment, at which the system is considered non-deformable; in the case of a uniform gravitational field, it coincides with the centre of mass of the very same system. In the case of a continuous mechanical system, one can make analogous considerations. If the mechanical system is non-deformable, then the "instantaneous" centre of gravity becomes "permanent".

In general, in the case of a *non-uniform gravitational field*, the centre of gravity of a mechanical system is the point of application of the resultant of the gravity forces acting upon the points of this system (if the resultant moment vanishes, and this point of application is independent of the position of the mechanical system; for instance, a mechanical system with central symmetry in a gravitational field with axial symmetry).

We notice that the centre of mass has a more general significance, the centre of gravity being put in evidence only in the presence of a gravitational field (for example, the gravitational field of the Earth). The centre of gravity of a non-deformable mechanical system is the same in any place on the surface of the Earth, because it coincides with the centre of mass. The point C is called also *centre of inertia* if we take in consideration the inertial property of the mass. In the case of a non-deformable mechanical system, the centre of mass C is a point rigidly linked to this system (the distance from C to any point of the system is constant in time).

In the case of a homogeneous continuous mechanical system, the centre of gravity is given by the relation (3.1.4); we are led thus to the notion of *geometric centre of gravity* (which has a purely geometric character). Analogously, in the case of a discrete mechanical system for which all the particles P_i have the same mass, the centre of gravity C is specified by the position vector

$$\boldsymbol{\rho} = \frac{1}{n} \sum_{i=1}^{n} \mathbf{r}_i, \tag{3.1.2"}$$

corresponding to the formula (3.1.2); this point is called also *the barycentre* of the system of points P_i.

Because the position of the centre of a system of parallel vectors does not depend on the frame of reference used, it follows that the centre of gravity represents an intrinsic characteristic of the considered mechanical system, the centre of mass having the same property.

1.1.2 Static moments

We define *the polar static moment* of the mechanical system $\mathscr{S}$, of geometric support Ω, with respect to the pole O in the form

$$\mathbf{S}_O = \int_\Omega \mathbf{r}\,\mathrm{d}m, \qquad (3.1.6)$$

where we have introduced a Stieltjes integral, the mass $m = m(\mathbf{r})$ being a distribution; its components are *the planar static moments* (with respect to the planes of co-ordinates)

$$S_{Oj} = S_{Ox_k x_l} = \int_\Omega x_j\,\mathrm{d}m, \quad j \neq k \neq l \neq j, \quad j,k,l = 1,2,3. \qquad (3.1.6')$$

In the case of a discrete mechanical system, considered at the previous subsection, we can write

$$\mathbf{S}_O = \sum_{i=1}^n m_i \mathbf{r}_i, \quad S_{Oj} = S_{Ox_k x_l} = \sum_{i=1}^n m_i x_j^{(i)}, \quad j \neq k \neq l \neq j, \quad j,k,l = 1,2,3, \qquad (3.1.7)$$

and in the case of a continuous mechanical system we have

$$\mathbf{S}_O = \int_V \mathbf{r}\mu(\mathbf{r})\,\mathrm{d}V, \quad S_{Oj} = S_{Ox_k x_l} = \int_\Omega x_j \mu(\mathbf{r})\,\mathrm{d}V, \quad j \neq k \neq l \neq j, \quad j,k,l = 1,2,3;$$

$$(3.1.7')$$

to a homogeneous continuous mechanical system will correspond *the geometric static moment*

$$\mathbf{S}_O = \int_V \mathbf{r}\,\mathrm{d}V, \quad S_{Oj} = S_{Ox_k x_l} = \int_\Omega x_j\,\mathrm{d}V, \quad j \neq k \neq l \neq j, \quad j,k,l = 1,2,3, \qquad (3.1.8)$$

while for a homogeneous discrete mechanical system we obtain

$$\mathbf{S}_O = \sum_{i=1}^n \mathbf{r}_i, \quad S_{Oj} = S_{Ox_k x_l} = \sum_{i=1}^n x_j^{(i)}, \quad j \neq k \neq l \neq j, \quad j,k,l = 1,2,3. \qquad (3.1.8')$$

Taking into account the relation of definition (3.1.1), one observes easily that

$$\mathbf{S}_O = M\boldsymbol{\rho} \qquad (3.1.9)$$

Mass geometry. Displacements. Constraints

or

$$S_{Oj} = M\rho_j, \quad j = 1, 2, 3; \qquad (3.1.9')$$

hence, we state

Theorem 3.1.1. *The polar (planar) static moment of a mechanical system with respect to a pole (plane) is equal to the static moment of the centre of mass, at which is considered to be concentrated the mass of the whole mechanical system, with respect to the same pole (plane).*

The relations (3.1.9), (3.1.9') show that, in a certain manner, the centre of mass can replace the whole given mechanical system; this observation holds also for other mechanical quantities which we will define, the usefulness of the centre of mass introduced above being thus put into evidence. If the pole O (a plane) coincides with (passes through) the centre of mass ($\boldsymbol{\rho} = \mathbf{0}$ or $\rho_j = 0$, $j = 1, 2, 3$), then the polar (planar) static moment with respect to this pole (plane) vanishes and reciprocally. Hence, the centre of mass of a mechanical system is characterized by the vanishing of the polar (planar) static moment with respect to it (to a plane passing through it); one can thus affirm once more that the centre of mass constitutes an *intrinsic characteristic* of the considered mechanical system (there is only one point C defined by the relation (3.1.1)).

We notice that we can write the relation (3.1.9) also in the form

$$\mathbf{S}_O(\mathscr{S}) = \mathbf{S}_O(\mathscr{S}_C), \qquad (3.1.9'')$$

where $\mathscr{S}_C$ is the mechanical system formed by only one material point, the centre of mass C, at which we consider concentrated the mass of the whole mechanical system $\mathscr{S}$. If we write the relation (3.1.9) for the poles O and O', respectively, and subtract one relation from the other, then we obtain the relation

$$\mathbf{S}_{O'} = \mathbf{S}_O + M \overrightarrow{O'O}, \qquad (3.1.10)$$

which may be written in the form

$$\mathbf{S}_{O'}(\mathscr{S}) = \mathbf{S}_O(\mathscr{S}) + \mathbf{S}_{O'}(\mathscr{S}_O) \qquad (3.1.10')$$

too, the notations being analogous to the above ones. It is thus put into evidence the variation of the polar static moment of a mechanical system $\mathscr{S}$ if one passes from a pole O to another pole O'; projecting on the co-ordinate axes, one obtains corresponding relations for the planar static moments.

In the case of a plane mechanical system (for which the geometric support Ω belongs, e.g., to the plane $x_3 = 0$), the components of the static moment $\mathbf{S}_O$ with respect to a pole in this plane are axial static moments (with respect to the co-ordinate axes)

$$S_{O_\alpha} = S_{Ox_\alpha} = \int_\Omega x_\beta \, \mathrm{d}m, \quad \alpha \neq \beta, \quad \alpha, \beta = 1, 2; \qquad (3.1.6'')$$

hence, the axial static moment of a plane mechanical system with respect to an axis contained in the plane of the system is equal to the static moment of the centre of mass, at which one considers concentrated the mass of the whole mechanical system, with respect to the very same axis. We can state also that the centre of mass of a plane mechanical system is characterized by the vanishing of the axial static moment with respect to an axis contained in the plane and passing through it.

In the case of a linear mechanical system (the geometric support Ω of which is a straight line), one has to do only with the polar static moment (with only one component) with respect to a pole belonging to the straight line.

1.1.3 Properties. Applications

We notice that the centre of mass is placed in the interior of any convex closed surface Σ which contains in its interior the geometric support Ω of the considered mechanical system. Indeed, taking the plane of co-ordinates Ox_1x_2 tangent to the surface Σ at an arbitrary point of it, taken as pole O, and admitting the direction of the axis Ox_3 towards the interior of Σ, it follows that for all the points of the geometric support Ω we have $x_3 \geq 0$; the formula (3.1.1') allows us to affirm that $\rho_3 \geq 0$ (the equality takes place if Ω is a plane geometric support). Hence, the centre C is situated in the same part of the considered plane as the surface Σ; because $O \in \Sigma$ is arbitrarily chosen, it follows that the centre C is situated in the interior of the surface Σ.

If the support Ω is a straight line or a plane, then the centre C is on the straight line or is contained in the plane. In the first case, if we choose this line as the Ox_1–axis, then we find easily that $\rho_2 = \rho_3 = 0$; in the second case we choose the respective plane as plane Ox_1x_2 and obtain $\rho_3 = 0$.

If the mechanical system admits a plane of geometric (the geometric support Ω admits a plane of symmetry) and mechanical (the symmetric points of the geometric support Ω have the same unit mass, in the case of a continuous mechanical system, or the same finite mass, in the case of a discrete mechanical system) symmetry, then we take this plane as plane Ox_1x_2; applying the formula (3.1.1'), we notice that some points of the support Ω belong to the plane $x_3 = 0$, while the other points are pairs of symmetric points, obtaining thus $\rho_3 = 0$. The centre of mass belongs thus to the plane of symmetry. If there exist two (or three) planes of geometric and mechanical symmetry, then the centre C belongs to the straight line (centre) of intersection of these planes, which – obviously – will be an axis (a centre) of geometric and mechanical symmetry. If the mechanical system admits a plane Π diametrically conjugate from geometric and mechanical point of view to a direction Δ (to each point P_i of the mechanical system there corresponds a point P_j of the same system, having the same mass, so that the segment P_iP_j has the direction of Δ and the middle in the plane Π), then the point C will belong to the plane Π; in particular, if $\Delta \perp \Pi$, then this plane is a plane of geometric and mechanical symmetry.

Mass geometry. Displacements. Constraints

For any division $\mathscr{S}_i$, $i = 1,2,...,n$, of the mechanical system $\mathscr{S}$ in n disjoint subsystems,

$$\mathscr{S} = \sum_{i=1}^{n} \mathscr{S}_i ,$$

we may write (see Theorem 3.1.1)

$$\boldsymbol{\rho} = \frac{1}{M} \sum_{i=1}^{n} M_i \boldsymbol{\rho}_i , \qquad (3.1.11)$$

where M_i and $\boldsymbol{\rho}_i$ are the mass and the position vector of the centre of mass C_i, respectively, corresponding to the subsystem $\mathscr{S}_i$, and M and $\boldsymbol{\rho}$ are the mass and the position vector of the centre of mass C, respectively, corresponding to the given mechanical system $\mathscr{S}$; by means of the formulae of definition and the property of associativity (additivity) of the finite sum (of the integral), the proof is obvious. If a mechanical system $\mathscr{S}$ may be considered as resulting by taking off a mechanical system $\mathscr{S}_2$ from a mechanical system $\mathscr{S}_1$, then the formula (3.1.11) allows us to write

$$\boldsymbol{\rho} = \frac{M_1 \boldsymbol{\rho}_1 - M_2 \boldsymbol{\rho}_2}{M_1 - M_2} , \qquad (3.1.12)$$

the notations being analogous to the above ones.

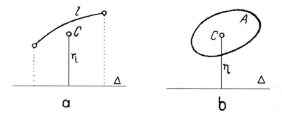

Figure 3.2. Pappus-Guldin theorems: for a surface (a) and for a volume (b).

The notion of centre of gravity allows us to state
Theorem 3.1.2 (*Pappus-Guldin*). *Let be an arc of a rectifiable plane curve of length l, coplanar with an axis Δ, at the same part of this one, which rotates by an angle $\alpha \leq 2\pi$ about the axis; the area of the surface thus obtained is given by*

$$S = \alpha \eta l , \qquad (3.1.13)$$

where η is the distance from the centre of gravity C of the arc of curve to the axis (Fig.3.2,a).

In the particular case of a surface of rotation ($\alpha = 2\pi$), we obtain

$$S = 2\pi \eta l , \qquad (3.1.13')$$

where the product of the perimeter of the circle described by the point C by the length l is put into evidence. We may also state

Theorem 3.1.3 (*Pappus-Guldin*). *Let be a plane figure of area A, coplanar with an axis Δ, at the same part of the latter one, which rotates by an angle $\alpha \leq 2\pi$ about the axis; the volume of the domain thus generated is given by*

$$V = \alpha \eta A, \qquad (3.1.14)$$

where η is the distance from the centre of gravity C of the plane figure to the axis Δ (Fig.3.2,b).

In particular, in the case of a body of rotation ($\alpha = 2\pi$), we can write

$$V = 2\pi \eta A, \qquad (3.1.14')$$

where the product of the perimeter of the circle described by the point C by the area A is put into evidence.

The proof of these theorems is obvious if one takes into account the definition and the properties of the Riemann integral.

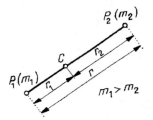

Figure 3.3. Centre of mass of a system of two particles.

For a discrete mechanical system, formed by two particles P_1 and P_2, of masses m_1 and m_2, respectively, the centre of mass is on the segment $P_1 P_2$, at the distances r_1 and r_2 from the extremities of it, so that (Fig.3.3)

$$r_1 = \frac{r}{m} m_2, \quad r_2 = \frac{r}{m} m_1, \quad r = r_1 + r_2, \quad m = m_1 + m_2, \quad m_1 \geq m_2; \qquad (3.1.15)$$

we notice that the centre C is closer to the particle of the greatest mass (in the considered case, closer to the particle P_1).

Taking into account the properties mentioned above, one observes easily that the centre of gravity of a discrete mechanical system formed by three particles of equal masses, or of a triangular line (or figure) is at the piercing point of its median lines; if the three masses are not equal, one obtains the famous theorem enounced in 1678 by Giovanni Ceva.

The centre of gravity of a contour (or figure) in form of a parallelogram (in particular, rectangle) is at the piercing point of its diagonals, the centre of gravity of a

Mass geometry. Displacements. Constraints

contour (or figure) having a circular form is at its centre a.s.o. Analogously, the centre of gravity of a discrete mechanical system formed by four non-coplanar particles of equal masses, or of a trihedral surface (or domain) is the piercing point of its median lines. Obviously, we can admit that all these lines, figures and domains correspond to homogeneous bodies.

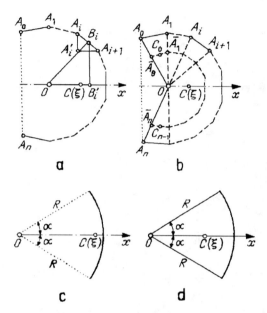

Figure 3.4. Centre of gravity of a regular polygonal line (a) or sector (b).
Centre of gravity of an arc (c) or sector (d) of circle.

In the case of a regular polygonal line $A_0 A_1 \ldots A_n$, formed by n segments of length l, the centre of gravity is on the symmetry axis, which passes through a vertex (n even) or through the middle of a segment (n odd). The similar triangles $A_i A_{i+1} A_i'$ and $OB_i B_i'$ (Fig.3.4,a) allow us to write $\overline{OB_i} \cdot \overline{A_i A_i'} = \overline{OB_i'} \cdot \overline{A_i A_{i+1}}$, so that the centre of gravity is given by

$$\xi = \frac{\sum_{i=0}^{n-1} \overline{A_i A_{i+1}} \cdot \overline{OB_i'}}{\sum_{i=0}^{n-1} \overline{A_i A_{i+1}}} = \frac{\overline{OB_i} \cdot \sum_{i=0}^{n-1} \overline{A_i A_i'}}{\sum_{i=0}^{n-1} \overline{A_i A_{i+1}}} = \frac{\overline{OB_i} \cdot \overline{A_0 A_n}}{\sum_{i=0}^{n-1} \overline{A_i A_{i+1}}};$$

if $\overline{OB_i} = a$ is the short radius, $\overline{A_i A_{i+1}} = l$ the side, $\overline{A_0 A_n} = c$ the chord and $p = nl$ the perimeter, then we can write

$$\xi = \frac{ac}{p}. \qquad (3.1.16)$$

The centre of gravity of the polygonal sector $A_0 A_1 ... A_n$ (Fig.3.4,b), formed by the triangles $OA_i A_{i+1}$, $i = 0,1,2,...,n-1$, coincides with the centre of gravity of the polygonal line $\overline{A_0 A_1} ... \overline{A_n}$, which passes through the centres of gravity $C_0, C_1, C_2, ..., C_{n-1}$ of the n triangles; in this case, it is obvious that

$$\xi = \frac{2}{3}\frac{ac}{p}. \qquad (3.1.17)$$

In the case of an arc of circle of angle $2\alpha \leq 2\pi$ and of radius R (Fig.3.4,c) we obtain, by a process of passing to the limit ($p = 2R\alpha$, $c = 2R\sin\alpha$, $a = R$),

$$\xi = \frac{c}{p}R = \frac{\sin\alpha}{\alpha}R; \qquad (3.1.16')$$

as well, for a circular sector of angle $2\alpha \leq 2\pi$ and radius R (Fig.3.4,d) we write

$$\xi = \frac{2}{3}\frac{c}{p}R = \frac{2}{3}\frac{\sin\alpha}{\alpha}R. \qquad (3.1.17')$$

In particular, for a semicircular line ($\alpha = \pi/2$) it results

$$\xi = \frac{2}{\pi}R, \qquad (3.1.16'')$$

while for the figure in form of a semicircle we obtain

$$\xi = \frac{4}{3\pi}R. \qquad (3.1.17'')$$

Analogous considerations can be made for three-dimensional bodies; for example, in the case of a homogeneous semisphere of radius R we have

$$\xi = \frac{3}{8}R. \qquad (3.1.18)$$

Let be a compound figure formed by two adjoining rectangles (Fig.3.5). We obtain thus the centres C_1 and C_2; the centre C of the compound figure is on the segment $C_1 C_2$, its position being determined by formulae of the form (3.1.15), where the masses, considered concentrated at those centres, are proportional to the areas of the component rectangles. Dividing the compound figure in other two rectangles, we are led to the centres C_1' and C_2'; the centre C will belong to the segment $C_1' C_2'$, hence it is at the piercing point of this segment with the segment $C_1 C_2$. Thus, the formula (3.1.11) leads to a graphic construction of the geometric centre of gravity C. We notice that the above-considered figure may also be obtained by taking off a rectangle from another one, applying thus the formula (3.1.12); obviously, a graphic construction can be used too.

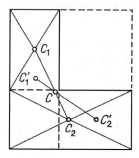

Figure 3.5. Centre of gravity of a compound figure formed by two adjoining rectangles.

1.2 Moments of inertia

In mechanics and physics can appear moments of the form

$$\int_\Omega x_1^{\alpha_1} x_2^{\alpha_2} x_3^{\alpha_3} \, \mathrm{d}m, \quad \alpha_1, \alpha_2, \alpha_3 \geq 0, \tag{3.1.19}$$

of order $\alpha = \alpha_1 + \alpha_2 + \alpha_3$, where the mass $m = m(\mathbf{r})$ is a distribution and we have introduced the Stieltjes integral. In the case $\alpha_1 = \alpha_2 = \alpha_3 = \alpha = 0$ one obtains the mass M, while the case $\alpha_i = 1$, $\alpha_j = \alpha_k = 0$, $i \neq j \neq k \neq i$, $i,j,k = 1,2,3$, $\alpha = 1$ corresponds to a planar static moment. The moments for which $\alpha > 2$ are not of a particular interest. In what follows, we will deal with moments of second order ($\alpha = 2$), called *moments of inertia*, which play an important rôle in the dynamics of mechanical systems. We mention also some considerations concerning tensors of *n*th order and, in particular, tensors of second order, useful in the study of the moment of inertia tensor.

1.2.1 Definitions. Properties

Let $\mathscr{S}$ be a mechanical system of geometric support Ω. We define *the polar moment of inertia* with respect to the pole O

$$I_O = \int_\Omega r^2 \, \mathrm{d}m, \tag{3.1.20}$$

the axial moments of inertia

$$I_{x_l} = \int_\Omega (x_j^2 + x_k^2) \, \mathrm{d}m, \quad j \neq k \neq l \neq j, \quad j,k,l = 1,2,3, \tag{3.1.21}$$

and *the planar moments of inertia*

$$I_{Ol} = I_{Ox_j x_k} = \int_\Omega x_l^2 \, \mathrm{d}m, \quad j \neq k \neq l \neq j, \quad j,k,l = 1,2,3, \tag{3.1.22}$$

where we have introduced the Stieltjes integral, $m = m(\mathbf{r})$ being a distribution.

In the case of a discrete mechanical system of particles of finite masses m_i, $i = 1,2,...,n$, we may write

$$I_O = \sum_{i=1}^{n} m_i r_i^2, \qquad (3.1.20')$$

$$I_{x_l} = \sum_{i=1}^{n} m_i \left[\left(x_j^{(i)}\right)^2 + \left(x_k^{(i)}\right)^2 \right], \quad j \neq k \neq l \neq j, \quad j,k,l = 1,2,3, \qquad (3.1.21')$$

$$I_{Ox_jx_k} = \sum_{i=1}^{n} m_i \left(x_l^{(i)}\right)^2, \quad j \neq k \neq l \neq j, \quad j,k,l = 1,2,3, \qquad (3.1.22')$$

while, in the case of a continuous mechanical system for which the geometric support Ω occupies the volume V, there result

$$I_O = \int_V r^2 \mu(\mathbf{r}) \, \mathrm{d}V, \qquad (3.1.20'')$$

$$I_{x_l} = \int_V \left(x_j^2 + x_k^2\right) \mu(\mathbf{r}) \, \mathrm{d}V, \quad j \neq k \neq l \neq j, \quad j,k,l = 1,2,3, \qquad (3.1.21'')$$

$$I_{Ox_jx_k} = \int_V x_l^2 \mu(\mathbf{r}) \, \mathrm{d}V, \quad j \neq k \neq l \neq j, \quad j,k,l = 1,2,3. \qquad (3.1.22'')$$

To a homogeneous continuous mechanical system there correspond *the geometric moments of inertia* (for the sake of simplicity, we use the same notation)

$$I_O = \int_V r^2 \, \mathrm{d}V, \qquad (3.1.20''')$$

$$I_{x_l} = \int_V \left(x_j^2 + x_k^2\right) \mathrm{d}V, \quad j \neq k \neq l \neq j, \quad j,k,l = 1,2,3, \qquad (3.1.21''')$$

$$I_{Ox_jx_k} = \int_V x_l^2 \, \mathrm{d}V, \quad j \neq k \neq l \neq j, \quad j,k,l = 1,2,3, \qquad (3.1.22''')$$

while for a homogeneous discrete mechanical system we can write

$$I_O = \sum_{i=1}^{n} r_i^2, \qquad (3.1.20^{\mathrm{iv}})$$

$$I_{x_l} = \sum_{i=1}^{n} \left[\left(x_j^{(i)}\right)^2 + \left(x_k^{(i)}\right)^2 \right], \quad j \neq k \neq l \neq j, \quad j,k,l = 1,2,3, \qquad (3.1.21^{\mathrm{iv}})$$

$$I_{Ox_jx_k} = \sum_{i=1}^{n} \left(x_l^{(i)}\right)^2, \quad j \neq k \neq l \neq j, \quad j,k,l = 1,2,3. \qquad (3.1.22^{\mathrm{iv}})$$

The above moments of inertia verify the relations

$$I_O = I_{Ox_2x_3} + I_{Ox_3x_1} + I_{Ox_1x_2} = \frac{1}{2}(I_{x_1} + I_{x_2} + I_{x_3}), \qquad (3.1.23)$$

$$I_{x_l} = I_{Ox_jx_l} + I_{Ox_kx_l}, \quad I_O = I_{Ox_jx_k} + I_{x_l}, \quad j \neq k \neq l \neq j, \quad j,k,l = 1,2,3, \qquad (3.1.23')$$

from which

$$I_{Ox_jx_k} = \frac{1}{2}(I_{x_j} + I_{x_k} - I_{x_l}), \quad j \neq k \neq l \neq j, \; j,k,l = 1,2,3; \qquad (3.1.23'')$$

hence, the polar moment of inertia and the planar moments of inertia can be expressed by means of the axial moments of inertia, so that it is sufficient to consider the latter ones. All these moments of inertia are non-negative.

We are led to the relations

$$I_{x_j} - I_{x_k} \leq I_{x_l} \leq I_{x_j} + I_{x_k}, \quad I_{x_j} \geq I_{x_k}, \quad j \neq k \neq l \neq j, \; j,k,l = 1,2,3, \qquad (3.1.24)$$

of the type of the triangle relations; the equalities take place if the geometric support Ω is contained in a manifold of the three-dimensional space (a plane or a straight line).

If Π_1, Π_2 and Π_3 are two by two orthogonal planes, then we can write, in general,

$$I_\Delta = I_{\Pi_1} + I_{\Pi_2}, \quad \Delta \equiv \Pi_1 \cap \Pi_2, \quad I_O = I_{\Pi_1} + I_{\Pi_2} + I_{\Pi_3}, \quad O \equiv \Pi_1 \cap \Pi_2 \cap \Pi_3. \qquad (3.1.25)$$

If Δ is an axis of geometric and mechanical symmetry, then the moments of inertia with respect to any plane Π passing through Δ are equal; taking into account the first relation (3.1.25), we obtain

$$I_\Delta = 2I_\Pi. \qquad (3.1.26)$$

If O is a centre of geometric and mechanical symmetry, then the moments of inertia with respect to any plane Π passing through O are equal; the second relation (3.1.25) and the relation (3.1.26) lead to

$$I_O = 3I_\Pi = \frac{3}{2}I_\Delta, \qquad (3.1.26')$$

where Δ is a straight line passing through O. For instance, in the case of a homogeneous whole sphere of radius R and mass M, we can write

$$I_O = \frac{3}{5}MR^2, \quad I_\Pi = \frac{1}{5}MR^2, \quad I_\Delta = \frac{2}{5}MR^2. \qquad (3.1.27)$$

We introduce *the centrifugal moments of inertia (products of inertia)*

$$I_{x_jx_k} = \int_\Omega x_j x_k \, \mathrm{d}m, \quad j \neq k, \; j,k = 1,2,3, \qquad (3.1.28)$$

too, which may be of the form

$$I_{x_j x_k} = \sum_{i=1}^{n} m_i x_j^{(i)} x_k^{(i)}, \quad I_{x_j x_k} = \int_V x_j x_k \mu(\mathbf{r}) \mathrm{d}V, \quad j \neq k, \quad j,k = 1,2,3, \qquad (3.1.28')$$

or of the form

$$I_{x_j x_k} = \sum_{i=1}^{n} x_j^{(i)} x_k^{(i)}, \quad I_{x_j x_k} = \int_V x_j x_k \, \mathrm{d}V, \quad j \neq k, \quad j,k = 1,2,3; \qquad (3.1.28'')$$

these moments of inertia can be also negative. We notice that

$$x_j x_k = \frac{1}{4}\left[(x_j + x_k)^2 - (x_j - x_k)^2\right] = \frac{1}{2}(d_{jk}^2 - d_{jk}'^2), \quad j \neq k, \quad j,k = 1,2,3,$$

where d_{jk}, d_{jk}' are the distances of a point of the mechanical system to the bisector planes Π_{jk}, Π_{jk}' of the dihedron formed by the planes $x_j = 0$, $x_k = 0$, respectively; we obtain

$$I_{x_j x_k} = \frac{1}{2}\left(I_{\Pi_{jk}} - I_{\Pi_{jk}'}\right) = \left(I_{\Delta_{jk}} - I_{\Delta_{jk}'}\right), \quad j \neq k, \quad j,k = 1,2,3, \qquad (3.1.29)$$

where Δ_{jk}, Δ_{jk}' are the bisectrices of the angle formed by the axes Ox_j, Ox_k, respectively, and we used formulae of the form (3.1.23'').

If the plane $Ox_j x_k$ is a plane of geometric and mechanical symmetry, (the points of co-ordinates x_l and $-x_l$ of the mechanical system have the same contribution in the computation of the integrals of the form (3.1.28)) or if the axis Ox_l is an axis of geometric and mechanical symmetry, then we obtain $I_{x_j x_l} = I_{x_k x_l} = 0$, $j \neq k \neq l \neq j$, $j,k,l = 1,2,3$.

The *gyration radius* (*radius of inertia*) with respect to the axis Δ is defined by the relation

$$i_\Delta = \sqrt{\frac{I_\Delta}{M}} \qquad (3.1.30)$$

and represents the distance of a material point at which is concentrated the mass M of the whole mechanical system to the axis Δ; in the case of a geometric moment of inertia of a homogeneous mechanical system of volume V, we may write

$$i_\Delta = \sqrt{\frac{I_\Delta}{V}}. \qquad (3.1.30')$$

If the geometric support Ω is contained in the plane $Ox_1 x_2$, then we may define the polar moment of inertia

$$I_O = \int_\Omega r^2 \, \mathrm{d}m = \int_\Omega (x_1^2 + x_2^2) \, \mathrm{d}m, \qquad (3.1.31)$$

Mass geometry. Displacements. Constraints

the axial moments of inertia

$$I_{x_\alpha} = \int_\Omega x_\beta^2 \, dm, \quad \alpha \neq \beta, \quad \alpha,\beta = 1,2, \quad (3.1.32)$$

and the products of inertia

$$I_{x_1 x_2} = \int_\Omega x_1 x_2 \, dm; \quad (3.1.33)$$

for discrete or continuous mechanical systems one can put in evidence analogous formulae. We notice the relation

$$I_O = I_{x_1} + I_{x_2}. \quad (3.1.34)$$

If the geometric support Ω is a circle, of centre O and radius R, then we obtain (for a homogeneous mechanical system)

$$I_O = \frac{\pi}{2} R^4, \quad I_{x_1} = I_{x_2} = \frac{\pi}{4} R^4. \quad (3.1.34')$$

In the case of an axis or of a centre of geometric and mechanical symmetry, respectively, one can emphasize properties analogous to those mentioned in the three-dimensional case. If the geometric support Ω is on a straight line, then one can define only polar moments of inertia (in the one-dimensional space).

1.2.2 Tensors of nth order

We will deal with some elements of tensor algebra and analysis, considering scalars, tensors of first order as well as of nth order, and we will put into evidence various particular cases; we will consider only tensors in E_3.

One can pass from a positive orthonormed basis $\mathcal{B}$ of unit vectors $\mathbf{i}_j$ to another positive orthonormed basis $\mathcal{B}'$ of unit vectors $\mathbf{i}'_k$ by means of formulae (2.1.8)-(2.1.11), given at Chap. 2, Subsec. 1.1.2.

The cosines α_{kj} which are introduced verify the relations

$$\alpha_{ij}\alpha_{ik} = \delta_{jk}, \quad \alpha_{ji}\alpha_{ki} = \delta_{jk}, \quad j,k = 1,2,3. \quad (3.1.35)$$

As well, one can pass from a system of co-ordinate axes to another one by means of the formulae (2.1.11) or of the formulae

$$\frac{\partial x'_k}{\partial x_j} = \alpha_{kj}, \quad \frac{\partial x_j}{\partial x'_k} = \alpha_{kj}, \quad k,j = 1,2,3. \quad (3.1.36)$$

Let be a function $U = U(x_1, x_2, x_3)$; by a change of co-ordinates of the form (2.1.11), we obtain the function $U' = U'(x'_1, x'_2, x'_3)$. If the condition

$$U' = U \tag{3.1.37}$$

is fulfilled, hence if the considered function is invariant to a change of orthogonal Cartesian co-ordinates, then we say that U is a *scalar* (or a *tensor of zeroth order*).

In the case of a scalar considered as a constant function (or of a scalar defined only at one point) we obtain a mathematical entity characterized by modulus and sign.

Let be a vector $\mathbf{V} = \mathbf{V}(\mathbf{r})$ (see Fig.2.2); by means of the two systems of co-ordinates, we may write

$$\mathbf{V} = V_j \mathbf{i}_j = V_k' \mathbf{i}_k', \tag{3.1.38}$$

where $V_j = V_j(x_1, x_2, x_3)$, $V_k' = V_k'(x_1, x_2, x_3)$, $j,k = 1,2,3$. Taking into account the relations (2.1.8) and (2.1.8'), respectively, we find the relations with the aid of which one may pass from the components of the vector $\mathbf{V}$ in the basis $\mathscr{B}$ to its components with respect to the basis $\mathscr{B}'$ and inversely, in the form

$$V_k' = V_j \alpha_{kj}, \quad V_j = V_k' \alpha_{kj}, \quad j,k = 1,2,3. \tag{3.1.39}$$

In general, three functions V_i, $i = 1,2,3$, which are transformed accordingly to the formulae (3.1.39) by a change of orthogonal Cartesian co-ordinates of the form (2.1.11), are the components of a *vector* (or of a *tensor of first order*) with respect to the considered frame of reference $\mathscr{B}$. A tensor of first order can be represented by its components with respect to a certain basis in the form of a row or column matrix

$$[V_i] \equiv \begin{bmatrix} V_1 & V_2 & V_3 \end{bmatrix} = \begin{bmatrix} V_1 \\ V_2 \\ V_3 \end{bmatrix}^{\mathrm{T}}, \tag{3.1.40}$$

where T indicates the transpose matrix.

Taking into account (3.1.36), we can express the transformation relations (3.1.39) also in the form

$$V_k' = V_j \frac{\partial x_k'}{\partial x_j}, \quad V_j = V_k' \frac{\partial x_j}{\partial x_k'}, \quad j,k = 1,2,3; \tag{3.1.41}$$

this form has a more general character, because it can be used also in the case of an arbitrary (non-linear) co-ordinate transformation (not only in the case of a linear one).

Extending these results, we call *tensor of second order* a mathematical entity $\mathbf{a}$ which is represented by its components with respect to a positive orthonormed basis $\mathscr{B}$ in the form of a square matrix

Mass geometry. Displacements. Constraints

$$[a_{ij}] \equiv \begin{bmatrix} a_{11} & a_{12} & a_{13} \\ a_{21} & a_{22} & a_{23} \\ a_{31} & a_{32} & a_{33} \end{bmatrix}; \qquad (3.1.42)$$

the elements of this matrix form a set of $3^2 = 9$ quantities a_{ij}, $i,j = 1,2,3$, which, by a change of orthogonal Cartesian co-ordinates of the form (2.1.11), behave corresponding to the relations

$$a'_{kl} = a_{ij}\alpha_{ki}\alpha_{lj}, \quad a_{ij} = a'_{kl}\alpha_{ki}\alpha_{lj}, \quad i,j,k,l = 1,2,3. \qquad (3.1.43)$$

Similarly, a *tensor of third order* is a mathematical entity **a**, represented by its components a_{ijk}, $i,j,k = 1,2,3$, with respect to a positive orthonormed basis $\mathcal{B}$; these components form a set of $3^3 = 27$ quantities which are transformed by a change of orthogonal Cartesian co-ordinates of the form (2.1.11) by means of the relations

$$a'_{lmn} = a_{ijk}\alpha_{li}\alpha_{mj}\alpha_{nk}, \quad a_{ijk} = a'_{lmn}\alpha_{li}\alpha_{mj}\alpha_{nk}, \quad i,j,k,l,m,n = 1,2,3. \qquad (3.1.44)$$

In general, a *tensor of nth order* is a mathematical entity **a**, represented by its components $a_{i_1 i_2 \ldots i_n}$, $i_k = 1,2,3$, $k = 1,2,\ldots,n$, with respect to a positive orthonormed basis $\mathcal{B}$; these components form a set of 3^n quantities which, by a change of orthogonal Cartesian co-ordinates of the form (2.1.11), are transformed corresponding to the relations

$$a'_{j_1 j_2 \ldots j_n} = a_{i_1 i_2 \ldots i_n}\alpha_{j_1 i_1}\alpha_{j_2 i_2}\ldots\alpha_{j_n i_n}, \quad a_{i_1 i_2 \ldots i_n} = a'_{j_1 j_2 \ldots j_n}\alpha_{j_1 i_1}\alpha_{j_2 i_2}\ldots\alpha_{j_n i_n},$$
$$i_k, j_k = 1,2,3, \quad k = 1,2,\ldots,n. \qquad (3.1.45)$$

The tensors thus defined are *Euclidean tensors*.
Let $\mathbf{a} \equiv [a_{i_1 i_2 \ldots i_n}]$ be a tensor of *n*th order; if the relation

$$a_{i_1 i_2 \ldots i_j \ldots i_k \ldots i_n} = a_{i_1 i_2 \ldots i_k \ldots i_j \ldots i_n} \qquad (3.1.46)$$

takes place in a basis $\mathcal{B}$, then we say that the tensor is *symmetric* with respect to the indices i_j and i_k; if the property (3.1.46) holds for any indices i_j and i_k, then the tensor **a** is *totally symmetric*. If the components of the tensor **a** verify the relation

$$a_{i_1 i_2 \ldots i_j \ldots i_k \ldots i_n} = -a_{i_1 i_2 \ldots i_k \ldots i_j \ldots i_n} \qquad (3.1.46')$$

in a basis $\mathcal{B}$, we say that this tensor is *antisymmetric* (*sqew-symmetric*) with respect to the indices i_j and i_k; if the property (3.1.46') takes place for all the indices i_j and i_k, then the tensor is *totally antisymmetric*. A tensor **a** which has not one of the properties mentioned above is an *asymmetric* tensor.

A tensor is, in general, definite at a point of the space; as well, we may consider a *tensor mapping* $(x_1, x_2, x_3) \to a_{i_1 i_2 \ldots i_n}(x_1, x_2, x_3)$, defining thus a *tensor field*.

If we use the upper index "prime" for the quantities which are obtained from the relation (2.1.9'), by a change of co-ordinate axes, it follows that $\alpha'_{lm} = \mathbf{i}'_l \cdot \mathbf{i}_m$, $l, m = 1, 2, 3$, the relation of definition of these cosines remaining the same; indeed, the scalar products (2.1.9') are invariant with respect to any frame of reference, in particular with respect to the frames $\mathcal{B}$ and $\mathcal{B}'$. We see that a relation of the form (3.1.43) holds, because

$$\alpha'_{lm} = \alpha_{kj}\alpha_{lk}\alpha_{mj} = \alpha_{lk}\delta_{km} = \alpha_{lm}, \quad l, m = 1, 2, 3,$$

where we have introduced Kronecker's symbol. Hence α_{ij}, $i, j = 1, 2, 3$, are the components of a tensor of second order

$$[\alpha_{ij}] \equiv \begin{bmatrix} \alpha_{11} & \alpha_{12} & \alpha_{13} \\ \alpha_{21} & \alpha_{22} & \alpha_{23} \\ \alpha_{31} & \alpha_{32} & \alpha_{33} \end{bmatrix}, \tag{3.1.47}$$

which makes clear the position of the orthonormed basis $\mathcal{B}'$ with respect to the basis $\mathcal{B}$ and inversely.

We can define Kronecker's symbol with respect to the basis $\mathcal{B}$ in the form

$$\delta'_{kl} = \begin{cases} 0, & k \neq l, \\ 1, & k = l, \end{cases} \quad k, l = 1, 2, 3;$$

also in this case, a relation of the form (3.1.43) is verified, because

$$\delta'_{kl} = \delta_{ij}\alpha_{ki}\alpha_{lj} = \alpha_{ki}\alpha_{li} = \delta_{kl}, \quad k, l = 1, 2, 3.$$

Hence, Kronecker's symbols are the components of a tensor of second order too

$$[\delta_{ij}] = \begin{bmatrix} 1 & 0 & 0 \\ 0 & 1 & 0 \\ 0 & 0 & 1 \end{bmatrix}, \tag{3.1.48}$$

which is just *the unit tensor* ($\boldsymbol{\delta} = \mathbf{1}$); this tensor is symmetric ($\delta_{ij} = \delta_{ji}$, $i, j = 1, 2, 3$).

Let us consider also the permutation symbol, defined by the formula (2.1.29). Taking into account the relations $(\mathbf{i}'_1, \mathbf{i}'_2, \mathbf{i}'_3) = \det[\alpha_{pq}] = 1$ and $(\mathbf{i}'_l, \mathbf{i}'_m, \mathbf{i}'_n) = \epsilon'_{lmn}$, $l, m, n = 1, 2, 3$, and the relations (2.1.37), it follows

$$\epsilon'_{lmn} = \epsilon_{ijk}\, \alpha_{li}\alpha_{mj}\alpha_{nk}\,, \quad l,m,n = 1,2,3\,.$$

Hence, the permutation symbol is a totally antisymmetric tensor of third order ϵ, for which the relations (2.1.45)-(2.1.46") take place.

Two tensors **a** and **b** of the same order n are *equal* (**a** = **b**) if they have the same components

$$a_{i_1 i_2 \ldots i_n} = b_{i_1 i_2 \ldots i_n}\,, \quad i_j = 1,2,3\,, \quad j = 1,2,\ldots,n\,, \tag{3.1.49}$$

in an arbitrary frame $\mathscr{B}$; this relation has the well known properties of *reflexivity*, *symmetry* and *transitivity*.

The *sum* of two tensors (**a** and **b**) of order n is a tensor **c** = **a** + **b** of the same order; hence, in a frame $\mathscr{B}$ we have

$$a_{i_1 i_2 \ldots i_n} + b_{i_1 i_2 \ldots i_n} = c_{i_1 i_2 \ldots i_n}\,. \tag{3.1.50}$$

The addition of tensors is *commutative* and *associative*.

Multiplying all the components of a tensor **a** of nth order in a frame $\mathscr{B}$ *by the same scalar* λ, one obtains a new tensor **b** = λ**a**, of the same order and of components $\lambda a_{i_1 i_2 \ldots i_n}$; this operation is *distributive* with respect to the addition of tensors, as well as of scalars. In particular, we can have $\lambda = -1$; hence, by *subtracting* two tensors of nth order one obtains a tensor of the same order. If **b** = **0** (all the components of a *null tensor* in an arbitrary basis $\mathscr{B}$ vanish), then we have $\lambda = 0$ or **a** = **0**.

Starting from the observations made at the previous subsection, one can show that an asymmetric tensor of nth order can be univocally decomposed in a sum of two tensors, the first one symmetric with respect to the indices i_j and i_k and the second one antisymmetric with respect to the same indices; we can thus write

$$a_{i_1 i_2 \ldots i_j \ldots i_k \ldots i_n} = \frac{1}{2}\left(a_{i_1 i_2 \ldots i_j \ldots i_k \ldots i_n} + a_{i_1 i_2 \ldots i_k \ldots i_j \ldots i_n}\right)$$
$$+ \frac{1}{2}\left(a_{i_1 i_2 \ldots i_j \ldots i_k \ldots i_n} - a_{i_1 i_2 \ldots i_k \ldots i_j \ldots i_n}\right). \tag{3.1.51}$$

The *tensor product* (*external product*) of two tensors **a** and **b** of nth and mth order, respectively, is a tensor **c** = **a** $\otimes$ **b** of $(n+m)$th order; in a frame $\mathscr{B}$, we can write

$$a_{i_1 i_2 \ldots i_n} b_{j_1 j_2 \ldots j_m} = c_{k_1 k_2 \ldots k_{n+m}}\,, \tag{3.1.52}$$

where the indices k_l, $l = 1,2,\ldots,n+m$ are the indices i_p, $p = 1,2,\ldots,n$, and j_q, $q = 1,2,\ldots,m$. For instance, by *the dyadic product* of two tensors of first order one obtains a tensor of second order

$$a_i b_j = c_{ij}\,, \quad i,j = 1,2,3\,. \tag{3.1.52'}$$

The external product of tensors is not commutative, but it is associative and distributive with respect to their addition.

If we make $j_k = j_l$ in the relation of definition (3.1.45) and if we take into account the formulae (3.1.35), then we obtain, in a basis $\mathscr{B}$,

$$a'_{j_1 j_2 \ldots j_{k-1} j_k j_{k+1} \ldots j_{l-1} j_k j_{l+1} \ldots j_n} = a_{i_1 i_2 \ldots i_{k-1} i_k i_{k+1} \ldots i_{l-1} i_k i_{l+1} \ldots i_n} \alpha_{j_1 i_1} \alpha_{j_2 i_2} \ldots$$
$$\ldots \alpha_{j_{k-1} i_{k-1}} \alpha_{j_{k+1} i_{k+1}} \ldots \alpha_{j_{l-1} i_{l-1}} \alpha_{j_{l+1} i_{l+1}} \ldots \alpha_{j_n i_n} ;$$

this operation is called *the contraction of the tensor*. Hence, by the contraction of two indices of a tensor of nth order one obtains a tensor of $(n-2)$th order. For instance, by the contraction of a tensor **a** of second order, of components a_{ij} in a basis $\mathscr{B}$, one obtains a scalar, called *the trace* of the tensor **a** and denoted

$$\operatorname{tr} \mathbf{a} = a_{ii} . \tag{3.1.53}$$

In particular, by the contraction of Kronecker's tensor one obtains $\operatorname{tr} \mathbf{1} = \delta_{ii} = 3$.

The internal product (*the contracted tensor product*) of two tensors **a** and **b** of nth and mth order, respectively, is a tensor **ab** of $(n + m - 2p)$th order, where p is the number of effected contractions. For instance, *the scalar product* of two vectors **a** and **b**, of components a_i and b_i, respectively, will be given by the contracted product $\mathbf{a} \cdot \mathbf{b} = \mathbf{ab} = a_i b_i$. If $\mathbf{c} = \mathbf{a} \times \mathbf{b}$ is *the vector product* of the two vectors, we may write $c_i = \in_{ijk} a_j b_k$, $i = 1,2,3$. We notice that *the triple scalar product* of three vectors **a**, **b** and **c** can be expressed also by means of a contracted product in the form $(\mathbf{a}, \mathbf{b}, \mathbf{c}) = \in_{ijk} a_i b_j c_k$. If **a** is a tensor of second order, of components a_{ij} in the basis $\mathscr{B}$, while **b** is a vector of components b_j in the same basis, then the contracted product $\mathbf{c} = \mathbf{ab}$ is a vector of components $c_i = a_{ij} b_j$; if **a** is a tensor of components a_{ijk}, while **b** is a tensor of components b_{kl}, then we obtain the contracted product of components $c_{ijl} = a_{ijk} b_{kl}$. The sum is obtained by contracting the last index of the first tensor by the first index of the second tensor; eventually, one can contract also the last but one index of the first tensor by the second index of the second tensor a.s.o. By means of two permutation tensors one obtains the external product $\in_{ijk} \in_{lmn}$, given by the formula (2.1.45); by contraction, one is led to various internal products of the form (2.1.46)-(2.1.46'').

We have seen that by algebraic operations of addition and of external or internal products one obtains new tensors. To identify if a quantity is a tensor, *the quotient law* is frequently used; so, if we have a relation of the form

$$f(i_1, i_2, \ldots, i_n) b_{j_1 j_2 \ldots j_m} = c_{k_1 k_2 \ldots k_{n+m}} , \tag{3.1.54}$$

where the indices k_l, $l = 1,2,\ldots,n+m$, are just the indices i_p, $p = 1,2,\ldots,n$, and j_q, $q = 1,2,\ldots,m$, then the function f is of the form

Mass geometry. Displacements. Constraints 135

$$f(i_1, i_2, \ldots, i_n) = a_{i_1 i_2 \ldots i_n},$$ (3.1.54')

obtaining thus the components of a tensor of nth order in a basis $\mathscr{B}$.

Let $U = U(x_1, x_2, x_3)$ be a scalar field, defined on a domain D, with $U \in C^1(D)$, and let be $\partial U / \partial x_i$, $i = 1,2,3$, the components of the conservative vector field which is thus obtained; by a change of co-ordinates, we obtain $U' = U'(x_1', x_2', x_3')$ and we may write

$$\frac{\partial U'}{\partial x_j'} = \frac{\partial U}{\partial x_i} \frac{\partial x_i}{\partial x_j'} = \frac{\partial U}{\partial x_i} \alpha_{ji}, \quad j = 1,2,3,$$

where we took into account the relations (3.1.36), (3.1.37). Hence, *the derivatives* of a scalar field with respect to the independent variables lead to a tensor field of first order; we denote $\partial U / \partial x_i = \partial_i U = U_{,i}$, $i = 1,2,3$. Analogously, starting from the vector field $V_i = V_i(x_1, x_2, x_3)$, $i = 1,2,3$, with $V_i \in C^1(D)$, and by a change of co-ordinates of the form (2.1.11), we can write

$$\frac{\partial V_l'}{\partial x_j'} = \frac{\partial (V_k \alpha_{lk})}{\partial x_i} \frac{\partial x_i}{\partial x_j'} = \frac{\partial V_k}{\partial x_i} \alpha_{lk} \alpha_{ji} \quad l, j = 1,2,3;$$

hence, the derivatives of first order of a tensor field of first order are the components of a tensor field of second order. We write $\partial V_i / \partial x_j = \partial_j V_i = V_{i,j}$, $i, j = 1,2,3$. In general, the derivatives of first order of a tensor field of nth order are the components of a tensor field of $(n + 1)$th order.

One can define, analogously, *derivatives of higher order* with respect to the independent variables of a tensor field. If $a_i \in C^2(D)$, the we can write

$$a_{i,jk} = a_{i,kj}, \quad i, j, k = 1,2,3,$$ (3.1.55)

the derivatives of second order being immaterial of the order of differentiation (Schwartz's theorem); we notice thus that the tensor $a_{i,jk}$ is symmetric with respect to the indices j and k.

Besides the operators grad, div and curl introduced in App. §2, for a vector field, we can introduce also the operators Grad, Div and Curl for the tensors of higher order.

1.2.3 Tensors of second order

Let **a** be an asymmetric tensor of second order, expressed by its components a_{ij} in the form of a square matrix (3.1.42). If we denote the symmetric part of this tensor by

$$\mathbf{a}_s = \frac{1}{2}(\mathbf{a} + \mathbf{a}^{\mathrm{T}}), \quad a_{(ij)} = \frac{1}{2}(a_{ij} + a_{ji}),$$ (3.1.56)

and the antisymmetric one by

$$\mathbf{a}_a = \frac{1}{2}(\mathbf{a} - \mathbf{a}^T), \quad a_{[ij]} = \frac{1}{2}(a_{ij} - a_{ji}), \tag{3.1.56'}$$

then we can write univocally

$$\mathbf{a} = \mathbf{a}_s + \mathbf{a}_a, \quad a_{ij} = a_{(ij)} + a_{[ij]}. \tag{3.1.56"}$$

Here $\mathbf{a}^T$ is the *transpose tensor*, represented by the *transpose matrix*

$$[a_{ji}] \equiv \begin{bmatrix} a_{11} & a_{21} & a_{31} \\ a_{12} & a_{22} & a_{32} \\ a_{13} & a_{23} & a_{33} \end{bmatrix}; \tag{3.1.57}$$

we have $(\mathbf{a}^T)^T = \mathbf{a}$, $\mathbf{a}_s^T = \mathbf{a}_s$ and $\mathbf{a}_a^T = -\mathbf{a}_a$.

A *symmetric tensor* of second order is represented by the matrix

$$[a_{ij}] \equiv \begin{bmatrix} a_{11} & a_{12} & a_{13} \\ a_{12} & a_{22} & a_{23} \\ a_{13} & a_{23} & a_{33} \end{bmatrix}, \tag{3.1.58}$$

which has only six distinct components, while an *antisymmetric tensor* of second order is represented in the form

$$[a_{ij}] \equiv \begin{bmatrix} 0 & a_{12} & -a_{13} \\ -a_{12} & 0 & a_{23} \\ a_{13} & -a_{23} & 0 \end{bmatrix}, \tag{3.1.58'}$$

and has only three distinct components.

Let us consider the contraction

$$\operatorname{tr} \mathbf{a} = a_{ii} = a_{11} + a_{22} + a_{33} \tag{3.1.59}$$

of the symmetric tensor of components a_{ij}; it is an invariant by the transformations of co-ordinates, being a scalar. The tensor $\mathbf{a}^0 = (\operatorname{tr} \mathbf{a}/3)\mathbf{1}$ of components $a_{ll}\delta_{ij}/3$, $i,j = 1,2,3$, and matrix

$$\frac{1}{3} a_{ll} \begin{bmatrix} 1 & 0 & 0 \\ 0 & 1 & 0 \\ 0 & 0 & 1 \end{bmatrix} \tag{3.1.60}$$

constitutes *the spheric tensor* of the symmetric tensor **a** ; the tensor $\mathbf{a}' = \mathbf{a} - (\operatorname{tr}\mathbf{a}/3)\mathbf{1}$ of components $a'_{ij} = a_{ij} - a_{ll}\delta_{ij}/3$, and matrix

$$\begin{bmatrix} a_{11} - \dfrac{1}{3}a_{ll} & a_{12} & a_{13} \\ a_{12} & a_{22} - \dfrac{1}{3}a_{ll} & a_{23} \\ a_{13} & a_{23} & a_{33} - \dfrac{1}{3}a_{ll} \end{bmatrix} \qquad (3.1.60')$$

is called *the deviator of the* considered *symmetric tensor*. We obtain thus *the canonical decomposition of a symmetric tensor of second order*

$$\mathbf{a} = \mathbf{a}^0 + \mathbf{a}'. \qquad (3.1.61)$$

An antisymmetric tensor **a** of second order is a degenerate one, equivalent to a tensor of first order; we can associate *the axial vector*

$$a_k = \frac{1}{2}\epsilon_{ijk}\, a_{ij}, \quad k = 1,2,3, \qquad (3.1.62)$$

being thus led to $a_1 = a_{23}$, $a_2 = a_{31}$, $a_3 = a_{12}$. Inversely, multiplying both members by ϵ_{klm} and taking into account (2.1.46), we find

$$\epsilon_{klm}\, a_k = \frac{1}{2}\epsilon_{klm}\epsilon_{ijk}\, a_{ij} = \frac{1}{2}\big(\delta_{li}\delta_{mj} - \delta_{lj}\delta_{mi}\big)a_{ij} = \frac{1}{2}(a_{lm} - a_{ml});$$

hence, one can write

$$a_{ij} = \epsilon_{ijk}\, a_k, \quad i,j = 1,2,3. \qquad (3.1.62')$$

For instance, the vector product **c** of two vectors **a** and **b** ($\mathbf{c} = \mathbf{a} \times \mathbf{b}$) leads to an antisymmetric tensor of second order

$$a_i b_j = 2\epsilon_{ijk}\, c_k, \quad i,j = 1,2,3. \qquad (3.1.63)$$

Let **a** be a symmetric tensor of second order, of components a_{ij}, represented by the matrix (3.1.58), and let be *the associate quadratic form*

$$\Phi = a_{ij} x_i x_j; \qquad (3.1.64)$$

this form being a scalar, we can write as well

$$\Phi = a'_{ij} x'_i x'_j, \qquad (3.1.64')$$

in a new system of orthonormed Cartesian co-ordinates. Considering the tensor **a** defined at a point, which we choose as the origin of the co-ordinate axes, the equation

$$\Phi = \pm 1 \tag{3.1.65}$$

represents a *quadric* Γ (one chooses the sign so as to obtain a real quadric); this one allows us to give a geometric image to the variation of the tensor components, by a change of the system of co-ordinate axes. To reduce the matrix (3.1.58) to a diagonal one, corresponds to the reduction of the quadratic form (3.1.64) to a sum of squares (eventual with a sign minus), as well as to the representation of the quadric Γ in the canonical form (with respect to its axes).

Let us consider the components for which the indices k and l are equal (without summation) in the relation (3.1.43). In this case, α_{ki} and α_{kj} are direction cosines of an axis Δ with respect to the frame $Ox_1x_2x_3$; denoting them by α_i, $i = 1,2,3$, we may write

$$a_\Delta = a_{ij}\alpha_i\alpha_j, \tag{3.1.66}$$

where a_Δ is a component of the principal diagonal of the tensor **a** in the new system $Ox_1'x_2'x_3'$, Δ being one of the new axes of co-ordinates. The directions for which this component has extreme values are called principal directions and correspond to the quadric's Γ axes; we have $a_{kl}' = 0$, $k \neq l$, $k,l = 1,2,3$, in this system of co-ordinates, and quadric's equation can be written in the canonical form, while the quadratic form (3.1.64') is expressed as a sum of squares.

We notice that

$$\alpha_i\alpha_i = 1; \tag{3.1.67}$$

we have thus to solve a problem of extremum with constraints. Using *the method of Lagrange's multipliers*, we consider the function

$$F(\alpha_i) = a_{ij}\alpha_i\alpha_j + \lambda(1 - \alpha_i\alpha_i);$$

we calculate the partial derivatives

$$\frac{\partial F}{\partial \alpha_i} = 2a_{ij}\alpha_j - 2\lambda\alpha_i = 2(a_{ij} - \lambda\delta_{ij})\alpha_j, \quad i = 1,2,3$$

and are led to the equations

$$(a_{ij} - \lambda\delta_{ij})\alpha_j = 0, \quad i = 1,2,3. \tag{3.1.68}$$

This system of three equations with three unknowns is homogeneous; because, by virtue of the relation (3.1.67), the system allows only non-zero solutions, its determinant must vanish, so that

Mass geometry. Displacements. Constraints

$$\det[a_{ij} - \lambda\delta_{ij}] = \begin{vmatrix} a_{11} - \lambda & a_{12} & a_{13} \\ a_{12} & a_{22} - \lambda & a_{23} \\ a_{13} & a_{23} & a_{33} - \lambda \end{vmatrix} = 0. \qquad (3.1.69)$$

Developing this equation, we get

$$\lambda^3 - \mathscr{I}_1\lambda^2 + \mathscr{I}_2\lambda - \mathscr{I}_3 = 0, \qquad (3.1.70)$$

where

$$\mathscr{I}_1 = \frac{1}{2}\epsilon_{ijk}\epsilon_{ljk}\,a_{il} = \delta_{il}a_{il} = a_{ii} = a_{11} + a_{22} + a_{33}, \qquad (3.1.70')$$

$$\mathscr{I}_2 = \frac{1}{2}\epsilon_{ijk}\epsilon_{lmk}\,a_{il}a_{jm} = \frac{1}{2}(a_{ii}a_{jj} - a_{ij}a_{ij}) = a_{22}a_{33} + a_{33}a_{11} + a_{11}a_{22}$$
$$- (a_{23}^2 + a_{31}^2 + a_{12}^2), \qquad (3.1.70'')$$

$$\mathscr{I}_3 = \frac{1}{6}\epsilon_{ijk}\epsilon_{lmn}\,a_{il}a_{jm}a_{kn} = \det[a_{pq}] = a_{11}a_{22}a_{33}$$
$$- (a_{11}a_{23}^2 + a_{22}a_{31}^2 + a_{33}a_{12}^2) + 2a_{23}a_{31}a_{12}. \qquad (3.1.70''')$$

The equation (3.1.70) is of the third degree and has three roots, at least one of them being real; the system (3.1.68) together with the condition (3.1.67) leads, for each root, to a principal direction for which

$$a_{ij}\alpha_j = \lambda\alpha_i, \quad i = 1,2,3. \qquad (3.1.71)$$

We may write the relations

$$a_{ij}\alpha_j' = \lambda_1\alpha_i', \quad a_{ij}\alpha_j'' = \lambda_2\alpha_i'', \quad i = 1,2,3,$$

corresponding to two roots λ_1 and λ_2; multiplying the first relation by α_i'' and the second one by α_i', we get

$$a_{ij}\alpha_j'\alpha_i'' = \lambda_1\alpha_i'\alpha_i'', \quad a_{ij}\alpha_j''\alpha_i' = \lambda_2\alpha_i''\alpha_i'.$$

Interchanging the dummy indices in the first member of the latter equation, taking into account that tensor **a** is symmetric, and subtracting the second equation from the first one, one obtain

$$(\lambda_1 - \lambda_2)\alpha_j'\alpha_j'' = 0. \qquad (3.1.71')$$

If $\lambda_1 = \lambda_2$, then the two roots are real; if $\lambda_1 \neq \lambda_2$, then we obtain $\alpha_j'\alpha_j'' = 0$; hence, the corresponding principal directions are orthogonal. If the two roots are complex conjugate ($\lambda_{1,2} = \alpha \pm i\beta$), then we have $\alpha'_j = \gamma_j + i\delta_j$, $\alpha''_j = \gamma_j - i\delta_j$, $j = 1,2,3$; hence, $\alpha'_j \alpha''_j = \gamma_j \gamma_j + \delta_j \delta_j = 0$, a relation which takes place only if all the terms vanish. In conclusion, all the roots of the equation (3.1.70) are real; hence, *there exist three principal directions* which, in the case of distinct roots, are *orthogonal*. These roots represent *the eigenvalues* of the given matrix, while the vectors along the corresponding principal directions are *the eigenvectors*.

If two roots are equal, then the corresponding quadric is *of rotation* (we have one principal direction, all the directions contained in the plane normal to this one being principal directions), while if all the roots are equal, then the quadric is a *sphere* (all the directions are principal directions).

The considered quadric is *Cauchy's quadric*, corresponding to the symmetric tensor **a**.

Because the tensor is a mathematical entity for which the principal directions do not depend on the system of co-ordinates, it follows that the coefficients $\mathscr{I}_1$, $\mathscr{I}_2$, $\mathscr{I}_3$ are invariant by a change of co-ordinates. These are the three invariants of the symmetric tensor **a**. An antisymmetric tensor **a** has only one invariant

$$\mathscr{I}_2 = a_{23}^2 + a_{31}^2 + a_{12}^2, \tag{3.1.72}$$

which corresponds to the modulus of its associate vector. If **a**' is the deviator (of associate matrix (3.1.60')) of the symmetric tensor, then we get

$$\mathscr{I}'_1 = 0. \tag{3.1.73}$$

Let us suppose that the system of axes $Ox_1x_2x_3$ corresponds to the principal directions. We may write, e.g., $\alpha_1 = 1$, $\alpha_2 = \alpha_3 = 0$ for $\lambda = \lambda_1$, and the relation (3.1.71) leads to $a_{11} = \lambda_1$, $a_{12} = a_{13} = 0$ for $i = 1,2,3$. Hence, λ_i, $i = 1,2,3$, represent the extreme values of the components a_Δ of the principal diagonal, corresponding to the reduction of the matrix (3.1.58) to a diagonal one and to the reduction of the quadratic form (3.1.64) to a sum of squares; obviously, we have

$$\mathscr{I}_1 = \lambda_1 + \lambda_2 + \lambda_3, \tag{3.1.74}$$
$$\mathscr{I}_2 = \lambda_2\lambda_3 + \lambda_3\lambda_1 + \lambda_1\lambda_2, \tag{3.1.74'}$$
$$\mathscr{I}_3 = \lambda_1\lambda_2\lambda_3. \tag{3.1.74''}$$

Using the direction cosines of the axis Δ with respect to the principal axes, we may write

$$a_\Delta = \lambda_1 \alpha_1^2 + \lambda_2 \alpha_2^2 + \lambda_3 \alpha_3^2. \tag{3.1.75}$$

Let be a system of axes $Ox_1x_2x_3$ and a point P of position vector $\mathbf{r} = x_j \mathbf{i}_j$ on the axis Δ (Fig.3.6); it results

$$\alpha_i = \frac{1}{r}x_i, \quad i = 1,2,3,$$

so that the relation (3.1.66) leads to

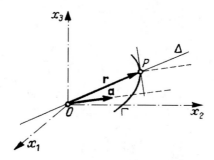

Figure 3.6. The Γ quadric associated to a symmetric tensor of second order.

$$a_\Delta = a_{ij}\frac{x_i x_j}{r^2}.$$

Let us choose the point P so as to fulfill the condition

$$r = \frac{1}{\sqrt{a_\Delta \operatorname{sign} a_\Delta}}; \qquad (3.1.76)$$

the locus of the point P, corresponding to all the positions which can be taken by the axis Δ, will be just the quadric (3.1.65), obtaining thus a geometric image of the variation of the component a_Δ. The equation of the quadric will be of the form

$$\lambda_1 x_1'^2 + \lambda_2 x_2'^2 + \lambda_3 x_3'^2 = \operatorname{sign} a_\Delta, \qquad (3.1.77)$$

where the function "sign" has been introduced to get always a real quadric. By convention, we denote the principal values so as to have

$$\lambda_1 \geq \lambda_2 \geq \lambda_3. \qquad (3.1.78)$$

If $\lambda_3 > 0$ or if $\lambda_1 < 0$, then we get an ellipsoid; otherwise, one obtains an one-sheet or a two-sheet hyperboloid (we notice that both hyperboloids, which are conjugate, form the considered locus). The relation (3.1.76) shows that the major and the minor principal axes correspond to $a_{\Delta\min}$ and $a_{\Delta\max}$, respectively. If we introduce the vector $\mathbf{a}$ of components $a_i = a_{ij}\alpha_j$, $i = 1,2,3$, then we obtain a direction which coincides with that of the normal to the surface $\Phi = \operatorname{const}$ at the point P (direction parameters $\Phi_{,i}$); hence, the vector $\mathbf{a}$ is normal to the plane tangent to the quadric at the point in which this one is pierced by the axis Δ (Fig.3.6). We notice that the vector $\mathbf{a}$ is along the axis Δ if the latter one coincides with a principal direction. The equation (3.1.77) allows us also to express the quadratic form (3.1.64) in the form of a sum of squares.

The three principal directions form a three-orthogonal principal trihedron of reference $O123$. One can prove that the components a_{ij}, $i \neq j$, $i,j = 1,2,3$, get their extreme values with respect to the trihedron formed by the bisector planes of the dihedral angles of the principal trihedron, i.e.:

$$\pm \frac{1}{2}(\lambda_2 - \lambda_3), \quad \pm \frac{1}{2}(\lambda_3 - \lambda_1), \quad \pm \frac{1}{2}(\lambda_1 - \lambda_2); \tag{3.1.79}$$

the two signs correspond to the two bisectors: the internal and the external one. Obviously, the greatest value is ($i \neq j$)

$$a_{ij\mathrm{max}} = \frac{1}{2}(\lambda_1 - \lambda_3). \tag{3.1.79'}$$

The components a_Δ corresponding to the normals to these bisector planes are equal to

$$\frac{1}{2}(\lambda_2 + \lambda_3), \quad \frac{1}{2}(\lambda_3 + \lambda_1), \quad \frac{1}{2}(\lambda_1 + \lambda_2), \tag{3.1.79''}$$

respectively.

1.2.4 The moment of inertia tensor

Let be the matrix

$$\begin{bmatrix} I_{x_1} & -I_{x_1 x_2} & -I_{x_1 x_3} \\ -I_{x_2 x_1} & I_{x_2} & -I_{x_2 x_3} \\ -I_{x_3 x_1} & -I_{x_3 x_2} & I_{x_3} \end{bmatrix}, \tag{3.1.80}$$

formed by means of the axial and the centrifugal moments of inertia. We notice that the moments of inertia may be expressed in a unitary form, i.e.

$$I_{jk} = \epsilon_{jnq} \epsilon_{kpq} \int_\Omega x_n x_p \mathrm{d}m = (\delta_{jk} \delta_{np} - \delta_{jp} \delta_{kn}) \int_\Omega x_n x_p \mathrm{d}m$$
$$= \int_\Omega (x_l x_l \delta_{jk} - x_j x_k) \mathrm{d}m, \quad j,k = 1,2,3, \tag{3.1.81}$$

where

$$I_{jk} = \begin{cases} I_{x_j}, & j = k, \\ -I_{x_j x_k}, & j \neq k; \end{cases} \tag{3.1.81'}$$

one puts thus into evidence *the moment of inertia tensor* **I** of components I_{jk}, which is a symmetric tensor of second order.

Mass geometry. Displacements. Constraints

Let us suppose that through the point O passes an axis Δ, the direction of which is given by the unit vector **n** of direction cosines n_i, $i = 1, 2, 3$ (Fig.3.7). Taking into account the transformation relations of the components I_{jk} by a rotation of the right-handed orthonormed frame of reference, we get

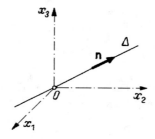

Figure 3.7. The axial moment of inertia with respect to a given axis.

$$I_\Delta = I_{jk} n_j n_k = I_{x_1} n_1^2 + I_{x_2} n_2^2 + I_{x_3} n_3^2 - 2(I_{x_2 x_3} n_2 n_3 + I_{x_3 x_1} n_3 n_1 + I_{x_1 x_2} n_1 n_2); \quad (3.1.82)$$

in a compact form, we may write

$$I_\Delta = (\mathbf{In}) \cdot \mathbf{n}, \quad (3.1.82')$$

where **In** represents a contracted tensor product.

Noting that

$$\mathbf{r} \times (\mathbf{u} \times \mathbf{r}) = r^2 \mathbf{u} - (\mathbf{u} \cdot \mathbf{r})\mathbf{r} = (u_j x_k - u_k x_j) x_k \mathbf{i}_j = (x_l x_l \delta_{jk} - x_j x_k) u_k \mathbf{i}_j,$$

where **u** is a given unit vector, we obtain the remarkable relation

$$\int_\Omega \mathbf{r} \times (\mathbf{u} \times \mathbf{r}) \mathrm{d}m = \mathbf{Iu}, \quad (3.1.83)$$

Iu being a contracted tensor product too.

Using the results of the previous subsection, we search the directions for which the axial moments of inertia get their extreme values. These values (the principal moments of inertia $I_1 \geq I_2 \geq I_3$) are the roots of the third degree equation in Lagrange's multiplier I

$$\det[I_{jk} - I\delta_{jk}] = \begin{vmatrix} I_{11} - I & I_{12} & I_{13} \\ I_{21} & I_{22} - I & I_{23} \\ I_{31} & I_{32} & I_{33} - I \end{vmatrix} = I^3 - \mathscr{I}_1 I^2 + \mathscr{I}_2 I - \mathscr{I}_3 = 0, \quad (3.1.84)$$

corresponding to three directions, orthogonal two by two (the principal directions $O1, O2, O3$), the direction cosines of which are given by the system of homogeneous equations with non-vanishing solutions

$$(I_{jk} - I\delta_{jk})n_k = 0, \quad j = 1,2,3, \qquad (3.1.85)$$

to which we associate the supplementary condition

$$n_k n_k = 1. \qquad (3.1.85')$$

The matrix (3.1.80) is reduced, in this case, to its diagonal form, the corresponding centrifugal moments of inertia being equal to zero. The coefficients of the equation (3.1.84) are the invariants of the tensor $\mathbf{I}$ and are given by

$$\mathscr{I}_1 = I_O = \frac{1}{2}\epsilon_{ijk}\epsilon_{ljk}I_{il} = \delta_{il}I_{il} = I_{jj} = I_{11} + I_{22} + I_{33} = I_1 + I_2 + I_3, \qquad (3.1.86)$$

$$\mathscr{I}_2 = \frac{1}{2}\epsilon_{ijk}\epsilon_{lmk}I_{il}I_{jm} = \frac{1}{2}(I_{ii}I_{jj} - I_{ij}I_{ij}) = I_{22}I_{33} + I_{33}I_{11} + I_{11}I_{22}$$
$$- I_{23}^2 - I_{31}^2 - I_{12}^2 = I_2 I_3 + I_3 I_1 + I_1 I_2, \qquad (3.1.86')$$

$$\mathscr{I}_3 = \frac{1}{6}\epsilon_{ijk}\epsilon_{lmn}I_{il}I_{jm}I_{kn} = \det[I_{ij}] = I_{11}I_{22}I_{33} - I_{11}I_{23}^2 - I_{22}I_{31}^2$$
$$- I_{33}I_{12}^2 + 2I_{23}I_{31}I_{12} = I_1 I_2 I_3. \qquad (3.1.86'')$$

If n_1, n_2, n_3 are the direction cosines of the axis Δ with respect to the principal directions, then we may write

$$I_\Delta = I_1 n_1^2 + I_2 n_2^2 + I_3 n_3^2. \qquad (3.1.82'')$$

1.2.5 The two dimensional case

If the support Ω is contained in the plane $Ox_1 x_2$, then we obtain $n_3 = 0$ and $I_{33} = I_{31} = I_{32} = 0$; the formula (3.1.82) leads to

$$I_\Delta = I_{x_1} n_1^2 + I_{x_2} n_2^2 - 2I_{x_1 x_2} n_1 n_2 = \frac{1}{2}(I_{x_1} + I_{x_2})$$
$$+ \frac{1}{2}(I_{x_1} - I_{x_2})\cos 2(\mathbf{n}, x_1) - I_{x_1 x_2}\sin 2(\mathbf{n}, x_1), \qquad (3.1.87)$$

where we put into evidence the angle formed by the unit vector $\mathbf{n}$ of the axis Δ with the co-ordinate axis Ox_1; for an axis Δ' normal to the axis Δ, we may write

$$I_{\Delta'} = \frac{1}{2}(I_{x_1} + I_{x_2}) - \frac{1}{2}(I_{x_1} - I_{x_2})\cos 2(\mathbf{n}, x_1) + I_{x_1 x_2}\sin 2(\mathbf{n}, x_1). \qquad (3.1.87')$$

The centrifugal moment of inertia is given by

$$I_{\Delta\Delta'} = \frac{1}{2}(I_{x_1} - I_{x_2})\sin 2(\mathbf{n}, x_1) + I_{x_1 x_2}\cos 2(\mathbf{n}, x_1); \qquad (3.1.87'')$$

we notice that we have also

$$I_{\Delta\Delta'} = \frac{1}{2}(I_{\Delta_1} - I_{\Delta_2}), \qquad (3.1.87''')$$

where Δ_1 and Δ_2 are the two bisectrices of the angle formed by the axes Δ and Δ'.

Equating to zero the derivative of the function I_Δ with respect to the argument $2(\mathbf{n}, x_1)$, we get

$$\tan 2(\mathbf{n}, x_1) = \frac{2 I_{x_1 x_2}}{I_{x_2} - I_{x_1}}, \qquad (3.1.88)$$

obtaining thus the directions for which the axial moment of inertia I_Δ attains its extreme (maximal or minimal) values; in this case, the axial moment of inertia $I_{\Delta'}$ attains its extreme (minimal or maximal) values too. It is easy to verify that for the angles $(\mathbf{n}, x_1)$ given by (3.1.88) (and only for those angles) the centrifugal moments of inertia vanish.

The relation (3.1.88) determines two angles $2(\mathbf{n}, x_1)$, differing by π; hence, the searched angles $(\mathbf{n}, x_1)$ differ by $\pi/2$. One obtains thus two principal directions, normal one to the other, denoted by $O1$ and $O2$, to which correspond the principal moments of inertia I_1 and I_2 ($I_1 \geq I_2$), respectively.

To see which of those directions corresponds to the maximal value of I_Δ (which will be denoted by I_1) and which to its minimal value (denoted by I_2), we calculate the second derivative

$$\frac{d^2 I_\Delta}{d(\mathbf{n}, x_1)^2} = -\frac{2\cos 2(\mathbf{n}, x_1)}{I_{x_1} - I_{x_2}}\left[(I_{x_1} - I_{x_2})^2 + 4 I_{x_1 x_2}^2\right]$$
$$= \frac{\sin 2(\mathbf{n}, x_1)}{I_{x_1 x_2}}\left[(I_{x_1} - I_{x_2})^2 + 4 I_{x_1 x_2}^2\right],$$

where we took into account also the relation (3.1.88). We suppose that $I_{x_1 x_2} \neq 0$. If $I_{x_1 x_2} > 0$ and $(\mathbf{n}, x_1)_{\min} < \pi/4$ and $I_{x_2} > I_{x_1}$ or $\pi/4 < (\mathbf{n}, x_1)_{\min} < \pi/2$ and $I_{x_2} < I_{x_1}$, then the second derivative is positive and we get $I_{\Delta\min} = I_2$; otherwise, if $I_{x_1 x_2} < 0$ and $(\mathbf{n}, x_1)_{\min} < \pi/4$ and $I_{x_1} > I_{x_2}$ or $\pi/4 < (\mathbf{n}, x_1)_{\min} < \pi/2$ and $I_{x_1} < I_{x_2}$, then the second derivative is negative and we obtain $I_{\Delta\max} = I_1$. The angle $(\mathbf{n}, x_1)_{\min}$ is the smallest positive angle. In conclusion, the principal axis $O1$ is that one which makes *the smallest angle* with the axis Ox_1 if $I_{x_1} > I_{x_2}$ or with the axis

Ox_2 if $I_{x_2} > I_{x_1}$, hence *with the axis with respect to which the moment of inertia is the greatest*. If $I_{x_1} = I_{x_2}$, then we get $(\mathbf{n}, x_1)_{\min} = \pi/4$; in this case

$$\frac{\mathrm{d}^2 I_\Delta}{\mathrm{d}(\mathbf{n}, x_1)^2} = 4 I_{x_1 x_2} \sin 2(\mathbf{n}, x_1),$$

and the corresponding direction is $O1$ or $O2$ as we have $I_{x_1 x_2} < 0$ or $I_{x_1 x_2} > 0$, respectively. If $I_{x_1 x_2} = 0$ and $I_{x_1} \neq I_{x_2}$, then we obtain

$$\frac{\mathrm{d}^2 I_\Delta}{\mathrm{d}(\mathbf{n}, x_1)^2} = -2(I_{x_1} - I_{x_2}) \cos 2(\mathbf{n}, x_1),$$

and $(\mathbf{n}, x_1)_{\min} = 0$ or $(\mathbf{n}, x_1)_{\min} = \pi/2$; if $I_{x_1} > I_{x_2}$, then $Ox_1 \equiv O1$ and $Ox_2 \equiv O2$, while if $I_{x_2} > I_{x_1}$, then $Ox_1 \equiv O2$ and $Ox_2 \equiv O1$. Finally, if $I_{x_1 x_2} = 0$ and $I_{x_1} = I_{x_2}$, then the expression (3.1.88) is indeterminate; *any direction passing through the respective point is a principal direction*, while the common magnitude of the axial moments of inertia is the magnitude of the principal moment of inertia.

We associate thus the diagonal matrix

$$I = \begin{bmatrix} I_1 & 0 \\ 0 & I_2 \end{bmatrix} \tag{3.1.89}$$

to the moment of inertia tensor. The equation (3.1.84) becomes

$$I^2 - \mathscr{I}_1 I + \mathscr{I}_2 = 0, \tag{3.1.90}$$

with the invariants

$$\mathscr{I}_1 = I_O = I_{x_1} + I_{x_2} = I_1 + I_2, \tag{3.1.90'}$$
$$\mathscr{I}_2 = I_{x_1} I_{x_2} - I_{x_1 x_2}^2 = I_1 I_2; \tag{3.1.90''}$$

the principal moments of inertia are thus given by

$$I_{1,2} = \frac{1}{2}(I_{x_1} + I_{x_2}) \pm \sqrt{(I_{x_1} - I_{x_2})^2 + 4 I_{x_1 x_2}^2}. \tag{3.1.91}$$

Taking into account the relations (3.1.88) and (3.1.91), the angles made by the principal directions with the axis Ox_1 are given also by one of the relations

$$\tan(\mathbf{n}, x_1) = \frac{I_{x_1} - I_2}{I_{x_1 x_2}} = \frac{I_{x_1 x_2}}{I_{x_2} - I_2} = \frac{I_1 - I_{x_2}}{I_{x_1 x_2}} = \frac{I_{x_1 x_2}}{I_1 - I_{x_1}}. \tag{3.1.92}$$

If the relation (3.1.87) is written with respect to the principal axes, then we get

Mass geometry. Displacements. Constraints

$$I_\Delta = I_1 \cos^2(\mathbf{n},1) + I_2 \sin^2(\mathbf{n},1) = \frac{1}{2}(I_1 + I_2) + \frac{1}{2}(I_1 - I_2)\cos 2(\mathbf{n},1);$$
(3.1.93)

the relations (3.1.87') and (3.1.87") lead to

$$I_{\Delta'} = \frac{1}{2}(I_1 + I_2) - \frac{1}{2}(I_1 - I_2)\cos 2(\mathbf{n},1), \qquad (3.1.93')$$

$$I_{\Delta\Delta'} = \frac{1}{2}(I_1 - I_2)\sin 2(\mathbf{n},1). \qquad (3.1.93'')$$

We notice that we have

$$I_\Delta - I_{\Delta'} = (I_1 - I_2)\cos 2(\mathbf{n},1) \qquad (3.1.93''')$$

too.

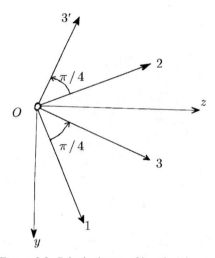

Figure 3.8. Principal axes of inertia (plane case).

It is easy to see that *the centrifugal moment of inertia* has an *extreme value* for the angles $(\mathbf{n},1) = \pi/4$ and $(\mathbf{n},1) = 3\pi/4$, hence for the bisectrices of the angles formed by the principal directions (Fig.3.8). The value of the extreme centrifugal moments of inertia are given by

$$I_{33'} = \pm\frac{1}{2}(I_1 - I_2), \qquad (3.1.94)$$

while the axial moment of inertia corresponding to the bisectrices $O3$ and $O3'$ is

$$I_{3'} = \frac{1}{2}(I_1 + I_2). \qquad (3.1.94')$$

Starting from the gyration radii (3.1.30) and (3.1.30'), we may write

$$i_1 = \sqrt{\frac{I_1}{M}}, \quad i_2 = \sqrt{\frac{I_2}{M}} \qquad (3.1.95)$$

and

$$i_1 = \sqrt{\frac{I_1}{A}}, \quad i_2 = \sqrt{\frac{I_2}{A}}, \qquad (3.1.95')$$

respectively, with respect to the principal axes of inertia, obtaining thus the principal gyration radii; with the aid of (3.1.93), we get

$$i_\Delta^2 = i_1^2 \cos^2(\mathbf{n},1) + i_2^2 \sin^2(\mathbf{n},1) = \frac{1}{2}(i_1^2 + i_2^2) + \frac{1}{2}(i_1^2 - i_2^2)\cos 2(\mathbf{n},1). \qquad (3.1.96)$$

In the case of a rectangle of sides a and b we may write

$$I_1 = \frac{1}{12}a^3 b, \quad I_2 = \frac{1}{12}ab^3, \quad I_O = \frac{1}{12}ab(a^2 + b^2), \quad a \geq b, \qquad (3.1.97)$$

with respect to the two axes of symmetry. For an ellipse of semiaxes a and b we get

$$I_1 = \frac{\pi}{4}a^3 b, \quad I_2 = \frac{\pi}{4}ab^3, \quad I_O = \frac{\pi}{4}ab(a^2 + b^2), \quad a \geq b, \qquad (3.1.98)$$

too; if $a = b = R$, we obtain

$$I_O = 2I_1 = 2I_2 = \frac{\pi}{2}R^4, \qquad (3.1.98')$$

for a circle of radius R. For an annulus of internal and external radii R_i and R_e, respectively, we get

$$I_O = 2I_1 = 2I_2 = \frac{\pi}{2}(R_e^4 - R_i^4). \qquad (3.1.99)$$

1.2.6 Three-dimensional geometric representations

Choosing a point P of co-ordinates x_i, $i = 1,2,3$, on the axis Δ, we introduce a vector $\overrightarrow{OP}$, the extremity of which is given by

$$\left|\overrightarrow{OP}\right| = \frac{K}{\sqrt{I_\Delta}}, \quad x_i = \frac{Kn_i}{\sqrt{I_\Delta}}, \quad i = 1,2,3, \qquad (3.1.100)$$

where $K > 0$ is a constant which determines the units; replacing in (3.1.82), we get

$$I_{x_1} x_1^2 + I_{x_2} x_2^2 + I_{x_3} x_3^2 - 2I_{x_2 x_3} x_2 x_3 - 2I_{x_3 x_1} x_3 x_1 - 2I_{x_1 x_2} x_1 x_2 = K^2 ; \qquad (3.1.101)$$

hence, the locus of the point P is a quadric. If $I_\Delta = 0$ (the geometric support Ω belongs to the axis Δ), then the quadric is a circular cylinder with the generatrices parallel to Δ; if $I_\Delta \neq 0$, then the quadric has all its points at a finite distance, hence it is an ellipsoid (*Poinsot's ellipsoid of inertia*). Taking into account the relation (3.1.100), this ellipsoid allows a geometric study of the variation of the moment of inertia I_Δ when the axis Δ rotates about the pole O (determination of the principal axes, of the principal moments of inertia etc.). If the ellipsoid of inertia is expressed with respect to the principal axes (the co-ordinates x_1, x_2, x_3 are considered to be with respect to these axes) and if we use the semiaxes a_1, a_2, a_3 given by the relations

$$I_1 a_1^2 = I_2 a_2^2 = I_3 a_3^2 = K^2, \qquad (3.1.102)$$

then we may express this ellipsoid in the form ($a_1 \leq a_2 \leq a_3$)

$$\frac{x_1^2}{a_1^2} + \frac{x_2^2}{a_2^2} + \frac{x_3^2}{a_3^2} = 1; \qquad (3.1.101')$$

hence, the principal moments of inertia are in inverse proportion to the squares of the semiaxes of the ellipsoid of inertia. Taking into account the relations (3.1.24) and (3.1.102), we get the conditions

$$\frac{1}{a_j^2} - \frac{1}{a_k^2} \leq \frac{1}{a_l^2} \leq \frac{1}{a_j^2} + \frac{1}{a_k^2}, \quad a_j \leq a_k, \quad j \neq k \neq l \neq j, \quad j,k,l = 1,2,3, \qquad (3.1.103)$$

which must be verified by the semiaxes of the ellipsoid (3.1.101') so as to be an ellipsoid of inertia; these conditions are superabundant, the inequalities on the right side being sufficient. To the semi-minor axis of the ellipsoid corresponds the maximal moment of inertia, while to the semi-major one corresponds the minimal moment of inertia. If two of the principal moments of inertia are equal (for instance $I_1 = I_2$), then the ellipsoid of inertia is an *ellipsoid of rotation* ($a_1 = a_2$); any axis passing through the pole O and situated in the plane $O12$ is a principal axis of inertia. If all the principal moments of inertia are equal ($I_1 = I_2 = I_3$), then the ellipsoid of inertia is a *sphere*. If the mechanical system admits three three-orthogonal planes of geometric and mechanical symmetry, then their intersection lines are principal axes of inertia. Corresponding to the properties mentioned in Subsec. 1.2.4 concerning the centrifugal moments of inertia, we may state that any axis normal to a plane of geometric and mechanical symmetry of a mechanical system is a principal axis of inertia for the point of piercing of the plane by this axis. As well, if a mechanical system has an axis of geometric and mechanical symmetry, then this one is a principal axis of inertia for all its points; the corresponding ellipsoid of inertia is an ellipsoid of rotation with respect to this axis.

Let be the equation of the ellipsoid of inertia written in the form

$$f(x_1, x_2, x_3) = I_1 x_1^2 + I_2 x_2^2 + I_3 x_3^2 - K^2 = 0$$

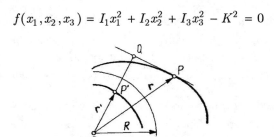

Figure 3.9. The ellipsoids of inertia and of gyration.

and an arbitrary point P of position vector $\mathbf{r}$ on this surface (Fig.3.9); the normal from O to the plane tangent at P to the ellipsoid of inertia pierces this plane at Q, so that $|\overrightarrow{OQ}| = h$. We construct the inverse P' of the point Q with respect to a sphere of centre O and arbitrary radius R; hence,

$$\overrightarrow{OP'} = \mathbf{r}' = r' \operatorname{vers} \overrightarrow{OQ}, \quad \operatorname{vers} \overrightarrow{OQ} = \frac{\operatorname{grad} f}{|\operatorname{grad} f|},$$

$$h = |\overrightarrow{OQ}| = \mathbf{r} \cdot \operatorname{vers} \overrightarrow{OQ}, \quad r'h = R^2,$$

the point P' being of co-ordinates x_1', x_2', x_3'. It follows that

$$\mathbf{r}' = \frac{R^2 \operatorname{grad} f}{\mathbf{r} \cdot \operatorname{grad} f}, \qquad (3.1.104)$$

wherefrom

$$\frac{x_1'}{I_1 x_1} = \frac{x_2'}{I_2 x_2} = \frac{x_3'}{I_3 x_3} = \frac{R^2}{K^2}. \qquad (3.1.104')$$

The locus of the point P' is an ellipsoid too ($K^2 = MR^4$)

$$\frac{x_1'^2}{i_1^2} + \frac{x_2'^2}{i_2^2} + \frac{x_3'^2}{i_3^2} = 1, \qquad (3.1.105)$$

called *the ellipsoid of gyration*, where we have introduced the gyration radii. One may pass from P' to P by a similar graphic construction and by a formula of the form (3.1.104), obtaining thus the ellipsoid of inertia; the two ellipsoids are *reciprocal*.

Let us consider a plane of axes of co-ordinates OI_Δ and $OI_{\Delta\Delta'}$ (Fig.3.10) and three circles (*Mohr's circles*): C_1 of centre $O_1((I_2 + I_3)/2, 0)$ and radius

$R_1 = (I_2 - I_3)/2$, C_2 of centre $O_2((I_3 + I_1)/2, 0)$ and radius $R_2 = (I_1 - I_3)/2$ and C_3 of centre $O_3((I_1 + I_2)/2, 0)$ and radius $R_3 = (I_1 - I_2)/2$. The co-ordinates of a point P in the hatched domain are an axial moment of inertia I_Δ and a centrifugal moment of inertia $I_{\Delta\Delta'}$. One obtains a plane geometric representation for a three-dimensional moment of inertia tensor; we have thus the possibility to get – in a simple way – the extreme values of the axial and centrifugal moments of inertia. For instance, the extreme centrifugal moments of inertia are the radii of the three circles, that is

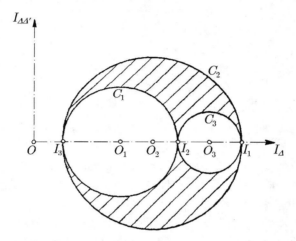

Figure 3.10. Mohr's circles for axial moments of inertia (three-dimensional case).

$$I'_{\Delta\Delta'} = \pm\frac{1}{2}(I_2 - I_3), \quad I''_{\Delta\Delta'} = \pm\frac{1}{2}(I_3 - I_1), \quad I'''_{\Delta\Delta'} = \pm\frac{1}{2}(I_1 - I_2), \qquad (3.1.106)$$

while the corresponding axial moments of inertia are the abscissae of the centres of the very same circles, (see also Subsec. 1.2.3), i.e.,

$$I'_\Delta = \frac{1}{2}(I_2 + I_3), \quad I''_\Delta = \frac{1}{2}(I_3 + I_1), \quad I'''_\Delta = \frac{1}{2}(I_1 + I_2). \qquad (3.1.106')$$

Once the principal moments of inertia specified, the three circles are easily obtained. Let be a direction of unit vector $\mathbf{n}$; one can thus build up three arcs of circle of centres O_1, O_2 and O_3, and of radii given by (Fig.3.11,a)

$$r_1^2 = \frac{1}{4}(I_2 - I_3)^2 + n_1^2(I_1 - I_2)(I_1 - I_3),$$
$$r_2^2 = \frac{1}{4}(I_3 - I_1)^2 + n_2^2(I_2 - I_3)(I_2 - I_1), \qquad (3.1.107)$$
$$r_3^2 = \frac{1}{4}(I_1 - I_2)^2 + n_3^2(I_3 - I_1)(I_3 - I_2);$$

152   MECHANICAL SYSTEMS, CLASSICAL MODELS

the point P of co-ordinates I_Δ and $I_{\Delta\Delta'}$ is the piercing point of these arcs of circle.

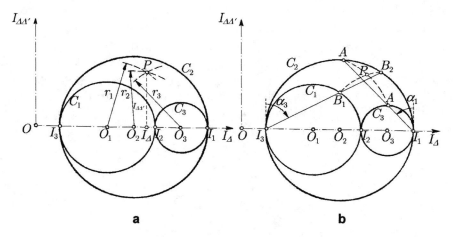

Figure 3.11. Mohr's circles. Determination of the components of the moments of inertia tensor for the direction **n** : analytical (a) and graphical (b) method.

One can use also a graphic method. Let be the tangents to the circle C_2, at the point in which the latter one intersects the axis OI_Δ and let us set up straight semi-lines inclined by the angles α_1 and α_3 with respect to those tangents, respectively (Fig.3.11,b); these semi-lines pierce the circles C_3 and C_2 at the points A_3 and A_2, respectively, and the circles C_1 and C_2 at the points B_1 and B_2, respectively. One can easily prove that the arc of circle of radius r_1 and centre O_1 passes through the points A_3 and A_2, while the arc of circle of radius r_3 and centre O_3 passes through the points B_1 and B_2; their intersection is just the point P. The construction is thus completely specified.

An important rôle is played by the approximation methods of computation of the integrals, by the graphical methods, by the methods using apparatuses for graphical determinations, by the experimental methods a.s.o.

1.2.7 Two-dimensional geometric representations

We may represent the variation of the axial moments of inertia I_Δ and $I_{\Delta'}$ given by (3.1.87) and (3.1.87'), respectively, by a rotation of the axes $O1$, $O2$ about the point O, as a function of the angle $(\mathbf{n},1)$ (Fig.3.12,a); the curve of fourth degree thus obtained is symmetric with respect to the two principal axes of inertia. The moments of inertia corresponding to the bisectrices of the angles formed by the principal axes of inertia are also put into evidence.

As well, the variation of the centrifugal moment of inertia $I_{\Delta\Delta'}$ by the angle $(\mathbf{n},1)$ is represented in Fig.3.12,b also by a curve of fourth degree, symmetric both with respect to the principal axes of inertia and to the bisectrices of the angles formed by those axes (with respect to which the considered moments of inertia have extreme values).

If the geometric support Ω is contained in the plane Ox_1x_2, then the relations (3.1.102) become

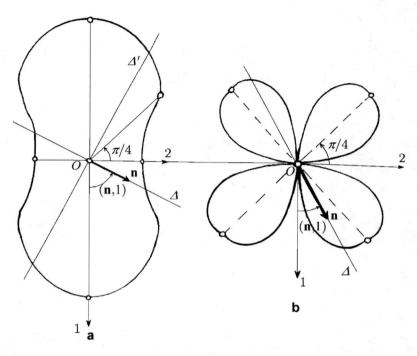

Figure 3.12. Variation of the axial (a) and centrifugal (b) moments of inertia.

$$I_1 a_1^2 = I_2 a_2^2 = K^2 \qquad (3.1.108)$$

and one obtains *the ellipse of inertia* ($a_1 \leq a_2$)

$$\frac{x_1^2}{a_1^2} + \frac{x_2^2}{a_2^2} = 1 ; \qquad (3.1.109)$$

hence, the principal moments of inertia are in inverse proportion to the squares of the semiaxes of the ellipse of inertia. To the semi-minor axis of the ellipse corresponds the maximal moment of inertia, while to the semi-major one corresponds the minimal moment of inertia. If the two principal moments of inertia are equal ($I_1 = I_2$), the ellipse of inertia is a circle ($a_1 = a_2$), and any axis passing through the pole O is a principal axis of inertia.

For instance, in the homogeneous case, we can take $K^2 = I_1 I_2 / A$, obtaining thus $a_1 = i_2$, $a_2 = i_1$, so that the equation of the ellipse becomes (Fig.3.13)

$$\frac{x_1^2}{i_2^2} + \frac{x_2^2}{i_1^2} = 1 \tag{3.1.109'}$$

We notice that we have $K^2 = A i_1^2 i_2^2$ too; on the other hand, taking into account (3.1.30') and (3.1.100), we obtain

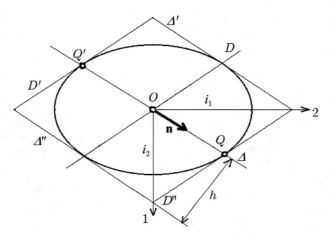

Figure 3.13. Properties of the ellipse of inertia.

$$\left| \overrightarrow{OQ} \right| = \frac{i_1 i_2}{i_\Delta}, \tag{3.1.110}$$

a remarkable relation, specifying the radius vector of the point Q.

Let be Δ' and Δ'' the tangents to the ellipse, parallel to the axis Δ, and D' and D'' the tangents parallel to the axis D, at the points Q and Q', respectively, Δ and D being two conjugate diameters of the ellipse; one obtains thus a parallelogram, the area of which is an invariant equal to $4h\left|\overrightarrow{OQ}\right|$, h being the distance from the centre O to one of the tangents Δ', Δ''. If the axis Δ coincides with one of the principal axes of inertia, the parallelogram becomes a rectangle of area $4i_1 i_2$; taking into account (3.1.110), we get

$$i_\Delta = h, \tag{3.1.110'}$$

hence a graphical evaluation of the gyration radius with respect to an arbitrary axis Δ.

Obviously, also in this case one can introduce an ellipse of gyration of the form (analogous to (3.1.105))

$$\frac{x_1'^2}{i_1^2} + \frac{x_2'^2}{i_2^2} = 1. \tag{3.1.111}$$

Mass geometry. Displacements. Constraints

By eliminating the angle $(\mathbf{n}, x_1)$ between one of the relations (3.1.87), (3.1.87') and the relation (3.1.87''), we get

$$\left[I_\Delta - \frac{1}{2}(I_{x_1} + I_{x_2})\right]^2 + I_{\Delta\Delta'}^2 = \left[\frac{1}{2}(I_{x_1} - I_{x_2})\right]^2 + I_{x_1 x_2}^2. \qquad (3.1.112)$$

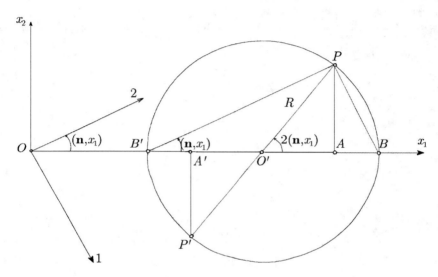

Figure 3.14. Mohr's circle. Principal axes and moments of inertia.

The relation remains valid if we replace I_Δ by $I_{\Delta'}$. Taking the axial moments of inertia I_Δ and $I_{\Delta'}$ along the axis Ox_1 and the centrifugal moment of inertia $I_{\Delta\Delta'}$ along the axis Ox_2, we notice that the equation (3.1.112) corresponds to a circle of centre $O'((I_{x_1} + I_{x_2})/2, 0)$ and radius $R = \sqrt{[(I_{x_1} - I_{x_2})/2]^2 + I_{x_1 x_2}^2}$ (Fig.3.14). Supposing that $I_{x_2} > I_{x_1}$ and $I_{x_1 x_2} > 0$, we obtain the points $P(I_{x_2}, I_{x_1 x_2})$ and $P'(I_{x_1}, -I_{x_1 x_2})$ on the circle; indeed, $\overline{A'O'} = \overline{O'A} = (I_{x_2} - I_{x_1})/2$ and $\overline{O'P} = \overline{O'P'} = R$, the affirmation made above being thus justified.

We notice that

$$\tan \widehat{AO'P} = \frac{\overline{PA}}{\overline{O'A}} = \frac{I_{x_1 x_2}}{(I_{x_2} - I_{x_1})/2}.$$

Taking into account the relation (3.1.88), it results $\widehat{AO'P} = 2(\mathbf{n}, x_1)$; we find also that $\widehat{AB'P} = (\mathbf{n}, x_1)$ (Fig.3.14). We are in the case in which $(\mathbf{n}, x_1)_{\min} < \pi/4$, $I_{x_2} > I_{x_1}$ and $I_{x_1 x_2} > 0$, so that, corresponding to the results of Subsec. 1.2.5, the direction of the axis $O2$ is given by $B'P$; analogously, the direction of the axis $O1$ corresponds to the straight line PB.

We see that $\overline{OB'} = \overline{OO'} - R = I_2$, $\overline{OB} = \overline{OO'} + R = I_1$, so that the extremities of the diameter BB' specify the magnitudes of the principal moments of inertia $I_1 > I_2$, corresponding to the formulae (3.1.91).

This circle, called *O.Mohr's circle* (initially introduced by Culmann and Rankine), puts into evidence – in a concise way – the properties of extremum of the moments of inertia with respect to axes rotating about a point.

By eliminating the angle $(\mathbf{n},1)$ between the relations (3.1.93) and (3.1.93"), one obtains

$$\left[I_\Delta - \frac{1}{2}(I_1 + I_2)\right]^2 + I_{\Delta\Delta'}^2 = \left[\frac{1}{2}(I_1 - I_2)\right]^2. \qquad (3.1.112')$$

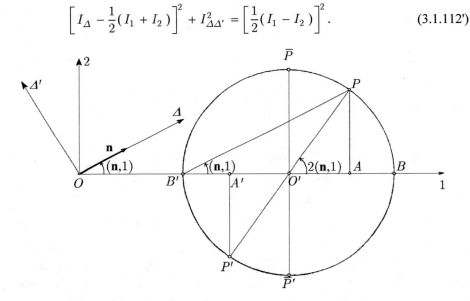

Figure 3.15. Mohr's circle. Determination of moments of inertia with respect to two orthogonal axes.

The equation holds also if we replace I_Δ by $I_{\Delta'}$. We get thus a circle of Mohr, drawn with the aid of the principal moments of inertia, in the plane $O12$, which allows us to determine the axial moments of inertia I_Δ and $I_{\Delta'}$, as well as the centrifugal moment of inertia $I_{\Delta\Delta'}$, corresponding to a given angle $(\mathbf{n},1)$ (Fig.3.15). Starting from the centre $O'((I_1 + I_2)/2, 0)$ of the circle of radius $R = (I_1 - I_2)/2$ and taking into account the relations (3.1.93)-(3.1.93"), we find the points $P(I_\Delta, I_{\Delta\Delta'})$ and $P'(I_{\Delta'}, -I_{\Delta\Delta'})$, $I_\Delta > I_{\Delta'}$, $I_{\Delta\Delta'} > 0$, corresponding to the angle $2(\mathbf{n},1)$ or to the angle $(\mathbf{n},1)$, respectively; the searched moments of inertia are thus obtained. We notice also that for $(\mathbf{n},1) = \pi/4$ we get the extreme centrifugal moments of inertia, which have the properties mentioned in Subsec. 1.2.5, corresponding to the points $\overline{P}$ and $\overline{P'}$ on the circle.

Mass geometry. Displacements. Constraints 157

Another geometric construction was proposed by Mohr too and was improved by R. Land; it is known as *Land's circle*. As in the previous case, we suppose that the axial moments of inertia $I_{x_1} < I_{x_2}$ and the centrifugal moment of inertia $I_{x_1 x_2}$ are known.

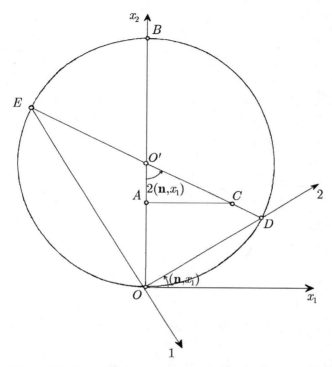

Figure 3.16. Land's circle. Principal axes and moments of inertia.

With respect to the frame of reference Ox_1x_2, we take the segments $\overline{OA} = I_{x_1}$ and $\overline{AB} = I_{x_2}$ on the axis Ox_2; on OB as diameter, we construct a circle with the centre at $O'(0,(I_{x_1} + I_{x_2})/2)$, of radius $R = (I_{x_1} + I_{x_2})/2$ and tangent at O to the axis Ox_1. We build up a segment AC, parallel to the axis Ox_1, and having the same direction, so as to have $\overline{AC} = I_{x_1 x_2}$ (if $I_{x_1 x_2} < 0$, then the point C must be in the opposite direction); the straight line $O'C$ pierces the circle at the points D and E (Fig.3.16).

If $(\mathbf{n}, x_1)$ is the angle formed by OD with the axis Ox_1, then it follows that $\widehat{OO'D} = 2(\mathbf{n}, x_1)$. We notice that $\overline{O'A} = (I_{x_1} + I_{x_2})/2 - I_{x_1} = (I_{x_2} - I_{x_1})/2$, hence

$$\overline{O'C} = \sqrt{[(I_{x_2} - I_{x_1})/2]^2 + I_{x_1 x_2}^2}.$$

Taking into account the formulae (3.1.91), we get $\overline{EC} = I_1$, $\overline{CD} = I_2$; on the other hand, $\tan 2(\mathbf{n}, x_1) = \overline{AC}/\overline{O'A} = 2I_{x_1 x_2}/(I_{x_2} - I_{x_1})$, so that OD is along the axis $O2$, while OE is along the axis $O1$.

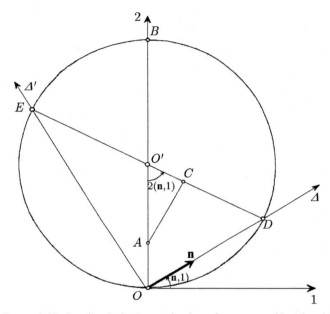

Figure 3.17. Land's circle. Determination of moments of inertia with respect to two orthogonal axes.

With respect to the principal axes $O1$, $O2$, one can construct a circle of diameter $\overline{OB} = \overline{OA} + \overline{AB}$, $\overline{OA} = I_2$, $\overline{AB} = I_1$ and of radius $R = (I_1 + I_2)/2$, tangent to $O1$ at O (Fig.3.17). By means of the angle $(\mathbf{n}, 1)$, we draw the straight lines OD and OE along the axes Δ and Δ' ($\Delta' \perp \Delta$), respectively; the points D and E are the extremities of a diameter. The point C is the foot of the normal from A to DE. It is easy to verify that $\overline{DC} = I_{\Delta'}$, $\overline{CE} = I_\Delta$ and $\overline{AC} = I_{\Delta \Delta'}$, obtaining thus the axial moments of inertia I_Δ, $I_{\Delta'}$ and the centrifugal moment of inertia $I_{\Delta\Delta'}$, corresponding to the direction $(\mathbf{n}, 1)$ with respect to the principal axis $O1$.

1.2.8 Huygens-Steiner theorems

Let be an axis Δ and an axis Δ_C, parallel to the first one and passing through the centre of mass C of the mechanical system. A point P of the system is projected at P' on a plane Π, normal to both axes Δ and Δ_C; this plane is pierced by the axes at the points P_Δ and P_{Δ_C}, respectively. Denoting $\overrightarrow{P'P_\Delta} = \mathbf{r}_\Delta$, $\overrightarrow{P'P_{\Delta_C}} = \mathbf{r}_{\Delta_C}$ and $\overrightarrow{P_{\Delta_C} P_\Delta} = \mathbf{d}$, and noting that we have to do with equipollent vectors, contained in the plane Π, we may write (Fig.3.18)

$$I_\Delta = \int_\Omega r_\Delta^2 \, dm = \int_\Omega \left(\mathbf{r}_{\Delta_C} + \mathbf{d}\right)^2 dm = \int_\Omega r_{\Delta_C}^2 \, dm + 2\mathbf{d} \cdot \int_\Omega \mathbf{r}_{\Delta_C} \, dm + d^2 \int_\Omega dm \,;$$

observing that the static moment with respect to the centre of mass vanishes, we have

$$I_\Delta = I_{\Delta_C} + Md^2 \qquad (3.1.113)$$

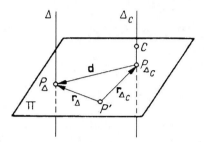

Figure 3.18. Huygens-Steiner theorem.

and we can state

Theorem 3.1.4 (*Huygens-Steiner*). *The moment of inertia of a mechanical system with respect to an axis Δ is equal to the sum of the moment of inertia of the same system with respect to an axis Δ_C parallel to the first one, passing through the centre of mass, and the moment of inertia of the centre of mass, at which we consider concentrated the mass of the whole mechanical system, with respect to the axis Δ.*

It follows that, being given all axes which have the same direction, the moment of inertia of a mechanical system is minimal for that axis which passes through the centre of mass. The moments of inertia with respect to the *central axes* (the axes which pass through the centre of mass) are called *central moments of inertia*; the moments of inertia which correspond to the central principal axes of inertia are called *central principal moments of inertia*. Corresponding to what was related before, from all the axes passing through the point C, there exists one (or at least one if we have equal moments of inertia) with respect to which the axial moment of inertia admits a *minimum minimorum*. We notice also that the locus of the parallel axes which have the same moment of inertia is a circular cylinder of radius d, the axis of the cylinder passing through C; as well, the variation of the moments of inertia with respect to axes of the same direction may be represented by a paraboloid of rotation, the axis of which has the same direction and passes through C.

Let Δ' be another axis, parallel to the axis Δ; if we write a formula of the form (3.1.113) for this axis too, and if we subtract the two formulae, then we get

$$I_{\Delta'} = I_\Delta + M\left(d'^2 - d^2\right); \qquad (3.1.114)$$

this result allows us to pass from an axis of a given direction to an axis parallel to the latter one, taking into account the distances d and d' from these axes to an axis which has the same direction and passes through C.

With the aid of the radii of gyration (3.1.30), the formulae (3.1.113), (3.1.114) become

$$i_\Delta^2 = i_{\Delta_C}^2 + d^2, \qquad (3.1.113')$$

$$i_{\Delta'}^2 = i_\Delta^2 + d'^2 - d^2, \qquad (3.1.114')$$

the notations corresponding to those used above.

For the polar and the planar moments of inertia, respectively, we may write also the formulae

$$I_O = I_C + M\rho^2, \qquad (3.1.115)$$

$$I_\Pi = I_{\Pi_C} + Md^2, \qquad (3.1.115')$$

corresponding to theorems of Huygens-Steiner type, where ρ is the position vector of C with respect to O, while d is the distance between the parallel planes Π and Π_C. The properties previously established can be adapted to those formulae too. Analogously, for the products of inertia we obtain a *theorem of Huygens-Steiner type* of the form

$$I_{\Delta\Delta'} = I_{\Delta_C \Delta'_C} + Mdd', \qquad (3.1.113'')$$

where Δ and Δ' are axes corresponding to orthogonal directions and which pass through the point O, while d and d' represent the co-ordinates of the centre of mass C with respect to those axes. These products of inertia may be also negative, so that for the central axes we cannot have minimal values. As well, starting from axes which pass through C, one can find axes parallel to latter ones for which $I_{\Delta\Delta'} = 0$; in the plane case, the locus of the pole O which has this property is an equilateral hyperbola.

Analogously, if $x_j = \rho_j + x_j^C$ (obvious notations) and if we take into account the characteristic property of the centre of mass, then the relation (3.1.81) leads to a synthesis of the Huygens-Steiner theorem in the form

$$I_{jk}^O = I_{jk}^C + M\left(\rho_l \rho_l \delta_{jk} - \rho_j \rho_k\right), \quad j,k = 1,2,3, \qquad (3.1.116)$$

or in the form

$$\mathbf{I}_O(\mathscr{S}) = \mathbf{I}_C(\mathscr{S}) + \mathbf{I}_O(\mathscr{S}_C); \qquad (3.1.116')$$

hence, we may state

Theorem 3.1.5. *The moment of inertia tensor of a mechanical system with respect to the point O is equal to the sum of a moment of inertia tensor of the same system with respect to the centre of mass C and the moment of inertia tensor of the centre of mass, at which we consider concentrated the mass of the whole mechanical system, with respect to the point O.*

Mass geometry. Displacements. Constraints

If we write the relation (3.1.116') for the poles O and O', respectively, and if we subtract the two relations, then we obtain the relation

$$\mathbf{I}_{O'}(\mathscr{S}) = \mathbf{I}_{O}(\mathscr{S}) + \mathbf{I}_{O'}(\mathscr{S}_C) - \mathbf{I}_{O}(\mathscr{S}_C); \qquad (3.1.117)$$

we put thus in evidence the variation of the moment of inertia tensor of a mechanical system $\mathscr{S}$ by passing from a pole O to a pole O'. Denoting by ρ'_j, ρ_j and η_j the components of the vectors $\overrightarrow{O'C}$, $\overrightarrow{OC}$ and $\overrightarrow{O'O}$, respectively ($\rho'_j = \rho_j + \eta_j$), we may write too

$$I^{O'}_{jk} = I^{O}_{jk} + M\left(\eta_l \eta_l \delta_{jk} - \eta_j \eta_k\right) - 2M\left[\frac{1}{2}(\rho_j \eta_k + \rho_k \eta_j) - \rho_l \eta_l \delta_{jk}\right],$$

wherefrom

$$\mathbf{I}_{O'}(\mathscr{S}) = \mathbf{I}_{O}(\mathscr{S}) + \mathbf{I}_{O'}(\mathscr{S}_C) - \overline{\mathbf{J}}_{OO'}(\mathscr{S}_C;\mathscr{S}_O), \qquad (3.1.117')$$

with

$$\overline{\mathbf{J}}_{OO'}(\mathscr{S}_C;\mathscr{S}_O) = \mathbf{J}_{OO'}(\mathscr{S}_C;\mathscr{S}_O) - \operatorname{tr}\mathbf{J}_{OO'}(\mathscr{S}_C;\mathscr{S}_O)\mathbf{1}, \qquad (3.1.117'')$$

where

$$\mathbf{J}_{OO'}(\mathscr{S}_C;\mathscr{S}_O) = \frac{2}{M}(\mathbf{S}_O(\mathscr{S}_C) \otimes \mathbf{S}_{O'}(\mathscr{S}_O)) \qquad (3.1.117''')$$

represents the symmetric part of the dyadic product of the two vectors.

In the case of a discrete mechanical system of particles P_i of position vectors $\mathbf{r}_i$, $i = 1, 2, \ldots, n$, we may write the identity

$$\sum_i m_i \sum_j m_j r_j^2 = \sum_i m_i^2 r_i^2 + \sum_i \sum_j m_i m_j r_j^2 = \left(\sum_i m_i \mathbf{r}_i\right)^2$$
$$-2\sum_i \sum_j m_i m_j \mathbf{r}_i \cdot \mathbf{r}_j + \sum_i \sum_j m_i m_j \left(r_i^2 + r_j^2\right), \quad i > j.$$

Denoting by r_{ij} the distance between the particles P_i and P_j ($\mathbf{r}_{ij} = \mathbf{r}_j - \mathbf{r}_i$), we get

$$MI_O = S_O^2 + \sum_{i=j+1}^{n}\sum_{j=1}^{n-1} m_i m_j r_{ij}^2 ; \qquad (3.1.118)$$

in particular, if $O \equiv C$, then the central polar moment of inertia is given by

$$I_C = \frac{1}{M}\sum_{i=j+1}^{n}\sum_{j=1}^{n-1} m_i m_j r_{ij}^2 = \frac{1}{2M}\sum_{i=1}^{n}\sum_{j=1}^{n} m_i m_j r_{ij}^2. \qquad (3.1.118')$$

2. Displacements. Constraints

A mechanical system can be free or subjected to constraints (internal constraints or constraints due to the interactions with other systems). We are thus led to a study of the constraints, which is based on the notion of geometric displacement.

2.1 Displacements

We put into evidence the real, possible and virtual displacements, which play an important rôle in the representation of constraints as well as in the study of equilibrium and motion of the mechanical systems.

2.1.1 Real, possible and virtual displacements

Let P be a point of a mechanical system $\mathscr{S}$ subjected to the action of a system of given forces. We suppose that the displacement of the point P from the given position of position vector $\mathbf{r}$ (at the moment t) to a neighbouring position P' of position vector $\mathbf{r}' = \mathbf{r} + \mathrm{d}\mathbf{r}$ (at the moment $t' = t + \mathrm{d}t$) is a differential quantity, compatible

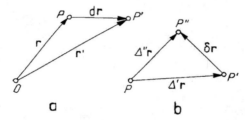

Figure 3.19. Real (a), possible and virtual (b) displacements.

with the constraints of the system (Fig.3.19,a). We call *real displacement* that one which depends on the time t and is determined by the forces acting upon the mechanical system. The real displacement is univocally determined for a given system of forces and for certain given conditions depending on the constraints of the mechanical system; it corresponds to the real motion of the mechanical system, supposing that the solution of the considered mechanical problem is unique. The displacements $\Delta \mathbf{r} = \mathbf{r}' - \mathbf{r}$, which depend on the time t and take place in a time interval Δt, but which are not determined by the forces which act upon the mechanical system, are called *possible displacements*; obviously, they are differential too and compatible with the constraints of the system (sometimes, to be more specific, they are called *infinitesimal possible displacements*). Possible displacements are all the differential displacements which correspond to a possible position of a mechanical system, at a given moment t; in general, they are not unique and can be also in an infinite number. The real displacement is one of the possible displacements (it is obtained if supplementary conditions are put to the latter ones, i.e., imposing the dependence on the forces which act upon the mechanical system). By means of relations

$$\Delta \mathbf{r} = \overline{\mathbf{v}} \Delta t, \quad \mathrm{d}\mathbf{r} = \mathbf{v} \mathrm{d}t, \tag{3.2.1}$$

Mass geometry. Displacements. Constraints 163

we may – analogously – introduce the *possible velocities* $\bar{\mathbf{v}}$, to the set of which belongs the *real velocity* $\mathbf{v}$ too. We notice that, in the case of a free mechanical system, any displacement is a possible displacement and any velocity is a possible velocity.

Let $\Delta'\mathbf{r}$ and $\Delta''\mathbf{r}$ be two possible displacements of the point P, which convey this point to the neighbouring positions P' and P'', respectively (Fig.3.19,b). We introduce the differential displacement $\delta\mathbf{r}$, which must be imposed to the point P, so that this one do change its position from P' to the neighbouring one P''; this displacement, equal to the difference of the two possible displacements

$$\delta\mathbf{r} = \Delta''\mathbf{r} - \Delta'\mathbf{r} \tag{3.2.2}$$

is called *virtual displacement*. Hence, by the summation of a possible displacement of a point and a virtual one of the very same point, we obtain a new possible displacement of the considered point. The relation (3.2.2) establishes a *geometric link* (and not a kinematic one) between two possible displacements of a point P and a virtual displacement of the very same point; it gives thus the possibility of passage from a possible displacement to another one. Thus, the virtual displacements (which are differential quantities too) do not take place in time, but are compatible with the constraints of the mechanical system at the time t; neither these displacements (as well as the possible displacements) are not determined by the given forces. The virtual displacements are thus displacements in the hypothesis in which the time t is fixed (the constraints of the system are "frozen", hence they do not depend on time); this point of view allows us to write – easily – constraint relations with the aid of virtual displacements (differentiating the geometric constraints with the assumption that $t = $ const). We notice that, although the virtual displacements do not take place in time, they depend – in a certain manner – on this variable, because they are different from a moment to another one; indeed, the virtual displacements represent the displacements of the points of a mechanical system from a possible position at the moment t to a neighbouring possible position at the very same moment. In a *stationary case*, the set of virtual displacements coincides with the set of possible displacements; thus, fixing the time t, we may pass from relations written by means of possible displacements $\Delta\mathbf{r}$ to relations expressed with the aid of virtual displacements $\delta\mathbf{r}$. Obviously, we admit that the virtual displacements $\delta\mathbf{r}$ are applied at the point P of position vector $\mathbf{r}$, being – in fact – variations of this vector. V. Vâlcovici has introduced the virtual displacements in the form

$$\delta\mathbf{r} = \Delta\mathbf{r} - \mathbf{v}\Delta t = \left(\Delta - \Delta t \frac{\mathrm{d}}{\mathrm{d}t}\right)\mathbf{r}, \quad \delta = \Delta - \Delta t \frac{\mathrm{d}}{\mathrm{d}t}, \tag{3.2.2'}$$

where $\mathbf{v}$ is the real velocity; the set of these displacements coincides with the set of the displacements defined by the relation (3.2.2). From (3.2.1), (3.2.2), it follows that

$$\delta\mathbf{r} = \mathbf{v}^*\Delta t, \quad \mathbf{v}^* = \bar{\mathbf{v}}'' - \bar{\mathbf{v}}'; \tag{3.2.1'}$$

we can thus introduce the *virtual velocities* $\mathbf{v}^*$ as a difference of possible velocities.

For instance, in the case of a particle P constrained to stay on a fixed curve C, the real displacement $d\mathbf{r}$ is tangent to the curve, having a specified direction; the possible and the virtual displacements are also tangent to the curve, but can be oriented in any of the two possible directions (there may exist two such displacements) (Fig.3.20,a). In the case of a particle constrained to stay on a fixed surface S, the real displacement $d\mathbf{r}$ is tangent to the surface and univocally directed; but the possible and the virtual displacements, the sets of which coincide (we are in a stationary case, as above), and which are also tangent to the surface, can be anyone in the tangent plane (Fig.3.20,b). If, besides the virtual displacement $\delta\mathbf{r}$, there exists the virtual displacement $-\delta\mathbf{r}$ too, then the virtual displacement is called *reversible*; otherwise, the virtual displacement is called *irreversible*.

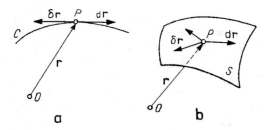

Figure 3.20. Real and virtual displacements on a fixed curve (a) or surface (b).

In both cases considered above, the curve and the surface are fixed, so that the real displacement belongs to the set of virtual displacements. If a particle P is, for instance, on a movable rigid surface S, the displacement of which is of translation velocity $\mathbf{u}$, and if we introduce the relative velocities $\mathbf{v}'$ and $\mathbf{v}''$, contained in the plane tangent to the surface at the point P, then we obtain the possible displacements

$$\Delta'\mathbf{r} = \mathbf{v}'\Delta t + \mathbf{u}\Delta t, \quad \Delta''\mathbf{r} = \mathbf{v}''\Delta t + \mathbf{u}\Delta t;$$

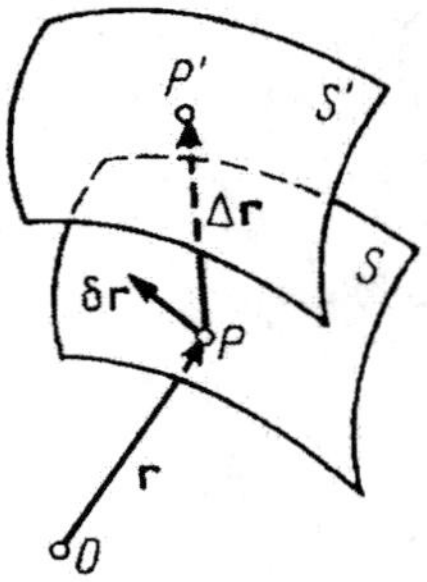

Figure 3.21. Real and virtual displacements on a movable rigid surface.

the virtual displacements of the form (3.2.1') thus obtained take place in the mentioned tangent plane, at the moment t. But the possible displacements connect points of the surface S to points of the surface S' (corresponding to the moment $t + \Delta t$) (Fig.3.21). Hence, in this case, the real displacement does not belong to the set of virtual displacements. In the case of a deformable surface, one may obtain analogous conclusions.

2.1.2 Real and virtual work

Let $\mathscr{S}$ be a mechanical system and P_i points of it of position vectors $\mathbf{r}_i$, $i = 1,2,...,n$. *The real elementary work* of the forces $\mathbf{F}_i$, acting at the mentioned n points of the system, is given by

$$\mathrm{d}W = \sum_{i=1}^{n} \mathbf{F}_i \cdot \mathrm{d}\mathbf{r}_i, \qquad (3.2.3)$$

where we used the definition formula (A.1.28'); analogously, we introduce *the virtual work* (the specification "elementary" is not necessary)

$$\delta W = \sum_{i=1}^{n} \mathbf{F}_i \cdot \delta \mathbf{r}_i. \qquad (3.2.3')$$

For internal forces which verify the relation (1.1.81), we may write (Fig.1.18)

$$\mathbf{F}_{ij} \cdot \mathrm{d}\mathbf{r}_i + \mathbf{F}_{ji} \cdot \mathrm{d}\mathbf{r}_j = \left(\mathbf{F}_{ij} + \mathbf{F}_{ji} \right) \cdot \mathrm{d}\mathbf{r}_i + \mathbf{F}_{ji} \cdot \mathrm{d}\mathbf{r}_{ij} = F_{ji}\mathbf{u} \cdot \left(r_{ij}\mathrm{d}\mathbf{u} + \mathbf{u}\mathrm{d}r_{ij} \right) = F_{ij}\mathrm{d}r_{ij},$$

where $\mathbf{u} = \mathrm{vers}\,\mathbf{r}_{ij}$, while $F_{ij} = F_{ji}$ are positive quantities, in the case of repulsive forces (which have the tendency to move off the points P_i and P_j one of the other), or negative ones, in the case of attractive forces; one obtains thus the real elementary work of internal forces in the remarkable form

$$\mathrm{d}W_{\mathrm{int}} = \sum_{i=j+1}^{n}\sum_{j=1}^{n-1} F_{ij}\mathrm{d}r_{ij} = \frac{1}{2}\sum_{i=1}^{n}\sum_{j=1}^{n} F_{ij}\mathrm{d}r_{ij}, \quad i \neq j. \qquad (3.2.4)$$

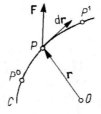

Figure 3.22. Real work along a curve C.

Noting that $\mathrm{d}r_{ij} \gtrless 0$ for $F_{ij} \gtrless 0$, it results that the real elementary work of internal forces is a strict positive quantity ($\mathrm{d}W_{\mathrm{int}} > 0$). In the case of a non-deformable mechanical system (discrete non-deformable system or rigid solid) and only in the case of such a system, we have $\mathrm{d}r_{ij} = 0$, so that $\mathrm{d}W_{\mathrm{int}} = 0$. In general, we can state that $\mathrm{d}W_{\mathrm{int}}$ is a non-negative quantity.

In the case of a force **F**, the point P of application of which describes a curve C between the points P^0 and P^1, we may write (Fig.3.22)

$$W_{\widehat{P^0 P^1}} = \int_{\widehat{P^0 P^1}} \mathbf{F} \cdot \mathrm{d}\mathbf{r} ; \qquad (3.2.5)$$

if the force is conservative, of the form (1.1.82), then we obtain

$$\mathrm{d}W = \mathrm{d}U, \quad W_{\widehat{P^0 P^1}} = U(P^1) - U(P^0) = U(\mathbf{r}_1) - U(\mathbf{r}_0), \qquad (3.2.5')$$

the real work depending only on the extreme positions of the point P. In the case of a closed curve C, we have

$$W_C = \oint_C \mathbf{F} \cdot \mathrm{d}\mathbf{r} = 0, \qquad (3.2.5'')$$

and the corresponding real work vanishes.

If there exists a function U_{ij} so that

$$\mathrm{d}U_{ij} = F_{ij} \mathrm{d}r_{ij} \qquad (3.2.6)$$

(for instance, if the internal forces depend only on the distances, $F_{ij} = F_{ij}(r_{ij})$, then $U_{ij} = \int F_{ij}(r_{ij}) \mathrm{d}r_{ij}$), then there exists the potential

$$U = \sum_{i=j+1}^{n} \sum_{j=1}^{n-1} U_{ij} = \frac{1}{2} \sum_{i=1}^{n} \sum_{j=1}^{n} U_{ij}, \quad i \neq j, \qquad (3.2.6')$$

too, and the mechanical system is *conservative*. Thus, the elastic solids in adiabatic or isothermic regime as well as the compressible fluids, in certain conditions (e.g., the perfect gases), are *conservative mechanical systems*.

2.2 Constraints

In the following, we deal with the notion of constraint, which will be characterized from a geometric, as well as from a kinematic point of view and for which we put in evidence various classifications. A special attention is given to ideal constraints and to constraints with friction, as well as to the constraints of a rigid solid; the constraint forces which arise are thus put into evidence.

2.2.1 Classification of constraints. Axiom of liberation from constraints

A mechanical system (discrete or continuous) can be free or can be subjected to some *restrictions* of *geometric* or *kinematic* nature; these restrictions, which represent a limitation of the positions of the points of the system, will be expressed by relations between the co-ordinates, or between the co-ordinates and the displacements (velocities) of the respective points. We say that the mechanical system is subjected to

Mass geometry. Displacements. Constraints 167

constraints, which can be due to internal causes (*internal constraints*) or may be due to interactions with other systems (*external constraints*). The constraints can be *unilateral* (if, for instance in E_3, a point P of the mechanical system is in one part with respect to a surface S or on the respective surface), being mathematically expressed by inequalities (strict, if the point P cannot belong to the surface, or not, otherwise), or can be *bilateral* (if the point P is on the surface S), being expressed by equalities. As well, the constraints can be *of contact* (if the mechanical systems – for instance, continuous – are lying one on the other) or *at distance* (if the distance between two points is invariant in time or depends on a certain law). Another classification of the constraints puts into evidence the *finite constraints* (*holonomic*, of *geometric* nature) and the *infinitesimal* (*differential*) *constraints*, of *kinematic* nature (in general, *non-holonomic*); thus, we reach a mathematical representation of the constraints. We notice also another classification, very important from the point of view of computation, i.e.: *ideal constraints* (perfect, smooth) and *constraints with friction* (real); analytical mechanics was developed just for mechanical systems subjected to ideal constraints. The constraints which do not change in time (do not depend explicitly on time) are called *stationary* (*scleronomic*) *constraints*; the constraints which vary in time are called *non-stationary* (*rheonomic*) *constraints*. We mention also the *critical constraints*; such constraints allow infinitesimal displacements which have not any correspondence in finite displacements.

We can pass from the study of a mechanical system with constraints to the study of a free one, using *the axiom of liberation* from *constraints*. This axiom allows us to replace the constraints by *constraint forces* applied to certain points of the mechanical system; in this case, the system may be considered as being free, but subjected to the action of constraint forces too. We are thus led to a new classification of the forces, i.e.: *given* (known) *forces* and *constraint* (unknown) *forces*; this classification is independent of the previous classifications (in external and internal forces or in conservative and non-conservative ones).

In these conditions, we may consider that the formulae (3.2.3), (3.2.3') give *the real elementary work* and *the virtual work of the given forces*, respectively. Analogously, *the real elementary work of the constraint forces* $\mathbf{R}_i$ (the letter $\mathbf{R}$ corresponds to the word "reaction"), applied at the same points P_i, is expressed in the form

$$\mathrm{d}W_R = \sum_{i=1}^{n} \mathbf{R}_i \cdot \mathrm{d}\mathbf{r}_i, \qquad (3.2.7)$$

while the corresponding *virtual work* is given by

$$\delta W_R = \sum_{i=1}^{n} \mathbf{R}_i \cdot \delta\mathbf{r}_i. \qquad (3.2.7')$$

In the case of *internal constraint forces* (corresponding to internal constraints and verifying relations of the form (1.1.81) and (2.2.50)) one can establish a formula of the form (3.2.4) for the corresponding work

$$\mathrm{d}W_{R\mathrm{int}} = \sum_{i=j+1}^{n}\sum_{j=1}^{n-1} R_{ij}\mathrm{d}r_{ij} = \frac{1}{2}\sum_{i=1}^{n}\sum_{j=1}^{n} R_{ij}\mathrm{d}r_{ij}, \quad i \neq j. \tag{3.2.4'}$$

The external constraint forces correspond to external constraints.

2.2.2 Geometric characterization of constraints

In an inertial frame of reference, the finite constraints of geometric nature are expressed by relations between the position vectors of the points of the mechanical system $\mathscr{S}$ and the time t. In the case of a discrete mechanical system of n particles P_i of position vectors $\mathbf{r}_i$, $i = 1,2,...,n$, we may write these relations in the form (we introduce a shortened notation, so that $\mathbf{r}_j$ represents the set of all position vectors)

$$f_l(\mathbf{r}_j;t) \equiv f_l(\mathbf{r}_1,\mathbf{r}_2,...,\mathbf{r}_n;t) = 0, \quad l = 1,2,...,p, \tag{3.2.8}$$

where p is the number of the respective constraints; thus, the mechanical system cannot occupy any position in the space, but only the positions allowed by the restrictions (3.2.8). We suppose that the functions f_l are of the class C^2, so that all p constraints be *distinct* (no one of the constraints can be a consequence of the other ones or of the differential equations of motion). We notice that $p < 3n$, so that the motion may take place; the difference $s = 3n - p$ represents *the number of geometric degrees of freedom* of the mechanical system $\mathscr{S}$. If $p = 3n$, then the mechanical system has not any geometric degree of freedom, being at rest with respect to the considered frame of reference; solving a system of $3n$ equations with $3n$ unknowns (the co-ordinates of the n particles), one obtains the position of the mechanical system (eventually, this position is not univocally determinate). We cannot have $p > 3n$, because in such a case the constraints are no more distinct. In the above considerations, we have admitted tacitly that the constraints are bilateral, in any expression being involved all the particles of the mechanical system $\mathscr{S}$; so that we can no more make considerations for each particle (point of the mechanical system), as in the case of a single particle. In the case of unilateral constraints of geometric nature of a mechanical system $\mathscr{S}$, we are led to relations of the form

$$f_l(\mathbf{r}_j;t) \geq 0, \quad l = 1,2,...,p. \tag{3.2.8'}$$

If the equality cannot take place for one of the constraints, then the respective constraint is called *strict*. Obviously, the condition $p < 3n$ must further hold.

We introduce the notations

$$X_{3(i-1)+j} = x_j^{(i)}, \quad i = 1,2,...,n, \quad j = 1,2,3, \tag{3.2.9}$$

where $x_j^{(i)}$ are the components of the position vectors $\mathbf{r}_i$; we may thus pass from the geometric support Ω of the discrete mechanical system $\mathscr{S}$ in the space E_3 (formed by

Mass geometry. Displacements. Constraints 169

the geometric points P_i) to a *representative geometric point* P (of generalized co-ordinates $X_1, X_2, ..., X_{3n}$) in a *representative space* E_{3n}. The relations (3.2.8), which express bilateral constraints, will be written in the form

$$f_l(X_j;t) \equiv f_l(X_1, X_2, ..., X_{3n};t) = 0, \quad l = 1, 2, ..., p, \qquad (3.2.8'')$$

while the relations (3.2.8'), corresponding to unilateral constraints, will take the form

$$f_l(X_j;t) \geq 0, \quad l = 1, 2, ..., p. \qquad (3.2.8''')$$

The relations (3.2.8'') represent thus the conditions that the representative point P be at the intersection of p hypersurfaces in the representative space E_{3n}, hence on a manifold of dimension $s = 3n - p$ (equal to the number of geometric degrees of freedom) of this space. Analogously, the relations (3.2.8''') represent the conditions that the representative point P be only in one side of this manifold or, eventually, on the manifold itself, if the constraint is not strict.

For instance, if a particle P can be only in the interior of a sphere of variable radius $R - vt$ (v is a velocity), we impose the condition (we take the origin O at the centre of the sphere)

$$(R - vt)^2 - (x_1^2 + x_2^2 + x_3^2) > 0. \qquad (3.2.10)$$

If the mechanical system $\mathscr{S}$ is formed by the particles P_1 and P_2, linked by an inextensible thread of length l, then we impose the condition

$$l^2 - (x_1^{(1)} - x_1^{(2)})^2 - (x_2^{(1)} - x_2^{(2)})^2 - (x_3^{(1)} - x_3^{(2)})^2 \geq 0; \qquad (3.2.10')$$

denoting $X_1 = x_1^{(1)}$, $X_2 = x_2^{(1)}$, $X_3 = x_3^{(1)}$, $X_4 = x_1^{(2)}$, $X_5 = x_2^{(2)}$, $X_6 = x_3^{(2)}$, the representative point P must be in a part of the hyperquadric (which corresponds to the equality)

$$l^2 - (X_1^2 + X_2^2 + X_3^2 + X_4^2 + X_5^2 + X_6^2) + 2(X_1X_4 + X_2X_5 + X_3X_6) \geq 0 \qquad (3.2.10'')$$

or even on this one.

2.2.3 Degrees of freedom of a non-deformable mechanical system. Euler's angles

Let $\mathscr{S}$ be a discrete mechanical system of n free particles P_i, $i = 1, 2, ..., n$. The position of a particle is determined by three arbitrary parameters (for instance, its co-ordinates); we say, as in the previous subsection, that the system $\mathscr{S}$ has $3n$ *degrees of freedom*, because there are $3n$ necessary arbitrary parameters to determine its position.

If the discrete mechanical system is non-deformable (between the particles take place geometric constraints (finite internal constraints), which maintain invariant the mutual distances), then we notice that for $n = 1$ it has 3 degrees of freedom, for $n = 2$ it has

$3 \cdot 2 - 1 = 5$ degrees of freedom, while for $n = 3$ it has $3 \cdot 3 - 3 = 6$ degrees of freedom. If we suppose that there are still 6 degrees of freedom for n particles, then for $n + 1$ particles we have $3(n + 1) - 3(n - 1) = 6$ degrees of freedom too (we take into account that the intervention of a supplementary particle introduces only 3 distinct constraints (Fig.3.23)); we have thus proved, by complete induction, that *a non-deformable discrete mechanical system, without external constraints, has* 6 *degrees of freedom for* $n \geq 3$.

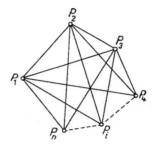

Figure 3.23. Degrees of freedom of a non-deformable discrete mechanical system.

This result holds also in the case of a non-deformable continuous mechanical system, hence in the case of a *rigid solid*. It is sufficient to show that the position of a rigid solid is univocally determined with respect to a fixed frame of reference with the aid of six independent parameters. Let thus be a fixed frame $O'x_1'x_2'x_3'$ (having a known position in space) and a movable frame $Ox_1x_2x_3$, rigidly connected to the rigid solid (eventually, the pole O is taken at the centre of mass of the solid); the pole O is determined by three parameters (the co-ordinates x_1^0, x_2^0, x_3^0 of its position vector). To determine the position of the rigid solid with respect to the fixed frame of reference, it is sufficient to specify the position of the movable frame with respect to the fixed one or with respect to a frame $O\bar{x}_1\bar{x}_2\bar{x}_3$ with the pole at O and the axes parallel to the corresponding axes of the fixed frame (Fig.3.24). We denote by ON (*the line of nodes*) the intersection of the planes $Ox_1'x_2'$ and Ox_1x_2. We give a positive rotation of angle $0 \leq \psi < 2\pi$ to the axis Ox_1' about the axis $O\bar{x}_3$ so as to coincide with ON and a positive rotation of angle $0 \leq \varphi < 2\pi$ to the ON-line about the axis Ox_3, so as to be superposed on Ox_1; as well, we give a positive rotation of angle $0 \leq \theta \leq 2\pi$ to the axis $O\bar{x}_3$ about ON so as to coincide with Ox_3. The axis Ox_2 is immediately obtained, observing that the frame $Ox_1x_2x_3$ must be a right-handed one. Starting from the frame $O\bar{x}_1\bar{x}_2\bar{x}_3$, the angles ψ, θ and φ, set up in the mentioned order (one determines firstly the line ON, than the axis Ox_3 and the axis Ox_1, and finally the axis Ox_2), specify univocally the movable frame, hence they represent three independent parameters. The angles ψ, θ and φ are called *Euler's angles*; the angle ψ is *the precession of the movable frame*, the angle θ is *the nutation of the movable frame*, while the angle φ is *the proper rotation of the movable frame*, by analogy with the

Mass geometry. Displacements. Constraints

parameters used in celestial mechanics to determine the position of a planet. Introducing the column matrices

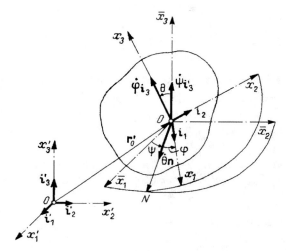

Figure 3.24. Degrees of freedom of a rigid solid. Euler's angles ψ, θ, φ.

$$\mathbf{i} = \begin{bmatrix} \mathbf{i}_1 \\ \mathbf{i}_2 \\ \mathbf{i}_3 \end{bmatrix}, \quad \mathbf{i}' = \begin{bmatrix} \mathbf{i}'_1 \\ \mathbf{i}'_2 \\ \mathbf{i}'_3 \end{bmatrix} \tag{3.2.11}$$

of the unit vectors of the co-ordinate axes and the matrices

$$\boldsymbol{\Psi} = \begin{bmatrix} \cos\psi & \sin\psi & 0 \\ -\sin\psi & \cos\psi & 0 \\ 0 & 0 & 1 \end{bmatrix}, \quad \boldsymbol{\Theta} = \begin{bmatrix} 1 & 0 & 0 \\ 0 & \cos\theta & \sin\theta \\ 0 & -\sin\theta & \cos\theta \end{bmatrix},$$

$$\boldsymbol{\Phi} = \begin{bmatrix} \cos\varphi & \sin\varphi & 0 \\ -\sin\varphi & \cos\varphi & 0 \\ 0 & 0 & 1 \end{bmatrix}, \tag{3.2.11'}$$

which give the direction cosines of the movable frame axes after a rotation of angle ψ, θ or φ, respectively, we may write

$$\mathbf{i} = \boldsymbol{\Phi}\boldsymbol{\Theta}\boldsymbol{\Psi}\mathbf{i}'. \tag{3.2.11''}$$

The matrix of the direction cosines of the movable frame axes ($\alpha_{jk} = \mathbf{i}'_j \cdot \mathbf{i}_k$) will be of the form

$$[\alpha_{jk}] = \begin{bmatrix} \cos\varphi\cos\psi - \cos\theta\sin\varphi\sin\psi & \cos\varphi\sin\psi + \cos\theta\sin\varphi\cos\psi & \sin\theta\sin\varphi \\ -\sin\varphi\cos\psi - \cos\theta\cos\varphi\sin\psi & -\sin\varphi\sin\psi + \cos\theta\cos\varphi\cos\psi & \sin\theta\cos\varphi \\ \sin\theta\sin\psi & -\sin\theta\cos\psi & \cos\theta \end{bmatrix},$$

(3.2.11''')

corresponding to a linear representation of the rotation group SO(3). Thus, the position of a rigid solid is univocally determined by six independent parameters (the co-ordinates x_1^0, x_2^0 and x_3^0 of the pole O and Euler's angles ψ, θ and φ); hence, *the rigid solid has six degrees of freedom*.

We can state

Theorem 3.2.1. *Any free non-deformable mechanical system* (*without external constraints*) *has six degrees of freedom* (*excepting the cases of one particle* (*three degrees*) *or of two particles* (*five degrees*)).

2.2.4 Kinematic characterization of constraints

In an inertial frame of reference, the infinitesimal constraints are expressed by relations between real displacements and the time interval dt in which these displacements take place. We may write such a relation in the form

$$f(d\mathbf{r}_j; dt) \equiv f(d\mathbf{r}_1, d\mathbf{r}_2, ..., d\mathbf{r}_n; dt) = 0,$$ (3.2.12)

using real displacements, in the case of a discrete mechanical system $\mathscr{S}$. Developing into a power series, we have

$$f = f_0 + f_1 + f_2 + ... = 0,$$ (3.2.12')

where f_k, $k = 0,1,2,...$, are homogeneous polynomials of degree k in $dx_1^{(j)}$, $dx_2^{(j)}$, $dx_3^{(j)}$, $j = 1,2,...,n$, and dt (the dimensional homogeneity being ensured). If $f_0 \neq 0$, then f_k, $k = 1,2,...$, can be neglected with respect to f_0, so that we get $f_0 \equiv 0$; in this case, the constraint is *finite*. If $f_0 \equiv 0$ but $f_1 \neq 0$, then f_k, $k = 2,3,...$, can be neglected with respect to f_1, and one obtains

$$f_1 = 0.$$ (3.2.12'')

If $f_0 \equiv 0$, $f_1 \equiv 0,...$, $f_{j-1} \equiv 0$ and $f_j \neq 0$, then we may neglect f_k, $k = j+1, j+2,...$, with respect to f_j, so that the relation (3.2.12') becomes $f_j = 0$. In nature, we do not encounter constraints corresponding to $j > 1$, so that we will admit only constraints of the form (3.2.12''). In the case of a discrete mechanical system $\mathscr{S}$, we express these constraints with the aid of a *Pfaff differential form* as follows

$$\sum_{i=1}^{n} \boldsymbol{\alpha}_{ki} \cdot \mathrm{d}\mathbf{r}_i + \alpha_{k0}\mathrm{d}t = 0, \quad k = 1,2,...,m, \tag{3.2.13}$$

assuming that we have m constraints; by the above reasoning, the coefficients $\boldsymbol{\alpha}_{ki}$ and α_{k0} cannot depend on the velocities, so that $\boldsymbol{\alpha}_{ki} = \boldsymbol{\alpha}_{ki}(\mathbf{r}_j;t)$, $\alpha_{k0} = \alpha_{k0}(\mathbf{r}_j;t)$, being functions of class C^1. We notice that the relations (3.2.13) can be expressed in the form

$$\sum_{i=1}^{n} \boldsymbol{\alpha}_{ki} \cdot \mathbf{v}_i + \alpha_{k0} = 0, \quad k = 1,2,...,m, \tag{3.2.13'}$$

too, by introducing the real velocities $\mathbf{v}_i = \mathrm{d}\mathbf{r}_i/\mathrm{d}t$. We put thus in evidence the *kinematic nature* of these constraints; the mechanical system $\mathscr{S}$ cannot have arbitrary velocities, but only those allowed by the velocity restrictions (3.2.13'). We admit that the constraints of kinematic nature are distinct, as well as these of geometric nature. In this case too, we must have $m < 3n$, so that the motion be possible; the difference $r = 3n - m$ constitutes *the number of kinematic degrees of freedom of the mechanical system $\mathscr{S}$*. This number vanishes if $m = 3n$, and the velocities of the n particles may be determined (eventually, not univocally), by solving a system of $3n$ equations with $3n$ unknowns (the components of the velocities of the n particles). If $m > 3n$, then the constraints are no more distinct, what was excluded from the very beginning. As in the case of finite constraints, the constraints considered above are bilateral ones; the unilateral constraints of a kinematic nature of the mechanical system $\mathscr{S}$ may be expressed in the form

$$\sum_{i=1}^{n} \boldsymbol{\alpha}_{ki} \cdot \mathrm{d}\mathbf{r}_i + \alpha_{k0}\mathrm{d}t \geq 0, \quad k = 1,2,...,m, \tag{3.2.14}$$

or in the form

$$\sum_{i=1}^{n} \boldsymbol{\alpha}_{ki} \cdot \mathbf{v}_i + \alpha_{k0} \geq 0, \quad k = 1,2,...,m. \tag{3.2.14'}$$

If the equality cannot take place for one of the constraints, then that one is called a *strict constraint*; but the condition $m < 3n$ must still hold.

Passing to the representative space E_{3n}, we use the notation (3.2.9), as well as

$$\alpha_{k0} = b_{k0}, \quad \boldsymbol{\alpha}_{ki} = \alpha_{kl}^{(i)}\mathbf{i}_l, \quad \alpha_{kl}^{(i)} = b_{k,3(i-1)+l}, \quad i = 1,2,...,n, \\ l = 1,2,3, \quad k = 1,2,...,m. \tag{3.2.9'}$$

The representative point P verifies the conditions

$$\sum_{j=1}^{3n} b_{kj}\mathrm{d}X_j + b_{k0}\mathrm{d}t = 0, \quad k = 1,2,...,m, \tag{3.2.13''}$$

or the conditions

$$\sum_{j=1}^{3n} b_{kj} V_j + b_{k0} = 0, \quad k = 1, 2, ..., m, \qquad (3.2.13''')$$

where dX_j are *the generalized real displacements*, while $V_j = dX_j/dt$ are the components of the velocity of the point P (*the generalized velocity*) in the representative space E_{3n}, in the case of bilateral constraints. The unilateral constraints (3.2.14), (3.2.14') are expressed in the form

$$\sum_{j=1}^{3n} b_{kj} dX_j + b_{k0} dt \geq 0, \quad k = 1, 2, ..., m, \qquad (3.2.14'')$$

or in the form

$$\sum_{j=1}^{3n} b_{kj} V_j + b_{k0} \geq 0, \quad k = 1, 2, ..., m. \qquad (3.2.14''')$$

Let $\Delta'\mathbf{r}_i$ and $\Delta''\mathbf{r}_i$ be two possible displacements which take place in the time interval Δt; the constraint relations may be written in the form

$$\sum_{i=1}^{n} \boldsymbol{\alpha}_{ki} \cdot \Delta'\mathbf{r}_i + \alpha_{k0} dt = 0, \quad \sum_{i=1}^{n} \boldsymbol{\alpha}_{ki} \cdot \Delta''\mathbf{r}_i + \alpha_{k0} dt = 0, \quad k = 1, 2, ..., m.$$

Subtracting and taking into account the relation (3.2.2) which defines the virtual displacements, we state that the latter ones verify the relations

$$\sum_{i=1}^{n} \boldsymbol{\alpha}_{ki} \cdot \delta \mathbf{r}_i = 0, \quad k = 1, 2, ..., m. \qquad (3.2.15)$$

Comparing with relations (3.2.13), we see once more that the virtual displacements correspond to a certain moment t and don't take place in time. Introducing the virtual velocities (3.2.1'), one may write these conditions also in the form

$$\sum_{i=1}^{n} \boldsymbol{\alpha}_{ki} \cdot \mathbf{v}_i^* = 0, \quad k = 1, 2, ..., m. \qquad (3.2.15')$$

If we start from possible displacements which satisfy unilateral constraints

$$\sum_{i=1}^{n} \boldsymbol{\alpha}_{ki} \cdot \Delta'\mathbf{r}_i + \alpha_{k0} dt \geq 0, \quad \sum_{i=1}^{n} \boldsymbol{\alpha}_{ki} \cdot \Delta''\mathbf{r}_i + \alpha_{k0} dt \geq 0, \quad k = 1, 2, ..., m,$$

we see that the virtual displacements do not satisfy any constraint relation (by subtraction, one cannot obtain any conclusion concerning the inequalities), being

Mass geometry. Displacements. Constraints 175

reversible. If in the relations concerning the possible displacements $\Delta''\mathbf{r}_i$ there is only the sign "equal" (the constraints are bilateral), then the virtual displacements which are obtained verify the relations

$$\sum_{i=1}^{n} \boldsymbol{\alpha}_{ki} \cdot \delta\mathbf{r}_i \geq 0, \quad k = 1, 2, \ldots, m, \qquad (3.2.16)$$

corresponding to unilateral constraints; one observes thus that, only in such a case, one may obtain unilateral constraints for virtual displacements which are irreversible. Analogously, one can express such constraints also in the form

$$\sum_{i=1}^{n} \boldsymbol{\alpha}_{ki} \cdot \mathbf{v}_i^* \geq 0, \quad k = 1, 2, \ldots, m. \qquad (3.2.16')$$

If δX_j are *generalized virtual displacements* (obtained as differences of *generalized possible displacements* $\Delta' X_j$ and $\Delta'' X_j$), we may write, for the representative point P in the space E_{3n}, the constraint relations in the form (bilateral constraints)

$$\sum_{j=1}^{3n} b_{kj} \delta X_j = 0, \quad k = 1, 2, \ldots, m, \qquad (3.2.15'')$$

or in the form

$$\sum_{j=1}^{3n} b_{kj} V_j^* = 0, \quad k = 1, 2, \ldots, m, \qquad (3.2.15''')$$

where V_j^* are *generalized virtual velocities* (which can be introduced as differences of *generalized possible velocities* $\overline{V}_j'$ and $\overline{V}_j''$); in the case of unilateral constraints (obtained as we have seen above), we use the relations

$$\sum_{j=1}^{3n} b_{kj} \delta X_j \geq 0, \quad k = 1, 2, \ldots, m, \qquad (3.2.16'')$$

or the relations

$$\sum_{j=1}^{3n} b_{kj} V_j^* \geq 0, \quad k = 1, 2, \ldots, m. \qquad (3.2.16''')$$

2.2.5 Case of a particle subjected to finite constraints

Let be the case of a single particle P, of position vector $\mathbf{r}$ and co-ordinates x_1, x_2, x_3, subjected to finite constraints

$$f_k(\mathbf{r}; t) \equiv f_k(x_1, x_2, x_3; t) = 0, \quad k = 1, 2; \qquad (3.2.17)$$

hence, the particle can be on a surface (we have only $k = 1$) or on a curve (at the intersection of two surfaces). A total differentiation with respect to time leads to conditions imposed to the velocity of the particle, in the form

$$\frac{\mathrm{d} f_k}{\mathrm{d} t} = \mathrm{grad}\, f_k \cdot \mathbf{v} + \dot f_k = f_{k,j} v_j + \dot f_k = 0\,, \quad k = 1,2\,; \qquad (3.2.18)$$

it results

$$\mathbf{v} = -\frac{\dot f_k}{|\mathrm{grad}\, f_k|^2} \mathrm{grad}\, f_k + \mathbf{c}_k\,, \quad \mathbf{c}_k \perp \mathrm{grad}\, f_k\,, \quad k = 1,2\,. \qquad (3.2.18')$$

The component of the velocity along the gradient (along the normal to the corresponding surface) is given by

$$v_g^k = -\frac{\dot f_k}{|\mathrm{grad}\, f_k|}\,, \quad k = 1,2\,, \qquad (3.2.18'')$$

while its component in a plane normal to the gradient is arbitrary; one obtains thus two components of the velocity. The total derivative of the relation (3.2.18) with respect to time leads to conditions imposed to the acceleration of the particle, given by

$$\frac{\mathrm{d}^2 f_k}{\mathrm{d} t^2} = \mathrm{grad}\, f_k \cdot \mathbf{a} + \mathrm{D}_2 f_k = 0\,, \quad k = 1,2\,, \qquad (3.2.19)$$

where we have introduced the notation

$$\mathrm{D}_2 f_k = f_{k,jl} v_j v_l + \dot f_{k,j} v_j + \ddot f_k\,; \qquad (3.2.19')$$

we obtain thus the acceleration

$$\mathbf{a} = -\frac{\mathrm{D}_2 f_k}{|\mathrm{grad}\, f_k|^2} \mathrm{grad}\, f_k + \overline{\mathbf{c}}_k\,, \quad \overline{\mathbf{c}}_k \perp \mathrm{grad}\, f_k\,, \quad k = 1,2\,. \qquad (3.2.19'')$$

Analogously, the component of the acceleration along the gradient is

$$a_g^k = -\frac{\mathrm{D}_2 f_k}{|\mathrm{grad}\, f_k|}\,, \quad k = 1,2\,, \qquad (3.2.19''')$$

while its component in a plane normal to the gradient is arbitrary; we obtain thus two components of the acceleration.

In the case of a unilateral constraint

$$f_k(\mathbf{r};t) \geq 0\,, \quad k = 1,2\,, \qquad (3.2.20)$$

Mass geometry. Displacements. Constraints 177

the velocity and the acceleration are not subjected, in general, to any condition. If at a moment t the constraint is bilateral ($f_k(t) = 0$), and than it becomes strict ($f_k(t + \Delta t) > 0$, $\Delta t > 0$), then we can use a development into a Taylor series

$$f_k(t + \Delta t) = f_k(t) + \frac{1}{1!}\frac{df_k}{dt}\Delta t + \frac{1}{2!}\frac{d^2 f_k}{dt^2}(\Delta t)^2 + ...;$$

in the frame of this hypothesis, we obtain, for $\Delta t \to 0$, the conditions $df_k/dt \geq 0$, which may be expressed also in the form

$$\operatorname{grad} f_k \cdot \mathbf{v} + \dot{f}_k \geq 0, \quad k = 1,2. \tag{3.2.20'}$$

If the constraint is strict, then it acts only on the velocity by which the particle leaves the surface $f_k = 0$, and not on the velocity by which it reaches the surface (because we have admitted that $\Delta t > 0$); the acceleration of the particle remains arbitrary. If we have $df_k/dt = 0$ at a moment t, then – by an analogous reasoning – we obtain the constraints

$$\frac{d^2 f_k}{dt^2} = \operatorname{grad} f_k \cdot \mathbf{a} + D_2 f_k \geq 0, \quad k = 1,2. \tag{3.2.20''}$$

2.2.6 Holonomic and non-holonomic constraints

If the Pfaff differential form (3.2.13) is integrable, hence if the first member of the relation (3.2.13") is a total differential with respect to the variables X_j and t, then the constraints have been denominated *holonomic* by Hertz; otherwise, they are called *non-holonomic*. We see that the holonomic constraints expressed in the form (3.2.13) are not infinitesimal, but finite ones, and may be represented in the form (3.2.8); thus, a holonomic mechanical system is a system with finite constraints or with infinitesimal integrable constraints (which – in fact – are finite constraints).

The finite holonomic constraints (3.2.8) may be expressed also by means of infinitesimal displacements in the form (relations of the form (3.2.13)).

$$df_l = \sum_{i=1}^{n} \operatorname{grad}_i f_l \cdot d\mathbf{r}_i + \dot{f}_l dt = \sum_{i=1}^{n} \nabla_i f_l \cdot d\mathbf{r}_i + \dot{f}_l dt = 0, \tag{3.2.21}$$

wherefrom

$$\boldsymbol{\alpha}_{li} = \nabla_i f_l = \frac{\partial f_l}{\partial x_j^{(i)}} \mathbf{i}_j, \quad i = 1,2,...,n, \quad \alpha_{l0} = \dot{f}_l, \quad l = 1,2,...,p. \tag{3.2.21'}$$

The virtual displacements must verify the relations

$$\sum_{i=1}^{n} \nabla_i f_l \cdot \delta \mathbf{r}_i = 0, \quad l = 1,2,...,p. \tag{3.2.21''}$$

Let be two neighbouring, possible, simultaneous (at the moment t) positions $\mathbf{r}_i$ and $\mathbf{r}_i + \delta\mathbf{r}_i$, $i = 1,2,...,n$, of the same holonomic discrete mechanical system. The constraint relations are of the form $f_l(\mathbf{r}_j) = 0$, $f_l(\mathbf{r}_j + \delta\mathbf{r}_j) = 0$, $l = 1,2,...,p$. Developing into a Taylor series, we may write

$$f_l(\mathbf{r}_j) + \sum_{i=1}^{n} \nabla_i f_l \cdot \delta\mathbf{r}_i + ... = 0;$$

neglecting the terms of higher order and taking into account the constraint relations, one obtains the conditions (3.2.21"). One can see once more that $\delta\mathbf{r}_i$ represent the differential displacements which must be effected by the particles of the mechanical system to pass from a position to another one, at the same moment t, being thus virtual displacements, corresponding to the relation of definition (3.2.2).

The relations (3.2.21) lead to the conditions

$$\sum_{i=1}^{n} \nabla_i f_l \cdot \mathbf{v}_i + \dot{f}_l = 0, \quad l = 1,2,...,p, \tag{3.2.21'''}$$

which must be verified by the velocities of the particles of the system; obviously, in these relations only the components of the velocities along the corresponding gradients of the constraints are involved. For instance, let be the particles P_i and P_j at an invariable mutual distance (or two points of a rigid solid); the relation (see Fig.1.18)

$$(\mathbf{r}_j - \mathbf{r}_i)^2 = r_{ij}^2 = \text{const} \tag{3.2.22}$$

takes place. Differentiating with respect to time (or applying the formula (3.2.21''')), we get

$$\mathbf{r}_{ij} \cdot \mathbf{v}_i = \mathbf{r}_{ij} \cdot \mathbf{v}_j; \tag{3.2.22'}$$

hence, the projections of the velocities of two points of a non-deformable mechanical system along the straight line defined by them are equal.

In what concerns the conditions imposed to the accelerations, we obtain

$$\frac{d^2 f_l}{dt^2} = \sum_{i=1}^{n} \nabla_i f_l \cdot \mathbf{a}_i + D_2 f_l = 0, \quad l = 1,2,...,p, \tag{3.2.23}$$

where

$$D_2 f_l = \sum_{i=1}^{n} \sum_{j=1}^{n} f_{l,ij} v_i v_j + \sum_{i=1}^{n} \dot{f}_{l,i} v_i + \ddot{f}_l, \quad l = 1,2,...,p. \tag{3.2.23'}$$

Thus, in the case of the non-deformable two points system considered above, we may write

$$\mathbf{r}_{ij} \cdot (\mathbf{a}_j - \mathbf{a}_i) + (\mathbf{v}_j - \mathbf{v}_i)^2 = 0. \qquad (3.2.23'')$$

We notice that, in the case of holonomic constraints, the number of degrees of freedom s given by the finite displacements is equal to the number of degrees of freedom r given by the infinitesimal displacements ($s = r$).

The non-holonomic constraints are expressed by relations of the form (3.2.13'), which represent conditions which must be verified by the real velocities. The total derivative with respect to time puts into evidence the conditions which are to be verified by the accelerations of the non-holonomic discrete mechanical system, i.e.,

$$\sum_{i=1}^{n} \boldsymbol{\alpha}_{ki} \cdot \mathbf{a}_i + \sum_{i=1}^{n} \frac{\mathrm{d}\boldsymbol{\alpha}_{ki}}{\mathrm{d}t} \cdot \mathbf{v}_i + \frac{\mathrm{d}\alpha_{k0}}{\mathrm{d}t} = 0, \quad k = 1, 2, \ldots, m; \qquad (3.2.24)$$

as in the case of the velocities, only the components of the accelerations along the parameters $\boldsymbol{\alpha}_{ki}$ are involved in this case too.

One can make analogous considerations in the representative space E_{3n}. Let be

$$\omega_k = \sum_{j=1}^{3n} b_{kj}\,\mathrm{d}X_j + b_{k0}\,\mathrm{d}t, \quad k = 1, 2, \ldots, m, \quad m < 3n, \qquad (3.2.25)$$

a differential form of the first degree, corresponding to the constraint relations (3.2.13''); these relations lead to a differential equation of the form $\omega_k = 0$. The form $\omega = \omega_k$ (for a fixed k) is (locally) integrable if there exist two functions $f \neq 0$ and g so that $\omega = f\mathrm{d}g$; the problem is thus reduced to the existence of an *integrating factor* for the considered differential equation. One may prove that the functions f and g do exist (in a sufficiently small neighbourhood) if and only if there exists a form of the first degree θ, so that

$$\mathrm{d}\omega = \theta \wedge \omega, \quad \omega \wedge \mathrm{d}\omega = 0, \qquad (3.2.26)$$

where we used the external product defined in the App., Subsec. 1.2.1. We introduce the matrices

$$\boldsymbol{\Omega} = \begin{bmatrix} \omega_1 \\ \omega_2 \\ \vdots \\ \omega_m \end{bmatrix}, \quad \mathbf{F} = \begin{bmatrix} f_{11} & f_{12} & \cdots & f_{1m} \\ f_{21} & f_{22} & \cdots & f_{2m} \\ \vdots & \vdots & \cdots & \vdots \\ f_{m1} & f_{m2} & \cdots & f_{mm} \end{bmatrix}, \quad \mathbf{G} = \begin{bmatrix} g_1 \\ g_2 \\ \vdots \\ g_m \end{bmatrix},$$

$$\boldsymbol{\Theta} = \begin{bmatrix} \theta_{11} & \theta_{12} & \cdots & \theta_{1m} \\ \theta_{21} & \theta_{22} & \cdots & \theta_{2m} \\ \vdots & \vdots & \cdots & \vdots \\ \theta_{m1} & \theta_{m2} & \cdots & \theta_{mm} \end{bmatrix}, \qquad (3.2.27)$$

where θ_{kj} are forms of the first degree; the system of differential equations $\omega_k = 0$, $k = 1,2,...,m$, is completely integrable if there exist the functions f_{kj} and g_j, so that

$$\omega_k = \sum_{j=1}^{m} f_{kj} \mathrm{d} g_j \qquad (3.2.28)$$

or

$$\mathbf{\Omega} = \mathbf{F}\mathrm{d}\mathbf{G}, \qquad (3.2.28')$$

F being a non-singular matrix. One proves
Theorem 3.2.2 (*Frobenius*). *The system of forms* (3.2.28) *is completely integrable if and only if there exists a system of forms of the first degree* θ_{kj}, *so that*

$$\mathrm{d}\mathbf{\Omega} = \mathbf{\Theta} \wedge \mathbf{\Omega}, \quad \mathrm{d}\omega_k = \sum_{j=1}^{m} \theta_{kj} \wedge \omega_j, \quad k = 1,2,...,m. \qquad (3.2.29)$$

The conditions (3.2.29) may be expressed also in an equivalent form (particularly convenient for applications; the condition (3.2.26) is a particular case)

$$\omega_1 \wedge \omega_2 \wedge ... \wedge \omega_m \wedge \mathrm{d}\omega_k = 0, \quad k = 1,2,...,m. \qquad (3.2.29')$$

In the particular case of the form

$$\omega = a_1(x_1,x_2,x_3)\mathrm{d}x_1 + a_2(x_1,x_2,x_3)\mathrm{d}x_2 + a_3(x_1,x_2,x_3)\mathrm{d}x_3, \qquad (3.2.30)$$

the condition of integrability becomes

$$\epsilon_{ijk}\, a_i a_{k,j} \mathrm{d}x_1 \wedge \mathrm{d}x_2 \wedge \mathrm{d}x_3 = 0. \qquad (3.2.30')$$

Indeed, the equation $\omega = 0$ is integrable if there exists an integrating factor $\lambda = \lambda(x_1,x_2,x_3)$, so that $\lambda \mathbf{a}$ be a gradient, hence so that $\mathrm{curl}(\lambda \mathbf{a}) = \lambda\, \mathrm{curl}\,\mathbf{a} + \mathrm{grad}\,\lambda \times \mathbf{a} = \mathbf{0}$; a scalar product by **a** leads to

$$\mathbf{a} \cdot \mathrm{curl}\,\mathbf{a} = \epsilon_{ijk}\, a_i a_{k,j} = 0. \qquad (3.2.30'')$$

By analogy to the considerations of Subsec. 2.2.2 for the holonomic case, we can state that the relations (3.2.13''') represent the conditions for the representative point P of a non-holonomic mechanical system to be at the intersection of m non-holonomic hypersurfaces (hypersurfaces studied by Gh. Vrănceanu) in the representative space E_{3n}, hence on a non-holonomic manifold of dimension $r = 3n - m$ (equal to the number of degrees of kinematic freedom) of this space.

In general, a mechanical system may be subjected to p holonomic and to m non-holonomic constraints; in the case of a discrete mechanical system we must have

$p + m < 3n$. The number of degrees of freedom will be $3n - (p + m)$ (obviously, kinematic degrees of freedom, because the holonomic constraints have the same number of geometric and kinematic degrees of freedom). In the case of a non-deformable mechanical system, one must have $p + m < 6$, and the number of degrees of freedom will be $6 - (p + m)$, the holonomic and non-holonomic constraints being – obviously – external.

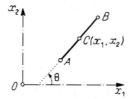

Figure 3.25. Motion of a rigid skate AB.

Let us consider, for instance, the motion of a rigid skate AB on the ice plane, considered to be the plane Ox_1x_2; let C be the middle of the segment AB (the theoretic point of contact between the curved sole of the skate and the ice). Obviously, the position of this segment is given by the co-ordinates x_1, x_2 of the point C and by the angle θ made by the segment AB with the axis Ox_1 (Fig.3.25). The parameters x_1, x_2 and θ are independent at a moment t, hence the skate has three degrees of freedom in finite displacements. We notice that the trajectory of the point C is tangent to AB; because the displacement of the point C (as well as its velocity) can take place only along the direction AB (to avoid a skipping), we may write the relation

$$\dot{x}_2 = \dot{x}_1 \tan \theta, \qquad (3.2.31)$$

linking the components of the velocity of the point C. Starting from the form

$$\omega = \sin\theta dx_1 - \cos\theta dx_2 \qquad (3.2.31')$$

and using the formula (A.1.55), we can write

$$d\omega = \cos\theta d\theta \wedge dx_1 + \sin\theta d\theta \wedge dx_2; \qquad (3.2.31'')$$

hence, with the aid of the properties emphasized in App., Subsec. 1.2.1, one obtains

$$\omega \wedge d\omega = -\sin^2\theta dx_1 \wedge dx_2 \wedge d\theta - \cos^2\theta dx_1 \wedge dx_2 \wedge d\theta$$
$$= -dx_1 \wedge dx_2 \wedge d\theta \neq 0, \qquad (3.2.31''')$$

so that the constraint (3.2.31) is non-integrable, and the considered mechanical system is non-holonomic. We may also suppose that the plane Ox_1x_2 is inclined with respect to a horizontal one, taking the axis Ox_1 along the direction of maximal inclination.

A sphere or a plane disk (in a vertical plane), which is rolling without sliding on a fixed horizontal plane may be other examples of mechanical systems subjected to non-holonomic constraints. We notice that in such a case the number s of degrees of freedom given by the finite displacements is greater than the number r of degrees of freedom given by the infinitesimal ones ($s > r$).

The constraints considered above are bilateral ones; in the case of unilateral constraints, one can make analogous considerations, using – for instance – relations of the form (3.2.16) or of the form

$$\sum_{i=1}^{n} \nabla_i f_l \cdot \delta \mathbf{r}_i \geq 0, \quad l = 1, 2, ..., p. \tag{3.2.16^{iv}}$$

2.2.7 Scleronomic and rheonomic constraints. Catastatic constraints

If the temporal variable t does not appear explicitly in the infinitesimal constraints (3.2.13) or (3.2.13') (*stationary case*), hence if

$$\dot{\boldsymbol{\alpha}}_{ki} = \frac{\partial \boldsymbol{\alpha}_{ki}}{\partial t} = \mathbf{0}, \quad \alpha_{k0} = 0, \quad i = 1,2,...,n, \quad k = 1,2,...,m, \tag{3.2.32}$$

then these constraints are called *scleronomic*; otherwise they are called *rheonomic* (*non-stationary case*). These conditions become

$$\dot{b}_{kj} = \frac{\partial b_{kj}}{\partial t} = 0, \quad b_{k0} = 0, \quad j = 1,2,...,3n, \quad k = 1,2,...,m, \tag{3.2.32'}$$

in the representative space E_{3n}.

Taking into account (3.2.21'), we observe that it is sufficient to have

$$\dot{f}_l = 0, \quad l = 1, 2, ..., p, \tag{3.2.33}$$

so that these holonomic constraints be scleronomic too; they must be of the form

$$f_l(\mathbf{r}_j) = 0, \quad l = 1,2,...,p. \tag{3.2.33'}$$

We may also write

$$\sum_{i=1}^{n} \nabla_i f_l \cdot \mathrm{d}\mathbf{r}_i = 0, \quad l = 1,2,...,p, \tag{3.2.33''}$$

as well as a relation of the form (3.2.21").

We notice that a constraint of the form (3.2.33') represents a fixed hypersurface; but a rheonomic constraint of the form (3.2.8) may represent a rigid surface, moving with respect to a rigid frame of reference, or a deformable surface.

We mention that the two classifications of the constraints (holonomic or non-holonomic and scleronomic or rheonomic) are independent; we may have, for instance, non-holonomic, rheonomic constraints (in the most general case) and holonomic, scleronomic constraints (in the most particular case).

Mass geometry. Displacements. Constraints 183

Taking into account the relations (3.2.15) verified by the virtual displacements, we notice that – in the case of scleronomic constraints – the set of virtual displacements coincides with the set of possible displacements; as well, in this case the real displacements belong to the set of virtual displacements. But these properties take place also if one has only

$$\alpha_{k0} = b_{k0} = 0, \quad k = 1,2,...,m. \tag{3.2.32''}$$

These conditions are sufficient; it is not necessary that the constraints be scleronomic. The constraints for which conditions of the form (3.2.32'') hold are called *catastatic constraints*. These conditions do not impose with necessity $\dot{\alpha}_{kl} = 0$ (or $\dot{b}_{kj} = 0$) if the constraints are non-holonomic; holonomic, rheonomic constraints which have these properties do not exist (in this case, the above condition holds too).

One can make analogous considerations in the case of unilateral constraints.

As examples of mechanical systems subjected to holonomic, scleronomic constraints one can mention the rigid solid (or a non-deformable discrete mechanical system), the points of which verify conditions of the form (3.2.22), the particle being constrained to stay on a fixed curve or surface etc. If the curve (surface) is moving, then the constraint is holonomic and rheonomic (for instance, a constraint of the form (3.2.10)); in general, in the case of a relative motion we have to do – in fact – just with such a constraint. The rigid solid with a fixed point or axis represents a holonomic and scleronomic mechanical system. We may consider also systems of rigid solids, subjected to mutual constraints (internal constraints), as well as to various external constraints. In a certain manner, each rigid solid behaves in the respective mechanical system as a particle, which has not three but six degrees of proper freedom.

The non-holonomic constraints of the previous subsection (for instance, the constraint expressed by the relation (3.2.31)) are scleronomic constraints.

2.2.8 Critical constraints

If the number s of the degrees of freedom given by the finite displacements is less than the number r of the degrees of freedom given by the infinitesimal relations ($s < r$), then we have to do with *critical constraints*.

Let, for instance, be a mechanical system formed by two particles P_1 and P_2 linked by a rigid bar of length $2r$ and constraint to be on a circle of radius r (Fig.3.26,a). The constraint relations (holonomic and scleronomic) are of the form (in a fixed frame of reference Oxy)

$$x_1^2 + y_1^2 = x_2^2 + y_2^2 = r^2, \quad (x_1 - x_2)^2 + (y_1 - y_2)^2 = 4r^2. \tag{3.2.34}$$

Because the four co-ordinates must verify three finite independent relations, it follows that the mechanical system has one geometric degree of freedom ($s = 1$) (in finite displacements). Indeed, we see that these co-ordinates may be expressed by means of the angle θ, taken as an independent parameter, in the form

$$x_1 = -x_2 = r\cos\theta, \quad y_1 = -y_2 = r\sin\theta. \tag{3.2.34'}$$

By differentiation, one obtains

$$x_1 dx_1 + y_1 dy_1 = 0, \quad x_2 dx_2 + y_2 dy_2 = 0,$$
$$(x_1 - x_2)(dx_1 - dx_2) + (y_1 - y_2)(dy_1 - dy_2) = 0; \tag{3.2.34''}$$

taking into account (3.2.34'), we notice that the last relation (3.2.34") is a linear consequence of the first two relations. Between the four co-ordinates take thus place two differential relations, so that the considered mechanical system has two kinematic degrees of freedom ($r = 2$) (in infinitesimal displacements). The condition mentioned above ($1 < 2$) is thus fulfilled. One can put into evidence an infinitesimal rotation $d\theta$ (Fig.3.26,b), corresponding to a finite rotation θ, as well as an infinitesimal translation $d\lambda$ (Fig.3.26,c), which has not any correspondent in finite displacements.

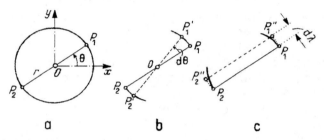

Figure 3.26. Critical system formed by two particles linked by a rigid bar and constrained to be on a circle (a). An infinitesimal rotation $d\theta$ (b) and an infinitesimal translation $d\lambda$ (c).

The fact that in case of critical constraints can take place infinitesimal displacements which do not have correspondence in finite displacements characterizes the respective constraints; indeed, this property corresponds to the above given definition.

2.2.9 Virtual work of constraint forces. Ideal constraints

We call *system of possible accelerations* of the particles P_i, $i = 1, 2, ..., n$, of the considered mechanical system $\mathscr{S}$ any system of accelerations $\mathbf{a}_i$ which satisfies, at a moment t, the relations (3.2.23), (3.2.23') and (3.2.24) (obtained from the holonomic and non-holonomic constraints, respectively, to which may be subjected the system $\mathscr{S}$), supposing that the position (the position vectors $\mathbf{r}_i$) and the velocities $\mathbf{v}_i$ of the system $\mathscr{S}$ verify the relations (3.2.8) and (3.2.13'), respectively. A system of accelerations which does not satisfy all these conditions is called a *system of impossible accelerations*. In the case of unilateral constraints analogous definitions can be given.

If the mechanical system $\mathscr{S}$ is free, then any particle P_i must satisfy an equation of the form

$$m_i \mathbf{a}_i = \mathbf{F}_i, \quad i = 1, 2, ..., n, \tag{3.2.35}$$

Mass geometry. Displacements. Constraints   185

corresponding to Newton's law (1.1.89), where $\mathbf{F}_i$ is the resultant of all the external and internal given *forces* which act upon the respective particle. In these conditions, if the accelerations $\mathbf{a}_i$ form a system of possible accelerations for the considered mechanical system $\mathcal{S}$ with constraints, then the constraint relations represent particular integrals of the equations of motion, having to do with a particular case of motion of this mechanical system. In the case of a system of impossible accelerations, one must introduce also *the constraint forces* $\mathbf{R}_i$, unknown a priori and acting upon the particles P_i, $i = 1, 2, ..., n$; the equations of motion become

$$m_i \mathbf{a}_i = \mathbf{F}_i + \mathbf{R}_i, \quad i = 1, 2, ..., n. \tag{3.2.35'}$$

To obtain possible accelerations, one must determine the constraint forces correspondingly; the mechanical system $\mathcal{S}$ becomes thus a free one, subjected to both given and constraint forces; in fact, this corresponds to the axiom of liberation from constraints.

We notice that one must determine $6n$ unknowns ($3n$ co-ordinates of the particles of the system $\mathcal{S}$ and $3n$ components of the constraint forces), with the aid of $3n$ scalar equations of motion (projections of the equations (3.2.35')) and $p + m$ constraint relations (3.2.8), (3.2.13'); there are still necessary $6n - (3n + p + m) = 3n - (p + m)$ scalar equations (a number of relations equal to the number of degrees of freedom) to solve the problem. To do this, we introduce an important class of constraints: *the class of ideal constraints*. Thus, we call ideal constraints these ones for which *the virtual work of the constraint forces*, given by (3.2.7'), vanishes

$$\sum_{i=1}^{n} \mathbf{R}_i \cdot \delta \mathbf{r}_i = 0 \tag{3.2.36}$$

for any system of virtual displacements of the considered mechanical system $\mathcal{S}$.

If the virtual displacements $\delta \mathbf{r}_i$ are arbitrary (we have not constraints), the relation (3.2.36) holds only if $\mathbf{R}_i = \mathbf{0}$, $i = 1, 2, ..., n$. Indeed, because the virtual displacements are arbitrary, we may equate them to zero, unless one, let be $\delta \mathbf{r}_j$; the relation (3.2.36) is reduced to $\mathbf{R}_j \cdot \delta \mathbf{r}_j = 0$. But the direction of $\delta \mathbf{r}_j$ is arbitrary, so that $\mathbf{R}_j = \mathbf{0}$; taking $j = 1, 2, ..., n$, one is led to the above conclusion. In this case, the system of equations of motion is sufficient to solve the problem.

But if p constraints of the form (3.2.21") and m constraints of the form (3.2.15) take place, we will use *the method of Lagrange's multipliers*. Starting from the relation (3.2.36) and with the aid of the relations (3.2.21") and (3.2.15), we may write

$$\sum_{i=1}^{n} \left(\mathbf{R}_i - \sum_{l=1}^{p} \lambda_l \nabla_i f_l - \sum_{k=1}^{m} \mu_k \boldsymbol{\alpha}_{ki} \right) \cdot \delta \mathbf{r}_i = 0,$$

where λ_l, $l = 1,2,...,p$, and μ_k, $k = 1,2,...,m$, are non-determined scalars (*Lagrange's multipliers*) and where we took into account that in a double finite sum one can invert the summation order; in components, we have

$$\sum_{i=1}^{n}\left(R_j^{(i)} - \sum_{l=1}^{p}\lambda_l\frac{\partial f_l}{\partial x_j^{(i)}} - \sum_{k=1}^{m}\mu_k\alpha_{kj}^{(i)}\right)\delta x_j^{(i)} = 0.$$

We introduce the notations (3.2.9) and we put thus into evidence $3n$ virtual displacements δX_j, $j = 1,2,...,3n$, which verify $p + m$ linear and distinct constraint relations (3.2.21''), (3.2.15), the matrix of the coefficients being of rank $p + m$ (otherwise the constraint relations could not be distinct). We express the virtual displacements $\delta X_1, \delta X_2,...,\delta X_{p+m}$ (we always may choose them so as the determinant Δ_{p+m} of the respective coefficients of the constraint relations be non-zero) with the aid of the other $3n - (p + m)$ virtual displacements, so that the latter ones may be considered as being independent. If we equate to zero the parentheses multiplying the first $p + m$ virtual displacements, then $p + m$ multipliers λ_l and μ_k are univocally determined (these multipliers are given by a system of $p + m$ linear algebraic equations with $p + m$ unknowns, of determinant $\Delta_{p+m} \neq 0$). As in the case previously considered, the independent virtual displacements $\delta X_{p+m+1}, \delta X_{p+m+2},...,\delta X_{3n}$ may all vanish, excepting only one, denoted by δX_j; because this non-zero displacement is arbitrary, it follows that the parenthesis multiplying it must vanish. If successively $j = p + m + 1, p + m + 2,...,3n$, and if we take into account the previous result, then all the parentheses multiplying the virtual displacements must vanish; one obtains thus the $3n - (p + m)$ supplementary relations searched to may solve the problem, so that the supplementary condition (3.2.36) introduced for the ideal constraints is sufficient. Finally, we may write

$$\mathbf{R}_i = \sum_{l=1}^{p}\lambda_l\nabla_i f_l + \sum_{k=1}^{m}\mu_k\boldsymbol{\alpha}_{ki}, \quad i = 1,2,...,n. \tag{3.2.37}$$

The motion of the mechanical system $\mathscr{S}$ as well as the constraint forces are completely determined.

By means of the expression (3.2.37) of the constraint forces and of the constraint relations (3.2.21), (3.2.13), we may write *the real elementary work of the constraint forces* in the form

$$\mathrm{d}W_R = -\sum_{l=1}^{p}\lambda_l\dot{f}_l\mathrm{d}t - \sum_{k=1}^{m}\mu_k\alpha_{k0}\mathrm{d}t. \tag{3.2.37'}$$

If $\dot{f}_l = 0$, $l = 1,2,...,p$ and $\alpha_{k0} = 0$, $k = 1,2,...,m$ (which holds, in general, in the case of catastatic constraints or, in particular, in the case of scleronomic constraints, as

Mass geometry. Displacements. Constraints 187

we have seen in Subsec. 2.2.7), then the real elementary work of the constraint forces vanishes; this result is justified also because the real displacements belong to the set of virtual displacements, the relation (3.2.36) implying $dW_R = 0$.

If, in the case of unilateral constraints of the form (3.2.16^{iv}) or (3.2.16), the virtual work of the constraint forces given by the relations (3.2.37) verifies the inequality

$$\sum_{i=1}^{n} \mathbf{R}_i \cdot \delta \mathbf{r}_i \geq 0, \qquad (3.2.36')$$

then these constraints are ideal ones; in this case one can make similar considerations to the above ones too.

To put into evidence the importance of the class of ideal constraints, we will show that a great number of mechanical systems belongs to this class. Let us consider a particle P constrained to stay on a fixed curve C (Fig.3.20,a) or on a fixed surface S (Fig.3.20,b); if we assume that the constraint is without friction, then the constraint force $\mathbf{R}$ is normal to the curve or to the surface, respectively (a tangential component would correspond to a sliding friction). Because the virtual displacement $\delta \mathbf{r}$ takes place along the tangent to C or in a plane tangent to S, it follows that $\mathbf{R} \cdot \delta \mathbf{r} = 0$; taking into account (3.2.37), we notice that this relation is of the form (3.2.21"), corresponding to the condition to which is subjected a particle constrained to be on a fixed curve or surface, respectively. This affirmation holds also in the case of a movable or deformable curve or surface (non-stationary case), because the constraint force $\mathbf{R}$ and the virtual displacement $\delta \mathbf{r}$ correspond to a fixed moment t; we mention that, in this case, $\mathbf{R} \cdot \Delta \mathbf{r} \neq 0$, and the necessity to use virtual displacements instead of possible ones is put into evidence. The constraints considered above are *constraints of contact*.

In the case of *constraints at distance*, for instance in the case in which the distance between two particles P_i and P_j is a function only on time ($r_{ij} = r_{ij}(t)$), we have (Fig.1.18)

$$\delta r_{ij}^2 = \delta (\mathbf{r}_j - \mathbf{r}_i)^2 = 2(\mathbf{r}_j - \mathbf{r}_i) \cdot (\delta \mathbf{r}_j - \delta \mathbf{r}_i) = 2\mathbf{r}_{ij} \cdot (\delta \mathbf{r}_j - \delta \mathbf{r}_i) = 0,$$

because the virtual displacements do not take place in time; if the internal constraint forces are of the form $\mathbf{R}_{ij} = \lambda \mathbf{r}_{ij}$, $\mathbf{R}_{ji} = \lambda \mathbf{r}_{ji}$, $\mathbf{R}_{ij} + \mathbf{R}_{ij} = \mathbf{0}$, λ being an indeterminate scalar, then the virtual work is given by $\mathbf{R}_{ij} \cdot \delta \mathbf{r}_i + \mathbf{R}_{ji} \cdot \delta \mathbf{r}_j = 0$. In the particular case in which $r_{ij} = \mathrm{const}$, we can state that a non-deformable discrete mechanical system is subjected to ideal internal constraints.

The ideal constraints may be introduced axiomatically with the aid of the relations (3.2.36), (3.2.36') in the case of a continuous mechanical system too, where the constraint forces $\mathbf{R}_i$ are applied at the points P_i. As a consequence, a rigid solid is also subjected to ideal internal constraints. In what concerns the external constraints, we may consider various cases of such ones. Thus, a rigid solid with a fixed point leads to a constraint force $\mathbf{R}$ applied at the very same point of position vector $\mathbf{r}$, hence to $\mathbf{R} \cdot \delta \mathbf{r} = 0$ (because $\delta \mathbf{r} = \mathbf{0}$). In the case of two fixed points P and P' of position vectors **r** and **r'**, respectively (a rigid solid with a fixed axis), we may write, analogously, $\mathbf{R} \cdot \delta \mathbf{r} + \mathbf{R}' \cdot \delta \mathbf{r}' = 0$. As well, in the case of a rigid solid sliding without friction on a fixed or movable curve (or surface), constraint forces normal to the curve (or surface) arise at the contact points of the rigid solid, while the corresponding virtual displacements are along the tangent (or in the tangent plane); the condition of ideal constraints is thus verified. The rigid solid which is rolling or pivoting without sliding on a fixed surface (it is subjected to a rotation about an instantaneous axis of rotation parallel or normal, respectively, to the tangent plane at the contact point), enjoys the same property; we suppose in these cases that the rolling and pivoting friction, respectively, vanishes.

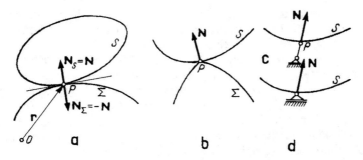

Figure 3.27. Two rigid solids S and Σ tangent along perfectly smooth surfaces (a). Case of an angular point (b). Representation of the simple support by a pendulum (c) or by a small cart (d).

Let be a mechanical system formed by two rigid solids S and Σ, constrained to remain tangent (one supposes that the surfaces in contact are perfectly smooth, the solids sliding one on the other); at the contact point $P(\mathbf{r})$ arise the constraint forces $\mathbf{N}_S = \mathbf{N}$ and $\mathbf{N}_\Sigma = -\mathbf{N}$ (on the basis of the principle of action and reaction, $\mathbf{N}_S + \mathbf{N}_\Sigma = \mathbf{0}$), which are normal to the considered surfaces (Fig.3.27,a). The relative velocity of the two rigid solids at P is $\mathbf{v}_\Sigma - \mathbf{v}_S$ (the difference between their velocities), so that the difference of two possible displacements (which are virtual displacements too) $\Delta \mathbf{r}_\Sigma - \Delta \mathbf{r}_S = (\mathbf{v}_\Sigma - \mathbf{v}_S)\Delta t$ lies in the common tangent plane; it follows that $\mathbf{N}_S \cdot \delta \mathbf{r}_S + \mathbf{N}_\Sigma \cdot \delta \mathbf{r}_\Sigma = \mathbf{N}_S \cdot \Delta \mathbf{r}_S + \mathbf{N}_\Sigma \cdot \Delta \mathbf{r}_\Sigma = \mathbf{N}_\Sigma \cdot (\Delta \mathbf{r}_\Sigma - \Delta \mathbf{r}_S)$ $= 0$. If the surfaces in contact are rough, the rigid solids rolling one on the other without sliding, then the relative velocity at the contact point vanishes ($\mathbf{v}_\Sigma - \mathbf{v}_S = \mathbf{0}$); we also get $\mathbf{R}_S \cdot \delta \mathbf{r}_S + \mathbf{R}_\Sigma \cdot \delta \mathbf{r}_\Sigma = \mathbf{R}_\Sigma (\Delta \mathbf{r}_\Sigma - \Delta \mathbf{r}_S) = \mathbf{R}_\Sigma \cdot (\mathbf{v}_\Sigma - \mathbf{v}_S)\Delta t = 0$, because the constraint forces verify the relation $\mathbf{R}_S + \mathbf{R}_\Sigma = \mathbf{0}$. Let be also the case in which the two rigid solids are linked by a hinge at $P(\mathbf{r})$ (Fig.3.28,a). If we neglect the frictions, then the action of the rigid solid Σ upon the rigid solid S is reduce to a force $\mathbf{R}_S$ applied at P, while the action of the rigid solid S upon the rigid solid Σ is reduced to a force $\mathbf{R}_\Sigma$, applied at P too; obviously, $\mathbf{R}_S + \mathbf{R}_\Sigma = \mathbf{0}$. It follows that

Mass geometry. Displacements. Constraints

$\mathbf{R}_S \cdot \delta \mathbf{r}_S + \mathbf{R}_\Sigma \cdot \delta \mathbf{r}_\Sigma = (\mathbf{R}_S + \mathbf{R}_\Sigma) \cdot \delta \mathbf{r} = 0$, because the point P has the same virtual displacement $\delta \mathbf{r}_S = \delta \mathbf{r}_\Sigma = \delta \mathbf{r}$, immaterial to which rigid solid it belongs.

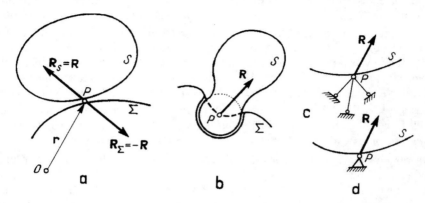

Figure 3.28. Two rigid solids S and Σ linked by a hinge (a). Spherical hinge (b) represented by three non-coplanar pendulums (c) or by an idealized fixed support (d).

In general, a *mechanism* is formed by a system of rigid solids, which are linked by hinges or by supports on perfectly smooth or rough surfaces; because there are not frictions, one may consider that this mechanical system is subjected to ideal constraints.

In the case in which arises also a sliding, a rolling or a pivoting friction, then one must introduce the corresponding components of the constraint forces and one must make supplementary hypotheses to allow the determination of these components; the mathematical model of the system of constraints must be completed. One may thus consider the motion with friction of a particle along a curve or a surface, the motion with friction of a rigid solid on another one etc.

We notice that the relation $\delta W_R = 0$, corresponding to ideal constraints, may be written also in the equivalent form

$$\sum_{i=1}^{n} \mathbf{R}_i \cdot \mathbf{v}_i^* = 0, \qquad (3.2.36'')$$

where we have put into evidence the virtual velocities of the points at which the constraint forces are applied.

2.2.10 Ideal constraints of the rigid solid

As we have seen, a free rigid solid has six degrees of freedom. If some external constraints appear, then the number s of these degrees of freedom becomes smaller (there is no more necessary the same number of independent parameters to specify the position of the rigid solid); otherwise, applying the axiom of liberation of constraints, one must determine the unknown constraint forces which are introduced, hence p unknown scalars. If in a given rigid solid problem one has $s + p = 6$, then this one is, in general, determinate (excepting some particular cases in which it could be indeterminate or impossible); we say that the respective mechanical system (in particular, the rigid solid) is *statically determinate* (*isostatic*). If $s + p > 6$, then the problem is indeterminate, the mechanical system being *statically indeterminate* (*hyperstatic*, with $s + p - 6$ *degrees of statical indeterminacy*). The unknowns of the problem cannot be determined in the case of a rigid solid, because of the limits of the mathematical model chosen for the solid; if we pass from a rigid to a deformable solid, closer to the physical reality, completing thus the considered mathematical model, then there appear supplementary relations which allow the complete solution of the problem. Finally, if $s + p < 6$, then the problem is impossible from the point of view of the rest with respect to a fixed frame of reference (in some particular cases, for special systems of given forces, the problem could be determinate); in this case, the mechanical system is a *mechanism* for which one has $6 - (s + p)$ *degrees of freedom*. In what follows, we pass in review some external constraints without friction, which are important in the case of a rigid solid.

Let S be a rigid solid, one of the points $P(\mathbf{r})$ of which is constrained to stay on a perfect smooth fixed surface Σ (we may suppose that this surface bounds another rigid solid which – for the sake of simplicity – will be denoted by Σ too); we say that the rigid solid has a *simple support* (*movable support*) at P (Fig.3.27,a). To state the position of the rigid solid, there are necessary only five scalar parameters (e.g., the co-ordinates u and v of the point P on the surface Σ and the three Euler's angles); hence, a simple support leaves out one degree of freedom of the rigid solid and can be replaced by a constraint force (a reaction) $\mathbf{N}$, normal to the surface Σ (as in the case of a particle constrained to stay on a given surface). If the surface Σ has at P a singular point (for instance, an angular point), then the direction of the constraint force is normal to the surface S bounding the rigid solid S (Fig.3.27,b); indeed, supposing that there are two rigid solids S and Σ, simply leaning one on the other, there arise two constraint forces $\mathbf{N}_S = \mathbf{N}$ and $\mathbf{N}_\Sigma = -\mathbf{N}$, in conformity to the principle of action and reaction, the force $\mathbf{N}_\Sigma$ being normal to the surface S (hence the force $\mathbf{N}_S$ too).

Besides the constraints at a contact surface-surface or surface-point considered above, we mention the constraints at a contact surface-curve, curve-curve, curve-point or point-point. As well, we can conceive the constraints on a curve in the contact surface-surface, surface-curve or curve-curve and the constraints on a surface in the contact surface-surface. One can make analogous considerations in all these cases.

The point of application as well as the support of the constraint force $\mathbf{N}$ are known; one must determine only its magnitude and its direction (a scalar unknown N, corresponding to a left out degree of liberty; the unknown N is obtained with the sign + or –, as the direction arbitrarily chosen at the beginning is or not the correct one); in general, the problem is thus determinate. Hence, to fix a rigid solid there are necessary six simple supports; the six degrees of freedom are thus vanishing and one must introduce six unknown constraint forces. A simple support may be graphically represented by a *pendulum*, indicating the direction in which the possibility of displacement is suppressed (Fig.3.27,c) or by an idealized support (schematized by a small cart, Fig.3.27,d), which puts into evidence the directions in which the displacement is possible. We notice that the directions in which the possibility of displacement is suppressed must verify certain conditions so that the rigid solid be fixed. Indeed, the constraint forces (reactions) N_j, $j = 1,2,...,6$, will be given by a linear system of equations of the form

$$\sum_{j=1}^{6} a_{ij} N_j + a_{i0} = 0, \quad i = 1,2,...,6; \tag{3.2.38}$$

this system has a unique solution if and only if $\det[a_{ij}] \neq 0$. To fulfill this condition, it is necessary that: i) the supports of any two reactions do not coincide; ii) the supports of any three reactions do not be coplanar and concurrent or parallel; iii) the supports of any four reactions do not be concurrent, parallel or do not belong to the same family of generatrices of an one-sheet hyperboloid; iv) the supports of any five reactions do not intersect two straight lines or do not intersect a straight line and be parallel to a plane or – in general – do not belong to a congruence of the first degree; v) the supports of the six reactions do not intersect the same straight line or do not be parallel to the same plane or – in general – do not belong to the same linear complex of the first degree. If one of these conditions is not fulfilled, then the rigid solid is no more fixed (the system (3.2.38) is no more compatible). In this case too, the rigid solid may be fixed for certain systems of forces, but the reactions are no more univocally determined (the rank of the matrix $[a_{ij}]$ remains smaller than six). If in a neighbouring position the rank of the matrix remains smaller than six, then we say that the rigid is not fixed; but if in such a position the rank is six, then the fixity takes place only for infinitesimal displacements, not for finite ones (the rigid solid is no more strictly fixed, having to do with *critical constraints*).

The simple support may be a bilateral or a unilateral constraint, as the displacement is hindered or not in both directions of the normal to the surface in contact, respectively.

Let S be a rigid solid for which one of the points, $P(\mathbf{r})$, is a fixed point of the space (with respect to a fixed frame of reference; eventually, the fixed point may belong to another rigid solid Σ); we say that the rigid solid S has a *spherical hinge* (*articulation, fixed support*) at P (Fig.3.28,a). To determine the position of the rigid solid, there are necessary only three scalar parameters (e.g., the three Euler's angles); hence, a spherical hinge suppresses three degrees of freedom of the rigid solid and may be replaced by a reaction $\mathbf{R}$ of unknown direction and magnitude, applied at the point P (in case of two rigid solids, the reactions $\mathbf{R}_S = \mathbf{R}$, $\mathbf{R}_\Sigma = -\mathbf{R}$, $\mathbf{R}_S + \mathbf{R}_\Sigma = \mathbf{0}$, arise). Hence, the constraint force has three unknown scalars (the components R_1, R_2, R_3 of $\mathbf{R}$ along three co-ordinate axes). Such a hinge may be obtained by a sphere of centre P (the theoretical point of support), with which the rigid solid S penetrates in a spherical cavity of the rigid solid Σ; hence, the possibility of displacement is suppressed, but the rotation is free (Fig.3.28,b). A spherical hinge is equivalent, from a mechanical point of view, to three simple supports at the same point, so that the directions in which the displacement is suppressed do not be coplanar; hence, a spherical hinge may be represented by three non-coplanar pendulums

(Fig.3.28,c) or by an idealized fixed support, denoting the impossibility of displacement (Fig.3.28,d).

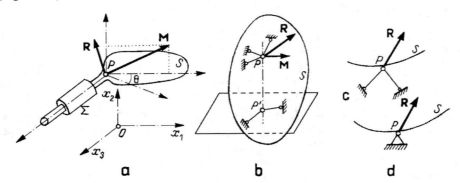

Figure 3.29. Cylindrical hinge (a). Representation by a fixed axis (b). Plane case: two concurrent pendulums (c) or an idealized fixed support (d).

Let us also consider a rigid solid S for which a straight line passing through a fixed point P of it (with respect to a fixed frame of reference) represents a fixed axis; in this case, the rigid solid S has a *cylindrical hinge (articulation)* at P. Such a hinge may be obtained by a cylinder passing through the theoretical point of support P, with which the rigid solid S penetrates in a cylindrical cavity of the rigid solid Σ; hence, besides the possibility of displacement, the possibility of rotation about two axes normal to the fixed axis is hindered too. Remains only the possibility of rotation about the fixed axis; hence, only a scalar parameter is necessary to determine the position of the rigid solid (e.g., the rotation angle θ, Fig.3.29,a). So that a cylindrical hinge suppresses five degrees of freedom and can be replaced by a reaction $\mathbf{R}$ of unknown direction and magnitude, applied at P, and by a couple of moment $\mathbf{M}$, applied at P too, and contained in a plane normal to the fixed axis (the rotation about the fixed axis is not hindered by not one component of the couple); there are thus introduced five scalar unknowns (the components R_1, R_2, R_3 of the reaction $\mathbf{R}$ and the components M_1, M_2 of the moment, supposing that the fixed axis is parallel to the axis Ox_3). A cylindrical hinge is equivalent, from a mechanical point of view, to a spherical hinge at P (three non-coplanar pendulums at the same point) and a support at a point P' of the fixed axis, formed by two pendulums contained in a plane normal to the axis (Fig.3.29,b); the reactions at P and P' are reduced at P to the torsor $\{\mathbf{R}, \mathbf{M}\}$ considered above (with $\mathbf{M}$ normal to the fixed axis).

If the given system of forces is plane, then a spherical hinge contained in this plane or a cylindrical hinge the axis of which is normal to this plane are replaced by a reaction $\mathbf{R}$, which has only two components in the plane; in this case, the spherical hinge and the cylindrical one are equivalent, and we may denote the respective support a *plane hinge (articulation)*. A plane hinge can be represented by two concurrent pendulums (Fig.3.29,c) or by an idealized fixed support (Fig.3.29,d).

Taking into account the observations made for the simple supports, we may affirm that a rigid solid cannot be fixed by means of two spherical hinges (the supports of six equivalent pendulums form two stars of concurrent straight lines); as well, the rigid solid cannot be fixed by a spherical hinge and three simple supports if the supports of the corresponding pendulums are concurrent or parallel or if they intersect the same straight line passing through the hinge, or if one of these supports passes through the hinge. We notice also that, in the case of a given system of forces contained in a plane, the rigid solid is not fixed by a hinge and a simple support if the support of the latter pendulum passes through the hinge (obviously plane).

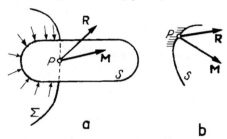

Figure 3.30. Built-in support (a). Graphical representation (b).

A support which suppresses all six degrees of freedom of a rigid solid S is called a *built-in (embedded) support* (a *rigid fixing*). A built –in mounting may be obtained by an extremity of a rigid solid S which penetrates in a rigid solid Σ, the latter one ensuring its fixity (Fig.3.30,a). On the whole surface of contact between the rigid solids S and Σ there appear constraint forces which cannot be determined in the frame of the rigid model considered; but the effect of this support may be replaced by the torsor of these forces at the theoretical point P (in fact, an arbitrary point): a reaction **R** and a couple of moment **M** (six unknown scalars R_i, M_i, $i = 1, 2, 3$, corresponding to the three axes of co-ordinates). A built-in support is equivalent, from a mechanical point of view, to six simple supports on the contact surface, which verify the necessary conditions mentioned in this case to obtain the fixity of the rigid solid, hence it can be represented by six pendulums which satisfy these conditions. Graphically, one can represent a built-in support (Fig.3.30,b). In the case of a given system of forces contained in a plane $x_3 = \text{const}$, by suppressing the six degrees of freedom which remain one must introduce only three scalar unknowns (R_1, R_2 and $M_3 = M$).

Besides the basic supports of the rigid solid, considered above, one may conceive also other ones, which are obtained – in general – starting from those previously considered. All the supports represent ideal external constraints of the rigid solid, which verify the relation (3.2.36).

But the rigid solid is a holonomic and scleronomic mechanical system, the external constraints being scleronomic too; in this case, the real elementary work of the constraint forces will vanish as well. We may write

$$\mathrm{d}W_R = \sum_j \mathbf{R}_j \cdot \mathrm{d}\mathbf{r}_{P_j} = \sum_j \mathbf{R}_j \cdot \mathbf{v}_{P_j} \mathrm{d}t = \sum_j \mathbf{R}_j \cdot \left(\mathbf{v}_P + \boldsymbol{\omega} \times \overrightarrow{PP_j}\right) \mathrm{d}t$$

$$= \left(\mathbf{v}_P \cdot \sum_j \mathbf{R}_j + \boldsymbol{\omega} \cdot \sum_j \overrightarrow{PP_j} \times \mathbf{R}_j\right) \mathrm{d}t = (\mathbf{R} \cdot \mathbf{v}_P + \mathbf{M}_P \cdot \boldsymbol{\omega}) \mathrm{d}t,$$

where we took into account the formula (5.2.3') (see Chap. 5, Subsec. 2.1.1), which links the velocities of two points P and P_j of the rigid solid, the sums (eventually, integrals) corresponding to all the points of the solid upon which act constraint forces; we put thus into evidence the resultant of the constraint forces, as well as their moment with respect to the theoretical point P of support (the torsor of these forces at P). Hence, the condition of ideal external constraints at the point P is of the form ($\mathbf{v}_P = \mathbf{v}, \mathbf{M}_P = \mathbf{M}$)

$$\mathbf{R} \cdot \mathbf{v} + \mathbf{M} \cdot \boldsymbol{\omega} = R_i v_i + M_i \omega_i = 0 \tag{3.2.39}$$

and may be verified in the considered particular cases. For example, in the case of a support on the plane $x_3 = \mathrm{const}$ we have $v_3 = 0$ and $R_1 = R_2 = M_1 = M_2 = M_3 = 0$; hence, the condition (3.2.39) holds and the constraint force $R_3 \neq 0$ is put in evidence. In the case of the coupling screw-nut, the advance being along the axis Ox_3, we have $\omega_1 = \omega_2 = 0$, $v_1 = v_2 = 0$ and $v_3 = (p/2\pi)\omega$, where p is the screw pitch, while $\omega = \omega_3 \neq 0$ is *the angular velocity* (as one can see in Chap. 5, Subsec. 1.3.3); we get thus $R_3(p/2\pi)\omega + M_3\omega = [(p/2\pi)R_3 + M_3]\omega = 0$, and the constraint forces R_1, R_2, R_3, M_1, M_2 and $M_3 = -(p/2\pi)R_3$ are put into evidence.

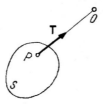

Figure 3.31. A thread acted upon by a tension **T**.

All the constraints considered above are holonomic ones, of geometric nature, and the results obtained are useful in the static as well as in the dynamic case; various examples of non-holonomic constraints, of kinematic nature, have been presented in Subsec. 2.2.6. All these constraints have been bilateral ones; but we can imagine unilateral constraints too, obtained with the aid of *threads*. A thread is considered to be perfectly flexible and inextensible; hence, the distance between the ends of a thread may diminish, but cannot grow. If a rigid solid is linked to a fixed point by a thread perfectly stretched (threads passing over pulleys, rigid solids oscillating at the end of a thread etc.), then a constraint force **T** (always of traction, it draws the rigid solid), arises along the latter one (Fig.3.31); this force is called *tension*, while $-\mathbf{T}$ stretches the thread. The corresponding constraint suppresses one degree of freedom of the rigid solid and introduces only one scalar unknown (the tension of modulus T), if the direction of the thread is fixed.

2.2.11 Constraints with friction

As we have seen in the case of a particle P constrained to stay on a fixed smooth surface S of equation $f(x_1, x_2, x_3) = 0$, a normal constraint force $\mathbf{N} = \lambda \operatorname{grad} f$,

Mass geometry. Displacements. Constraints 195

where λ is a scalar to be determined, arises. If the surface is rough, then the constraint is no more ideal, while the constraint force **R** is no more normal to the surface; there appears a component **T** of this force, contained in the plane tangent at P to the surface and which has an opposite direction with respect to the component in this plane of the resultant of the given forces to which is subjected the particle P. To can determine *the force of sliding friction* (it hinders the sliding of the particle P on the surface, hence the displacement of the particle in the frame of this constraint) one must make supplementary hypotheses. *The* most usual *model* which corresponds in many cases to the physical reality is that *due to Coulomb*; in this case, the modulus of the friction component is given by (experimentally established for the first time by Amonton in 1699)

$$T \leq fN, \quad (3.2.40)$$

the equality taking place in the *limit case of rest* (if the inequality takes place, then the particle P is *at rest* on the surface). The numerical coefficient f, which depends only on the nature and the state (dry or wet) of the rough surface, and does not depend on the velocity **v** of the particle if it begins to move, is called *coefficient of friction* (coefficient of *static* friction, unlike a *dynamic* one, which is of the form $f = f(v)$). We introduce also *the angle of friction* φ, defined by the relation

$$f = \tan \varphi. \quad (3.2.41)$$

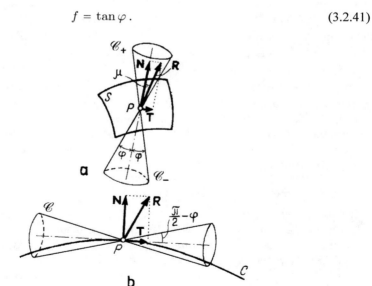

Figure 3.32. Particle P on a surface S (a) or on a curve C (b) with friction.

The relation $T/N = \tan \mu \leq \tan \varphi$ leads to $\mu \leq \varphi$, in this case (Fig.3.32,a); hence, for rest, the angle μ between the constraint force **R** and its normal component **N** must be smaller or at the most equal to the angle of friction φ. We are thus led to construct a circular cone $\mathscr{C}$ with the vertex at P and the axis normal to the surface S, the vertex angle being 2φ (*the cone of friction*). For rest, in the case of a bilateral constraint, the support of the constraint force **R** must be contained in the interior of the zone $\mathscr{C}_+$ or of the zone $\mathscr{C}_-$ of the cone $\mathscr{C}$, or on the cone itself; if the constraint is unilateral, then we take into consideration only the zone $\mathscr{C}_+$ (or $\mathscr{C}_-$) of the cone of friction, situated in the part of the surface S in which the particle P may be.

In the case of a particle P constraint to stay on a smooth curve C, of equations $f_i(x_1, x_2, x_3) = 0$, $i = 1, 2$, arises a constraint force $\mathbf{N} = \lambda_1 \operatorname{grad} f_1 + \lambda_2 \operatorname{grad} f_2$, contained in the plane normal to the curve at P, where λ_1, λ_2 are scalars which must be determined. As in the previous case, if the curve is rough, then we introduce a force of sliding friction **T** of the form (3.2.40), along the tangent to the curve at P. For equilibrium, it is necessary and sufficient that the angle between the support of the constraint force **R** and the tangent to the curve be greater or at least equal to $\pi/2 - \varphi$ (Fig.3.32,b). We are thus led to consider a circular *cone of friction* $\mathscr{C}$, the axis of which is tangent to the curve $\mathscr{C}$ at P, the vertex angle being $\pi - 2\varphi$. For rest, it is necessary and sufficient that the support of the reaction **R** be situated in the exterior of the cone $\mathscr{C}$ or on it.

In both cases considered above, the problem of the position of rest is indeterminate; in fact, we specify only the corresponding limit positions. In the case of a discrete mechanical system $\mathscr{S}$ of particles, subjected to constraints with friction, the degree of indetermination is greater, because neither the directions of the forces of friction are not known. To solve the problem, we must know the motion by which the system $\mathscr{S}$ reached the position of rest, to can determine the forces of friction sufficient to maintain it in this position, or we must search the forces of friction by which the system $\mathscr{S}$ can pass once more from the state of rest in a state of motion.

If the friction does not take place between a particle and a rigid surface (or curve), but between a particle and a fluid (liquid or gas), then we have to do with a *viscous friction*, and a *coefficient of viscosity* appears too. The constraint force is – in this case – a *force of resistance*.

2.2.12 Constraints with friction of a rigid solid

Let be the rigid solids S and Σ simply supported at P, considered at Subsec. 2.2.10 (Fig.3.27,a). In reality, the solids are deformed in the neighbourhood of the point P, so that the contact is obtained on a small surface on which appear constraint forces (Fig.3.33,a); by reducing this system of forces at the point P, we obtain the corresponding torsor (the constraint force **R** and the constraint couple of moment **M**), which replaces the action of the solid Σ upon the solid S (Fig.3.33,b). The *normal constraint force* **N** (*ideal constraint*), the constraint force **T**, contained in the tangent plane (corresponding to *the sliding friction*), the couple of moment $\mathbf{M}_p$, normal to the tangent plane (corresponding to *the pivoting friction*, which is opposed to the rotation about this normal), and the couple of moment $\mathbf{M}_r$, contained in the tangent plane (corresponding to *the rolling friction*, which is opposed to the rotation about an axis in the tangent plane, passing through P) (Fig.3.33,c) are thus put into evidence. Hence, in the case of the rigid solid, the friction phenomenon has a more complex aspect.

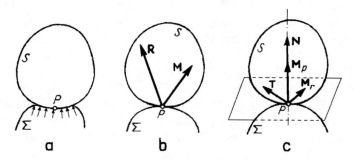

Figure 3.33. Zone of contact between the rigid solids S and Σ (a). The constraint force **R** and the constraint couple of moment **M** (b) and their components (c).

Let us consider the rigid solids S and Σ in contact; as well, let **Q** be the component of the given forces along the normal to the common tangent plane and **N** = −**Q** the corresponding ideal constraint force (Fig.3.34,a); besides the component **N** of the constraint force **R**, arises a tangential component too, which is opposing to the tendency of sliding of the solid S with respect to the solid Σ. This component appears even if one of the solids does not slide with respect to the other one (because of the roughness of the surfaces in contact), having the tendency to equate to zero an

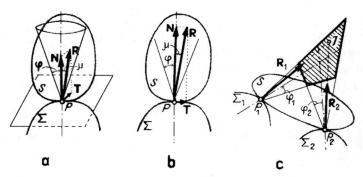

Figure 3.34. Two rigid solids S' and Σ in sliding friction contact: the case of friction (a); the angle of friction (b); the quadrangle of friction (c).

eventual given tangential force which could produce such a sliding; in this case, we have to do with a *friction of adherence* (*of adhesion*), which is opposed to the tangential component of the given forces. In the limit case in which the solid S begins to slide on the solid Σ, one obtains *the force of sliding friction* **T** (*the friction of motion*); we suppose that this force is of *Coulombian nature* and is given by a formula of the form (3.2.40), where *the coefficient f of sliding friction* (3.2.41) does not depend – in general – on the relative velocity of sliding of the two solids or on the magnitude of the surfaces in contact, but only on their nature (roughness). Numerical values of the coefficient f are given in Table 3.1. An example of dependence of the velocity (for greater values) is given in Table 3.2 for metal on metal. Also in this case, a *cone of friction* of vertex angle 2φ is introduced, and the resultant of the constraint forces must be in its interior or on it. If the given forces are contained in a plane passing through the point of contact P, then the problem has a *plane* character, the cone of friction being reduced to an *angle of friction* (the trace of the cone of friction on the considered plane) (Fig.3.34,b). Let us consider a rigid solid acted upon by coplanar forces and leaning with friction at the points P_1 and P_2 on the surfaces Σ_1 and Σ_2, respectively; the conditions $T_1 \leq f_1 N_1$, $T_2 \leq f_2 N_2$ are put. The angles of friction constructed at these points determine a quadrangle called *the quadrangle of friction* (Fig.3.34,c); the state of rest is possible if the point of piercing I of the supports of the constraint forces $\mathbf{R}_1$ and $\mathbf{R}_2$ is in the interior of the quadrangle or on it. If the system of given forces is not plane, then one can take into consideration the three-dimensional domain obtained by the intersection of the two cones of friction.

Table 3.1

wood on wood, dry	0.25-0.50	leather on metals, dry	0.56
wood on wood, soapy	0.20	leather on metals, wet	0.36
metals on oak, dry	0.50-0.60	leather on metals, greasy	0.23
metals on oak, wet	0.24-0.26	leather on metals, oily	0.15
metals on oak, soapy	0.20	steel on agate, dry	0.20
metals on elm, dry	0.20-0.25	steel on agate, oily	0.107
hemp on oak, dry	0.53	iron on stone	0.30-0.70
hemp on oak, wet	0.33	wood on stone	0.40
leather on oak	0.27-0.38	earth on earth	0.25-1.00
metals on metals, dry	0.30	earth on earth, wet clay	0.31
metals on metals, wet	0.15-0.20	metals on ice	0.01-0.03
smooth surfaces, best results ...	0.03-0.036	smooth surfaces, occasionally greased	0.07-0.08

Table 3.2

v (km/h)	0	10.93	21.08	43.5	65.8	87.6	96.48
f	0.242	0.088	0.072	0.07	0.057	0.038	0.027

To put into evidence *the rolling friction* we will consider the contact between a circular wheel of radius R and a horizontal plane. The wheel can be acted upon by the afferent vertical given force $\mathbf{Q}$, by a horizontal given force $\mathbf{F}$ and by a *turning moment* $\mathbf{M}$. If the local deformation is not taken into consideration, then there appears only the normal constraint force $\mathbf{N} = -\mathbf{Q}$ (Fig.3.35,a). In reality, in the neighbourhood of the point P arise local deformations, hence constraint forces on a relatively small but finite surface (Fig.3.35,b); besides the normal constraint force, one can thus put into evidence the tangential constraint force $\mathbf{T} = -\mathbf{F}$ and *the moment of rolling friction* $\mathbf{M}_r$ of modulus $M_r = Fr = Ne$ (in the absence of the turning moment) (Fig.3.35,c,d). This is the case of *the drawn wheel*. Because the position of rest can take place only for

Mass geometry. Displacements. Constraints 199

limited values of the modulus of the force **F**, it follows that – in fact – M_r and e take limit values; if $e_{\max} = s$, then – with necessity – arises the condition

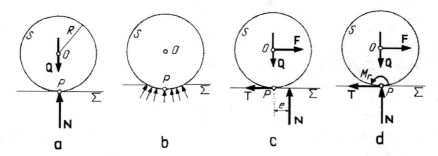

Figure 3.35. The case of rolling friction: without (a) and with (b) local deformations; the moment of rolling friction (c,d).

$$M_r \leq sN, \qquad (3.2.42)$$

where s is a *coefficient of rolling friction* and has the dimension of a length (the maximal parallel displacement of the support of the reaction **N** with respect to the point of contact P, so as to oppose to the motion of rolling; it depends only on the nature of the solids in contact). But besides the friction of rolling arises a friction of sliding too, so that: if $T \leq fN$, $M_r \leq sN$ (hence if $M_r/N \leq fR \leq s$ or if $M_r/N \leq s \leq fR$), then the solid is *in rest*, if $T \leq fN$, $M_r > sN$ (hence if $s < M_r/N \leq fR$), then the solid is *rolling without sliding*, if $T > fN$, $M_r \leq sN$ (hence if $fR < M_r/N \leq s$), then the solid is *sliding without rolling*, while if $T > fN$, $M_r > sN$ (hence if $fR < s < M_r/N$ or if $s < fR < M_r/N$), then the solid is *sliding and rolling* at the same time. In the case of a *motive wheel* intervenes a turning moment **M** too; one can prove that, if the horizontal plane is too smooth (the coefficient f is too small), then the wheel does not move, neither for a very great turning moment **M**. The problem of the contact of a circular wheel with an inclined plane may be studied analogously.

Let us consider two solids S and Σ in contact; because of their local deformation, the contact takes place on a surface σ (Fig.3.36,a). We suppose that the solid S is acted upon by a vertical force **Q** and by a couple of moment **M** along the common normal; this solid has the tendency to rotate about this normal, maintaining unchanged the surface of contact, hence the tendency of pivoting if M is greater than a minimal value. In fact the *pivoting* is a sliding on the surface of contact σ; on an element of area $d\sigma$ arises a force of sliding friction of magnitude $dT = fn d\sigma$ ($\mathbf{n} d\sigma$ is the reaction normal to the element of area $d\sigma$, while f is the coefficient of sliding friction), at a distance r from P, in a direction normal to the vector radius (Fig.3.36,b); the moment of pivoting friction $\mathbf{M}_p$, of modulus M_p, must verify the condition

$$M_p \leq \int_\sigma fnr \, d\sigma; \qquad (3.2.43)$$

if the force **Q** is uniformly distributed on the surface of contact, then we have $n = Q/\sigma = N/\sigma$, so that

$$M_p \leq f \frac{N}{\sigma} \int_\sigma r \, d\sigma, \qquad (3.2.43')$$

where the integral is of the nature of a scalar geometric static moment. This condition is of the form

$$M_p \leq aN, \qquad (3.2.43'')$$

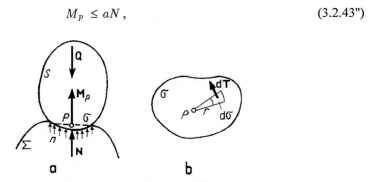

Figure 3.36. The case of pivoting friction: zone of contact (a); the local constraint (b).

where a is a *coefficient of pivoting friction*, having the dimension of a length (it depends only on the geometry and the nature of the surface of contact and not on the magnitude of this surface or on the velocity of the motion of pivoting). If the surface of contact is circular, of radius R, then one obtains

$$a = \frac{2}{3} fR; \qquad (3.2.44)$$

as well,

$$a = \frac{2}{3} f \frac{R_e^3 - R_i^3}{R_e^2 - R_i^2}, \qquad (3.2.44')$$

in the case of an annular surface of contact of external radius R_e and internal radius R_i.

Other cases of supporting with friction (e.g., friction in articulations, eye joints, bearings etc.) may be studied in an analogous manner, using the basic results considered above.

Chapter 4

STATICS

As we have seen, *statics deals with the equivalence of the systems of forces* (in particular, *the equivalence to zero*) which act upon a mechanical system; it is thus studied a particular case of motion, that is the case in which a mechanical system, subjected to the action of a system of forces, remains *at rest* with respect to a given (inertial) frame of reference. We will thus consider the statics of discrete or continuous mechanical systems, especially the problems which arise in the statics of rigid solids.

1. Statics of discrete mechanical systems

In the study of the problems of discrete mechanical systems, we start from the case of a single particle; the results thus obtained may be extended to the case of other mechanical systems.

1.1 Statics of the particle

We deal successively with the case of a free particle and with the case of a particle subjected to constraints; as well, we make some considerations concerning the stability of the equilibrium of a particle.

1.1.1 The free particle

By *free particle* (*material point*) we understand that one which may take any position in the space; this position depends only on the forces acting upon the particle and is independent of any geometric or kinematic restriction. A free particle (in E_3) has *three degrees of freedom*, its position being specified by three independent parameters (e.g., the co-ordinates x_j, $j = 1,2,3$, the components of the position vector $\mathbf{r}$ of the corresponding geometric point).

We consider a body which may be modelled as a particle. The forces $\mathbf{F}_i$, $i = 1,2,...,n$, which act upon this body are – in this case – concurrent forces, and can be modelled as *bound vectors* (Fig.4.1,a). Applying to this system elementary operations of equivalence, we obtain *the resultant*

$$\mathbf{F} = \sum_{i=1}^{n} \mathbf{F}_i , \qquad (4.1.1)$$

which is an *invariant* of the respective system; hence, two systems of forces, which are acting upon the same particle, are equivalent if they have the same resultant **F**.

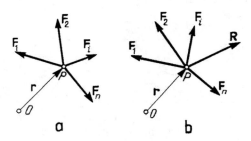

Figure 4.1. Statics of a free (a) or constraint (b) particle.

Let be a free particle P which is not acted upon by any force, at rest with respect to an inertial frame of reference. Let us suppose that this particle is then subjected to the action of a system of concurrent forces. The principle of the parallelogram of forces allows us to replace the given system of forces by its resultant, while the principles of inertia and of initial conditions allow us to affirm that the state of rest is maintained if to the particle is no more applied any force (or is applied a force equal to zero). If the system of concurrent forces has a *zero resultant* (it is *equivalent to zero*), then we say that this system of forces is in *static equilibrium* (if no confusion may occur, then it is sufficient to say that the system of forces is in *equilibrium*); in this case, the particle remains at *rest* with respect to the considered inertial frame of reference (sometimes we say that the particle is in *equilibrium*). In conclusion, the necessary and sufficient condition of equilibrium (of rest; here and in what follows we exclude the possibility of a rectilinear and uniform motion, foreseen by the principle of inertia) of a particle P, with respect to a given frame of reference, is written in the form

$$\mathbf{F} = \mathbf{0} \qquad (4.1.2)$$

or, in components, in the form

$$F_j = 0, \quad j = 1, 2, 3. \qquad (4.1.2')$$

In the case of a system of forces which have the same support remains only one scalar condition, while in the case of a coplanar system of forces, two scalar conditions must be fulfilled.

The first basic problem is that in which there are given the forces which act upon the particle, and one must search the position of equilibrium; the unknowns are – in this case – the parameters specifying the position of the particle (one, two or three parameters, as we are in a particular case – in E_1 or in E_2 – in what concerns the system of forces or, in the general case, in E_3). If the system of equations of equilibrium is indeterminate, then there exists an infinity of possible positions of equilibrium, while if this system is impossible there is not one position of equilibrium.

In *the second basic problem*, the position of equilibrium of the particle is given, while the forces which act upon the latter one to maintain this position must be

Statics

obtained. In this case, the unknowns are the magnitudes and the directions of the forces, the corresponding solution being – in general – *indeterminate*. If certain conditions are imposed to those forces in what concerns their number, directions or magnitudes, then it is possible that the solutions be *determinate*; it is necessary that the number of the scalar unknowns thus introduced be at the most equal to three (or to two in the case of coplanar forces).

We notice also the case of *the basic mixed problem*, in which a part of the three unknowns are elements specifying the system of forces, the other unknowns being parameters which determine the position of equilibrium.

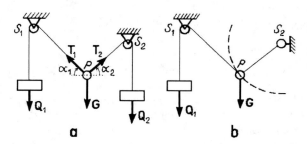

Figure 4.2. Heavy particle in equilibrium: free (a) or constraint (b) particle.

To illustrate the considerations made above, let be, in a vertical plane, a particle P (materialized by a small annulus) of weight $\mathbf{G}$, linked by two inextensible threads which pass over two small pulleys S_1 and S_2 and are tensioned by the weights $\mathbf{Q}_1$ and $\mathbf{Q}_2$, respectively (Fig.4.2,a). Because the particle is free and subjected to a system of coplanar forces, we may choose as unknown parameters (we are in the case of the first basic problem) the angles α_1 and α_2 made by the two threads with the horizontal line, respectively. The threads are extended by the tensions $\mathbf{T}_1$ and $\mathbf{T}_2$, respectively, for which we have $|\mathbf{T}_1| = |\mathbf{Q}_1| = Q_1$ and $|\mathbf{T}_2| = |\mathbf{Q}_2| = Q_2$, respectively, so that the particle is acted upon by the forces $\mathbf{G}$, $\mathbf{T}_1$ and $\mathbf{T}_2$, which must be in equilibrium. Projecting on the horizontal and the vertical, we get

$$Q_1 \cos \alpha_1 - Q_2 \cos \alpha_2 = 0, \quad Q_1 \sin \alpha_1 + Q_2 \sin \alpha_2 - G = 0,$$

wherefrom

$$\sin \alpha_1 = \frac{G^2 + Q_1^2 - Q_2^2}{2GQ_1}, \quad \sin \alpha_2 = \frac{G^2 + Q_2^2 - Q_1^2}{2GQ_2}. \tag{4.1.3}$$

We notice that the conditions $|G^2 + Q_1^2 - Q_2^2| < 2GQ_1$, $|G^2 + Q_2^2 - Q_1^2| < 2GQ_2$ must hold; the equalities never take place, because the pulleys S_1 and S_2 cannot be on the same vertical. These conditions may be written in the form

$$|G - Q_1| < Q_2 < |G + Q_1|, \quad |G - Q_2| < Q_1 < |G + Q_2| \tag{4.1.3'}$$

too, corresponding to the conditions verified by the sides of a triangle (the polygon of the forces **G**, **T**$_1$ and **T**$_2$); otherwise, the equilibrium is impossible, and the particle cannot remain at rest. The position of equilibrium depends, obviously, on the position of the pulleys S_1 and S_2; for instance, if the pulleys S_1 and S_2 are on the same horizontal, then the conditions which must take place become $0 < G^2 + Q_1^2 - Q_2^2 < 2GQ_1$, $0 < G^2 + Q_2^2 - Q_1^2 < 2GQ_2$.

The intervening threads do not introduce any restriction of geometric nature so that the above considered problem is that of a free particle. If the particle would be linked by one of the threads (considered inextensible) to a fixed point (a ring), then it should be on a circle, the centre of which is at this fixed point (Fig.4.2,b).

1.1.2 The constraint particle

If a particle P is subjected to constraints, which will be considered scleronomic, then the latter ones have an influence on the conditions of equilibrium, hence on the position of rest, diminishing the number of degrees of freedom of the particle. As it was shown in Chap. 3, Subsec. 2.2.1, *the axiom of liberation from constraints* (*the axiom of liberation, the axiom of constraints, the axiom of constraint forces*) allows us to replace these constraints by constraint forces (reactions); in this case, the particle P may be considered to be a free particle, subjected to the action of the given as well as of the constraint forces, so that one can use the considerations of the previous subsection. If **R** is the constraint force applied to the particle P (in fact, the resultant of all constraint forces acting upon this particle), which is subjected also to the action of the system of given forces **F**$_i$, $i = 1, 2, ..., n$, then the necessary and sufficient condition of equilibrium (of rest with respect to a given frame of reference) is of the form (Fig.4.1,b)

$$\mathbf{F} + \mathbf{R} = 0 \tag{4.1.4}$$

or, in components, of the form

$$F_j + R_j = 0, \quad j = 1, 2, 3. \tag{4.1.4'}$$

In the case in which the given and the constraint forces are coplanar, only two scalar conditions must be fulfilled, while if these forces have the same support, then only one scalar condition remains.

The unknowns of the problem are of two kinds: unknowns corresponding to the parameters which specify the position of equilibrium and unknowns which determine the constraint force (the basic problem is thus a *mixed* one); some conditions which must be verified by the given forces so as the particle be at rest with respect to a given frame of reference there arise. If no condition is imposed to the constraint force, then the problem is *indeterminate* (the number of the components of the constraint force is equal to the number of the equations of equilibrium, but there intervene also the parameters specifying the position of equilibrium), excepting the case in which the particle is constrained to be at a certain fixed position in the space; in the latter case, we have

Statics

$$\mathbf{R} = -\mathbf{F}, \quad R_j = -F_j, \quad j = 1,2,3. \tag{4.1.5}$$

To point out the restrictions which may be imposed to the constraint force $\mathbf{R}$, we will effect the decomposition

$$\mathbf{R} = \mathbf{N} + \mathbf{T}, \tag{4.1.6}$$

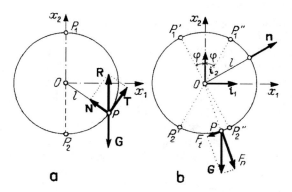

Figure 4.3. Heavy particle constraint to stay on a circle in a vertical plane: without friction (a), with friction (b).

where the components $\mathbf{N}$ and $\mathbf{T}$ do not allow the particle to leave the constraint or to move in the frame of it, respectively. Let us consider, for instance, a particle P of weight $\mathbf{G}$, which may remain in equilibrium on a fixed circle, in a vertical plane (Fig.4.3,a); the component $\mathbf{N}$ of the constraint force is in the direction of the normal to the circle, being the normal reaction, while the component $\mathbf{T}$ is tangent to the circle, representing a constraint force with friction (the force of friction). The problem will be studied in Subsecs. 1.1.5 and 1.1.8.

1.1.3 Geometry of a curve. Frenet's trihedron

Let be a curve C defined by the *parametric equations*

$$x_j = x_j(q), \quad q \in Q \equiv [q_0, q_1], \quad j = 1,2,3, \tag{4.1.7}$$

where $x_j \in C^3(Q)$; a point P is specified by the position vector $\mathbf{r}(q) = x_j(q)\mathbf{i}_j$ (Fig.4.4). Introducing *the derivative* $\mathbf{r}'(q)$ and *the differential* $\mathrm{d}\mathbf{r}$, and taking into account the metrics of the space E_3, we may write

$$|\mathrm{d}\mathbf{r}| = \sqrt{x_1'^2 + x_2'^2 + x_3'^2}\,\mathrm{d}q = \sqrt{x_j' x_j'}\,\mathrm{d}q = \mathrm{d}s, \tag{4.1.8}$$

where *the curvilinear abscissa* represents the length of *the arc of curve* $\widehat{P_0 P}$, given by

$$s(q) = \int_{q_0}^{q} \sqrt{x_1'^2(u) + x_2'^2(u) + x_3'^2(u)}\,\mathrm{d}u = \int_{q_0}^{q} \sqrt{x_j'(u)x_j'(u)}\,\mathrm{d}u. \tag{4.1.8'}$$

A point of the curve C for which the derivative of the position vector is non-zero is called an *ordinary point*; a non-ordinary point is called *singular*. In what follows we consider only ordinary points. An *osculating circle* at an ordinary point P on the curve C is the limit to which tends a circle defined by the point P and two other neighbouring points if the latter ones tend to the point P; the osculating circle is tangent to the curve at this very point and is an approach of the curve in the neighbourhood of the respective point. The plane normal to the tangent at this point is called *normal plane*; the straight lines which pass through the point P and belong to this plane are *normals* to the curve C. The plane in which lays the osculating circle is called *the osculating plane*; this plane contains the tangent to the curve too. The centre of the osculating circle (called also the *circle of curvature*) lies on the normal to the curve at the point P, contained in the osculating plane; this normal is called *the principal normal*. The plane normal to the principal normal (which contains also the tangent to the curve) is called *the rectifying plane*; the normal contained in this plane is called *binormal*. Hence, the osculating plane is formed by the tangent and the principal normal, the normal plane is formed by the principal normal and the binormal, while the rectifying plane is defined by the binormal and the tangent. In the case of a plane curve, the osculating plane coincides with the plane of the curve. At the point P, we introduce a *local frame of reference* (a *movable frame*), that is *the intrinsic Frenet's trihedron* (sometimes denoted as *the Serret-Frenet trihedron*); this one is a three-orthogonal trihedron, the unit vectors of the corresponding co-ordinate axes being defined as follows: the unit vector $\boldsymbol{\tau}$ is *tangent* to the curve C and has the direction in which the parameter q (as well as the curvilinear abscissa) increases; the unit vector $\boldsymbol{\nu}$ is along *the principal normal* and is directed towards the interior of the curve; the unit vector $\boldsymbol{\beta}$ is along *the binormal* and is directed so that the scalar triple product $(\boldsymbol{\tau},\boldsymbol{\nu},\boldsymbol{\beta}) > 0$ (to have a right-handed trihedron) (Fig.4.4).

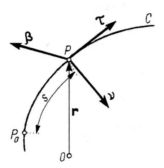

Figure 4.4. Frenet's trihedron.

Taking into account (4.1.8) and noting that $d\mathbf{r}$ has the direction of the tangent to the curve C at the point P, we get

$$\boldsymbol{\tau} = \frac{d\mathbf{r}}{ds} = \mathbf{r}'(s) = \frac{x'_j(q)\mathbf{i}_j}{\sqrt{x'_k(q)x'_k(q)}}. \qquad (4.1.9)$$

Noting that the derivative of a vector of constant modulus is normal to it and that the unit vector $\boldsymbol{\tau}$ is contained in the osculating plane to the curve at the point P, we may write

$$\frac{d\boldsymbol{\tau}}{ds} = \boldsymbol{\tau}'(s) = \frac{1}{\rho}\mathbf{v}; \qquad (4.1.10)$$

the scalar $1/\rho > 0$ is *the curvature* of the curve at the point P, while ρ is the *radius of curvature* (*the radius of the osculating circle* to the curve at the very same point P). We have thus $\boldsymbol{\beta} \cdot (d\boldsymbol{\beta}/ds) = 0$, because $\boldsymbol{\beta}^2 = 1$, while the orthogonality relation $\boldsymbol{\tau} \cdot \boldsymbol{\beta} = 0$ leads to $\boldsymbol{\tau} \cdot (d\boldsymbol{\beta}/ds) = 0$, where we took into account the relation (4.1.10) and the condition of orthogonality $\mathbf{v} \cdot \boldsymbol{\beta} = 0$. We may write

$$\frac{d\boldsymbol{\beta}}{ds} = \boldsymbol{\beta}'(s) = \frac{1}{\rho'}\mathbf{v}; \qquad (4.1.10')$$

the scalar $1/\rho'$ is *the torsion* of the curve at the point P, while ρ' is the *radius of torsion* at the very same point. Differentiating $\mathbf{v} = \boldsymbol{\beta} \times \boldsymbol{\tau}$ with respect to the curvilinear abscissa s and taking into account (4.1.10), (4.1.10'), as well as the vector products $\boldsymbol{\tau} = \mathbf{v} \times \boldsymbol{\beta}$ and $\boldsymbol{\beta} = \boldsymbol{\tau} \times \mathbf{v}$, we may write

$$\frac{d\mathbf{v}}{ds} = \mathbf{v}'(s) = -\frac{1}{\rho}\boldsymbol{\tau} - \frac{1}{\rho'}\boldsymbol{\beta}. \qquad (4.1.10'')$$

The formulae (4.1.10)-(4.1.10'') are called *Frenet's formulae* (sometimes, *the Serret-Frenet formulae*).

Starting from the formula (4.1.10), we get $(1/\rho)\boldsymbol{\tau} \times \mathbf{v} = (1/\rho)\boldsymbol{\beta} = \boldsymbol{\tau} \times (d\boldsymbol{\tau}/ds)$; taking into account (4.1.9) and the modulus in both members, we obtain

$$\frac{1}{\rho} = |\mathbf{r}'(s) \times \mathbf{r}''(s)| = \frac{1}{ds^3}|d\mathbf{r} \times d^2\mathbf{r}|. \qquad (4.1.11)$$

The scalar product of the relation (4.1.10'') by $\boldsymbol{\beta}$, leads to $1/\rho' = -\boldsymbol{\beta} \cdot (d\mathbf{v}/ds)$; taking into account (4.1.10), we may compute $\mathbf{v}'(s)$, while the vector product $\boldsymbol{\beta} = \boldsymbol{\tau} \times \mathbf{v}$ leads to two mixed products, one of which vanishes. Using once more the relation (4.1.10), as well as the relation (4.1.11), we may write

$$\frac{1}{\rho'} = -\rho^2\left(\boldsymbol{\tau}(s), \boldsymbol{\tau}'(s), \boldsymbol{\tau}''(s)\right) = -\rho^2\left(\mathbf{r}'(s), \mathbf{r}''(s), \mathbf{r}'''(s)\right)$$

$$= -\frac{\left(\mathbf{r}'(s), \mathbf{r}''(s), \mathbf{r}'''(s)\right)}{\left[\mathbf{r}'(s) \times \mathbf{r}''(s)\right]^2} = -\frac{\left(d\mathbf{r}, d^2\mathbf{r}, d^3\mathbf{r}\right)}{\left(d\mathbf{r} \times d^2\mathbf{r}\right)^2}. \qquad (4.1.12)$$

1.1.4 Geometry of a surface. Darboux's trihedron

Let be a surface S, defined by *the parametric equations*

$$x_j = x_j(u,v), \quad (u,v) \in D, \quad j = 1,2,3, \qquad (4.1.13)$$

where $x_j \in C^2(D)$, D being a given two-dimensional domain; a point P is specified by the position vector $\mathbf{r}(u,v) = x_j(u,v)\mathbf{i}_j$, where u and v are curvilinear co-ordinates on the surface (Fig.4.5,a). If one or the other one of the two curvilinear co-ordinates is constant, then one obtains two families of *co-ordinate lines*

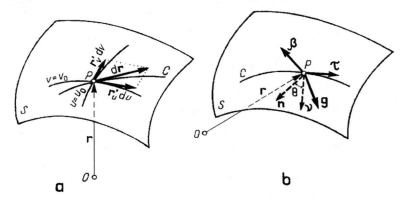

Figure 4.5. Co-ordinate lines on a surface (a). Darboux's trihedron (b).

$$\mathbf{r} = \mathbf{r}(u,v_0), \quad \mathbf{r} = \mathbf{r}(u_0,v), \quad u_0,v_0 = \mathrm{const}, \qquad (4.1.14)$$

where the vectors $\mathbf{r}'_u = \mathrm{d}\mathbf{r}/\mathrm{d}u$, $\mathbf{r}'_v = \mathrm{d}\mathbf{r}/\mathrm{d}v$ have the same directions as the tangents to these curves. Through a point (u_0,v_0) of this surface passes only one co-ordinate line of each family. The directions specified by the derivatives $\mathbf{r}'_u$, $\mathbf{r}'_v$ are distinct if $\mathbf{r}'_u \times \mathbf{r}'_v \neq \mathbf{0}$, and the parametric representation (4.1.13) effectively defines a surface. In this case, if $\mathbf{r}'_u \cdot \mathbf{r}'_v = 0$, then the co-ordinate lines form an *orthogonal system*. The vectors $\mathbf{r}'_u$ and $\mathbf{r}'_v$ define *the tangent plane* to the surface at the point (u_0,v_0). A point (u_0,v_0) which is an ordinary point for each of the two co-ordinate lines passing through it and for which the condition $\mathbf{r}'_u \times \mathbf{r}'_v \neq \mathbf{0}$ holds is called an *ordinary point* of the surface; a point which is not ordinary is called *singular*.

In this case too, we introduce a *local system of reference* (a *movable frame*), that is *the intrinsic Darboux's trihedron* (called also *the Darboux-Ribaucourt trihedron*); the unit vectors of the co-ordinate axes of this three-orthogonal trihedron are defined as follows: the unit vector $\boldsymbol{\tau}$ is *tangent* to a curve C on the surface and is directed so that its curvilinear abscissa increases; the unit vector $\mathbf{g}$ is along *the tangential normal* (normal to C at P, belonging to the tangent plane to S at P); the unit vector $\mathbf{n}$ is along *the normal to the surface*, its direction being so that $(\mathbf{r}'_u,\mathbf{r}'_v,\mathbf{n}) > 0$. The direction of the unit vector $\mathbf{g}$ is chosen so that the trihedron be right-handed $((\boldsymbol{\tau},\mathbf{g},\mathbf{n}) > 0)$ (Fig.4.5,b). We suppose that P is an ordinary point.

If we introduce the mappings $t \to u(t)$, $t \to v(t)$, $t \in [t_0, t_1]$, then the point P describes a curve C on the surface S; the tangent to this curve is given by $\mathrm{d}\mathbf{r} = \mathbf{r}'_u \mathrm{d}u + \mathbf{r}'_v \mathrm{d}v$. Starting from $\mathrm{d}\mathbf{r} \cdot \mathrm{d}\mathbf{r} = |\mathrm{d}\mathbf{r}|^2 = \mathrm{d}s^2$, we may write *the element of arc* in the form

$$\mathrm{d}s^2 = E\mathrm{d}u^2 + 2F\mathrm{d}u\mathrm{d}v + G\mathrm{d}v^2, \qquad (4.1.15)$$

where

$$E = \mathbf{r}'^2_u, \quad F = \mathbf{r}'_u \cdot \mathbf{r}'_v, \quad G = \mathbf{r}'^2_v; \qquad (4.1.15')$$

this expression represents *the first basic quadratic form of the surface* and was introduced by Gauss. Taking into account Lagrange's identity (2.1.33) and noting that P is an ordinary point, we find the condition

$$H^2 = (\mathbf{r}'_u \times \mathbf{r}'_v)^2 = EG - F^2 > 0; \qquad (4.1.15'')$$

hence, this quadratic form is *positive definite* in all the ordinary points of the surface. The properties of a surface S which depend only on the coefficients of the first quadratic form (and of the partial derivatives of these coefficients) are called *intrinsic properties of the surface*.

The element of area, that is the area of the curvilinear parallelogram obtained with the vectors $\mathbf{r}'_u \mathrm{d}u$ and $\mathbf{r}'_v \mathrm{d}v$, is given by

$$\mathrm{d}S = |\mathbf{r}'_u \times \mathbf{r}'_v|\mathrm{d}u\mathrm{d}v = H\mathrm{d}u\mathrm{d}v. \qquad (4.1.16)$$

We notice that the tangents to all the curves passing through P are contained in the plane defined by $\mathbf{r}'_u$ and $\mathbf{r}'_v$; the unit vector $\mathbf{n}$ of the normal to the surface at the point mentioned above is – in this case – given by

$$\mathbf{n} = \frac{\mathbf{r}'_u \times \mathbf{r}'_v}{|\mathbf{r}'_u \times \mathbf{r}'_v|} = \frac{1}{H}\mathbf{r}'_u \times \mathbf{r}'_v = \mathbf{r}'_u \times \mathbf{r}'_v \frac{\mathrm{d}u\mathrm{d}v}{\mathrm{d}S}, \qquad (4.1.17)$$

its direction being specified by $(\mathbf{r}'_u, \mathbf{r}'_v, \mathbf{n}) > 0$.

Projecting the first formula of Frenet (4.1.10) on the unit vector $\mathbf{n}$, we get

$$\frac{\cos\theta}{\rho}\mathrm{d}s^2 = L\mathrm{d}u^2 + 2M\mathrm{d}u\mathrm{d}v + N\mathrm{d}v^2, \qquad (4.1.18)$$

where $\theta = \sphericalangle(\mathbf{n}, \mathbf{v})$, and

$$L = -\mathbf{r}'_u \cdot \mathbf{n}'_u = \mathbf{n} \cdot \mathbf{r}''_{uu} = \frac{1}{H}(\mathbf{r}'_u, \mathbf{r}'_v, \mathbf{r}''_{uu}),$$

$$M = -\mathbf{r}'_u \cdot \mathbf{n}'_v = -\mathbf{r}'_v \cdot \mathbf{n}'_u = \mathbf{n} \cdot \mathbf{r}''_{uv} = \frac{1}{H}(\mathbf{r}'_u, \mathbf{r}'_v, \mathbf{r}''_{uv}), \qquad (4.1.18')$$

$$N = -\mathbf{r}'_v \cdot \mathbf{n}'_v = \mathbf{n} \cdot \mathbf{r}''_{vv} = \frac{1}{H}(\mathbf{r}'_u, \mathbf{r}'_v, \mathbf{r}''_{vv}),$$

with the notations $\mathbf{n}'_u = \partial \mathbf{n}/\partial u$, $\mathbf{n}'_v = \partial \mathbf{n}/\partial v$, $\mathbf{r}''_{uu} = \partial^2 \mathbf{r}/\partial u^2$, $\mathbf{r}''_{uv} = \partial^2 \mathbf{r}/\partial u \partial v$, $\mathbf{r}''_{vv} = \partial^2 \mathbf{r}/\partial v^2$; the expression in the second member of the relation (4.1.18) *represents the second basic quadratic form of the surface.*

Let be a curve C on the smooth surface S and a point P on this curve; *the radius of curvature* of this curve at the point P will be denoted by ρ. The plane determined by the unit vectors $\boldsymbol{\tau}$ and $\mathbf{n}$ pierces the surface S by a curve C_n, called *the normal section* associated at P to the curve C, the unit vector of the corresponding principal normal being $\mathbf{v}_n$; the curvature $1/\rho_n$ of the normal section C_n associated at P to the curve C is called *the normal curvature of the curve* C at P and is positive or negative as $\sphericalangle(\mathbf{n}, \mathbf{v}_n) = 0$ or $\sphericalangle(\mathbf{n}, \mathbf{v}_n) = \pi$, respectively. We state thus

Theorem 4.1.1 (*Meusnier*). *The normal curvature of a curve C on a surface S is the projection of the curvature vector $\mathbf{v}/\rho$ on the unit vector $\mathbf{n}$ of the normal to the surface S*

$$\frac{1}{\rho_n} = \frac{\cos\theta}{\rho}. \qquad (4.1.19)$$

A unit vector $\boldsymbol{\tau}$ to which corresponds a zero normal curvature defines an *asymptotic direction*; the curves on the surface for which the tangent to each point is an asymptotic direction are called *asymptotic lines of the surface*. On a smooth surface S there exist two families of asymptotic lines determined by the differential equation

$$L\,du^2 + 2M\,du\,dv + N\,dv^2 = 0; \qquad (4.1.20)$$

hence, through a point $P(u,v)$ of the surface pass two asymptotic lines. If a straight line lays on a surface, then this one is obviously an asymptotic line. The projection of the curvature vector $\mathbf{v}/\rho$ on the tangent plane to a surface S at the point P is called, by definition, *the geodesic curvature* of the curve C at the point P and is given by

$$\frac{1}{\rho_g} = \frac{\sin\theta}{\rho} = \pm(\boldsymbol{\tau}(s), \boldsymbol{\tau}'(s), \mathbf{n}(s)) = \pm(\mathbf{r}'(s), \mathbf{r}''(s), \mathbf{n}(s)); \qquad (4.1.21)$$

one can prove that the geodesic curvature at a point P of a curve C laying on the surface S is equal to the curvature of the projection of the curve C on the tangent plane to the surface at the very same point P. The radius ρ_g is called *radius of geodesic curvature* (or *tangential*). The geodesic curvature of the curve C at a point P

Statics

is an intrinsic invariant of the surface S. Let be a curve C for which the principal normal $\mathbf{v}$ coincides with the normal $\mathbf{n}$ to the surface; in this case, the osculating plane to the curve at the point P is normal to the surface at this point, and we have $\sin\theta = 0$, hence the geodesic curvature vanishes. The curves on the surface for which, at any point, the geodesic curvature vanishes are called *geodesic lines*; the geodesic lines of a smooth surface S are determined by the differential equation

$$(\mathrm{d}\mathbf{r}, \mathrm{d}^2\mathbf{r}, \mathbf{n}) = 0, \qquad (4.1.22)$$

so that through an ordinary point P of it passes an infinity of geodesic lines. A geodesic line passing through two points of the surface S represents the shortest way between these two points; it is a property of variational nature. For instance, the geodesic lines on a sphere are its great circles. We introduce also *the geodesic torsion* of the curve C at the point P

$$\frac{1}{\rho'_g} = \frac{1}{\rho_g} - \frac{\mathrm{d}\theta}{\mathrm{d}s}, \qquad (4.1.23)$$

where ρ'_g is *the radius of geodesic torsion*; we notice that the geodesic torsion of a curve C at a point P of the surface is equal to the torsion of the geodesic line tangent to this curve at the point P.

Returning to Darboux's trihedron and introducing also Frenet's trihedron, which corresponds to a curve C on a surface S, we may write

$$\mathbf{n} = \mathbf{v}\cos\theta + \boldsymbol{\beta}\sin\theta, \quad \mathbf{g} = \mathbf{v}\sin\theta - \boldsymbol{\beta}\cos\theta, \qquad (4.1.24)$$

wherefrom

$$\mathbf{v} = \mathbf{g}\sin\theta + \mathbf{n}\cos\theta, \quad \boldsymbol{\beta} = -\mathbf{g}\cos\theta + \mathbf{n}\sin\theta. \qquad (4.1.24')$$

Starting form the Frenet's formulae (4.1.10)-(4.1.10"), using the formulae (4.1.24), (4.1.24'), and introducing the normal curvature (4.1.19), as well as the geodesic curvature (4.1.21) and the geodesic torsion (4.1.23), we may write the derivatives of the unit vectors of Darboux's trihedron with respect to the arc s in the form

$$\begin{aligned}
\frac{\mathrm{d}\boldsymbol{\tau}}{\mathrm{d}s} &= \boldsymbol{\tau}'(s) = \frac{1}{\rho_n}\mathbf{n} + \frac{1}{\rho_g}\mathbf{g}, \\
\frac{\mathrm{d}\mathbf{g}}{\mathrm{d}s} &= \mathbf{g}'(s) = -\frac{1}{\rho_g}\boldsymbol{\tau} - \frac{1}{\rho'_g}\mathbf{n}, \\
\frac{\mathrm{d}\mathbf{n}}{\mathrm{d}s} &= \mathbf{n}'(s) = -\frac{1}{\rho_n}\boldsymbol{\tau} + \frac{1}{\rho'_g}\mathbf{g}.
\end{aligned} \qquad (4.1.25)$$

1.1.5 Particle subjected to ideal constraints

The *ideal* (*smooth, frictionless*) *constraints* are these constraints for which $\mathbf{T} = \mathbf{0}$; in this case, the relation (3.2.36) takes place (the virtual work of the constraint forces vanishes). In reality, such constraints do not exist; but there exist curves and surfaces (constraints of contact) for which the force of friction can be neglected in a first approximation. In this case, the reaction

$$\mathbf{R} = \mathbf{N} \qquad (4.1.26)$$

is along the normal to the surface or is contained in the plane normal to the curve, respectively, at the point at which stays the particle. Taking into account the condition of equilibrium (4.1.4) and the relation (4.1.26), we may state

Theorem 4.1.2. *A particle constrained to stay on a fixed smooth surface* (*curve*) (*ideal constraints*) *is in equilibrium if and only if the resultant of the given forces acting upon it is directed along the normal to the surface or is contained in the plane normal to the curve, respectively, at the point which represents the position of equilibrium.*

In the case of the particle subjected to ideal constraints, there appear unknowns concerning the position of equilibrium and unknowns corresponding to the constraint forces. The system of equations (4.1.4') is – in general – sufficient to solve the equilibrium problem; but in some particular cases, this system can be indeterminate or impossible, thus existing an infinity of such positions or none.

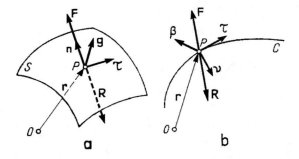

Figure 4.6. Particle in rest on a surface (a) or on a curve (b).

If the particle P is constrained to stay on a fixed smooth surface S (Fig.4.6,a), of equation (given in an implicit form)

$$f(\mathbf{r}) \equiv f(x_1, x_2, x_3) = 0, \qquad (4.1.27)$$

then the constraint force, normal to the surface, is of the form

$$\mathbf{R} = \lambda \operatorname{grad} f, \quad R_j = \lambda f_{,j}, \quad j = 1, 2, 3, \qquad (4.1.27')$$

where the scalar λ is a parameter which must be determined; we get the equations of equilibrium

$$\mathbf{F} + \lambda \operatorname{grad} f = \mathbf{0}, \quad F_j + \lambda f_{,j} = 0, \quad j = 1, 2, 3. \qquad (4.1.27'')$$

Statics

The system of equations (4.1.27"), considered in the unknown λ, is compatible if and only if

$$\frac{F_1}{f_{,1}} = \frac{F_2}{f_{,2}} = \frac{F_3}{f_{,3}};\qquad (4.1.27''')$$

these equations, together with (4.1.27), specify the position of equilibrium, while the parameter λ (hence, the constraint force (4.1.27')) is given by the system (4.1.27").

If the equation of the surface S is given in the explicit form

$$x_3 = \varphi(x_1, x_2),\qquad (4.1.28)$$

then we get the equations which determine the constraint force

$$F_1 + \lambda\varphi_{,1} = 0,\quad F_2 + \lambda\varphi_{,2} = 0,\quad F_3 + \lambda\varphi_{,3} = 0,\qquad (4.1.28')$$

while the conditions specifying the position of equilibrium are

$$\frac{F_1}{\varphi_{,1}} = \frac{F_2}{\varphi_{,2}} = -F_3.\qquad (4.1.28'')$$

Analogously, if the surface S is given by the parametric equations

$$x_i = x_i(u,v),\quad u \in [u_1, u_2],\quad v \in [v_1, v_2],\quad i = 1,2,3,\qquad (4.1.29)$$

then the conditions of equilibrium become

$$\frac{F_1}{\det\left[\dfrac{\partial(x_2, x_3)}{\partial(u,v)}\right]} = \frac{F_2}{\det\left[\dfrac{\partial(x_3, x_1)}{\partial(u,v)}\right]} = \frac{F_3}{\det\left[\dfrac{\partial(x_1, x_2)}{\partial(u,v)}\right]},\qquad (4.1.29')$$

where we have put into evidence the director parameters of the normal in the form of functional determinants, while the constraint force is given by the equations

$$F_i + \frac{1}{2}\lambda\,\epsilon_{ijk}\det\left[\frac{\partial(x_j, x_k)}{\partial(u,v)}\right] = 0,\quad i = 1,2,3,\qquad (4.1.29'')$$

where ϵ_{ijk} is Ricci's symbol. Using Darboux's local frame of reference, we notice that $R_\tau = R_g = 0$, $R_n = R$, so that

$$F_\tau = 0,\quad F_g = 0,\quad R = -F_n;\qquad (4.1.30)$$

the first two relations (4.1.30) state the position of equilibrium, while the last relation specifies the constraint force.

If the particle P lies on a fixed smooth curve C (Fig.4.6,b) of equations (in an implicit form)

$$f_k(\mathbf{r}) \equiv f_k(x_1, x_2, x_3) = 0, \quad k = 1, 2, \tag{4.1.31}$$

then one obtains the constraint force (in the plane normal to the curve C, specified by the normals to the surfaces passing through this curve)

$$\mathbf{R} = \lambda_1 \operatorname{grad} f_1 + \lambda_2 \operatorname{grad} f_2, \quad R_j = \lambda_1 f_{1,j} + \lambda_2 f_{2,j}, \quad j = 1, 2, 3, \tag{4.1.31'}$$

where the scalars λ_k, $k = 1, 2$, are parameters to be determined, and the equations of equilibrium

$$\mathbf{F} + \lambda_1 \operatorname{grad} f_1 + \lambda_2 \operatorname{grad} f_2 = \mathbf{0}, \quad F_j + \lambda_1 f_{1,j} + \lambda_2 f_{2,j} = 0, \quad j = 1, 2, 3. \tag{4.1.31''}$$

The system of linear equations in the unknowns λ_1, λ_2 is compatible if and only if

$$\begin{vmatrix} F_1 & f_{1,1} & f_{2,1} \\ F_2 & f_{1,2} & f_{2,2} \\ F_3 & f_{1,3} & f_{2,3} \end{vmatrix} = 0, \tag{4.1.31'''}$$

obtaining thus the condition which, together with (4.1.31), specifies the position of equilibrium; the parameters λ_1, λ_2 (hence, the constraint force (4.1.31')) are subsequently determined by the system (4.1.31'').

If the equations of the curve C are given in the explicit form

$$x_3 = \varphi(x_1, x_2), \quad x_3 = \psi(x_1, x_2), \tag{4.1.32}$$

then we get the equations which give the constraint force in the form

$$F_1 + \lambda_1 \varphi_{,1} + \lambda_2 \psi_{,1} = 0, \quad F_2 + \lambda_1 \varphi_{,2} + \lambda_2 \psi_{,2} = 0, \quad F_3 - \lambda_1 - \lambda_2 = 0, \tag{4.1.32'}$$

the condition specifying the position of equilibrium being

$$F_1(\psi_{,2} - \varphi_{,2}) + F_2(\varphi_{,1} - \psi_{,1}) + F_3(\varphi_{,1}\psi_{,2} - \varphi_{,2}\psi_{,1}) = 0. \tag{4.1.32''}$$

Analogously, for the parametric form of the curve C

$$x_i = x_i(q), \quad q \in [q_1, q_2], \quad i = 1, 2, 3, \tag{4.1.33}$$

the components of the constraint force must fulfil the condition (the constraint force is contained in the plane normal to the curve C, hence it is normal to the tangent of director parameters $x_i'(q)$, $i = 1, 2, 3$)

$$R_i x_i'(q) = 0. \tag{4.1.33'}$$

Statics

This relation is identically fulfilled if

$$R_i = \epsilon_{ijk} \lambda_j x'_k, \quad i = 1,2,3, \tag{4.1.34}$$

where λ_j, $j = 1,2,3$, are arbitrary parameters; the equations of equilibrium become

$$F_i + \epsilon_{ijk} \lambda_j x'_k = 0, \quad i = 1,2,3, \tag{4.1.33''}$$

and are compatible if

$$F_i x'_i(q) = 0. \tag{4.1.33'''}$$

The condition (4.1.33''') determines the values of the parameter $q \in [q_1, q_2]$, which correspond to the position of equilibrium. In this case, the system (4.1.33'') leads to

$$\lambda_i = k x'_i + \frac{1}{3} \epsilon_{ijk} \frac{F_j}{x'_k}, \quad i = 1,2,3, \tag{4.1.34'}$$

where k is an indeterminate parameter; replacing in (4.1.34), we find the constraint forces in the form

$$R_i = -F_i, \quad i = 1,2,3. \tag{4.1.34''}$$

Using Frenet's frame of reference, we notice that $R_\tau = 0$, so that

$$F_\tau = 0, \quad R_\nu = -F_\nu, \quad R_\beta = -F_\beta; \tag{4.1.35}$$

the first of these relations gives the position of equilibrium, while the other two relations specify the constraint force.

In the particular case of a heavy particle P, constrained to stay on a fixed smooth circle in a vertical plane (Fig.4.3,a), we have to do with an ideal constraint; the constraint force is reduced to the normal component ($\mathbf{T} = \mathbf{0}$), and the equation of equilibrium is of the form

$$\mathbf{G} + \mathbf{N} = \mathbf{0}. \tag{4.1.36}$$

Taking into account the Theorem 4.1.2, there results that the positions of equilibrium are the points P_1 and P_2; the normal reaction is given by

$$\mathbf{N} = -\mathbf{G}. \tag{4.1.36'}$$

In general, as it was shown in Chap. 3, Sec. 2.2, the particle P is subjected to holonomic, scleronomic constraints of the form

$$f_l(x_1, x_2, x_3) = 0, \quad l = 1,2, \tag{4.1.37}$$

or to scleronomic, non-holonomic ones of the form

$$\boldsymbol{\alpha}_k \cdot \mathrm{d}\mathbf{r} = 0, \quad k = 1,2; \tag{4.1.37'}$$

the total number of constraints may be at the most two. If we should have three constraints, then the particle would be at a fixed point. We notice also that the particle may be subjected only to holonomic constraints (as it was considered above) or only to non-holonomic constraints or to a holonomic constraint and to a non-holonomic one. In the case of ideal constraints, the virtual work of the constraint forces vanishes

$$\delta W_R = \mathbf{R} \cdot \delta \mathbf{r} = 0, \tag{4.1.38}$$

while the constraint force is given by the formula (3.2.37) in the form

$$\mathbf{R} = \sum_{l=1}^{2} \lambda_l \operatorname{grad} f_l + \sum_{k=1}^{2} \mu_k \boldsymbol{\alpha}_k . \tag{4.1.38'}$$

We notice that the first sum corresponds to the cases in which the particle is constrained to stay on a fixed smooth surface or curve; the constraint force is expressed only with the aid of the second sum if only non-holonomic constraints appear. The equations of equilibrium are of the form

$$\mathbf{F} + \lambda_1 \operatorname{grad} f_1 + \lambda_2 \operatorname{grad} f_2 + \mu_1 \boldsymbol{\alpha}_1 + \mu_2 \boldsymbol{\alpha}_2 = \mathbf{0}, \tag{4.1.39}$$

and only one or two of the four indetermined parameters λ_1, λ_2, μ_1, μ_2 (the vector coefficients of the other three or two parameters are equal to zero, because the corresponding constraints do not take place) are involved. In the general case, the problem is solved as it was shown above (the cases in which only one indeterminate parameter λ is involved or only two indeterminate parameters λ_1, λ_2 are involved).

1.1.6 Particle subjected to unilateral ideal constraints

We admitted, in the previous subsections, that the constraints are bilateral. Analogously, *the unilateral holonomic, scleronomic* constraints of a particle may be expressed in the form

$$f_l(x_1, x_2, x_3) \geq 0, \quad l = 1,2, \tag{4.1.40}$$

and the *unilateral non-holonomic, scleronomic* ones in the form

$$\boldsymbol{\alpha}_k \cdot \mathrm{d}\mathbf{r} \geq 0, \quad k = 1,2, \tag{4.1.40'}$$

their total number being at the most equal to two; if three constraints should be, then the particle would be at a fixed point, and all the relations would be equalities. As in the case of bilateral constraints, the particle may be subjected only to holonomic or only to non-holonomic constraints or to a holonomic constraint and to a non-holonomic one. As

Statics

it was shown in Chap. 3, Subsec. 2.2.9, in the case of unilateral constraints, the virtual work of the constraint forces given by the formula (4.1.38') is non-negative

$$\delta W_R = \mathbf{R} \cdot \delta \mathbf{r} \geq 0. \qquad (4.1.41)$$

To solve the problem of equilibrium of a particle subjected to unilateral ideal constraints, the positions of equilibrium are firstly determined, supposing that the constraints are bilateral (equality in relations (4.1.40), (4.1.40')); then, the direction of the resultant **F** is analysed for each of these positions. If this direction ensures the constraint, then the position of equilibrium is possible; otherwise, the positions of equilibrium thus obtained do not correspond to the imposed constraints.

In the case of a particle which verifies a condition of the form

$$f(x_1, x_2, x_3) \geq 0, \qquad (4.1.42)$$

the constraint force is given by

$$\mathbf{R} = \lambda \operatorname{grad} f, \quad \lambda \geq 0, \qquad (4.1.42')$$

and is directed along the normal to the surface $f(x_1, x_2, x_3) = 0$, in the direction in which the function f is increasing, as well as its gradient (to respect the imposed unilateral constraint). As it is shown by the condition (4.1.4), for equilibrium one must have $\mathbf{F} \cdot \mathbf{R} < 0$, so that

$$\mathbf{F} \cdot \operatorname{grad} f < 0 \qquad (4.1.42'')$$

or, in components,

$$F_i f_{,i} < 0. \qquad (4.1.42''')$$

In general, in the case of unilateral constraints (4.1.40), (4.1.40'), the equation of equilibrium (4.1.39) leads, on the same way, to the conditions

$$\begin{aligned}
(\mathbf{F} + \lambda_2 \operatorname{grad} f_2 + \mu_1 \boldsymbol{\alpha}_1 + \mu_2 \boldsymbol{\alpha}_2) \cdot \operatorname{grad} f_1 &< 0, \\
(\mathbf{F} + \lambda_1 \operatorname{grad} f_1 + \mu_1 \boldsymbol{\alpha}_1 + \mu_2 \boldsymbol{\alpha}_2) \cdot \operatorname{grad} f_2 &< 0,
\end{aligned} \qquad (4.1.43)$$

$$\begin{aligned}
(\mathbf{F} + \lambda_1 \operatorname{grad} f_1 + \lambda_2 \operatorname{grad} f_2 + \mu_2 \boldsymbol{\alpha}_2) \cdot \boldsymbol{\alpha}_1 &< 0, \\
(\mathbf{F} + \lambda_1 \operatorname{grad} f_1 + \lambda_2 \operatorname{grad} f_2 + \mu_1 \boldsymbol{\alpha}_1) \cdot \boldsymbol{\alpha}_2 &< 0,
\end{aligned} \qquad (4.1.43')$$

where $\lambda_1, \lambda_2, \mu_1, \mu_2 \geq 0$ (in fact, only two of these conditions take place, because one can have only two unilateral constraints).

Let be, for instance, the case of a particle of weight **G**, linked by a flexible and inextensible thread of length l to a fixed point O and constrained to stay on the vertical plane Ox_1x_2 (Fig.4.3,a); hence, the particle must be on a circle in this plane or in its interior, satisfying the condition

$$f(x_1, x_2) = l^2 - x_1^2 - x_2^2 \geq 0. \tag{4.1.44}$$

Noting that $F_1 = F_3 = 0$, $F_2 = -G$, assuming that the constraint is bilateral and associating the constraint relation $x_3 = 0$, the condition (4.1.31''') reads

$$\begin{vmatrix} 0 & -2x_1 & 0 \\ -G & -2x_2 & 0 \\ 0 & 0 & 1 \end{vmatrix} = -2Gx_1 = 0,$$

so that the positions of equilibrium are the points $P_1(0, l, 0)$ and $P_2(0, -l, 0)$ (the same positions as in the previous subsection); the condition (4.1.42''') becomes $Gx_2 < 0$, and is verified only at the point P_2, which is the only position of equilibrium corresponding to the unilateral constraint.

1.1.7 Notions concerning the stability of equilibrium

Let P be a particle of position vector $\mathbf{r} = \mathbf{r}(t)$. If $t = t_0$ is an initial moment, then we say that $\mathbf{r} = \mathbf{c} = \overrightarrow{\text{const}}$ is a *position of equilibrium* if $\mathbf{r}(t_0) = \mathbf{c}$, $\dot{\mathbf{r}}(t_0) = \mathbf{0} \Rightarrow \mathbf{r}(t) \equiv \mathbf{c}$; hence, if the particle P is at the mentioned position with zero velocity, then it remains at any moment at this position. Taking into account the form (1.1.95) of the force $\mathbf{F}$, Newton's equation (1.1.89) leads – in this case – to

$$\mathbf{F}(\mathbf{c}, \mathbf{0}; t) = \mathbf{0} \tag{4.1.45}$$

for a free particle. If this force depends on time, then the equation (4.1.45) has not, in general, a constant solution (the same for any t); if the force does not depend on time, then the equation (4.1.45) represents the necessary and sufficient condition of equilibrium and one can obtain the vector $\mathbf{c}$.

We say that the position of equilibrium is *stable* if $\forall \varepsilon > 0, \varepsilon' > 0$, $\exists \eta > 0, \eta' > 0$, so that $|\mathbf{r}(t_0) - \mathbf{c}| < \eta$, $|\dot{\mathbf{r}}(t_0)| < \eta' \Rightarrow |\mathbf{r}(t) - \mathbf{c}| < \varepsilon$, $|\dot{\mathbf{r}}(t)| < \varepsilon'$, $\forall t > t_0$; hence, perturbing the position of equilibrium in a sufficiently small neighbourhood with a sufficiently small velocity, this position remains at any moment in a previously given neighbourhood, and its velocity is not greater than a certain limit, previously given too. Otherwise, the position of equilibrium is *instable*, and can be *labile* or, at the limit, *critical*; in the latter case, the equilibrium can be *indifferent* (any position of the particle is a position of equilibrium).

For instance, in the particular case considered in Subsec. 1.1.2, one finds easily that P_1 represents a labile position of equilibrium, while P_2 is a stable one (Fig.4.3,a); if, passing to the limit ($l \to \infty$), the circle becomes a horizontal straight line, then any position is a position of equilibrium (the equilibrium is indifferent). We notice that, in the considered case, the particle has a minimal, maximal, or stationary x_2 – co-ordinate, as we are in a stable, labile, or indifferent case, respectively; we are thus led to

Theorem 4.1.3 (*E. Torricelli*). *The position of equilibrium of a particle subjected only to the action of a given uniform gravitational field is a stable, labile or indifferent position of equilibrium as this one has a minimal, maximal, or stationary applicate, respectively* (with respect to a frame of reference for which one of the axes is parallel to the considered field, opposite to its direction).

Let P be a free particle subjected to the action of a conservative force of the form (1.1.82). The condition of equilibrium (4.1.2) becomes

$$U_{,i} = 0, \quad i = 1,2,3, \qquad (4.1.46)$$

corresponding thus to the necessary conditions to have an extremum of the potential function U. When passing to the curvilinear co-ordinates q_i, $i = 1,2,3$, by relations of the form (A.1.32), we may write ($U(\mathbf{r}(q_1,q_2,q_3)) = \bar{U}(q_1,q_2,q_3)$)

$$\mathbf{F} \cdot \mathrm{d}\mathbf{r} = \mathrm{grad}\, U \cdot \mathrm{d}\mathbf{r} = \mathrm{d}U = U_{,i}\mathrm{d}x_i = \mathrm{grad}\, U \cdot \frac{\partial \mathbf{r}}{\partial q_j}\mathrm{d}q_j = U_{,i}\frac{\partial x_i}{\partial q_j}\mathrm{d}q_j = \frac{\partial \bar{U}}{\partial q_j}\mathrm{d}q_j;$$

hence, the conditions of equilibrium (4.1.46) are equivalent to the conditions

$$\frac{\partial \bar{U}}{\partial q_j} = 0, \quad j = 1,2,3, \qquad (4.1.46')$$

in curvilinear co-ordinates.

In the particular case considered above, the gravity force $\mathbf{G}$ is conservative and derives from the potential $U = -Gx_2$ (the additive constant is taken equal to zero). With the aid of Torricelli's theorem, we notice that the minimal applicate of the particle P corresponds to a maximum of the potential U, while the maximal one corresponds to a minimum of it; if the applicate of the particle is stationary, then the potential U enjoys this property too. We may state

Theorem 4.1.4 (*Lagrange-Dirichlet*). *The position of equilibrium P^0 of a particle P subjected to scleronomic, holonomic constraints, acted upon by a field of conservative forces, the potential of which has an isolated maximum at the point P^0, is a position of stable equilibrium.*

This theorem may be proved with the aid of the theorem of energy. Intuitively, let us suppose that the potential U of a free particle P has an isolated maximum equal to U_0 at the point P^0; $U < U_0$ in the neighbourhood of P^0, so that the equipotential surface $U = U_0 - \varepsilon$, $\varepsilon > 0$ sufficiently small, is a closed surface surrounding the point P^0 and reducing, by continuity, to this very point if $\varepsilon \to 0$ (Fig.4.7,a). The force $\mathbf{F} = \mathrm{grad}\, U$ is normal to the potential surface at each point of it and is directed in the growing direction of U, hence towards the interior of the surface; the considered force does not allow the particle P to move away from the position P^0, so that this one is a position of *stable* equilibrium.

Otherwise, if – at a point P^0 – U reaches an isolated minimum U_0, then the equipotential surface $U = U_0 + \varepsilon$, $\varepsilon > 0$ sufficiently small, surrounds this point, while the force $\mathbf{F}$ is normal to the mentioned surface, in the direction of the growing U, hence towards the exterior (Fig.4.7,b); this force has the tendency to move away the particle P form the point P^0, so that it represents a position of *labile* equilibrium.

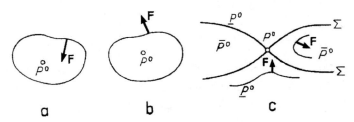

Figure 4.7. Stable (a) and labile (b) positions of equilibrium.
Conical point of a level surface (c).

Let us suppose that the three equations (4.1.46) or (4.1.46') are verified at the point P^0, but the potential U has neither an isolated maximum, nor an isolated minimum. In the neighbourhood of this point there exist two regions $\overline{P}^0$ and $\underline{P}^0$, so that in $\overline{P}^0$ the function U takes values greater that U_0 (the value of U at P^0), while in $\underline{P}^0$ it takes values smaller than U_0; these regions are separated by a level surface Σ for which $U = U_0$ and which, obviously, passes through P^0, where it has a conical point (Fig.4.7,c). The force $\mathbf{F}$ has the tendency to carry the particle P at the point P^0, in the region $\underline{P}^0$, while in the region $\overline{P}^0$ the respective force has the tendency to move away the particle from this position; because an arbitrary perturbation of the position of equilibrium P^0 may lead the particle P in the region $\overline{P}^0$, it follows that this one is an *instable* position of equilibrium (*labile* or *critic* equilibrium).

We notice that Torricelli's theorem is a particular case of the Lagrange-Dirichlet theorem.

Let be the case of a particle P constrained to stay on a fixed smooth surface S (Fig.4.6,a), given by the equations (4.1.29). The given force $\mathbf{F}$ must be normal to the surface at the respective point, hence to each of the co-ordinate curves $v = \text{const}$ and $u = \text{const}$; the conditions of equilibrium are thus of the form (Q_1 and Q_2 are called generalized forces)

$$Q_1(u,v) \equiv \mathbf{F} \cdot \frac{\partial \mathbf{r}}{\partial u} = F_i(u,v) \frac{\partial x_i}{\partial u} = 0,$$

$$Q_2(u,v) \equiv \mathbf{F} \cdot \frac{\partial \mathbf{r}}{\partial v} = F_i(u,v) \frac{\partial x_i}{\partial v} = 0,$$

(4.1.47)

obtaining thus the values of the parameters u and v corresponding to the searched positions. An interesting case is that in which $Q_1 \mathrm{d}u + Q_2 \mathrm{d}v$ is a total differential of a function $U(u,v)$; this function is thus given by

Statics

$$\bar{U}(u,v) = \int F_i(u,v)\mathrm{d}x_i(u,v). \tag{4.1.47'}$$

The positions of equilibrium correspond to the points for which the function U of two independent variables has an extremum, hence for which

$$\frac{\partial \bar{U}}{\partial u} = 0, \quad \frac{\partial \bar{U}}{\partial v} = 0. \tag{4.1.47''}$$

In particular, if the force $\mathbf{F}$ derives from a potential ($F_i = U_{,i}$, $i = 1,2,3$), then the latter one can be obtained in the form

$$U(x_1, x_2, x_3) = \int F_i(x_1, x_2, x_3)\mathrm{d}x_i; \tag{4.1.47'''}$$

the transformation relations (4.1.29) lead to $U(\mathbf{r}(u,v)) = \bar{U}(u,v)$, the conditions of equilibrium being of the form (4.1.47''). In general, the equipotential surface U passing through a position of equilibrium P^0 is tangent at the very same point to the given surface S, because the given force $\mathbf{F}$ must be normal both to this surface and to the equipotential surface. To justify – in this case – the Lagrange-Dirichlet theorem, we may study the form of the curves $\bar{U} = \bar{U}_0 \pm \varepsilon$, $\varepsilon > 0$ sufficiently small, on the surface S, $\bar{U}_0$ corresponding to the position of equilibrium.

Let us consider a particle P constrained to stay on a fixed smooth curve C (Fig.4.6,b), given by the parametric equations (4.1.33). The condition of equilibrium (4.1.33''') (the given force is normal to the curve at the point corresponding to the position of equilibrium) is of the form (Q is called generalized force)

$$Q(q) \equiv \mathbf{F} \cdot \frac{\mathrm{d}\mathbf{r}}{\mathrm{d}q} = F_i(q)x_i'(q) = 0, \tag{4.1.48}$$

obtaining thus the parameter q which corresponds to the searched positions. Taking into account the function

$$\bar{U}(q) = \int F_i(q)\mathrm{d}x_i(q) = \int Q(q)\mathrm{d}q, \tag{4.1.48'}$$

the positions of equilibrium correspond to the values of q for which the derivative vanishes

$$\frac{\mathrm{d}\bar{U}(q)}{\mathrm{d}q} = 0, \tag{4.1.48''}$$

hence, for the points of extremum of this function. If the force $\mathbf{F}$ is conservative, then the potential $U(\mathbf{r})$ is given by (4.1.47'''), while the transformation relations (4.1.33) lead to $U(\mathbf{r}(q)) = \bar{U}(q)$, the condition of equilibrium being of the form (4.1.48''). The study of the tendency of displacement of the particle on the curve in the neighbourhood of the position of equilibrium may justify the Lagrange-Dirichlet theorem.

1.1.8 Particle subjected to constraints with friction

As we have seen, in general, the constraint force which acts upon a particle can be decomposed in the form (4.1.6); in the case of an ideal constraint, the component **N** which hinders the particle to leave the constraint is sufficient. If the constraint is with friction, then the component $\mathbf{T} \neq \mathbf{0}$ hinders the particle to move along this constraint. In what follows, we use the Coulombian model introduced in Chap. 3, Subsec. 2.2.11 for the constraint force, supposing that the constraints are scleronomic and holonomic. We notice that this force is tangent to the rough surface or curve on which the particle is constrained to stay; its direction is opposite to the sliding tendency, while its modulus verifies the relation (3.2.40), the particle remaining in equilibrium.

In the case of constraints with friction, a supplementary unknown (the tangential component **T**), for the determination of which we dispose of the inequality (3.2.40), is thus introduced; in general, the corresponding problems are indeterminate (there are regions on the surface or on the curve in which the equilibrium is possible). The limit positions at which a particle remains in equilibrium may be determined in the case of a rough curve, the inequality (3.2.40) becoming an equality; but in the case of a rough surface, the limit positions of equilibrium are curves on this surface (in fact, the force **T** has two unknown components in this case).

If the particle P is subjected to constraints with friction, then the equation of equilibrium is written in the form

$$\mathbf{F} + \mathbf{N} + \mathbf{T} = \mathbf{0}; \qquad (4.1.49)$$

we associate to it the equation of the rough surface S (for a particle constrained to stay on this surface) and the inequality (3.2.40). We dispose thus of five scalar equations for the unknowns N, T and x_i^0, $i = 1,2,3$, which specify the constraint forces and the position of equilibrium P^0. Taking into account (3.2.40), the relation (4.1.49) leads to $T^2 = (\mathbf{F} + \mathbf{N})^2 = F^2 + 2\mathbf{F} \cdot \mathbf{N} + N^2 \leq f^2 N^2$; projecting on the external normal **n** to the surface, one obtains $F_n + N = 0$, as well as $\mathbf{F} \cdot \mathbf{N} = F_n N$. There results $N^2(1 + f^2) \geq F^2$; the region of equilibrium on the surface S is thus specified by the data of the problem in the form

$$F_n^2 (1 + f^2) \geq F^2. \qquad (4.1.50)$$

Noting that $F_n^2 = F^2 \cos^2 \mu$, where μ is the angle made by the force **F** (or the total constraint force **R**) with the normal to the surface at the position of equilibrium P^0, and introducing the angle of friction given by (3.2.41), we obtain the geometric condition $\mu \leq \varphi$; hence, the support of the force **F** (or of the constraint force **R**) must be in the interior or on the frontier of the cone of friction of vertex angle 2φ

(Fig.3.32,a). If the constraint is unilateral, of the form (4.1.42), then the cone of friction has only one sheet.

If the particle P is constrained to stay on a rough curve C, then we must associate the equations of the curve to the equation of equilibrium (4.1.49) and to the inequality (3.2.40); we dispose thus of six scalar relations for the unknowns N (equivalent to two unknowns, the components of $\mathbf{N}$ in the normal plane to the curve), T and x_i^0, $i = 1,2,3$, which give the constraint forces and the position of equilibrium P^0. Taking into account (3.2.40), from (4.1.49) we get $N^2 = (\mathbf{F} + \mathbf{T})^2 = F^2 + T^2 + 2\mathbf{F} \cdot \mathbf{T} \geq T^2 / f^2$; projecting on the tangent to the curve, we obtain $F_t + T = 0$, as well as $\mathbf{F} \cdot \mathbf{T} = F_t T$. The region of equilibrium on the curve C is thus specified by the data of the problem in the form

$$F_t^2 \left(1 + f^2\right) \leq f^2 F^2. \tag{4.1.51}$$

Because $F_t^2 = F^2 \cos^2 \mu$, where μ is the angle made by the force $\mathbf{F}$ (or by the total constraint force $\mathbf{R}$) with the tangent to the curve at the position of equilibrium P^0, we find the condition $\cos^2 \mu \leq \sin^2 \varphi = \cos^2(\pi/2 - \varphi)$, hence the geometric condition $\mu \geq \pi/2 - \varphi$, where φ is the angle of friction given by (3.2.41); hence, the support of the given force $\mathbf{F}$ (or of the constraint force $\mathbf{R}$) must be in the exterior or on the frontier of the cone of friction of vertex angle $\pi - 2\varphi$ (Fig.3.32,b). In the case of unilateral constraints, only one sheet of the mentioned zone corresponds to the positions of equilibrium.

A synthesis of the above results is given by

Theorem 4.1.5. *A particle constrained to stay on a fixed rough surface or curve (constraints with friction) is in equilibrium if and only if the resultant of the given forces which act upon it is contained in the interior of the cone of friction of vertex angle 2φ or in the exterior of the cone of friction of vertex angle $\pi - 2\varphi$, respectively, or on the frontier of the cone (the case of limit equilibrium). In the plane case, the cone of friction becomes an angle of friction.*

Let be a particle P of weight $\mathbf{G}$ subjected to stay on a circle of radius l, $x_1^2 + x_2^2 = l^2$, in a vertical plane $x_3 = 0$ (Fig.4.3,b). Noting that $l\mathbf{n} = x_1\mathbf{i}_1 + x_2\mathbf{i}_2$ and $\mathbf{G} = -G\mathbf{i}_2$, we get $F_n = \mathbf{G} \cdot \mathbf{n} = -Gx_2/l$; we may use the formula (4.1.50), which leads to $G^2 \left(x_2^2/l^2\right)\left(1 + f^2\right) \geq G^2$. Hence, the positions of equilibrium are on the arcs of circle $\widehat{P_1'P_1''}$ and $\widehat{P_2'P_2''}$, specified by the relations

$$-l \leq x_2 \leq -\frac{l}{\sqrt{1+f^2}}, \quad \frac{l}{\sqrt{1+f^2}} \leq x_2 \leq l, \tag{4.1.52}$$

respectively; because $f = \tan \varphi$, we may write

$$l \leq x_2 \leq -l\cos\varphi, \quad l\cos\varphi \leq x_2 \leq l \tag{4.1.52'}$$

too. The two arcs on which takes place the equilibrium are contained in an angle of vertex O and which is equal to 2φ.

1.2 Statics of discrete systems of particles

Let us consider, in what follows, free or constraint discrete mechanical systems, hence the case of a finite number of particles; We introduce the principle of virtual work in the case of ideal constraints. The general results thus obtained may be used in the case of continuous mechanical systems too.

1.2.1 Free discrete mechanical systems

Let $\mathscr{S}$ be a free discrete mechanical system, hence a finite system of n free particles $\mathscr{S} \equiv \{P_i, i = 1,2,...,n\}$. We suppose that a particle P_i of position vector $\mathbf{r}_i$ is acted upon by a given external force $\mathbf{F}_i$ (the resultant of all given external forces acting upon this particle) and by the given internal forces $\mathbf{F}_{ij}$, $j \neq i$, $i,j = 1,2,...,n$; we notice that the internal forces verify the axiomatic relation (1.1.81). The forces acting upon this mechanical system are modelled by bound vectors, hence we say that the system of particles is *at rest* with respect to a given frame of reference (the system of given forces is in equilibrium or the free discrete mechanical system is in equilibrium) if the system of bound vectors is equivalent to zero (using the principles of mechanics, as in the case of a single particle), hence if

$$\mathbf{F}_i + \sum_{j=1}^{n} \mathbf{F}_{ij} = \mathbf{0}, \quad j \neq i, \quad i = 1,2,...,n. \qquad (4.1.53)$$

A finite system of free particles is in equilibrium if any of its particles is in equilibrium; hence, any subsystem of the considered system will have this property (any particle which forms this subsystem is in equilibrium). We may thus state

Theorem 4.1.6 (*theorem of equilibrium of parts*). *If a free discrete mechanical system $\mathscr{S}$ is in equilibrium under the action of given external and internal forces, then any of its parts (any subsystem $S \subset \mathscr{S}$) will be in equilibrium too under the action of the given forces corresponding to the respective part.*

Computing the torsor of the given forces and noting that the torsor of the internal forces is equal to zero (as it was shown in Chap. 2, Subsec. 2.2.8, $\tau_O \{\mathbf{F}_{ij}\} = \mathbf{0}$), it follows that

$$\tau_O \{\mathbf{F}_i\} = \mathbf{0}; \qquad (4.1.54)$$

hence, a *necessary condition* of equilibrium is obtained by equating to zero the torsor of the given external forces with respect to an arbitrary pole. In Chap. 2, Subsec. 2.2.2 it was shown that, in the case of a non-deformable mechanical system, the forces are modelled with the aid of sliding vectors; taking into account the conditions in which a system of forces modelled by sliding vectors is equivalent to zero, it follows that, in the case of a non-deformable discrete mechanical system, the condition (4.1.54) is a *sufficient condition* of equilibrium too. This condition may be written in the form

$$\sum_{i=1}^{n} \mathbf{F}_i = \mathbf{0}, \quad \sum_{i=1}^{n} \mathbf{r}_i \times \mathbf{F}_i = \mathbf{0}. \tag{4.1.54'}$$

The condition (4.1.54) has a great advantage, i.e. it does not contain the internal forces. We may state

Theorem 4.1.7 (*theorem of rigidity*). *Supposing that a free discrete mechanical system $\mathscr{S}$ becomes rigid, the conditions of equilibrium of this new mechanical system represent necessary conditions of equilibrium for the initially given mechanical system.*

The first basic problem is that in which the forces acting upon the free discrete mechanical system $\mathscr{S}$ are given, and one must determine its position of equilibrium. In *the second basic problem*, the positions of equilibrium of the particles which form the free discrete mechanical system $\mathscr{S}$ are given, and one asks to determine the forces which act upon this system. In general, one may enunciate a *mixed basic problem* with respect to the above mentioned questions. The conditions of equilibrium (4.1.53) are equivalent to $3n$ scalar relations; in the case in which the considered mechanical system is plane (from the point of view of the positions of the particles, the forces which are acting being coplanar too), the number of these relations is reduced to $2n$, while if the mechanical system is linear (all the particles as well as the forces are on the same support), we may write only n scalar relations.

1.2.2 Constraint discrete mechanical systems

Let us consider a discrete mechanical system $\mathscr{S}$, hence a finite system of n particles $\mathscr{S} \equiv \{P_i, i = 1, 2, ..., n\}$ subjected to m scleronomic and holonomic or non-holonomic constraints. As in the previous case, we admit that a particle P_i of position vector $\mathbf{r}_i$ is acted upon by the external force $\mathbf{F}_i$ and by the internal forces $\mathbf{F}_{ij}$, $i \neq j$, $i, j = 1, 2, ..., n$; using the axiom of liberation from constraints, we introduce the external constraint force $\mathbf{R}_i$ (the resultant of all the external constraint forces which act upon this particle) and the internal constraint forces $\mathbf{R}_{ij}$, $i \neq j$, $i, j = 1, 2, ..., n$, which verify the axiomatic relation (1.1.81) too. All the forces which act upon the mechanical system are modelled by bound vectors; we may say (as in the case studied in the previous subsection) that the system of particles is *at rest* with respect to a given frame of reference (in equilibrium) if

$$\mathbf{F}_i + \mathbf{R}_i + \sum_{j=1}^{n} \left(\mathbf{F}_{ij} + \mathbf{R}_{ij} \right) = \mathbf{0}, \quad j \neq i, \quad i = 1, 2, ..., n. \tag{4.1.55}$$

Because each particle must be in equilibrium, we may state

Theorem 4.1.6' (*theorem of equilibrium of parts*). *If a constraint discrete mechanical system $\mathscr{S}$ is in equilibrium under the action of given and constraint forces, then any of* its parts (*any subsystem* $S \subset \mathscr{S}$) *will be in equilibrium too, under the action of the given and constraint forces corresponding to the respective part.*

Noting that the torsor of the internal given and constraint forces vanishes we obtain a *necessary condition* of equilibrium in the form

$$\tau_O \{ \mathbf{F}_i \} + \tau_O \{ \mathbf{R}_i \} = \mathbf{0} \tag{4.1.56}$$

or in the form

$$\sum_{i=1}^{n} (\mathbf{F}_i + \mathbf{R}_i) = \mathbf{0}, \quad \sum_{i=1}^{n} \mathbf{r}_i \times (\mathbf{F}_i + \mathbf{R}_i) = \mathbf{0}. \tag{4.1.56'}$$

In these conditions, which are also sufficient for a non-deformable discrete mechanical system, the internal forces do not intervene; this is an important advantage for computation. Hence, we state

Theorem 4.1.7' (*theorem of rigidity*). *Supposing that a constraint discrete mechanical system $\mathscr{S}$ becomes rigid, the conditions of equilibrium of the new mechanical system represent necessary conditions of equilibrium for the initially given mechanical system.*

The basic problem which arises is, in general, a *mixed problem*, in which the constraint forces acting upon the given discrete mechanical system must also be determinate. If the constraints are expressed by m distinct scalar relations, then the number of independent parameters which specify the position of equilibrium is equal to $3n - m$; these parameters may be obtained in an explicit form in the case of holonomic constraints, which are thus eliminated from the computation. The system of $3n$ scalar relations (4.1.55) allows to determine the position of equilibrium of the mechanical system as well as the constraint forces; but this system of equations is not always compatible and determinate.

A discrete mechanical system of n particles for which the equations (4.1.55) lead to a finite and determinate solution (we have $3n - m \geq 0$) is called a *statically determinate* (*isostatic*) *system*; in the case of equality, the mechanical system is *at rest*, whatever given forces are acting upon it. If the system of equations is indeterminate (the number of the unknowns of the problem is greater that the number of equations, $3n - m < 0$), then the mechanical system is *statically indeterminate* (*hyperstatic*).

1.2.3 Principle of virtual work

Let $\mathscr{S}$ be a discrete mechanical system subjected to ideal constraints (for which the virtual work of the constraint forces (3.2.36) vanishes). Starting from the necessary and sufficient conditions of equilibrium (4.1.55), written in the form ($\mathbf{F}_i$ and $\mathbf{R}_i$ are the resultants of all given and constraint forces, respectively, immaterial if they are external or internal)

$$\mathbf{F}_i + \mathbf{R}_i = \mathbf{0}, \quad i = 1,2,...,n, \tag{4.1.57}$$

performing a scalar product by the virtual displacements $\delta \mathbf{r}_i$, summing for all the particles of the system $\mathscr{S}$ and taking into account the relation of definition of the ideal constraints (3.2.36), we obtain the relation

$$\delta W = \sum_{i=1}^{n} \mathbf{F}_i \cdot \delta \mathbf{r}_i = 0, \qquad (4.1.58)$$

which represents a *necessary condition* of equilibrium.

Supposing that the condition (4.1.58) is fulfilled and that p constraints of the form (3.2.21") and m constraints of the form (3.2.15) take place, one can use the method of Lagrange's multipliers; we may thus write

$$\sum_{i=1}^{n} \left(\mathbf{F}_i + \sum_{l=1}^{p} \lambda_l \nabla_i f_l + \sum_{k=1}^{m} \mu_k \boldsymbol{\alpha}_{ki} \right) \cdot \delta \mathbf{r}_i = 0,$$

where λ_l, $l = 1,2,...,p$, μ_k, $k = 1,2,...,m$, are scalars to be determined (the Lagrange's multipliers) and where we notice that in a finite double sum one can invert the order of summation. By a reasoning analogous to that given in Chap. 3, Subsec. 2.2.9, we obtain

$$\mathbf{F}_i + \sum_{l=1}^{p} \lambda_l \nabla_i f_l + \sum_{k=1}^{m} \mu_k \boldsymbol{\alpha}_{ki} = \mathbf{0}, \quad i = 1,2,...,n. \qquad (4.1.59)$$

We find again the relations (3.2.37), which give the constraint forces; hence, the relations (4.1.59) are equivalent to the relations (4.1.57). We may state (the relation (4.1.58) is now a *sufficient condition* too)

Theorem 4.1.8 (*theorem of virtual work*). *The necessary and sufficient condition of equilibrium of a discrete mechanical system subjected to ideal constraints and acted upon by a system of given forces is obtained by equating to zero the virtual work of these forces for any system of virtual displacements.*

Taking into account the equivalence between the relation (4.1.58), which represents the theorem of virtual work, and the relations (4.1.57), which represent the form taken by *Newton's principle* (3.2.35') *in the static case* (equating to zero the accelerations $\mathbf{a}_i$), it follows that the theorem of virtual work may be considered as being a principle (*the principle of virtual work* or *the principle of virtual displacements*), because, starting from it, one can solve the basic problems of statics.

In contradistinction to the necessary condition (4.1.56) or to the necessary conditions (4.1.56'), where the internal forces do not appear, but the constraint forces do intervene, in the necessary and sufficient condition (4.1.58) are involved all the given forces (external and internal), but the constraint forces are absent; it is an advantage for the computation, because one can specify the position of equilibrium even if the constraint forces are not determined.

The equations (4.1.59) are called *Lagrange's equations of equilibrium of the first kind*.

Introducing the virtual displacements (3.2.1'), we may write the condition (4.1.58) in the form

$$\sum_{i=1}^{n} \mathbf{F}_i \cdot \mathbf{v}_i^* = 0, \qquad (4.1.58')$$

so that the above considered principle may be called *the principle of virtual velocities* too.

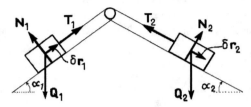

Figure 4.8. Equilibrium of two heavy bodies on inclined plains.

Let be, for instance, two solids of weights $\mathbf{Q}_1$ and $\mathbf{Q}_2$, respectively, staying on two planes inclined by the angles α_1 and α_2, respectively, with respect to a horizontal line, and linked by an inextensible thread, which passes over a small pulley (Fig.4.8). Between the two bodies (which may be modelled as two particles) and the inclined planes arise the external constraint forces $\mathbf{N}_1$ and $\mathbf{N}_2$, respectively, while in the thread appear the tensions $\mathbf{T}_1$ and $\mathbf{T}_2$, respectively (internal constraint forces for the considered mechanical system); we admit that do not appear frictions (the considered constraints are ideal). The principle of virtual work is written in the form

$$\mathbf{Q}_1 \cdot \delta \mathbf{r}_1 + \mathbf{Q}_2 \cdot \delta \mathbf{r}_2 = -Q_1 |\delta \mathbf{r}_1| \sin \alpha_1 + Q_2 |\delta \mathbf{r}_2| \sin \alpha_2 = 0;$$

noting that $|\delta \mathbf{r}_1| = |\delta \mathbf{r}_2|$, there results the necessary and sufficient condition of equilibrium

$$Q_1 \sin \alpha_1 = Q_2 \sin \alpha_2, \qquad (4.1.60)$$

which is independent of the constraint forces.

A particle P constrained to stay on a fixed smooth surface S is specified by the position vector $\mathbf{r} = \mathbf{r}(u,v)$, where u and v are co-ordinates on the surface; the principle of virtual work

$$\mathbf{F} \cdot \delta \mathbf{r} = \mathbf{F} \cdot \frac{\partial \mathbf{r}}{\partial u} \delta u + \mathbf{F} \cdot \frac{\partial \mathbf{r}}{\partial v} \delta v = Q_1(u,v)\delta u + Q_2(u,v)\delta v = 0$$

yields conditions of equilibrium of the form (4.1.47) for the generalized forces Q_1 and Q_2. If the particle P is constrained to stay on a fixed smooth curve C, being specified by the position vector $\mathbf{r} = \mathbf{r}(q)$, where q is a parameter, then we may write the principle of virtual work in the form

$$\mathbf{F} \cdot \delta \mathbf{r} = \mathbf{F} \cdot \frac{\partial \mathbf{r}}{\partial q} \delta q = Q(q) \delta q = 0 \, ;$$

we find thus again the condition of equilibrium (4.1.48) for the generalized force.

In the case of *unilateral ideal constraints* of the form (3.2.16^{iv}) or of the form (3.2.16), the virtual work of the constraint forces verifies the inequality (3.2.36'). The principle of virtual work will be expressed in the form

$$\delta W = \sum_{i=1}^{n} \mathbf{F} \cdot \delta \mathbf{r}_i \leq 0 \qquad (4.1.61)$$

for any system of virtual displacements, representing *the necessary and sufficient condition of equilibrium* of a discrete mechanical system subjected to unilateral ideal constraints; in this case too, one may make considerations analogous to those made above.

2. Statics of solids

Among the continuous mechanical systems, we consider – in what follows – only the solids, that is rigid solids and deformable ones. Concerning the latter ones, we deal only with perfect flexible, torsionable and inextensible threads, as well as with bars and systems of bars; the general study of deformable solids needs more complex mathematical models.

2.1 Statics of rigid solids

We start with the general problem of equilibrium of free and constraint rigid solids; the results thus obtained are then applied to various particular cases (the rigid solid with a fixed point or axis, the rigid solid subjected to constraints with friction etc.). The systems of rigid solids are taken into consideration too.

2.1.1 Statics of the free rigid solid

Let be a *free rigid solid*, hence a non-deformable continuous mechanical system subjected to a system of given forces $\{\mathbf{F}_i, i = 1, 2, ..., n\}$; these forces are external ones. The internal forces are due to the constraints (the cohesion forces between the particles which constitute the rigid solid) and do not intervene in computation; moreover, the work effected by these forces vanishes.

A free rigid system may have any position in space that depends only on the system of forces acting upon it. As we have seen in Chap. 2, Subsec. 2.2.2, the forces are modelled by means of sliding vectors in the case of a non-deformable mechanical system. The necessary and sufficient condition of equilibrium of a free rigid solid acted upon by the given forces $\{\mathbf{F}_i\}$ is written in the form

$$\tau_O\{\mathbf{F}_i\} = \mathbf{0}, \quad \mathbf{R} = \sum_{i=1}^{n} \mathbf{F}_i = \mathbf{0}, \quad \mathbf{M}_O = \sum_{i=1}^{n} \mathbf{r}_i \times \mathbf{F}_i = \mathbf{0}; \quad (4.2.1)$$

one can thus state

Theorem 4.2.1. *A free rigid solid subjected to the action of a system of given forces is at rest (in equilibrium) with respect to a fixed frame of reference if and only if the torsor of these forces with respect to an arbitrary pole vanishes.*

Projecting the condition (4.2.1) on the co-ordinate axes, we get six conditions of equilibrium

$$R_j = \sum_{i=1}^{n} F_{ij} = 0, \quad M_{Oj} = \sum_{i=1}^{n} \epsilon_{jkl} x_k^{(i)} F_{il} = 0, \quad j = 1,2,3. \quad (4.2.1')$$

We notice that these scalar conditions may be replaced by other equivalent scalar conditions. Thus, *a free rigid solid is in equilibrium if and only if the sum of the moments of all the given forces with respect to each of the edges of a non-degenerate tetrahedron* (the six straight lines do not belong to the same complex of first degree) *vanishes*. Obviously, these conditions are necessary. Let us suppose that they hold for a tetrahedron $OA_1A_2A_3$. Because the sum of the moments with respect to the edges OA_1, OA_2, OA_3 is zero, it follows that the moment with respect to the pole O is zero too; in this case, the given forces are in equilibrium or have a unique resultant, which passes through O. This reasoning may be repeated for all the vertices of the tetrahedron, so that the given forces must be in equilibrium, because there cannot exist a unique resultant passing through all four vertices; hence, the above mentioned conditions are sufficient too. But these equations are not independent, hence they are not conditions of equilibrium for the free rigid solid if: three of the straight lines are concurrent or parallel and coplanar at the same time (in particular, two of the straight lines are parallel, while a third one is the straight line at infinity of the plane defined by the first two ones or three of these straight lines are concurrent straight lines at infinity), four of the straight lines are generatrices of the same family of a ruled quadric (in particular, they can be concurrent or parallel), five of the straight lines intersect two other straight lines or intersect a same straight line and are parallel to the same plane (they belong to a linear congruence) or the six straight lines intersect the same straight line or are parallel to the same plane (they belong to a linear complex).

In particular, in the case of a system of coplanar forces acting upon the free rigid solid, three of the equations (4.2.1') are identically verified; there remain two equations for the resultant $\mathbf{R}$ (projections on two non-parallel axes in the plane of the forces) and an equation of moment with respect to an axis normal to the considered plane. Corresponding to the results in Chap. 2, Subsec. 2.2.6, we may replace these equations by three equations of moment with respect to three non-coplanar axes, normal to the plane of forces; as well, we may use two equations of moment with respect to two axes normal to the plane of forces and an equation of projection of their resultant on an axis contained in this plane and which is not normal to the plane of the other two axes. If the above mentioned conditions are not fulfilled, then one of the scalar equations is a linear consequence of the other two ones.

Corresponding to the results in Chap. 2, Subsec. 2.2.7, in the case of a system of parallel forces, one can use for the resultant an equation of projection on the common direction of the forces and one may write two equations of moment with respect to two non-parallel axes, normal to this direction; we can use also three equations of moment about three non-concurrent and non-parallel axes, contained in a plane normal to the direction of the forces.

If $n = 1, 2, ..., 6$, then one can put into evidence some necessary conditions of equilibrium of the free rigid solid, which must be verified a priori and which depend on the geometric configuration of the given system of forces. So, a system formed of only one non-zero force cannot be in equilibrium. A system of two forces can be in equilibrium only if the forces have the same support; as well, a system of three forces is in equilibrium only if their supports are concurrent or parallel and coplanar. For $n = 4$ it is necessary that the supports belong to the same linear series of straight lines (e.g., generatrices of the same family of a ruled quadric – in particular, concurrent or parallel) to be in equilibrium. A necessary condition of equilibrium for $n = 5$ is the belonging of the supports of the forces to the same congruence of the first degree (for instance, they intersect two straight lines or they intersect a straight line and are parallel to a plane). A system of six forces is in equilibrium (necessary condition) if their supports belong to the same complex of first degree (e.g., an intersection with the same straight line or the parallelism to a same plane).

In *the first basic problem*, the forces which act upon the free rigid solid are given, and one asks the position of equilibrium. As we have seen in Chap. 3, Subsec. 2.2.3, a free rigid solid has six degrees of freedom; in this case, the unknowns are the six parameters (eventually, the co-ordinates of a point of the rigid solid and the three Euler's angles), which specify the position of the rigid solid. If the system of six equations of equilibrium is indeterminate, then there exists an infinity of possible positions of equilibrium, while if this system of equations is impossible, then such a position does not exist. These observations may be put in connection with the considerations previously made, concerning the cases in which one cannot have equilibrium or in which some necessary conditions of equilibrium have been emphasized.

The second basic problem is that in which the position of equilibrium of the free rigid solid is given, and the forces which must act upon it to maintain this position are searched; obviously, one supposes that this system of forces depends on a certain number parameters, which are the unknowns of the problem (the magnitudes and the directions of the forces). The solution of the problem is, in general, indeterminate; if certain conditions, which limit the number of the unknowns to six, are imposed, then it is possible that the solution of the problem be determinate.

We mention *the mixed basic problem* too, in which the position of equilibrium of the rigid solid is partially known, as well as the system of forces; in this case, the position of equilibrium and the system of forces are searched.

2.1.2 Statics of the rigid solid with ideal constraints

The six degrees of freedom of a free rigid solid may be partially or totally annulled by the introduction of some *constraints*, which we suppose to be *ideal*. Applying the axiom of liberation from constraints, there appear constraint forces (reactions); supplementary unknowns are thus introduced, but less scalar parameters are necessary to determine the position of equilibrium. Let us suppose that the rigid solid with ideal constraints is acted upon by a system of given forces $\{\mathbf{F}_i, i = 1,2,...,n\}$ and a system of constraint forces $\{\mathbf{R}_j, j = 1,2,...,m\}$; in this case, *the necessary and sufficient condition of equilibrium* reads

$$\tau_O\{\mathbf{F}_i\} + \tau_O\{\mathbf{R}_i\} = \mathbf{0}, \quad \mathbf{R} + \mathbf{\bar{R}} = \mathbf{0}, \quad \mathbf{M}_O + \mathbf{\overline{M}}_O = \mathbf{0}, \tag{4.2.2}$$

where we have introduced the torsor of given forces in the form

$$\mathbf{R} = R_k \mathbf{i}_k = \sum_{i=1}^{n} \mathbf{F}_i, \quad \mathbf{M}_O = M_{Ok} \mathbf{i}_k = \sum_{i=1}^{n} \mathbf{r}_i \times \mathbf{F}_i, \tag{4.2.2'}$$

while the torsor of constraint forces is given by

$$\mathbf{\bar{R}} = \bar{R}_k \mathbf{i}_k = \sum_{j=1}^{m} \mathbf{R}_j, \quad \mathbf{\overline{M}}_O = \overline{M}_{Ok} \mathbf{i}_k = \sum_{j=1}^{m} \mathbf{r}_j \times \mathbf{R}_j. \tag{4.2.2''}$$

We may thus state

Theorem 4.2.1'. *A rigid solid subjected to ideal constraints is at rest (in equilibrium) with respect to a fixed frame of reference if and only if the sum of the torsors of given and constraint forces with respect to the same arbitrary pole vanishes.*

Projecting on the co-ordinate axes, we get six conditions of equilibrium

$$\sum_{i=1}^{n} F_{ik} + \sum_{j=1}^{m} R_{jk} = 0, \quad \sum_{i=1}^{n} \epsilon_{kpq} x_p^{(i)} F_{iq} + \sum_{j=1}^{m} \epsilon_{kpq} x_p^{(j)} R_{jq} = 0, \quad k = 1,2,3. \tag{4.2.2'''}$$

The basic problem is, in general, a *mixed problem* in which both the unknown position of equilibrium and the constraint forces are searched. Let p and q be the number of unknown scalars necessary to determine the constraint forces and the position of equilibrium, respectively. If in such a problem we have $p + q = 6$, then this one is, in general, determinate (however, it is possible that in some particular cases (critical cases) be indeterminate), and we say that the rigid solid is *statically determinate (isostatic)*. If $p + q > 6$, then the problem is indeterminate, the rigid solid being *statically indeterminate (hyperstatic)*. We are limited by the mathematical model chosen for the solid, so that the unknowns of the problem cannot be determined; if we consider a deformable solid, closer to physical reality, completing thus the mathematical model, then there arise supplementary relations which allow the complete solving of the problem. If $p + q < 6$, then the problem is, in general, impossible from the point of

Statics 233

view of the rest with respect to a fixed frame of reference, and we have to do with a *mechanism* (in some particular cases, for certain systems of forces, the equilibrium could be possible). In what follows, we deal only with statically determined rigid solids.

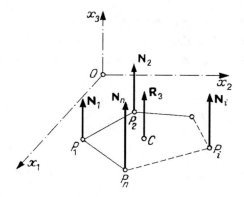

Figure 4.9. Polygon of sustentation.

Among the ideal constraints frequently encountered, we mention: *the simple support, the hinge* (equivalent to three or two simple supports) and *the built-in support* (equivalent to six simple supports), introduced in Chap. 3, Subsec. 2.2.10; taking into account the above mentioned equivalences, we may consider the case of *the rigid solid on several simple supports* too (in particular, the case of six simple supports), case considered in the same subsection. Let thus be a rigid solid leaning on the plane $x_3 = 0$ at the points $P_i\left(x_1^{(i)}, x_2^{(i)}, 0\right)$, $i = 1, 2, ..., n$ (Fig.4.9). The equations of equilibrium are of the form

$$R_1 = R_2 = 0, \quad R_3 + \sum_{i=1}^{n} N_i = 0,$$

$$M_{O1} + \sum_{i=1}^{n} N_i x_2^{(i)} = 0, \quad M_{O2} - \sum_{i=1}^{n} N_i x_1^{(i)} = 0, \quad M_{O3} = 0,$$

where $\mathbf{N}_i$, $i = 1, 2, ..., n$, are the unknown constraint forces. Hence, the external given forces must verify the conditions $R_1 = R_2 = M_{O3} = 0$, so that the rigid solid be in equilibrium; these forces must reduce to a resultant $\mathbf{R}_3$, normal to the plane $x_3 = 0$, because the scalar of the torsor vanishes ($R_i M_{Oi} = 0$). The other three equations determine the constraint forces, and the problem is indeterminate if the number of the points of support is greater than three. If $n = 3$, then the given force $\mathbf{R}_3$ must be decomposed in three components of supports parallel to this force. We can mention, e.g., the tripod of a painter or of a shoemaker; in case of a four-legged stool ($n = 4$), the problem is statically indeterminate (excepting the case in which the force $\mathbf{R}_3$ acts at the middle of the square $P_1 P_2 P_3 P_4$). The solution of the problem is determined if, for $n = 3$, the points P_1, P_2 and P_3 are not collinear; otherwise, the solution is indeterminate if the force $\mathbf{R}_3$ pierces the straight line on which are the three points or impossible if it does not pierce it. Hence, the rigid body simple supported on more than two points, the reactions of which have supports parallel and coplanar, constitutes a hyperstatic mechanical system. If the supports mentioned above are unilateral constraints (as it happens in most cases), then $N_i > 0$, $i = 1, 2, ..., n$; these reactions are modelled by a system of parallel sliding vectors, their resultant $\mathbf{R}_3$ being along the central axis, which pierces the plane $x_3 = 0$ at the point C of co-ordinates

$$\xi_1 = \frac{\sum_{i=1}^{n} N_i x_1^{(i)}}{\sum_{i=1}^{n} N_i}, \quad \xi_2 = \frac{\sum_{i=1}^{n} N_i x_2^{(i)}}{\sum_{i=1}^{n} N_i}.$$

There results that $x_{k\min}^{(i)} < \xi_k < x_{k\max}^{(i)}$, $k = 1, 2$; hence, the point C is in the interior of a convex polygon, which contains in its interior or on its contour the points P_i, $i = 1, 2, ..., n$. The polygon of minimal area which fulfils these conditions is called *polygon of sustentation*. Hence, to have equilibrium, the resultant of the external given forces must pierce the plane of support (case of unilateral constraints) in the interior of the polygon of sustentation.

Another important constraint, put in evidence in Chap. 3, Subsec. 2.2.10, is *the constraint by threads*. This constraint is unilateral, introducing only one unknown (*the tension* in the thread) if the direction of the thread is fixed. If the direction of the thread may be anyone, then the constraint force has three unknown components; in this case, it is possible that the thread does not introduce geometric restrictions (no constraints) for the rigid solid (Fig.4.2,a) or may constrain the point of fixing of the same rigid solid to stay on a curve or on a surface (Fig.4.2,b).

In the case of a *system of coplanar given forces* (e.g., contained in a plane $x_3 = $ const) remain only three conditions of equilibrium

$$\sum_{i=1}^{n} F_{ik} + \sum_{j=1}^{m} R_{jk} = 0, \quad k = 1, 2,$$

$$\sum_{i=1}^{n} \left(x_1^{(i)} F_{i2} - x_2^{(i)} F_{i1} \right) + \sum_{j=1}^{m} \left(x_1^{(j)} R_{j2} - x_2^{(j)} R_{j1} \right) = 0. \quad (4.2.3)$$

Analogously, if the rigid solid is acted upon by a *system of parallel given forces* (e.g., with supports parallel to the axis Ox_3), then the equations of equilibrium read

$$\sum_{i=1}^{n} F_{i3} + \sum_{j=1}^{m} R_{j3} = 0, \quad \sum_{i=1}^{n} \epsilon_{kpq} \, x_p^{(i)} F_{iq} + \sum_{j=1}^{m} \epsilon_{kpq} \, x_p^{(j)} R_{jq} = 0, \quad k = 1, 2. \quad (4.2.4)$$

As in the case of a free rigid solid, the conditions of equilibrium may be expressed also in other forms, equivalent to those above.

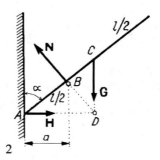

Figure 4.10. Equilibrium of a homogeneous heavy straight bar, which leans on a vertical wall and at a fixed point.

Let be, for instance, a homogeneous straight bar of length $2l$, which leans at A on a vertical wall and at a fixed point B, at a distance from the wall, which is acted upon only by its own weight **G** (Fig.4.10). We introduce the constraint forces **N** and **H**, so that the bar be in equilibrium under the action of the given and constraint forces; for equilibrium, the supports of the three forces must be concurrent. We write two equations of moments with respect to the points A and B and an equation of projection of the forces on the vertical (we eliminate thus a constraint force from each equation), in the form

$$N\frac{a}{\sin\alpha} - Gl\sin\alpha = 0, \quad Ha\cot\alpha - G(l\sin\alpha - a) = 0, \quad N\sin\alpha - G = 0,$$

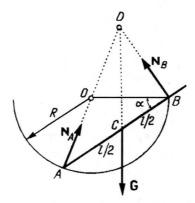

Figure 4.11. Equilibrium of a homogeneous heavy straight bar, leaned on a body bounded by a semispherical surface.

where α is the angle made by the bar with the horizontal (the generalized co-ordinate, which determines the position of equilibrium); we get

$$N = G\sqrt[3]{\frac{l}{a}}, \quad H = G\sqrt[3]{\frac{l}{a}}\left[1 - \sqrt[3]{\left(\frac{a}{l}\right)^2}\right], \quad \sin^3\alpha = \frac{a}{l}, \quad (4.2.5)$$

obtaining thus the constraint forces (because $N, H \geq 0$, it results that the directions of these forces have been correctly chosen) and the position of equilibrium. We notice that the condition $a \leq l$ must hold; in the limit case, the equilibrium is labile, and $N = G$, $H = 0$.

We suggest to the reader the solving of the problem (Fig.4.11) (the equilibrium of a homogeneous heavy straight bar leaned on a body bound by a semispherical surface).

In the general case of a rigid solid subjected to constraints without friction takes place a relation of the form (3.2.39); interesting particular cases have been considered in Chap. 3, Subsec. 2.2.10.

2.1.3 Statics of a rigid solid with a fixed point or axis

Let be a *rigid solid with a fixed point* (which may be a spherical hinge), subjected to the action of given forces of torsor $\tau_O \{ \mathbf{F}_i \} = \{ \mathbf{R}, \mathbf{M}_O \}$; without losing anything from the generality, we may assume that the pole O is at the fixed point (Fig.4.12). One has only one constraint force $\bar{\mathbf{R}}$ at this point, so that the equations (4.2.2) lead to the conditions

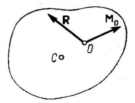

Figure 4.12. Equilibrium of a rigid solid with a fixed point.

$$\mathbf{M}_O = \mathbf{0}, \qquad (4.2.6)$$

which must be fulfilled by the given forces and which determines the position of equilibrium, and to the constraint force

$$\bar{\mathbf{R}} = -\mathbf{R}. \qquad (4.2.6')$$

If the fixed point is just the centre of gravity ($C \equiv O$), while the rigid solid is subjected only to the action of its own weight $\mathbf{G}$, then the condition (4.2.6) is identically fulfilled, and we have $\bar{\mathbf{R}} = -\mathbf{G}$; the rigid solid is thus in equilibrium in any position, the respective property being characteristic for the centre of gravity.

Let us consider a rigid solid which, besides the fixed point O, admits another fixed point O'; in this case, the axis OO' is a fixed one, and we have to do with a *rigid solid with a fixed axis* (taken as axis Ox_3) (Fig.4.13). The two fixed points at which, due to the system of given forces, appear the constraint forces $\bar{\mathbf{R}}$ and $\bar{\mathbf{R}}'$, may be spherical hinges. The equations (4.2.2) read

$$\mathbf{R} + \bar{\mathbf{R}} + \bar{\mathbf{R}}' = \mathbf{0}, \quad \mathbf{M}_O + \mathbf{h} \times \bar{\mathbf{R}}' = \mathbf{0}, \qquad (4.2.7)$$

where $\overrightarrow{OO'} = \mathbf{h}$, h being the distance between the two fixed points; in a developed form, we may write

$$R_1 + \overline{R}_1 + \overline{R}'_1 = 0, \quad R_2 + \overline{R}_2 + \overline{R}'_2 = 0, \quad R_3 + \overline{R}_3 + \overline{R}'_3 = 0,$$
$$M_{O1} - h\overline{R}'_2 = 0, \quad M_{O2} + h\overline{R}'_1 = 0, \quad M_{O3} = 0.$$
(4.2.7')

For the equilibrium of a rigid solid with a fixed axis it is necessary and sufficient that the resultant moment of the system of given forces with respect to this axis be equal to zero ($M_{O3} = 0$). As the corresponding equation of condition is compatible (determinate or indeterminate) or incompatible, the rigid solid admits positions of equilibrium (determinate or indeterminate – equilibrium in any position) or does not allow such a position. Such a condition is fulfilled if the supports of all the given forces intersect the fixed axis (e.g., the rigid solid with a fixed axis, the centre of gravity of which is on this axis ($C \in OO'$), if it is acted upon by its own weight).

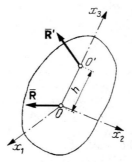

Figure 4.13. Equilibrium of a rigid solid with a fixed axis.

The components of the constraint forces are given by

$$\overline{R}_1 = \frac{M_{O2}}{h} - R_1, \quad \overline{R}_2 = -\frac{M_{O1}}{h} - R_2, \quad \overline{R}'_1 = -\frac{M_{O2}}{h}, \quad \overline{R}'_2 = \frac{M_{O1}}{h};$$
(4.2.7")

the other components ($\overline{R}_3$ and $\overline{R}'_3$) are linked by the third relation (4.2.7') and cannot be obtained independently, but only by renouncing to the hypothesis of rigidity (hence, using another mathematical model of the solid). Indeed, one can apply at O a force $\mathbf{F}$ along the fixed axis and at O' an analogous force $-\mathbf{F}$, without any influence on the mechanical phenomenon (in the case of the rigid solid, the forces are modelled by sliding vectors); consequently, the components of the constraint forces along the Ox_3–axis are modified. Hence, the considered mechanical system is hyperstatic. To can determine the constraint forces $\overline{R}_3$ and $\overline{R}'_3$, a supplementary relation is necessary, which may be obtained only by considerations concerning the deformation of the solid, hence admitting another mathematical model of it. On the other hand, if at one of the fixed points, for instance at O', we replace the spherical hinge by a cylindrical one, along the fixed axis, then the problem becomes statically determinate (indeed, we have

$\overline{R}'_3 = 0$, hence $\overline{R}_3 = -R_3$). If we impose the condition that the constraint force at O' be equal to zero for any position of equilibrium, then we get $M_{O1} = M_{O2} = 0$; hence, the system of given forces must be reduced to a resultant passing through the fixed point O (in this case, the rigid solid with a fixed axis behaves as a rigid solid with a fixed point).

2.1.4 Statics of the rigid solid with constraints with friction

We have seen in Chap. 3, Subsec. 2.2.12 that, in reality, the solids are deformed in the vicinity of the theoretical point of contact, so that the constraint forces which arise have also tangential components, appearing moments (couples) too; thus, the constraints with friction are put into evidence. The general case (Fig.3.33) leads to *the sliding friction* (which hinders the displacement in the tangential plane), *the pivoting friction* (which hinders the rotation about the normal to the tangential plane) and *the rolling friction* (which hinders a rotation about an axis in the tangent plane).

The conditions of equilibrium will be, in general, of the form (4.2.2)-(4.2.2"), but as in the case of one particle, the sign "=" is replaced by the sign "≤", so that the equalities become inequalities in the formulae (4.2.2). Thus, a certain zone of equilibrium is emphasized, as well as a domain of variation of the given forces for these positions of equilibrium. Practically, one considers firstly the case of the limit equilibrium; then, one passes to the case indicated by the inequality which is modelling the mechanical phenomenon. The three types of friction mentioned above have many applications in technique; we consider some of these ones in what follows, especially those which have interesting theoretical implications.

Concerning the sliding friction, we mention – especially – the possibility to use a quadrangle of friction to study the equilibrium of a rigid solid leaning at two points on other two solids and acted upon by forces coplanar with these points. In what concerns the pivoting friction, we have considered the fundamental case of a vertical shaft of circular or annular section (a rigid solid of cylindrical form) on a bearing. One may thus study the case of an axially symmetric axle tree, the supports of which have analogous properties; this last problem is much more difficult, its solution requiring some supplementary hypotheses (a certain mathematical modelling).

A fundamental case of rolling friction is that of *the drawn wheel* (of radius R and weight **G**) of a vehicle by the horizontal force **F** (Fig.4.14,a). The imposed conditions of equilibrium lead to $N = G$, $T = F$, $M_r = FR$, $-fN \leq T \leq fN$, $-sN \leq M_r \leq sN$, where f is the coefficient of sliding friction, while s is the coefficient of rolling friction; hence, we have determined the normal constraint force N, the sliding constraint force T and the rolling moment M_r, and we have emphasized the possibility of sliding and rolling in both directions. The conditions of equilibrium will be thus of the form $-fG \leq F \leq fG$ and $-G(s/R) \leq F \leq G(s/R)$; as a matter of fact, one must fulfil the condition for which the limits are the closest. The wheel begins to move by sliding or by rolling if the force **F** does not verify the first of those conditions, and we have $f < s/R$, or the second one, and we have $f > s/R$, respectively.

Analogously, we may consider *the motive wheel* (of radius R and weight **G**), acted upon, besides the traction force **F**, by the turning couple of moment **M** (Fig.4.14,b). From the conditions of equilibrium, we get $N = G$, $T = F$, $M_r = M - FR$, $F \le fN$, $-sN \le M_r \le sN$, where we have emphasized the two tendencies of rolling and only one of sliding, in an opposite direction with respect to the turning couple. We obtain thus $F \le fG$ and $-sG \le M - FR \le sR$ or $[F - G(s/R)]R \le M \le [F + G(s/R)]R$. Therefrom, we get the minimal turning couple $M = [F + G(s/R)]R$ necessary to put the wheel in motion; the maximal traction force is $F = fG$. If F is greater than this magnitude (hence, if the horizontal plane is too smooth – the coefficient too small), then the traction is not possible, no matter how much greater is the turning couple M.

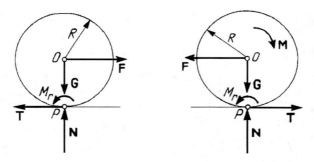

Figure 4.14. The drawn (a) and the motive (b) wheel on a horizontal plane.

The case of the drawn wheel of a vehicle on a *plane inclined* by the angle α with respect to the horizontal (Fig.4.15,a) may be studied in the same manner. Thus, one can show that if $f < s/R$ or $f > s/R$, $f = \tan\varphi$, then the wheel begins to move by sliding or by rolling, respectively; but the force **F** must not verify the condition of equilibrium

$$G\frac{\sin(\alpha - \varphi)}{\cos\varphi} \le F \le G\frac{\sin(\alpha + \varphi)}{\cos\varphi},$$

or

$$G\left(\sin\alpha - \frac{s}{R}\cos\alpha\right) \le F \le G\left(\sin\alpha + \frac{s}{R}\cos\alpha\right),$$

respectively. If a turning couple of moment **M** is introduced (Fig.4.15,b), then the conditions of equilibrium are

$$F \le G\frac{\sin(\varphi - \alpha)}{\cos\varphi}$$

and

$$\left[F + G\left(\sin\alpha - \frac{s}{R}\cos\varphi\right)\right]R \le M \le \left[F + G\left(\sin\alpha + \frac{s}{R}\cos\varphi\right)\right]R\,;$$

the latter inequality specifies the minimal turning couple for which the towage is possible, while from the first inequality one obtains the maximal traction force.

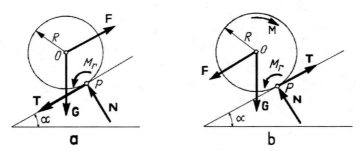

Figure 4.15. The drawn (a) and the motive (b) wheel on an inclined plane.

A *hinge* allows a rigid solid to effect a motion of rotation about an axis which passes through the centre of this constraint; experimentally, one states that, applying to the rigid solid a couple in a plane normal to the axis of rotation, the solid begins to rotate only if this couple attains a certain limit value, because of the forces of friction which arise. One emphasizes thus a *couple of friction in the hinge*, $\mathbf{M}_f$, which must verify the condition of equilibrium

$$M_f \le f'rR\,, \tag{4.2.8}$$

where r is the radius of the hinge journal (we assume a cylindrical hinge), $\mathbf{R}$ is the total reaction in the hinge, while f' is a dimensionless *coefficient of friction in the hinge*. The circle of radius $f'r$, the centre of which is at the centre of the hinge, is called *circle of friction*; for equilibrium, it is necessary that the reaction $\mathbf{R}$ do pierce the circle of friction or, at the limit, be tangent to it.

Cylindrical hinges intervene in technique especially in the form of bearings. Thus, the wheels of a machine are fixed on *axle trees*, which lean on *bearings*. The parts of the axle trees which are in contact with bearings are called *journals* and are special disposed. Because of the friction in bearings appears a moment $\mathbf{M}_f$, which is difficult to be determined; indeed, the respective mechanical phenomenon is particularly complex. Adopting the simplified hypotheses of a dry friction, one may find formulae of the form (4.2.8), which are upper limits of the moment M_f; here, r is the radius of the journal, $\mathbf{R}$ is the frictionless reaction in the axle tree, while f' is a *coefficient of friction* which depends on the type of the axle tree.

In the case of a *sliding bearing* we distinguish between a *bearing without clearance* (Fig.4.16,a) or *with clearance* (Fig.4.16,b). In the first case, if we write the equation of moment of the tangential friction forces with respect to a point of the journal axis, then we get

Statics 241

$$f' = \frac{f}{R}\sum_i N_i,\qquad(4.2.9)$$

where N_i is the normal reaction at the arbitrary point P_i of contact between the journal and the bearing, while f is the corresponding coefficient of sliding friction. In the second case, from equilibrium conditions too, we get

$$f' = f + \frac{s}{r},\qquad(4.2.9')$$

where s is the coefficient of rolling friction between the journal and the bearing. In the case of a *ball contact bearing* (Fig.4.16,c) one obtains a coefficient of friction f' smaller that those obtained above; analogously, one can show that

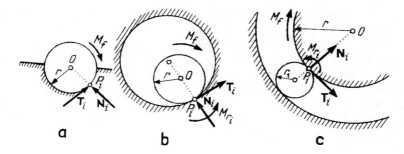

Figure 4.16. Sliding bearing without (a) or with (b) clearance. Ball contact bearing (c).

$$f' = \left(\frac{s+s_1}{2r_1} + \frac{s}{r}\right)\sum_i \frac{N_i}{R},\qquad(4.2.9'')$$

where N_i is the normal reaction at a point P_i of contact between the journal and the ball, r and r_1 are the radii of the journal and of the ball, respectively, while s and s_1 represent the coefficients of rolling friction between the journal and the ball and between the ball and the bearing, respectively.

Obviously, other cases of constraints with friction, imposed by the practice, may be put in evidence; but all these cases reduce to a sliding, pivoting, or rolling friction or to a combination of them.

2.1.5 Statics of systems of rigid solids

A mechanical system $\mathscr{S}$ constituted by a finite number of rigid solids represents a *system of rigid solids* (which are considered to be n); we assume – in general – that the system is subjected to constraints. Thus, the system of rigid solids may be acted upon by given and constraint external forces (we apply the axiom of liberation from constraints), as well as by given and constraint internal forces (between the various component rigid solids). In general, the constraints between the rigid solids can be simple supports, hinges, built-in supports and constraints with friction.

If l, $0 \leq l \leq 6n$, is the number of degrees of freedom of the system, n_e and n_i, $0 \leq n_e + n_i \leq 6n$, are the number of the unknowns introduced by the external and internal constraint forces, respectively, while e, $0 \leq e \leq 6n$, is the number of equations of equilibrium which may be written, then a *necessary condition* to determine the statical or the instantaneous (for a mechanism) equilibrium, respectively, is of the form

$$l + n_e + n_i = e. \qquad (4.2.10)$$

To specify the number of degrees of freedom of a *mechanism* is, in general, a sufficiently difficult problem. This is a basic problem in the theory of mechanisms, but we do not deal with it in what follows; we will suppose that the system of rigid solids is *statically determinate* (the equations of static equilibrium are sufficient to determine the constraint forces and the position of equilibrium). If the number e of equations is not sufficiently great, then the mechanical system is *hyperstatic*, and we cannot determine all the unknowns of the problem; as for the rest, the mathematical model of rigid bodies is – in this case – no more sufficient, and we must use a model of deformable body.

We notice that the notion of system of rigid solids is a conventional one; indeed, a system is constituted of subsystems (parts of the system), and a subsystem $S \subset \mathscr{S}$ may be considered as an independent system, if only this one is of interest. Thus, the notions of "external force" and "internal force" are relative; indeed, an internal force of a system may be external one for a subsystem. If upon a system of rigid solids do not act external forces, then we say that the system is *isolated*.

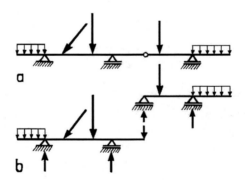

Figure 4.17. Gerber bar.

A system of rigid solids is at rest (in equilibrium) with respect to a fixed frame of reference if any constituent rigid solid of the system is in equilibrium under the action of all the given and constraint forces (all these forces are, for a particular rigid solid, external forces) which act upon it. For example, let be a *bar with hinges* (a *Gerber bar*) (Fig.4.17,a), which may be considered as being formed by two simply supported bars (Fig.4.17,b); the Gerber bar is in equilibrium only if each of the simply supported bars is in equilibrium. Obviously, this result is valid for any subsystem of an arbitrary system, so that we may state

Statics

Theorem 4.2.2 (*theorem of equilibrium of parts*). *If a system $\mathscr{S}$ of rigid solids is in equilibrium under the action of a system of given and constraint forces, then any of its parts (any subsystem $S \subset \mathscr{S}$) is in equilibrium too under the action of the given and constraint forces corresponding to the respective part.*

Noting that the torsor of the internal given and constraint forces vanishes, we obtain a *necessary condition of equilibrium* of the whole mechanical system in the form

$$\tau_O \{ \mathbf{F}_i \} + \tau_O \{ \mathbf{R}_i \} = \mathbf{0}, \qquad (4.2.11)$$

where the torsor operator is applied to all the external forces which are acting upon this system. In this condition, which is also *sufficient* in the case of a non-deformable mechanical system, the internal forces do not intervene, and that is an important advantage for the computation. We may thus state

Theorem 4.2.3 (*theorem of rigidity*). *Assuming that a given system of rigid solids with constraints becomes rigid, the conditions of equilibrium of the new mechanical system represent necessary conditions of equilibrium for the given mechanical system.*

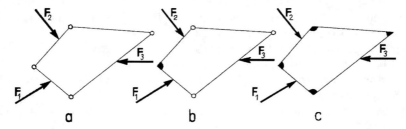

Figure 4.18. System of four bars: with four articulations (a), with one vertex built-in (b) or with all vertices built-in (c).

It results that a deformable or non-deformable system of rigid solids, which is in rest under the action of a given system of forces, remains further at rest if it becomes rigid by introducing *supplementary internal constraints*. Thus, e.g., if a system of four articulated bars (Fig.4.18,a) is in equilibrium under the action of the forces $\mathbf{F}_1, \mathbf{F}_2, \mathbf{F}_3$, then it remains in equilibrium if the articulation of one vertex is replaced by a built-in support (Fig.4.18,b), or if all four vertices are built-in (Fig.4.18,c).

The basic problem is – in general – a *mixed problem*, in which the position of equilibrium and the constraint forces which act upon the mechanical system are searched. If the constraints are specified by $n_e + n_i = m$ independent parameters (corresponding to m degrees of freedom which are cancelled), then the number of independent parameters which specify the position of equilibrium will be equal to $6n - m$; in general, in the case of a statically determinate mechanical system, these parameters may be obtained explicitly.

In the case of a system of rigid solids subjected to constraints with friction, other problems may arise too. Let thus be the case of a system constituted of n rigid solids, which has only one degree of freedom and is subjected to the action of two active given forces: the *motive force* $\mathbf{F}_m$ and the *resistent force* $\mathbf{F}_r$ (one may have couples of moments $\mathbf{M}_m$ and $\mathbf{M}_r$, respectively). There exist, in general, two tendencies of displacement, corresponding to the increasing or the decreasing of the parameter which is fixing the position of the mechanical system. We call *direct motion* that one which corresponds to the direction of displacement due to the motive force and *inverse motion* that one which corresponds to the direction of displacement of the resistent force. One determines the magnitudes of the motive force corresponding to the two tendencies of displacement from the position of equilibrium; these two limit magnitudes depend on the resistent force $\mathbf{F}_r$ (or on the couple $\mathbf{M}_r$), on the geometry of the static position of equilibrium of the mechanism (distances or angles), on the fixed geometric elements (independent of the configuration of the mechanism) and on the coefficients of friction. The two limit magnitudes coincide if we do not take into account the phenomenon of friction.

We say that a system of rigid solids is subjected to a phenomenon of *self-fixing* (or *self-braking*), the position of equilibrium being maintained if the motive mechanical element is no more acting ($\mathbf{F}_m$ or $\mathbf{M}_m$), but the resistent mechanical element ($\mathbf{F}_r$ or $\mathbf{M}_r$) still acts. We can say that the system is in equilibrium under the limit (or at the limit) of sliding, rolling or pivoting, in the opposite tendency of a direct motion; to have a motion in this case, $\mathbf{F}_m$ (or $\mathbf{M}_m$) must change of direction. Analogously, a system of rigid solids is subjected to a phenomenon of *self-locking* if, to obtain a tendency of direct motion, in a certain configuration of the system, the motive mechanical element ($\mathbf{F}_m$ or $\mathbf{M}_m$) must tend to infinity. The first of these phenomena may be useful in practice, but the second one must be avoid; thus, the study of those phenomena has a particular importance.

Using the above exposure, we may emphasize three important methods of computation. Thus, in the *method of isolating the solids*, each rigid solid of the system is isolated by introducing the corresponding constraint forces and the conditions of equilibrium (the torsor of the system of given and constraint forces with respect to an arbitrary pole vanishes); there are obtained $6n$ equations of equilibrium for the $6n$ unknowns of the problem (corresponding to the position of equilibrium and to the constraint forces). In the plane case, there remain only $3n$ equations of equilibrium. Taking into account the principle of action and reaction, some of the unknowns may affect two solids in linkage. The solving of the system of $6n$ equations may – sometimes – require a very arduous computation.

In the *method of equilibrium of parts*, subsystems of the considered system are isolated, introducing the corresponding external and internal given and constraint forces, and necessary conditions of global equilibrium (the torsor of the external given and constraint forces with respect to an arbitrary pole vanishes) are written for each subsystem. Choosing conveniently the subsystems, one can obtain thus some constraint forces (selecting the forces which we wish to determine) from a system of equations with a smaller number of unknowns. In the *method of rigidity* (which is, in fact, a particular case of the previous method, corresponding to the case in which the part is the whole system), only six equations of equilibrium are written, which may be sufficient to obtain the external constraint forces of the given mechanical system; the application of this method as a first attempt to compute is thus justified. We notice that the equations which are obtained by the method of equilibrium of parts or, in particular, by the method of rigidity are linear combinations of the equations obtained by the method of isolating the solids; the mechanical interpretations given above may be very useful in computation. We notice too that the equations obtained by the method of rigidity may be equations of verification after the method of isolating the solids has been applied.

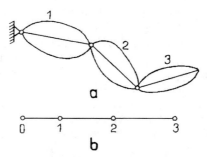

Figure 4.19. System of rigid solids (a) having an arborescent graph (b).

In practice, it is convenient to determine first the parameters defining the configuration of equilibrium of the system of rigid solids; this depends on the structure of the considered system. Conventionally, we may represent a solid of the system by a point, while the linkage between two solids is represented by a segment of straight line, which joins the corresponding points. A *graph* is thus obtained – a schema corresponding to the structure of the system; hence, *the method of graphs* may be used too.

To the system of rigid solids in Fig.4.19,a there corresponds the graph in Fig.4.19,b, while to the system in Fig.4.20,a does correspond the graph in Fig.4.20,b; the fixed element has been denoted by 0. A *cycle* of a graph is a succession of its lines, which forms a closed polygon. A graph which has not one cycle is called *arborescent*. Thus, the graphs are of two classes: arborescent graphs (as that in Fig.4.19,b) and *graphs with cycles* (as that in Fig.4.20,b, which has three cycles). In the first case, one may write the equations which have as unknowns only independent parameters, specifying the

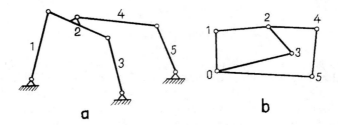

Figure 4.20. System of rigid solids (a) having a graph with cycles (b).

configuration of equilibrium (in the considered case, four parameters). In the second case, the problem can be reduced to the previous one if one transforms the given system in a system of rigid solids with arborescent graphs, by removing some internal connections, each of them interrupting a cycle; because there exist several possibilities, it is – obviously – recommendable to remove the linkages which introduce the smallest number of scalar unknowns (for instance, between a simple support and a hinge, it is preferable to eliminate the simple support).

2.1.6 Applications. Simple devices

In what follows, we deal – shortly – with applications of statics of rigid solids to some well known mechanical devices, called – usually – *simple devices*; these ones are used in their simplest form or in the construction of various machines or equipments. They are rigid solids or systems of rigid solids and are subjected to two categories of forces: *motive forces*, which try to put the system in motion, and *resistent forces*, which are opposed to the motion. The systems of forces which act upon these devices must be in equilibrium. In general, a simple device allows to overcome a resistent force of greater intensity by means of a motive force of smaller intensity. Sometimes, if the aim of the mechanical device is to change the direction of the motive force or a better equilibration from statical point of view, then it is possible to have the same intensity for the resistent force as for the motive one (or the intensity of the latter one may be smaller). We deal only with simple devices with a single degree of freedom, characterized by only one parameter of geometric nature, which specifies the position of equilibrium. One must find the relation between the modulus of the motive force and the modulus of the resistent force for equilibrium. In general, the own weight of these devices is neglected, because it is small with respect to the magnitude of the considered forces.

Although they are numerous, the simple devices can be divided in two great classes: simple devices of *the class of the inclined plane* (the inclined plane, the wedge, the screw) and simple devices of *the class of the lever* (the lever, the systems of articulated levers, the apparatuses of weight, the hoists and the pulleys).

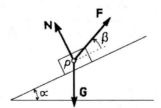

Figure 4.21. Inclined plane without friction.

The inclined plane is a simple device which allows to move up or down the solid bodies, through the agency of a force materialized, for instance, by a cable. Let be such a body, modelled by a particle P of weight $\mathbf{G}$ which leans *frictionless* on an inclined plane of angle α (Fig. 4.21); one must determine the force $\mathbf{F}$ which must be applied at the point P, so that this one be at rest. Writing the equations of equilibrium along the inclined plane and the normal to it, we get

$$F\cos\beta = G\sin\alpha, \quad F\sin\beta + N = G\cos\alpha, \quad N \geq 0,$$

where β is the angle made by the force **F** with the inclined plane, while **N** is the normal reaction. The second equation determines the constraint force, while the first one specifies the force **F**; we notice that the solution is not unique (indeed, we have obtained only a relation between the modulus of the force **F** and the direction, characterized by the angle β). Because $0 \le \alpha \le \pi/2$, there results $0 \le \cos\beta \le 1$ and $-\pi/2 \le \beta \le \pi/2$, as well as the condition $F \ge G\sin\alpha$. We may also write $N = G\cos(\alpha+\beta)\sec\beta \ge 0$, wherefrom $-\pi/2 \le \beta \le \pi/2 - \alpha$. In particular, for $\beta = 0$ we obtain $F = G\sin\alpha$, $N = G\cos\alpha$, for $\beta = \pi/2 - \alpha$ we have $N = 0$, $F = G$, while for $\beta = -\alpha$ there results $F = G\tan\alpha$, $N = G\sec\alpha$. In the limit case $\alpha = 0$ (horizontal plane) we get $\beta = -\pi/2$, $N = G + F$, the modulus F being arbitrary, or $\beta = \pi/2$, $N = G - F$, $F \le G$; in the limit case $\alpha = \pi/2$ we obtain $F = G\sec\beta$, $N = -G\tan\beta$, $-\pi/2 < \beta \le 0$. Between the motive force of modulus $F_m = F$ and the resistent force of modulus $F_r = G$ the relation

$$F_m = F_r \frac{\sin\alpha}{\cos\beta} \qquad (4.2.12)$$

takes place; if $\beta = 0$, we notice that $F_m < F_r$, and the motive force necessary to move up a body of weight **G** is smaller than the resistent force. For $\beta \ne 0$, this relation of order takes further place only if

$$\alpha + \beta < \frac{\pi}{2}. \qquad (4.2.12')$$

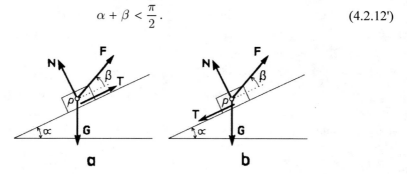

Figure 4.22. Inclined plane with friction, which hinders the particle to "move down" (a) or to "move up" (b).

If the material point P stays *with friction* on the inclined plane, the coefficient of friction being $f = \tan\varphi$ (Fig.4.22), we distinguish two cases, as the force of sliding friction has a direction or another one, contrary to the tendency of sliding of the considered body. The equations of equilibrium will be (along the inclined plane and the normal to it)

$$F\cos\beta \pm T = G\sin\alpha, \quad F\sin\beta + N = G\cos\alpha, \quad 0 \le T \le fN,$$

N being the normal reaction (the condition $N > 0$ is included above). The first two equations determine the constraint forces; by replacing in the third relation, one obtains two inequalities, which put into evidence a relation between the modulus of the force **F** and its direction, characterized by the angle β. If the force **T** hinders the particle to "move down" along the inclined plane (Fig.4.22,a), then we obtain the condition $0 \leq G\sin\alpha - F\cos\beta \leq f(G\cos\alpha - F\sin\beta)$ or $G\sin\alpha \geq F\cos\beta$, as well as $G\sin(\alpha - \varphi) \leq F\cos(\beta + \varphi)$; if the force **T** hinders the particle to "move up" along the inclined plane (Fig.4.22,b), then we obtain the condition $0 \leq F\cos\beta - G\sin\alpha \leq f(G\cos\alpha - F\sin\beta)$ or $F\cos\beta \geq G\sin\alpha$, as well as $G\sin(\alpha + \varphi) \geq F\cos(\beta - \varphi)$. We consider now the particular case $0 < \alpha < \varphi$. If the particle tends to slide down along the inclined plane, then we are in the first case. Noting that $\sin(\alpha - \varphi) < 0$, the second condition is fulfilled for $\cos(\beta + \varphi) \geq 0$ or $-\pi/2 - \varphi \leq \beta \leq \pi/2 - \varphi$; analogously, the first condition is fulfilled for $\cos\beta \leq 0$, hence for $-\pi \leq \beta \leq -\pi/2$ or $\pi/2 \leq \beta \leq \pi$. It follows that for $-\pi/2 - \varphi \leq \beta \leq -\pi/2$, F arbitrary, the equilibrium takes place. If $-\pi/2 < \beta \leq \pi/2 - \varphi$, then the condition $F \leq G\sin\alpha\sec\beta$ must hold too. For $-\pi \leq \beta < -\pi/2 - \varphi$ and $\pi/2 \leq \beta \leq \pi$ we have $\cos(\beta + \varphi) < 0$, and $\cos\beta \leq 0$, hence the condition $F \leq G\sin(\alpha - \varphi) \cdot \sec(\beta + \varphi)$ is sufficient. For $\pi/2 - \varphi < \beta < \pi/2$ we have $\cos\beta > 0$, so that we associate also the condition $F \leq G\sin\alpha\sec\beta$. Noting that $\sin(\alpha - \varphi)\sec(\beta + \varphi) \lessgtr \sin\alpha\sec\beta$, hence $\sin(\alpha - \varphi)\cos\beta \gtrless \sin\alpha\cos(\beta + \varphi)$, as we have $\cos(\alpha + \beta) \lessgtr 0$, there result two subcases (we remember that $\alpha < \varphi$): if $\pi/2 - \varphi < \beta \leq \pi/2 - \alpha$ then we have $F \leq G\sin\alpha\sec\beta$, while if $\pi/2 - \alpha \leq \beta < \pi/2$, then the condition $F \leq G\sin(\alpha - \varphi)\sec(\beta + \varphi)$ is imposed. In conclusion, the equilibrium takes place for $-\pi \leq \beta < -\pi/2 - \varphi$ or $\pi/2 - \alpha \leq \beta \leq \pi$, $F \leq G\sin(\alpha - \varphi)\sec(\beta + \varphi)$ for $-\pi/2 - \varphi \leq \beta \leq -\pi/2$, F arbitrary, and for $-\pi/2 < \beta \leq \pi/2 - \alpha$, $F \leq G\sin\alpha\sec\beta$. If the particle tends to slide up along the inclined plane, then we are in the second case. Noting that we cannot have $\cos\beta \leq 0$ and that the second condition is verified if $\cos(\beta - \varphi) \leq 0$, we have equilibrium for $-\pi/2 < \beta \leq -\pi/2 + \varphi$, $F \geq G\sin\alpha\sec\beta$. As well, for $-\pi/2 + \varphi < \beta < \pi/2$, $G\sin\alpha\sec\beta \leq F \leq G\sin(\alpha + \varphi)\sec(\beta - \varphi)$ we have equilibrium too; because $\sin\alpha\sec\beta \leq \sin(\alpha + \varphi)\sec(\beta - \varphi)$ only if $\cos(\alpha + \beta) \geq 0$, it follows that this condition holds only for $-\pi/2 + \varphi < \beta \leq \pi/2 - \alpha$. We have thus obtained all the possibilities of equilibrium for $0 < \alpha < \varphi$, immaterial of the tendency of sliding of the particle. Effecting an analogous study for all the possible values of the angle α, the following conditions of equilibrium are obtained (immaterial of the tendency of sliding)

$$0 \leq \alpha \leq \varphi: \quad -\pi \leq \beta \leq -\frac{\pi}{2} - \varphi \text{ or } \frac{\pi}{2} - \alpha \leq \beta \leq \pi, \quad F \leq G\frac{\sin(\alpha - \varphi)}{\cos(\beta + \varphi)},$$

Statics

$$-\frac{\pi}{2} - \varphi \leq \beta \leq -\frac{\pi}{2} + \varphi, \quad F \text{ arbitrary,}$$

$$\varphi \leq \alpha \leq \frac{\pi}{2}: \quad \begin{aligned} -\frac{\pi}{2} + \varphi \leq \beta \leq \frac{\pi}{2} - \alpha, & \quad F \leq G\frac{\sin(\alpha + \varphi)}{\cos(\beta - \varphi)}; \\ -\frac{\pi}{2} - \varphi \leq \beta \leq -\frac{\pi}{2} + \varphi, & \quad F \geq G\frac{\sin(\alpha - \varphi)}{\cos(\beta + \varphi)}, \\ -\frac{\pi}{2} + \varphi \leq \beta \leq \frac{\pi}{2} - \alpha, & \quad G\frac{\sin(\alpha - \varphi)}{\cos(\beta + \varphi)} \leq F \leq G\frac{\sin(\alpha + \varphi)}{\cos(\beta - \varphi)}. \end{aligned}$$

If the material point is in equilibrium for $F \leq 0$ (hence if it is acted upon only by the force **G**), then one obtains the phenomenon of *self-fixation*. If the equilibrium takes place for an arbitrary F (no matter how great), then the phenomenon of *self-locking* is obtained.

If the inclined plane is used *to move up* a body of weight **G**, then the relation

$$F_m = F_r \frac{\sin(\alpha + \varphi)}{\cos(\beta - \varphi)} \tag{4.2.13}$$

takes place. We notice that $F_m < F_r$ if

$$\alpha - \beta + 2\varphi < \frac{\pi}{2}, \quad \beta \leq \varphi, \tag{4.2.13'}$$

or if the relation (4.2.12') holds for $\beta \geq \varphi$. If $\beta = 0$, then the force which moves up the body along the inclined plane has a modulus less than the force necessary to move it along the vertical if

$$\alpha + 2\varphi < \frac{\pi}{2}. \tag{4.2.13''}$$

If the inclined plane is used *to move down* the heavy bodies, then the tendency of sliding is the descending one. The formula (4.2.12) remains, further, valid; in the case of a sliding friction, the relation

$$F_m = F_r \frac{\sin(\alpha - \varphi)}{\cos(\beta + \varphi)} \tag{4.2.14}$$

takes place. We notice that for $\alpha \leq \varphi$, $0 \leq \beta + \varphi < \pi/2$, we have $F_m \leq 0$, while the body is self-fixed on the plane; the forces of friction are sufficient to ensure this equilibrium.

The wedge is a *dismountable* simple device (a *device of fixed joining*), having the form of a triangular or trapezoidal prism, which is introduced between two pieces, acting upon them by pressure and friction forces. In the case of a *symmetric wedge* (with *double inclination*), of vertex angle 2α (Fig.4.23), we may write

$$P = 2(\sin\alpha + f\cos\alpha)\sum_i N_i ,$$

$$Q = (\cos\alpha - f\sin\alpha)\sum_i N_i ,$$

where $\mathbf{N}_i$ is the normal reaction at the point P_i of the lateral surface, $\mathbf{P}$ is the force by which the wedge is beaten, $\mathbf{Q}$ ($\mathbf{Q}\perp\mathbf{P}$) is the force by which the wedge pushes laterally each piece, while $f = \tan\varphi$ is the coefficient of sliding friction. Noting that $F_m = P$, $F_r = Q$, there results

$$F_m = 2\frac{\sin\alpha + f\cos\alpha}{\cos\alpha - f\sin\alpha} F_r = 2 F_r \tan(\alpha + \varphi). \qquad (4.2.15)$$

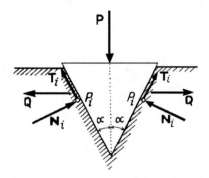

Figure 4.23. The equilibrium of a wedge.

We observe that one obtains a great pressure Q with the aid of a small force P, if α and φ are small angles. If the wedge is pulled out, then the direction of the friction force is changing and one obtains the relation

$$F_m = 2 F_r \tan(\alpha - \varphi). \qquad (4.2.15')$$

If $F_m \leq 0$, hence if $\alpha \leq \varphi$, then the wedge remains *self-fixed*.

The *asymmetric wedge* with a *simple* or a *double inclination* can be studied analogously.

The screw is a simple device used for detachable installings with clamping, for the transmission of motion (by transforming the motion of rotation in a motion of translation and inversely), for the adjustement of the relative position of two pieces or for the elimination of wear plays, as well as for the measuring of the lengths. On the lateral surface of a right circular cylinder is cut a screw thread in the form of a circular helix; developing the lateral surface of the cylinder, it is easy to verify that the slope of the helix is given by the relation

$$\tan\alpha = \frac{p}{2\pi r}, \qquad (4.2.16)$$

where p is the pitch of the helix, while r is the radius of the cylinder. The screw is thread in the nut, its relative motion with respect to it being a particular helical motion, called a screw motion. It is possible that the nut be fixed, the screw having a helical motion or the screw be fixed and the nut having such a motion or the nut have a motion

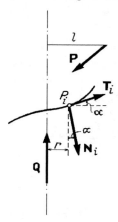

Figure 4.24. The equilibrium of a screw.

of rotation and the screw a motion of translation or – finally – the screw have a motion of rotation and the nut a motion of translation. In the first of these cases, at one of the extremities acts upon the screw a turning (motive) couple of moment $M_m = Pl$, where $F_m = P$ is the intensity of the force which acts at the end of a lever arm of length l; at the other extremity is acting a resistent force $F_r = Q$ along the screw axis, with a direction opposite to that of the screw driving. In any point P_i of contact with the nut fillet arises a force representing its action of component of modulus N_i, normal to the fillet, and of component of modulus $T_i = fN_i$, tangent to this one, guided in a direction opposite to the direction of displacement of the screw fillet with respect to that of the nut (Fig.4.24).

The equations of equilibrium (projection on the screw axis and moment about the same axis) lead to

$$Q = (\cos\alpha - f\sin\alpha)\sum_i N_i, \quad M_m = r(\sin\alpha + f\cos\alpha)\sum_i N_i,$$

wherefrom

$$M_m = Qr\frac{\sin\alpha + f\cos\alpha}{\cos\alpha - f\sin\alpha} = Qr\tan(\alpha + \varphi), \qquad (4.2.17)$$

where $f = \tan\varphi$ is the coefficient of sliding friction. By unscrewing, the direction of the friction forces is changing, so that one must have

$$Qr\tan(\alpha - \varphi) \leq M_m \leq Qr\tan(\alpha + \varphi) \qquad (4.2.17')$$

for equilibrium. We notice that for $\alpha + \varphi = \pi/2$ the turning moment M_m (hence, the motive force F_m) must be sufficiently great to can obtain the clamping of the screw; it is the case of *self-locking*. As well, for $\alpha < \varphi$ it is necessary to act with a turning moment $\mathbf{M}_m$ of a direction opposite to that of the clamping for screwing, independent of the force $\mathbf{Q}$; in this case, the screw is *self-fixed*.

The level is a simple device formed by a rigid solid with a fixed point or axis, acted upon by two forces: a motive force of intensity $F_m = P$ and a resistent force of intensity $F_r = Q$. The supports of these forces are contained in a plane normal to the axis of rotation of the rigid solid and do not pierce this axis.

Neglecting the frictions, the equation of moments with respect to the fixed point or axis leads to *Archimedes' relation*

$$F_m = \frac{b}{a} F_r, \qquad (4.2.18)$$

where a is the level arm of the motive force, while b is the level arm of the resistent force.

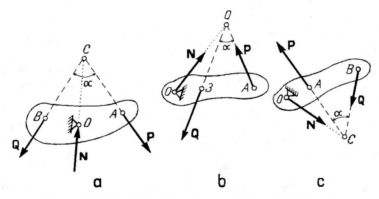

Figure 4.25. The level of first (a), second (b) or third (c) order.

As a function of the position of the articulation O with respect to the points of application A and B of the motive and resistent force, respectively, the levels may be of three kinds. At *the level of first order*, the hinge O is between the points A and B (Fig.4.25,a); if $a = b$, then it results $F_m = F_r$ (the case of the balance with equal arms), if $a > b$, then it results $F_m < F_r$ (case in which motive force is saved), and if $a < b$, then one obtains $F_m > F_r$ (non-economic case). *The levels of second order* are those for which the point B of application of the resistent force is between the points O and A (Fig.4.25,b); in this case, $a > b$, hence $F_m < F_r$ (motive force is saved). In the case of *the levels of third order* the point A of application of the motive force is between the points O and B (Fig.4.25,c); in this case $a < b$, hence $F_m > F_r$ (non-economic case).

Taking into account the friction in the bearing, we may use the formula (4.2.8), obtaining the equation of equilibrium

$$Pa - Qb = f'Nr, \qquad (4.2.19)$$

where r is the hinge journal radius, f' is the corresponding coefficient of friction, while N is the modulus of the reaction, given by

$$N = \sqrt{P^2 + Q^2 + 2PQ\cos\alpha}, \qquad (4.2.19')$$

α being the angle formed by the forces **P** and **Q**. There result the extreme values of the motive force F_m (for equilibrium)

$$F_m = \frac{F_r}{a^2 - f'^2 r^2}\left(ab + f'^2 r^2 \cos\alpha \pm f'r\sqrt{a^2 + b^2 + 2ab\cos\alpha - f'^2 r^2 \sin^2\alpha}\right).$$
$$(4.2.20)$$

In the particular case $\alpha = 0$ (**P** $\parallel$ **Q**), we get

$$F_m = k\frac{b}{a}F_r; \qquad (4.2.21)$$

the ratio b/a in the formula (4.2.18) is thus multiplied by the coefficient

$$k = \frac{1 + \dfrac{f'r}{b}}{1 - \dfrac{f'r}{a}} > 1. \qquad (4.2.21')$$

Hence, a motive force greater than that used if there are not frictions is necessary.

To obtain scale ratios greater than a/b (corresponding to a single level), one may use *systems of articulated levels*.

The level is the basic element for *weighting apparatuses*. We mention thus *the balance with equal arms, the Roman balance, the decimal balance, the Roberval balance* etc.

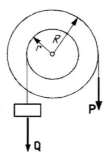

Figure 4.26. Equilibrium of a cable hoist.

254   MECHANICAL SYSTEMS, CLASSICAL MODELS

The cable hoist is a simple device used to raise the weights. The simple hoist is formed by a cylindrical drum of radius r, on which a cable is rapped up; an extremity of the cable is fixed to the drum, while by the other extremity the weight (the resistent force) $\mathbf{F}_r = \mathbf{Q}$, which must be raised, is hanging down. On the drum is fixed a wheel of radius R, on which the motive force $\mathbf{F}_m = \mathbf{P}$ is tangentially applied (Fig.4.26). The equation of moment with respect to the rotation axis leads to

$$F_m = \frac{r}{R} F_r. \qquad (4.2.22)$$

To maintain $F_m = \mathrm{const}$ when taking into account the own weight of the cable too, one may use a truncated cone drum (*regulator cable hoist*). Sometimes, a hoist with a vertical axis is called a *capstan*. In the case of a *differential hoist*, the drum is constituted of two cylindrical sections of different radii, on which a cable is rapped up, in distinct directions, raising a weight $\mathbf{Q}$ with the aid of a pulley.

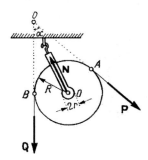

Figure 4.27. Equilibrium of a pulley.

The pulley is a simple device constituted of a circular disk of radius R, on the circumference of which passes a cable (chain); the axle of the pulley is fastened by a fork with a hook. *The fixed pulley* has a fixed axle, while the movable *pulley* has a shifting one.

At the two extremities of the cable of a fixed pulley (Fig.4.27), the forces $\mathbf{F}_m = \mathbf{P}$ and $\mathbf{F}_r = \mathbf{Q}$ are acting so that, in the absence of frictions, we have

$$F_m = F_r. \qquad (4.2.23)$$

Taking into account the friction in the bearing, we may write

$$(P - Q)R = f'Nr, \qquad (4.2.24)$$

where r is the radius of the pulley journal, f' is the coefficient of friction of the bearing, while N is the modulus of the reaction, given by a formula of the form (4.2.19'). We get

$$F_m = \frac{F_r}{R^2 - f'^2 r^2} \left(R^2 + f'^2 r^2 \cos^2 \alpha \pm 2 f'r \cos \frac{\alpha}{2} \sqrt{R^2 - f'^2 r^2 \sin^2 \frac{\alpha}{2}} \right). \qquad (4.2.25)$$

Statics

In the particular case $\alpha = 0$ ($\mathbf{P} \parallel \mathbf{Q}$), there results $N = P + Q$, and we may write

$$F_m = kF_r,\qquad (4.2.26)$$

the multiplicative coefficient k being given by

$$k = \frac{1 + f'\dfrac{r}{R}}{1 - f'\dfrac{r}{R}} > 1. \qquad (4.2.26')$$

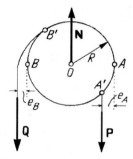

Figure 4.28. Influence of the rigidity of the cable in the equilibrium of a pulley.

In general, the cable is supposed to be perfectly flexible; in reality, the cable has a certain *rigidity*, so that in the zones AA' and BB' it nears by e_A the pulley axle or moves to a distance e_B from this one, the curvature having a continuous variation (Fig.4.28). The equation of moments yields a relation of the form (4.2.26) too, where

$$k = 1 + \frac{e_A + e_B + 2f'r}{R - e_A - f'r}; \qquad (4.2.26'')$$

neglecting e_A with respect to R, we can also write

$$k = 1 + \lambda + \frac{2f'r}{R - f'r}, \qquad (4.2.27)$$

where the influence of the rigidity is given by

$$\lambda = \frac{e_A + e_B}{R - f'r}. \qquad (4.2.27')$$

Because $k > 1$, we have $F_m > F_r$, and the simple fixed pulley has now the rôle to change the direction of transmission of the force (in fact, its support).

The movable pulley allows to raise a weight $\mathbf{Q}$ using a force $\mathbf{P}$ of a smaller intensity ($F_m < F_r$). For instance, in the case of the movable pulley for which $\mathbf{P} \parallel \mathbf{Q}$ (Fig.4.29) we have

$$F_m = \frac{1}{2} F_r, \tag{4.2.28}$$

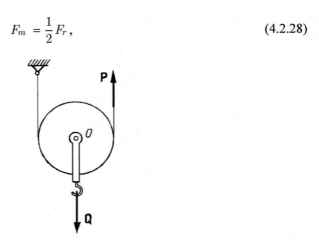

Figure 4.29. Equilibrium of a movable pulley.

the tension in the cable being $T = P/2$. If we take into account the frictions and the rigidity of the cable, we may use the coefficient k introduced above, so that

$$F_m = kT, \quad F_m + T = F_r,$$

wherefrom

$$F_m = \frac{k}{1+k} F_r. \tag{4.2.29}$$

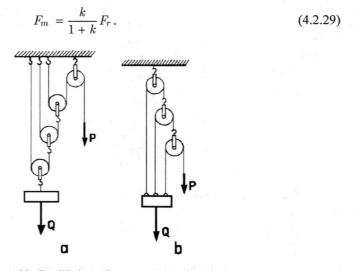

Figure 4.30. Equilibrium of exponential pulley blocks.

By means of fixed and movable pulleys, we may constitute *systems of pulleys*. We mention thus the *exponential pulley block* (Fig.4.30,a), formed by a fixed pulley and n mobile ones, for which we have

Statics

$$F_m = \frac{k^{n+1}}{(1+k)^n} F_r \qquad (4.2.30)$$

or, neglecting the frictions and the rigidity of the cables,

$$F_m = \frac{1}{2^n} F_r. \qquad (4.2.30')$$

Analogously, for another exponential pulley block (Fig.4.30,b) we obtain

$$F_m = \frac{1}{\left(1+\frac{1}{k}\right)^{n+1} - 1} F_r \qquad (4.2.31)$$

or

$$F_m = \frac{1}{2^{n+1} - 1} F_r. \qquad (4.2.31')$$

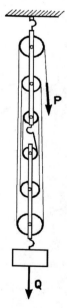

Figure 4.31. Equilibrium of a pulley block with n fixed and n movable pulleys.

In the case of the pulley block with n fixed and n movable pulleys (Fig.4.31), there results

$$F_m = \frac{k^{2n}(k-1)}{k^{2n} - 1} F_r \qquad (4.2.32)$$

or
$$F_m = \frac{1}{2n} F_r. \qquad (4.2.32')$$

With the aid of a hoist and of a mobile pulley we may obtain a *differential pulley block* too.

2.1.7 Efforts in bars

The notion of bar has been introduced in Chap. 1, Subsec. 1.1.10. Let thus be a *curved bar* (the axis of which must be – in general – a skew curve) in equilibrium under the action of a system $\mathscr{F}$ of given and constraint external forces. A cross section (plane and normal to the bar axis) divides the bar in two parts, and the internal forces which arise on the two faces thus obtained are put into evidence (Fig.4.32,a,b). Thus, the part I is acted upon by the subsystem $\mathscr{F}_1$ of given forces and reactions and by the system $\mathscr{F}_{12}$ of forces with which the part II is acting upon this one; analogously, the part II is acted upon by the subsystem $\mathscr{F}_2$ of given forces and reactions, as well as by the system $\mathscr{F}_{21}$ of forces with which the part I is acting upon it. The global condition of equilibrium and the theorem of equilibrium of parts allow us to write

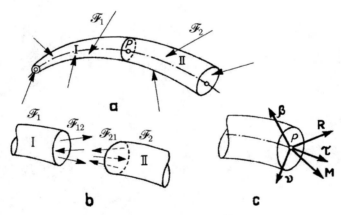

Figure 4.32. Equilibrium of a curved bar (a). Influence of the internal forces (b). Resultant efforts on a cross section (c).

$$\mathscr{F} \sim \mathscr{F}_1 + \mathscr{F}_2 \sim \{0\}, \quad \mathscr{F}_1 + \mathscr{F}_{12} \sim \{0\}, \quad \mathscr{F}_2 + \mathscr{F}_{21} \sim \{0\}, \qquad (4.2.33)$$

wherefrom

$$\mathscr{F}_{12} + \mathscr{F}_{21} \sim \{0\}, \qquad (4.2.33')$$

corresponding to the principle of action and reaction; as well, we get

$$\mathscr{F}_{21} \sim \mathscr{F}_1, \quad \mathscr{F}_{12} \sim \mathscr{F}_2 \qquad (4.2.33'')$$

Statics

too, so that we may state

Theorem 4.2.4. *The system of internal forces by which a part of a bar acts upon the other part is equivalent to the system of external (given and constraint) forces, corresponding to the first part.*

Applying the operator torsor (usually, at the centre of gravity of the cross section) to the relations (4.2.33"), we get

$$\tau\{\mathscr{F}_{21}\} = \tau\{\mathscr{F}_1\}, \quad \tau\{\mathscr{F}_{12}\} = \tau\{\mathscr{F}_2\}; \qquad (4.2.34)$$

the pole with respect to which has been made the computation was considered to be the same for the two systems of forces (before the detachment of the two parts), so that we did not put it in evidence. It is suitable to consider the torsor of the system of internal forces corresponding to the face which is encountered the first by getting over the bar axis. Frequently, one does it from left to right (part I), so that one has to do with the torsor $\tau\{\mathscr{F}_{12}\}$, corresponding to the left face. The components of the resultant **R** and of the resultant moment **M** are applied at the centre of gravity of this cross section (Fig.4.32,c) and are called *efforts* (three forces and three moments); usually, Frenet's intrinsic trihedron is considered, the direction of the unit vector $\boldsymbol{\tau}$ of the tangent coinciding with the direction of getting over the bar axis. The components of the resultant **R** are: *The axial force* **N**, along the tangent $\boldsymbol{\tau}$, and *the shearing (cross, transverse) force* **T**, contained in the normal plane (of components $\mathbf{T}_\nu$ and $\mathbf{T}_\beta$ along the principal normal and the binormal, respectively) (Fig.4.33,a). As well, the components of the resultant moment **M** are: *the moment of torsion (twisting moment)* $\mathbf{M}_t$, along the tangent, and *the bending moment* $\mathbf{M}_b$, contained in the normal plane (of components $\mathbf{M}_\nu$ and $\mathbf{M}_\beta$ along the corresponding unit vectors) (Fig.4.33,b). By

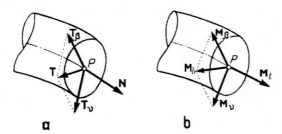

Figure 4.33. Efforts on a cross section of a curved bar: force components (a) and moment components (b).

convention, the scalars of these vector components are positive if they have the same directions as the unit vectors of the axes of the intrinsic trihedron. We may thus write

$$\mathbf{R} = N\boldsymbol{\tau} + T_\nu \mathbf{v} + T_\beta \boldsymbol{\beta}, \quad \mathbf{M} = M_t \boldsymbol{\tau} + M_\nu \mathbf{v} + M_\beta \boldsymbol{\beta}. \qquad (4.2.35)$$

The variation of the six efforts $N, T_\nu, T_\beta, M_t, M_\nu, M_\beta$ along the bar axis may be represented by *diagrams of efforts*, which put in evidence their values in each section;

obviously, it is necessary to specify also some laws of variation of those efforts. To do this, we consider an element of the bar axis of length ds, reducing thus all the forces with respect to the axis' points, corresponding to the cross sections on which act these forces; as well, we suppose that the bar is acted upon only by *distributed forces* $\mathbf{p}ds$ and by *distributed moments* $\mathbf{m}ds$, for which

$$\mathbf{p} = p_\tau \boldsymbol{\tau} + p_\nu \mathbf{v} + p_\beta \boldsymbol{\beta}, \quad \mathbf{m} = m_t \boldsymbol{\tau} + m_\nu \mathbf{v} + m_\beta \boldsymbol{\beta}. \tag{4.2.35'}$$

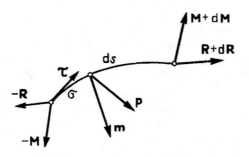

Figure 4.34. Equilibrium of an element of bar.

Equating to zero the torsor of all the forces which act upon this bar element (Fig.4.34), we obtain

$$-\mathbf{R} + (\mathbf{R} + d\mathbf{R}) + \mathbf{p}ds = \mathbf{0},$$

$$-\mathbf{M} + (\mathbf{M} + d\mathbf{M}) + \mathbf{m}ds + (\boldsymbol{\tau}ds) \times (\mathbf{R} + d\mathbf{R}) + \int_0^{ds} (\boldsymbol{\tau}\sigma) \times \mathbf{p}(\sigma) d\sigma = \mathbf{0}.$$

Applying a mean value formula to the above integral and neglecting the terms of higher order, there results

$$\frac{d\mathbf{R}}{ds} + \mathbf{p} = \mathbf{0}, \quad \frac{d\mathbf{M}}{ds} + \boldsymbol{\tau} \times \mathbf{R} + \mathbf{m} = \mathbf{0}. \tag{4.2.36}$$

Taking into account Frenet's formulae given in Subsec. 1.1.3, we get

$$\frac{dN}{ds} - \frac{1}{\rho}T_\nu + p_\tau = 0, \quad \frac{dT_\nu}{ds} + \frac{1}{\rho}N + \frac{1}{\rho'}T_\beta + p_\nu = 0,$$

$$\frac{dT_\beta}{ds} - \frac{1}{\rho'}T_\nu + p_\beta = 0, \tag{4.2.37}$$

$$\frac{dM_t}{ds} - \frac{1}{\rho}M_\nu + m_\tau = 0, \quad \frac{dM_\nu}{ds} + \frac{1}{\rho}M_t + \frac{1}{\rho'}M_\beta - T_\beta + m_\nu = 0,$$

$$\frac{dM_\beta}{ds} - \frac{1}{\rho'}M_\nu + T_\nu + m_\beta = 0. \tag{4.2.37'}$$

With the aid of the first two equations (4.2.37) one can express the efforts T_ν and T_β as functions of the given external loads and the normal force N; replacing in the third

Statics

equation, we get a differential equation of the third order in N, which may determine the latter effort. Analogously, the equations (4.2.37') lead to a differential equation of the same order for the moment of torsion M_t which contains the axial force N too, now known; we mat determine thus all six efforts.

In the case of a *plane curved bar* $1/\rho' \to 0$, $\boldsymbol{\beta}$ being the unit vector normal to the plane of the bar; we get

$$\frac{dN}{ds} - \frac{1}{\rho}T_\nu + p_\tau = 0, \quad \frac{dT_\nu}{ds} + \frac{1}{\rho}N + p_\nu = 0, \quad \frac{dT_\beta}{ds} + p_\beta = 0, \quad (4.2.38)$$

$$\frac{dM_t}{ds} - \frac{1}{\rho}M_\nu + m_\tau = 0, \quad \frac{dM_\nu}{ds} + \frac{1}{\rho}M_t - T_\beta + m_\nu = 0,$$

$$\frac{dM_\beta}{ds} + T_\nu + m_\beta = 0. \quad (4.2.38')$$

One of the efforts N and T_ν may be eliminated between the first two equations (4.2.38), obtaining thus a differential equation of the second order in the other effort; the third equation allows the computation of the shearing force T_β. These efforts being obtained, one can make analogous considerations for the subsystem (4.2.38').

If the plane curved bar is acted upon by *forces contained in its plane*, then one has $p_\beta = 0$, $m_\tau = m_\nu = 0$, and, taking into account the Theorem 4.2.4 (the resultant of the external forces is contained in the plane of the bar axis, while the resultant moment of these forces is normal to this plane) one gets $T_\beta = 0$, $M_t = M_\nu = 0$ too; the remaining equations are

$$\frac{dN}{ds} - \frac{1}{\rho}T_\nu + p_\tau = 0, \quad \frac{dT_\nu}{ds} + \frac{1}{\rho}N + p_\nu = 0, \quad \frac{dM_\beta}{ds} + T_\nu + m_\beta = 0, \quad (4.2.39)$$

and one makes observations analogous to those above.

If the plane curved bar is acted upon by *forces normal to its plane*, then we have $p_\tau = p_\nu = 0$, $m_\beta = 0$, and, on the basis of the same theorem (the external forces constitute a system of parallel forces), we obtain $N = 0$, $T_\nu = 0$, $M_\beta = 0$; the system of equations becomes

$$\frac{dT_\beta}{ds} + p_\beta = 0, \quad \frac{dM_t}{ds} - \frac{1}{\rho}M_\nu + m_\tau = 0, \quad \frac{dM_\nu}{ds} + \frac{1}{\rho}M_t - T_\beta + m_\nu = 0, \quad (4.2.40)$$

and may be analogously studied.

Starting from the two particular cases considered above, we obtain the solution corresponding to the general case, by superposing the effects.

In the case of *straight bars*, the direction of the binormal and the direction of the principal normal are not determined; it is convenient to use a right-handed fixed orthonormed frame of reference $Ox_1x_2x_3$, the axis Ox_3 being along the bar axis

$(1/\rho \to 0$ and $\mathrm{d}s = \mathrm{d}x_3$). Corresponding to the conventions in technical mechanics of deformable solids (strength of materials), the components of the torsor of internal forces are of the form (Fig.4.35,a,b)

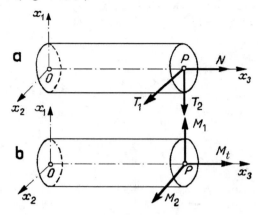

Figure 4.35. Efforts on a cross section of a straight bar: force components (a) and moment components (b).

$$\mathbf{R} = -T_2\mathbf{i}_1 + T_1\mathbf{i}_2 + N\mathbf{i}_3, \quad \mathbf{M} = M_1\mathbf{i}_1 + M_2\mathbf{i}_2 + M_3\mathbf{i}_3, \qquad (4.2.41)$$

and the external load is given by

$$\mathbf{p} = p_j\mathbf{i}_j, \quad \mathbf{m} = m_j\mathbf{i}_j, \qquad (4.2.41')$$

where $\mathbf{i}_j$, $j = 1,2,3$, are the unit vectors of the co-ordinate axes. We obtain the relations

$$\mathbf{R}_{,3} + \mathbf{p} = \mathbf{0}, \quad \mathbf{M}_{,3} + \mathbf{i}_3 \times \mathbf{R} + \mathbf{m} = \mathbf{0} \qquad (4.2.42)$$

or, in components,

$$T_{2,3} - p_1 = 0, \quad T_{1,3} + p_2 = 0, \quad N_{,3} + p_3 = 0, \qquad (4.2.42')$$
$$M_{1,3} - T_1 + m_1 = 0, \quad M_{2,3} - T_2 + m_2 = 0, \quad M_{t,3} + m_3 = 0. \qquad (4.2.42'')$$

Eliminating the shearing forces, we get

$$M_{1,33} + m_{1,3} + p_2 = 0, \quad M_{2,33} + m_{2,3} - p_1 = 0. \qquad (4.2.43)$$

In the case of external loads *contained in the plane* Ox_1x_3, we have $p_2 = 0$, $m_1 = m_3 = 0$; there results

$$T_{1,3} = 0, \quad M_{1,3} - T_1 = 0, \quad M_{t,3} = 0,$$

wherefrom

$$T_1 = T_1^0 = \text{const}, \quad M_1 = T_1^0 x_3 + M_1^0, \quad M_1^0 = \text{const}, \quad M_t = M_t^0 = \text{const} \tag{4.2.44}$$

and

$$T_{2,3} = p_1, \quad M_{2,3} = T_2 - m_2, \quad M_{2,33} = p_1 - m_{2,3}. \tag{4.2.44'}$$

If we have too $m_2 = 0$, then we may write

$$T_{2,3} = p_1, \quad M_{2,3} = T_2, \tag{4.2.45}$$

so that

$$M_{2,33} = p_1. \tag{4.2.45'}$$

In this case, the diagrams of efforts are, e.g., (Fig.4.36), where we have put into evidence the correspondence between the point of extremum for the bending moment and the point at which the shearing force vanishes; we notice also the correspondence between the inflexion point of the diagram of bending moments, the point of extremum of the diagram of shearing forces and the point at which the diagram of normal loadings vanishes, as well as the correspondence between the point of inflexion of the diagram of shearing forces and the point of extremum of the diagram of normal loadings. Sometimes, the positive part of the diagram of bending moments is plotted under the axis (to be in concordance with the deflection line of the bar axis).

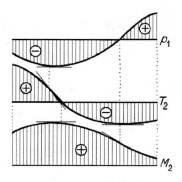

Figure 4.36. Diagrams of efforts in a straight bar acted upon by loads in a plane.

In the case of loads contained in the plane Ox_2x_3 ($p_1 = 0$, $m_2 = m_3 = 0$) we can make analogous observations.

We notice that, in all considerations made in this subsection, the bar is reduced to its axis; as a matter of fact, so it is for all problems concerning the efforts on the cross section. This is a mathematical model, which is used currently in technical mechanics of solids.

In the case of *external concentrated loads* (e.g., concentrated forces), the above relations remain valid if the usual functions are replaced by distributions, the operations of differentiation being effected in the same sense too. For instance, in the case of a straight bar acted upon by a concentrated force

$$p_1(x_3) = P\delta(x_3 - x_3^0) \tag{4.2.46}$$

at the point of co-ordinate x_3^0, the equations (4.2.45) lead to

$$T_2(x_3) = P\theta(x_3 - x_3^0) + T_2^0,$$
$$M_2(x_3) = P(x_3 - x_3^0)\theta(x_3 - x_3^0) + T_2^0 x_3 + M_2^0; \tag{4.2.46'}$$

the shearing force and the bending moment at the cross section $x_3 = 0$ are thus put in evidence.

2.1.8 Articulated systems

An *articulated system* is a structure of bars (a system of rigid solids) linked by articulations, so as to form a *non-deformable system from a geometric point of view* (not a mechanism); the loads are supposed to be applied at the nodes.

By definition, for a bar between two articulations, all the given and constraint forces, can be reduced to resultants at each extremity of it; for equilibrium, the two resultants must be directed along the straight line connecting the articulations, having the same modulus and opposite directions. Hence, from a statical point of view, a bar requires only one unknown: the corresponding *axial force*.

The theoretical articulations at the nodes cannot be practically realized. The nodes are more or less rigid in the case of constructions in concrete or metal; these rigidities introduce secondary efforts, which may be computed separately. As well, if the bars are acted also transversally (for instance, by their own weight), one calculates the resultants at the nodes, while the effect of the other efforts (for instance of bending) are separately computed.

Starting from the most simple non-deformable geometric construction formed by articulated bars (the triangle, which has three nodes and three bars), one can constitute a plane articulated system. By complete induction, one can show that between the number b of bars and the number n of nodes takes place the relation

$$2n = b + 3, \tag{4.2.47}$$

which represents a *necessary condition* for the geometric non-deformability of the plane articulated system; thus, a plane articulated system 1-2-...-6, constructed starting form a basic triangle 1-2-3, is shown in Fig.4.37. On the other hand, a system must be analysed, from case to case, to see if it is not a *system of critical form* (at least partially), the form of which may be lost under the action of a particular system of external loads; we will not deal with such articulated systems.

A free articulated system, as that considered above, can be fixed in its plane with the aid of three simple constraints (a hinge and a simple support or three simple supports of non-concurrent directions at the nodes. Because a simple support is equivalent to a pendulum (a bar), the relation (4.2.47) has the more general form

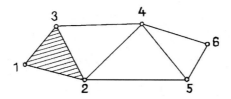

Figure 4.37. The setting up of a plane articulated system.

$$2n = b + s, \quad s \geq 3, \qquad (4.2.47')$$

where s is the number of simple supports. If $s = 3$, then the plane articulated system (structure) constitutes a single geometrically non-deformable body, by suppressing the supports, and is called a *free articulated system* (structure).

If we are isolating each node (sectioning all the bars around the node), then the given and the constraint forces, as well as the efforts in bars form a system of concurrent forces; writing two equations of equilibrium for each node we obtain $2n$ equations. If the relation (4.2.47') holds, then the articulated system is *statically determinate* (on the basis of the Kronecker-Capelli theorem for systems of linear algebraic equations); if $2n < b + s$, then the articulated system is *statically indeterminate*, while if $2n > b + s$, then the system is a *mechanism* (geometrically deformable).

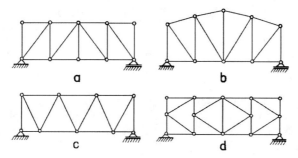

Figure 4.38. Rectangular system (a), trapezoidal system (b), triangular system (c) and K lattice (d) framework.

A plane articulated system may – often – play, in its totality, the rôle of a bar, which is called *truss* (*framework*). The bars of the contour (excepting the lateral ones) are called *flanges*; according to their positions, we distinguish between *superior* and *inferior flanges*. The bars linking the flanges are called *members* (*lattices*); the vertical lattices are called *vertical members*, while the inclined ones are called *diagonals*. We denote the axial force in the bar of the framework by S, I, V or D if this one is a superior or inferior flange, a vertical member or a diagonal, respectively. According to the form of the contour, the frameworks may be *frameworks with parallel flanges* (Fig.4.38,a) or *frameworks with polygonal flanges* (Fig.4.38,b).

Among the simple articulated systems, we mention: *rectangular system trusses* (Fig.4.38,a), *trapezoidal system trusses* (Fig.4.38,b), *triangular system trusses* (Fig.4.38,c) and *K lattice trusses* (Fig.4.38,d). If over a *primitive system* of bars is introduced a supplementary system, then one obtains a *compound articulated system*. The double rectangular system trusses, the double triangular system trusses or the trusses with multiple lattices form *complex articulated systems*.

To determine the efforts in the bars of a framework, one may use various methods of computation. Thus, in *the method of sections* (*Ritter's method*) a complete section is made, and the conditions of equilibrium of one of the two parts are put. Only three unknowns may appear, because one can write only three equations of equilibrium; if the reactions have been previously obtained, then the unknowns may be only efforts in the sectioned bars, so that these ones cannot be more than three. We notice that these bars must not be all concurrent or parallel; in such a case, the three equations are no more linearly independent.

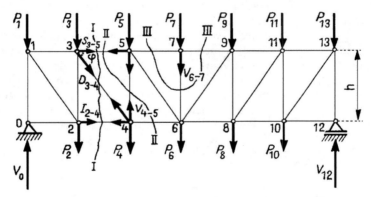

Figure 4.39. Equilibrium of a framework with parallel flanges.

Let be a framework with parallel flanges (Fig.4.39), of height h, acted upon by vertical concentrated forces at the nodes. Because the truss is simply supported, the reactions can be obtained from its global equilibrium, as in the case of a usual straight bar. Let thus be a section I-I; we replace the sectioned bars (3-5, 2-4 and 3-4) by the corresponding efforts (S_{3-5}, I_{2-4} and D_{3-4}), with a positive sign. We consider the equilibrium of the left part of the truss. The equation of moment with respect to the node 3 leads to

$$I_{2-4} = \frac{M_3}{h}, \qquad (4.2.48)$$

where M_3 is the moment of the given and constraint forces acting upon the left part of the bar with respect to the node 3 (positive if the corresponding couple leads to a clockwise rotation in the plane); analogously, the equation of moment about the node 4 allows to write

$$S_{3-5} = -\frac{M_4}{h}, \tag{4.2.48'}$$

where M_4 has a similar signification. The equation of projection on the normal common to the two parallel flanges yields

$$D_{3-4} = \frac{T_1}{\sin \varphi}, \tag{4.2.48''}$$

where T_1 is the projection on the considered direction of all the given and constraint forces at the left of section I-I, while φ is the angle formed by the diagonal with one of the flanges. We notice that M and T have the significance of a bending moment and of a shearing force in a straight bar, respectively, their sign being analogously established. In the case of a section II-II, the equation of projection on the normal common to the parallel flanges leads to

$$V_{4-5} = T_{II}, \tag{4.2.49}$$

where T_{II} has an analogous significance. One obtains thus the efforts in all the bars of the framework. Only for the vertical member 6-7 one must make a section of the form III-III (we isolate the node 7), wherefrom

$$V_{6-7} = -P_7. \tag{4.2.50}$$

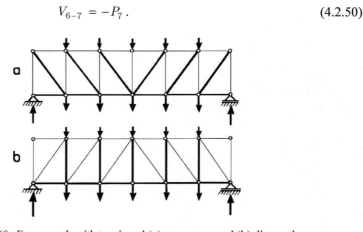

Figure 4.40. Framework with tensioned (a) or compressed (b) diagonals.

Such a bar for which the effort depends only on a local load, being independent of the form of the beam and of its loading, is called a *supplementary bar*.

As we can see, it follows that in the bars of the superior flanges and in the vertical beams appear only efforts of compression, while in the bars of the inferior flanges and in the diagonals we have efforts of tension. The compressed bars are traced by thin lines, while the tensioned ones are traced by thick lines (Fig.4.40,a). If the diagonals of each pane would be ascendent towards the middle of the truss (instead to be descendent, as in the previous case), then they would be compressed and the vertical members tensioned, but the bars of the flanges would have efforts of the same sign as above. These results are rendered by the same graphical convention in Fig.4.40,b; we notice that the efforts in the superior flanges of the panes and in the end vertical members vanish. Indeed, these affirmations are justified because between the efforts in the bars and the forces acting at the nodes take place relations of the type of those between the internal forces (see, e.g., the nodes 3 and 4 and the bar 3-4, Fig.4.41).

This method of computation may be applied analogously in the case of a truss with polygonal flanges; we mention that it can be used in a graphical variant too.

Another method of computation often used is *the method of isolation of nodes*, which is – in fact – a particular case of the method of sections; in the frame of this method, each node is isolated, by sectioning the bars which start form this one and by replacing them by the corresponding efforts. One obtains thus, in each node, a system

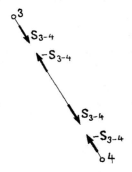

Figure 4.41. Efforts in the members of a framework.

of forces which must be in equilibrium. Two equations can be thus written for each node; this method of computation is convenient if we can solve these equations for each node separately. We start from a node where intervene only two unknowns, passing than to neighbouring nodes where intervene only two unknowns too. For instance, in the case of the truss with parallel flanges considered in Fig.4.39 one starts from the

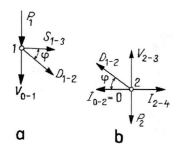

Figure 4.42. Method of separation of nodes.

node O; previously, the reactions have been determined by a global equilibrium (the method of rigidity). One obtains thus $I_{0-2} = 0$ and $V_{0-1} = -V_0$ (Fig.4.39). Then, one passes – successively – to the equilibrium of the node 1 (Fig.4.42,a), to the equilibrium of node 2 (Fig.4.42,b) a.s.o. Sometimes, it is useful to combine the method of isolation of nodes with the method of sections, as in the case of K lattice trusses (Fig.4.38,d).

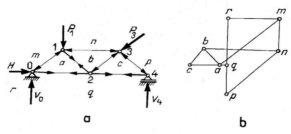

Figure 4.43. A triangular truss (a) and the corresponding Maxwell-Cremona draught (b).

The method of isolation of nodes is graphically concretized in *the Maxwell-Cremona draught*. Let be the truss 0-1-2-3-4, acted upon by the forces P_1 and P_3 (Fig.4.43,a). We use *Bow's notation*, denoting by m, n, p, q, r the zones in which is divided the exterior of the framework by the given and constraint forces which act upon it; as well, we denote by a, b, c the zones in the interior of the truss, separated by bars or by the corresponding efforts. Each given and external force and each effort will be expressed in the form $N_{\alpha\beta}$, where the indices α, β correspond to the zones m, n, p, q, r, a, b, c on one and the other part of the respective force. First of all, the polygon of given and constraint forces, which is a closed one, is drawn. Then, one constructs the polygon of forces for the equilibrium of each node, taking the forces clockwise around the node, in the order in which they are met. We notice that to each node of the truss (Fig.4.43,a) there corresponds a closed polygon in the Maxwell-Cremona draught (Fig.4.43,b), while to each closed polygon of the truss (the polygons at the exterior of the truss are closed at infinity) there corresponds a point of intersection of the respective sides on the draught. Such figures are called *reciprocal*, while the method is called *the method of reciprocal polygons* too. Denoting the intervals on the framework, the signs of the efforts are determined on the figure, taking into account the direction along a polygon according the clockwise direction around the corresponding node on the framework; if the force stretches the node, then the effort is of tension (as in Fig.4.41), otherwise, it is of compression.

The compound structures may be studied analogously, by the methods indicated above. In the case of *complex structures* in which there is not one node with only two unknowns and not one section cutting only three non-concurrent members or for which one cannot make a decomposition in simple structures to which the application of these methods be possible, one must make – in general – a study of the whole framework; such a study involves the solution of a system of $2n$ linear algebraic equations for the n nodes of the truss. But, from case to case, one can apply *the method of bars replacing* of Henneberg or *the two sections method* of S.A. Tsaplin.

The spatial articulated systems are also systems of bars linked by hinges so as to form a non-deformable structure from a geometric point of view, and where the loads are applied at the nodes; but the members of the truss are no more coplanar.

The simple articulated systems are those obtained by joining tetrahedra formed by bars; such a tetrahedron has six bars, corresponding to its sides. By complete induction, one can show that

$$3n = b + 6,\tag{4.2.51}$$

where b is the number of bars, while n is the number of nodes; this is a *necessary condition* of non-deformability of the articulated system.

If the articulated system is not free, then it is necessary to introduce constraints equivalent to six simple supports, which may be materialized by pendulums; besides *simple supports* (which allow the rotation in any direction and the displacement in a plane normal to its direction, Fig.4.44,a), such constraints may be *plane hinges* (*double supports*, materialized by two concurrent pendulums, which allow the rotation in any direction and a displacement along a direction normal to the plane of the support, Fig.4.44,b) and *spherical hinges* (*triple supports* which allow the rotation in any direction, but not one displacement is allowed, Fig.4.44,c).

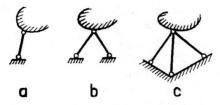

Figure 4.44. Simple (a), double (plane hinge) (b) and triple (spherical hinge) (c) supports.

The relation (4.2.51) becomes a more general form

$$3n = b + s, \quad s \geq 6,\tag{4.2.51'}$$

where s is the number of simple supports.

With the three types of supports mentioned above one can obtain various supporting systems; for instance, a simple support, a double support and a triple support ensure the fixity in space of the articulated system. In general, the six pendulums (the six bars) of the supporting system must fulfil some conditions to ensure a correct fixity of the articulated system, hence to avoid a critical case. These critical forms may be identified – in general – on a static way (by equations of projection and of moment) or on a kinematic way (analysing all the possibilities of motion). Thus, not one straight line about which the resultant moment of all the constraint forces be zero must exist, because the resultant moment of all the given forces about this line (which – in general – does not vanish) could no more be equilibrated; in this case, the articulated system could rotate about this straight line. As well, there must not exist some straight line so that the sum of the projections along it of all the six constraint forces be zero, because the projection along the very same line of the resultant of the given forces would no more be equilibrated; in this case, the rigid solid would have a motion of translation. We mention also that a certain number of pendulums of the six ones must not constitute a plane critical form (for instance, three coplanar pendulums do not be concurrent).

Among *the critical forms* which are the most encountered, depending on the directions of the supports' pendulums, one can mention the following ones: the case in which all the six directions pierce the same straight line; the case in which two sets of three directions are parallel or concurrent; the case in which at least four directions are parallel or concurrent or are coplanar; the case in which two directions are in a plane which contains also the piercing point of other three directions; the case in which three directions are coplanar, the plane containing also the piercing point of other two directions; the case in which five directions are in two planes, their intersection of which is coplanar with the sixth direction; the case in which all the six directions are in parallel planes. These results may be easily verified. We consider that a thorough examination of the supporting system is very important, to can avoid the critical cases.

If the supporting is correct, then the six unknown reactions may be determined by a system of six scalar equations with six unknowns; in some particular case, one can make various observations, simplifying thus the computation. The condition (4.2.51') is *only a necessary condition* of geometric non-deformability; besides this condition, one must verify if the articulated system is not a *critical form*. In this case, to very small variations of the lengths of the bars correspond very great displacements of the nodes; expressing the conditions of equilibrium on the deformed form of the articulated system, to an arbitrary loading may correspond very great values of the efforts in bars or some particular load may lead to indeterminate efforts.

The efforts in the members of the articulated system are given by a system of $3n$ equations with $3n$ unknowns (the equations of equilibrium in each node), so that the condition of non-deformability of such a system is given by

$$\Delta_{3n} \neq 0, \qquad (4.2.52)$$

where Δ_{3n} is the determinant of the coefficients of the system of equations.

Because it is rather difficult to express such conditions for a great n, in particular cases one may use some special methods of investigation. Thus, if no one force is applied at the nodes (hence, if one applies null loads), then all the free terms of the system of equations vanish and, if we take into account the condition (4.2.52), the system has only zero solutions. Hence, in *the method of null loading*, if one succeeds to show that, for zero loads at the nodes, all the efforts in the bars vanish, then it follows that the articulated system is not a critical form; otherwise, this system is a critical form.

On the other hand, the relation (4.2.51') ensures us that the articulated system is *statically determinate*; if $b + s < 3n$, then the system is a mechanism, while if $b + s > 3n$, then the system is *statically indeterminate*. If we have $s = 6$ in the relation (4.2.51'), then the system is a *free articulated system*, the geometric non-deformability of which does not depend on the constraints.

Besides *the simple articulated systems*, we mention also the compound articulated systems, obtained by the composition of various simple systems with the aid of some bars of connection. *The complex articulated systems* are those which cannot be reduced to simple articulated ones.

An important case is that of articulated spatial systems which form a polyhedron without internal diagonals. One may thus use polyhedra the faces of which are constituted by plane trusses, their nodes being on the edges of the polyhedra. We notice that the conditions of geometric non-deformability are fulfilled. To prove this assertion, we start from *Euler's relation*

$$m - n = f - 2, \qquad (4.2.53)$$

valid for a closed polyhedron, where n is the number of vertices, m is the number of edges and f is the number of faces. Assuming that each face is a triangle, each edge being common to two faces, and each face having three edges, we find the supplementary relation

$$2m = 3f;\qquad(4.2.54)$$

eliminating f between the last two relations, we may write

$$m = 3n - 6.\qquad(4.2.53')$$

Noting that the edges are just the bars of the truss ($m = b$), it follows that the relation (4.2.53') is equivalent to the relation (4.2.51); the non-deformability of this articulated system is thus emphasized. If each face of the polyhedron is constituted by a plane truss, non-deformable from a geometric point of view, having nodes only on the edges, so that at each node be at least three coplanar bars, then the above reasoning is valid; such a spatial framework is geometrically non-deformable.

We notice that the hypothesis of perfect hinges at the nodes has a greater importance in the spatial case than in the plane one. Indeed, in this case, the rigidity of the nodes may have a great influence on the values of the efforts in bars.

To state the efforts in bars, we use – in general – the same methods of computation as in the plane case. We mention thus *the method of isolation of nodes*, which can be applied analytically, as well as graphically; one must have at the most three unknown efforts (the reactions are obtained – previously – from conditions of global equilibrium) at each node. In *the method of sections*, each section must cut at the most six bars with unknown efforts. Eventually, one can combine the two methods. These methods are no more sufficient – in general – in the case of complex structures; then one must use *the method of bars replacing* of Henneberg or *the two sections method* of Tsaplin. If it is possible, then one can make also a decomposition of the spatial articulated system in several plane articulated systems, which are separately studied.

2.1.9 Open articulated systems

We call *open articulated system* (*polygonal articulated line, chain*) a system of articulated bars having the form of a polygonal line; if such a system is constituted of more than two bars, then it is *movable*. Assuming that the system of given and constraint external forces $\mathbf{F}_0, \mathbf{F}_1, \mathbf{F}_2, \ldots, \mathbf{F}_n$, equivalent to zero, is acting at the nodes $P_0, P_1, P_2, \ldots, P_n$ (Fig.4.45,a), one must find *the form of the polygonal line*, as well as *the efforts* $\mathbf{N}_{01} = -\mathbf{N}_{10}$, $\mathbf{N}_{12} = -\mathbf{N}_{21}, \ldots, \mathbf{N}_{n-1,n} = -\mathbf{N}_{n,n-1}$ in the bars (according to the considerations synthetized in Fig.4.41, we assume that these forces are applied at the respective nodes). An *open* articulated system is called *simply connected* because, by suppressing an intermediary bar, it loses its unity, unlike the *closed* systems, considered in the previous subsection, which are called *multiply connected* (they are not losing their unity by suppressing an intermediary bar).

For the equilibrium of an open articulated system, it is necessary and sufficient that each node of it be in equilibrium under the action of the external loads (given and constraints), as well as of the internal ones (efforts in bars). To the node P_1 corresponds the polygon of forces OA_1A_2, to the node P_2 the polygon of forces OA_2A_3 a.s.o., while to the node P_{n-1} the polygon of forces $OA_{n-1}A_n$; we construct these polygons in a cumulative *polygon of forces* $OA_1A_2...A_n$, where the internal forces corresponding to two consecutive nodes equilibrate one the other (Fig.4.45,b). The polygons corresponding to the nodes P_0 and P_n are reduced to segments of straight lines. The point O is called *pole*, $OA_1, OA_2,...,OA_n$ being *polar radii*. The geometric figure formed by the sides of the articulated polygonal line, parallel to the polar radii, is called funicular (link) polygon. Firstly, one constructs the polygon of forces $OA_1A_2...A_n$ (the forces $\mathbf{F}_0$ and $\mathbf{F}_n$ may be external constraint forces); the funicular polygon sides are then drawn, by parallels to the polar radii (the length of which is proportional to the magnitude of the efforts in bars, at a certain scale, the directions of these efforts being so as to close the polygons of forces). Thus, the position of equilibrium of the open articulated system, as well as the constraint forces and the efforts in bars are specified.

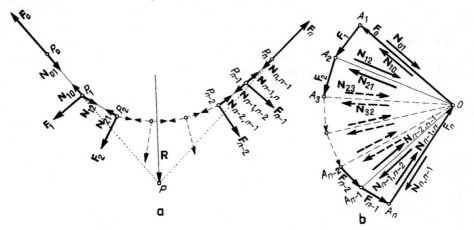

Figure 4.45. Funicular polygon (a). Polygon of forces (b).

Because the external force which acts at a node forms a triangle with the efforts in the contiguous bars, there results that this force and the two adjacent bars are coplanar; thus, an open articulated system acted upon by coplanar external forces is contained in the respective plane too. If in all the bars arise efforts of tension, then the polygonal articulated system may be replaced by a perfect flexible and inextensible thread, acted upon analogously, obtaining the same results.

The graphic construction of the polygon of forces and of the funicular polygon is particularly useful in many plane problems. Thus, the resultant **R** of the external forces $\mathbf{F}_1, \mathbf{F}_2, \ldots, \mathbf{F}_{n-1}$ is specified in direction and magnitude by the oriented segment $\overrightarrow{A_1 A_n}$; one can show that the resultant passes through the piercing point P of the extreme sides of the funicular polygon (Fig.4.45,a). If the polygon of forces is closing, then two cases may take place: if the two sides of the funicular polygon are parallel, then the system of forces is reduced to a couple, while if these sides coincide, then the system of forces is equivalent to zero. We may state

Theorem 4.2.5. *The necessary and sufficient condition of equilibrium of a system of coplanar forces which act upon a rigid solid* (forces modelled by sliding forces) *consists in the closing of the polygon of forces, as well as of the funicular polygon.*

Let be n coplanar forces $\mathbf{F}_1, \mathbf{F}_2, \ldots, \mathbf{F}_n$. One can show that the $n+1$ sides of a funicular polygon, corresponding to a pole O_1, pierce the corresponding sides of another funicular polygon, corresponding to a pole O_2, in $n+1$ points on a straight line parallel to $O_1 O_2$; this one is called *the Culmann straight line*. Starting from this property, one can easily see that, for a system of given forces, all the funicular polygons which pass through two fixed points P_0 and P_n have their poles on a straight line parallel to $P_0 P_n$. Finally, one proves that there is only one funicular polygon which passes through three non-collinear given points. These basic properties allow the graphical study of a system of forces which act upon a rigid solid or upon a system of rigid solids and constitute the basis of the methods of *graphical statics*. For instance, one can decompose a force along three non-concurrent coplanar supports, one can construct the moment of a system of forces, one can determine reactions in a graphical way, one can calculate graphically (with a certain approximation) static moments (obtaining thus the position of centres of gravity), moments of inertia etc.

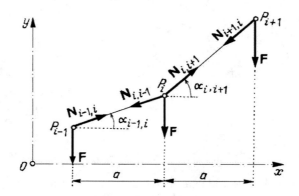

Figure 4.46. Polygonal articulated line. Analytical method.

Analytical methods imply – in general – a great volume of computation. Let us consider the particular case of an open articulated system, the extremities of which are fixed at two fixed points, and which is acted upon by equidistant equal parallel forces; such a system may be encountered in the case of suspension bridges. We consider thus

Statics 275

the articulated polygonal line $P_0 P_1 P_2 ... P_n$, acted upon by the forces $\mathbf{F}_1 = \mathbf{F}_2 = ... = \mathbf{F}_{n-1} = \mathbf{F}$ with respect to a Cartesian right-handed frame of reference Oxy. We use the notations in Fig.4.46, assuming – for the sake of simplicity – that $O \equiv P_0$. We may thus write

$$N_{i,i-1} \cos \alpha_{i-1,i} = N_{i,i+1} \cos \alpha_{i,i+1} = H = \text{const},$$
$$N_{i,i-1} \sin \alpha_{i-1,i} + F = N_{i,i+1} \sin \alpha_{i,i+1},$$

where $N_{i-1,i} = N_{i,i-1}$ is the modulus of the effort in the bar $P_{i-1} P_i$, while H is the constant modulus of the projection of this effort on the Ox - axis; eliminating the efforts in bars, we get

$$\tan \alpha_{i-1,i} = \tan \alpha_{01} + (i-1) \frac{F}{H}.$$

The ordinate of the point P_i will be

$$y_i = a\left(\tan \alpha_{01} + \tan \alpha_{12} + ... + \tan \alpha_{i-1,i} \right)$$
$$= a\left[i \tan \alpha_{01} + \frac{F}{H} + 2\frac{F}{H} + ... + (i-1)\frac{F}{H} \right] = a\left[i \tan \alpha_{01} + \frac{i(i-2)}{2} \frac{F}{H} \right];$$

noting that $x_i = ia$, we have

$$y_i = x_i \tan \alpha_{01} + \frac{F}{2H} x_i \left(\frac{x}{a} - 1 \right).$$

Hence, the open articulated system considered above can be inscribed in a parabola of equation

$$y = x \tan \alpha_{01} + \frac{F}{2H} x \left(\frac{x}{a} - 1 \right). \tag{4.2.55}$$

Knowing the co-ordinates of the point P_n, one can determine the angle α_{01}, as well as the modulus H.

2.2 Statics of threads

As we have seen in Chap. 1, Subsec. 1.1.10, a *thread* is a deformable solid (a bar) for which two dimensions (of the cross section) are completely negligible with respect to the third dimension (the length); one considers that the threads are *perfect flexible and torsionable* (they cannot take over efforts of bending and of torsion), even if – in reality – such ideal models do not exist, as it was shown in Subsec. 2.1.6 (Fig.4.28). We suppose also that the threads to study (materialized by cables, chains, ropes etc.) are *inextensible*. In reality, such threads do not exist; to take into consideration their extensibility (as in the case of rope bridges) passes beyond the frame of this chapter.

After deducing the equation of equilibrium of threads, together with a study of them, we will consider some particular cases of equilibrium under the action of distributed or concentrated loads; as well, we will emphasize the problems which arise in the case of threads constrained to stay on a surface.

2.2.1 Equilibrium equations of threads

Because its thickness is negligible, a thread may be modelled by its axis, the points of application, both of the external forces and of the efforts being thus situated on this axis. If in the equations of equilibrium of a bar (4.2.36) we neglect the moments on the cross section as well as the moment external loads (because of the perfect flexibility and torsionability), we may express the equilibrium of threads in the form

$$\frac{d\mathbf{R}(s)}{ds} + \mathbf{p}(s) = \mathbf{0}, \quad \mathbf{\tau}(s) \times \mathbf{R}(s) = \mathbf{0}, \qquad (4.2.56)$$

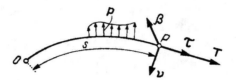

Figure 4.47. Equilibrium of a thread.

where $\mathbf{p}(s)$ is the external load (considered distributed on the unit length of the thread), applied at the point of curvilinear abscissa s, while $\mathbf{R}(s)$ is the resultant of the efforts on the corresponding cross section. Taking into account the model assumed for the thread, as well as the second equation (4.2.56), $\mathbf{R}(s)$ is reduced to the axial force of *tension* $\mathbf{T}(s)$ (one uses the letter $\mathbf{T}$, from the word tension), along the unit vector $\mathbf{\tau}$, tangent to the thread (Fig.4.47); the equation of equilibrium is thus reduced to

$$\frac{d\mathbf{T}(s)}{ds} + \mathbf{p}(s) = \mathbf{0}. \qquad (4.2.57)$$

Noting that the effort $\mathbf{T}(s) = T(s)\mathbf{\tau}(s)$, where $T(s) \geq 0$ is the tension in the thread, at the section s, we may write the equations of equilibrium with respect to Frenet's trihedron in the form

$$\frac{dT}{ds} + p_\tau = 0, \quad \frac{T}{\rho} + p_\nu = 0, \quad p_\beta = 0, \qquad (4.2.57')$$

where we took into account Frenet's formula (4.1.10). Extending the considerations made in the previous subsection concerning articulated polygonal lines and passing to the limit, one obtains the threads; hence, their form of equilibrium is a *funicular curve*.

The third equation (4.2.57') shows that, for equilibrium, the external load $\mathbf{p}$ must be contained in the corresponding osculating plane, in any point of the funicular curve. It results that, in the case of external coplanar loads, the funicular curve is a plane one; the same conclusion is obtained if the load **p** is of constant direction ($\mathbf{p} = p(s)\mathbf{u}$, $\mathbf{u} = \overrightarrow{\mathrm{const}}$). Finally, if the distributed load **p** has the same support, the funicular curve becomes a straight line, which coincides with the common support; this happens also if the thread is not acted upon by some forces ($\mathbf{p} = \mathbf{0}$) or is acted upon only by tangential forces ($p_\nu = 0$). The first two equations (4.2.57') show that the external load **p** is directed towards the convex part of the funicular curve (because $p_\nu = -T/\rho < 0$), exactly in the direction of the decreasing tensions ($p_\tau = -\mathrm{d}T/\mathrm{d}s > 0$, for T decreasing). Moreover, the tangential component p_τ of the external load specifies the tension T; then, the normal component p_ν determines the curvature $1/\rho$, hence the funicular curve. If the external load **p** is normal to the thread for any s (hence if $p_\tau = 0$), then the tension T is constant; one can thus explain why the tension T remains constant along a thread, even if this one passes frictionless over a pulley. This observation remains valid also in the case of a thread of negligible weight, on a smooth surface ($T = \mathrm{const}$), if the funicular curve is a geodesic line of the surface (**v** is along the normal to the surface).

An interesting particular case is that in which the external force is conservative, deriving from a potential $U = U(s)$, hence the case in which

$$\mathbf{p} = \mathrm{grad}\, U, \quad p_\tau = \frac{\mathrm{d}U}{\mathrm{d}s}. \tag{4.2.58}$$

The first equation (4.2.57') leads to

$$T + U = \mathrm{const}; \tag{4.2.58'}$$

by choosing in a convenient form the origin O, the constant can be taken equal to zero, the tension in the thread being thus specified. We notice that the tension T is constant at all the points of a thread on an equipotential surface.

Referring the equation (4.2.57) to the orthonormed frame Ox_i, $i = 1,2,3$, we get

$$\frac{\mathrm{d}\left(T\frac{\mathrm{d}x_i}{\mathrm{d}s}\right)}{\mathrm{d}s} + p_i = 0, \quad i = 1,2,3; \tag{4.2.59}$$

associating the relation

$$\frac{\mathrm{d}x_i}{\mathrm{d}s}\frac{\mathrm{d}x_i}{\mathrm{d}s} = 1 \tag{4.2.59'}$$

too, we can determine the unknowns $T(s)$ and $x_i(s)$, $i = 1,2,3$. The four differential equations (4.2.59), (4.2.59') are of first order with respect to T and of second order with respect to the co-ordinates; the six constants of integration thus introduced are determined by boundary conditions (for instance, bilocal conditions at the two extremities of the thread).

2.2.2 Particular configurations of equilibrium

Let be a thread fixed at the points P_1 and P_2, acted upon by a distributed load $\mathbf{p}(s)$ of constant direction, proportional to the element of arc (for instance, the own weight). According to an observation in the previous subsection, the funicular curve will be a plane one. Referring to a right-handed orthogonal Cartesian system Oxy, we may write

$$\frac{\mathrm{d}}{\mathrm{d}s}\left(T\frac{\mathrm{d}x}{\mathrm{d}s}\right) = 0, \quad \frac{\mathrm{d}}{\mathrm{d}s}\left(T\frac{\mathrm{d}y}{\mathrm{d}s}\right) = p,$$

wherefrom

$$T\frac{\mathrm{d}x}{\mathrm{d}s} = H = \mathrm{const}, \tag{4.2.60}$$

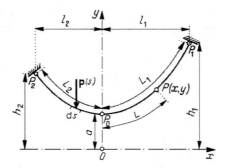

Figure 4.48. Catenary curve.

the tension at the point P_0 being thus constant and equal to H (Fig.4.48). This result has a more general character; if the external load is of the form $\mathbf{p} = p(s)\mathbf{u}$, $\mathbf{u} = \overrightarrow{\mathrm{const}}$, then the projection of the tension on a direction normal to the unit vector $\mathbf{u}$ is constant.

Eliminating the tension T, we get

$$a\frac{\mathrm{d}}{\mathrm{d}s}\left(\frac{\mathrm{d}y}{\mathrm{d}s}\right) = 1, \quad a = \frac{H}{p}, \tag{4.2.61}$$

wherefrom, taking into account $\mathrm{d}s = \sqrt{1 + y'^2}\,\mathrm{d}x$ and integrating, we obtain

$$y' = \sinh\frac{x - x_0}{a}, \quad y - y_0 = a\cosh\frac{x - x_0}{a};$$

hence, the funicular curve, called *catenary curve* in this case, is a hyperbolic line. Effecting a translation of the co-ordinate axes towards the position in Fig.4.48, we find, finally (without losing anything of the generality)

Statics

$$y = a\cosh\frac{x}{a}, \quad y' = \sinh\frac{x}{a}, \quad T = H\cosh\frac{x}{a} = py; \qquad (4.2.62)$$

the last relation corresponds to the formula (4.2.58'), the considered external force **p** being conservative ($\mathbf{p} = \mathrm{grad}(-py + \mathrm{const})$). We notice that the length of the arc of curve between the points P_0 and P is given by

$$L_x = \int_0^x \mathrm{d}s = a\sinh\frac{x}{a}, \quad L_x^2 = y^2 - a^2, \qquad (4.2.62')$$

and the curvature at a current point is expressed in the form

$$\frac{1}{\rho} = \frac{a}{y^2}. \qquad (4.2.62'')$$

Taking into account the position of the points of suspension P_1 and P_2 (Fig.4.48), we introduce the notations $2l = l_1 + l_2$, $2h = h_1 - h_2$; the total length of the catenary curve $L = L_1 + L_2$ and the difference of level $2h$ are given by

$$2L = a\left(\sinh\frac{l_1}{a} + \sinh\frac{l_2}{a}\right), \quad 2h = a\left(\cosh\frac{l_1}{a} - \cosh\frac{l_2}{a}\right),$$

so that

$$L^2 - h^2 = a^2 \sinh^2\frac{l}{a}. \qquad (4.2.62''')$$

Noting that

$$\frac{\sqrt{L^2 - h^2}}{l} = \frac{\sinh\frac{l}{a}}{\frac{l}{a}} = 1 + \frac{1}{3!}\left(\frac{l}{a}\right)^2 + \frac{1}{5!}\left(\frac{l}{a}\right)^4 + \ldots$$

and taking into account that the function in the second member is increasing for $l/a \in (0,\infty)$, it results that this equation has only one positive root a, function of the known parameters l, h and L; the condition $l^2 + h^2 < L^2$ must be fulfilled, hence the length of the thread must be greater than the distance $\overline{P_1 P_2}$.

If such a thread has a large span and a very small deflection with respect to it, then the tension is increasing very much. In this case, the projection H of the tension is greater than the own weight of the thread; one can thus neglect the powers greater than 3 of the ratio $2pL/H$ and – obviously – of the ratio $px/H = x/a$. Hence,

$$\sinh\frac{x}{a} \cong \frac{x}{a} + \frac{x^3}{6a^3}, \quad \cosh\frac{x}{a} \cong 1 + \frac{x^2}{2a^2}.$$

Replacing in (4.2.62), (4.2.62') and translating the axis Ox so that the origin does coincide with the vertex of the catenary curve (which is now a parabola), we obtain (Fig.4.49)

$$y = \frac{p}{2H}x^2, \quad L_x = x + \frac{p^2}{6H^2}x^3 = x\left(1 + \frac{2}{3}\frac{y^2}{x^2}\right), \quad T = H + \frac{p^2}{2H}x^2 \cong H.$$
(4.2.63)

Assuming that the points P_1 and P_2 have the same ordinate ($h = 0$), equal to the deflection f of the thread, and noting that $l_1 = l_2 = l$, we may also write ($a = l^2/2f$)

$$y = \frac{f}{l^2}x^2, \quad 2L = 2l + \frac{4f^2}{3l}, \quad T \cong H = \frac{pl^2}{2f}.$$
(4.2.63')

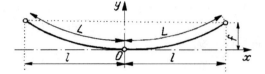

Figure 4.49. Parabolic curve.

We remark the great influence of a small variation of the length $2L$ of the thread on the deflection f (due – for instance – to dilatation or contraction), the distance $2l$ between the points of suspension remaining constant. If $L = l + \Delta l$, then the second relation (4.2.63') leads to

$$f = \sqrt{\frac{3}{2}}\sqrt{\frac{\Delta l}{l}}\,l \cong 1,225\sqrt{\frac{\Delta l}{l}}\,l\,;$$
(4.2.64)

thus, for a variation of the length $2L$ given by $\Delta l/l = 0,0001$ (e.g., $2\Delta l = 1$ cm for $2l = 100$ m), we obtain $f = 0,01225l$ (hence, $f \cong 61$ cm).

Analogously, one can study the case of a thread fixed at the extremities and acted upon by a distributed load of constant direction, proportional to the length of the projection of the element of thread on a plane normal to the direction of the force; using the same frame of reference as in the previous case (Fig.4.49), we may introduce a uniformly distributed load of intensity q by the relation $p\,\mathrm{d}s = q\,\mathrm{d}x$. The equation (4.2.60) holds further and becomes

$$H\frac{\mathrm{d}^2 y}{\mathrm{d}x^2} = q,$$

wherefrom, by integration,

Statics

$$\frac{dy}{dx} = \frac{q}{H}(x - x_0), \quad y - y_0 = \frac{q}{2H}(x - x_0)^2.$$

By a translation of the co-ordinate axes, we find again the results of the preceding case of the catenary curve with a relatively small deflection, in the form

$$y = \frac{q}{2H}x^2, \quad f = \frac{q}{2H}l^2, \quad H = \frac{ql^2}{2f}, \quad T(x) = \sqrt{H^2 + q^2x^2}. \tag{4.2.65}$$

Hence, the funicular curve is a parabola; these results are used – for instance – in the case of threads acted upon by loads laid down by the snow.

2.2.3 Threads acted upon by concentrated loads

If the external load $\mathbf{p}(s)$ is no more distributed, the thread being acted upon by one or several *concentrated loads*, the equation of equilibrium (4.2.57) can no more be applied in the usual manner. In this case, noting that a thread acted upon by only two concentrated forces at two distinct points of it is in equilibrium if and only if the two forces have the same modulus, the same support and opposite directions; the configuration of equilibrium of this portion of thread is a segment of a line which coincides with the common support of the two forces. Thus, the thread may be assimilated to an open articulated system, the problem of equilibrium being thus reduced to the problem of determination of the funicular polygon of the external given forces; various conditions which may be imposed to this polygon, function of the given limit conditions, have been discussed in Subsec. 2.1.9. Analytically, the study of such a thread can be made writing the equations of equilibrium for each point of application of a concentrated load.

To pass over these difficulties of computation, as it was shown by W. Kecs and P.P. Teodorescu, one may use *the methods of the theory of distributions*. Thus, we can admit that the equation (4.2.57) maintains its form in distributions, $\mathbf{T}(s)$ and $\mathbf{p}(s)$ being distributions, while the differentiation is made in the sense of the theory of distributions; besides, this equation may be obtained directly in distributions, in an analogous manner. Using the formulae (1.1.50), (1.1.51) and assuming that the tension $\mathbf{T}(s)$ is a regular distribution, we may write

$$\frac{d\mathbf{T}(s)}{ds} = \frac{\tilde{d}\mathbf{T}(s)}{ds} + \sum_{i=1}^{n-1}(\Delta\mathbf{T})_i\,\delta(s - s_i), \tag{4.2.66}$$

where "tilde" indicates the derivatives in the usual sense, while $(\Delta\mathbf{T})_i$ are the jumps of the tension at the points of discontinuity $s = s_i$, $i = 1, 2, ..., n - 1$. We decompose the external load in the form

$$\mathbf{p}(s) = \tilde{\mathbf{p}}(s) + \sum_{i=1}^{n-1}\mathbf{F}_i\,\delta(s - s_i), \tag{4.2.66'}$$

where $\tilde{\mathbf{p}}(s)$ are forces in the usual sense (distributed loads), while $\mathbf{F}_i$ are concentrated forces, applied at the points $s = s_i$, $i = 1,2,...,n-1$. In this case, the equation (4.2.57) can be decomposed in the form

$$\frac{\tilde{\mathrm{d}}\mathbf{T}(s)}{\mathrm{d}s} + \tilde{\mathbf{p}}(s) = \mathbf{0}, \quad (4.2.67)$$

$$(\Delta \mathbf{T})_i + \mathbf{F}_i = \mathbf{0}, \quad i = 1,2,...,n-1; \quad (4.2.67')$$

the equation (4.2.67) coincides with the classical one, while the $n-1$ relations (4.2.67') must take place for the points of application of the concentrated forces.

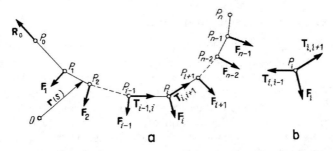

Figure 4.50. Thread acted upon by concentrated forces (a). Local equilibrium at a vertex (b).

Let thus be a thread acted upon only by the concentrated forces $\mathbf{F}_i$, $i = 1,2,...,n-1$, applied at the points $s_1 < s_2 < ... < s_{n-1}$; the equation (4.2.67) becomes

$$\frac{\tilde{\mathrm{d}}\mathbf{T}(s)}{\mathrm{d}s} = \mathbf{0},$$

wherefrom we see that the tension $\mathbf{T}(s)$ is piecewise constant, so that the funicular curve is a funicular polygon (Fig.4.50,a). On the other hand, we notice that

$$(\Delta \mathbf{T})_i = \mathbf{T}_{i,i+1} - \mathbf{T}_{i-1,i}, \quad \mathbf{T}_{i-1,i} + \mathbf{T}_{i,i-1} = \mathbf{0};$$

hence, the equation (4.2.67') corresponds to the equilibrium of the point P_i, subjected to the action of the concentrated force $\mathbf{F}_i$ and of the tensions $\mathbf{T}_{i,i-1}$ and $\mathbf{T}_{i,i+1}$ (Fig.4.50,b). Integrating the equation

$$\frac{\mathrm{d}\mathbf{T}(s)}{\mathrm{d}s} + \sum_{i=1}^{n-1} \mathbf{F}_i \delta(s - s_i) = \mathbf{0},$$

we get

$$\mathbf{T}(s) = -\sum_{i=1}^{n-1} \mathbf{F}_i \theta(s - s_i) - \mathbf{R}_0; \quad (4.2.68)$$

Statics

we put thus in evidence the tension at each point of the thread, excepting the points of discontinuity, but at these points appear the corresponding jumps. We denote by $\mathbf{R}_0 = \mathbf{T}_{01}$ the constraint force applied at the point P_0. If $\mathbf{r}(s)$ is the position vector of a point of the thread and if we take into account the formula (4.1.9), then we obtain

$$\boldsymbol{\tau}(s) = \frac{\mathrm{d}\mathbf{r}(s)}{\mathrm{d}s} = \frac{1}{T}\mathbf{T}(s) = -\frac{1}{T}\left[\sum_{i=1}^{n-1}\mathbf{F}_i\theta(s-s_i) + \mathbf{R}_0\right],$$

where

$$T = |\mathbf{T}| = \sqrt{\left[\mathbf{R}_0 + \sum_{i=1}^{n-1}\mathbf{F}_i\theta(s-s_i)\right]^2}. \qquad (4.2.68')$$

Noting that the unit vector of the tangent to the curve of equilibrium has only discontinuities of the first species and taking the origin at the point P_0, we get, by integration,

$$\mathbf{r}(s) = \sum_{\substack{j=0 \\ (k\geq 1)}}^{k-1} \frac{\sum_{i=0}^{j}\mathbf{F}_i}{\sqrt{\left[\sum_{i=0}^{j}\mathbf{F}_i\right]^2}}(s_{j+1}-s_j) + \frac{\sum_{i=0}^{k}\mathbf{F}_i}{\sqrt{\left[\sum_{i=0}^{j}\mathbf{F}_i\right]^2}}(s-s_k), \qquad (4.2.68'')$$

for $s_k \leq s \leq s_{k+1}$, $k = 0,1,2,\ldots,n-1$, with $s_0 = 0$ and $s_n = L$, where L is the total length of the thread; as well, it was denoted $\mathbf{F}_0 = \mathbf{R}_0$. We have thus obtained the vector equation of the searched funicular polygon.

2.2.4 Threads constrained to stay on a surface

We consider first of all a thread C staying on a *smooth surface* S, supposing that the external forces which act on it are distributed; in the case of concentrated forces, one can adapt the results of the preceding subsection. Let be

$$f(x_1, x_2, x_3) = 0 \qquad (4.2.69)$$

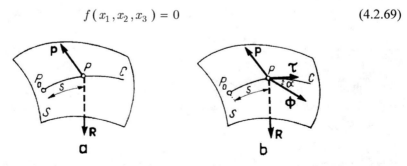

Figure 4.51. Equilibrium of a thread on a frictionless surface (a) or on a surface with friction (b).

the equation of the surface, with respect to a right-handed orthonormed frame of reference Ox_i, $i = 1,2,3$. An element ds of the thread is acted upon by a given external force $\mathbf{p}(s)ds$ and by the constraint force $\mathbf{R}(s)ds$, where $\mathbf{R} = \lambda \operatorname{grad} f$ ($\mathbf{R}$ is along the normal to the surface), λ being an indeterminate scalar (Fig.4.51,a). The equation of equilibrium is written in the form

$$\frac{d\mathbf{T}(s)}{ds} + \mathbf{p}(s) + \mathbf{R}(s) = \mathbf{0} \tag{4.2.70}$$

or, in projection on the co-ordinate axes,

$$\frac{d}{ds}\left(T\frac{dx_i}{ds}\right) + p_i + \lambda f_{,i} = 0, \quad i = 1,2,3. \tag{4.2.70'}$$

Thus, the equations (4.2.69), (4.2.70') and the relation (4.2.59') allow the determination of the unknowns T and λ, and of the co-ordinates x_1, x_2, x_3 as functions of s.

Projecting the equation

$$\frac{dT(s)}{ds}\boldsymbol{\tau}(s) + T(s)\frac{d\boldsymbol{\tau}(s)}{ds} + \mathbf{p}(s) + \mathbf{R}(s) = \mathbf{0}$$

on the axes of the intrinsic Darboux's trihedron and taking into account the first formula (4.1.25), we get the equations

$$\frac{dT}{ds} + p_\tau = 0, \quad \frac{T}{\rho_g} + p_g = 0, \quad \frac{T}{\rho_n} + p_n + R = 0. \tag{4.2.71}$$

In particular, if the thread is not acted upon by an external given load ($\mathbf{p} = \mathbf{0}$), then $T = \text{const}$, $1/\rho = 0$ and $R = -T/\rho_n$, the thread staying along a geodesic line of the surface (case mentioned in Subsec. 2.2.1); if $T = 0$, then the thread is not extended, its form being arbitrary.

If the *surface* S is *rough*, then its reaction on an element of thread will have, besides the normal component $\mathbf{R}(s)ds$, a tangential component $\boldsymbol{\Phi}(s)ds$ too, called *force of sliding friction*. The equations of equilibrium in Darboux's trihedron become

$$\frac{dT}{ds} + p_\tau + \Phi\cos\alpha = 0, \quad \frac{T}{\rho_g} + p_g + \Phi\sin\alpha = 0, \quad \frac{T}{\rho_n} + p_n + R = 0, \tag{4.2.72}$$

where $\Phi = |\boldsymbol{\Phi}|$, while α is the angle made by the force $\boldsymbol{\Phi}$ with the unit vector $\boldsymbol{\tau}$ (Fig.4.51,b), tangent to the thread; according to Chap. 3, Subsec. 2.2.12, one must have

$$\Phi \leq fR \tag{4.2.72'}$$

for equilibrium, where f is a *Coulombian coefficient of friction* (of sliding). It follows that, for the same external load $\mathbf{p}(s)$, the configurations of equilibrium C of the thread on the surface S will be contained between two limit curves C_1 and C_2. In the limit case (the sign "=" in the relation (4.2.72')), assuming that the thread is not acted upon by external loads ($\mathbf{p} = \mathbf{0}$), we may write

$$\frac{\mathrm{d}T}{\mathrm{d}s} + fR\cos\alpha = 0, \quad \frac{T}{\rho_g} + fR\sin\alpha = 0, \quad \frac{T}{\rho_n} + R = 0. \tag{4.2.73}$$

Eliminating R and α between these relations, we obtain the differential equations

$$\frac{\mathrm{d}T_{\mathrm{ext}}}{T_{\mathrm{ext}}} = \pm\frac{1}{\rho_n}\sqrt{f^2 - \tan^2\theta}\mathrm{d}s$$

for the extreme values of the tension T, where we took into account the formulae (4.1.19), (4.1.21), θ being the angle formed by the normal $\mathbf{n}$ to the surface S with the principal normal $\mathbf{v}$ to the line C. We obtain thus *Euler's inequalities*

$$T_{\min} \leq T(s) \leq T_{\max}, \tag{4.2.74}$$

with

$$T_{\min} = T_0 e^{-\chi}, \quad T_{\max} = T_0 e^{\chi}, \quad \chi = \int_0^s \frac{1}{\rho_n(\sigma)}\sqrt{f^2 - \tan^2\theta(\sigma)}\mathrm{d}\sigma, \tag{4.2.74'}$$

where T_0 is the tension at the point P_0, origin of the considered co-ordinates.

Denoting $f = \tan\varphi$, where φ is *the angle of sliding friction*, we notice that, for equilibrium, one must have $\theta \leq \varphi$. If the thread is along a geodesic line of the surface, then $\theta = 0$, so that

$$\chi = f\int_0^s \frac{\mathrm{d}\sigma}{\rho(\sigma)}, \tag{4.2.75}$$

where, on the basis of Meusnier's formula (4.1.19), $\rho_n = \rho$, ρ being the radius of curvature of the curve C.

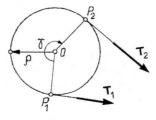

Figure 4.52. Thread wrapped up a circular cylinder along a director curve of it.

In particular, we consider the case of a thread wrapped up a *circular cylinder*, along a director curve of it (a geodesic line of the cylinder), case which appears often in practice. Let P_1 and P_2 be two extreme points of the thread on the rough cylinder (Fig.4.52). We may calculate

$$\chi = f \int_0^s \frac{d\sigma}{\rho(\sigma)} = f \int_0^\gamma d\psi = f\gamma,$$

so that *Euler's relation*

$$e^{-f\gamma} \leq \frac{T_2}{T_1} \leq e^{f\gamma}, \qquad (4.2.76)$$

where γ is the arc, measured in radians, covered by the thread, must hold for equilibrium. Because $e^{f\gamma}$ increases rapidly with γ, for great values of this argument, one may equilibrate a force of great intensity, which acts at an extremity of the thread, by a force of a relative small intensity, applied at the other extremity; for instance, for $f = 0.25$, a contact of the thread on half of the circumference ($\gamma = \pi$) or on a whole circumference or on two or four circumferences, we are led to an amplification of the tension in the thread 2.19 times or 4.81 times or 23.14 times or 535.49 times, respectively.

A geodesic line of the cylinder is – in general – a *helix*, and we can make an analogous study in this case too.

Chapter 5

KINEMATICS

Kinematics deals with the motion of mechanical systems in time, without taking into account their masses and the forces that act upon them; thus, its object of study is the *geometry of motion*. We consider the kinematics of the particle, developing the notions of velocity and acceleration; as well, the kinematics of rigid solids and the kinematics of mechanical system – in general – are dealt with, emphasizing the relative motion too.

1. Kinematics of the particle

We consider the motion of a *particle* (*material point*) with respect to a fixed frame of reference, emphasizing thus its trajectory, velocity and acceleration; the results thus obtained are particularized for some important cases.

1.1 Trajectory and velocity of the particle

In what follows, we define the trajectory and the velocity of a particle, as well as the horary equation of motion; we specify then the velocity in curvilinear co-ordinates and in some particular systems of co-ordinates.

1.1.1 Trajectory. Horary equation of motion. Velocity

We have introduced in Chap. 1, Subsec. 1.1.4 the notion of *frame of reference* with respect to which the motion is studied, using *arbitrary curvilinear co-ordinates* or, in particular, *spherical co-ordinates, cylindrical co-ordinates, or orthogonal Cartesian co-ordinates*. As well, in Chap. 1, Subsec. 1.1.5, we have shown that a particle P describes a *trajectory* C (Fig.1.5), of *vector equation* (1.1.6) or of *parametric equations* (1.1.16')-(1.1.16^{iv}), which define *the law of motion*. The functions which are involved must be *continuous* and *bounded in modulus* for $t \in T \equiv [t_0, t_1]$; they must be differentiable too, excepting – eventually – a finite number of moments, distinguishing thus between a *continuous motion* (the functions are everywhere of class C^2) and a *discontinuous motion*. If the trajectory is a *rectifying curve*, the mapping (1.1.17) (or (1.1.18')) is *the horary equation of motion*.

By analogy with the relation (1.1.18'), a particle P verifies the *law of motion*

$$\mathbf{r} = \mathbf{r}(t; \mathbf{r}_0), \quad x_i = x_i(t; x_i^0), \quad i = 1, 2, 3, \tag{5.1.1}$$

where $\mathbf{r}_0 = x_j^0 \mathbf{i}_j$ corresponds to the position P_0 of the particle P at *the initial moment* t_0.

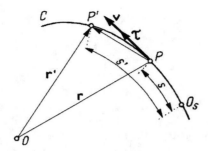

Figure 5.1. Trajectory and velocity of a particle.

It was also shown that the first derivative $\dot{\mathbf{r}}$ is an invariant with respect to a change of fixed frame of reference. To emphasize the mechanical significance of this derivative, we will study how the particle P moves on the trajectory. Let P and P' be the positions of the particle at time t and time t', respectively; *the mean velocity* of the particle in the interval of time $t' - t$ is defined in the form (Fig.5.1)

$$\mathbf{v}_{\text{mean}} = \frac{\overrightarrow{PP'}}{t' - t} = \frac{\mathbf{r}' - \mathbf{r}}{t' - t} = \frac{\mathbf{r}(t') - \mathbf{r}(t)}{t' - t}. \tag{5.1.2}$$

Analogously, *the mean magnitude of the velocity* is given by

$$v_{\text{mean}} = \frac{\widehat{PP'}}{t' - t} = \frac{s' - s}{t' - t} = \frac{s(t') - s(t)}{t' - t}; \tag{5.1.2'}$$

we notice that this magnitude coincides with the modulus of the mean velocity only by identifying the arc $\widehat{PP'}$ with the chord $\overline{PP'}$. Passing to the limit, we obtain *the instantaneous velocity* (*the velocity at the point P*)

$$\mathbf{v}_P = \lim_{t' \to t} \mathbf{v}_{\text{mean}}, \quad v_P = \lim_{t' \to t} v_{\text{mean}}, \tag{5.1.3}$$

supposing that these limits exist; in this case, $|\mathbf{v}| = v$. It follows that *the velocity* of a particle P is given by (for the sake of simplicity, we renounce to the index P)

$$\mathbf{v} = \dot{\mathbf{r}}, \tag{5.1.4}$$

and is expressed by *the derivative of the position vector with respect to time* (corresponding to the formulae (1.1.20')). We notice that the instantaneous velocity (which, as a matter of fact, will be used) is an ideal notion, because only mean velocities can be practically measured; the approximation is the best if one considers, in the series of the mean velocities, terms which correspond to as small as possible

Kinematics 289

difference $t' - t$, tending to the instantaneous velocity. *The components of the velocity in orthogonal Cartesian co-ordinates* are given by

$$v_i = \dot{x}_i, \quad i = 1,2,3 \tag{5.1.4'}$$

$\dot{x}_i$ being the velocities of some particles situated in the projections of the point P on the three axes of co-ordinates. It results that *the modulus of the velocity* is given by

$$v = \sqrt{v_i v_i} = \sqrt{\dot{x}_i \dot{x}_i} = \dot{s}, \tag{5.1.5}$$

where we took into consideration the relations (1.1.18). The velocity v of the particle P on the trajectory (which is a scalar) is thus the modulus of the velocity $\mathbf{v}$ (vector quantity); *the horary equation of motion* is obtained by the integration of the differential equation

$$ds = v(t)dt, \tag{5.1.5'}$$

obtaining the formula (1.1.18'). If $v(t) = v_0 = \text{const}$, the motion is called *uniform*; in this case

$$s = v_0 t + s_0. \tag{5.1.6}$$

Passing to the limit in formula (5.1.3) we notice that the oriented segment $\overrightarrow{PP'}$ situated along the chord $\overline{PP'}$ tends to the tangent to the trajectory at the point P, the velocity $\mathbf{v}$ enjoying this property. In *intrinsic co-ordinates* (after Frenet's trihedron) we may thus write

$$v_\tau = v, \quad v_\nu = v_\beta = 0, \tag{5.1.7}$$

so that

$$\mathbf{v} = v\boldsymbol{\tau}. \tag{5.1.7'}$$

We notice that the velocity vector is applied at the point P, its direction corresponding to the direction of the motion.

1.1.2 Velocity of a particle in curvilinear co-ordinates

Let be a *system of curvilinear co-ordinates* q_1, q_2, q_3, linked to the position vector and to the orthogonal Cartesian co-ordinates by relations of the form (A.1.32), (A.1.33). Taking into account the formulae (A.1.35) and the notation (A.1.36), the velocity of a particle is expressed in the form

$$\mathbf{v} = \frac{\partial \mathbf{r}}{\partial q_i} \dot{q}_i = \dot{q}_i \mathbf{e}_i = v_i \mathbf{e}_i, \tag{5.1.8}$$

where $v_i = \dot{q}_i$ are *the components of the velocity* (as a matter of fact, they are *the contravariant components*, but we do not use this notion, neither the corresponding notations) in the frame of reference specified by the basis' vectors $\mathbf{e}_i$. The modulus of the velocity is given by

$$v^2 = g_{ij}\dot{q}_i\dot{q}_j, \qquad (5.1.8')$$

where we use the formula (A.1.37) of the element of arc.

Taking into account the notations (A.1.40), we may write

$$\mathbf{v} = \sum_{i=1}^{3} H_i v_i \mathbf{i}_i, \quad \mathbf{i}_i = \operatorname{vers}\mathbf{e}_i, \qquad (5.1.9)$$

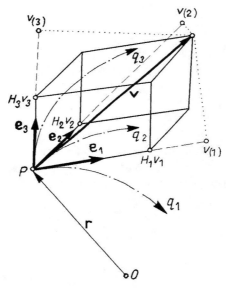

Figure 5.2. Velocity of a particle in curvilinear co-ordinates.

$H_1 v_1$, $H_2 v_2$, $H_3 v_3$ being *the physical components* of the velocity $\mathbf{v}$ (Fig.5.2). Analogously, *the orthogonal projections of the velocity* $\mathbf{v}$ on the basis' vectors $\mathbf{e}_i$ will be given by $v_{(i)} = \mathbf{v} \cdot \mathbf{i}_i$, hence by (without summation for $i = 1,2,3$)

$$v_{(i)} = \sum_{j=1}^{3} \frac{g_{ij}}{H_j} v_i. \qquad (5.1.10)$$

In the case of *orthogonal curvilinear co-ordinates* $g_{ij} = 0$, $i \neq j$, so that (without summation for $i = 1,2,3$)

$$v_{(i)} = H_i v_i; \qquad (5.1.11)$$

Kinematics

hence, the orthogonal projections of the velocity are the physical components of it. If the frame of reference is orthonormed, then $H_1 = H_2 = H_3 = 1$, obtaining again the components (5.1.4').

1.1.3 Important particular cases

In *spherical co-ordinates*, we obtain

$$\mathbf{v} = v_r \mathbf{i}_r + v_\theta \mathbf{i}_\theta + v_\varphi \mathbf{i}_\varphi, \qquad (5.1.12)$$

where

$$v_r = \dot{r}, \quad v_\theta = r\dot{\theta}, \quad v_\varphi = r\sin\theta\dot{\varphi}, \qquad (5.1.12')$$

taking into account the results in App., Subsec. 1.1.5; the modulus of the velocity is given by

$$v^2 = \dot{r}^2 + r^2\dot{\theta}^2 + r^2\sin^2\theta\dot{\varphi}^2. \qquad (5.1.12'')$$

Analogously, in *cylindrical co-ordinates* we may write

$$\mathbf{v} = v_r \mathbf{i}_r + v_\theta \mathbf{i}_\theta + v_z \mathbf{i}_z, \qquad (5.1.13)$$

where

$$v_r = \dot{r}, \quad v_\theta = r\dot{\theta}, \quad v_z = \dot{z}, \qquad (5.1.13')$$

the modulus of the velocity being given by

$$v^2 = \dot{r}^2 + r^2\dot{\theta}^2 + \dot{z}^2; \qquad (5.1.13'')$$

in particular, in *polar co-ordinates* (in the plane Ox_1x_2), we have

$$\mathbf{v} = v_r \mathbf{i}_r + v_\theta \mathbf{i}_\theta, \quad v_r = \dot{r}, \quad v_\theta = r\dot{\theta} \qquad (5.1.14)$$

and

$$v^2 = \dot{r}^2 + r^2\dot{\theta}^2, \qquad (5.1.14')$$

corresponding to the results in App., Subsec. 1.1.2.

1.1.4 Areal velocity

In the case of a plane trajectory C, we may define a quantity of the nature of a velocity, which characterizes the variation of the sectorial area between two vector radii and the arc of the trajectory, when the particle P describes the latter one. Let thus be the sectorial areas A and A' at the moments t and t', respectively. Noting that (Fig.5.3)

$$\frac{r^2(\theta'-\theta)}{2(t'-t)} < \frac{A'-A}{t'-t} < \frac{r'^2(\theta'-\theta)}{2(t'-t)}$$

and passing to the limit for $t' \to t$, we obtain *the areal velocity* $\Omega = \dot{A}$ (in the plane Ox_1x_2), given by

$$\Omega = \frac{1}{2}r^2\dot{\theta} = \frac{1}{2}rv_\theta. \qquad (5.1.15)$$

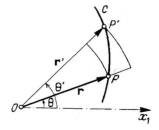

Figure 5.3. Areal velocity of a particle.

Vectorially, we have

$$\mathbf{\Omega} = \Omega \mathbf{i}_3 = \frac{1}{2}\mathbf{r} \times \mathbf{v}, \qquad (5.1.15')$$

where we took into account $\mathbf{r} = r\mathbf{i}_r$, $\mathbf{i}_r \times \mathbf{i}_\theta = \mathbf{i}_3$ and the first formula (5.1.14); it follows also that

$$\Omega = \frac{1}{2}(x_1\dot{x}_2 - \dot{x}_1x_2). \qquad (5.1.15'')$$

If the trajectory C is a tortuous curve, we may consider formulae of the form (5.1.15'), (5.1.15'') for the projections of the particle P on the three co-ordinate planes. We introduce thus *the areal velocity*

$$\mathbf{\Omega} = \frac{1}{2}\mathbf{r} \times \mathbf{v} = \frac{1}{2}\mathbf{r} \times \dot{\mathbf{r}} = \Omega_i \mathbf{i}_i, \quad \Omega_i = \frac{1}{2}\epsilon_{ijk}\, x_j \dot{x}_k, \qquad (5.1.16)$$

which characterizes the variation of the area of the sector between two vector radii on the lateral surface of the cone of vertex O and directrix C.

In particular, if $\mathbf{\Omega} = \mathbf{C} = \overrightarrow{\text{const}}$, then we may write

$$\frac{1}{2}\mathbf{r}\cdot(\mathbf{r}\times\mathbf{v}) = \mathbf{C}\cdot\mathbf{r} = 0;$$

hence, the trajectory C is a plane curve, which passes through the origin O of the co-ordinate axes. In the case of a vanishing areal velocity ($\mathbf{C} = \mathbf{0}$), the velocity $\mathbf{v}$ has the same direction as the position vector $\mathbf{r}$, the trajectory being rectilinear.

Kinematics 293

We notice that one can introduce the torsor of the velocity **v** with respect to the pole O in the form

$$\tau_O(\mathbf{v}) = \{\mathbf{v}, 2\mathbf{\Omega}\}. \qquad (5.1.16')$$

1.2 Acceleration of the particle

We define – in the following – the acceleration of a particle, by introducing the velocity hodograph; we calculate the acceleration in curvilinear co-ordinates and in some particular cases of co-ordinates. We introduce accelerations of higher order too, as well as the acceleration of a discontinuous motion.

1.2.1 Velocity hodograph. Acceleration

Let M be an arbitrary pole at which we apply the vector **V**, equipollent to the velocity vector **v**; if the particle P describes the trajectory C, then the extremity Q of the vector **V** describes a curve Γ, called *the velocity hodograph* (Fig.5.4). Observing that $\overrightarrow{MQ} = \mathbf{V}$ plays the rôle of a vector radius, it results that the velocity by which the point Q is moving on the curve Γ is the velocity of the velocity of the particle P, equipollent to the acceleration **a** (introduced in Chap. 1, Subsec.1.1.5), which is invariant with respect to a change of a fixed frame of reference.

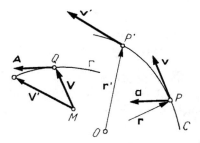

Figure 5.4. Hodograph of velocity.

Let P and P' be two positions of the particle on the trajectory C. We introduce the mean acceleration

$$\mathbf{a}_{\mathrm{mean}} = \frac{\mathbf{v}' - \mathbf{v}}{t' - t}; \qquad (5.1.17)$$

the instantaneous acceleration (acceleration at the point P) becomes

$$\mathbf{a} = \lim_{t' \to t} \mathbf{a}_{\mathrm{mean}}. \qquad (5.1.17')$$

Taking into account the velocity hodograph, we may write

$$\mathbf{a} = \dot{\mathbf{v}} = \ddot{\mathbf{r}}; \qquad (5.1.18)$$

the acceleration, applied at the point P, is thus *the derivative of the velocity* with respect to time or *the second derivative of the position vector* with respect to time.

The components of the acceleration in orthogonal Cartesian co-ordinates are expressed in the form

$$a_i = \dot{v}_i = \ddot{x}_i, \quad i = 1,2,3, \tag{5.1.18'}$$

$\ddot{x}_i$ being the accelerations of the projections of the point P along the three axes of co-ordinates. The modulus of the acceleration is given by

$$a = \sqrt{a_i a_i} = \sqrt{\dot{v}_i \dot{v}_i} = \sqrt{\ddot{x}_i \ddot{x}_i}. \tag{5.1.18''}$$

Starting from (4.1.7'), we obtain, by differentiation,

$$\mathbf{a} = \dot{v}\boldsymbol{\tau} + v\dot{\boldsymbol{\tau}} = \dot{v}\boldsymbol{\tau} + v\frac{d\boldsymbol{\tau}}{ds}\dot{s},$$

wherefrom

$$a_\tau = \dot{v}, \quad a_\nu = \frac{v^2}{\rho}, \quad a_\beta = 0, \tag{5.1.19}$$

so that

$$\mathbf{a} = a_\tau \boldsymbol{\tau} + a_\nu \boldsymbol{\nu}, \tag{5.1.19'}$$

taking into account the first formula of Frenet (4.1.10). The component of the acceleration along the binormal vanishes, hence the acceleration is contained in the osculating plane. The modulus of the acceleration may be written in the form

$$a = \sqrt{\dot{v}^2 + \frac{v^4}{\rho^2}}. \tag{5.1.19''}$$

We notice that, in the case of a *uniform* motion ($v = $ const), the tangential acceleration vanishes; if $a_\tau = $ const, then the motion is *uniformly varied* (*uniformly accelerated* or *uniformly decelerated* as a_τ and v have the same sign or are of opposite signs). The normal acceleration (the acceleration $\mathbf{a}$ too) is directed always towards the interior of the trajectory (towards the centre of curvature), being *centripetal* ($a_\nu \geq 0$); it vanishes only at the inflection points of the trajectory or in the case of a rectilinear motion ($1/\rho = 0$). If the acceleration vanishes ($\mathbf{a} = \mathbf{0}$, $a_\tau = a_\nu = 0$), the motion of the particle is *rectilinear* and *uniform*. Starting form the areal velocity (5.1.16), we may define the areal acceleration in the form

$$\dot{\boldsymbol{\Omega}} = \frac{1}{2}\mathbf{r} \times \mathbf{a} = \frac{1}{2}\mathbf{r} \times \ddot{\mathbf{r}} = \dot{\Omega}_i \mathbf{i}_i, \quad \dot{\Omega}_i = \frac{1}{2}\in_{ijk} x_j \ddot{x}_k; \tag{5.1.20}$$

Kinematics 295

in this case, the torsor of the acceleration with respect to the pole O is given by

$$\tau_O(\mathbf{a}) = \dot{\tau}_O(\mathbf{v}) = \{\mathbf{a}, 2\mathbf{\Omega}\}. \tag{5.1.20'}$$

1.2.2 Approximation of the motion of a particle in the neighbourhood of a given position. Deviation

Let P and P' be two positions of a particle, corresponding to the moments t and t', respectively. Assuming that the function $\mathbf{r}(t)$ is of class C^2, we may use the formula (A.1.20) of Taylor type in the form

$$\mathbf{r}(t') = \mathbf{r}(t) + (t' - t)\dot{\mathbf{r}}(t) + \frac{1}{2}(t' - t)^2 \ddot{\mathbf{r}}(t) + \mathbf{\eta}(t, t' - t),$$

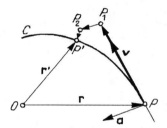

Figure 5.5. Deviation of a particle.

where $\mathbf{\eta}(t, t' - t)$ represents the rest. We can thus approximate the motion of the particle P in the neighbourhood of the point P by $\overrightarrow{PP'} = \overrightarrow{PP_1} + \overrightarrow{P_1P_2} + \overrightarrow{P_2P'}$ (Fig.5.5), with

$$\overrightarrow{PP_1} = (t' - t)\dot{\mathbf{r}}(t), \quad \overrightarrow{P_1P_2} = \frac{1}{2}(t' - t)^2 \ddot{\mathbf{r}}(t), \quad \overrightarrow{P_2P'} = \mathbf{\eta}(t, t' - t); \tag{5.1.21}$$

for a fixed t, we may write

$$\frac{\mathrm{d}\overrightarrow{PP_1}}{\mathrm{d}t'} = \dot{\mathbf{r}}(t), \quad \frac{\mathrm{d}\overrightarrow{P_1P_2}}{\mathrm{d}t'} = (t' - t)\ddot{\mathbf{r}}(t), \quad \frac{\mathrm{d}^2 \overrightarrow{P_1P_2}}{\mathrm{d}t'^2} = \ddot{\mathbf{r}}(t). \tag{5.1.21'}$$

We can thus state

Theorem 5.1.1. *The continuous motion of a particle P may be approximated, at a given moment t, in a neighbourhood of first order* (the segment $\overrightarrow{PP'}$), *by a rectilinear and uniform motion along the tangent at P to the trajectory, where the velocity is equal to the velocity of the particle at P. As well, in a neighbourhood of second order* (the segments $\overrightarrow{PP_1}$ and $\overrightarrow{P_1P_2}$), *the motion can be approximated by a succession of two rectilinear motions: the motion previously presented, to which a uniform varied motion with the acceleration of the particle at P is added.*

The time interval $t' - t$ is very small in a vicinity of first order; in a vicinity of second order, this interval is somewhat greater, but still small, so that the segment $\overrightarrow{P_2P'} = \mathbf{\eta}$ is negligible. The vector $\overrightarrow{P_1P_2}$ is called *deviation*; its rôle is to bring back the particle P from the tangent (on which it moves if it is not acted by a force, according to the principle of inertia) on the trajectory C.

1.2.3 Acceleration of a particle in curvilinear co-ordinates

Starting from the expression (5.1.8) of the velocity, by differentiation, one obtains

$$\mathbf{a} = \frac{\partial^2 \mathbf{r}}{\partial q_j \partial q_k} \dot{q}_j \dot{q}_k + \frac{\partial \mathbf{r}}{\partial q_j} \ddot{q}_j = \mathbf{e}_{j,k} \dot{q}_j \dot{q}_k + \mathbf{e}_j \ddot{q}_j. \tag{5.1.22}$$

Noting that

$$\mathbf{a} \cdot \mathbf{e}_j = a_k \mathbf{e}_k \cdot \mathbf{e}_j = a_k g_{kj}, \quad g_{kj} g^{ij} = \delta_k^i,$$

we may write

$$a_i = g^{il} \mathbf{a} \cdot \mathbf{e}_l = \mathbf{e}_l \cdot \mathbf{e}_{j,k} g^{il} \dot{q}_j \dot{q}_k + \mathbf{e}_j \cdot \mathbf{e}_l g^{il} \ddot{q}_j = \ddot{q}_i + [jk,l] g^{il} \dot{q}_j \dot{q}_k,$$

using the notations introduced in App., Subsec.1.1.5, where $[jk,l]$ is Christoffel's symbol of first species. With the aid of Christoffel's symbol of second species (A.1.45), we may write the components (*contravariant components*) of the acceleration in the frame $\mathbf{e}_i$ in the form

$$a_i = \ddot{q}_i + \begin{Bmatrix} i \\ j\ k \end{Bmatrix} \dot{q}_j \dot{q}_k, \quad i = 1,2,3. \tag{5.1.22'}$$

The physical components of the acceleration are $a_1 H_1$, $a_2 H_2$, $a_3 H_3$, while *the orthogonal projections of the acceleration* on the basis' vectors $\mathbf{e}_i$ are written in the form (without summation with respect to $i = 1,2,3$)

$$a_{(i)} = \sum_{j=1}^{3} \frac{g_{ij}}{H_i} a_j. \tag{5.1.23}$$

In the case of *orthogonal curvilinear co-ordinates*, we have (without summation with respect to $i = 1,2,3$)

$$a_{(i)} = H_i a_i, \tag{5.1.23'}$$

so that the orthogonal projections of the acceleration are its physical components. For an orthonormed frame of reference ($H_1 = H_2 = H_3 = 1$) we find again the components (5.1.18').

Kinematics

Using a method due to Lagrange, we can calculate the components of the acceleration also in a movable system of curvilinear co-ordinates, given by

$$\mathbf{r} = \mathbf{r}(q_1, q_2, q_3; t).\tag{5.1.24}$$

We have

$$\mathbf{v} = \frac{\partial \mathbf{r}}{\partial q_j}\dot{q}_j + \frac{\partial \mathbf{r}}{\partial t} = \mathbf{e}_j \dot{q}_j + \dot{\mathbf{r}},$$

wherefrom

$$\frac{\partial \mathbf{v}}{\partial \dot{q}_j} = \frac{\partial \mathbf{r}}{\partial q_j} = \mathbf{e}_j;$$

then

$$\mathbf{a} \cdot \mathbf{e}_i = \frac{d\mathbf{v}}{dt} \cdot \frac{\partial \mathbf{r}}{\partial q_i} = \frac{d}{dt}\left(\mathbf{v} \cdot \frac{\partial \mathbf{r}}{\partial q_i}\right) - \mathbf{v} \cdot \frac{d}{dt}\left(\frac{\partial \mathbf{r}}{\partial q_i}\right).$$

We notice that

$$\frac{d}{dt}\left(\frac{\partial \mathbf{r}}{\partial q_i}\right) = \frac{\partial^2 \mathbf{r}}{\partial q_j \partial q_i}\dot{q}_j + \frac{\partial^2 \mathbf{r}}{\partial t \partial q_i}, \quad \frac{\partial \mathbf{v}}{\partial q_i} = \frac{\partial^2 \mathbf{r}}{\partial q_i \partial q_j}\dot{q}_j + \frac{\partial^2 \mathbf{r}}{\partial q_i \partial t};$$

because the vector function $\mathbf{r}$ is of class C^2, it follows

$$\frac{d}{dt}\left(\frac{\partial \mathbf{r}}{\partial q_i}\right) = \frac{\partial \mathbf{v}}{\partial q_i} = \frac{\partial}{\partial q_i}\left(\frac{d\mathbf{r}}{dt}\right),$$

so that the operators d/dt and $\partial/\partial q_i$ are permutable. Finally, we get (*covariant components* of the acceleration)

$$\mathbf{a} \cdot \mathbf{e}_i = \frac{d}{dt}\left(\mathbf{v} \cdot \frac{\partial \mathbf{v}}{\partial \dot{q}_i}\right) - \mathbf{v} \cdot \frac{\partial \mathbf{v}}{\partial q_i} = \frac{1}{2}\frac{d}{dt}\left(\frac{\partial v^2}{\partial \dot{q}_i}\right) - \frac{1}{2}\frac{\partial v^2}{\partial q_i},$$

wherefrom

$$a_i = \frac{1}{2}g^{ij}\left[\frac{d}{dt}\left(\frac{\partial v^2}{\partial \dot{q}_j}\right) - \frac{\partial v^2}{\partial q_j}\right], \quad i = 1, 2, 3,\tag{5.1.25}$$

obtaining thus *Lagrange's formula*.

1.2.4 Important particular cases

Taking into account the results in App., Subsec. 1.1.5, we obtain, in *spherical co-ordinates*,

$$\mathbf{a} = a_r \mathbf{i}_r + a_\theta \mathbf{i}_\theta + a_\varphi \mathbf{i}_\varphi, \tag{5.1.26}$$

with

$$a_r = \ddot{r} - r\dot{\theta}^2 - r\sin^2\theta\dot{\varphi}^2,$$
$$a_\theta = r\ddot{\theta} + 2\dot{r}\dot{\theta} - r\sin\theta\cos\theta\dot{\varphi}^2 = \frac{1}{r}\frac{\mathrm{d}}{\mathrm{d}t}(r^2\dot{\theta}) - \frac{r}{2}\sin 2\theta\dot{\varphi}^2, \tag{5.1.26'}$$
$$a_\varphi = r\sin\theta\ddot{\varphi} + 2\dot{r}\sin\theta\dot{\varphi} + 2r\cos\theta\dot{\theta}\dot{\varphi} = \frac{1}{r\sin\theta}\frac{\mathrm{d}}{\mathrm{d}t}(r^2\sin^2\theta\dot{\varphi}).$$

Analogously, in *cylindrical co-ordinates*, we get

$$\mathbf{a} = a_r \mathbf{i}_r + a_\theta \mathbf{i}_\theta + a_z \mathbf{i}_z, \tag{5.1.27}$$

where

$$a_r = \ddot{r} - r\dot{\theta}^2, \quad a_\theta = r\ddot{\theta} + 2\dot{r}\dot{\theta}, \quad a_z = \ddot{z}. \tag{5.1.27'}$$

In particular, in *polar co-ordinates*, in the plane Ox_1x_2, we may write

$$\mathbf{a} = a_r \mathbf{i}_r + a_\theta \mathbf{i}_\theta, \quad a_r = \ddot{r} - r\dot{\theta}^2, \quad a_\theta = r\ddot{\theta} + 2\dot{r}\dot{\theta} = \frac{1}{r}\frac{\mathrm{d}}{\mathrm{d}t}(r^2\dot{\theta}). \tag{5.1.28}$$

1.2.5 Accelerations of higher order

As we have seen in Chap. 1, Subsec. 1.1.4, the derivatives of higher order of the position vector are invariant with respect to changes of a fixed frame of reference; these derivatives are identified with the derivatives of the acceleration $\mathbf{a}$, which will be called *acceleration of first order* ($\mathbf{a} = \mathbf{a}_{(1)}$). We obtain thus *accelerations of higher order*: the acceleration of second order ($\mathbf{a}_{(2)} = \dot{\mathbf{a}} = \dddot{\mathbf{r}}$), the acceleration of third order ($\mathbf{a}_{(3)} = \ddot{\mathbf{a}} = \ddddot{\mathbf{r}}$) and – in general – the acceleration of nth order, given by

$$\mathbf{a}_{(n)} = \frac{\mathrm{d}^{n-1}\mathbf{a}}{\mathrm{d}t^{n-1}} = \frac{\mathrm{d}^{n+1}\mathbf{r}}{\mathrm{d}t^{n+1}}. \tag{5.1.29}$$

Although the accelerations of higher order do not intervene directly in the Newtonian model of mechanics, which needs only the acceleration of first order, one considers that some mechanical phenomena (collisions, seismic phenomena, etc., which take place by a rapid variation of the intensity of the force) may lead to other mathematical modelling, in which these accelerations play an important rôle.

Starting from formulae (5.1.19), (5.1.19'), which give the acceleration of first order with the aid of its intrinsic components (along Frenet's trihedron axes), and using Frenet's formulae (4.1.10), (4.1.10'), we get

$$\mathbf{a}_{(2)} = a_\tau^{(2)}\boldsymbol{\tau} + a_\nu^{(2)}\boldsymbol{\nu} + a_\beta^{(2)}\boldsymbol{\beta}, \qquad (5.1.30)$$

where

$$a_\tau^{(2)} = \ddot{v} - \frac{v^3}{\rho^2}, \quad a_\nu^{(2)} = 3\frac{v\dot{v}}{\rho} - v^2\frac{\dot{\rho}}{\rho^2}, \quad a_\beta^{(2)} = -\frac{v^3}{\rho\rho'}. \qquad (5.1.30')$$

We notice that the acceleration of second order has a component along the binormal too, which vanishes only in the case of a rectilinear trajectory or, more general, of a torsionless trajectory (in the osculatory plane). The acceleration of second order along the principal normal vanishes (for a curvilinear trajectory) if

$$v^3 = \frac{v_0^3}{\rho_0}\rho, \qquad (5.1.31)$$

where ρ_0 and v_0 correspond to the moment $t = t_0$ (for instance, in the case of a uniform circular motion).

Obviously, the acceleration of second order may be introduced also with the aid of the hodograph of the acceleration of first order.

1.2.6 Acceleration in case of a discontinuous motion

In the case in which the position vector is a continuous function on $[t',t'']$, while the velocity and the acceleration are continuous on the same interval, excepting a finite number of moments $t_i \in [t',t'']$, $i = 1,2,...,n$ (piecewise continuous), to which correspond discontinuities of the first species, that is

$$\mathbf{v}(t_i - 0) \neq \mathbf{v}(t_i + 0), \quad \mathbf{a}(t_i - 0) \neq \mathbf{a}(t_i + 0), \qquad (5.1.32)$$

it is necessary to use methods of the theory of distributions. The integrals

$$\int_{t'}^{t''} \mathbf{v}(t)\mathrm{d}t, \quad \int_{t'}^{t''} \mathbf{a}(t)\mathrm{d}t \qquad (5.1.33)$$

do exist in these conditions.

Taking into account the formula (1.1.51), we may write

$$\frac{\mathrm{d}^2\mathbf{r}(t)}{\mathrm{d}t^2} = \frac{\tilde{\mathrm{d}}^2\mathbf{r}(t)}{\mathrm{d}t^2} + \sum_{i=1}^n \mathbf{V}_i\delta(t - t_i), \qquad (5.1.34)$$

where

$$\mathbf{V}_i = \mathbf{v}(t_i + 0) - \mathbf{v}(t_i - 0), \quad i = 1, 2, \ldots, n, \tag{5.1.34'}$$

represents *the jump of the velocity*, corresponding to the moment of discontinuity $t = t_i$, while the sign "tilde" is for the derivative in the usual sense; we may thus state

Theorem 5.1.2. *The acceleration of a particle in the sense of the theory of distributions is equal to the distribution defined by the acceleration of the particle in the usual sense, where the latter one exists, to which is added the sum of the products of the velocity jump, of the particle by the Dirac distribution.*

We introduce the notations

$$\mathbf{a}(t) = \frac{d^2 \mathbf{r}(t)}{dt^2}, \quad \tilde{\mathbf{a}}(t) = \frac{\tilde{d}^2 \mathbf{r}(t)}{dt^2}, \quad \mathbf{a}_c(t) = \sum_{i=1}^n \mathbf{V}_i \delta(t - t_i), \tag{5.1.34''}$$

where $\mathbf{a}(t)$ is *the acceleration in the sense of the theory of distributions*, $\tilde{\mathbf{a}}(t)$ is *the acceleration in the usual sense*, while $\mathbf{a}_c(t)$ is *the complementary acceleration* due to the discontinuities. With these notations, the relation (5.1.34) becomes

$$\mathbf{a}(t) = \tilde{\mathbf{a}}(t) + \mathbf{a}_c(t). \tag{5.1.34'''}$$

The acceleration in the sense of the theory of distributions will be called generalized acceleration too.

1.3 Particular cases of motion of a particle

We consider, in what follows, some particular cases of motion: the rectilinear motion, the circular motion, the parabolic motion, the helical motion, as well as the cycloidal motion.

1.3.1 Parabolic and rectilinear motion

We consider the particular case $\mathbf{a} = \mathbf{a}_0 = \overrightarrow{\text{const}}$; integrating, we get

$$\mathbf{v} = (t - t_0)\mathbf{a}_0 + \mathbf{v}_0, \quad \mathbf{r} = \frac{1}{2}(t - t_0)^2 \mathbf{a}_0 + (t - t_0)\mathbf{v}_0 + \mathbf{r}_0, \tag{5.1.35}$$

where we have supposed that $\mathbf{v} = \mathbf{v}_0$ and $\mathbf{r} = \mathbf{r}_0$ at the initial moment. We notice that, taking a new origin of the frame of reference at $\mathbf{r} = \mathbf{r}_0$, nothing of the generality of the trajectory is lost; the position vector $\mathbf{r}$ is thus a linear combination of the constant vectors $\mathbf{v}_0$ and $\mathbf{a}_0$, hence, it belongs to a fixed plane, so that the trajectory is a plane curve. We suppose that the trajectory is contained in the plane Ox_1x_2; we may thus write the parametric equations in the form (Fig.5.6)

$$x_1(t) = v_1^0(t - t_0), \quad x_2(t) = \frac{1}{2}a_2^0(t - t_0)^2 + v_2^0(t - t_0), \tag{5.1.35'}$$

where, for the sake of simplicity, we take the Ox_2-axis along the acceleration $\mathbf{a}_0$. Eliminating the time t, we get

$$x_2 = \frac{a_2^0}{2(v_1^0)^2} x_1^2 + \frac{v_2^0}{v_1^0} x_1; \qquad (5.1.36)$$

hence, the trajectory is a parabola, the axis of which is parallel to the acceleration $\mathbf{a}_0$; the vertex Q of the parabola is of co-ordinates (we notice that $v_1^0, v_2^0 > 0$, $a_2^0 < 0$)

$$x_1 = -\frac{v_1^0 v_2^0}{a_2^0}, \quad x_{2\max} = -\frac{(v_2^0)^2}{2a_2^0}, \qquad (5.1.36')$$

and is obtained by the particle at the moment $t = t_0 - v_2^0 / a_2^0$. We may write

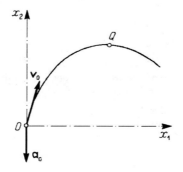

Figure 5.6. Parabolic motion of a particle.

$$v_2^0 = \sqrt{-2a_2^0 x_{2\max}}, \qquad (5.1.37)$$

obtaining thus *Torricelli's formula*, which gives the component v_2^0 of the initial velocity of the particle which attains the ordinate $x_{2\max}$, assuming an acceleration of modulus $|\mathbf{a}_0| = -a_2^0$. These results stay at the basis of studies on external ballistics.

In the particular case in which $v_1^0 = 0$, the motion is rectilinear; if we assume that the motion is along an Ox-axis, then the second relation (5.1.35') allows to write the equation of motion (which is – at the same time – the horary equation) in the form

$$x(t) = s(t) = \frac{1}{2} a_0 (t - t_0)^2 + v_0 (t - t_0) + x_0, \qquad (5.1.38)$$

where we put into evidence the initial position (eventually another one that the origin O); there results

$$v(t) = a_0 (t - t_0) + v_0. \qquad (5.1.38')$$

If $a_0 = 0$, then the motion is *uniform* and the diagram of motion is given in Fig.5.7,a. As a matter of fact, the reciprocal implication is also true; indeed, if $\mathbf{a}_0 = \mathbf{0}$, then

$$\mathbf{v} = \mathbf{v}_0, \quad \mathbf{r} = \mathbf{v}_0(t - t_0) + \mathbf{r}_0, \tag{5.1.39}$$

the trajectory being a straight line. If $a_0 \neq 0$, then the motion is *uniformly varied*; eliminating the time t between the relations (5.1.38), (5.1.38'), we obtain *Torricelli's formula* in the form

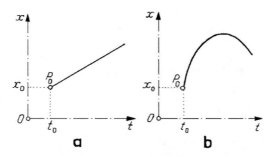

Figure 5.7. Diagram of a rectilinear motion of a particle: uniform (a) and uniformly varied (b).

$$v = \sqrt{v_0^2 + 2a_0(x - x_0)}; \tag{5.1.40}$$

in particular, if the particle has not initial velocity at the origin O, then we get

$$v = \sqrt{2a_0 x}. \tag{5.1.40'}$$

If, for a certain interval of time, the velocity and the acceleration have the same direction, then the motion is *accelerated*, on the contrary, the motion is *decelerated*. In the case of a varied motion, it is possible to have both phases (for instance, a motion the diagram of which is given in Fig.5.7,b) or only one of them.

1.3.2 Circular motion

If the trajectory is a circle of radius R ($\mathbf{r} = R\mathbf{i}_r$), then the motion is called *circular* (Fig.5.8). By means of the formulae (5.1.7), (5.1.19) or of the formulae (5.1.14), (5.1.28), we obtain

$$\mathbf{v} = \boldsymbol{\omega} \times \mathbf{r}, \quad v = v_\tau = v_\theta = \dot{s} = R\omega, \quad s = R\theta + s_0, \tag{5.1.41}$$

$$\mathbf{a} = \dot{\boldsymbol{\omega}} \times \mathbf{r} - \omega^2 \mathbf{r}, \quad a_\tau = a_\theta = R\dot{\omega}, \quad a_\nu = -a_r = R\omega^2, \quad a = R\sqrt{\omega^4 + \dot{\omega}^2}, \tag{5.1.41'}$$

where we have introduced the angular velocity ω and the angular acceleration ε, defined by the relations

Kinematics

$$\boldsymbol{\omega}(t) = \omega(t)\mathbf{i}_3, \quad \boldsymbol{\varepsilon}(t) = \dot{\boldsymbol{\omega}}(t) = \dot{\omega}(t)\mathbf{i}_3, \quad \omega(t) = \dot{\theta}(t), \quad \varepsilon(t) = \dot{\omega}(t) = \ddot{\theta}(t).$$
(5.1.42)

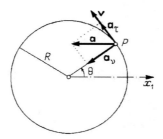

Figure 5.8. Circular motion of a particle.

If $\dot{\omega} = 0$, $\omega = \text{const}$ (and $v = \text{const}$), then the motion is uniform and we have

$$\theta = \omega t + \theta_0, \quad s = R(\omega t + \theta_0), \quad v = R\omega, \quad a_\tau = 0, \quad a = a_\nu = R\omega^2.$$
(5.1.43)

1.3.3 Helical motion

Let be a particle in a uniform motion on a *helix* of pitch $p = 2\pi R \tan\alpha$, situated on a circular cylinder of radius R (Fig.5.9). The trajectory is given by

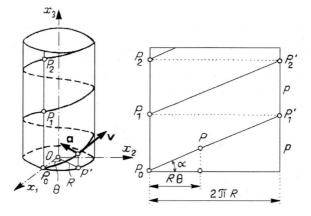

Figure 5.9. Helical motion of a particle.

$$x_1 = R\cos\omega t, \quad x_2 = R\sin\omega t, \quad x_3 = \frac{p\omega}{2\pi}t.$$
(5.1.44)

Hence,

$$v_1 = -R\omega\sin\omega t, \quad v_2 = R\omega\cos\omega t, \quad v_3 = \frac{p\omega}{2\pi},$$
(5.1.45)

so that

$$v = \omega\sqrt{R^2 + \frac{p^2}{4\pi^2}} = \frac{R\omega}{\cos\alpha};$$ (5.1.45')

then

$$a_1 = -R\omega^2\cos\omega t,\quad a_2 = -R\omega^2\sin\omega t,\quad a_3 = 0$$ (5.1.46)

and

$$a = R\omega^2.$$ (5.1.46')

Thus, the motion is uniform and the acceleration is reduced to the normal acceleration given by (5.1.20). Comparing to (5.1.46') and taking into account (5.1.45'), we obtain the radius of curvature

$$\rho = R + \frac{p^2}{4\pi^2 R} = \frac{R}{\cos^2\alpha};$$ (5.1.47)

the principal normal is parallel to the plane Ox_1x_2 and passes through the Ox_3-axis; hence, the acceleration **a** enjoys the same property (as it results from the formulae (5.1.46)).

1.3.4 Cycloidal motion

Another interesting particular case of motion is that of a particle P on a *cycloid*; that one is the locus of a point of a circle (for instance, on the peripheral of a wheel in a vertical plane) of radius R, which is rolling without sliding on a straight line (horizontal) (Fig.5.10). The imposed condition leads to

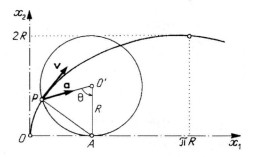

Figure 5.10. Cycloidal motion of a particle.

$$\theta = \omega t,\quad \omega = \frac{v_0}{R}$$ (5.1.48)

where $\mathbf{v}_0 = \overrightarrow{\text{const}}$ is the velocity of the centre O' of the circle, supposed to have a uniform motion. The parametric equations of the trajectory are

Kinematics

$$x_1 = R(\omega t - \sin \omega t), \quad x_2 = R(1 - \cos \omega t), \tag{5.1.49}$$

wherefrom one obtains the components of the velocity and of the acceleration

$$v_1 = R\omega(1 - \cos \omega t), \quad v_2 = R\omega \sin \omega t, \tag{5.1.49'}$$

$$a_1 = R\omega^2 \sin \omega t, \quad a_2 = R\omega^2 \cos \omega t, \tag{5.1.49''}$$

respectively.

One can easily prove that the modulus of the velocity and of the acceleration are expressed in the form

$$v = 2R\omega \sin \frac{\omega t}{2} = 2v_0 \sin \frac{\omega t}{2} = \overline{AP}\omega, \quad a = R\omega^2 = \overline{O'P}\omega^2, \tag{5.1.49'''}$$

respectively; the acceleration **a** is directed towards the centre O' of the circle. Hence, from the point of view of the velocities, the particle behaves as in a uniform motion of angular velocity ω around the point A; what concerns the accelerations, it behaves as in a rotation around the point O'.

2. Kinematics of the rigid solid

In the study of the motion of a rigid solid, it is necessary to study the motion of an arbitrary point P of it with respect to a fixed frame of reference $\mathscr{R}'$ of orthogonal Cartesian co-ordinates $O'x_1'x_2'x_3'$. By means of some basic kinematics formulae which are deduced, one considers some particular cases of motion; hence, one can pass to the general case of motion of the rigid solid.

2.1 Kinematical formulae in the motion of a rigid solid

We put in evidence, in what follows, some results concerning the determination of the velocity and the acceleration in the motion of a rigid solid.

2.1.1 Velocity in the motion of the rigid solid

Let be *a movable frame of reference* $\mathscr{R}$ *of orthogonal Cartesian co-ordinates* $Ox_1x_2x_3$, in a rigid linkage with the rigid solid (hence, in motion with respect to *the fixed frame of reference* $\mathscr{R}'$, specified by the constant unit vectors $\mathbf{i}_j'$, $j = 1,2,3$); obviously, we use right orthonormed frames of reference. We denote by $\mathbf{r}_0'$ the position vector of the pole O with respect to the pole O' and by $\mathbf{r}' = x_j'\mathbf{i}_j'$ the position vector of the point P with respect to the frame $\mathscr{R}'$; analogously, the position vector of the same point P with respect to the movable frame of reference is $\mathbf{r} = x_j\mathbf{i}_j$, where $\mathbf{i}_j$, $j = 1,2,3$, are the unit vectors of the frame $\mathscr{R}$ (Fig.5.11). We mention that, during the motion, $\mathbf{r} = \overrightarrow{\text{const}}$ with respect to the moving frame $\mathscr{R}$. We notice that the latter frame can be determined by the vector $\mathbf{r}_0$ and the unit vectors $\mathbf{i}_j$, $j = 1,2,3$ (as a matter of fact, one of the unit vectors and the plane formed by the other two unit vectors are sufficient), that is 2+1=3 independent scalar quantities; this result corresponds to the six degrees of freedom of the rigid solid. The position of the point P with respect to the frame $\mathcal{R}'$ is given by

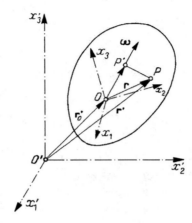

Figure 5.11. Motion of the rigid solid with respect to a fixed and a movable frame of reference.

$$\mathbf{r}' = \mathbf{r}'_0 + \mathbf{r}. \tag{5.2.1}$$

Using the formula (A.2.37) which links the absolute derivative (with respect to the frame $\mathcal{R}'$) to the relative derivative (with respect to the frame $\mathcal{R}$), we may write

$$\dot{\mathbf{r}} = \frac{\mathrm{d}\mathbf{r}}{\mathrm{d}t} = \boldsymbol{\omega} \times \mathbf{r}, \tag{5.2.2}$$

where $\boldsymbol{\omega}$ is a vector specified by (A.2.36), the same for all the points of the rigid solid (hence, an invariant); we took into account that the derivative of $\mathbf{r}$ with respect to the movable frame of reference is equal to zero ($\partial \mathbf{r}/\partial t = \mathbf{0}$). For the velocity $\mathbf{v}' = \dot{\mathbf{r}}'$ of the point P with respect to the fixed frame of reference we obtain *Euler's formula*, which gives the distribution of the velocities in a rigid solid

$$\mathbf{v}' = \mathbf{v}'_0 + \boldsymbol{\omega} \times \mathbf{r}, \tag{5.2.3}$$

where we have introduced the velocity $\mathbf{v}'_0 = \dot{\mathbf{r}}'_0$ of the pole O of the frame $\mathcal{R}$ with respect to the same frame $\mathcal{R}'$. We notice that the relations (5.2.2), (5.2.3) may be written in the form

$$\frac{\mathrm{d}\overrightarrow{OP}}{\mathrm{d}t} = \boldsymbol{\omega} \times \overrightarrow{OP}, \tag{5.2.2'}$$

$$\mathbf{v}'_0 - \mathbf{v}'_P + \boldsymbol{\omega} \times \overrightarrow{OP} = \mathbf{0} \tag{5.2.3'}$$

Kinematics

too, where O and P are two arbitrary points of the rigid solid. Scalarly, we have

$$v_i = v_i^O + \epsilon_{ijk}\, \omega_j x_k, \quad i = 1,2,3\,. \tag{5.2.3''}$$

A scalar product of the relation (5.2.3) by vers **r** leads to a mixed product which vanishes, so that

$$\mathbf{v}' \cdot \text{vers}\,\mathbf{r} = \mathbf{v}'_0 \cdot \text{vers}\,\mathbf{r}\,; \tag{5.2.4}$$

hence, *the projections of the velocities of two points of a rigid solid on the straight line which links these points are equal.* We notice that the relation (5.2.4) corresponds to the relation (3.2.22'), a consequence of the rigidity condition (3.2.22), and represents *the condition of compatibility of the velocities in the motion of the rigid solid* (*relation of holonomic constraint*).

Let P_1 and P_2 be two points of the rigid solid. From the relation $\overrightarrow{P_1 P_2} = \overrightarrow{OP_2} - \overrightarrow{OP_1}$, it results $\overrightarrow{P_1 P_2}^2 = \overrightarrow{OP_1}^2 + \overrightarrow{OP_2}^2 - 2\overrightarrow{OP_1} \cdot \overrightarrow{OP_2}$; hence, the condition of rigidity of the solid ($\left|\overrightarrow{P_1 P_2}\right| = \text{const}$, $\left|\overrightarrow{OP_1}\right| = \text{const}$, $\left|\overrightarrow{OP_2}\right| = \text{const}$) leads to $\sphericalangle\left(\overrightarrow{OP_1}, \overrightarrow{OP_2}\right) = \text{const}$. We may thus state that *the angle of two arbitrary segments of a rigid solid is conserved in a general motion of it.*

Analogously, effecting the scalar product of the relation (5.2.3) by the vector $\boldsymbol{\omega}$, we get

$$\mathbf{v}' \cdot \boldsymbol{\omega} = \mathbf{v}'_0 \cdot \boldsymbol{\omega} \tag{5.2.5}$$

and may state

Theorem 5.2.1. *The scalar product of the velocity of a point of the rigid solid by the vector* $\boldsymbol{\omega}$ *is an invariant* (the same for all the points of the rigid solid).

We can state also that *the projection of the velocity of a point of the rigid solid on the vector* $\boldsymbol{\omega}$ *is a constant* (the same for all the points of the rigid solid). It follows that, in the case of a general motion (for $\mathbf{v}'_0$ and $\boldsymbol{\omega}$ arbitrary vectors), there are not points of vanishing velocity (for which $\mathbf{v}' = \mathbf{0}$). We obtain this result also by observing that the equation $\mathbf{v}'_0 + \boldsymbol{\omega} \times \mathbf{r} = \mathbf{0}$ has a solution only if $\mathbf{v}'_0 \cdot \boldsymbol{\omega} = 0$; as well, if the vectors $\mathbf{v}'_0$ and $\boldsymbol{\omega}$ are orthogonal at a point, then they are orthogonal at any other point.

Hence, in the motion of the rigid solid do appear two *kinematic invariants*: the vector $\boldsymbol{\omega}$ as we will see in Subsec. 2.2.2, it is an angular velocity) and the scalar product $\mathbf{v} \cdot \boldsymbol{\omega}$ (or the projection of the vector $\mathbf{v}$ on the direction of the vector $\boldsymbol{\omega}$, that is $\mathbf{v} \cdot \boldsymbol{\omega}/\omega$).

2.1.2 Acceleration in the motion of a rigid solid

Differentiating the relation (5.2.3) with respect to time in the frame $\mathcal{R}'$, we obtain the acceleration with respect to the same frame

$$\mathbf{a}' = \mathbf{a}'_0 + \dot{\boldsymbol{\omega}} \times \mathbf{r} + \boldsymbol{\omega} \times (\boldsymbol{\omega} \times \mathbf{r}) \qquad (5.2.6)$$

where we have introduced the acceleration $\mathbf{a}'_0 = \dot{\mathbf{v}}'_0$ of the pole O of the movable frame of reference with respect to the same fixed frame; we took into account the formula (5.2.2) and the relation $\dot{\boldsymbol{\omega}} = \mathrm{d}\boldsymbol{\omega}/\mathrm{d}t = \partial\boldsymbol{\omega}/\partial t$ (on the basis of the formula (A.2.37)). Using the basic formula of the triple vector product (2.1.49), we may write

$$\mathbf{a}' = \mathbf{a}'_0 + \dot{\boldsymbol{\omega}} \times \mathbf{r} + (\boldsymbol{\omega} \cdot \mathbf{r})\boldsymbol{\omega} - \omega^2 \mathbf{r} \qquad (5.2.6')$$

too; in components, we have

$$a'_i = a'^O_i + \left(\omega_i \omega_j - \omega^2 \delta_{ij} - \epsilon_{ijk} \dot{\omega}_k \right) x_j, \quad i = 1,2,3. \qquad (5.2.6'')$$

Denoting by P' the projection of the point P on the vector $\boldsymbol{\omega}$ (we have $\mathbf{r} = \overrightarrow{OP} = \overrightarrow{OP'} + \overrightarrow{P'P}$) and noting that $\boldsymbol{\omega} \times \overrightarrow{OP'} = \mathbf{0}$ in (5.2.6) and $\boldsymbol{\omega} \cdot \overrightarrow{P'P} = 0$ in (5.2.6'), we obtain *Rivals' formula*

$$\mathbf{a}' = \mathbf{a}'_0 + \dot{\boldsymbol{\omega}} \times \overrightarrow{OP} - \omega^2 \overrightarrow{P'P}. \qquad (5.2.6''')$$

Imposing the condition $a'_i = 0$, $i = 1,2,3$, we get a system of linear equations in the co-ordinates x_j, obtaining thus the points in which the acceleration of the rigid solid vanishes. If we use the formula (2.1.36"), then we may write the determinant of the coefficients of the unknowns of this system in the form

$$\Delta = \det\left[\omega_i \omega_j - \omega^2 \delta_{ij} - \epsilon_{ijk} \dot{\omega}_k \right] = \frac{1}{6} \epsilon_{ijk} \epsilon_{lmn} \left(\omega_i \omega_l - \omega^2 \delta_{il} \right.$$
$$\left. - \epsilon_{ilp} \dot{\omega}_p \right)\left(\omega_j \omega_m - \omega^2 \delta_{jm} - \epsilon_{jmq} \dot{\omega}_q \right)\left(\omega_k \omega_n - \omega^2 \delta_{kn} - \epsilon_{knr} \dot{\omega}_r \right);$$

developing, we obtain 27 sums of products, seven of them being equal to zero, because they correspond to products of symmetric tensors by antisymmetric ones with respect to the same indices. Taking into account the formulae (2.1.46)-(2.1.46") and the above observation, we get

$$\Delta = \omega_i \omega_j \dot{\omega}_i \dot{\omega}_j - \omega_i \omega_i \dot{\omega}_j \dot{\omega}_j = (\boldsymbol{\omega} \cdot \dot{\boldsymbol{\omega}})^2 - \omega^2 \dot{\omega}^2 = -(\boldsymbol{\omega} \times \dot{\boldsymbol{\omega}})^2, \qquad (5.2.7)$$

where we have used Lagrange's identity (2.1.33).

If $\boldsymbol{\omega} \times \dot{\boldsymbol{\omega}} \neq \mathbf{0}$, then there exists a point (and only one), called *the pole of accelerations*, for which the acceleration vanishes at a given moment; hence, this pole is moving with respect to both frames of reference (fixed and movable). One may thus state

Theorem 5.2.2. *In the general motion of a rigid solid, the instantaneous distribution of the accelerations is the same as in the case of a rigid solid with a fixed point* (the pole of accelerations) *at the respective moment.*

Kinematics

If $\boldsymbol{\omega} \times \dot{\boldsymbol{\omega}} = \mathbf{0}$, then the system of linear equations may be impossible, so that points of null acceleration do not exist (the case of a translation or of a helical motion), or may be indetermined, existing a straight line (an instantaneous axis of rotation, support of the vector $\boldsymbol{\omega}$) for which all the points have a vanishing acceleration (the case of a rotation or of a plane-parallel motion). All these particular cases will be considered in Secs 2.2 and 2.3.

The scalar product of relation (5.2.6) by the vector $\boldsymbol{\omega}$, where $\mathbf{a}' = \mathbf{0}$, leads to a scalar equation of the form (2.1.53), hence

$$\boldsymbol{\omega} \cdot \mathbf{a}'_0 + (\boldsymbol{\omega} \times \dot{\boldsymbol{\omega}}) \cdot \mathbf{r} = 0, \qquad (5.2.8)$$

where we took into account the relation (2.1.53) and the conditions in which a triple scalar product vanishes. The solution of this equation is of the form (2.1.53') and we may write

$$\mathbf{r} = \mathbf{p} \times (\boldsymbol{\omega} \times \dot{\boldsymbol{\omega}}) - \frac{\boldsymbol{\omega} \cdot \mathbf{a}'_0}{(\boldsymbol{\omega} \times \dot{\boldsymbol{\omega}})^2} \boldsymbol{\omega} \times \dot{\boldsymbol{\omega}}. \qquad (5.2.8')$$

The arbitrary vector $\mathbf{p}$ is determined so that the vector equation obtained from (5.2.6), where we make $\mathbf{a}' = \mathbf{0}$, be verified; one observes once more the important rôle played by the vector product $\boldsymbol{\omega} \times \dot{\boldsymbol{\omega}}$.

In general, $(\boldsymbol{\omega}, \dot{\boldsymbol{\omega}}, \boldsymbol{\omega} \times \dot{\boldsymbol{\omega}}) \neq 0$, so that we may represent the vectors $\mathbf{r}$ and $\mathbf{a}'_0$ in a frame of reference defined by the three factors of this mixed product, in the form

$$\mathbf{r} = \lambda \boldsymbol{\omega} + \mu \dot{\boldsymbol{\omega}} + \nu \boldsymbol{\omega} \times \dot{\boldsymbol{\omega}}, \quad \mathbf{a}'_0 = a'^O_\lambda \boldsymbol{\omega} + a'^O_\mu \dot{\boldsymbol{\omega}} + a'^O_\nu \boldsymbol{\omega} \times \dot{\boldsymbol{\omega}}; \qquad (5.2.9)$$

replacing in the vector equation obtained from (5.2.6), making $\mathbf{a}' = \mathbf{0}$, and noting that basis' vectors are arbitrary, we obtain the scalar system

$$\begin{aligned} a'_\lambda + (\boldsymbol{\omega} \cdot \dot{\boldsymbol{\omega}})\mu + \dot{\omega}^2 \nu &= 0, \\ a'_\mu - \omega^2 \mu - (\boldsymbol{\omega} \cdot \dot{\boldsymbol{\omega}})\nu &= 0, \\ a'_\nu - \lambda - \omega^2 \nu &= 0; \end{aligned} \qquad (5.2.9')$$

We see thus that the determinant of the coefficients of the unknowns is given by the same formula (5.2.7), similar conclusions being obtained.

The condition (5.2.4) of compatibility of the velocities may be written also in the form

$$(\mathbf{v}'_O - \mathbf{v}'_P) \cdot \overrightarrow{OP} = 0; \qquad (5.2.4')$$

hence, *the difference of the projections of the velocities of two points of the rigid solid on the straight line which links them vanishes* (or *the difference of the velocities of the two points is normal to the straight line which links them*). Analogously, *the condition (3.2.23") of compatibility of the accelerations* becomes

$$(\mathbf{a}'_O - \mathbf{a}'_P) \cdot \overrightarrow{OP} = (\mathbf{v}'_O - \mathbf{v}'_P)^2 = (\boldsymbol{\omega} \times \overrightarrow{OP})^2. \qquad (5.2.10)$$

We may thus state that *the product of the modulus of the difference of the projections of the accelerations of two points of a rigid solid on the straight line which links them by the distance between these points is equal to the square of the difference of the velocities of the respective points.*

2.2 Particular cases of motion of the rigid solid

In what follows, we consider some particular cases of motion of the rigid solid, i.e.: the motion of translation, the motion of rotation and the motion of rototranslation. As well, we emphasize the motion of the local frames of Frenet and Darboux.

2.2.1 Motion of translation

We say that a rigid solid is subjected to a *motion of translation* if an arbitrary straight bar of it remains parallel to itself at any moment of the motion. In particular, the axes Ox_i, $i = 1, 2, 3$, of the movable frame of reference $\mathcal{R}$ must enjoy this property; as a matter of fact, it is sufficient that the three axes do remain parallel to themselves (hence, the moving frame of reference must have a displacement parallel to itself), in order that any other straight line do enjoy the same property (because any straight line of the rigid solid is rigidly connected to the frame $\mathcal{R}$). The position vector $\mathbf{r}$ remains constant in the relation (5.2.1) (it has a displacement parallel to itself), in this case; the trajectory C of the point P is obtained from the trajectory Γ of the point O by a translation of vector $\mathbf{r} = \mathbf{r}' - \mathbf{r}'_0$ (Fig.5.12). If the point O attains the point O_1 on the curve Γ, then the point P attains the point P_1 on the curve C, while the vector $\overrightarrow{O_1P_1} = \mathbf{r}_1$ is equipollent to the vector $\mathbf{r}$.

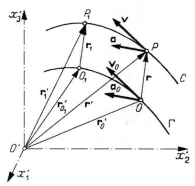

Figure 5.12. Motion of translation of a rigid solid.

Taking into account Poisson's formula (A.238), it results that $\mathbf{i}_j = \mathbf{0}$, $j = 1, 2, 3$, if and only if $\boldsymbol{\omega} = \mathbf{0}$; in this case, the distribution laws of the velocities and accelerations are given by

Kinematics

$$\mathbf{v}' = \mathbf{v}'_O, \quad \mathbf{a}' = \mathbf{a}'_O. \tag{5.2.11}$$

Hence, at a given moment, all the points of a rigid solid in a motion of translation have the same velocity and the same acceleration; the respective vectors may be modelled by free vectors. If the velocity $\mathbf{v}'_O$ is given, then the equation of the trajectory becomes

$$\mathbf{r}'(t) = \mathbf{r}'(t_0) + \int_{t_0}^{t} \mathbf{v}'(\tau)\, d\tau \tag{5.2.12}$$

If $\text{vers}\,\overrightarrow{\mathbf{v}'_O} = \text{const}$, then the motion of translation is rectilinear; otherwise, it is curvilinear. If $|\mathbf{v}'_O| = \text{const}$, then the motion of translation is uniform.

2.2.2 Motion of rotation

We say that a rigid solid is subjected to a *motion of rotation* if two points of it remain fixed during the motion; the straight line which links the two points (called *axis of rotation*) is fixed too, in this case. We assume (without losing from the generality of the study) that $O \equiv O'$ and we choose as axis of rotation $O'x'_3 \equiv Ox_3$. The rigid solid remains with only one degree of freedom, which is the angle

$$\theta = \theta(t) \tag{5.2.13}$$

between the axes $O'x'_\alpha$ and Ox_α, $\alpha = 1,2$ (Fig.5.13). Noting that the unit vectors of the movable frame may be expressed with the aid of the unit vectors of the fixed frame of reference in the form

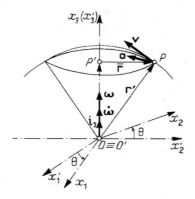

Figure 5.13. Motion of rotation of a rigid solid.

$$\mathbf{i}_1 = \cos\theta\, \mathbf{i}'_1 + \sin\theta\, \mathbf{i}'_2, \quad \mathbf{i}_2 = -\sin\theta\, \mathbf{i}'_1 + \cos\theta\, \mathbf{i}'_2, \quad \mathbf{i}_3 = \mathbf{i}'_3$$

the formula (A.2.36) leads to

$$\omega_1 = \omega_2 = 0, \quad \omega_3(t) = \omega(t) = \dot\theta(t), \tag{5.2.13'}$$

hence, the vector $\boldsymbol{\omega}$ is a sliding vector situated along the axis of rotation ($\boldsymbol{\omega}(t) = \omega(t)\mathbf{i}_3$), being *the angular velocity vector*. Thus, we have put in evidence the mechanical significance of the vector $\boldsymbol{\omega}$, formally introduced in the formula (5.2.3).

The velocity of a point of the rigid solid is thus given by

$$\mathbf{v} = \dot{\mathbf{r}} = \boldsymbol{\omega} \times \mathbf{r} = \overrightarrow{PO} \times \boldsymbol{\omega} \tag{5.2.14}$$

and can be considered to be the moment of the vector $\boldsymbol{\omega}$ with respect to the point P. Multiplying scalarly this relation by the vector $\mathbf{r}$ and by the vector $\mathbf{i}_3$, respectively, and noting that one obtains null mixed products, one gets

$$\mathbf{r} \cdot \dot{\mathbf{r}} = \frac{1}{2}\frac{\mathrm{d}}{\mathrm{d}t}r^2 = 0, \quad \mathbf{i}_3 \cdot \dot{\mathbf{r}} = \frac{\mathrm{d}}{\mathrm{d}t}(\mathbf{i}_3 \cdot \mathbf{r}) = 0,$$

wherefrom $r^2 = \mathrm{const}$ and $\sphericalangle(\mathbf{i}_3, \mathbf{r}) = \mathrm{const}$. It follows that the trajectory of a point P is the intersection of a sphere of centre O and r radius and a cone, the vertex of which is at O too; hence, these trajectories are circles situated in planes normal to the axis of rotation, having the centres in points P' on the very same axis. Because $\mathbf{r} = \overrightarrow{OP} = \overrightarrow{OP'} + \overrightarrow{P'P} = \overrightarrow{OP'} + \overline{\mathbf{r}}$, where P' is the projection of the point P on the axis of rotation, one can write the relation (5.2.14) also in the form

$$\mathbf{v} = \boldsymbol{\omega} \times \overline{\mathbf{r}}, \quad v = \omega|\overline{\mathbf{r}}|; \tag{5.2.14'}$$

comparing this formula with the formula (5.1.41), which corresponds to the circular motion of a particle, we see that the velocity vector is tangent to the trajectory in the normal plane to the axis of rotation. Noting that the vectors $\overline{\mathbf{r}}$ and $\mathbf{v}$ are polar vectors, it results that $\boldsymbol{\omega}$ is an axial vector; besides, the above considerations justify the denomination given to this vector. Taking into account the considerations in Chap. 3, Subsec. 1.2.3, we may attach to the vector $\boldsymbol{\omega}$ an antisymmetric tensor of second order, given by the formula (3.1.62'); we write

$$\omega_{ij} = \epsilon_{ijk}\omega_k, \quad \omega_i = \frac{1}{2}\epsilon_{ijk}\omega_{jk}, \tag{5.2.15}$$

and the relations (5.2.14), (5.2.14') become

$$v_i = \epsilon_{ijk}\omega_j x_k = \omega_{ji}x_j, \quad v_1 = -\omega x_2, \quad v_2 = \omega x_1, \quad v_3 = 0, \tag{5.2.14''}$$

so that all the points of the rigid solid have the same angular velocity. We notice that the points on the axis of rotation are the only ones of vanishing velocity, while the velocity vectors of the points of a straight line parallel to this axis are equipollent; as well, the velocities are parallel and their moduli have a linear variation for the points of a normal to the axis of rotation. Taking into account (A.2.31'), (A.2.31''), the relations

$$\mathrm{curl}\,\mathbf{v} = 2\boldsymbol{\omega}, \quad \mathrm{div}\,v = 0 \tag{5.2.16}$$

Kinematics

hold, the field of velocities being thus *solenoidal*.

Rivals formula (5.2.6''') leads to

$$\mathbf{a} = \dot{\boldsymbol{\omega}} \times \mathbf{r} - \omega^2 \overline{\mathbf{r}} = \dot{\boldsymbol{\omega}} \times \overline{\mathbf{r}} - \omega^2 \overline{\mathbf{r}}, \quad a = |\overline{\mathbf{r}}|\sqrt{\omega^4 + \dot{\omega}^2}, \tag{5.2.17}$$

corresponding to the formula (5.1.41'), which gives the distribution of accelerations in the circular motion of particle; we have

$$a_i = \epsilon_{ijk}\,\dot{\omega}_j x_k - \omega^2 x_i, \quad i = 1,2,3,$$
$$a_1 = -\dot{\omega}x_2 - \omega^2 x_1, \quad a_2 = \dot{\omega}x_1 - \omega^2 x_2, \quad a_3 = 0, \tag{5.2.17'}$$

so that the accelerations are contained in a plane normal to the axis of rotation and directed towards the interior of the circular trajectory. The determinant of the homogeneous system which gives the points of null acceleration is $\omega^4 + \dot{\omega}^2 \ne 0$ and allows to affirm that only the points of the axis of rotation enjoy this property (it is one of the cases mentioned in Subsec.2.1.1, the angular acceleration vector $\boldsymbol{\varepsilon} = \dot{\boldsymbol{\omega}}$ being a sliding vector too, the support of which is the same axis of rotation. As in the case of the velocities, the accelerations are equipollent vectors for the points of a straight line parallel to the axis of rotation; as well, for the points of a straight line normal to the axis of rotation, the velocities are parallel, their modulus having a linear variation. If $\dot{\boldsymbol{\omega}} = \mathbf{0}$, then the motion of rotation is uniform. If the vectors $\boldsymbol{\omega}$ and $\dot{\boldsymbol{\omega}}$ have the same direction, then the motion is accelerated; otherwise, it is decelerated.

2.2.3 Helical motion. Motion of rototranslation

We say that a rigid solid has a helical *motion* if two of its points remain on a fixed straight line during the motion; we may also say that a straight line rigidly linked to the rigid solid (*axis of rotation and sliding*) maintains its support fixed (or slides along a fixed support). We choose the axis of rotation and sliding as axis $O'x_3'$ and the point O on this axis (in general, $O \ne O'$), so that the axis Ox_3 coincides with it. This motion is defined by the scalar functions (Fig.5.14)

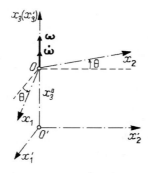

Figure 5.14. Helical motion of a rigid solid.

$$x_3'^O = x_3'^O(t), \quad \theta = \theta(t), \tag{5.2.18}$$

which specifies the position of the rigid solid at a given moment (the rigid solid has two degrees of freedom). The distribution of the velocities is given by

$$\mathbf{v}' = \mathbf{v}'_O + \boldsymbol{\omega} \times \overline{\mathbf{r}}, \quad v' = \sqrt{v'^2_O + \omega^2 \overline{r}^2},$$
$$v'_1 = -\omega x_2, \quad v'_2 = \omega x_1, \quad v'_3 = v'^O_3 = v'_O,$$
(5.2.19)

where $\omega(t) = \dot{\theta}(t)$ has the significance given at the previous subsection. The formulae (5.2.19) show that the modulus of the velocity is minimal (equal to v'_O) for the points of the axis of rotation and sliding; in fact, the velocity $v'_O = v'_O(t)$ is a function of time, so that one may have points of null velocity only – eventually – for some particular cases. As in the case of a motion of rotation, the velocities of the points of a straight line parallel to the axis of rotation and sliding are equipollent vectors. In a plane normal to the axis of rotation and sliding we obtain a component of the velocity which has a behaviour analogous to that in case of a motion of rotation; the component of the velocity along the direction of the same axis behaves as in the case of a motion of translation.

The distribution of the accelerations is given by

$$\mathbf{a}' = \mathbf{a}'_O + \dot{\boldsymbol{\omega}} \times \overline{\mathbf{r}} - \omega^2 \overline{\mathbf{r}}, \quad a' = \sqrt{a'^2_O + (\omega^4 + \dot{\omega}^2)\overline{r}^2},$$
(5.2.20)

leading to the same components a'_1, a'_2 as those given by the formula (5.2.17') and to the component $a'_3 = a'_0$. As in the case of the velocities, the modulus of the acceleration is minimal (equal to a'_0) for the points of the axis of rotation and sliding; but there cannot be points of null acceleration, excepting – eventually – for some particular moments (case mentioned in Subsec. 2.1.1). We notice that $\boldsymbol{\varepsilon}(t) = \dot{\boldsymbol{\omega}}(t) = \dot{\omega}(t)\mathbf{i}_3 = \ddot{\theta}(t)\mathbf{i}_3$ is the angular acceleration, the same for any point of the rigid solid. The accelerations are equipollent vectors for the points of a straight line parallel to the axis of rotation and sliding. The component of the acceleration which is contained in a plane normal to the axis of rotation and sliding has a behaviour analogous to that in case of a motion of rotation; while the component of the acceleration along the same axis has a behaviour analogous to that of a motion of translation.

From the above considerations, it results that the helical motion can be obtained, from the point of view of the distribution of the velocities and accelerations, by the composition of a motion of rotation with a motion of translation along the axis of rotation. We notice that the vectors $\mathbf{v}'_O(t)$ and $\boldsymbol{\omega}(t)$ have the same constant direction ($\mathbf{v}'_O(t) \times \boldsymbol{\omega} = \mathbf{0}$).

The considered motion is called helical, the trajectory of a point of the rigid solid being situated on a circular cylinder. If the first of the scalar functions (5.2.18) which defines the motion verifies a relation of the form $x'^O_3 = k\theta$, k being a constant the dimension of which is a length, then the trajectories are *helices* and we have to do with a *screw motion*; the rigid solid has – in this case – only one degree of freedom. We obtain $v'_O = k\omega$, $a'_O = k\dot{\omega}$, so that

Kinematics

$$v = \omega\sqrt{\bar{r}^2 + k^2}, \quad a = \sqrt{\bar{r}^2\omega^4 + (\bar{r}^2 + k^2)\dot\omega^2}. \tag{5.2.21}$$

The pitch of the helix is $p = 2\pi k$, so that

$$v'_O = \frac{p}{2\pi}\omega, \quad a'_O = \frac{p}{2\pi}\dot\omega. \tag{5.2.22}$$

We notice that the helical motion is a particular case of a motion of rototranslation, i.e., the case in which the vectors $\boldsymbol{\omega}$ and $\mathbf{v}'_O$ have the same support. In the general case, $\boldsymbol{\omega} \times \mathbf{v}'_O \neq \mathbf{0}$, $\boldsymbol{\omega}$ and $\mathbf{v}'_O$ having fixed directions, and we obtain an arbitrary *motion of rototranslation* (called also *motion of finite rototranslation*), which is not a helical motion; this case is basic for the study of the general motion of the rigid solid.

The relations (5.2.16) remain still valid in the case of a motion of rototranslation.

2.2.4 Motion of Frenet's and Darboux's trihedra

If we know the trajectory of a point of the rigid solid, then we may choose as movable frame of reference *Frenet's trihedron*. Taking into account Poisson's formulae (A.2.38) and Frenet's formulae (4.1.10)-(4.1.10"), and noting that

$$\frac{d\boldsymbol{\tau}}{dt} = \frac{d\boldsymbol{\tau}}{ds}\frac{ds}{dt}, \quad \frac{d\boldsymbol{\nu}}{dt} = \frac{d\boldsymbol{\nu}}{ds}\frac{ds}{dt}, \quad \frac{d\boldsymbol{\beta}}{dt} = \frac{d\boldsymbol{\beta}}{ds}\frac{ds}{dt},$$

we may write

$$\frac{d\boldsymbol{\tau}}{dt} = \boldsymbol{\omega} \times \boldsymbol{\tau} = \frac{v}{\rho}\boldsymbol{\nu},$$

$$\frac{d\boldsymbol{\nu}}{dt} = \boldsymbol{\omega} \times \boldsymbol{\nu} = -\frac{v}{\rho}\boldsymbol{\tau} - \frac{\rho}{v'}\boldsymbol{\beta}, \tag{5.2.23}$$

$$\frac{d\boldsymbol{\beta}}{dt} = \boldsymbol{\omega} \times \boldsymbol{\beta} = \frac{v}{\rho'}\boldsymbol{\nu},$$

which leads to interesting kinematic interpretations for the radii of curvature and torsion. In this case, the angular velocity vector is given by

$$\boldsymbol{\omega} = \omega_\tau\boldsymbol{\tau} + \omega_\nu\boldsymbol{\nu} + \omega_\beta\boldsymbol{\beta}, \quad \omega_\tau = -\frac{v}{\rho'}, \quad \omega_\nu = 0, \quad \omega_\beta = \frac{v}{\rho} \tag{5.2.24}$$

and we may write

$$\omega = v\sqrt{\frac{1}{\rho^2} + \frac{1}{\rho'^2}}. \tag{5.2.24'}$$

In the case of a torsionless motion, hence in the case in which the trajectory of a point of the rigid solid is a plane curve, we have

$$v = \rho\omega \,. \tag{5.2.24''}$$

Analogously, choosing as a movable frame of reference *Darboux's trihedron* (assuming that we know a surface on which moves a point of the rigid solid, for instance a cylinder in the case of a motion of rototranslation) and using the formulae (A.2.38) and (4.1.25), as well as observations analogous to those above, we obtain

$$\frac{d\boldsymbol{\tau}}{dt} = \boldsymbol{\omega} \times \boldsymbol{\tau} = \frac{v}{\rho_g}\mathbf{g} + \frac{v}{\rho_n}\mathbf{n}\,,$$
$$\frac{d\mathbf{g}}{dt} = \boldsymbol{\omega} \times \mathbf{g} = -\frac{v}{\rho_g}\boldsymbol{\tau} - \frac{v}{\rho'_g}\mathbf{n}\,, \tag{5.2.25}$$
$$\frac{d\mathbf{n}}{dt} = \boldsymbol{\omega} \times \mathbf{n} = -\frac{v}{\rho_n}\boldsymbol{\tau} + \frac{v}{\rho'_g}\mathbf{g}\,;$$

we are thus led to interesting kinematic interpretations for the radii of normal curvature, as well as of geodesic curvature and torsion. We may express the angular velocity vector in the form

$$\boldsymbol{\omega} = \omega_\tau \boldsymbol{\tau} + \omega_g \mathbf{g} + \omega_n \mathbf{n}\,, \quad \omega_\tau = -\frac{v}{\rho'_g}\,, \quad \omega_g = -\frac{v}{\rho_n}\,, \quad \omega_n = \frac{v}{\rho_g}\,, \tag{5.2.26}$$

wherefrom

$$\omega = v\sqrt{\frac{1}{\rho_g^2} + \frac{1}{\rho_g'^2} + \frac{1}{\rho_n^2}}\,. \tag{5.2.26'}$$

2.3 General motion of the rigid solid

We consider, in the following, the general motion of the rigid solid as an instantaneous helical motion, introducing the fixed and the movable axoids; according to the form of the axoids (conical or cylindrical), we obtain the motion of a rigid solid with a fixed point or the plane-parallel motion of the rigid solid, respectively.

2.3.1 Instantaneous helical motion. Static-kinematic analogy

In the general case of motion of a rigid solid we use the results given in Subsecs 2.1.1 and 2.1.2 concerning the distribution of velocities and accelerations. Using the formula (5.2.3), which gives the velocity $\mathbf{v}'$ of a point P of the rigid solid with respect to a fixed frame of reference, and the formula (5.2.14), we observe that $\{\boldsymbol{\omega}, \mathbf{v}'_O\} = \tau_O\{\boldsymbol{\omega}\}$, $\{\boldsymbol{\omega}, \mathbf{v}'_P\} = \tau_P\{\boldsymbol{\omega}\}$, with

$$\mathbf{v}'_P = \mathbf{v}'_O + \overrightarrow{PO} \times \boldsymbol{\omega} = \mathbf{v}'_O + \boldsymbol{\omega} \times \overrightarrow{OP} = \mathbf{v}'_O + \boldsymbol{\omega} \times \mathbf{r}\,. \tag{5.2.27}$$

We obtain thus a torsor of the angular velocities, $\boldsymbol{\omega}$ playing the rôle of the resultant vector (an axial, sliding vector), while $\mathbf{v}'$ is the moment resultant vector (a polar,

Kinematics 317

bounded vector). This observation stays at the basis of *the static-kinematic analogy*, which allows the kinematic study of the motion of the rigid solid with the aid of the static methods. Analysing the distribution of the velocities, we see that – at a given moment – the general motion of a rigid solid may be identified with an *instantaneous motion of rototranslation*, characterized by an *instantaneous torsor of the angular velocities*.

Hence, the general motion of a rigid solid is characterized by the vectors $\boldsymbol{\omega}$ and $\mathbf{v}'_O$ (the components of the torsor $\tau_O\{\boldsymbol{\omega}\}$ of the angular velocities with respect to the pole O, rigidly linked to the rigid solid); in general, these vectors are not collinear. There exist points (which can belong to the rigid solid or not) for which $\boldsymbol{\omega} \parallel \mathbf{v}'_O$, the general motion of the rigid solid becoming thus an *instantaneous helical motion*. Indeed, taking into account the scalar of the torsor

$$\boldsymbol{\omega} \cdot \mathbf{v}'_P = \boldsymbol{\omega} \cdot \mathbf{v}'_O, \qquad (5.2.28)$$

which is an invariant, it follows that $\mathbf{v}'^{\parallel}_P = \mathbf{v}'^{\parallel}_O$, where we have emphasized the components parallel to the vector $\boldsymbol{\omega}$ of these vectors. The formula (5.2.27) becomes

$$\mathbf{v}'^{\perp}_P = \mathbf{v}'^{\perp}_O + \boldsymbol{\omega} \times \mathbf{r}, \qquad (5.2.29)$$

where the components of the velocities normal to the vector $\boldsymbol{\omega}$ have been introduced. Using the method of determination of the central axis (emphasized in Chap. 2, Subsec. 2.2.5), we put the condition $\mathbf{v}'^{\perp}_P = \mathbf{0}$, wherefrom

$$\mathbf{r} = \lambda\boldsymbol{\omega} + \frac{\boldsymbol{\omega} \times \mathbf{v}'_O}{\omega^2}; \qquad (5.2.30)$$

we obtain thus the instantaneous axis (the locus of the searched points) with respect to which we have an instantaneous helical motion. This axis is parallel to the vector $\boldsymbol{\omega}$; a vector equipollent to the latter one, the support of which is the instantaneous axis, is called *Chasles' vector*. We may write the equation (5.2.30) also in the form (a vector product by $\boldsymbol{\omega}$ eliminates the scalar λ)

$$\mathbf{v}'_O - \mathbf{r} \times \boldsymbol{\omega} = \frac{\boldsymbol{\omega} \cdot \mathbf{v}'_O}{\omega^2}\boldsymbol{\omega}; \qquad (5.2.30')$$

in components one has

$$\frac{1}{\omega_1}\left(v'^O_1 - x_2\omega_3 + x_3\omega_2\right) = \frac{1}{\omega_2}\left(v'^O_2 - x_3\omega_1 + x_1\omega_3\right)$$

$$= \frac{1}{\omega_3}\left(v'^O_3 - x_1\omega_2 + x_2\omega_1\right) = \frac{1}{\omega^2}\left(\omega_1 v'^O_1 + \omega_2 v'^O_2 + \omega_3 v'^O_3\right). \qquad (5.2.30'')$$

We notice that the instantaneous axis, defined by the equations (5.2.30)-(5.2.30"), is the locus of the points for which, at a given moment, the modulus of the velocity is minimal. The above considerations allow us to state

Theorem 5.2.3 (*Chasles*). *The general motion of a rigid solid at a given moment may be identified with an instantaneous helical motion about an instantaneous axis of rotation and sliding.*

We can say that this helical motion is *tangent* to the general motion of the rigid solid at a given moment.

From the above analysis it follows that if $\boldsymbol{\omega} = \mathbf{0}$, $\mathbf{v}'_O = \mathbf{0}$, then the rigid solid is in rest. If $\boldsymbol{\omega} = \mathbf{0}$, $\mathbf{v}'_O \neq \mathbf{0}$, then a motion of translation takes place; if $\boldsymbol{\omega} \neq \mathbf{0}$, $\mathbf{v}'_O = \mathbf{0}$, then the motion of the rigid solid can be identified with a motion of rotation about an instantaneous axis of rotation passing through the fixed point O (in particular, a finite rotation), hence with a rigid solid with a fixed point. If $\boldsymbol{\omega} \neq \mathbf{0}$, $\mathbf{v}'_O \neq \mathbf{0}$, then two cases may occur: if the torsor's scalar vanishes ($\boldsymbol{\omega} \cdot \mathbf{v}'_O = 0$), then the motion of the rigid solid is an instantaneous motion of rotation about an instantaneous axis of rotation (which does not pass through a fixed point); if the scalar does not vanish ($\boldsymbol{\omega} \cdot \mathbf{v}'_O \neq 0$), then a general motion of the rigid solid takes place.

In the general motion of the rigid solid, the distribution of accelerations is that emphasized in Subsec. 2.1.2.

2.3.2 The fixed and the movable axoids

We notice that the instantaneous axis of rotation and sliding varies both with respect to the fixed and the movable frames of reference, because $\boldsymbol{\omega}$ and $\mathbf{v}'_O$ are functions of time. The locus of the instantaneous axes of rotation and sliding with respect to the frame $\mathcal{R}'$ is a ruled surface $\mathcal{A}_f$, called *the fixed axoid*, while the locus of the same axis with respect to the frame $\mathcal{R}$ is a ruled surface $\mathcal{A}_m$ too, called *the movable axoid* (Fig.5.15); these two surfaces play an important rôle in the general motion of a rigid solid, as it was emphasized by Poncelet.

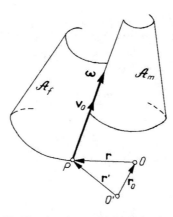

Figure 5.15. Fixed and movable axoids of a rigid solid.

Let P be a point of the rigid solid on the instantaneous axis of rotation and sliding, rigidly linked to this axis, for which a relation of the form (5.2.1) takes place.

Kinematics

Differentiating with respect to the frame $\mathscr{R}'$ and using the formula (A.2.37), we obtain

$$\frac{d\mathbf{r}'}{dt} = \frac{d\mathbf{r}_O}{dt} + \frac{\partial \mathbf{r}}{\partial t} + \boldsymbol{\omega} \times \mathbf{r} ; \qquad (5.2.31)$$

$\mathbf{v}'_P = d\mathbf{r}'/dt$ is the velocity of the point P in the motion of the instantaneous axis of rotation and sliding with respect to the frame $\mathscr{R}'$, while $\mathbf{v}_P = \partial \mathbf{r}/\partial t$ is the velocity of the same point in the motion of this axis with respect to the frame $\mathscr{R}$. Noting that $\mathbf{v}'_O = d\mathbf{r}'_O/dt$ and $\mathbf{v}'_O + \boldsymbol{\omega} \times \mathbf{r} = \lambda \boldsymbol{\omega}$, where λ is a scalar, because the translation velocity of the point P and the angular velocity $\boldsymbol{\omega}$ are collinear vectors for the points of the instantaneous axis of rotation and sliding, we get the relation

$$\mathbf{v}'_P - \mathbf{v}_P = \lambda \boldsymbol{\omega}. \qquad (5.2.31')$$

If the velocities $\mathbf{v}_P$ and $\mathbf{v}'_P$ would be equal, then the axoids would have a motion of rolling without sliding; because of the vector difference $\lambda \boldsymbol{\omega}$, occurs also a sliding of the movable axoid on the fixed one along the direction of the vector $\boldsymbol{\omega}$, with the velocity $\lambda \boldsymbol{\omega}$. Projecting the relation (5.2.31) on a direction normal to the vector $\boldsymbol{\omega}$, we obtain

$$\mathbf{v}'^{\perp}_P = \mathbf{v}^{\perp}_P, \qquad (5.2.32)$$

the components normal to the instantaneous axis of rotation and sliding of the velocities of a point P of this axis with respect to the fixed and movable frames of reference, respectively, being equal. We may thus state

Theorem 5.2.4. *The general motion of a rigid solid takes place so that the movable axoid be tangent to the fixed axoid along the instantaneous axis of rotation and sliding* (the common generatrix of the two axoids), *its motion being a rolling about this axis, over the fixed axoid, together with a sliding along the same instantaneous axis.*

We notice that the two axoids must be both developable or warped surfaces.

The motion of the movable axoid characterizes thus completely the general motion of the rigid solid.

2.3.3 Motion of the rigid solid with a fixed point

In the case of a rigid solid with a fixed point both origins may coincide with this point, without losing anything of the generality ($O \equiv O'$); we have thus $\mathbf{v}'_O = \mathbf{v}_O = \mathbf{0}$, $\mathbf{a}'_O = \mathbf{a}_O = \mathbf{0}$, the vectors $\boldsymbol{\omega}$ and $\dot{\boldsymbol{\omega}}$ being arbitrary. The rigid solid remains with three degrees of freedom and its position at a given moment may be specified with the aid of Euler's angles ψ, θ, φ, emphasized in Chap. 3, Subsec. 2.2.3.

The distribution of the velocities is given by the relation

$$\mathbf{v} = \boldsymbol{\omega} \times \mathbf{r}, \qquad (5.2.33)$$

the points situated on the support of the angular velocity vector being of null velocity; this support passes through the fixed point and is just the instantaneous axis of rotation (in this case is no sliding). The axoids are two tangent cones, having the vertices at the fixed point (*Poinsot's cones*); the motion of a rigid solid with a fixed point is thus characterized by *the rolling without sliding of the polhodic cone* $\mathscr{C}_p$ (movable) *over the herpolhodic cone* $\mathscr{C}_h$ (fixed) (Fig.5.16).

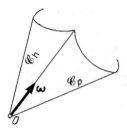

Figure 5.16. Herpolodic and polodic cones of a rigid solid.

Introducing the components $\dot\psi$, $\dot\varphi$ and $\dot\theta$ of the angular velocity with respect to the axes Ox_3', Ox_3 and ON, respectively, (Fig.3.15), we may write

$$\boldsymbol{\omega} = \dot\psi \mathbf{i}_3' + \dot\varphi \mathbf{i}_3 + \dot\theta \mathbf{n}, \qquad (5.2.34)$$

where $\mathbf{n} = \operatorname{vers} \overrightarrow{ON}$. Projecting on the axes of the frame $\mathscr{R}$, it results

$$\begin{aligned}\omega_1 &= \dot\theta \cos\varphi + \dot\psi \sin\theta \sin\varphi, \\ \omega_2 &= -\dot\theta \sin\varphi + \dot\psi \sin\theta \cos\varphi, \\ \omega_3 &= \dot\varphi + \dot\psi \cos\theta, \end{aligned} \qquad (5.2.35)$$

while, in projection on the axes of the frame $\mathscr{R}'$, we may write

$$\begin{aligned}\omega_1' &= \dot\theta \cos\psi + \dot\varphi \sin\theta \sin\psi, \\ \omega_2' &= \dot\theta \sin\psi - \dot\varphi \sin\theta \cos\psi, \\ \omega_3' &= \dot\psi + \dot\varphi \cos\theta. \end{aligned} \qquad (5.2.35')$$

If we put $x_i' = \lambda \omega_i'$ and $x_i = \lambda \omega_i$, $i = 1,2,3$, then we obtain the parametric equations of the herpolhodic and polhodic cones, respectively, the parameters being λ and t.

To pass from $\omega_i(t)$, $i = 1,2,3$, to Euler's angles $\psi(t)$, $\theta(t)$ and $\varphi(t)$, hence to integrate the system (5.2.35) with respect to the latter unknown functions, it is useful to introduce the intermediate unknown functions $\alpha_i(t)$, $i = 1,2,3$, which represent the direction cosines of the axis Ox_3' with respect to the frame $\mathscr{R}$. We have the relations

$$\alpha_1 = \sin\theta \sin\varphi, \quad \alpha_2 = \sin\theta \cos\varphi, \quad \alpha_3 = \cos\theta, \qquad (5.2.36)$$

Kinematics 321

which allow the determination of Euler's angles φ, θ, if one knows α_i; the angle ψ is then obtained by a quadrature from the third relation (5.2.35). The link between the functions α_i and ω_i is obtained writing that the derivative of the unit vector $\mathbf{i}'_3$ with respect to the fixed frame of reference vanishes; we have

$$\dot{\mathbf{i}}'_3 = \boldsymbol{\omega} \times \mathbf{i}'_3 = \mathbf{0}, \quad \mathbf{i}'^2_3 = 1 \tag{5.2.37}$$

or, in components,

$$\dot{\alpha}_i + \epsilon_{ijk}\,\omega_j \alpha_k = 0, \quad i = 1,2,3, \quad \alpha_i \alpha_i = 1. \tag{5.2.37'}$$

The distribution of accelerations is of the form

$$\mathbf{a} = \dot{\boldsymbol{\omega}} \times \mathbf{r} + \boldsymbol{\omega} \times (\boldsymbol{\omega} \times \mathbf{r}) = \dot{\boldsymbol{\omega}} \times \mathbf{r} + (\boldsymbol{\omega} \cdot \mathbf{r})\boldsymbol{\omega} - \omega^2 \mathbf{r}; \tag{5.2.38}$$

taking into account the results in Subsec. 2.1.2, we can state that – in general – in the motion of a rigid solid with a fixed point, excepting the fixed point, there are not other points of null acceleration. This motion is reducible to a motion of rotation only in the case in which the vectors $\boldsymbol{\omega}$ and $\dot{\boldsymbol{\omega}}$ are collinear or one of them vanishes.

2.3.4 Plane-parallel motion

We say that a rigid solid has a *plane-parallel motion* if three non-collinear points of it are contained, during the motion, in a fixed plane (hence, *if a plane section of the rigid slides on a fixed plane*). We notice that the motion of rotation and the motion of translation, the trajectory of which is plane, are plane-parallel motions. Because each point of a normal to the considered plane section has the same translated trajectory, we may refer only to this plane section in the fixed plane. If the rigid solid is reduced to a plate of small thickness (negligible), the median plane of which is just the fixed plane, then the motion is called *plane*. We choose the two frames of reference so that the axes $O'x'_\alpha$ and Ox_α, $\alpha = 1,2$, of the fixed and movable frames, respectively, be contained in the fixed plane. In this case, the rigid solid has three degrees of freedom, and its position at a moment t can be specified by the scalar functions

$$x'^O_\alpha = x'^O_\alpha(t), \quad \alpha = 1,2, \quad \theta = \theta(t), \tag{5.2.39}$$

where x'^O_α are the co-ordinates of the pole O with respect to the fixed frame of reference, while θ is the angle made by the Ox_α-axis with the $O'x'_\alpha$-axis (Fig.5.17). The trajectories of the points of the rigid solid are, obviously, plane curves, while the points situated on a parallel to the $O'x'_3$-axis describe identical curves; hence, it is sufficient to study the motion in the plane $O'x'_1x'_2$, considered as a fixed plane. In this case,

$$\mathbf{v}'_O = v'^O_\alpha \mathbf{i}'_\alpha, \quad \mathbf{a}'_O = a'^O_\alpha \mathbf{i}'_\alpha, \quad \boldsymbol{\omega} = \omega \mathbf{i}'_3, \quad \dot{\boldsymbol{\omega}} = \dot{\omega} \mathbf{i}'_3, \tag{5.2.39'}$$

Where, in the summation, the Greek indices take only the values 1 and 2.

The formula (5.2.3) allows to write the distribution of the velocities in the form

$$v'_1 = v'^O_1 - \omega x_2, \quad v'_2 = v'^O_2 + \omega x_1, \quad v'_3 = 0. \qquad (5.2.40)$$

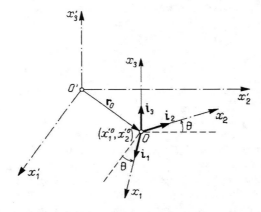

Figure 5.17. Plane-parallel motion of a rigid solid.

The points of null velocity of the rigid solid are given by

$$\xi_1 = -\frac{v'^O_2}{\omega}, \quad \xi_2 = \frac{v'^O_1}{\omega}, \quad \xi_3 \text{ arbitrary,} \qquad (5.2.41)$$

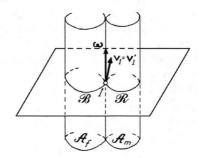

Figure 5.18. Fixed and movable centrodes in a plane-parallel motion.

with respect to the movable frame of reference, these points being situated on a straight line normal to the fixed plane, which is the instantaneous axis of rotation (from the given definition of the plane-parallel motion, it results that there is no sliding along this axis). Hence, the rigid solid will be in instantaneous motion of rotation about an instantaneous axis of rotation which is of constant direction and the trace of which on the fixed plane is a point I, called *instantaneous centre of rotation*; indeed, in this plane (and in parallel planes too), an instantaneous motion of rotation of the points of the rigid solid about the point I (of null velocity) takes place, as it was shown by Euler. The two axoids are, in this case, two cylinders, the traces of which on the fixed plane are two curves: the *basis $\mathcal{B}$* (*the fixed (space) centrode*) and *the rolling curve $\mathcal{R}$*

Kinematics 323

(*the movable (body) centrode*), tangent at the instantaneous centre of rotation I (Fig.5.18). The velocities of the point I in its motion with respect to the fixed and the movable frames of reference are $\mathbf{v}'_I$ and $\mathbf{v}_I$, respectively; taking into account the relation (5.2.32), it results

$$\mathbf{v}'_I = \mathbf{v}_I, \tag{5.2.42}$$

so that the two velocities are directed along the common tangent of the two centroids, their elements of arc being equal. The plane-parallel motion of the rigid solid is thus characterized by *the rolling without sliding of the rolling curve over the basis*.

The parametric equations of the rolling curve are, obviously, given by (5.2.41); by a change of axes of co-ordinates, we obtain the parametric equations of the basis in the form

$$\xi'_1 = x^O_1 + \xi_1 \cos\theta - \xi_2 \sin\theta, \quad \xi'_2 = x^O_2 + \xi_1 \sin\theta + \xi_2 \cos\theta. \tag{5.2.43}$$

We notice that, at a moment t, the velocity vector is normal to a radius starting from the instantaneous centre of rotation and its modulus is proportional to the length of the radius (an instantaneous motion of rotation about this centre takes place); hence, we may use *geometric methods* to determine the point I and to draw the two centroids. Thus, being given the velocities of two points P and Q of a rigid solid in a plane-parallel motion (their non-parallel supports are sufficient), the centre I will be at the intersection of the normals to $\mathbf{v}'_P$ and $\mathbf{v}'_Q$ at these points (Fig.5.19,a); it follows

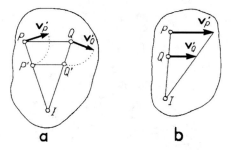

Figure 5.19. Geometric determination of the instantaneous centre of rotation if one knows the velocities of two points: concurrent velocities (a) or parallel velocities (b).

$$\omega = \frac{v'_P}{\overline{IP}} = \frac{v'_Q}{\overline{IQ}}, \quad v'_Q = \frac{\overline{IQ}}{\overline{IP}} v'_P. \tag{5.2.44}$$

If the supports of the two velocities are parallel, then the centre I is thrown to infinity, $\mathbf{v}'_P = \mathbf{v}'_Q$, and we have to do with a motion of translation. Finally, if the velocities $\mathbf{v}'_P$ and $\mathbf{v}'_Q$ are both normal to the straight line PQ, then the centre I will be on this line at the point of intersection of it with the straight line which links the extremities of the two velocities (Fig.5.19,b). By turning down the vectors $\mathbf{v}'_P$ and $\mathbf{v}'_Q$ in the same direction of a rigid angle, we obtain the vectors $\overrightarrow{PP'}$ and $\overrightarrow{QQ'}$, the points P' and Q' being on the straight lines PI and QI, respectively (Fig.5.19,a); noting that $v'_P = \overrightarrow{PP'}$, $v'_Q = \overrightarrow{QQ'}$, and taking into account the relations (5.2.44) there results that the straight line $P'Q'$ is parallel to the straight line PQ. Thus, *the velocities turning down method* allows to construct graphically the velocity $\mathbf{v}'_Q$ of a point of a rigid motion in a plane-parallel motion if we know the instantaneous centre of rotation I, as well as the velocity $\mathbf{v}'_P$ of another point P; if $\mathbf{v}'_P \perp \overrightarrow{PQ}$, then we use the graphic of Fig.5.19,b.

Let us consider the points $P, Q, R, ...$ in the plane section of the fixed plane, where the velocity vectors $\mathbf{v}'_P$, $\mathbf{v}'_Q$, $\mathbf{v}'_R, ...$ are applied; we construct the equipollent vectors $\overrightarrow{OP'} = \mathbf{v}'_P$, $\overrightarrow{OQ'} = \mathbf{v}'_Q$, $\overrightarrow{OR'} = \mathbf{v}'_R, ...$ at a pole $\overline{O}$. We may introduce the relative velocities $\mathbf{v}_{QP} = \overrightarrow{P'Q'}$, $\mathbf{v}_{RQ} = \overrightarrow{Q'R'}$, $\mathbf{v}_{PR} = \overrightarrow{R'P'}, ...$ too; the figure $\overline{O}P'Q'R'...$ forms the velocities plane. Taking into account the relation (5.2.3'), one obtains (Fig.5.20)

$$\mathbf{v}'_Q = \mathbf{v}'_P + \mathbf{v}_{QP}, \quad \mathbf{v}_{QP} = \boldsymbol{\omega} \times \overrightarrow{PQ}. \tag{5.2.45}$$

We notice also that $\mathbf{v}_{QP} \perp \overrightarrow{PQ}$ or $\overrightarrow{P'Q'} \perp \overrightarrow{PQ}$ and other analogous relations (Fig.5.20). Introducing the moduli in the second relation (5.2.45) and proceeding analogously with similar relations, we get

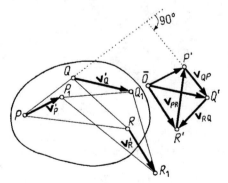

Figure 5.20. Velocities' similarity theorem.

$$\frac{\overline{P'Q'}}{\overline{PQ}} = \frac{\overline{Q'R'}}{\overline{QR}} = \frac{\overline{R'P'}}{\overline{RP}} = \omega. \tag{5.2.45'}$$

Hence, the triangles $P'Q'R'$ and PQR are similar, their similarity ratio being ω; we notice that the sides of the triangle $P'Q'R'$ are normal to the homologous sides of the triangle PQR. Because a polygon may be decomposed in a certain number of triangles, we may state

Theorem 5.2.5 (*the velocities' similarity theorem*; *Burmester*). *If a polygon $PQR\ldots$, formed by the points of a rigid solid in plane-parallel motion is given, then the polygon $P'Q'R'\ldots$, formed by the homologous points of the velocities plane, is similar to the first polygon and is rotated through an angle of 90^0 with respect to this one in the direction of the angular velocity.*

Let be, e.g., a rigid bar AB of length $2l$, which is moving so that its extremities do remain on the axes $O'x_1'$ and $O'x_2'$, respectively, hence on two fixed orthogonal straight lines (*Cardan's problem*) (Fig.5.21). We choose the pole of the movable frame of reference in $O \equiv A$ and the axis Ox_1 normal to AB; the rigid solid has only one degree of freedom, its position being specified by the angle $\theta(t)$. The velocities of the points A and B will be along the fixed axes of co-ordinates, so that we may easily determine the centre I ($2l\sin\theta$, $2l\cos\theta$). We notice that the basis is a circle of centre in O' and radius $\overline{O'I} = 2l$; the rolling curve is a circle too, of diameter $\overline{AB} = 2l$, passing through the points O' and I. If the bar can stay only in the first quadrant, then the basis is a quarter of a circle, while the rolling curve is a semicircle.

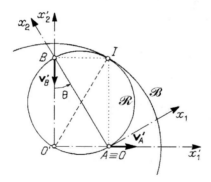

Figure 5.21. Cardan's problem.

Let us consider also the case of a wheel of radius R, which is moving on a horizontal straight line (taken as axis $O'x_1'$); we assume that the centre O of the wheel moves with a horizontal velocity $\mathbf{v}_O'$ while this one rotates with an angular velocity ω_O. The position of the wheel is given by two parameters; the abscissa x_1' of the point O and the angle θ specify a point of it, corresponding thus to two degrees of freedom (the third one is annihilated by the imposed condition). The basis $\mathcal{B}$ is a straight line, parallel to the horizontal $O'x_1'$ and situated under the point O, while the rolling curve is a circle $\mathcal{R}$ of centre O and radius $\overline{OI} = v_O'/\omega_O$. If $v_O' = R\omega_O$, then $\overline{OI} = R$, and the motion of the wheel is a rolling without sliding (Fig.5.22,a). If $v_O' > R\omega_O$, then we have $\overline{OI} > R$, and the wheel slides in the direction in which it advances (the velocities of all its points have the same direction; the case of *the drawn wheel*) (Fig.5.22,b); as well, if $v'_O < R\omega_O$, then we see that $\overline{OI} < R$ and the wheel slides in the opposite direction to that in which it advances (there are points which have a direction opposite to that of $\mathbf{v}'_O$; the case of *the driving wheel*) (Fig.5.22,c).

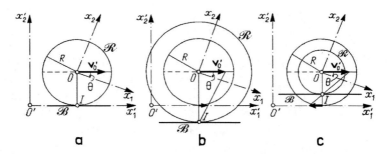

Figure 5.22. Motion of a wheel of a horizontal line: case $v'_0 = R\omega_0$ (a); case $v'_0 > R\omega_0$ (b); case $v'_0 < R\omega_0$ (c).

The distribution of accelerations in the plane-parallel motion is given by the formula (5.2.6) in the form ($\boldsymbol{\omega} \cdot \mathbf{r} = 0$)

$$\mathbf{a}' = \mathbf{a}'_O + \dot{\boldsymbol{\omega}} \times \mathbf{r} - \omega^2 \mathbf{r}, \tag{5.2.46}$$

$$a'_1 = a'^O_1 - \dot{\omega} x_2 - \omega^2 x_1, \quad a'_2 = a'^O_2 + \dot{\omega} x_1 - \omega^2 x_2, \quad a'_3 = 0. \tag{5.2.46'}$$

There exists a straight line parallel to $O'x'_3$ for which the accelerations vanish ($\dot{\boldsymbol{\omega}} \parallel \boldsymbol{\omega}$, corresponding to the considerations in Subsec. 2.1.2). The point J at which the acceleration vanishes is called *the centre (pole) of the accelerations*, of co-ordinates

$$\eta_1 = \frac{\omega^2 a'^O_1 - \dot{\omega} a'^O_2}{\omega^4 + \dot{\omega}^2}, \quad \eta_2 = \frac{\dot{\omega} a'^O_1 + \omega^2 a'^O_2}{\omega^4 + \dot{\omega}^2}, \tag{5.2.47}$$

with respect to the movable frame of reference. The instantaneous distribution of the accelerations is identical to that of an instantaneous rotation about the centre J, as one can see effecting a change of axes in this pole.

We notice that the instantaneous centre I and the pole of accelerations J are distinct points, which do not coincide in the particular case of a motion of rotation. Hence, in general, $\mathbf{v}'_I = \mathbf{0}$, $\mathbf{a}'_I \neq \mathbf{0}$ and $\mathbf{v}'_J \neq \mathbf{0}$, $\mathbf{a}'_J = \mathbf{0}$.

We have seen that the instantaneous centre may be used to determine the velocities of some points of the rigid solid, when one of these velocities is known; we notice that the pole of accelerations J can play an analogous rôle to determine the accelerations. In *the method of the accelerations' pole* we suppose known the acceleration a'_P of the point P, the angular velocity $\boldsymbol{\omega}$ and the angular acceleration $\boldsymbol{\varepsilon} = \dot{\boldsymbol{\omega}}$. Thus, starting from the point P, we draw the segment PJ, which makes the angle given by $\tan \varphi = \dot{\omega}/\omega^2$, in the direction indicated by the angular acceleration, with the

Kinematics

acceleration $\mathbf{a}'_P$, the point J being specified by $\overline{PJ} = |\mathbf{a}'_P|/\sqrt{\omega^4 + \dot\omega^2}$ (Fig.5.23); indeed, if $O \equiv P$ and one takes the Ox_1-axis along the vector $\mathbf{a}'_P$, then the formulae (5.2.47) justify the above affirmation. Noting that

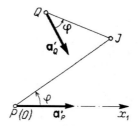

Figure 5.23. Method of the acceleration's pole.

$$\frac{|\mathbf{a}'_P|}{|\mathbf{a}'_Q|} = \frac{\overline{JP}}{\overline{JQ}}, \qquad (5.2.48)$$

we may easily draw the vector $|\mathbf{a}'_Q|$ applied at the point Q, which makes an angle φ (in the same direction as that indicated by the angular acceleration) with JQ.

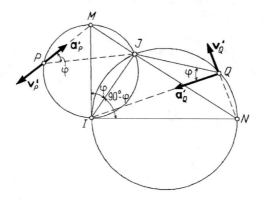

Figure 5.24. Inflections and turning back circles.

In a plane-parallel motion, the locus of the points P for which $\mathbf{v}'_P \times \mathbf{a}'_P = \mathbf{0}$ is a circle (called *the inflections circle*), while the locus of the points Q for which $\mathbf{v}'_Q \cdot \mathbf{a}'_Q = 0$ is a circle too (called *the turning back circle*). If we draw from I the orthogonal straight lines which make the angles φ and $90° - \varphi$ with IJ and meet the normal to IJ at J in M and N, respectively, then the circles of diameters IM and IN, respectively, are the searched loci (*the circles of Bresse*); the drawing easily justifies this assertion (Fig.5.24).

To the points $P, Q, R, ...$, of accelerations $\mathbf{a}'_P$, $\mathbf{a}'_Q$, $\mathbf{a}'_R$, ..., correspond the equipollent vectors $\overrightarrow{O'P'} = \mathbf{a}'_P$, $\overrightarrow{O'Q'} = \mathbf{a}'_Q$, $\overrightarrow{O'R'} = \mathbf{a}'_R$, ..., applied at the point O';

the figure $O'P'Q'R'\ldots$ constitutes the *plane of accelerations* (Fig.5.25). Introducing the *relative accelerations*

$$\mathbf{a}_{QP} = \overrightarrow{P'Q'} = \mathbf{a}'_Q - \mathbf{a}'_P = \dot{\boldsymbol{\omega}} \times \overrightarrow{PQ} + \boldsymbol{\omega} \times (\boldsymbol{\omega} \times \overrightarrow{PQ}),\ldots, \quad (5.2.49)$$

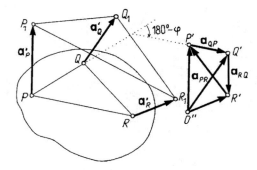

Figure 5.25. Accelerations' similarity theorem.

as sums of two terms which have the same expressions as in case of the circular motion, we may state (as in the case of velocities)

Theorem 5.2.6 (*the accelerations' similarity theorem*). *If a polygon* $PQR\ldots$, *formed by points of a rigid solid in plane-parallel motion is given, then the polygon* $P'Q'R'\ldots$, *formed by the homologous points of the accelerations plane is similar to the first polygon and is rotated through an angle of* $180° - \varphi$ *with respect to this one in the direction of the angular acceleration.*

The relations

$$\frac{\overline{P'Q'}}{\overline{PQ}} = \frac{\overline{Q'R'}}{\overline{QR}} = \frac{\overline{R'P'}}{\overline{RP}} \quad (5.2.50)$$

take place for the triangles PQR and $P'Q'R'$. We mention that the similarity theorem and the relation (5.2.50) remain valid also in the case of collinear points P,Q,R (to these points correspond the collinear points P',Q',R' in the accelerations plane).

The velocities plane and the accelerations plane lead to methods useful for the graphical computation to determine the corresponding kinematic quantities.

Let $P_1, Q_1, R_1,\ldots$ be the extremities of the velocity vectors $\mathbf{v}'_P, \mathbf{v}'_Q, \mathbf{v}'_R,\ldots$, applied at the points $P,Q,R,\ldots$, respectively (Fig.5.20). We may write

$$\overrightarrow{P_1 Q_1} = \overrightarrow{PQ} + \mathbf{v}'_Q - \mathbf{v}'_P = \overrightarrow{PQ} + \mathbf{v}_{QP} = \overrightarrow{PQ} + \boldsymbol{\omega} \times \overrightarrow{PQ};$$

noting that the vector product $\boldsymbol{\omega} \times \overrightarrow{PQ}$ is a polar vector normal to $\overrightarrow{PQ}$ and contained in the plane of motion, and taking into account that the vector $\boldsymbol{\omega}$ is normal to this plane, it results

Kinematics

$$\overrightarrow{P_1Q_1}^2 = \overrightarrow{PQ}^2 + \left(\boldsymbol{\omega} \times \overrightarrow{PQ}\right)^2 = \left(1 + \omega^2\right)\overrightarrow{PQ}^2.$$

Completing the similarity theorem (for velocities) we may state that the polygons $PQR\ldots$ and $P_1Q_1R_1\ldots$ are similar too, the similarity ratio being $\sqrt{1+\omega^2}$ (as well with the polygon $P'Q'R'\ldots$).

If, analogously, $P_1, Q_1, R_1 \ldots$ are the extremities of the acceleration vectors $\mathbf{a}'_P$, $\mathbf{a}'_Q$, $\mathbf{a}'_R, \ldots$, applied at the points $P, Q, R, \ldots$, respectively (Fig.5.25), we may write

$$\overrightarrow{P_1Q_1} = \overrightarrow{PQ} + \mathbf{a}'_Q - \mathbf{a}'_P = \overrightarrow{PQ} + \mathbf{a}_{QP} = \overrightarrow{PQ} + \dot{\boldsymbol{\omega}} \times \overrightarrow{PQ} - \omega^2 \overrightarrow{PQ},$$

where we took into account the formula (5.2.46); as above, we notice that the vector product $\dot{\boldsymbol{\omega}} \times \overrightarrow{PQ}$ is a polar vector, normal to $\overrightarrow{PQ}$, contained in the plane of motion, and taking into account that the vector $\dot{\boldsymbol{\omega}}$ is normal to this plane, we obtain

$$\overrightarrow{P_1Q_1}^2 = \left(1 - \omega^2\right)^2 \overrightarrow{PQ}^2 + \left(\dot{\boldsymbol{\omega}} \times \overrightarrow{PQ}\right)^2 = \left[(1-\omega)^2 + \dot{\omega}^2\right]\overrightarrow{PQ}^2.$$

Hence, completing the similarity theorem (for the accelerations), we may state that the polygons $PQR\ldots$ and $P_1Q_1R_1\ldots$ are similar too, the similarity ratio being $\sqrt{\left(1-\omega^2\right)^2 + \dot{\omega}^2}$ (the same statement with the polygon $P'Q'R'\ldots$).

We can emphasize also some interesting properties concerning the displacement of a segment of straight line PQ in a plane-parallel motion. Let be the segments PQ and RS in the positions P_1Q_1 and R_1S_1, at the moment t_1, and in the positions P_2Q_2 and R_2S_2, at the moment t_2, respectively, with respect to a fixed frame of reference. Taking into account a property emphasized in Subsec. 2.1.1, we may write $\sphericalangle\left(\overrightarrow{P_1Q_1}, \overrightarrow{R_1S_1}\right) = \sphericalangle\left(\overrightarrow{P_2Q_2}, \overrightarrow{R_2S_2}\right)$; noting that $\sphericalangle\left(\overrightarrow{P_1Q_1}, \overrightarrow{P_2Q_2}\right) = \sphericalangle\left(\overrightarrow{P_1Q_1}, \overrightarrow{R_1S_1}\right) + \sphericalangle\left(\overrightarrow{R_1S_1}, \overrightarrow{R_2S_2}\right) + \sphericalangle\left(\overrightarrow{R_2S_2}, \overrightarrow{P_2Q_2}\right)$ and taking into account the previous relation, we get

$$\theta_{12} = \sphericalangle\left(\overrightarrow{P_1Q_1}, \overrightarrow{P_2Q_2}\right) = \sphericalangle\left(\overrightarrow{R_1S_1}, \overrightarrow{R_2S_2}\right). \quad (5.2.51)$$

We may thus state

Theorem 5.2.7. *In a plane-parallel motion, the angle θ_{12} formed by two homologous segments of straight line at the moments t_1 and t_2 depends only on the two moments, but not on the two considered segments* (being thus equal to the angle formed by any other two homologous segments of straight line at the same moments).

If $\theta_{12} = 0$, then we have $\overrightarrow{P_1Q_1} \parallel \overrightarrow{P_2Q_2}$; but $\left|\overrightarrow{P_1Q_1}\right| = \left|\overrightarrow{P_2Q_2}\right|$ so that $\overrightarrow{P_1P_2} = \overrightarrow{Q_1Q_2}$, these vectors defining a motion of finite translation. If $\theta_{12} \neq 0$ we construct a point I_0

at the intersection of the midperpendiculars of the segments P_1P_2 and Q_1Q_2 (Fig.5.26). Noting that the triangles $I_0P_1Q_1$ and $I_0P_2Q_2$ are equal (their sides are equal, by construction), it results $\widehat{P_1I_0Q_1} = \widehat{P_2I_0Q_2}$; we may thus write $\widehat{P_1I_0P_2} = \widehat{Q_1I_0Q_2}$ too, so that the segment of straight line P_1Q_1 can be superposed over the segment of straight line P_2Q_2 by a finite rotation about the point I_0. We may state

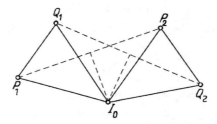

Figure 5.26. Euler's theorem for a plane-parallel motion.

Theorem 5.2.8 (*Euler*). *In a plane-parallel motion of a rigid solid one may pass from a position corresponding to a moment t_1 to a position corresponding to a moment t_2 by a finite translation or by a finite rotation.*

If $t_2 \to t_1$, then the point I_0 becomes the instantaneous centre of rotation.

In a general motion of the rigid solid, we construct the vectors $\overrightarrow{O'P'} = \mathbf{v}'_P$, $\overrightarrow{O'Q'} = \mathbf{v}'_Q$ and $\overrightarrow{O'R'} = \mathbf{v}'_R$ equipollent to the velocities of the points P, Q, R at an arbitrary point O. The points P', Q', R' determine a plane Π on which we project the points P, Q, and R; the projections of the corresponding velocities $\mathbf{v}'_P = \overrightarrow{PP_1}$, $\mathbf{v}'_Q = \overrightarrow{QQ_1}$ and $\mathbf{v}'_R = \overrightarrow{RR_1}$ are the vectors $\overrightarrow{pp_1}$, $\overrightarrow{qq_1}$, $\overrightarrow{rr_1}$. As it was shown by Poncelet, the normals at the points p, q, and r to these projections on the plane Π, respectively, are concurrent; the instantaneous axis of rotation and sliding passes through this point and is normal to the considered plane.

3. Relative motion. Kinematics of mechanical systems

Starting from the results concerning the relative motion of a particle, we pass to the relative motion of a mechanical system, in particular of a rigid solid. We give then some results concerning the systems of rigid solids.

3.1 Relative motion of a particle

We analysed till now the motion of a particle P with respect to a fixed frame of reference $\mathcal{R}'$ of axes $O'x'_1x'_2x'_3$; sometimes, it is useful to consider its motion with respect to a frame of reference of axes $Ox_1x_2x_3$ in motion with respect to the fixed frame (a movable frame $\mathcal{R}$) (Fig.5.27). It is thus put the problem to determine the kinematic quantities which characterize the motion of the particle with respect to the

Kinematics

frame $\mathcal{R}'$ if the quantities corresponding to the motion of the particle with respect to the frame $\mathcal{R}$, as well as the motion of $\mathcal{R}$ with respect to $\mathcal{R}'$ are known. Such a problem is important – for instance – in the study of the motion of a particle on the surface of the Earth.

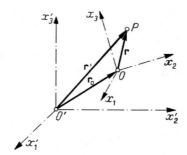

Figure 5.27. Relative motion of a rigid solid.

The motion of the particle with respect to the fixed frame of reference is called *absolute motion*, its motion with respect to the movable frame of reference being *the relative motion*; *the motion of transport* characterizes the motion of $\mathcal{R}$ with respect to $\mathcal{R}'$ and can be specified by the motion of its pole and of its axes. The velocities and accelerations are called *absolute*, *relative* and *of transport*, after the motions to which these quantities correspond. The mechanical phenomenon in its totality is called *relative motion*. We assume that the time t is the same in the two frames of reference (the Newtonian model). Eventually, we may have $t' = t + \text{const}$; the constant can be taken equal to zero, without losing something of the generality of the phenomenon.

In what follows, we consider the composition laws of velocities and accelerations in the case of a particle, emphasizing thus the basic kinematic laws of the relative motion.

3.1.1 Composition of velocities

Starting from the relation (5.2.1) which links the position vectors **r** and **r**′ of the particle P with respect to the frames $\mathcal{R}$ and $\mathcal{R}'$, respectively, we differentiate with respect to time in the fixed frame of reference. Applying the formula (A.2.37) which links *the absolute derivative* (with respect to $\mathcal{R}'$) to *the relative derivative* (with respect to $\mathcal{R}$), we may write

$$\frac{d\mathbf{r}}{dt} = \frac{\partial \mathbf{r}}{\partial t} + \boldsymbol{\omega} \times \mathbf{r}. \tag{5.3.1}$$

Noting

$$\mathbf{v}_a = \frac{d\mathbf{r}'}{dt}, \quad \mathbf{v}_r = \frac{\partial \mathbf{r}}{\partial t}, \quad \mathbf{v}_t = \mathbf{v}'_O + \boldsymbol{\omega} \times \mathbf{r}, \tag{5.3.2}$$

where $\mathbf{v}_a$, $\mathbf{v}_r$ and $\mathbf{v}_t$ are *the absolute velocity, the relative velocity* and *the velocity of transport* of the particle, respectively; we obtain

$$\mathbf{v}_a = \mathbf{v}_t + \mathbf{v}_r \qquad (5.3.3)$$

and may state

Theorem 5.3.1. *The absolute velocity of a particle is obtained by the vector composition of the velocity of transport with the relative velocity of it.*

Equating to zero the relative motion ($\mathbf{v}_r = \mathbf{0}$), we notice that the velocity in the motion of transport is – as a matter of fact – the velocity of a point of the rigid solid, rigidly linked to the movable frame of reference.

3.1.2 Composition of accelerations

Differentiating the relation (5.3.3) with respect to time in the frame $\mathcal{R}'$ and taking into account (5.3.2) and (A.2.37), we may write

$$\frac{d\mathbf{v}_a}{dt} = \frac{d\mathbf{v}_t}{dt} + \frac{d\mathbf{v}_r}{dt},$$

$$\frac{d\mathbf{v}_t}{dt} = \frac{d\mathbf{v}'_O}{dt} + \frac{d\boldsymbol{\omega}}{dt} \times \mathbf{r} + \boldsymbol{\omega} \times \frac{d\mathbf{r}}{dt}, \quad \frac{d\mathbf{v}_r}{dt} = \frac{\partial \mathbf{v}_r}{\partial t} + \boldsymbol{\omega} \times \mathbf{v}_r.$$

Using the formula (5.3.1) and the notations

$$\mathbf{a}_a = \frac{d\mathbf{v}_a}{dt}, \quad \mathbf{a}_r = \frac{\partial \mathbf{v}_r}{\partial t}, \quad \mathbf{a}_C = 2\boldsymbol{\omega} \times \mathbf{v}_r,$$

$$\mathbf{a}_t = \mathbf{a}'_O + \dot{\boldsymbol{\omega}} \times \mathbf{r} + \boldsymbol{\omega} \times (\boldsymbol{\omega} \times \mathbf{r}), \quad \mathbf{a}'_O = \frac{d\mathbf{v}'_O}{dt}, \qquad (5.3.4)$$

where $\mathbf{a}_a$, $\mathbf{a}_r$, $\mathbf{a}_t$ and $\mathbf{a}_C$ are *the absolute acceleration, the relative acceleration, the acceleration of transport* and *the acceleration of Coriolis* (*the complementary acceleration*) of the particle, respectively, we obtain

$$\mathbf{a}_a = \mathbf{a}_t + \mathbf{a}_r + \mathbf{a}_C; \qquad (5.3.5)$$

noting that the absolute acceleration is not obtained by summing vectorially the acceleration of transport with the relative acceleration (by vanishing the motion of transport, a complementary term, due to Coriolis, is added to the relative acceleration), so that we state

Theorem 5.3.2. *The absolute acceleration of a particle is obtained by the vector composition of the acceleration of transport with the relative and Coriolis' accelerations.*

Excepting the trivial case in which $\mathbf{v}_r = \mathbf{0}$, the acceleration of Coriolis vanishes if $\boldsymbol{\omega} = \mathbf{0}$, hence if the movable frame of reference has a motion of translation with respect to the fixed frame of reference (the movable frame is not rotating, its axes having a displacement parallel to themselves), or if $\mathbf{v}_r \parallel \boldsymbol{\omega}$ (e.g., the case of a particle in motion on the generatrix of a right circular cylinder, which rotates about its axis).

Kinematics 333

As it can be easily verified, after Resal, if $\omega = \omega' + \omega''$ or if $\mathbf{v}_r = \mathbf{v}'_r + \mathbf{v}''_r$, corresponding to two successive rotations or relative motions, then the resultant acceleration of Coriolis is the sum of the component Coriolis' accelerations ($\mathbf{a}_C = \mathbf{a}'_C + \mathbf{a}''_C$).

As in the case of velocities, by equating to zero the relative motion ($\mathbf{v}_r = \mathbf{a}_r = \mathbf{0}$), we notice that the acceleration in the motion of transport is the acceleration of a point of the rigid solid, rigidly linked to the moving frame of reference.

We may apply the above results – for instance – to the computation of the velocities and of the accelerations in polar, cylindrical, or spherical co-ordinates, by the composition of rectilinear motions or of motions of rotation, hence by the composition of a motion of transport with a relative one.

3.2 Relative motion of the rigid solid

The results obtained at the preceding section may be used in the study of the motion of a point of an arbitrary mechanical system, hence in the study of the motion of the respective mechanical system. In particular, we consider the relative motion of the rigid solid, emphasizing the corresponding composition of the velocities and of the accelerations.

3.2.1 Composition of velocities

Let us consider a fixed frame of reference $\mathscr{R}_0$ of pole O, a moving frame of reference $\mathscr{R}_1$ of pole O_1 and another moving frame of reference $\mathscr{R}_2$ of pole O_2, rigidly linked to the rigid solid. In this case, the relative velocity (with respect to the movable frame of reference) of a point P of the rigid solid is given by $\mathbf{v}_r = \mathbf{v}_{21} + \boldsymbol{\omega}_{21} \times \mathbf{r}_2$, where $\mathbf{v}_{21}$ is the velocity of the pole O_2 with respect to the pole O_1, $\boldsymbol{\omega}_{21}$ is the angular velocity of the frame $\mathscr{R}_2$ with respect to the frame $\mathscr{R}_1$, while $\mathbf{r}_2$ is the position vector of the point P with respect to the frame $\mathscr{R}_2$; using analogous notations, we may write the velocity of transport (of the moving frame with respect to the fixed one) in the form $\mathbf{v}_t = \mathbf{v}_{10} + \boldsymbol{\omega}_{10} \times \mathbf{r}_1$. The formula (5.3.3) of composition of velocities allows to write the absolute velocity (of the point P with respect to the frame $\mathscr{R}_0$) in the form

$$\mathbf{v}_P = \mathbf{v}_{10} + \mathbf{v}_{21} + \boldsymbol{\omega}_{10} \times \mathbf{r}_1 + \boldsymbol{\omega}_{21} \times \mathbf{r}_2 . \tag{5.3.6}$$

In the case of $n-1$ motions of transport, corresponding to $n-1$ frames $\mathscr{R}_1$, $\mathscr{R}_2$,..., $\mathscr{R}_{n-1}$, the frame $\mathscr{R}_0$ being fixed, while the frame $\mathscr{R}_n$ is rigidly linked to the rigid solid, we may write

$$\mathbf{v}_P = \mathbf{v}_{n,0} + \boldsymbol{\omega}_{n,0} \times \mathbf{r}_n = \sum_{i=1}^{n} \mathbf{v}_{i,i-1} + \sum_{i=1}^{n} \boldsymbol{\omega}_{i,i-1} \times \mathbf{r}_i , \tag{5.3.7}$$

the formula being proved by complete induction, taking into account (5.3.6) too; analogously, the angular velocity of the rigid solid with respect to the fixed frame $\mathcal{R}_0$ is given by

$$\boldsymbol{\omega} = \boldsymbol{\omega}_{n,0} = \sum_{i=1}^{n} \boldsymbol{\omega}_{i,i-1} \, . \qquad (5.3.7')$$

In particular, if all the component motions (both relative and of transport) are translations, then we obtain ($\boldsymbol{\omega}_{10} = \boldsymbol{\omega}_{21} = ... = \boldsymbol{\omega}_{n,n-1} = \mathbf{0}$)

$$\mathbf{v}_P = \sum_{i=1}^{n} \mathbf{v}_{i,i-1} \, , \qquad (5.3.8)$$

all the points of the rigid solid having the same velocity. We may state

Theorem 5.3.3. *By the composition of n motions of translation of a rigid solid one obtains a resultant motion which is a motion of translation too, the velocity of a point of the rigid solid being equal to the vector sum of the velocities of translation of the component motions.*

If all the component motions are instantaneous rotations and if the origins of the corresponding frames are on the instantaneous axes of rotation and coincide, then we have ($\mathbf{v}_{10} = \mathbf{v}_{21} = ... = \mathbf{v}_{n,n-1} = \mathbf{0}, \mathbf{r}_1 = \mathbf{r}_2 = ... = \mathbf{r}_n = \mathbf{r}$)

$$\mathbf{v}_P = \boldsymbol{\omega} \times \mathbf{r}, \quad \boldsymbol{\omega} = \sum_{i=1}^{n} \boldsymbol{\omega}_{i,i-1} \qquad (5.3.9)$$

and we may state

Theorem 5.3.4. *By the composition of n instantaneous motions of rotation, the instantaneous axes of rotation of which are concurrent at a fixed point O, one obtains an instantaneous resultant motion which is also an instantaneous motion of rotation about an instantaneous axis of rotation which passes through the same point O and the angular velocity of which is the vector sum of the angular velocities of the component motions.*

This result allows the study of the motion of a rigid solid with a fixed point, by the composition of three instantaneous motions of rotation about three instantaneous axes of rotation, passing through the fixed point.

In the case of instantaneous motions of rotation about some instantaneous parallel axes of rotation, we may write ($\mathbf{v}_{i,i-1} = \mathbf{0}, \boldsymbol{\omega}_{i,i-1} = \omega_{i,i-1}\mathbf{u}, |\mathbf{u}| = 1, i = 1,2,...,n$)

$$\mathbf{v}_P = \boldsymbol{\omega} \times \boldsymbol{\rho}, \quad \boldsymbol{\omega} = \omega\mathbf{u}, \quad \mathbf{u} = \sum_{i=1}^{n} \boldsymbol{\omega}_{i,i-1}, \quad \boldsymbol{\rho} = \frac{\sum_{i=1}^{n} \omega_{i,i-1}\mathbf{r}_i}{\omega}, \qquad (5.3.10)$$

assuming that $\boldsymbol{\omega} \neq \mathbf{0}$. We may pass from a point P to a point Q by the relation $\overrightarrow{O_iQ} = \overrightarrow{O_iP} + \overrightarrow{PQ}$, wherefrom

Kinematics

$$\mathbf{v}_Q = \mathbf{u} \times \sum_{i=1}^{n} \boldsymbol{\omega}_{i,i-1}\left(\mathbf{r}_i + \overrightarrow{PQ}\right) = \mathbf{v}_P + (\omega \mathbf{u}) \times \overrightarrow{PQ};$$

if $\boldsymbol{\omega} = \mathbf{0}$, then we obtain $\mathbf{v}_Q = \mathbf{v}_P$, hence a motion of translation. We may thus state

Theorem 5.3.5. *By the composition of n instantaneous motions of rotation, the instantaneous axes of rotation of which are parallel, of angular velocities $\boldsymbol{\omega}_{i,i-1}$, one obtains an instantaneous resultant motion which is an instantaneous motion of rotation about an instantaneous axis of rotation passing through the centre of the parallel vectors $\boldsymbol{\omega}_{i,i-1}$ of angular velocity*

$$\boldsymbol{\omega} = \sum_{i=1}^{n} \boldsymbol{\omega}_{i,i-1},$$

or a motion of translation, as we have $\boldsymbol{\omega} \neq \mathbf{0}$ or $\boldsymbol{\omega} = \mathbf{0}$, respectively.

If, for $n = 2$, we have $\boldsymbol{\omega}_{10} + \boldsymbol{\omega}_{21} = \mathbf{0}$, then we can say that a couple of instantaneous rotations is equivalent to a translation.

In the general case, by passing from a point P to a point Q of the rigid solid, we obtain a relation of the same form as that above, i.e.

$$\mathbf{v}_Q = \mathbf{v}_P + \boldsymbol{\omega} \times \overrightarrow{PQ} \tag{5.3.11}$$

(of the form (5.2.27)). We attach the relation

$$\mathbf{v}_P = \sum_{i=1}^{n} \mathbf{v}_{i,i-1} + \sum_{i=1}^{n} \overrightarrow{PO_i} \times \boldsymbol{\omega}_{i,i-1} \tag{5.3.7''}$$

to the relation (5.3.7'), so that we can introduce the torsor of the angular velocities, applying the static-kinematic analogy. The considerations made in Subsec. 2.3.1 remain valid, and we may classify the resultant instantaneous motions (obtained by the composition of some instantaneous motions), as in the case of only one such motion.

In particular, for rest with respect to a fixed frame of reference we must have

$$\boldsymbol{\omega} = \mathbf{0}, \quad \mathbf{v}_P = \mathbf{0}, \tag{5.3.12}$$

the point P being arbitrarily chosen.

Observing that

$$\sum_{i=1}^{n} \boldsymbol{\omega}_{i,i-1} \times \mathbf{r}_i = \sum_{i=1}^{n} \boldsymbol{\omega}_{i,i-1} \times \overrightarrow{O_iP} = \sum_{i=1}^{n} \boldsymbol{\omega}_{i,i-1} \times \left(\overrightarrow{O_1O_{i+1}} + \overrightarrow{O_{i+1}O_{i+2}} \ldots \right.$$
$$\left. \ldots + \overrightarrow{O_{n-1}O_n} + \overrightarrow{O_nP}\right) = \boldsymbol{\omega}_{10} \times \overrightarrow{O_1O_2} + (\boldsymbol{\omega}_{10} + \boldsymbol{\omega}_{21}) \times \overrightarrow{O_2O_3} \ldots$$
$$\ldots + \left(\sum_{i=1}^{n} \boldsymbol{\omega}_{i,i-1}\right) \times \overrightarrow{O_nP}, \quad O_{n+1} \equiv P,$$

we may also write

$$\mathbf{v}_P = \sum_{i=1}^{n} \mathbf{v}_{i,i-1} + \sum_{i=1}^{n-1} \boldsymbol{\omega}_0 \times \overrightarrow{O_i O_{i+1}} + \boldsymbol{\omega}_{n,0} \times \overrightarrow{O_n P}, \qquad (5.3.13)$$

an useful relation for practical applications. If we take into account a formula of the form (5.3.11), then we get

$$\mathbf{v}_{n,0} = \sum_{i=1}^{n} \mathbf{v}_{i,i-1} + \sum_{i=1}^{n-1} \boldsymbol{\omega}_{i,0} \times \overrightarrow{O_i O_{i+1}}. \qquad (5.3.13')$$

The general motion of a rigid solid is completely characterized by the vectors $\mathbf{v}'_0(t)$ and $\boldsymbol{\omega}(t)$, corresponding to a motion of translation and of rotation, respectively; as we have seen, the vector $\mathbf{v}'_0(t)$ may be replaced by a couple of vectors $\boldsymbol{\omega}_0(t)$, $-\boldsymbol{\omega}_0(t)$. Using the general results given in Chap. 2, Subsec. 2.2.4, concerning the systems of sliding vectors, we can state

Theorem 5.3.6. *The general motion of a rigid solid at a given moment may be obtained by the composition of three instantaneous motions of rotation about three instantaneous axes of rotation passing through three given points or by the composition of two instantaneous motions of rotation about two instantaneous axes of rotation, one of them passing through a given point.*

We notice that the relation (5.2.31) established for a point of the instantaneous axis of rotation and sliding corresponds, in fact, to the basic formula (5.3.3) in a relative motion. We come back to the respective problem in the particular case of a plane-parallel motion, considered in Subsec. 2.3.4. Let thus be the instantaneous centre of rotation I ($I(t)$ at the moment t), which describes the curve $\mathcal{B}$ (basis) in the absolute motion, and the curve $\mathcal{R}$ (rolling curve) in the relative motion, respectively; obviously, its velocity vanishes ($\mathbf{v}_t = \mathbf{0}$) in the motion of transport. The formula (5.3.3) leads thus to $\mathbf{v}_a = \mathbf{v}_r$, so that the two centrodes are tangent at $I(t)$ at the moment t. In modulus, we have $\dot{s}' = \dot{s}$ too, so that $s'(t) = s(t)$ on the curves $\mathcal{B}$ and $\mathcal{R}$, respectively; we start from the initial moment $t = t_0$ at which the two centrodes are tangent at $I(t_0)$ and which is considered to be the origin for the corresponding curvilinear co-ordinates. We may thus state

Theorem 5.3.7. *In the plane-parallel motion of a rigid solid, the basis $\mathcal{B}$ and the rolling curve $\mathcal{R}$ are centrodes tangent at $I(t)$ at the moment t; during the motion, the rolling curve is rolling without sliding over the basis.*

We may also state

Theorem 5.3.7' (*reciprocal*). *If in a plane-parallel motion a smooth curve $\mathcal{R}$, rigidly linked to the rigid solid, is rolling without sliding over a fixed smooth curve $\mathcal{B}$, then the point of contact $I(t)$ of the two curves at the moment t is the instantaneous centre of rotation, $\mathcal{B}$ and $\mathcal{R}$ being the basis and the rolling curve, respectively.*

Indeed, at the point $I(t)$ we have $\mathbf{v}_a = \lambda \mathbf{v}_r$, λ scalar, and $\dot{s}' = \dot{s}$, so that $\mathbf{v}_a = \mathbf{v}_r$; the relation (5.3.3) leads – in this case – to $\mathbf{v}_t = \mathbf{0}$, condition which

Kinematics 337

characterizes the centre $I(t)$. If we enlarge somewhat the conditions of this reciprocal theorem so that, in general, $\dot{s}' \neq \dot{s}$, then $\lambda \neq 1$; on the basis of the relation (5.3.3), the velocity $\mathbf{v}_t \neq \mathbf{0}$ is along the common direction of the velocities $\mathbf{v}_a$ and $\mathbf{v}_r$, hence it is tangent to the two curves at the point M, while $I(t)$ is on their common normal (Fig.5.28,a). We thus state

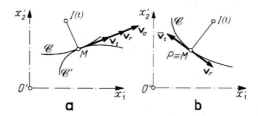

Figure 5.28. Properties of a smooth curve rigidly linked to a rigid solid in a plane-parallel motion.

Theorem 5.3.8. *If, in a plane-parallel motion, a smooth curve $\mathscr{C}$, rigidly linked to the rigid solid, remains all the time tangent to a smooth fixed curve $\mathscr{C}'$, then the instantaneous centre of rotation $I(t)$ is, at any moment t, on the common normal at the point M of contact.*

Let also be a curve $\mathscr{C}$, rigidly linked to a rigid solid, which passes through a fixed point M, and let be a point P on this curve, which coincides all the time with the point M; the point P describes the curve $\mathscr{C}$ with respect to the movable frame of reference, remaining at a fixed point with respect to the fixed frame of reference. Hence, at this point $\mathbf{v}_a = \mathbf{0}$ and $\mathbf{v}_t + \mathbf{v}_r = \mathbf{0}$; but $\mathbf{v}_r$ is along the tangent at P to the curve $\mathscr{C}$, so that $\mathbf{v}_t$ enjoys the same property (Fig.5.28,b). It results that the centre I is on the normal to this tangent and we may state

Theorem 5.3.9. *If, in a plane-parallel motion, a smooth curve $\mathscr{C}$, rigidly linked to the rigid solid, passes all the time through a fixed point M, then the instantaneous centre of rotation $I(t)$ is at any moment t on the normal at M to the curve $\mathscr{C}$.*

3.2.2 Composition of accelerations

As in the case of the distribution of velocities, we consider a fixed frame of reference $\mathscr{R}_0$ of pole O, a movable frame $\mathscr{R}_1$ of pole O_1 and a frame $\mathscr{R}_2$ of pole O_2, rigidly linked to the rigid solid. The relative acceleration (with respect to the movable frame) of a point P of the rigid solid is given by $\mathbf{a}_r = \mathbf{a}_{21} + \dot{\boldsymbol{\omega}}_{21} \times \mathbf{r}_2 + \boldsymbol{\omega}_{21} \times (\boldsymbol{\omega}_{21} \times \mathbf{r}_2)$, where $\mathbf{a}_{21}$ is the acceleration of the pole O_2 with respect to the pole O_1, $\boldsymbol{\omega}_{21}$ and $\dot{\boldsymbol{\omega}}_{21}$ are the angular velocity and acceleration, respectively, of the frame $\mathscr{R}_2$ with respect to the frame $\mathscr{R}_1$, while $\mathbf{r}_2$ is the position vector of the point P with respect to the frame $\mathscr{R}_2$; using analogous notations, we may write the acceleration of transport (of the movable frame with respect to the fixed one) in the form $\mathbf{a}_t = \mathbf{a}_{10} + \dot{\boldsymbol{\omega}}_{10} \times \mathbf{r}_1$

$+\boldsymbol{\omega}_{10} \times (\boldsymbol{\omega}_{10} \times \mathbf{r}_1)$. The acceleration of Coriolis is given by $\mathbf{a}_C = 2\boldsymbol{\omega}_{10} \times \mathbf{v}_r = 2\boldsymbol{\omega}_{10} \times (\mathbf{v}_{21} + \boldsymbol{\omega}_{21} \times \mathbf{r}_2)$. The formula (5.3.5) of composition of accelerations allows to write the absolute acceleration (of the point P with respect to $\mathscr{R}_0$) in the form

$$\begin{aligned}\mathbf{a}_P &= \mathbf{a}_{10} + \mathbf{a}_{21} + \dot{\boldsymbol{\omega}}_{10} \times \mathbf{r}_1 + \dot{\boldsymbol{\omega}}_{21} \times \mathbf{r}_2 + \boldsymbol{\omega}_{10} \times (\boldsymbol{\omega}_{10} \times \mathbf{r}_1)\\ &\quad + \boldsymbol{\omega}_{21} \times (\boldsymbol{\omega}_{21} \times \mathbf{r}_2) + 2\boldsymbol{\omega}_{10} \times (\mathbf{v}_{21} + \boldsymbol{\omega}_{21} \times \mathbf{r}_2).\end{aligned} \qquad (5.3.14)$$

In the case of $n-1$ motions of transport, corresponding to $n-1$ frames $\mathscr{R}_1$, $\mathscr{R}_2, \ldots, \mathscr{R}_{n-1}$, the frame of reference $\mathscr{R}_0$ being fixed, while the frame $\mathscr{R}_n$ is rigidly linked to the rigid solid, we have

$$\begin{aligned}\mathbf{a}_P &= \mathbf{a}_{n,0} + \dot{\boldsymbol{\omega}}_{n,0} \times \mathbf{r}_n + \boldsymbol{\omega}_{n,0} \times (\boldsymbol{\omega}_{n,0} \times \mathbf{r}_n) = \sum_{i=1}^{n} \mathbf{a}_{i,i-1} + \sum_{i=1}^{n} \dot{\boldsymbol{\omega}}_{i,i-1} \times \mathbf{r}_i \\ &\quad + \sum_{i=1}^{n} \boldsymbol{\omega}_{i,i-1} \times (\boldsymbol{\omega}_{i,i-1} \times \mathbf{r}_i) + 2\sum_{i=2}^{n}\sum_{j=1}^{i-1} \boldsymbol{\omega}_{j,j-1} \times (\mathbf{v}_{i,i-1} + \boldsymbol{\omega}_{i,i-1} \times \mathbf{r}_i);\end{aligned} \qquad (5.3.15)$$

the formula may be proved by complete induction, taking into account the relation (5.3.14) and the sum (5.3.7'). Differentiating the last formula with respect to time in the fixed frame of reference, we may write

$$\begin{aligned}\frac{d\boldsymbol{\omega}}{dt} &= \frac{d\boldsymbol{\omega}_{n,0}}{dt} = \sum_{i=1}^{n} \frac{d\boldsymbol{\omega}_{i,i-1}}{dt} = \sum_{i=1}^{n}\left(\frac{\partial \boldsymbol{\omega}_{i,i-1}}{\partial t} + \boldsymbol{\omega}_{i,0} \times \boldsymbol{\omega}_{i,i-1}\right) \\ &= \sum_{i=1}^{n} \dot{\boldsymbol{\omega}}_{i,i-1} + \sum_{i=1}^{n}\sum_{j=1}^{i} \boldsymbol{\omega}_{j,j-1} \times \boldsymbol{\omega}_{i,i-1},\end{aligned}$$

so that the angular acceleration of the rigid solid with respect to the fixed frame $\mathscr{R}_0$ will be given by

$$\dot{\boldsymbol{\omega}} = \dot{\boldsymbol{\omega}}_{n,0} = \sum_{i=1}^{n}\dot{\boldsymbol{\omega}}_{i,i-1} + \sum_{i=2}^{n}\sum_{j=1}^{i-1}\boldsymbol{\omega}_{j,j-1} \times \boldsymbol{\omega}_{i,i-1} = \sum_{i=1}^{n}\dot{\boldsymbol{\omega}}_{i,i-1} + \sum_{i=2}^{n}\boldsymbol{\omega}_{i-1,0} \times \boldsymbol{\omega}_{i,i-1}. \qquad (5.3.15')$$

In particular, if all the component motions (both of transport and relative) are translations, we obtain ($\boldsymbol{\omega}_{i,i-1} = \dot{\boldsymbol{\omega}}_{i,i-1} = \mathbf{0}$, $i = 1,2,\ldots,n$)

$$\mathbf{a}_P = \sum_{i=1}^{n} \mathbf{a}_{i,i-1}, \qquad (5.3.16)$$

all the points of the rigid solid having the same acceleration, and we may state

Theorem 5.3.10. *By the composition of n motions of translation of a rigid solid, one obtains a resultant motion which is a motion of translation too, the acceleration of a*

Kinematics 339

point of the rigid solid being equal to the vector sum of the accelerations of translation of the component motions.

If all the component motions are instantaneous rotations, the origins of the corresponding frames of reference being on the instantaneous axes of rotation and being coincident, we get ($\mathbf{v}_{i,i-1} = \mathbf{0}, \mathbf{a}_{i,i-1} = \mathbf{0}, \mathbf{r}_i = \mathbf{r}, i = 1,2,...,n$)

$$\mathbf{a}_P = \left(\sum_{i=1}^{n} \dot{\boldsymbol{\omega}}_{i,i-1}\right) \times \mathbf{r} + \sum_{i=1}^{n} \boldsymbol{\omega}_{i,i-1} \times (\boldsymbol{\omega}_{i,i-1} \times \mathbf{r}) + 2\sum_{i=2}^{n}\sum_{j=1}^{i-1} \boldsymbol{\omega}_{j,j-1} \times (\boldsymbol{\omega}_{i,i-1} \times \mathbf{r}) ;$$

(5.3.17)

noting that, in general, $\dot{\boldsymbol{\omega}} \neq \mathbf{0}$, even if the component rotations are uniform ($\dot{\boldsymbol{\omega}}_{i,i-1} = \mathbf{0}$, $i = 1,2,...,n$), we may state that, by the composition of n instantaneous motions of rotation, the instantaneous axes of which are concurrent at a point O, one obtains a motion with a distribution of accelerations characteristic to a rigid solid with a fixed point (the point O).

In the case of some instantaneous motions of rotation about some parallel instantaneous axes of rotation, we can write ($\boldsymbol{\omega}_{j,j-1} \times \boldsymbol{\omega}_{i,i-1} = \mathbf{0}$, $i,j = 1,2,...,n$)

$$\dot{\boldsymbol{\omega}} = \sum_{i=1}^{n} \dot{\boldsymbol{\omega}}_{i,i-1} , \qquad (5.3.18)$$

the vector $\dot{\boldsymbol{\omega}}$ being parallel to the vector $\boldsymbol{\omega}$; we obtain as resultant motion an instantaneous rotation about an instantaneous axis of rotation or a translation, the distribution of accelerations being a corresponding one.

Noting that $\mathbf{r}_i = \overrightarrow{O_i O_{i+1}} + \overrightarrow{O_{i+1} O_{i+2}} + ... + \overrightarrow{O_{n-1} O_n} + \overrightarrow{O_n P}$, taking into account (5.3.7') and (5.3.15'), and using the relation

$$(\boldsymbol{\omega}_{i-1,0} \times \boldsymbol{\omega}_{i,i-1}) + \mathbf{r}_i = \boldsymbol{\omega}_{i-1,0} \times (\boldsymbol{\omega}_{i,i-1} \times \mathbf{r}_i) - \boldsymbol{\omega}_{i,i-1} \times (\boldsymbol{\omega}_{i-1,0} \times \mathbf{r}_i),$$

a consequence of the relation (2.1.50'), we may write the law of composition (5.3.15) also in the form

$$\mathbf{a}_P = \sum_{i=1}^{n} \mathbf{a}_{i,i-1} + \sum_{i=1}^{n} \dot{\boldsymbol{\omega}}_{i,0} \times \overrightarrow{O_i O_{i+1}} + \dot{\boldsymbol{\omega}}_{n,0} \times \overrightarrow{O_n P}$$
$$+ \sum_{i=1}^{n-1} \boldsymbol{\omega}_{i,0} \times (\boldsymbol{\omega}_{i,0} \times \overrightarrow{O_i O_{i+1}}) + \boldsymbol{\omega}_{n,0} \times (\boldsymbol{\omega}_{n,0} \times \overrightarrow{O_n P}) + 2\sum_{i=2}^{n} \boldsymbol{\omega}_{i-1,0} \times \mathbf{v}_{i,i-1} \qquad (5.3.19)$$

useful in many applications.

In the general case, passing from a point P to a point Q of the rigid solid, we obtain a relation of the form

$$\mathbf{a}_Q = \mathbf{a}_P + \dot{\boldsymbol{\omega}} \times \overrightarrow{PQ} + \boldsymbol{\omega} \times (\boldsymbol{\omega} \times \overrightarrow{PQ}), \qquad (5.3.20)$$

where the angular velocity and acceleration are given by the formulae (5.3.7'), (5.3.15'), respectively.

We mention that the formula (5.3.19) may lead to a torsor of the angular accelerations, less useful that the torsor of the angular velocities, because of its complicated form.

3.3 Kinematics of systems of rigid solids

After some general considerations concerning the systems of rigid solids (called also multibody systems, where it is understood that the bodies are rigid ones), we introduce the notion of mechanism, as an example of such systems; the results thus obtained allow to present some applications concerning the transmission of displacements, velocities and accelerations.

3.3.1 Systems of rigid solids

The results obtained for the relative motion of a rigid solid may be used also for systems of rigid solids. Let be thus two rigid solids $\mathscr{S}$ and $\mathscr{S}'$, bounded by two convex surfaces S and S', respectively, which at any moment t, have the same tangent in an ordinary common point $P \equiv P'$, $P \in S$, $P' \in S'$ (Fig.5.29); we assume that the solid $\mathscr{S}'$ is fixed, while the solid $\mathscr{S}$ is movable, remaining at any moment in contact with the solid $\mathscr{S}'$. Let us suppose that a movable point Q coincides with $P \equiv P'$ at any moment t; the locus of the point Q with respect to the movable surface S is a curve C, its velocity $\mathbf{v}_P$ being a relative velocity, and with respect to the fixed surface S' a curve C', its velocity $\mathbf{v}_{P'}$ being an absolute velocity. Taking into account the relation (5.3.3), we obtain the velocity of the point Q as a velocity of transport with respect to the surface S' in the form

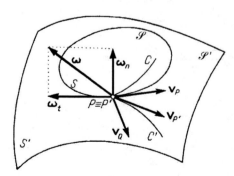

Figure 5.29. Motion of a rigid solid over another rigid solid.

$$\mathbf{v}_Q = \mathbf{v}_{P'} - \mathbf{v}_P. \tag{5.3.21}$$

Because the velocities $\mathbf{v}_{P'}$ and $\mathbf{v}_P$ are contained in the common tangent plane, the velocity $\mathbf{v}_Q$ which characterizes the sliding of the surface S over the surface S' will belong to the same plane. The distribution of the velocities is known if, besides the velocity $\mathbf{v}_Q(t)$, the angular velocity $\boldsymbol{\omega}(t)$, which passes through the point P, is also given. The motion of the rigid solid $\mathscr{S}$ with respect to the rigid solid $\mathscr{S}'$ is thus characterized by a translation of velocity $\mathbf{v}_Q(t)$ and by a rotation of angular velocity $\boldsymbol{\omega}(t)$. The vector $\boldsymbol{\omega}(t)$ may be decomposed in two components: an angular velocity $\boldsymbol{\omega}_n(t)$, along the normal to the tangent plane, which characterizes a pivoting about the respective axis, and an angular velocity $\boldsymbol{\omega}_t(t)$, contained in the tangent plane, which characterizes a rolling about the corresponding axis. We can say that the general motion of the rigid solid $\mathscr{S}$ over the rigid solid $\mathscr{S}'$ takes place so that the surface S is *rolling and pivoting with sliding* over the surface S'. If $\mathbf{v}_Q(t) = \mathbf{0}$, then the motion of the rigid solid $\mathscr{S}$ over the rigid solid $\mathscr{S}'$ is an *instantaneous rotation* (*pivoting and rolling*) about an instantaneous axis of rotation passing through the point of contact. The fixed axoid intersects the surface S' after the curve C', while the movable axoid intersects the surface S after the curve C; in this case, $\mathbf{v}_{P'} = \mathbf{v}_P$, so that the curve C is rolling without sliding over the curve C' during the motion. If $\mathbf{v}_Q \neq \mathbf{0}$ but $\boldsymbol{\omega}_n = \mathbf{0}$ ($\boldsymbol{\omega} = \boldsymbol{\omega}_t$), then the surface S is rolling with sliding over the surface S', along an instantaneous axis of rotation and sliding; if $\mathbf{v}_Q = \mathbf{0}$ too, then the motion is only a rotation without sliding.

Let be the frames $\mathscr{R}_1, \mathscr{R}_2, \ldots, \mathscr{R}_n$. rigidly linked to the rigid solids $\mathscr{S}_1, \mathscr{S}_2, \ldots, \mathscr{S}_n$, respectively. Denoting by $\boldsymbol{\omega}_{jk}$ the angular velocity in the motion of the frame $\mathscr{R}_j$ with respect to the frame $\mathscr{R}_k$ and by $(\partial/\partial_j)\mathbf{V}$ the derivative with respect to time of a vector $\mathbf{V}$ in the frame $\mathscr{R}_j$, we may write

$$\frac{\partial}{\partial_i}\mathbf{V} = \frac{\partial}{\partial_{i+1}}\mathbf{V} + \boldsymbol{\omega}_{i+1,i} \times \mathbf{V}, \quad i = 1, 2, \ldots, n-1,$$

$$\frac{\partial}{\partial_n}\mathbf{V} = \frac{\partial}{\partial_1}\mathbf{V} + \boldsymbol{\omega}_{1,n} \times \mathbf{V};$$

summing with respect to i and noting that the vector $\mathbf{V}$ is arbitrary, we obtain the relation (equivalent to the relation (5.3.7'))

$$\boldsymbol{\omega}_{21} + \boldsymbol{\omega}_{32} + \ldots \boldsymbol{\omega}_{n,n-1} + \boldsymbol{\omega}_{1,n} = \mathbf{0}, \tag{5.3.22}$$

which links the relative rotations of n rigid solids. In particular, one obtains the remarkable relations

$$\boldsymbol{\omega}_{ij} + \boldsymbol{\omega}_{ji} = \mathbf{0}, \quad \boldsymbol{\omega}_{ij} = \boldsymbol{\omega}_{ik} + \boldsymbol{\omega}_{kj}, \quad i \neq j \neq k \neq i, \quad i,j,k = 1,2,\ldots,n. \tag{5.3.22'}$$

Starting from

$$\frac{\partial}{\partial_i}\overrightarrow{O_iO_j} = \frac{\partial}{\partial_j}\overrightarrow{O_iO_j} + \boldsymbol{\omega}_{ji} \times \overrightarrow{O_iO_j}, \quad \overrightarrow{O_iO_j} = \mathbf{r}_i - \mathbf{r}_j,$$

we find also the relation

$$\mathbf{v}_{ij} + \mathbf{v}_{ji} + \boldsymbol{\omega}_{ij} \times \mathbf{r}_i + \boldsymbol{\omega}_{ji} \times \mathbf{r}_j = \mathbf{0}, \quad i \neq j, \quad i,j = 1,2,\dots,n, \tag{5.3.23}$$

complementary to the first relation (5.3.22'), where $\mathbf{v}_{ij}$ is the velocity of the pole O_i of the frame $\mathcal{R}_i$ with respect to the pole O_j of the frame $\mathcal{R}_j$. Starting from the relation (5.3.7) and using the relation (5.3.23), we get the relation

$$\sum_{i=1}^{n-1} \mathbf{v}_{i+1,i} + \mathbf{v}_{1,n} + \sum_{i=1}^{n-1} \boldsymbol{\omega}_{i+1,i} \times \mathbf{r}_{i+1} + \boldsymbol{\omega}_{1,n} \times \mathbf{r}_1 = \mathbf{0}, \quad n \geq 2, \tag{5.3.24}$$

corresponding to the frames $\mathcal{R}_1, \mathcal{R}_2,\dots, \mathcal{R}_n$, the position vectors being those of a point P of the rigid solid with respect to the above mentioned frames. We mention that one may write the relation (5.3.24) also in the form

$$\sum_{i=1}^{n-1} \mathbf{v}_{i+1,i} + \mathbf{v}_{1,n} + \sum_{i=1}^{n-1} \overrightarrow{PO_{i+1}} \times \boldsymbol{\omega}_{i+1,i} + \overrightarrow{PO_1} \times \boldsymbol{\omega}_{1,n} = \mathbf{0}, \quad n \geq 2; \tag{5.3.24'}$$

the formulae (5.3.22) and (5.3.24') allow thus to use the static-kinematic analogy for the sliding vectors $\boldsymbol{\omega}_{i+1,i}$, $i = 1,2,\dots,n-1$, and $\boldsymbol{\omega}_{1,n}$, the velocities $\mathbf{v}_{i+1,i}$, $i = 1,2,\dots,n-1$, and $\mathbf{v}_{1,n}$ playing the rôle of moments of those vectors.

The relation (5.3.13) may lead, on the same way, to some interesting results. Noting that

$$\frac{\partial}{\partial_i}\boldsymbol{\omega}_{ji} = \frac{\partial}{\partial_j}\boldsymbol{\omega}_{ji} + \boldsymbol{\omega}_{ji} \times \boldsymbol{\omega}_{ji} = -\frac{\partial}{\partial_j}\boldsymbol{\omega}_{ij},$$

we obtain the relation

$$\dot{\boldsymbol{\omega}}_{ij} + \dot{\boldsymbol{\omega}}_{ji} = \mathbf{0}. \tag{5.3.25}$$

The relation (5.3.15') allows to write

$$\sum_{i=1}^{n-1} \dot{\boldsymbol{\omega}}_{i+1,i} + \dot{\boldsymbol{\omega}}_{1,n} + \sum_{i=1}^{n-2} \boldsymbol{\omega}_{i+1,1} \times \boldsymbol{\omega}_{i+2,i+1} \tag{5.3.26}$$

in this case; in particular, one obtains

$$\dot{\boldsymbol{\omega}}_{ij} = \dot{\boldsymbol{\omega}}_{ik} + \dot{\boldsymbol{\omega}}_{kj} - \boldsymbol{\omega}_{ik} \times \boldsymbol{\omega}_{kj}. \tag{5.3.25'}$$

Using a relation of the form (2.1.50'), as for the distribution of velocities, and starting from the relation (5.3.15), we get a relation of the form

$$\sum_{i=1}^{n}\mathbf{a}_{i+1,i} + \mathbf{a}_{1,n} + \sum_{i=1}^{n}\mathbf{a}_{i+1,i}^{t} + \mathbf{a}_{1,n}^{t} + \sum_{i=1}^{n-1}\mathbf{a}_{i+1,i}^{C} + \mathbf{a}_{1,n}^{C}$$
$$+ \sum_{i=1}^{n}\overrightarrow{PO_{i+1}} \times \dot{\boldsymbol{\omega}}_{i+1,i} + \overrightarrow{PO_{1}} \times \dot{\boldsymbol{\omega}}_{1,n} + \sum_{i=1}^{n-2}\overrightarrow{PO_{i+2}} \times (\boldsymbol{\omega}_{i+1,1} \times \boldsymbol{\omega}_{i+2,i+1}) = \mathbf{0}; \quad (5.3.27)$$

thus, the formulae (5.3.26), (5.3.27) lead also to the static-kinematic analogy for the sliding vectors of the nature of angular accelerations, where the relative accelerations, the accelerations of transport and the accelerations of Coriolis play the rôle of moments of those vectors.

Analogous results can be obtained starting from the relation (5.3.19) too.

We consider also a mechanical system made up by three rigid solids $\mathscr{S}_i$, $\mathscr{S}_j$ and $\mathscr{S}_k$, which have a plane-parallel motion. The rigid solid $\mathscr{S}_i$ has a motion of rotation with respect to the rigid solid $\mathscr{S}_j$ about an instantaneous axis of rotation, normal to the plane of motion at the instantaneous centre of rotation I_{ij}, with an angular velocity $\boldsymbol{\omega}_{ij}$; analogously, one introduce the instantaneous centres of rotation I_{jk} and I_{ki}, as well as the corresponding angular velocities $\boldsymbol{\omega}_{jk}$ and $\boldsymbol{\omega}_{ki}$. Taking into account the second relation (5.3.22'), there results that the three parallel vectors are also coplanar, so that one may state

Theorem 5.3.11 (*theorem of the three instantaneous centres of rotation*). *If a mechanical system made up by three rigid solids* $\mathscr{S}_i$, $\mathscr{S}_j$ *and* $\mathscr{S}_k$ *has a plane-parallel motion, then the three instantaneous centres of rotation* I_{ij}, I_{jk} *and* I_{ki} *corresponding to their relative motions are collinear.*

We notice that $I_{ij} \equiv I_{ji}$, hence the instantaneous centre of rotation of the rigid solid $\mathscr{S}_i$ with respect to the rigid solid $\mathscr{S}_j$ coincides with the instantaneous centre of rotation of the rigid solid $\mathscr{S}_j$ with respect to the rigid solid $\mathscr{S}_i$.

3.3.2 Kinematic chains. Mechanisms

The rigid solids which constitute a system of rigid solids are called *elements*; one of those elements may be considered fixed, the other elements being – in general – movable. The link which restricts the motion of an element with respect to another one is called *kinematic couple* relative to the two elements; one may say also that this is the possibility *to transmit the motion from one element to another one*. Among the kinematic couples we mention: *the articulation* (it allows the rotations), *the coulisse* or *the slideway* (it allows the displacement in a given direction) and *the simple support* (it does not allow displacement in a given direction). An element of a kinematic couple is considered fixed, studying – in fact – the relative motion of the second element with respect to the first one. The conditions of linkage (the restrictions of the relative motion) diminish the number of degrees of freedom of the free element (which is six). We denote by c *the number of the conditions of linkage*; if $c = 0$, then the elements are *free one with respect to the other*, while if $c = 6$, then the two elements are *built in one into the other* (*rigid constraint*). Hence $1 \le c \le 5$; if N is *the number of degrees of freedom* which remain, then we have $c = 6 - N$. We may thus classify (after Malyshev) the kinematic couples in *five classes*, after the number of the conditions of linkage. We mention thus the plane-sphere and the plane-plane couple (of class I and class II, respectively), the spherical and the plane couple (of class III), the annular and the cylindrical couple (of class IV), the couple of rotation, of translation and the helical couple (of class V); for instance, the plane-sphere couple allows two translations and three rotations, while the cylindrical couple allows only a translation and a rotation. If the relative motion of the two elements is a plane-parallel motion, then we have to do with a *plane couple*; otherwise, the couple is a *space* one. In another classification, we have: inferior kinematic couples, if the contact zone is a surface, and superior kinematic couples, if the contact zone is a line or a point.

A system of rigid solids (elements) linked by kinematic couples, which allow the motion from one element to the other, by successive transformations, is called a *kinematic chain*. If there is at least a *singular element*, which belongs to only one kinematic couple, then the kinematic chain is called *open* (Fig.5.30,a); otherwise (if each element of it belongs at least to two kinematic couples), the kinematic chain is called *closed* (Fig.5.30,b). A kinematic chain is *simple* if each element of it belongs at the most to two kinematic couples (Fig.5.30). Otherwise, the kinematic chains are

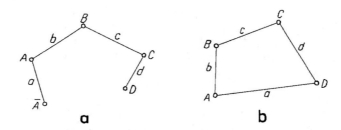

Figure 5.30. Kinematic chains: open (a) and closed (b).

complex, having at least an element involved in more than two kinematic couples; these chains may be open or closed too (Fig.5.31,a,b). If all the couples are plane, then the kinematic chain is *plane*; otherwise, it is *spatial*. We call *basis* of the kinematic

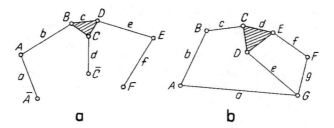

Figure 5.31. Complex kinematic chains: open (a) and closed (b).

chain an element of it which is considered fixed; the other elements are *leading elements* (if they induce the motion coming from the exterior to the other elements) or *followers* (if they receive the motion from the first elements). A kinematic chain for which at any

Kinematics 345

position of one or several leading elements of it corresponds a unique position for all the other followers (with respect to the considered basis) is called *desmodromous*; otherwise, the kinematic chain is called *non-desmodromous*. For instance, the articulated quadrangle of vertices A, B, C, D and sides a, b, c, d (Fig.5.32), for which a is the basis, while the element b is leading is a desmodromous kinematic chain, the elements c and d being followers. In analogous conditions, an articulated pentagon, e.g., is a non-desmodromous kinematic chain.

The *number of degrees of freedom* of a kinematic chain, constituted of n elements, linked by n_c couples of class c, $c = 1, 2, ..., 5$, is given by

$$N = 6n - \sum_{c=1}^{5} c n_c . \tag{5.3.28}$$

If one of the elements of the kinematic chain is a basis, then its *degree of mobility* is specified by *the Somov-Malyshev formula* (*the structural formula of the kinematic chains*)

$$M = 6(n-1) - \sum_{c=1}^{5} c n_c , \tag{5.3.28'}$$

the kinematic couples being independent. If there exist l restrictions of motion (due, for instance, to some anterior common linkages), then we have

$$N = (6-l)n - \sum_{c=l+1}^{5} (c-l) n_c , \tag{5.3.29}$$

the degree of mobility being given by *Dobrovolski's formula*

$$M_l = (6-l)(n-1) - \sum_{c=l+1}^{5} (c-l) n_c . \tag{5.3.29'}$$

In the plane case ($l = 3$), we get $N = 3n - n_4 - 2n_5$ and $M_3 = 3(n-1) - n_4 - 2n_5$, the latter formula being due to Chebyshev.

A closed kinematic chain which has a basis and is subjected to a desmodromous motion is called a *mechanism*; the number of its leading elements is – in general – equal to the number of its degrees of mobility. In the case of *the articulated quadrangle* ($n = 4$, $n_4 = 0$, $n_5 = 4$) (Fig.5.32), Chebyshev's formula gives $M = 3(4-1) - 2 \cdot 4 = 1$. In this particular case, we distinguish six centres of rotation, i.e.: four *centres of permanent rotation* (the fixed centres of rotation $I_{ba} \equiv A$ and $I_{da} \equiv D$ and the movable centres of rotation $I_{cb} \equiv B$ and $I_{dc} \equiv C$) and two *instantaneous centres of rotation* I_{ca} and I_{db}; corresponding to the Theorem 5.3.11, we notice that the above mentioned centres of rotation are three abreast on the four sides of the quadrangle. In general, a mechanism made up of n elements has $n(n-1)/2$ centres of rotation; if the mechanism is a polygon, then n centres are permanent (two of them being fixed), while $n(n-3)/2$ are instantaneous ones.

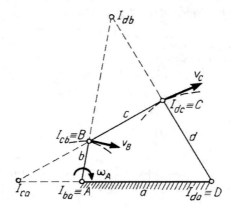

Figure 5.32. Desmodromous articulated quadrangle.

3.3.3 Applications to the transmission of displacements, velocities and accelerations

The determination of the kinematic characteristics of a mechanism (i.e., the positions, the velocities and the accelerations of its points) is of particular importance in the study of it; the methods of computation used are analytical, graphical, or graphoanalytical ones.

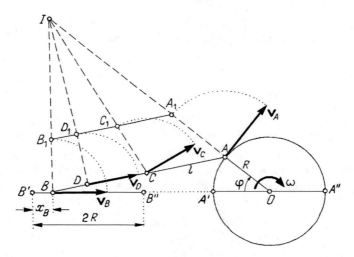

Figure 5.33. Crank and connecting-rod mechanism.

Let be, for instance, *the crank and connecting-rod mechanism* BAO (one of the most usual mechanisms, which transforms a rectilinear motion in a circular one) (Fig.5.33). the leading element is *the crank* AO, articulated at O and of length R,

Kinematics

the position of which is specified by the angle φ with respect to the position OA', corresponding to *the dead-point* (*dead-position*); in the latter case, the extremity B of the connecting-rod BA, which glides at B and is of length l, is at the point B'. Taking the point B' as origin for the displacements, the position of the point B at a given moment will be

$$x_B = \overline{B'B} = R(1 - \cos\varphi) + l\left(1 - \sqrt{1 - \lambda^2 \sin^2\varphi}\right), \quad \lambda = \frac{R}{l}.$$

The ratio λ being subunitary (in general, $\lambda < 1/3$), we may use binomial's formula for the radical, so that

$$x_B \cong R(1 - \cos\varphi) + \frac{R^2}{2l}\sin^2\varphi = 2\sin^2\frac{\varphi}{2}\left(1 + \lambda\cos^2\frac{\varphi}{2}\right). \tag{5.3.30}$$

Hence, one has the velocity

$$v_B = v_A\left(\sin\varphi + \frac{R}{2l}\sin 2\varphi\right) = v_A\sin\varphi(1 + \lambda\cos\varphi), \quad v_A = \omega R, \quad \omega = \dot\varphi. \tag{5.3.31}$$

We must have $\cos\varphi + \lambda\cos 2\varphi = 0$ for $v_{B\max}$, wherefrom

$$\cos\varphi = \frac{1}{4\lambda}\left(-1 + \sqrt{1 + 8\lambda^2}\right) \cong \lambda, \quad \varphi < 90°,$$

using the same approximation method as above; consequently,

$$v_{B\max} \cong \left(1 - \frac{\lambda^2}{2}\right)(1 + \lambda^2)v_A \cong \left(1 + \frac{\lambda^2}{2}\right)v_A \tag{5.3.31'}$$

For $\lambda = 1/5$ we obtain $\varphi \cong 78°27'47''$ and $v_{B\max} \cong 1.02 v_A$.

The acceleration is given by

$$a_B = \omega^2 R(\cos\varphi + \lambda\cos 2\varphi) + R\dot\omega\sin\varphi(1 + \lambda\cos\varphi). \tag{5.3.32}$$

In case of a *constant regime* ($\omega = $ const), it results

$$a_B = a_A(\cos\varphi + \lambda\cos 2\varphi), \quad a_A = \omega^2 R. \tag{5.3.32'}$$

We get $a_{B\max} = a_A(1 + \lambda)$ for $\varphi = 0°$ and $a_{B\min} = -a_A(1 - \lambda)$ for $\varphi = 180°$; we can also have $a_{B\min} = -a_A(1 + \lambda)/8\lambda$ too for $\cos\varphi = -1/4\lambda$.

Another analytical method is *the method of independent cycles*, at the basis of which stay the formulae (5.3.22) and (5.3.24'), allowing to write two vector equations of equilibrium for each independent cycle of the considered mechanism; in the case of n

independent cycles, one may write $2n$ vector equations to determine the velocities $\mathbf{v}_{1,n}$, $\mathbf{v}_{i+1,i}$ and the angular velocities $\boldsymbol{\omega}_{1,n}$, $\boldsymbol{\omega}_{i+1,i}$, $i = 1,2,...,n-1$. The formulae (5.3.26), (5.3.27) lead to an analogous method of computation for the distribution of accelerations. But the analytical methods are difficult to use in the case of more intricate mechanisms.

Among *the graphical* or *grapho-analytical methods* we mention – first of all – *the method of connecting-rod curves* to determine the displacements of the points of elements of this nature.

To determine the velocities, one can use – sometimes – *the method of the instantaneous centre of rotation*; for instance, in the case of the just considered crank and connecting-rod mechanism (Fig.5.33), the centre I can be easily obtained, so that

$$\omega_I = \frac{\overline{OA}}{\overline{IA}}\omega, \quad \omega_C = \overline{IC}\omega_I, \tag{5.3.33}$$

for an arbitrary point C of the driving rod. *The velocities' turning down method*, emphasized in Subsec. 2.3.4, allows to determine graphically the magnitude and the direction of velocities; taking into account Fig.5.19, we may obtain the drawing of Fig.5.33. The point D, the foot of the normal from I to the connecting-rod AB, is the point of minimal magnitude of the velocity, which has the direction of the connecting-rod, and is its *characteristic point*. We mention that it is not necessary to obtain previously the point I; that is an advantage of the latter method. But if this point is specified and a particular scale for the velocities is chosen, so that the magnitude of the velocity of a point be equal to the corresponding instantaneous radius (in our case $v_A = \overline{IA}$), then the velocities of all the points will have the moduli equal to the corresponding instantaneous radii; this is *the method of the normal velocities*. Euler's formula (5.2.3') leads to *the method of the relative motion* (*the method of the vector equations*). As well, the formulae (5.2.4), (5.2.4') stay at the basis of *the method of velocities' projections*; thus, if we know the velocities of the points A and B, which correspond to two elements of a mechanism, then we may obtain the projections of the velocity of a point C, rigidly linked to each of the points A and B and non-collinear with these points, on the straight lines AC and BC, hence obtaining the velocity of the point C. The drawing of the velocities' plane and the similarity theorem (for velocities) lead to *the method of velocities' polygon* for a plane-parallel motion. Using the theorem of the three instantaneous centres of rotation (see Subsec. 3.3.1), one may often determine the distribution of the velocities in the case of a plane mechanism (*the method of collinearity of the instantaneous centres of rotation*).

To determine the accelerations of an element, we may use the graphical representation of the accelerations of its points. As in the case of velocities, we mention *the method of relative motion* (*the method of vector equations*), based on the formulae (5.2.6)-(5.2.6"), *the projections' method*, based on the formula (5.2.10), and *the method of accelerations' polygon*, based on the introduction of the accelerations' plane and on the similarity theorem (for accelerations), introduced in Subsec. 2.3.4.

We notice that a mechanism realizes a transformation of an *input quantity* (displacement, velocity, acceleration) into an *output quantity* of the same nature; thus,

Kinematics

the output quantity will be equal to the input one, amplified by a coefficient, which is a *transmission ratio* (or a *transfer function*), hence $d_o = \lambda_d d_i$, $v_o = \lambda_v v_i$, d being a displacement and v a velocity, where the indices o and i stay for output and input, respectively, while λ_d, λ_v are the corresponding transmission ratios. We mention that, in the case of accelerations, we cannot speak about a transmission ratio, but only if $\lambda_v = \text{const}$, case in which $a_o = \lambda_v a_i$ with analogous notations; otherwise, the ratio λ_a of transmission of accelerations depends not only on the position of the leading element, but also on its velocity and acceleration.

Among the mechanisms which realize these transformations, we mention – first of all – *the mechanisms with articulated levels*. One of the simplest mechanisms of this type is *the articulated quadrangle*, which may be – in particular – an *articulated parallelogram*. *The link mechanisms* may be with a *translation, oscillating* or *rotation link*; to the first category belongs also *the crank and connecting-rod mechanism*, just considered. We mention also the case in which the axle of the link does not pass through the articulation of the crank, as well as the case of *eccentric gears* (for which the ratio $\lambda = R/l$ is very small). Another mechanism of this kind is *the shaping mechanism*. *The director mechanisms* may be used to obtain a rectilinear or a curvilinear trajectory; they may be *reversers (exact director mechanisms)* or *approximate mechanisms*. *The mechanisms with a Cardan joint* are based on a *Cardan (universal) coupling*.

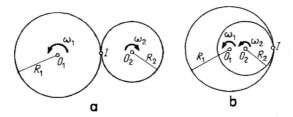

Figure 5.34. Wheel mechanisms with (exterior) (a) or interior (b) wheels.

In the case of *wheel mechanisms*, one must have in view the relative position of the axes of rotation and the angular velocities; we mention that the wheels may be *exterior* (Fig.5.34,a) (the rotation direction is changed) or *interior* (Fig.5.34,b) (the rotation direction is maintained). The transmission may be obtained by *friction wheels* or by *gear wheels* (*trains of gears*). In the case in which the axes of rotation are *parallel* (Fig.5.34), having a rolling without sliding, the velocity of their point of contact (which is an instantaneous centre of rotation too) is the same, so that the transmission ratio is given by

$$\lambda_\omega = \frac{\omega_1}{\omega_2} = \pm \frac{R_2}{R_1}, \qquad (5.3.34)$$

where one takes the sign + or – as the direction of rotation is maintained or not; if the motions of rotation are uniform, then we may write

$$\lambda_\omega = \pm \frac{n_1}{n_2}, \qquad (5.3.34')$$

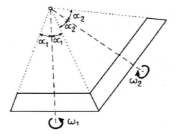

Figure 5.35. Gear wheels with concurrent rotation axes.

where n_1 and n_2 represent the number of rotations per unit time. In the case in which the rotation axes are *concurrent*, the wheels are truncated cones of vertex angles α_1 and α_2 (Fig.5.35), respectively, and we get

$$\lambda_\omega = \frac{\omega_1}{\omega_2} = \pm \frac{\sin \alpha_2}{\sin \alpha_1}; \qquad (5.3.35)$$

if we have to do with trains of gears, then R_1 and R_2 of the formula (5.3.34) are the radii of the two centrodes, and the formula may be written also in the form given by Willis

$$\lambda_\omega = \frac{\omega_1}{\omega_2} = \pm \frac{N_1}{N_2}, \qquad (5.3.34'')$$

where N_1, N_2 represent the number of teeth of the two wheels, respectively. We mention also *the trains of gear rack wheels*, as well as *the helical gear wheels* (for which the rotation axes have arbitrary relative positions). A more complex character have *the planetary* and *differential trains of gears*. We mention also *the worm-spiral wheel trains* (*endless screw-helical wheel trains*).

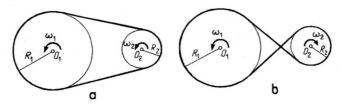

Figure 5.36. Mechanisms with flexible elements (a); using of crossed belts.

The mechanisms with flexible elements may use belts, cables, chains etc. In this case, the direction of rotation is maintained (Fig.5.36,a); the direction may be changed

Kinematics 351

if *crossed belts* are used (Fig.5.36,b). The formulae (5.3.34), (5.3.34,b) remain still valid.

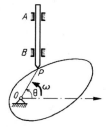

Figure 5.37. Cam gears.

The cam gears are used to transform the motions. *The cam* plays the rôle of a leading element, having the pole O fixed, while the *draw bar* AB is the follower, P being the point of contact (Fig.5.37). The point P is moving after the law $\overline{OP} = r = f(t)$, where $f(t)$ is a periodic function; if the cam has a rotation of angle $\theta = \omega t + \theta_0$, then the shape of the cam is expressed in polar co-ordinates in the form

$$r = f\left(\frac{\theta - \theta_0}{\omega}\right). \tag{5.3.36}$$

Chapter 6

DYNAMICS OF THE PARTICLE WITH RESPECT TO AN INERTIAL FRAME OF REFERENCE

Dynamics deals with the motion of mechanical systems subjected to the action of given forces. We begin this study with a single particle in motion with respect to an *inertial frame of reference*; as it was shown in Chap. 1, Subsec. 1.1.4, it is the frame with respect to which the basic laws of mechanics are verified. If these laws hold with respect to a certain frame, in a Newtonian model, then they are verified in any other frame in rectilinear and uniform motion with respect to the first one, obtaining thus a *set* of inertial frames of reference. To study the motion in such a frame, we emphasize the corresponding general theorems, both for the free and the constraint (frictionless or with friction) particle.

1. Introductory notions. General theorems

After introducing mechanical quantities which play an essential rôle in the frame of the Newtonian model, we formulate the problem of the free particle, emphasizing the methods of solving it; the theorems of existence and uniqueness are thus presented and stress is put on the notion of first integral. The principle of relativity allows to establish the Galileo-Newton transformations group. Starting from the general theorems corresponding to the motion of the particle, one obtains the conservation theorems, hence the first integrals of the equations of motion; the first integral of areas leads then to the notion of central force.

1.1 Introductory notions

In what follows, we introduce the notions of momentum, moment of momentum, work, kinetic and potential energy, power and mechanical efficiency; the conservative and non-conservative forces are then considered. We mention also the formulation of the problem of the free particle in motion and the presentation of the equations of motion in curvilinear co-ordinates.

1.1.1 Momentum. Moment of momentum. Torsor of momentum

We have introduced the notion of momentum in Chap. 1, Subsec. 1.1.6; thus *the momentum (linear momentum) of a particle* P of position vector $\mathbf{r}$ with respect to a given fixed frame of reference (which is supposed to be inertial), is expressed in the form

$$\mathbf{H} = m\mathbf{v} = m\dot{\mathbf{r}} \tag{6.1.1}$$

and is a vector collinear with the velocity $\mathbf{v}$ of the respective particle.

The moment of the momentum with respect to the pole O of the frame of reference is called *moment of momentum* (*angular momentum*) *of the particle*, with respect to this pole, and is given by

$$\mathbf{K}_O = \mathbf{r} \times \mathbf{H} = m\mathbf{r} \times \mathbf{v} = m\mathbf{r} \times \dot{\mathbf{r}} = 2m\boldsymbol{\Omega}_O, \tag{6.1.2}$$

where we have introduced the areal velocity (5.1.16). In general, we may consider a moment of momentum with respect to any given fixed point (fixed with respect to an inertial frame of reference).

In components, we have

$$H_i = mv_i = m\dot{x}_i, \quad K_{Oi} = 2m\Omega_{Oi} = \epsilon_{ijk} m x_j \dot{x}_k, \quad i = 1,2,3. \tag{6.1.3}$$

The notion of torsor, introduced in Chap. 2, Subsec. 2.1.3, allows to write

$$\tau_O(\mathbf{H}) = \{\mathbf{H}, \mathbf{K}_O\}; \tag{6.1.4}$$

hence, the set formed by the linear and angular momentum of a particle represents *the torsor of the momentum of the* respective *particle* with respect to the considered pole. We notice that the torsor (6.1.4) may be obtained from the torsor (5.1.16') multiplying it by the mass m. The notion of torsor plays thus – in the problems of dynamics – a rôle analogous to that performed in statics.

1.1.2 Work. Kinetic and potential energy. Conservative forces

The notion of work (mechanical work) has been introduced in Chap. 3, Subsec. 2.1.2, in the form of real elementary work (3.2.3); considering only real displacements, we may omit the adjective "real" in what follows. In this case, *the elementary work of the given forces* is of the form

$$dW = \mathbf{F} \cdot d\mathbf{r} = F_i dx_i, \tag{6.1.5}$$

where $\mathbf{F}$ is the resultant of the given forces which act upon the particle P, which effects a real displacement $d\mathbf{r}$; analogously, *the elementary work of the constraint forces* is given by

$$dW_R = \mathbf{R} \cdot d\mathbf{r} = R_i dx_i, \tag{6.1.5'}$$

where $\mathbf{R}$ is the resultant of the constraint forces applied at the point P. Using the property of distributivity of the scalar product with respect to the addition of vectors, it results that the elementary work of a resultant force is equal to the sum of the elementary works of the component forces which act upon the same particle. The same property allows us to state that the elementary work of a resultant displacement is equal to the sum of the elementary works corresponding to successive component displacements; in this case, if the particle P describes a trajectory C between the points P^0 and P^1, the work of the given force is expressed by

$$W_{\widehat{P^0P^1}} = \int_{\widehat{P^0P^1}} \mathbf{F} \cdot d\mathbf{r} = \int_{\widehat{P^0P^1}} F_i dx_i \,. \tag{6.1.6}$$

The work is a scalar quantity, expressed in *ergs* in the CGS-system or in *joules* in the SI-system (see also Table 1.1).

We notice that we may express the elementary work in the form

$$dW = \mathbf{F} \cdot \mathbf{v} dt = dr \operatorname{pr}_{d\mathbf{r}} \mathbf{F} = F \operatorname{pr}_{\mathbf{F}} d\mathbf{r} \tag{6.1.5''}$$

too, obtaining – in general – a Pfaff form (it is not an exact differential).

The given force may be of the form $\mathbf{F} = \mathbf{F}(\mathbf{r},\dot{\mathbf{r}};t)$; if we know the trajectory $\mathbf{r} = \mathbf{r}(t)$, then the elementary work depends only on time, and the curvilinear integral (6.1.6) becomes a simple one. If the position of the particle and the force $\mathbf{F}$ are specified by a parameter q ($\mathbf{r} = \mathbf{r}(q), \mathbf{F} = \mathbf{F}(q)$), then

$$dW = \mathbf{F}(q) \cdot \mathbf{r}'(q) dq = F_i(q) x_i'(q) dq = Q(q) dq \,, \tag{6.1.5'''}$$

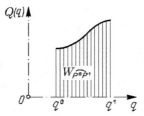

Figure 6.1. Work diagram.

and the work of the given forces is expressed with the aid of a simple integral too; in this case, we may introduce *the work diagram* (Fig.6.1), sometimes useful in practice. If the force $\mathbf{F}$ depends only on the position of the particle ($\mathbf{F} = \mathbf{F}(\mathbf{r};t)$), then the work depends only on the trajectory and is independent on the velocity by which that one is travelled through.

In the case of a *conservative force* $\mathbf{F}$ (which derives from a *simple potential*, being of the form $\mathbf{F} = \operatorname{grad} U$, $U = U(\mathbf{r})$), the elementary work becomes an exact differential

$$dW = dU\,, \tag{6.1.7}$$

the work W depending only on the extreme positions of the particle ($W_{\widehat{P^0P^1}} = U(\mathbf{r}_1) - U(\mathbf{r}_0)$). If the trajectory C is a closed curve, then the corresponding work vanishes ($W_C = 0$).

In the case of a *quasi-conservative force*, which derives from a *simple quasi-potential* $U = U(\mathbf{r};t)$, the elementary work is no more an exact differential, but is of the form

$$dW = \operatorname{grad} U \cdot d\mathbf{r} = dU - \dot{U}dt, \qquad (6.1.8)$$

where we emphasized the partial derivative with respect to time.

In Chap. 1, Subsec. 1.1.12, we have introduced *the generalized potential*

$$U = \mathbf{U} \cdot \mathbf{v} + U_0 = U_j v_j + U_0, \qquad (6.1.9)$$

where $U_0 = U_0(\mathbf{r})$ and $\mathbf{U} = \mathbf{U}(\mathbf{r})$ are *scalar* and *vector potentials*, respectively. The components of *the conservative force*, defined by means of this potential, are of the form

$$F_j = [U]_j, \quad j = 1,2,3, \qquad (6.1.10)$$

where we have put in evidence *the Euler-Lagrange derivative*, corresponding to the index j, given by

$$[U]_j = U_{,j} - \frac{d}{dt}\left(\frac{\partial U}{\partial \dot{x}_j}\right); \qquad (6.1.10')$$

this derivative corresponds to the formula (1.1.88) and allows to express these components by a formula analogous to (1.1.82). Taking into account (6.1.9) and noting that

$$\frac{\partial U}{\partial \dot{x}_j} = \frac{\partial U}{\partial v_j} = U_j,$$

we obtain

$$\mathbf{F} = \operatorname{grad} U - \frac{d\mathbf{U}}{dt} = \operatorname{grad}(\mathbf{U} \cdot \mathbf{v} + U_0) - \frac{d\mathbf{U}}{dt}. \qquad (6.1.11)$$

Replacing the vector potential $\mathbf{U}$ by the vector potential $\overline{\mathbf{U}} = \mathbf{U} + \operatorname{grad}\varphi$, $\varphi = \varphi(\mathbf{r})$, and taking into account the relation $d\varphi/dt = \mathbf{v} \cdot \operatorname{grad}\varphi$, as well as that one may invert the order of application of the operators $(\operatorname{grad}(d/dt) = (d/dt)\operatorname{grad})$ for a function φ of class C^2, we find the same force $\mathbf{F}$; hence, the vector potential is determined abstraction of a field of gradients. The velocity $\mathbf{v}$ being a function only on time, the formula (A.2.31) allows us to write

$$\operatorname{grad}(\mathbf{U} \cdot \mathbf{v}) = (\mathbf{v} \cdot \nabla)\mathbf{U} + \mathbf{v} \times \operatorname{curl} \mathbf{U} = \frac{d\mathbf{U}}{dt} + \mathbf{v} \times \operatorname{curl} \mathbf{U};$$

taking into account (6.1.9), we may express the conservative force **F** in the form

$$\mathbf{F} = \operatorname{grad} U_0 + \mathbf{v} \times \operatorname{curl} \mathbf{U} = \left[\left(U_{k,j} - U_{j,k} \right) v_k + U_{0,j} \right] \mathbf{i}_j. \tag{6.1.11'}$$

The above observation concerning the transformation of a vector potential $\mathbf{U} \to \bar{\mathbf{U}}$ is easily verified ($\operatorname{curl}\operatorname{grad}\varphi = \mathbf{0}$).

In the case of a *generalized quasi-potential*, the two quasi-potentials are of the form $U_0 = U_0(\mathbf{r};t)$ and $\mathbf{U} = \mathbf{U}(\mathbf{r};t)$, respectively; the expression (6.1.11) of *the quasi-conservative force* remains still valid. Noting that $\mathrm{d}\mathbf{U}/\mathrm{d}t = (\mathbf{v} \cdot \nabla)\mathbf{U} + \dot{\mathbf{U}}$, where the point indicates the partial derivative with respect to time, we may express the force **F** in the form

$$\mathbf{F} = \operatorname{grad} U_0 + \mathbf{v} \times \operatorname{curl} \mathbf{U} - \dot{\mathbf{U}}. \tag{6.1.11''}$$

If to the transformation of vector quasi-potential $\mathbf{U} \to \bar{\mathbf{U}}$, considered above, we associate the transformation of scalar quasi-potential $U_0 \to \bar{U}_0 = U_0 + \dot{\varphi}$, hence if we effect the transformation $U \to \bar{U} = U + \mathrm{d}\varphi/\mathrm{d}t$, then the form (6.1.11) (or the form (6.1.11')) of the quasi-conservative force remains invariant; that is a *gauge transformation*.

In the case of a conservative force which derives from a generalized potential, the elementary work is given by

$$\mathrm{d}W = \mathbf{F} \cdot \mathrm{d}\mathbf{r} = \operatorname{grad} U_0 \cdot \mathrm{d}\mathbf{r} + (\mathbf{v}, \operatorname{curl}\mathbf{U}, \mathrm{d}\mathbf{r}),$$

hence it is a total differential

$$\mathrm{d}W = \mathrm{d}U_0. \tag{6.1.12}$$

Thus, the formulae (6.1.11') and (6.1.12) show that the scalar potential U_0 plays the rôle of the simple potential in the frame of the generalized potential (6.1.9), and the vector potential has no contribution in what concerns the elementary work; all the considerations made for the simple potential remain still valid. If the force is quasi-conservative, deriving from a generalized quasi-potential, we obtain

$$\begin{aligned}\mathrm{d}W &= \mathrm{d}U_0 - \dot{U}_0 \mathrm{d}t - \dot{\mathbf{U}} \cdot \mathrm{d}\mathbf{r} = \mathrm{d}U_0 - (\dot{\mathbf{U}} \cdot \mathbf{v} + \dot{U}_0)\mathrm{d}t \\ &= (\operatorname{grad} U_0 - \dot{\mathbf{U}}) \cdot \mathrm{d}\mathbf{r} = \left(U_{0,j} - \dot{U}_j \right) \mathrm{d}x_j;\end{aligned} \tag{6.1.12'}$$

one can easily see that the gauge transformation mentioned above has not one influence on this elementary work too.

We notice also that a conservative force derives always from a generalized or a simple potential as it depends or not explicitly on the velocity of the particle.

The notion of work appeared in XIXth century with the occasion of experiments concerning the transformation of mechanical motion into heat (a non-mechanical form of motion of the matter); one observes that in this transformation there is a constant ratio between the effected work and the obtained heat. The work represents thus a measure of *the mechanical motion* of a particle, which is transformed in a *non-mechanical motion* of it.

As well, the work allows a dynamic measurement of the action of a force. The positive work is a *motive work* W_m, which puts the particle in motion, while the negative work is a *resistent work* W_r for which it is necessary to use up external energy. The work vanishes if the force or the displacement vanish or if they are perpendicular one to the other.

We notice that one may use a quantity of state to characterize the mechanical motion of a particle. Indeed, taking into account Newton's basic law (1.1.89"), we have, successively,

$$\mathrm{d}W = \mathbf{F} \cdot \mathrm{d}\mathbf{r} = \frac{\mathrm{d}}{\mathrm{d}t}(m\mathbf{v}) \cdot \mathrm{d}\mathbf{r} = m\frac{\mathrm{d}\mathbf{v}}{\mathrm{d}t} \cdot \mathrm{d}\mathbf{r} = m\frac{\mathrm{d}\mathbf{r}}{\mathrm{d}t} \cdot \mathrm{d}\mathbf{v} = m\mathbf{v} \cdot \mathrm{d}\mathbf{v}$$

$$= \frac{\mathrm{d}}{\mathrm{d}t}\left(\frac{1}{2}m\mathbf{v}^2\right) = \frac{\mathrm{d}}{\mathrm{d}t}\left(\frac{1}{2}mv^2\right) = \mathrm{d}T,$$

introducing the scalar quantity

$$T = \frac{1}{2}m\mathbf{v}^2 = \frac{1}{2}mv^2, \qquad (6.1.13)$$

called *kinetic energy of the particle* P; this quantity depends on the mass and the velocity of the particle. Taking into account the above considerations concerning the work, it follows that the kinetic energy is a quantity of state of the particle, which measures its mechanical motion and its capacity to be transformed into a non-mechanical motion.

Taking into account the modality of introducing the kinetic energy, we notice that the simple potential U (or the scalar potential U_0) is a quantity of energetical nature; hence, we introduce the function

$$V = -U \qquad (6.1.14)$$

or

$$V = -U_0, \qquad (6.1.14')$$

where $V = V(\mathbf{r})$ is called *potential energy*.

The sum

$$E = T + V \qquad (6.1.15)$$

is called *total mechanical energy* (or *mechanical energy*).

Dynamics of the particle with respect to an inertial frame of reference 359

1.1.3 Power. Mechanical efficiency. Power of non-conservative forces

The quantity

$$P = \frac{dW}{dt} \qquad (6.1.16)$$

is called *power* (*mechanical power*) and is measured in erg/s in the CGS-system or in J/s=W in the SI-system (see Table 1.1); in practice, 1 HP=75 kgf · m/s (HP means horse-power) is used too. Taking into account (6.1.5), we may also write

$$P = \mathbf{F} \cdot \frac{d\mathbf{r}}{dt} = \mathbf{F} \cdot \mathbf{v}. \qquad (6.1.16')$$

This quantity is used to the calibration of motors, engines, apparatuses etc.

Assuming that an engine may be modelled as a particle, the motive work is equal to the resistent one ($W_m = W_r$), in a regime working of it; here and in what follows we consider the work in absolute value (always positive). We notice that the resistent work is formed by the *useful work* W_u, realized by the engine for the goal for which it was built up, and by the *passive* (*lost*) *work* W_p, used up by the passive forces (frictions, various resistent forces etc.); we have thus $W_m = W_u + W_p$, the engine playing the rôle of a transformer of work. To the motive, the useful and the passive work correspond *the motive power* P_m, *the useful power* P_u and *the passive power* P_p ($P_m = P_u + P_p$), respectively, taken – in what follows – also in absolute value (always positive).

We call *mechanical efficiency* the ratio

$$\eta = \frac{W_u}{W_m} = 1 - \frac{W_p}{W_m} = \frac{P_u}{P_m} = 1 - \frac{P_p}{P_m} < 1, \qquad (6.1.17)$$

which is always subunitary (if not, the engine would be a "perpetuum mobile"), being a transfer function of the power; the ratio $W_p / W_m = P_p / P_m$ put in evidence is the *loss factor*.

If we introduce the *force transmission factor* and the *velocity transmission factor* (used in Chap. 5, Subsec. 3.3.3) by relations

$$\lambda_F = \frac{F_m}{F_u}, \quad \lambda_v = \frac{v_m}{v_u}, \qquad (6.1.18)$$

where v_m and v_u are the components of the velocity (in modulus) along the direction of the motive force F_m and of the useful force F_u, respectively, we may write

$$\eta \lambda_F \lambda_v = \frac{P_u}{P_m} \frac{F_m}{F_u} \frac{v_m}{v_u};$$

noting that $P_m = F_m v_m$ and $P_u = F_u v_u$, we get

Theorem 6.1.1 (*the black box law*). *The product of the force transmission factor by the velocity transmission factor and by the mechanical efficiency of an engine is equal to unity*

$$\lambda_F \lambda_v \eta = 1. \tag{6.1.19}$$

Because the transmission factors may be obtained experimentally, this law determines the mechanical efficiency, so that it is not necessary to disassemble the engine for this (as it would be in the interior of a black box). In the case of an *ideal engine* (for which $\eta = 1$, the friction being so small that it can be neglected), we obtain

$$\lambda_F \lambda_v = 1; \tag{6.1.19'}$$

in this case, *the force transmission factor is the inverse of the velocity transmission factor*.

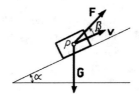

Figure 6.2. Case of inclined plane.

For instance, in the case of the inclined plane (which makes an angle α with the horizontal line (Fig.6.2)) with friction (coefficient of sliding friction $f = \tan\varphi$), considered in Chap. 4, Subsec. 2.1.6, which allows to move up a rigid solid (modelled by a material point P), of weight $\mathbf{G}$, with the aid of a force $\mathbf{F}$ (which makes the angle β with the inclined plane), we have ($F_m = F$, $F_u = G$, $v_m = v\cos\beta$, $v_u = v\sin\alpha$)

$$\eta = \frac{1}{\lambda_F \lambda_v} = \frac{G \sin\alpha}{F \cos\beta};$$

if the inclined plane is used to move up a rigid solid, then we use the relation (4.2.13) and obtain the mechanical efficiency

$$\eta = \frac{\sin\alpha \cos(\beta - \varphi)}{\cos\beta \sin(\alpha + \varphi)} = \frac{\sin\alpha(\cos\beta + f\sin\beta)}{\cos\beta(\sin\alpha + f\cos\alpha)} = \frac{1 + f\tan\beta}{1 + f\cot\alpha}, \tag{6.1.20}$$

while if it is used to move down the heavy bodies, then the relation (4.2.14) leads to

$$\eta = \frac{\sin\alpha \cos(\beta + \varphi)}{\cos\beta \sin(\alpha - \varphi)} = \frac{\sin\alpha(\cos\beta - f\sin\beta)}{\cos\beta(\sin\alpha - f\cos\alpha)} = \frac{1 - f\tan\beta}{1 - f\cot\alpha}. \tag{6.1.20'}$$

If $\beta = 0$, then we obtain

Dynamics of the particle with respect to an inertial frame of reference

$$\eta = \frac{1}{1 \pm f \cot \alpha}, \qquad (6.1.20'')$$

corresponding to the moving up or down of the rigid solid along the inclined plane.

A force which does not derive from a simple quasi-potential may be always written in the form

$$\mathbf{F} = \operatorname{grad} U + \overline{\mathbf{F}}, \quad U = U(\mathbf{r};t), \qquad (6.1.21)$$

where $\overline{\mathbf{F}}$ is a *non-potential force*; the elementary work is given by

$$\mathrm{d}W = \mathbf{F} \cdot \mathrm{d}\mathbf{r} = \operatorname{grad} U \cdot \mathrm{d}\mathbf{r} + \overline{\mathbf{F}} \cdot \mathrm{d}\mathbf{r} = \mathrm{d}U + (\overline{\mathbf{F}} \cdot \mathbf{v} - \dot{U})\mathrm{d}t.$$

In the case in which $\dot{U} = 0$ ($U = U(\mathbf{r})$ is a potential), the quantity $\overline{\mathbf{F}} \cdot \mathbf{v}$ is of the nature of a mechanical power and represents *the power of non-potential forces*. The *non-potential forces* $\overline{\mathbf{F}}$ of vanishing power ($\overline{\mathbf{F}} \cdot \mathbf{v} = 0$) are called *gyroscopic forces* and – obviously – depend on the distribution of velocities; the forces $\mathbf{F}$ are, in this case, *conservative*. If the power of non-potential forces is non-zero, then the forces are *non-conservative*. The non-potential forces of negative power ($\overline{\mathbf{F}} \cdot \mathbf{v} < 0$) are called *dissipative forces*, because – in this case – the energy diminishes (we notice that $U = -V$).

For instance, a force $\overline{\mathbf{F}} = \mathbf{v} \times \operatorname{curl} \mathbf{U}$ (corresponding to the vector potential of (6.1.11')) is a gyroscopic force; we may thus state that *a conservative force which derives from a generalized potential may be expressed in the form of a sum between a force which derives from a simple potential and a gyroscopic force*. As well, a force of the form $\mathbf{F} = \overline{\mathbf{F}} = -\lambda \mathbf{v}$, $\lambda > 0$, which arises in the motion of a particle in a resistent medium, is a dissipative force.

1.1.4 Formulation of problems of a particle in motion

As it is stated by the principle of action of forces, the motion of a free particle P, of mass m, is governed by the differential equation (1.1.89), which is of the form

$$m\ddot{\mathbf{r}} = \mathbf{F}, \qquad (6.1.22)$$

where $\mathbf{F}$ is the resultant of the given forces, applied at point P.

In *the first basic problem* (*the direct problem*) is given the force $\mathbf{F} = \mathbf{F}(\mathbf{r}, \dot{\mathbf{r}};t)$, hence of the form (1.1.93), and the determination of the trajectory, hence of the vector function $\mathbf{r} = \mathbf{r}(t)$, and of the velocity $\mathbf{v}(t) = \dot{\mathbf{r}}$ is asked. The problem is solved by integrating the vector equation (6.1.22) or the scalar equations

$$m\ddot{x}_i = F_i, \quad i = 1,2,3, \qquad (6.1.22')$$

with certain *boundary conditions*. The most times, *initial conditions* (at *the initial moment* $t = t_0$) are put; but one may put also other boundary conditions (e.g., *bilocal conditions*, at the moments $t = t_0$ and $t = t_1$). In certain conditions, the solution of the problem is unique.

In *the second basic problem* (*the inverse problem*) the motion of the particle is known (hence, the position vector $\mathbf{r} = \mathbf{r}(t)$ is given) and the determination of the force $\mathbf{F}$, which induces this motion, is asked. In general, the solution of the problem is not unique.

We mention *the mixed basic problem* too, in which some elements characterizing the motion and the force are given; other unknown elements which specify the motion of the particle and the force applied upon it are asked. Also, this problem has not a unique solution, but only in certain conditions.

From the point of view of mathematical modelling of mechanical phenomena the second basic problem is very important, because it allows to discover laws which are verified by various forces which are encountered in nature (for instance, the force of universal attraction). If a particular motion $\mathbf{r} = \mathbf{r}(t)$, which corresponds to certain given initial conditions, is considered, then we may set up the solution

$$\overline{\mathbf{F}} = \mathbf{F} + (\mathbf{r} - \overline{\mathbf{r}}) \times \mathbf{G}(\mathbf{r}, \dot{\mathbf{r}}, t), \qquad (6.1.23)$$

where $\overline{\mathbf{F}} = m\ddot{\overline{\mathbf{r}}}$, while $\mathbf{G}$ is an arbitrary vector function of class C^0; indeed, the boundary value problem

$$m\ddot{\mathbf{r}} = \overline{\mathbf{F}}, \quad \mathbf{r}(t_0) = \overline{\mathbf{r}}(t_0), \quad \dot{\mathbf{r}}(t_0) = \dot{\overline{\mathbf{r}}}(t_0)$$

is satisfied, while the given problem is indeterminate.

More exactly, the equations (6.1.22') allow to calculate the numerical values of the components F_i, $i = 1, 2, 3$, at any moment t. Hence, if the components of the acceleration are uniform functions of time, then the inverse problem has a unique solution from numerical point of view, but the vector function $\mathbf{F}$ is not univocally determined; indeed, in the expression of the function $\mathbf{F}(\mathbf{r}, \dot{\mathbf{r}}; t)$ as a function of time one may – partially or totally – replace t as function of the components of vectors $\mathbf{r}$ and $\dot{\mathbf{r}}$.

The problem becomes determinate if other conditions are imposed to the expression of the force, e.g.: $\mathbf{F} = \mathbf{F}(t)$, $\mathbf{F} = \mathbf{F}(\mathbf{r})$ or $\mathbf{F} = \mathbf{F}(\dot{\mathbf{r}})$. If the equation of the trajectory is given in the form $\mathbf{r} = \mathbf{r}(t; \mathbf{r}_0, \mathbf{v}_0)$, hence function of the position and the velocity at a given moment, then we may compute $\mathbf{v} = \dot{\mathbf{r}}(t; \mathbf{r}_0, \mathbf{v}_0)$, $\mathbf{F} = m\mathbf{a} = m\ddot{\mathbf{r}}(t; \mathbf{r}_0, \mathbf{v}_0)$; eliminating $\mathbf{r}_0$ and $\mathbf{v}_0$, we are led – in general – to a law of the form $\mathbf{F} = \mathbf{F}(\mathbf{r}, \mathbf{v}; t)$.

The mixed basic problem can be studied analogously.

1.1.5 Equations of motion in curvilinear co-ordinates

Using *the curvilinear co-ordinates* q_1, q_2, q_3, linked to the position vector and to orthogonal Cartesian co-ordinates by relations of the form (A.1.32), (A.1.33), and taking into account Lagrange's formula (5.1.25), which gives the components

Dynamics of the particle with respect to an inertial frame of reference 363

(contravariant components) of the acceleration in the frame specified by the basis' vectors $\mathbf{e}_i$, we may write the law of motion in the form

$$\frac{1}{2}mg^{ij}\left[\frac{\mathrm{d}}{\mathrm{d}t}\left(\frac{\partial v^2}{\partial \dot{q}_j}\right) - \frac{\partial v^2}{\partial q_j}\right] = F_i, \quad i = 1,2,3, \qquad (6.1.24)$$

where F_i are the components (contravariant components) of the given force $\mathbf{F}$.

The results in Chap. 5, Subsec. 1.2.4 lead to

$$m\left(\ddot{r} - r\dot{\theta}^2 - r\sin^2\theta\dot{\varphi}^2\right) = F_r,$$
$$m\left[\frac{1}{r}\frac{\mathrm{d}}{\mathrm{d}t}\left(r^2\dot{\theta}\right) - \frac{r}{2}\sin 2\theta\dot{\varphi}^2\right] = F_\theta, \qquad (6.1.25)$$
$$\frac{m}{r\sin\theta}\frac{\mathrm{d}}{\mathrm{d}t}\left(r^2\sin^2\theta\dot{\varphi}\right) = F_\varphi,$$

in *spherical co-ordinates*, with

$$\mathbf{F} = F_r\mathbf{i}_r + F_\theta\mathbf{i}_\theta + F_\varphi\mathbf{i}_\varphi. \qquad (6.1.25')$$

Analogously, in *cylindrical co-ordinates*, one obtains

$$m\left(\ddot{r} - r\dot{\theta}^2\right) = F_r, \quad \frac{m}{r}\frac{\mathrm{d}}{\mathrm{d}t}\left(r^2\dot{\theta}\right) = F_\theta, \quad m\ddot{z} = F_z, \qquad (6.1.26)$$

where

$$\mathbf{F} = F_r\mathbf{i}_r + F_\theta\mathbf{i}_\theta + F_z\mathbf{i}_z; \qquad (6.1.26')$$

in particular, in *polar co-ordinates* in the plane Ox_1x_2, we may write

$$\mathbf{F} = F_r\mathbf{i}_r + F_\theta\mathbf{i}_\theta, \quad m\left(\ddot{r} - r\dot{\theta}^2\right) = F_r, \quad \frac{m}{r}\frac{\mathrm{d}}{\mathrm{d}t}\left(r^2\dot{\theta}\right) = F_\theta. \qquad (6.1.26'')$$

Using the components (5.1.19) of the acceleration, we obtain *Euler's equations* of motion

$$m\dot{v} = F_\tau, \quad \frac{mv^2}{\rho} = F_\nu, \quad 0 = F_\beta, \qquad (6.1.27)$$

in *intrinsic co-ordinates*, where F_τ, F_ν, F_β are the components of the given force in Frenet's frame of reference. The third of these equations shows that – at any moment – the particle P moves subjected to the action of the force $\mathbf{F}$, so that this one is situated in the osculating plane of the trajectory, corresponding to the position of the particle at the respective moment.

Let us multiply the equations (6.1.24) by g_{ik} and let us sum, taking into account the relation $g_{ik}g^{ij} = \delta_k^j$; let us introduce also the components $Q_k = F_i g_{ik}$ (covariant components) of the force, which are called *generalized forces* and are given by

$$Q_k = \mathbf{F} \cdot \mathbf{e}_k = \mathbf{F} \cdot \frac{\partial \mathbf{r}}{\partial q_k}. \qquad (6.1.28)$$

By means of the kinetic energy (6.1.13), we may write the equations of motion of the particle in the form

$$\frac{\mathrm{d}}{\mathrm{d}t}\left(\frac{\partial T}{\partial \dot{q}_k}\right) - \frac{\partial T}{\partial q_k} = Q_k, \quad k = 1, 2, 3, \qquad (6.1.24')$$

obtaining thus the corresponding *Lagrange's equations*. In this context, the co-ordinates q_k, $k = 1, 2, 3$, are called *generalized co-ordinates* too. Using the Euler-Lagrange derivative, we may also write

$$[T]_k + Q_k = 0, \quad k = 1, 2, 3. \qquad (6.1.24'')$$

Taking into account the conditions in which the formula (5.1.25) has been established, we may state that the equations (6.1.24') are valid also for a movable system of curvilinear co-ordinates, hence for which $\mathbf{r} = \mathbf{r}(q_1, q_2, q_3; t)$. Noting that $\mathbf{v} = \mathbf{e}_i \dot{q}_i + \dot{\mathbf{r}}$, it results that the kinetic energy T may be expressed as a sum of a quadratic form, a linear form and a constant with respect to the quantities $\dot{q}_k$, called *generalized velocities*; in this case, in the equations (6.1.24'), the derivative of maximal order is $\ddot{q}_k$, hence *the generalized acceleration*. These equations are thus of second order, the unknown functions being the generalized co-ordinates $q_k = q_k(t)$, $k = 1, 2, 3$. The boundary conditions are – usually – initial conditions of the form

$$q_k(t_0) = q_k^0, \quad \dot{q}_k(t_0) = \dot{q}_k^0; \qquad (6.1.29)$$

such a boundary value problem in which the generalized co-ordinates and the generalized velocities (hence the position and the velocity of the particle) are given for a certain moment (usually, the initial moment) is called a *Cauchy type problem*. In the particular case in which $\mathbf{r}$ does not depend explicitly on time ($\dot{\mathbf{r}} = \mathbf{0}$), the formula (5.1.22') allows to write the equations of motion in the form

$$m\left(\ddot{q}_i + \begin{Bmatrix} i \\ j \ k \end{Bmatrix} \dot{q}_j \dot{q}_k\right) = F_i, \quad i = 1, 2, 3, \qquad (6.1.24''')$$

where we have introduced Christoffel's symbol of second kind.

1.2 General theorems

In what follows, the theorems of existence and uniqueness are given and stress is put on the notion of first integral. To may obtain first integrals of the equations of motion, one states the universal theorems in the motion of a particle, obtaining the corresponding conservation theorems too; the first integral of areas is put in connection with the notion of central force. We mention also the Galileo-Newton transformation group, which corresponds to the principle of relativity.

1.2.1 Theorems of existence and uniqueness

For the sake of simplicity, we consider the vector equation (6.1.22) or the system of three scalar equations (6.1.22') in orthogonal Cartesian co-ordinates; obviously, the results which will be obtained hold also for the equations of motion in curvilinear co-ordinates.

We replace the system of three differential equations of second order by a system of six differential equations of first order, written in the normal form

$$\dot{x}_i = v_i, \quad \dot{v}_i = \frac{F_i}{m}, \quad i = 1, 2, 3, \qquad (6.1.30)$$

where $v_i = v_i(x_1, x_2, x_3; t)$, $F_i = F_i(x_1, x_2, x_3, v_1, v_2, v_3; t)$. Such a system is called *non-autonomous*; if the time does not intervene explicitly in v_i and F_i, then the system is called *autonomous* (or *dynamic*). We associate to this system the initial conditions (the position and the velocity of the particle at the initial moment)

$$x_i(t_0) = x_i^0, \quad v_i(t_0) = v_i^0, \quad i = 1, 2, 3, \qquad (6.1.30')$$

the boundary value problem (6.1.30), (6.1.30') being thus a Cauchy type problem. The boundary value problem (6.1.22'), (6.1.30') is equivalent to the boundary value one (6.1.30), (6.1.30'); for this latter problem we may prove

Theorem 6.1.2 (*of existence and uniqueness; Cauchy-Lipschitz*). *If the functions* v_i *and* F_i, $i = 1, 2, 3$, *are continuous on the heptadimensional interval* $\mathcal{D}$, *specified by* $x_i^0 - X_i^0 \leq x_i \leq x_i^0 + X_i^0$, $v_i^0 - V_i^0 \leq v_i \leq v_i^0 + V_i^0$, $t_0 - T_0 \leq t \leq t_0 + T_0$, X_i^0, $V_i^0, T_0 = \text{const}$, $i = 1, 2, 3$, *and defined on the Cartesian product of the phase space* (*of canonical co-ordinates* $x_1, x_2, x_3, mv_1, mv_2, mv_3$) *by the time space* (*of co-ordinate* t), *and if Lipschitz's conditions*

$$|v_i(x_1, x_2, x_3; t) - v_i(\overline{x}_1, \overline{x}_2, \overline{x}_3; t)| \leq \frac{1}{\mathcal{T}} \sum_{j=1}^{3} |x_j - \overline{x}_j|,$$

$$|F_i(x_1, x_2, x_3, v_1, v_2, v_3; t) - F_i(\overline{x}_1, \overline{x}_2, \overline{x}_3, \overline{v}_1, \overline{v}_2, \overline{v}_3; t)|$$
$$\leq \frac{m}{\mathcal{T}} \sum_{j=1}^{3} \left(\frac{1}{\tau} |x_j - \overline{x}_j| + |v_j - \overline{v}_j|\right)$$

are verified for $i = 1,2,3$, where $\mathscr{T}$ is a time constant independent of v_i and t, while τ is a time constant equal to unity, then it exists a unique solution $x_i = x_i(t)$, $v_i = v_i(t)$ of the system (6.1.30) which satisfies the initial conditions (6.1.30') and is defined on the interval $t_0 - T \leq t \leq t_0 + T$, where

$$T \leq \min\left(T_0, \frac{X_i^0}{\mathscr{V}}, \tau\frac{V_i^0}{\mathscr{V}}, \mathscr{T}\right), \quad \mathscr{V} = \max\left(v_i, \tau\frac{F_i}{m} \in \mathscr{D}\right).$$

The continuity of functions v_i *and* F_i *on the interval* $\mathscr{D}$ *ensures* the existence *of the solution, after* the theorem of Peano. *For* the uniqueness *of the solution,* the conditions of Lipschitz *must be fulfilled too; the latter conditions may be replaced by other less restrictive conditions, in conformity to which the partial derivatives of first order of functions* v_i *and* F_i, $i = 1,2,3$, *must exist and be bounded in absolute value on the interval* $\mathscr{D}$. Besides, the conditions in Theorem 6.1.2 are *sufficient conditions* of existence and uniqueness which *are not necessary*.

We notice that the existence and the uniqueness of the solution have been put in evidence only on the time interval $[t_0 - T, t_0 + T]$, in the neighbourhood of the initial moment t_0 (besides, the moment t_0 must not be – necessarily – the initial moment, but may be a moment arbitrarily chosen); taking, for instance, $t_0 + T$ as initial moment, it is possible, respecting the above reasoning, to extend the solution on an interval of length $2T_1$ a.s.o., if – certainly – the sufficient conditions of existence and uniqueness of the theorem hold in the neighbourhood of the new initial moment. Thus, we can *prolong* the solution for $t \in [t_1, t_2]$, corresponding to an interval of time in which takes place the considered mechanical phenomenon (even for $t \in (-\infty, \infty)$). Often, even the solution of the boundary value problem is not unique from a mathematical point of view, the principle of inertia may bring the necessary precision for the searched solution, which becomes unique from a mechanical point of vies, as it was shown by V. Vâlcovici (see Chap. 1, Subsec. 1.2.1).

Other theorems emphasize some important properties of the solution; thus, we state **Theorem 6.1.3** (*on the continuous dependence of the solution on a parameter*). *If the functions* $v_i(x_1, x_2, x_3; t, \mu)$, $F_i(x_1, x_2, x_3, v_1, v_2, v_3; t, \mu)$ *are continuous with respect to the parameter* $\mu \in [\mu_1, \mu_2]$ *and satisfy the conditions of the existence and uniqueness theorem, while the constant* $\mathscr{T}$ *of Lipschitz does not depend on* μ, *then the solution* $x_i(t, \mu)$, $v_i(t, \mu)$, $i = 1,2,3$, *of the system* (6.1.30), *which satisfies the conditions* (6.1.30'), *depends continuously on* μ.

Analogously, one may state theorems concerning the continuous dependence of the solution on the initial conditions (that allows to get an approximate solution, fulfilling the initial conditions with a certain approximation) or on several parameters. Concerning the analyticity problem, we mention

Theorem 6.1.4 (*on the analytical dependence of the solution on a parameter*; Poincaré). *The solution* $x_i(t, \mu)$, $v_i(t, \mu)$, $i = 1,2,3$, *of the system* (6.1.30), *which*

satisfies the conditions (6.1.30'), depends analytically on the parameter $\mu \in [\mu_1, \mu_2]$ in the neighbourhood of the value $\mu = \mu_0$ if, in the interval $\mathscr{D} \times [\mu_1, \mu_2]$, the functions v_i and F_i are continuous with respect to t and analytical with respect to x_i, v_i, $i = 1, 2, 3$, and μ.

One may also state

Theorem 6.1.5 (*on the differentiability of the solution*). *If in a neighbourhood of a point* $\mathscr{P}(x_1^0, x_2^0, x_3^0, v_1^0, v_2^0, v_3^0; t_0)$ *the functions* $v_i(x_1, x_2, x_3; t)$ *and* $F_i(x_1, x_2, x_3, v_1, v_2, v_3; t)$, $i = 1, 2, 3$, *are of class* C^k, *the solutions* $x_i(t)$ *and* $v_i(t)$ *of the boundary value problem* (6.1.30), (6.1.30') *are of class* C^{k+1} *in the same neighbourhood.*

The points $\mathscr{P}$ in the neighbourhood of which the boundary value problem (6.1.30), (6.1.30') has not solution or even if the solution exist, this one is not unique, are called *singular points*; the integral curves constituted only of singular points are called *singular curves*, the respective solution being a *singular solution*. In the case of singular points, supplementary conditions are necessary, leading to the choice of one of *the branches of the many-valued solution.*

1.2.2 First integrals. General integral. Constants of integration

We call *integrable combination* of the system (6.1.30) a differential equation which is a consequence of this system, but which can be easily integrated, for instance an equation of the form

$$\mathrm{d}f(x_1, x_2, x_3, v_1, v_2, v_3; t) = 0. \tag{6.1.31}$$

One obtains thus the finite relation

$$f(x_1, x_2, x_3, v_1, v_2, v_3; t) = C, \quad C = \mathrm{const}, \tag{6.1.31'}$$

which links the co-ordinates x_1, x_2, x_3 and the components of the velocity v_1, v_2, v_3 to the time t; the function f which is reduced to a constant along the integral curves is called *first integral* of the system (6.1.30).

If we determine $k \leq 6$ first integrals, for which

$$f_j(x_1, x_2, x_3, v_1, v_2, v_3; t) = C_j, \quad C_j = \mathrm{const}, \quad j = 1, 2, \ldots, k, \tag{6.1.32}$$

the matrix

$$\mathbf{M} \equiv \left[\frac{\partial(f_1, f_2, \ldots, f_k)}{\partial(x_1, x_2, x_3, v_1, v_2, v_3)} \right]$$

being of rank k, then all the first integrals are functionally independent (for the sake of simplicity, further we say *independent first integrals*) and we may express k unknown functions of the system (6.1.32) with respect to the other ones; replacing in (6.1.30), the problem is reduced to the integration of a system of equations with only $6-k$ unknowns (hence, a smaller number of unknowns). If $k = 6$, then all the first integrals are independent, so that the system (6.1.32) of first integrals determines all the unknown functions. We notice that for $k > 6$ the first integrals (6.1.32) are no more independent; we may thus set up at the most six independent first integrals.

Solving the system (6.1.32) for $k = 6$, we obtain (the matrix $\mathbf{M}$ is a square matrix of sixth order for which $\det \mathbf{M} \neq 0$)

$$x_i = x_i(t; C_1, C_2, \ldots, C_6), \quad v_i = v_i(t; C_1, C_2, \ldots, C_6), \quad i = 1, 2, 3, \tag{6.1.33}$$

hence the general integral of the system of equations (6.1.30). Analogously, the general integral of the vector equation (6.1.22) is

$$\mathbf{r} = \mathbf{r}(t; C_1, C_2, \ldots, C_6); \tag{6.1.34}$$

eventually, we have

$$\mathbf{r} = \mathbf{r}(t; \mathbf{K}_1, \mathbf{K}_2). \tag{6.1.34'}$$

Thus, we put in evidence six scalar constants or two vector constants of integration. Because the vector functions (6.1.34) and (6.1.34') verify the equation (6.1.22) for any constants of integration, we may state that the same particle acted upon by the same force, has various possibilities of motion. Imposing the initial conditions (6.1.30'), which may be written also in the vector form

$$\mathbf{r}(t_0) = \mathbf{r}_0, \quad \mathbf{v}(t_0) = \dot{\mathbf{r}}(t_0) = \mathbf{v}_0, \tag{6.1.30''}$$

we get

$$x_i(t_0; C_1, C_2, \ldots, C_6) = x_i^0, \quad v_i(t_0; C_1, C_2, \ldots, C_6) = v_i^0, \quad i = 1, 2, 3;$$

the conditions (6.1.30') being independent, we can write

$$\det\left[\frac{\partial(x_1^0, x_2^0, x_3^0, v_1^0, v_2^0, v_3^0)}{\partial(C_1, C_2, C_3, C_4, C_5, C_6)}\right] \neq 0,$$

so that, on the basis of the theorem of implicit functions, we deduce

$$C_j = C_j(t_0; x_1^0, x_2^0, x_3^0, v_1^0, v_2^0, v_3^0), \quad j = 1, 2, \ldots, 6.$$

Thus, finally, we have

$$x_i = x_i(t; t_0, x_1^0, x_2^0, x_3^0, v_1^0, v_2^0, v_3^0), \quad v_i = v_i(t; t_0, x_1^0, x_2^0, x_3^0, v_1^0, v_2^0, v_3^0),$$
$$j = 1, 2, \ldots, 6, \tag{6.1.35}$$

Dynamics of the particle with respect to an inertial frame of reference

or

$$\mathbf{r} = \mathbf{r}(t; t_0, \mathbf{r}_0, \mathbf{v}_0), \quad \mathbf{v} = \mathbf{v}(t; t_0, \mathbf{r}_0, \mathbf{v}_0). \tag{6.1.35'}$$

Hence, in the frame of the conditions of the theorem of existence and uniqueness, the principle of action of forces (Newton's law) and the principle of initial conditions determine, univocally, the motion of the particle in a finite interval of time; by prolongation, the affirmation may become valid for any t. Thus, *the deterministic aspect of Newtonian mechanics* is put into evidence.

Sometimes, one may perform a computation in two steps. Starting from the system of differential equations of second order (6.1.22'), we can find three integrable combinations leading to three first integrals written in the form

$$\varphi_i(x_1, x_2, x_3, \dot{x}_1, \dot{x}_2, \dot{x}_3; t) = C_i, \quad i = 1, 2, 3; \tag{6.1.36}$$

if, starting from these relations, in a second step, we build up other three integrable combinations, leading to the first integrals

$$\psi_i(x_1, x_2, x_3, ; t; C_1, C_2, C_3) = C_{i+3}, \quad i = 1, 2, 3 \tag{6.1.36'}$$

the problem is solved. Indeed, noting that

$$\det\left[\frac{\partial(\psi_1, \psi_2, \psi_3)}{\partial(x_1, x_2, x_3)}\right] \neq 0,$$

we find the first group of relations (6.1.33).

In the case of a *two-dimensional problem* remain four scalar constants of integration, the solution (6.1.35) being of the form

$$x_\alpha = x_\alpha\left(t; t_0, x_1^0, x_2^0, v_1^0, v_2^0\right), \quad v_\alpha = v_\alpha\left(t; t_0, x_1^0, x_2^0, v_1^0, v_2^0\right), \quad \alpha = 1, 2, \tag{6.1.37}$$

corresponding to the system of equations

$$m\ddot{x}_\alpha = F_\alpha, \quad \alpha = 1, 2; \tag{6.1.37'}$$

as well, in the case of a *unidimensional problem* to which corresponds the equation ($Ox_1 \equiv Ox$)

$$m\ddot{x} = F, \tag{6.1.38}$$

we get the solution

$$x = x(t; t_0, x_0, v_0), \quad v = v(t; t_0, x_0, v_0), \tag{6.1.38'}$$

which introduces only two constants of integration.

1.2.3 Principle of relativity. Galileo-Newton group

We notice that the equation of motion (6.1.22) of the particle P is written in an inertial frame of reference $\mathcal{R}'$ with respect to which we suppose that the basic principles which set up the mathematical model of mechanics hold. Let be a frame $\mathcal{R}$ in a rectilinear and uniform motion of translation, specified by the velocity $\mathbf{v}'_0 = \overrightarrow{\mathrm{const}}$, with respect to the frame $\mathcal{R}'$. Assuming that the motion with respect to the frame $\mathcal{R}$ is a relative motion, while the motion with respect to $\mathcal{R}'$ is an absolute one, the velocity and the acceleration of transportation are given by $\mathbf{v}_t = \mathbf{v}'_0$, $\mathbf{a}_t = \mathbf{0}$, respectively. We notice that the frame $\mathcal{R}$ does not rotate ($\boldsymbol{\omega} = \mathbf{0}$) and use the formulae (5.3.3) and (5.3.5) for the composition of velocities and accelerations, respectively; we may thus write

$$\mathbf{v}' = \mathbf{v} + \mathbf{v}'_0, \quad \mathbf{a}' = \mathbf{a}, \quad \ddot{\mathbf{r}}' = \ddot{\mathbf{r}}, \quad \dot{\mathbf{r}}' = \dot{\mathbf{r}} + \mathbf{v}'_0. \qquad (6.1.39)$$

In the frame $\mathcal{R}$ one obtains the equation of motion

$$m\ddot{\mathbf{r}} = \mathbf{F},$$

which is of the same form as the equation (6.1.22), the acceleration $\ddot{\mathbf{r}}$ and the given force $\mathbf{F} = \mathbf{F}'(\mathbf{r}'(\mathbf{r}), \mathbf{v}'(\mathbf{r}); t) = \mathbf{F}(\mathbf{r}, \dot{\mathbf{r}}; t)$ being – obviously – expressed in the new system of co-ordinates. We may thus state

Theorem 6.1.6 (*of relativity; Galileo*). *If there exists an inertial frame of reference, then there exists an infinity of inertial frames, obtained one from the other by a rectilinear and uniform motion of translation.*

This theorem has been stated by Galileo as a principle and is known as the *principle of relativity* (corresponding to the classical model of mechanics). From the point of view of the mathematical modelling of mechanics, it results that – using only experiments of mechanical character – an observer linked to the frame $\mathcal{R}$ cannot put in evidence the motion with respect to the frame $\mathcal{R}'$. In other words, one cannot determine the absolute motion (with respect to an absolute frame, which – as a matter of fact – does not exist) of a particle, but only abstraction making of a rectilinear and uniform motion; hence, one can emphasize only the relative motion with respect to an inertial frame of reference.

The relation of passing from the frame $\mathcal{R}$ to the frame $\mathcal{R}'$ (hence the transformation $\mathcal{R} \to \mathcal{R}'$) is obtained from (6.1.39), and is given by

$$\mathbf{r}' = \mathbf{r} + \mathbf{v}'_0 t, \quad t' = t. \qquad (6.1.40)$$

Choosing conveniently the two frames $\mathcal{R}$ and $\mathcal{R}'$, without loss of generality of the mechanical phenomenon, we may express the relation of transformation (6.1.40) in the form (we use right-handed orthonormed frames for which the axes $O'x'_2$ and Ox_2, as well as $O'x'_3$ and Ox_3 are parallel, respectively, while the axes $O'x'_1$ and Ox_1 are collinear (Fig.6.3))

Dynamics of the particle with respect to an inertial frame of reference

$$x_1' = x_1 + v_0't, \quad x_2' = x_2, \quad x_3' = x_3; \tag{6.1.40'}$$

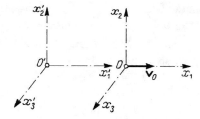

Figure 6.3. Principle of relativity. The Galileo group.

it results that the set of these transformations contains only one essential parameter, that is the magnitude v_0' of the velocity of the frame $\mathcal{R}$ with respect to the frame $\mathcal{R}'$. At the moment $t = 0$ (which can be taken as initial moment) the two frames coincide; we notice that the inverse transformation $\mathcal{R}' \to \mathcal{R}$ takes place too. As well, considering also a frame $\mathcal{R}''$ and the transformation $\mathcal{R}' \to \mathcal{R}''$, of the form $\mathbf{r}'' = \mathbf{r}' + \mathbf{v}_0't'$, $t'' = t'$, we observe that, by composition of the two transformations, we obtain the transformation $\mathcal{R} \to \mathcal{R}''$, of the form $\mathbf{r}'' = \mathbf{r} + \mathbf{v}_0''t$, $t'' = t$, hence a transformation of the same set of transformations, if we accept *the law of composition of velocities* of the relative motion

$$\mathbf{v}_0'' = \mathbf{v}_0 + \mathbf{v}_0'; \tag{6.1.41}$$

in the case of particular frames which have the properties of Fig.6.3, hence in the case of the transformation (6.1.40'), the law of composition of velocities is

$$v_0'' = v_0 + v_0'. \tag{6.1.41'}$$

It results that the set of transformations (6.1.40) forms a group, denoted by Γ and called *the Galileo group*; this group contains three parameters.

We may attach to this group *the group of time translations*, denoted by $\mathcal{T}$ and containing one parameter, specified by

$$t' = t + t_0, \tag{6.1.42}$$

which leads to a change of initial moment (which becomes t_0), as well as *the group of space translations* in E_3, denoted by T, containing three parameters and expressed in the form

$$\mathbf{r}' = \mathbf{r} + \mathbf{r}_0', \tag{6.1.42'}$$

so that the two frames can no more coincide. Combining these transformations, we may write

$$\mathbf{r}' = \mathbf{r} + \mathbf{v}'_0 t + \mathbf{r}'_0, \quad t' = t + t_0. \tag{6.1.42"}$$

We obtain thus a transformation which forms a group with seven parameters; in such a transformation, the axes of the right-handed orthonormed frames $\mathcal{R}$ and $\mathcal{R}'$ remain parallel to themselves. We obtain a rotation of the frame $\mathcal{R}$ with respect to the frame $\mathcal{R}'$ by the transformation (see also Chap. 2, Subsec. 1.1.2)

$$\mathbf{r}' = \boldsymbol{\alpha}\mathbf{r}, \quad x'_i = \alpha_{ij} x_j, \quad i = 1, 2, 3, \tag{6.1.43}$$

which forms *the group of proper (finite) rotations* in E_3, denoted by SO(3) (*the special orthogonal group* in E_3); this group contains only three distinct constants (which – eventually – may be Euler's angles), because the tensor $\boldsymbol{\alpha}$ verifies the six relations of orthogonality (3.1.35).

Finally, starting from the transformations (6.1.40), (6.1.42), (6.1.42') and (6.1.43), we may set up the transformation

$$\mathbf{r}' = \boldsymbol{\alpha}\mathbf{r} + \mathbf{v}_0 t + \mathbf{r}_0, \quad t' = t + t_0, \tag{6.1.44}$$

which forms a group with ten parameters, denoted by $\mathcal{G}$ and called *the Galileo-Newton group*; the groups Γ, $\mathcal{T}$, T and SO(3) are subgroups of the group $\mathcal{G}$. In particular, for $\boldsymbol{\alpha} = 1$ we find again the transformation (6.1.42").

1.2.4 General theorems

Starting from the equation of motion (6.1.22), written with respect to an inertial frame of reference $\mathcal{R}$, considered fixed, we may state some theorems with a general character, consequence of this equation, which are known as *the general (universal) theorems* of the dynamics of the particle.

Taking into account the momentum (6.1.1) and that the mass m of the particle is constant, we may write the equation (6.1.22) in the form

$$\dot{\mathbf{H}} = \frac{d\mathbf{H}}{dt} = \mathbf{F}, \quad \dot{H}_i = F_i, \quad i = 1, 2, 3, \tag{6.1.45}$$

corresponding to the second law of mechanics, so as it was stated by Newton (see also, Chap. 1, Subsec. 1.2.1); we may thus state

Theorem 6.1.7 (*theorem of momentum*). *The derivative with respect to time of the momentum of a free particle is equal to the resultant of the given forces which act upon it.*

This form of Newton's second law is the same as that stated by Einstein in the Special Theory of Relativity.

If we perform a left vector product of the relation (6.1.22) by $\mathbf{r}$ and notice that

$$\frac{d}{dt}[\mathbf{r} \times (m\mathbf{v})] = \dot{\mathbf{r}} \times (m\mathbf{v}) + \mathbf{r} \times (m\dot{\mathbf{v}}) = \mathbf{r} \times (m\ddot{\mathbf{r}}),$$

Dynamics of the particle with respect to an inertial frame of reference

introducing also the moment of momentum (6.1.2), then we may write

$$\dot{\mathbf{K}}_O = \frac{d\mathbf{K}_O}{dt} = \mathbf{r} \times \mathbf{F} = \mathbf{M}_O, \quad \dot{K}_{Oi} = M_{Oi}, \quad i = 1,2,3; \tag{6.1.46}$$

we obtain thus
Theorem 6.1.8 (*theorem of moment of momentum*). *The derivative with respect to time of the moment of momentum of a free particle with respect to a fixed pole is equal to the moment of the resultant of the given forces which act upon it, with respect to the same pole.*

By means of the notion of hodograph, introduced in Chap. 5, Subsec. 1.2.1, we may give a kinematic interpretation to the Theorems 6.1.7 and 6.1.8 too, stating:
Theorem 6.1.7'. *The velocity of a point which describes the hodograph of the momentum of a free particle with respect to a fixed pole is equipollent to the resultant of the given forces which act upon it.*
Theorem 6.1.8'. *The velocity of a point which describes the hodograph of the moment of momentum of a free particle with respect to a fixed pole is equipollent to the moment of the given forces which act upon it, with respect to the same pole.*

The torsor of the momentum, specified by the relation (6.1.4) allows to write

$$\dot{\tau}_O(\mathbf{H}) = \frac{d\tau_O(\mathbf{H})}{dt} = \tau_O(\mathbf{F}), \tag{6.1.47}$$

where we took into account the formulae (6.1.45), (6.1.46); therefore, we state
Theorem 6.1.9 (*torsor's theorem*). *The derivative with respect to time of the torsor of the momentum of a free particle with respect to a fixed pole is equal to the torsor of the resultant of the given forces which act upon it, with respect to the same pole.*

The relation (6.1.45) may be written also in the form

$$d\mathbf{H} = \mathbf{F}dt, \tag{6.1.45'}$$

wherefrom

$$\Delta \mathbf{H} = \mathbf{H}(t_2) - \mathbf{H}(t_1) = \int_{t_1}^{t_2} \mathbf{F}dt; \tag{6.1.45''}$$

the variation of the momentum of a free particle in a finite interval of time is thus emphasized. The quantity $\int_{t_1}^{t_2} \mathbf{F}dt$ represents *the impulse of the given force*, corresponding to the interval of time $[t_1, t_2]$.

As well, another form of the relation (6.1.46) is

$$d\mathbf{K}_O = \mathbf{M}_O dt, \tag{6.1.46'}$$

so that

$$\Delta \mathbf{K}_O = \mathbf{K}_O(t_2) - \mathbf{K}_O(t_1) = \int_{t_1}^{t_2} \mathbf{M}_O dt, \tag{6.1.46''}$$

and *the variation of the moment of momentum of a free particle in a finite interval of time* is put into evidence; the quantity $\int_{t_1}^{t_2} \mathbf{M}_O \mathrm{d}t$ represents *the impulse of the moment of the given force* with respect to the pole O, corresponding to the interval of time $[t_1, t_2]$. Analogously, we get *the variation of the torsor of the momentum of a free particle in a finite interval of time* in the form

$$\Delta \tau_O(\mathbf{H}) = \tau_O \left(\int_{t_1}^{t_2} \mathbf{F} \mathrm{d}t \right). \tag{6.1.47'}$$

The above considerations play an important rôle in the case of an interval of time and – particularly – in the case of discontinuous phenomena.

If we return to the kinetic energy T, introduced in Subsec. 1.1.2, we may write the relation

$$\mathrm{d}T = \mathrm{d}W = \mathbf{F} \cdot \mathrm{d}\mathbf{r} ; \tag{6.1.48}$$

hence, we state

Theorem 6.1.10 (*theorem of kinetic energy*). *The differential of the kinetic energy of a free particle is equal to the elementary work of the resultant of the given forces which act upon it.*

Dividing the relation (6.1.48) by $\mathrm{d}t$ and taking into account (6.1.16'), we can write this theorem in a form closer to that of the previous theorems, i.e.

$$\dot{T} = \frac{\mathrm{d}T}{\mathrm{d}t} = P, \tag{6.1.48'}$$

obtaining thus

Theorem 6.1.10' (*theorem of kinetic energy; second form*). *The derivative with respect to time of the kinetic energy of a free particle is equal to the power of the resultant of the given forces which act upon it.*

We notice that the elementary work is an exact differential only in the case of a conservative force (which derives from a simple or a generalized potential). In general, this work is not a total differential (it is a Pfaff form) and the theorem of kinetic energy is written in the form (for $t \in [t_1, t_2]$)

$$\Delta T = T(t_2) - T(t_1) = T_2 - T_1 = W_{\widehat{P_1 P_2}}$$
$$= \int_{\widehat{P_1 P_2}} \mathbf{F} \cdot \mathrm{d}\mathbf{r} = \int_{t_1}^{t_2} \mathbf{F} \cdot \mathbf{v} \mathrm{d}t = \int_{t_1}^{t_2} P \mathrm{d}t, \tag{6.1.48''}$$

integrating between P_1 and P_2; we may thus state

Theorem 6.1.10'' (*theorem of kinetic energy; finite form*). *The variation of the kinetic energy of a free particle in a finite interval of time is equal to the work of the resultant of the given forces which act upon it in that interval of time.*

The scalar product of relation (6.1.45'') by $\mathbf{v}_2$ leads to

Dynamics of the particle with respect to an inertial frame of reference

$$m\mathbf{v}_2^2 - m\mathbf{v}_1 \cdot \mathbf{v}_2 = \mathbf{v}_2 \cdot \int_{t_1}^{t_2} \mathbf{F} \, dt \, ;$$

introducing the notations

$$T_1 = \frac{1}{2} m\mathbf{v}_1^2, \quad T_2 = \frac{1}{2} m\mathbf{v}_2^2, \quad \Delta T = T_2 - T_1,$$
$$T_0 = \frac{1}{2} m\mathbf{v}_0^2, \quad \mathbf{v}_0 = \mathbf{v}_2 - \mathbf{v}_1, \tag{6.1.49}$$

we may write

$$\Delta T + T_0 = \mathbf{v}_2 \cdot \int_{t_1}^{t_2} \mathbf{F} \, dt, \tag{6.1.50}$$

wherefrom we state

Theorem 6.1.11. *The sum of the variation of the kinetic energy of a free particle in a finite interval of time and the kinetic energy of the variation of the velocity in the same interval of time is equal to the scalar product of the impulse of the resultant of the given forces corresponding to the considered interval of time by the velocity of the particle at the final moment.*

The scalar product of the relation (6.1.45") by $\mathbf{v}_1$ leads – analogously – to

$$\Delta T - T_0 = \mathbf{v}_1 \cdot \int_{t_1}^{t_2} \mathbf{F} \, dt \, ; \tag{6.1.50'}$$

we thus state

Theorem 6.1.11'. *The difference between the variation of the kinetic energy of a free particle in a finite interval of time and the kinetic energy of the variation of the velocity in the same interval of time is equal to the scalar product of the impulse of the resultant of the given forces, corresponding to the considered interval of time, by the velocity of the particle at the initial moment.*

Summing the relations (6.1.50) and (6.1.50') and taking into account the relation (6.1.48"), we get

$$\Delta T = W_{\widehat{P_1 P_2}} = \frac{1}{2} (\mathbf{v}_1 + \mathbf{v}_2) \cdot \int_{t_1}^{t_2} \mathbf{F} \, dt, \tag{6.1.51}$$

and we may state

Theorem 6.1.12 (*Kelvin*). *The work of the resultant of the given forces which act upon a free particle in a finite interval of time (the variation of the kinetic energy of the particle) is equal to the scalar product of the impulse of this resultant, corresponding to the considered interval of time, by the semisum of the velocities of the particle at the initial and the final moment.*

Subtracting the relations (6.1.50) and (6.1.50') one of the other, we may write

$$T_0 = \frac{1}{2} \mathbf{v}_0 \cdot \int_{t_1}^{t_2} \mathbf{F} \, dt \, ; \tag{6.1.51'}$$

we get

Theorem 6.1.12' (*analogous of Kelvin's theorem*). *The kinetic energy of the variation of the velocity of a free particle in a finite interval of time is equal to half of the scalar product of the impulse of the resultant of the given forces, corresponding to the considered interval of time, by the variation of the velocity in the same interval of time.*

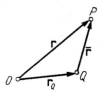

Figure 6.4. General theorems with respect to a movable pole.

The general theorems stated above take place in an inertial frame of reference $\mathscr{R}$, considered fixed; the theorems of moment of momentum and of torsor, which depend on the pole O, maintain their form also with respect to another pole Q, rigidly linked to the frame $\mathscr{R}$ (fixed with respect to this frame). If the pole Q is movable and the calculation is made with respect to the frame $\mathscr{R}$ too, the momentum $\mathbf{H}$ remains invariant, but the moment of momentum and the moment of the resultant of given forces become (Fig.6.4)

$$\dot{\mathbf{K}}_O = \mathbf{r} \times \mathbf{H} = (\mathbf{r}_Q + \bar{\mathbf{r}}) \times \mathbf{H} = \mathbf{K}_Q + \mathbf{r}_Q \times \mathbf{H},$$
$$\mathbf{M}_O = \mathbf{M}_Q + \mathbf{r}_Q \times \mathbf{F};$$

in this case, replacing in relation (6.1.46) and taking into account the relation (6.1.45), we may write

$$\dot{\mathbf{K}}_Q = \frac{\mathrm{d}\mathbf{K}_Q}{\mathrm{d}t} = \mathbf{M}_Q - \mathbf{v}_Q \times \mathbf{H}, \qquad (6.1.52)$$

obtaining thus a generalized form of the theorem of moment of momentum. As a consequence, the formula (6.1.47) is generalized in the form

$$\dot{\tau}_Q(\mathbf{H}) = \frac{\mathrm{d}\tau_Q(\mathbf{H})}{\mathrm{d}t} = \tau_Q(\mathbf{F}) - \{\mathbf{0}, \mathbf{v}_Q \times \mathbf{H}\}. \qquad (6.1.52')$$

1.2.5 Conservation theorems

If the resultant $\mathbf{F}$ of the given forces which act upon the particle P fulfils certain conditions, then the general theorems presented in the previous subsection allow to state some conservation theorems. Thus, if the force $\mathbf{F}$ is parallel to a fixed plane (is normal to a fixed direction of unit vector $\mathbf{u}$ with respect to the frame $\mathscr{R}$ or has a zero component), then the theorem of momentum allows to write

Dynamics of the particle with respect to an inertial frame of reference

$$\dot{\mathbf{H}} \cdot \mathbf{u} = \dot{\mathbf{H}} \cdot \mathbf{u} + \mathbf{H} \cdot \dot{\mathbf{u}} = \frac{\mathrm{d}(\mathbf{H} \cdot \mathbf{u})}{\mathrm{d}t} = \mathbf{F} \cdot \mathbf{u} = 0,$$

wherefrom

$$\mathbf{H} \cdot \mathbf{u} = (m\mathbf{v}) \cdot \mathbf{u} = H_i u_i = m v_i u_i = C, \quad C = \mathrm{const}; \tag{6.1.53}$$

we obtain thus a scalar first integral of the equations of motion. Hence, *if the force* **F** *is parallel to a fixed plane, then the projection of the velocity of the free particle P on the normal to this plane is conserved* (is constant) *in time*; associating a particle P' to the projection of the particle P on this normal, we may state that the particle P' has a uniform motion. Because

$$(m\mathbf{v}) \cdot \mathbf{u} = (m\dot{\mathbf{r}}) \cdot \mathbf{u} = m\frac{\mathrm{d}}{\mathrm{d}t}(\mathbf{r} \cdot \mathbf{u}) = C,$$

it results

$$(m\mathbf{r}) \cdot \mathbf{u} = m x_i u_i = Ct + C', \quad C, C' = \mathrm{const}, \tag{6.1.53'}$$

being thus led to a new scalar first integral, independent of the previous one; the mentioned condition allows us to set up two independent scalar first integrals. As well, *if the force* **F** *has a fixed direction* (is normal to two distinct fixed directions with respect to the frame $\mathscr{R}$ or has two zero components), *then* one obtains four independent scalar first integrals of the form (6.1.53), (6.1.53'), while *the projection of the velocity of the particle on a plane normal to the given force* (determined by the two fixed directions) *is conserved in time*. Eliminating the time between the two first integrals of the form (6.1.53'), we may state that – in this case – the trajectory of the particle P is a plane curve, the support of the force **F** being contained in the plane of the curve too. We may start also from the equation of motion

$$m\ddot{\mathbf{r}} = F(t)\mathbf{u}, \quad \mathbf{u} = \mathrm{vers}\,\mathbf{F} = \overrightarrow{\mathrm{const}},$$

wherefrom one obtains the vector

$$m\mathbf{r} - \mathbf{C}' = \mathbf{u}\int \mathrm{d}t \int F(t)\mathrm{d}t + \mathbf{C}t, \quad \mathbf{C}, \mathbf{C}' = \overrightarrow{\mathrm{const}},$$

as a linear combination of the constant vectors **u** and **C**, hence it is contained in the plane defined by these vectors. Associating a particle P' to the projection of the particle P on a normal to the direction of the force **F** in the considered plane, we notice that this particle has a uniform and rectilinear motion.

Finally, if the force **F** vanishes (is normal to three distinct fixed directions), then we may build up three independent scalar first integrals of the form (6.1.53). Besides, $\mathbf{F} = \mathbf{0}$ leads to $\dot{\mathbf{H}} = \mathbf{0}$, so that

$$\mathbf{H} = m\mathbf{v} = \mathbf{C}, \quad \mathbf{C} = \overrightarrow{\mathrm{const}}, \quad H_i = C_i, \quad i = 1,2,3; \tag{6.1.53''}$$

hence we get a vector first integral, equivalent to three independent scalar first integrals. We may state

Theorem 6.1.13 (*conservation theorem of momentum*). *The momentum (and the velocity) of a free particle is conserved in time if and only if the resultant of the given forces which act upon it vanishes.*

We notice that the relation $m\mathbf{v} = m\dot{\mathbf{r}} = \mathbf{C}$ leads to

$$m\mathbf{r} = \mathbf{C}t + \mathbf{C}', \quad \mathbf{C}, \mathbf{C}' = \overline{\text{const}}, \quad mx_i = C_i t + C_i', \quad i = 1,2,3; \qquad (6.1.53''')$$

we obtain thus a new vector first integral, equivalent to three scalar first integrals. The conservation theorem of momentum allows to set up two independent vector first integrals or six independent scalar first integrals (the maximal number of independent scalar first integrals which can be obtained). The motion of the particle P is thus rectilinear and uniform, being completely determined with respect to the frame $\mathcal{R}$. Besides, this result corresponds to the principle of inertia, which appears thus as a particular case of the principle of action of forces; however, this principle preserves its independence, because it is not necessary to introduce the notion of zero resultant ($\mathbf{F} = 0$), as well as because it can lead to a selection of the solution of the problem with initial conditions (of Cauchy type) if not all the conditions asked by the sufficient theorem of existence and uniqueness are fulfilled. As it was seen above, this principle is not in contradiction with the other principles. We notice also that this principle maintains its form in relativistic mechanics, even if it cannot be deduced by particularizing.

Analogously, if the moment $\mathbf{M}_O$ is contained in a fixed plane (is normal to a fixed axis Δ, $O \in \Delta$, of unit vector $\mathbf{u}$, with respect to a frame $\mathcal{R}$, or has a vanishing component), the theorem of moment of momentum leads to

$$\dot{\mathbf{K}}_O \cdot \mathbf{u} = \frac{\mathrm{d}}{\mathrm{d}t}(\mathbf{K}_O \cdot \mathbf{u}) = \mathbf{M}_O \cdot \mathbf{u} = 0,$$

so that

$$\mathbf{K}_O \cdot \mathbf{u} = (\mathbf{r}, m\mathbf{v}, \mathbf{u}) = K_{Oi} u_i = \epsilon_{ijk}\, m x_i v_j u_k = \overline{C}, \quad \overline{C} = \text{const}, \qquad (6.1.54)$$

resulting thus a new scalar first integral of the equations of motion. The mentioned condition holds *if and only if the force* $\mathbf{F}$ *is coplanar* (concurrent or parallel) *with the axis* Δ ($M_\Delta = 0$); in this case, *the projection of the moment of momentum* $\mathbf{K}_O$ *on the axis* Δ *is conserved in time*. If, in particular, the axis Δ coincides with the axis Ox_3, we have

$$m(x_1 v_2 - x_2 v_1) = m(x_1 \dot{x}_2 - x_2 \dot{x}_1) = \overline{C}, \quad \overline{C} = \text{const}. \qquad (6.1.54')$$

As well, if the moment $\mathbf{M}_O$ has a fixed support (is normal to two distinct fixed axes Δ_1 and Δ_2 with respect to the frame $\mathcal{R}$ or has two zero components), then we obtain two independent scalar first integrals of the form (6.1.54). This condition holds if and only if the force **F** is contained in a fixed plane Π (normal to the direction of the moment $\mathbf{M}_O$ and passing through $O \equiv \Delta_1 \cap \Delta_2$) or the support of the force **F** passes through the point O. Indeed, if the axes Δ_1 and Δ_2 coincide with the axes Ox_1 and Ox_2, respectively, the relations

$$M_{O1} \equiv x_2 F_3 - x_3 F_2 = 0, \quad M_{O2} \equiv x_3 F_1 - x_1 F_3 = 0$$

lead to

$$x_1 M_{O1} + x_2 M_{O2} = -x_3 (x_1 F_2 - x_2 F_1) = -x_3 M_{O3} = 0;$$

hence, we may have $x_3 = 0$ or $M_{O3} = 0$.

In the first case, the field of forces **F** being coplanar, the trajectory of the particle P is a plane curve contained in the plane Π, while the moment of momentum $\mathbf{K}_O$ is normal to this plane. In the second case, $\mathbf{M}_O = \mathbf{0}$ (the moment $\mathbf{M}_O$ is normal to three distinct axes, so that we may set up three independent first integrals of the form (6.1.54). Besides, $\mathbf{M}_O = \mathbf{0}$ leads to $\mathbf{K}_O = \mathbf{0}$, so that

$$\mathbf{K}_O = \mathbf{r} \times (m\mathbf{v}) = \overline{\mathbf{C}}, \quad \overline{\mathbf{C}} = \overrightarrow{\text{const}}, \quad K_{Oi} = \epsilon_{ijk} m x_j v_k = \overline{C}_i, \quad i = 1,2,3, \quad (6.1.54'')$$

hence a vector first integral, equivalent to three scalar first integrals; we may state

Theorem 6.1.14 (*conservation theorem of moment of momentum*). *The moment of momentum of a free particle with respect to a fixed pole is conserved in time if and only if the moment of the resultant of the given forces which act upon it, with respect to the same pole, vanishes.*

We notice that the conservation theorem of momentum entails the conservation theorem of moment of momentum; hence, the vector first integral or the three corresponding scalar first integrals, given by the moment of momentum conservation theorem, are not independent of the six scalar first integrals given by the momentum conservation theorem. In this case, we may write

$$\tau_O(\mathbf{H}) = \overrightarrow{\text{const}} \qquad (6.1.54''')$$

and we may state

Theorem 6.1.15 (*conservation theorem of torsor*). *The torsor of the momentum of a free particle with respect to a fixed pole is conserved in time if and only if the resultant of the given forces which act upon it vanishes.*

Considering only the vector first integrals (6.1.53'') and (6.1.54''), which form the first integral of the torsor (6.1.54'''), one obtains six scalar first integrals (we do not take into account the vector first integral (6.1.53''')), which are not independent, being linked by the relation $\mathbf{C} \cdot \overline{\mathbf{C}} = 0$ (consequence of the relation $\mathbf{H} \cdot \mathbf{K}_O = (m\mathbf{v}, \mathbf{r}, m\mathbf{v}) = 0$).

We notice that, in the frame of the Theorem 6.1.15, the torsor of the force **F** vanishes too. As a matter of fact, the torsor conservation theorem allows to write nine scalar first integrals, from which only six are independent.

If the force **F** is contained in the fixed plane Ox_1x_2, then we have $F_3 = 0$, $M_{O1} = M_{O2} = 0$ and we may build up four first integrals

$$f_1 \equiv v_3 = C_3, \quad f_2 \equiv x_3 - C_3 t = C_3', \quad f_3 \equiv x_2 v_3 - x_3 v_2 = \frac{\overline{C}_1}{m},$$

$$f_4 \equiv x_3 v_1 - x_1 v_3 = \frac{\overline{C}_2}{m}, \quad C_3, C_3', \overline{C}_1, \overline{C}_2 = \text{const},$$

which are independent, because the matrix

$$\left[\frac{\partial(f_1, f_2, f_3, f_4)}{\partial(x_1, x_2, x_3, v_1, v_2, v_3)} \right] = \begin{bmatrix} 0 & 0 & 0 & 0 & 0 & 1 \\ 0 & 0 & 1 & 0 & 0 & 0 \\ 0 & v_3 & -v_2 & 0 & -x_3 & x_2 \\ -v_3 & 0 & v_1 & x_3 & 0 & -x_1 \end{bmatrix}$$

is of rank four (e.g., the determinant of the fourth order formed by the last four columns does not vanish). If we put the initial conditions in the plane Ox_1x_2, then we notice that $C_3 = C_3' = 0$; the other first integrals lead to $\overline{C}_1 = \overline{C}_2 = 0$. The co-ordinates $x_1 = x_1(t)$ and $x_2 = x_2(t)$ and the components of the velocity $v_1 = v_1(t)$, $v_2 = v_2(t)$ remain to be determinate. Thus, the theorems of linear and angular momentum may give independent first integrals; but such a result cannot be obtained always. Thus, if the force **F** is parallel to the fixed axis Ox_3 (without having – necessarily – a fixed support), then we have $F_1 = F_2 = 0$, $M_{O3} = 0$ and we may set up five first integrals

$$f_1 \equiv v_1 = C_1, \quad f_2 \equiv x_1 - C_1 t = C_1', \quad f_3 \equiv v_2 = C_2, \quad f_4 \equiv x_2 - C_2 t = C_2',$$

$$f_5 \equiv x_1 v_2 - x_2 v_1 = \frac{\overline{C}_3}{m}, \quad C_1, C_1', C_2, C_2', \overline{C}_3 = \text{const};$$

one may easily see that the last first integral is a consequence of the four first integrals, having only four independent first integrals.

Using the considerations made in Subsec. 1.1.2, let us suppose that the resultant **F** of the given forces is a conservative force which derives from a simple or generalized potential; in this case, the elementary work is a total differential and the formula (6.1.48) leads to

$$T = U + h \tag{6.1.55}$$

or to

$$T = U_0 + h, \tag{6.1.55'}$$

where $h = \text{const}$, hence to a scalar first integral. Introducing the potential energy (6.1.14) or (6.1.14') and the mechanical energy (6.1.15), we may write

$$E = T + V = h, \quad h = \text{const} \tag{6.1.55''}$$

too, obtaining thus

Theorem 6.1.16 (*mechanical energy conservation theorem*). *The mechanical energy of a free particle is conserved in time if and only if the resultant of the given forces which act upon it is conservative.*

Thus, the denomination given to these forces, which form a conservative field, is justified; h is *the energy constant*.

The mechanical energy conservation theorem allows to determine the magnitude of the velocity of the particle without knowing its trajectory; we may thus write

$$v^2 = \frac{2}{m}[h - V(\mathbf{r})]. \tag{6.1.56}$$

We notice that – in particular – the motion of a particle constrained to stay on an equipotential surface and acted upon by the corresponding conservative force is uniform.

The constants corresponding to the first integrals introduced above may be determined observing that the latter ones are conserved in time (the respective functions have the same value at any moment t, inclusive at the initial moment $t = t_0$). Thus, in the case of the linear momentum conservation theorem it results $\mathbf{C} = m\mathbf{v}_0$, in the case of the angular momentum conservation theorem we have $\overline{\mathbf{C}} = \mathbf{r}_0 \times (m\mathbf{v}_0)$, while in the case of the mechanical energy conservation theorem we may write $h = mv_0^2/2 + V(\mathbf{r}_0)$.

1.2.6 Theorem of areas. Central forces

Starting from the velocities torsor (5.1.16'), multiplying by the mass m and taking into account (6.1.1), (6.1.2), we find again the torsor of momentum (6.1.4); multiplying – analogously – the accelerations torsor (5.1.20') by m and taking into account (6.1.45), (6.1.46'), we obtain the torsor theorem (6.1.47). We can thus see that the areal velocity and acceleration play for the angular momentum and its derivative a rôle similar to that played by the velocity and the acceleration for the linear momentum and its derivative, respectively. Thus, if the force $\mathbf{F}$ is coplanar (concurrent or parallel) with an axis Δ of unit vector $\mathbf{u}$ ($M_\Delta = 0$), then the projection of the areal velocity $\mathbf{\Omega}_O$ of the free particle P on the axis Δ is constant in time

$$\mathbf{\Omega}_O \cdot \mathbf{u} = \frac{1}{2}(\mathbf{r}, \mathbf{v}, \mathbf{u}) = \Omega_{Oi} u_i = \frac{1}{2} \epsilon_{ijk} x_i v_j u_k = C, \quad C = \text{const}, \tag{6.1.57}$$

obtaining a scalar first integral equivalent to (6.1.54); associating a particle P' to the projection of the particle P on a plane Π normal to the axis Δ, we may state that the areal velocity of the particle P' with respect to the trace of the axis Δ on the plane Π is conserved in time. Assuming that the axis Δ coincides with the axis Ox_3, we may write

$$\Omega_{O3} = \frac{1}{2}(x_1 v_2 - x_2 v_1) = \frac{1}{2}(x_1 \dot{x}_2 - x_2 \dot{x}_1) = C, \quad C = \text{const}. \tag{6.1.57'}$$

As well, if the moment $\mathbf{M}_O$ has a fixed support, then we obtain two independent scalar first integrals of the form (6.1.57). If the force $\mathbf{F}$ is contained in a fixed plane Π, then the trajectory of the particle P is a plane curve, while the moment of momentum is of the form $\mathbf{K}_O = K_O(t)\mathbf{u}$, where $\mathbf{u} = \text{vers}\,\mathbf{K}_O$ is normal to the plane Π; one can easily see that

$$\mathbf{r} \cdot \mathbf{K}_O = \mathbf{r} \cdot [\mathbf{r} \times (m\mathbf{v})] = K_O(t)\mathbf{u} \cdot \mathbf{r} = 0,$$

hence $\mathbf{u} \cdot \mathbf{r} = 0$, so that this is the equation of the plane Π, which passes through O.

The condition $\mathbf{M}_O = \mathbf{0}$ is verified if and only if the support of the given force $\mathbf{F}$ passes through the pole O; such a force is called *central force*. In such a case, the moment of momentum conservation theorem takes place; we also obtain

$$\mathbf{\Omega}_O = \frac{1}{2}\mathbf{r} \times \mathbf{v} = \frac{1}{2}\mathbf{C}, \quad \mathbf{C} = \overrightarrow{\text{const}}, \quad \Omega_{Oi} = \frac{1}{2}\epsilon_{ijk} x_j v_k = \frac{1}{2}C_i, \quad i = 1,2,3, \tag{6.1.57''}$$

hence, a vector first integral, equivalent to three scalar first integrals, and we may state **Theorem 6.1.17** (*areal velocity conservation theorem*). *The areal velocity of a free particle with respect to a fixed pole is conserved in time if and only if the resultant of the given forces which act upon it is a central force (its support passes – permanently – through the same pole).*

In the latter case $\mathbf{\Omega}_O \cdot \mathbf{r} = (1/2)(\mathbf{r},\mathbf{v},\mathbf{r}) = (1/2)\mathbf{C} \cdot \mathbf{r}$; hence, the trajectory of the particle P is a plane curve, contained in the plane

$$\mathbf{C} \cdot \mathbf{r} = 0, \quad \mathbf{C} = \overrightarrow{\text{const}} \neq \mathbf{0}, \tag{6.1.58}$$

which passes through the pole O. If $\mathbf{C} = \mathbf{0}$, then it results that $\mathbf{r}$ and $\mathbf{v}$ are collinear vectors; the equation $\dot{\mathbf{r}} = \lambda(t)\mathbf{r}$, λ scalar, leads to

$$\mathbf{r} = \mathbf{r}_0 e^{\int_{t_0}^{t} \lambda(\tau)\,d\tau}, \tag{6.1.58'}$$

the trajectory of the particle being rectilinear. We notice that $\mathbf{C} = \mathbf{r}_0 \times \mathbf{v}_0$. As a conclusion, if the moment $\mathbf{M}_O$ has a fixed support, then the trajectory of the particle P is rectilinear or a plane curve (contained in the plane Π) as the vectors $\mathbf{r}_0$ and $\mathbf{v}_0$ of the initial conditions have or not the same support.

Dynamics of the particle with respect to an inertial frame of reference

Assuming that the plane Π coincides with the plane Ox_1x_2, the particle P describes the curve C, the trajectory of the particle P' being the curve C', the projection of the curve C on the plane Π (Fig.6.5); if the trajectory C is a plane curve, in the conditions mentioned above, then $P' \equiv P$. We may write

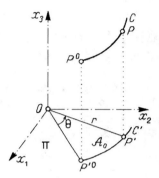

Figure 6.5. Areas theorem.

$$\Omega_{O3} = \frac{1}{2}(x_1\dot{x}_2 - x_2\dot{x}_1) = \frac{1}{2}r^2\dot{\theta} = \frac{1}{2}C, \qquad (6.1.59)$$

where we used polar co-ordinates in the plane Π; the first integral $r^2\dot{\theta} = C$ is called also the areas first integral. Noting that $\Omega_{O3} = \mathrm{d}\mathcal{A}_O/\mathrm{d}t$, where we have attached the sign +, to the area $\mathcal{A}_O$, corresponding to a positive rotation in the plane Π, we get

$$\mathcal{A}_O = \frac{1}{2}C(t - t_0), \qquad (6.1.59')$$

and are led to

Theorem 6.1.18 (*areas theorem*). *The area described by the vector radius of a free particle, beginning with its initial position, is proportional to the interval of time covered if and only if the resultant of the given forces which act upon it is a central force.*

We can also say that the vector radius describes equal areas in equal times. The constant C is called *the areas constant*. We mention that this theorem has been stated in the same conditions in which the areal velocity conservation theorem takes place; but it can be applied to the particle P' too (Fig.6.5), in the case in which we may write only a single scalar first integral of the form (6.1.57).

Taking into account the observations made at the beginning of this subsection, the angular momentum theorem allows to write

$$2m\dot{\mathbf{\Omega}}_O = \mathbf{M}_O \qquad (6.1.60)$$

and we may state

Theorem 6.1.19 (*areal acceleration theorem*). *The product of the double mass of a free particle by its areal acceleration with respect to a fixed pole is equal to the moment of the resultant of the given forces which act upon it, with respect to the same pole.*

Hence, the equation (6.1.60) plays – with respect to the moment $\mathbf{M}_O$ – the same rôle as that played by Newton's equation (1.1.89) with respect to the force $\mathbf{F}$.

2. Dynamics of the particle subjected to constraints

Assuming that the particle is subjected to ideal constraints or to constraints with friction, we complete – in what follows – the general and conservation theorems previously stated and emphasize the differential principles of mechanics for a particle. Using then the general results thus obtained, we study the motion of a particle constrained to stay on a curve or on a surface; the case of constraints with friction is also taken into consideration. By means of the generalized co-ordinates, we may study the motion of a particle with only one degree of freedom in the conservative case, as well as in a dissipative case.

2.1 General considerations

The case of one or two ideal holonomic (rheonomic or scleronomic) constraints is considered and the equation of motion, the general and the conservation theorems are completed by introducing the constraint forces. We mention also the presentation of other differential principles of mechanics (d'Alembert, d'Alembert-Lagrange), equivalent to Newton's differential principle.

2.1.1 Ideal constraints. Introductory notions

We introduced in Chap. 3, Sec. 2.2 the notion of constraint, together with its multiple implications. Using the considerations made on this occasion, we assume that a particle may be subjected to two holonomic (finite, of geometric nature) bilateral constraints of the form (3.2.17); if a third constraint, compatible with the first two ones, would appear, then the position of the particle would be specified from geometric point of view (uniqueness or not). The case of non-holonomic constraints of the form (4.1.37') will be studied subsequently by analytical methods. As well, we admit the existence of unilateral constraints of the form (3.2.8') in some particular problems.

If the particle is constrained to stay on a curve C, then its co-ordinates must verify the relations

$$f_1(x_1, x_2, x_3; t) = 0, \quad f_2(x_1, x_2, x_3; t) = 0 \tag{6.2.1}$$

or the relations

$$f_1(x_1, x_2, x_3) = 0, \quad f_2(x_1, x_2, x_3) = 0, \tag{6.2.1'}$$

as the constraint is rheonomic or scleronomic, respectively; in the first case, the curve is movable, while in the second one it is fixed. The particle remains with only one degree of freedom and its position may be specified by means of a single independent parameter q (obtained by eliminating the two considered constraints) in the form

$$\mathbf{r} = \mathbf{r}(q;t), \quad x_i = x_i(q;t), \quad i = 1,2,3, \quad q = q(t), \qquad (6.2.2)$$

in the general case of a rheonomic constraint. In particular, *the generalized co-ordinate* q may be the curvilinear co-ordinate s on the curve C or even the time t.

Analogously, to determine the positions of a particle constrained to stay on a surface S of equation

$$f(x_1, x_2, x_3; t) = 0 \qquad (6.2.3)$$

or

$$f(x_1, x_2, x_3) = 0, \qquad (6.2.3')$$

as the constraint is non-stationary (movable surface) or stationary (fixed surface), respectively, there are necessary two parameters q_1, q_2 (obtained by the elimination of the considered constraint), corresponding to two degrees of freedom; it results

$$\mathbf{r} = \mathbf{r}(q_1, q_2; t), \quad x_i = x_i(q_1, q_2; t), \quad i = 1,2,3, \quad q_\alpha = q_\alpha(t), \quad \alpha = 1,2. \qquad (6.2.4)$$

The generalized co-ordinates q_1, q_2, may – eventually – be the co-ordinates on the surface S.

Using the axiom of liberation from constraints, we introduce the constraint force $\mathbf{R}$ so that, corresponding to the formula (3.2.35'), the equation of motion (6.2.2) becomes

$$m\mathbf{a} = m\ddot{\mathbf{r}} = \mathbf{F} + \mathbf{R}, \quad ma_i = m\ddot{x}_i = F_i + R_i, \quad i = 1,2,3; \qquad (6.2.5)$$

in such conditions, the particle behaves as a free one. If the constraints are ideal, then one obtains the virtual work $\delta W_R = \mathbf{R} \cdot \delta \mathbf{r} = 0$; in the cases considered above, the constraint force $\mathbf{R}$ is normal to the surface S, hence it is of the form

$$\mathbf{R} = \lambda \operatorname{grad} f, \quad R_i = \lambda f_{,i}, \quad i = 1,2,3, \quad \lambda \text{ scalar}, \qquad (6.2.6)$$

or it belongs to the normal plane to the curve C, hence it is of the form

$$\mathbf{R} = \lambda_1 \operatorname{grad} f_1 + \lambda_2 \operatorname{grad} f_2, \quad R_i = \lambda_1 f_{1,i} + \lambda_2 f_{2,i}, \quad i = 1,2,3, \quad \lambda_1, \lambda_2 \text{ scalars}, \qquad (6.2.6')$$

at the point occupied by the particle (corresponding to the formula (3.2.37)). A constraint with friction implies a constraint force tangent to the surface S or to the curve C, which is determined by a supplementary modelling of the mechanical phenomenon.

We observe that all the results obtained in the case of a free particle may be used also for a particle subjected to constraints if to the resultant $\mathbf{F}$ of the given forces is added also the resultant **R** (unknown) of the constraint forces. In the first basic problem, besides the trajectory (the vector function $\mathbf{r} = \mathbf{r}(t)$) is asked also the constraint force **R** (which involves – in general – three unknown scalar components). In the second basic problem, one must determine the forces **F** and **R** ; as in the case $\mathbf{R} = \mathbf{0}$, the problem has not a unique solution. As well, neither in the case of the mixed basic problem the solution is not unique.

In what concerns the theorem of existence and uniqueness, they remain further valid if the functions (6.2.1') and (6.2.3') are of class C^1 and their derivatives of first order fulfil conditions of Lipschitz type.

2.1.2 General theorems

Corresponding to the results in Subsec. 1.2.4, we may write

$$\dot{\mathbf{H}} = \mathbf{F} + \mathbf{R}, \quad \dot{H}_i = F_i + R_i, \quad i = 1,2,3, \tag{6.2.7}$$

$$\dot{\mathbf{K}}_O = \mathbf{r} \times (\mathbf{F} + \mathbf{R}) = \mathbf{M}_O + \bar{\mathbf{M}}_O, \quad \dot{K}_{Oi} = M_{Oi} + \bar{M}_{Oi}, \quad i = 1,2,3, \tag{6.2.7'}$$

thus stating:

Theorem 6.2.1 (*theorem of momentum*). *The derivative with respect to time of the momentum of a particle subjected to constraints is equal to the resultant of the given and constraint forces which act upon it.*

Theorem 6.2.2 (*theorem of moment of momentum*). *The derivative with respect to time of the moment of momentum of a particle subjected to constraints, with respect to a fixed pole, is equal to the moment of the resultant of the given and constraint forces which act upon it, with respect to the same pole.*

Introducing the notion of hodograph, we may state theorems analogous to Theorems 6.1.7' and 6.1.8'.

Noting that

$$\dot{\tau}_O(\mathbf{H}) = \tau_O(\mathbf{F}) + \tau_O(\mathbf{R}), \tag{6.2.7''}$$

we may state

Theorem 6.2.3 (*theorem of torsor*). *The derivative with respect to time of the torsor of momentum of a particle subjected to constraints, with respect to a fixed pole, is equal to the torsor of the resultant of the given and constraint forces which act upon it, with respect to the same pole.*

Introducing *the impulse of the constraint force* $\int_{t_1}^{t_2} \mathbf{R} \mathrm{d}t$ and *the impulse of the moment of the constraint force* $\int_{t_1}^{t_2} \bar{\mathbf{M}}_O \mathrm{d}t$, corresponding to the interval of time $[t_1, t_2]$, we may write

$$\Delta \mathbf{H} = \int_{t_1}^{t_2} \mathbf{F} \mathrm{d}t + \int_{t_1}^{t_2} \mathbf{R} \mathrm{d}t, \tag{6.2.8}$$

$$\Delta \mathbf{K}_O = \int_{t_1}^{t_2} \mathbf{M}_O \mathrm{d}t + \int_{t_1}^{t_2} \bar{\mathbf{M}}_O \mathrm{d}t, \tag{6.2.8'}$$

$$\Delta\boldsymbol{\tau}_O(\mathbf{H}) = \boldsymbol{\tau}_O\left(\int_{t_1}^{t_2} \mathbf{F}\,\mathrm{d}t\right) + \boldsymbol{\tau}_O\left(\int_{t_1}^{t_2} \mathbf{R}\,\mathrm{d}t\right). \tag{6.2.8''}$$

The relation (6.1.48) is completed in the form

$$\mathrm{d}T = \mathrm{d}W + \mathrm{d}W_R = \mathbf{F}\cdot\mathrm{d}\mathbf{r} + \mathbf{R}\cdot\mathrm{d}\mathbf{r} \tag{6.2.9}$$

and we may state

Theorem 6.2.4 (*theorem of kinetic energy*). *The differential of the kinetic energy of a particle subjected to constraints is equal to the sum of the elementary works of the resultants of the given and constraint forces which act upon it.*

As it was seen in Chap. 3, Subsec. 2.2.9, in the case of scleronomic (or – more general – catastatic) constraints, we have $\mathrm{d}W_R = 0$; in this case, the Theorem 6.2.4, corresponding to a particle subjected to constraints, is of the same form as the Theorem 6.1.10, corresponding to a free particle. In what concerns Theorems 6.1.10' and 6.1.10", one can make analogous observations.

We notice that, if we take the moment of momentum with respect to a pole Q, movable with respect to the origin O, the formula (6.1.52) leads to

$$\dot{\mathbf{K}}_Q = \mathbf{M}_Q + \overline{\mathbf{M}}_Q - \mathbf{v}_Q \times \mathbf{H}. \tag{6.2.10}$$

2.1.3 Conservation theorems

Using the results given in Subsec. 1.2.5, we may build up first integrals, in certain conditions, in the case of a particle subjected to constraints too. Thus, if the sum $\mathbf{F} + \mathbf{R}$ is parallel to a fixed plane (is normal to a fixed direction of unit vector $\mathbf{u}$, $(\mathbf{F} + \mathbf{R})\cdot\mathbf{u} = 0$), then we may write the first integral (6.1.53); analogously, if the sum $\mathbf{M}_O + \overline{\mathbf{M}}_O$ is contained in a fixed plane (is normal to a fixed axis Δ, $O \in \Delta$, of unit vector $\mathbf{u}$, $(\mathbf{M}_O + \overline{\mathbf{M}}_O)\cdot\mathbf{u} = 0$), then one obtains the first integral (6.1.54).

If $\mathbf{F} + \mathbf{R} = \mathbf{0}$ (necessary and sufficient condition of static equilibrium), then we may state a *momentum conservation theorem*, while if $\mathbf{M}_O + \overline{\mathbf{M}}_O = \mathbf{0}$ (necessary condition of static equilibrium), then we may state a *moment of momentum conservation theorem*. The first condition mentioned above allows to state a *torsor conservation theorem* too.

In the case of scleronomic constraints and of a conservative force we may write also a *mechanical energy conservation theorem*.

The moment of momentum conservation theorem is equivalent to the areal velocity conservation theorem; these theorems take place if and only if the sum $\mathbf{F} + \mathbf{R}$ passes through the fixed pole O (it is a central force).

As in the case of a free particle, one can obtain only six independent first integrals; but these ones are sufficient to determine the motion. Even if the constraint force $\mathbf{R}$ is not known a priori, the conditions imposed above are often fulfilled, and we may set up first integrals in the case of the particle subjected to constraints too.

2.1.4 Differential principles of mechanics

The equation of motion of a particle P subjected to constraints is written in the form (6.2.5); this equation represents *Newton's principle* (the first differential principle of mechanics). But this basic principle can be expressed also in other equivalent forms, which are useful in various particular cases; if we consider these forms as consequences of Newton's principle, then they will be theorems.

Introducing *the force of inertia*

$$\overline{\mathbf{F}} = -m\ddot{\mathbf{r}}, \qquad (6.2.11)$$

the law of motion becomes

$$\mathbf{F} + \overline{\mathbf{F}} + \mathbf{R} = \mathbf{0} \qquad (6.2.12)$$

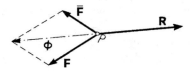

Figure 6.6. D'Alembert's theorems.

and we may state (Fig.6.6)

Theorem 6.2.5 (*d'Alembert*). *The motion of a particle subjected to constraints takes place so that – at any moment – it is in dynamic equilibrium under the action of the resultant of the given and constraint forces, as well as of the force of inertia.*

We introduce the force

$$\mathbf{\Phi} = \mathbf{F} + \overline{\mathbf{F}} = \mathbf{F} - m\ddot{\mathbf{r}}, \qquad (6.2.13)$$

which is called *the lost force of d'Alembert*; in this case, the equation becomes (Fig.6.6)

$$\mathbf{\Phi} + \mathbf{R} = \mathbf{0} \qquad (6.2.14)$$

and we can state

Theorem 6.2.6 (*d'Alembert*). *The motion of a particle subjected to constraints takes place so that the constraint force be equilibrated – at any moment – by the lost force of d'Alembert.*

We notice that the relation (6.2.13) may be written also in the form

$$\mathbf{F} = \mathbf{\Phi} + (-\overline{\mathbf{F}}) = \mathbf{\Phi} + m\ddot{\mathbf{r}};$$

hence, only the component $m\ddot{\mathbf{r}} = -\overline{\mathbf{F}}$ of the force $\mathbf{F}$ contributes to the motion of the particle, while the component $\mathbf{\Phi}$ is lost because it equilibrates the constraint force (justifying thus the given denomination).

Formally, the equation (6.2.14), which represents the necessary and sufficient condition for dynamic equilibrium (characterizing – entirely – the motion of the particle subjected to constraints), is not different from the relation (4.1.4), which represents the

Dynamics of the particle with respect to an inertial frame of reference 389

necessary and sufficient condition of static equilibrium. As a consequence, all the considerations made for static problems, starting form the relation (4.1.4), may be transposed for similar problems of dynamic character, replacing the given force **F** by the lost force of d'Alembert $\boldsymbol{\Phi}$; e.g., the condition (4.1.56), written for only one particle, leads to the theorem of torsor, characterized by the formula (6.2.7''). As well, we may use the results in Chap. 4, Subsecs. 1.1.5, 1.1.6, 1.1.8, corresponding to the particle constrained to stay on a surface or on a curve.

If $\delta W_R = 0$, then we may write the relation (4.1.58) for a single particle in the form

$$\delta W = \boldsymbol{\Phi} \cdot \delta \mathbf{r} = 0, \tag{6.2.15}$$

stating

Theorem 6.2.7 (*theorem of virtual work; d'Alembert-Lagrange*). *The motion of a particle subjected to ideal constraints takes place so that the virtual work of the lost force of d'Alembert, which acts upon the particle, vanishes for any virtual displacement of it.*

In the case of unilateral ideal constraints of the form (3.2.16^{iv}) or of the form (3.2.16), the virtual work of the lost force of d'Alembert verifies the relation

$$\delta W = \boldsymbol{\Phi} \cdot \delta \mathbf{r} \leq 0. \tag{6.2.15'}$$

We notice that each of the above theorems may stay at the basis of the Newtonian mathematical model of mechanics, representing thus a differential principle of mechanics.

2.2 Motion of the particle with one or two degrees of freedom

In what follows, we consider the motion of a particle (frictionless or with friction), constrained to stay on a curve or on a surface. A study is then made for the case in which the particle has only one degree of freedom (conservative or dissipative case).

2.2.1 Motion of a particle constrained to stay on a curve

Let P be a particle in motion on a smooth movable or fixed curve C (Fig.6.7) of equations (6.2.1) or (6.2.1'). The constraint force will be expressed in the form (6.2.6'), while the equation of motion will be

$$m\ddot{\mathbf{r}} = \mathbf{F} + \lambda_1 \operatorname{grad} f_1 + \lambda_2 \operatorname{grad} f_2; \tag{6.2.16}$$

in components, we may write

$$m\ddot{x}_i = F_i + \lambda_1 f_{1,i} + \lambda_2 f_{2,i}, \quad i = 1, 2, 3. \tag{6.2.16'}$$

The equations (6.2.1) and (6.2.16') form a system of five scalar equations for the unknown functions $x_i = x_i(t)$ and for the parameters λ_1 and λ_2, which specify the constraint force.

We notice that, in the case of the rheonomic constraints,

$$\operatorname{grad} f_\alpha \cdot \mathrm{d}\mathbf{r} + \dot{f}_\alpha \mathrm{d}t = 0, \quad \alpha = 1,2,$$

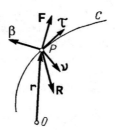

Figure 6.7. Motion of a particle constrained to stay on a curve C.

the real work of the constraint forces is written in the form

$$\mathrm{d}W_R = \mathbf{R} \cdot \mathrm{d}\mathbf{r} = \lambda_1 \operatorname{grad} f_1 \cdot \mathrm{d}\mathbf{r} + \lambda_2 \operatorname{grad} f_2 \cdot \mathrm{d}\mathbf{r} = -\left(\lambda_1 \dot{f}_1 + \lambda_2 \dot{f}_2\right) \mathrm{d}t,$$

being non-zero; indeed, taking into account the variation in time of the curve, the real displacement of the particle is not tangent to the curve frozen at the moment t, in the normal plane of which is the constraint force $\mathbf{R}$.

If the constraint is scleronomic or, at least, catastatic, then the work $\mathrm{d}W_R = 0$; in this case, the curve C is fixed and the trajectory coincides with it, the constraint force belonging to the normal plane to the trajectory at the point P. In the theorem of kinetic energy (6.2.9), the constraint force does no more appear, and we may specify the position of the particle on the trajectory. The parametric representation (6.2.2) becomes $x_i = x_i(q)$, $i = 1,2,3$, so that we obtain

$$\frac{\mathrm{d}}{\mathrm{d}t}\left(\frac{m}{2} x'_i x'_i \dot{q}^2\right) = Q(q,\dot{q};t)\dot{q}, \tag{6.2.17}$$

where $x'_i = \mathrm{d}x_i(q)/\mathrm{d}q$, while $Q(q,\dot{q};t)\dot{q} = \mathbf{F}(\mathbf{r},\dot{\mathbf{r}};t) \cdot \dot{\mathbf{r}} = F_i(q,\dot{q};t)\dot{x}_i(q)$, which determines the generalized co-ordinate $q = q(t)$. In particular, if the given force depends only on the position of the particle ($\mathbf{F} = \mathbf{F}(\mathbf{r})$), then we have $Q = Q(t)$, so that

$$\frac{m}{2}(v^2 - v_0^2) = \int_{q_0}^{q} Q(\eta)\mathrm{d}\eta.$$

Noting

$$\varphi(q) = \frac{1}{x'_i x'_i}\left[v_0^2 + \frac{2}{m}\int_{q_0}^{q} Q(\eta)\mathrm{d}\eta\right], \tag{6.2.18}$$

the equation (6.2.17) becomes

Dynamics of the particle with respect to an inertial frame of reference 391

$$\dot{q}^2 = \left(\frac{\mathrm{d}q}{\mathrm{d}t}\right)^2 = \varphi(q)$$

and we obtain

$$t - t_0 = \pm \int_{q_0}^{q} \frac{\mathrm{d}\eta}{\sqrt{\varphi(\eta)}}, \qquad (6.2.18')$$

where the position of the particle at the initial moment has been emphasized. The sign taken in this formula is that of the generalized velocity $\dot{q}$ (the sign of the initial velocity, which is preserved if $\dot{q}$ does not vanish); if $\dot{q}$ vanishes, then the velocity $\mathbf{v}$ vanishes too, while the sign is specified by the sense of the velocity $\mathbf{v}$, hence by the sense of the tangential component of the given force $\mathbf{F}$.

The trajectory and the velocity of the particle being determined, in the case of a fixed curve, as well as in the case of a movable one, we may put in evidence the constraint force $\mathbf{R}$, by calculating the scalars λ_1 and λ_2 from two of the equations (6.2.16') (the system (6.2.16') is – in this case – compatible).

Projecting the equation of motion (6.2.5) on the axes of Frenet's intrinsic frame, in case of stationary constraints, we obtain

$$m\dot{v} = m\ddot{s} = F_\tau(s,\dot{s};t), \quad m\frac{v^2}{\rho} = F_\nu + R_\nu, \quad F_\beta + R_\beta = 0. \qquad (6.2.19)$$

The first of these equations does not contain the constraint force, so that it determines the motion of the particle along the trajectory (as a matter of fact, multiplying both members of the equation by $\mathrm{d}s$, we get the theorem of kinetic energy), while the other two equations give the components of the constraint force in the plane normal to the trajectory (in a simpler form as that previously mentioned)

$$R_\nu = -F_\nu + m\frac{v^2}{\rho}, \quad R_\beta = -F_\beta. \qquad (6.2.20)$$

The first equation (6.2.19) shows that the law of motion does not change if the curve is deformed without changing its length or by modifying the force $\mathbf{F}$, but maintaining its tangential component (only the constraint force is changing); in particular, we may transform the curve into a straight line, reducing the problem to the study of a rectilinear motion.

If $\mathbf{F}$ is a conservative force, deriving from a simple potential $U = U(\mathbf{r})$ or from a generalized potential (the simple part of which is $U_0 = U_0(\mathbf{r})$), then we may write a conservation theorem of energy in the form (6.1.55) or (6.1.55'); the component of the constraint force along the principal normal is thus of the form

$$R_\nu = -F_\nu + \frac{2}{\rho}(U + h), \qquad (6.2.20')$$

and is obtained without knowing the motion of the particle on the trajectory (in case of a generalized potential we replace the function U by the function U_0).

In the general case of rheonomic constraints we may use Lagrange's equations (6.1.24'); because we have only one generalized co-ordinate q, we obtain the differential equation ($\mathbf{r} = \mathbf{r}(q;t)$)

$$\frac{d}{dt}\left(\frac{\partial T}{\partial \dot{q}}\right) - \frac{\partial T}{\partial q} = Q(q,\dot{q};t), \qquad (6.2.21)$$

where

$$T = \frac{1}{2}m\left(\frac{\partial \mathbf{r}}{\partial q}\dot{q} + \frac{\partial \mathbf{r}}{\partial t}\right)^2, \quad Q = \mathbf{F}(\mathbf{r},\dot{\mathbf{r}};t) \cdot \frac{\partial \mathbf{r}}{\partial q}, \qquad (6.2.21')$$

the unknown function being $q = q(t)$. In particular, we may consider $\mathbf{r} = \mathbf{r}(s;t)$, where s is the curvilinear co-ordinate along the curve C; a scalar product of the equation of motion (6.2.5) by the unit vector $\boldsymbol{\tau} = d\mathbf{r}/ds$, tangent to the curve C, leads to ($\mathbf{R} \cdot \boldsymbol{\tau} = 0$)

$$m\ddot{\mathbf{r}} \cdot \boldsymbol{\tau} = \mathbf{F} \cdot \boldsymbol{\tau} = F_\tau;$$

but

$$\dot{\mathbf{r}} = \dot{s}\boldsymbol{\tau} + \frac{\partial \mathbf{r}}{\partial t}, \quad \ddot{\mathbf{r}} = \ddot{s}\boldsymbol{\tau} + \dot{s}\dot{\boldsymbol{\tau}} + \frac{\partial^2 \mathbf{r}}{\partial t^2},$$

so that the equation of motion, which is a generalization of the equation (6.2.19), becomes

$$m\ddot{s} + \frac{\partial^2 \mathbf{r}}{\partial t^2} \cdot \boldsymbol{\tau} = F_\tau, \qquad (6.2.22)$$

where the unknown function is $s = s(t)$.

2.2.2 Motion of a particle constrained to stay on a surface

Let P be a particle in motion on a smooth movable or fixed surface S (Fig.6.8) of equation (6.2.3) or (6.2.3'). The constraint force is given by (6.2.6) and the equation of motion becomes

$$m\ddot{\mathbf{r}} = \mathbf{F} + \lambda \operatorname{grad} f; \qquad (6.2.23)$$

in components, we may write

$$m\ddot{x}_i = F_i + \lambda f_{,i}, \quad i = 1,2,3. \qquad (6.2.23')$$

The equations (6.2.3) and (6.2.23') form a system of four scalar equations for the three unknown functions $x_i = x_i(t)$ and for the parameter λ, which specifies the constraint force. Thus, the equations (6.2.23') allow to write

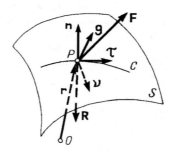

Figure 6.8. Motion of a particle constrained to stay on a surface S.

$$\frac{m\ddot{x}_1 - F_1}{f_{,1}} = \frac{m\ddot{x}_2 - F_2}{f_{,2}} = \frac{m\ddot{x}_3 - F_3}{f_{,3}}.$$

Adding also the equation (6.2.3), we may determine the trajectory of the particle; the parameter λ is then given by any of the equations (6.2.23').

In the case of a unilateral constraint (case in which the particle may leave the surface, remaining on one part of it), one must take into consideration the direction of the constraint force **R**. Thus, if the force **R** is directed towards the part in which the particle may leave the surface, that one will remain on the surface. We notice that the force **R** is directed towards the part of the surface for which $f \geqslant 0$ if $\lambda \geqslant 0$; hence, the particle remains on the surface only if the parameter λ maintains its sign. Consequently, if this parameter vanishes and changes of sign, then the particle leaves the surface and moves – further – as a free particle.

In the case of a rheonomic constraint, we may write

$$\operatorname{grad} f \cdot \mathrm{d}\mathbf{r} + \dot{f}\mathrm{d}t = 0,$$

and the real work of the constraint force is given by

$$\mathrm{d}W_R = \mathbf{R} \cdot \mathrm{d}\mathbf{r} = \lambda \operatorname{grad} f \cdot \mathrm{d}\mathbf{r} = -\lambda \dot{f}\mathrm{d}t;$$

indeed, the real displacement of the particle is not tangent to the surface frozen at the moment t, while the constraint force **R** is normal to this surface at the point P. If the constraint is scleronomic (or – at least – catastatic), then $\mathrm{d}W_R = 0$, and the theorem of kinetic energy can be written in the form (6.1.48); because the particle is constrained to stay on a surface, and has two degrees of freedom, it is necessary one equation more to can specify the motion.

Using Darboux's frame, we may write, in case of a fixed surface,

$$m\dot{v} = F_\tau, \quad m\frac{v^2}{\rho_n} = F_n + R, \quad m\frac{v^2}{\rho_g} = F_g; \quad (6.2.24)$$

the first and the last equation determine the motion corresponding to the two mentioned degrees of freedom, while the second equation gives the constraint force

$$R = -F_n + m\frac{v^2}{\rho_n}. \quad (6.2.25)$$

If the force **F** is conservative, deriving – for instance – from a simple potential $U = U(\mathbf{r})$, we may use the formula (6.1.55), and the constraint force is given by

$$R = -F_n + \frac{2}{\rho_n}(U + h). \quad (6.2.25')$$

In the case of a given zero force ($\mathbf{F} = \mathbf{0}$), the first equation (6.2.24) shows that the motion of the particle is uniform (the magnitude v of the velocity is constant in time). The component of the acceleration along the tangent to the trajectory vanishes, so that the acceleration is directed towards the principal normal $\mathbf{\nu}$ to the trajectory. The second equation (6.2.24) emphasizes ($F_n = 0$) that the acceleration vector is normal to the surface S; hence, at the point P, the principal normal to the trajectory of the particle has the same support as the normal to the surface S ($\mathbf{n} \times \mathbf{\nu} = \mathbf{0}$). The last equation (6.2.24) shows that ($F_g = 0$) the geodesic curvature vanishes; hence, the trajectory of the particle P is a geodesic of the surface S, the motion being uniform. This result may be put in connection with the principle of inertia, the straight line being a geodesic of the space E_3; in this case, $\rho_n \to \infty$, while $R = 0$. It results the constraint force in the form

$$R = m\frac{v_0^2}{\rho_n}, \quad (6.2.26)$$

where v_0 is the magnitude of the initial velocity. For instance, a particle which is constrained to stay on a fixed sphere and is acted upon by a zero force describes a great circle of it.

We notice that the trajectory is a geodesic if the necessary and sufficient condition $F_g = 0$ is fulfilled, hence if the force **F** belongs to the osculating plane of the trajectory (determined by the unit vectors $\mathbf{\tau}$ and $\mathbf{\nu}$, $\mathbf{\nu} \times \mathbf{n} = \mathbf{0}$); in this case, the motion is uniform only if $F_\tau = 0$, hence if the given force is normal to the surface S. If the particle has a uniform motion on a geodesic circle ($1/\rho_g = \text{const}$, in the sense of Darboux), then we have $F_g = \text{const}$, $F_\tau = 0$, so that the given force is normal to the trajectory, its projection on the tangent plane being constant during the motion. As well, if $R = -F_n$ during the motion, then the trajectory is an asymptotic line of the surface.

Dynamics of the particle with respect to an inertial frame of reference

As in the preceding subsection, in the case of rheonomic constraints we may use Lagrange's equations (6.1.24') in the form (we have two generalized co-ordinates q_1 and q_2, so that $\mathbf{r} = \mathbf{r}(q_1, q_2; t)$)

$$\frac{\mathrm{d}}{\mathrm{d}t}\left(\frac{\partial T}{\partial \dot{q}_\alpha}\right) - \frac{\partial T}{\partial q_\alpha} = Q_\alpha(q_1, q_2, \dot{q}_1, \dot{q}_2; t), \quad \alpha = 1, 2, \tag{6.2.27}$$

where

$$T = \frac{1}{2}m\left(\frac{\partial \mathbf{r}}{\partial q_1}\dot{q}_1 + \frac{\partial \mathbf{r}}{\partial q_2}\dot{q}_2 + \frac{\partial \mathbf{r}}{\partial t}\right)^2, \quad Q_\alpha = \mathbf{F}(\mathbf{r}, \dot{\mathbf{r}}; t) \cdot \frac{\partial \mathbf{r}}{\partial q_\alpha}; \tag{6.2.27'}$$

the unknown functions $q_1(t)$ and $q_2(t)$ may be – eventually – curvilinear co-ordinates s_1 and s_2 on the surface S. In this case, taking into account the formula (6.1.24''') and the results in Chap. 5, Subsec. 1.2.3, the equations of motion read

$$m\ddot{s}_\alpha + m\begin{Bmatrix}\alpha\\\beta\ \gamma\end{Bmatrix}\dot{s}_\beta\dot{s}_\gamma = g^{\alpha\beta}Q_\beta(s_1, s_2), \quad Q_\beta = \mathbf{F}\cdot\frac{\partial \mathbf{r}}{\partial s_\beta}, \quad \alpha, \beta = 1, 2, \tag{6.2.28}$$

and the unknown functions are $s_1 = s_1(t)$ and $s_2 = s_2(t)$. Making $Q_1 = Q_2 = 0$ and noting that one can take the time t proportional to the curvilinear co-ordinate s on the trajectory (the motion is uniform), we find the equations of the geodesic curves

$$s''_\alpha + \begin{Bmatrix}\alpha\\\beta\ \gamma\end{Bmatrix}s'_\beta s'_\gamma = 0, \quad \alpha = 1, 2, \tag{6.2.29}$$

where $s'_\alpha = \partial s_\alpha / \partial s$, $\alpha = 1, 2$.

We have seen that the conservation theorem of the mechanical energy allows – in the conditions in which this theorem takes place – to determine easily the constraint force, if the trajectory of the particle is known. In what concerns the conservation theorem of moment of momentum, the condition

$$\mathbf{r} \times (\mathbf{F} + \mathbf{R}) = \mathbf{r} \times \mathbf{F} + \lambda \mathbf{r} \times \operatorname{grad} f = \mathbf{0}$$

must hold; this condition is fulfilled for any λ if $\mathbf{r} \times \mathbf{F} = \mathbf{0}$ and $\mathbf{r} \times \operatorname{grad} f = \mathbf{0}$. We obtain a scalar first integral if the components of the two moments along one of the axes (for instance, Ox_3) vanish. Thus, we start from $x_1 f_2 - x_2 f_1 = 0$; the associate system of differential equations

$$\frac{\mathrm{d}x_1}{-x_2} = \frac{\mathrm{d}x_2}{x_1} = \frac{\mathrm{d}x_3}{0}$$

leads to the integrals $x_1^2 + x_2^2 = C_1$, $x_3 = C_2$, so that $f = f\left(x_1^2 + x_2^2, x_3\right) = 0$. Hence, the surface S must be a surface of rotation (with Ox_3 as symmetry axis);

indeed, the support of the constraint force intersects – at any moment – the rotation axis. If the particle P is constrained to stay on a surface of rotation, while the force $\mathbf{F}$ is coplanar with the symmetry axis of the surface, then we may write a first integral of the form (6.1.54').

2.2.3 Motion of a particle subjected to constraints with friction

In the case of a constraint with friction, the constraint force $\mathbf{R}$ has not only the normal component $\mathbf{N}$, which does not allow the particle to leave the constraint, but also a tangential one $\mathbf{T} \neq \mathbf{0}$, which hinders the particle to move in the frame of this constraint. As in the static case, considered in Chap. 4, Subsec. 1.1.8, in what follows we use the Coulombian model introduced in Chap. 3, Subsec. 2.2.11, for the force of friction; we assume that the constraints are holonomic and scleronomic. The force of friction is tangent to the rough surface or curve on which stays the particle, its direction being opposite to that of the motion.

In the case of a particle constrained to stay on a rigid fixed or movable curve C, of equations (6.2.1), the equation of motion reads

$$m\ddot{\mathbf{r}} = \mathbf{F} + \mathbf{N} + \mathbf{T}. \tag{6.2.30}$$

We introduce also the relations

$$|\mathbf{T}| = f|\mathbf{N}|, \quad \mathbf{T} = -|\mathbf{T}|\frac{\mathbf{v}_r}{v_r}, \tag{6.2.31}$$

where $\mathbf{v}_r$ is the relative velocity of the particle; we notice that $\operatorname{vers} \mathbf{T} + \operatorname{vers} \mathbf{v}_r = \mathbf{0}$. If the particle is constrained to stay on a rigid fixed or movable surface S, of equation (6.2.3), then the equations which determine the unknown quantities will be of the form (6.2.30) and (6.2.31) too. The relative velocity is given by $\mathbf{v}_r = \mathbf{v} - \mathbf{v}_t$, where $\mathbf{v}_t = \mathbf{v}_Q(t) + \boldsymbol{\omega}(t) \times \overrightarrow{QP}$ is the velocity of transportation of the surface (or of the curve); Q is a point of the surface (or curve), while $\boldsymbol{\omega}(t)$ is the instantaneous rotation.

Eliminating the force of friction $\mathbf{T}$, we obtain the vector equation

$$m\ddot{\mathbf{r}} = \mathbf{F} + N\mathbf{n} - fN\frac{\mathbf{v} - \mathbf{v}_t}{|\mathbf{v} - \mathbf{v}_t|}, \tag{6.2.32}$$

where $\mathbf{n} = \operatorname{vers} \operatorname{grad} f$, or the vector equation

$$m\ddot{\mathbf{r}} = \mathbf{F} + N_1\mathbf{n}_1 + N_2\mathbf{n}_2 - f|\mathbf{N}_1 + \mathbf{N}_2|\frac{\mathbf{v} - \mathbf{v}_t}{|\mathbf{v} - \mathbf{v}_t|}, \tag{6.2.32'}$$

where $\mathbf{n}_\alpha = \operatorname{vers} \operatorname{grad} f_\alpha$, $\alpha = 1,2$. If the surface S (or the curve C) is fixed, then the velocity of transportation vanishes ($\mathbf{v}_t = \mathbf{0}$).

If we consider the motion on a fixed curve with respect to Frenet's frame, then we get

Dynamics of the particle with respect to an inertial frame of reference

$$m\dot{v} = F_\tau - fN\,\text{sign}\,v\,, \quad m\frac{v^2}{\rho} = F_\nu + N_\nu\,, \quad 0 = F_\beta + N_\beta\,, \qquad (6.2.33)$$

with $N = \sqrt{N_\nu^2 + N_\beta^2}$; the equations (6.2.33) and (6.2.1') determine thus the unknown functions $x_i = x_i(t)$, $i = 1, 2, 3$, as well as the components N_ν, N_β of the constraint force. Eliminating the constraint force, we may write the equation

$$m\dot{v} = mv\frac{\mathrm{d}v}{\mathrm{d}s} = F_\tau - f\sqrt{F_\beta^2 + \left(m\frac{v^2}{\rho} - F_\nu\right)^2}\,\text{sign}\,v\,, \qquad (6.2.34)$$

which gives the velocity $v = v(s)$. In the case of a plane curve, the force **F** belonging to the respective plane, we have $F_\beta = 0$ and we get

$$mv\frac{\mathrm{d}v}{\mathrm{d}s} = \frac{1}{2}m\frac{\mathrm{d}(v^2)}{\mathrm{d}s} = F_\tau \mp f\left(m\frac{v^2}{\rho} - F_\nu\right)\,, \qquad (6.2.34')$$

where the sign corresponds to a direction of the constraint force opposite to the motion. Integrating, we have

$$v^2 = \varphi(s)\,, \qquad (6.2.35)$$

where

$$\varphi(s) = C\mathrm{e}^{\mp 2f\psi(s)} + \frac{2}{m}\mathrm{e}^{\mp 2f\psi(s)}\int_0^s \mathrm{e}^{2f\psi(\sigma)}\left[F_\tau(\sigma) + fF_\nu(\sigma)\right]\mathrm{d}\sigma\,, \qquad (6.2.35')$$

with

$$\psi(s) = \int\frac{\mathrm{d}s}{\rho}\,, \quad C = \text{const}\,. \qquad (6.2.35'')$$

Hence, we obtain

$$t - t_0 = \int_{s_0}^s \frac{\mathrm{d}\sigma}{\sqrt{\varphi(\sigma)}}\,, \qquad (6.2.35''')$$

the motion of the particle on the fixed curve C being thus specified; returning to the equations of motion (6.2.33), we emphasize the constraint force too.

In the case of a particle constrained to move on a fixed surface S, we introduce Darboux's trihedron and find the equations

$$m\dot{v} = F_\tau - fN\,, \quad m\frac{v^2}{\rho_n} = F_n + N\,, \quad m\frac{v^2}{\rho_g} = F_g\,; \qquad (6.2.36)$$

these three equations, together with the equation (6.2.3'), form a system of four equations for the unknown functions $x_i = x_i(t)$, $i = 1,2,3$, and the constraint force N. Eliminating the constraint force between the first two equations, we get a new equation which, together with the third equation (6.2.36), constitutes a system of two differential equations for the unknown curvilinear co-ordinates $s_\alpha = s_\alpha(t)$, $\alpha = 1,2$, on the surface S; thus, the relation (6.2.3') is eliminated, and the co-ordinates which we use are generalized co-ordinates. The constraint force may be easily determined from the second equation (6.2.36). If $F_g = 0$, then the motion takes place as in the case in which the surface is smooth; that motion is uniform if and only if $F_\tau = fN$, hence if and only if the tangential component of the given force equilibrates the force of sliding friction.

2.2.4 Motion of a particle with a single degree of freedom in the conservative case

In the case of a *particle* (or of a mechanical system) *with a single degree of freedom*, for which the equation of motion is of the form

$$\ddot{q} = f(q), \tag{6.2.37}$$

where q is the generalized co-ordinate, we may set up a first integral of energy of the form

$$\dot{q}^2 - \dot{q}_0^2 = 2[U(q) - U(q_0)], \quad U(q) = \int f(q)\mathrm{d}q, \tag{6.2.38}$$

by introducing the simple potential U (or the scalar potential U_0 of a generalized potential); hence, the corresponding mechanical system is a *conservative system*. As well, one can show that a *unidimensional conservative mechanical system* (with a single degree of freedom) or a pluridimensional one (if we succeed, by means of the first integrals, to eliminate the corresponding parameters, obtaining a unidimensional one) leads to an equation of motion of the form (6.2.37). We notice that this equation corresponds to a *non-linear free oscillation, without damping*; the function $f(q)$ is thus a *calling force*. The equation (6.2.38) can be integrated in the form (6.2.18'), using the notation

$$\varphi(q) = \dot{q}_0^2 + 2[U(q) - U(q_0)]. \tag{6.2.38'}$$

One takes the sign + or − in (6.2.18') as the function $q(t)$ is monotone increasing or decreasing, respectively. It is necessary that $\varphi(q) \geq 0$, so that the motion be real. Observing that $\varphi(q_0) = \dot{q}_0^2 \geq 0$, we may assume that the function $q(t)$ begins to increase together with t (corresponding to the direction of the initial velocity); the sign + is thus chosen. A study of the variation of the function $\varphi(q)$ and of its zeros leads to interesting conclusions concerning the motion of the particle (or of the mechanical system).

Noting $\dot{q} = p$, we may replace the equation (6.2.37) by the system

Dynamics of the particle with respect to an inertial frame of reference

$$\frac{dq}{dt} = p, \quad \frac{dp}{dt} = f(q), \tag{6.2.39}$$

which leads to

$$\frac{dp}{dq} = \frac{f(q)}{p}; \tag{6.2.39'}$$

the motion of the particle is thus equivalent to the motion of a *representative point* $\mathscr{P}$ in the *phase space* of co-ordinates q, p. The trajectory C in this space pierces the Oq-axis by a right angle, and the tangent to it is parallel to this axis for $f(q) = 0$, $p \neq 0$; if we have also $p = 0$, then one obtains a *singular point*, corresponding to a position of equilibrium, as it results from the system (6.2.39).

Expressing the first integral (6.2.38) in the form

$$p^2 + 2V(q) = h, \quad h = \dot{q}_0^2 + 2V(q_0), \quad V(q) = -U(q), \tag{6.2.40}$$

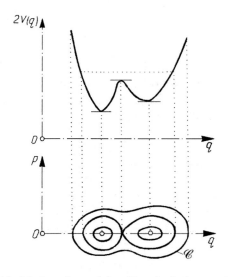

Figure 6.9. Motion of a particle with a single degree of freedom in the conservative case.

where h is *the energy constant*, we notice that the trajectory $\mathscr{C}$ is symmetric with respect to the Oq-axis and is situated in the domain $2V(q) \leq h$. On the basis of the Lagrange-Dirichlet theorem (see Chap. 4, Subsec. 1.1.7), to the points of local minimum of the potential energy $V(q)$ correspond *stable positions of equilibrium*, while to the points of maximum correspond *labile positions of equilibrium* (Fig.6.9). From the first equation (6.2.39), we see that – for $p > 0$ – q increases together with the time t, so that we may specify the direction of the trajectory. We obtain *the period of the motion* in the form

$$T = \oint \frac{\mathrm{d}p}{q}, \qquad (6.2.41)$$

integrating along a closed curve.

2.2.5 Study of a dissipative mechanical system with a single degree of freedom

If, during the motion, intervenes the friction too, then the mechanical energy is no more conserved, but diminishes in time, due to the phenomenon of dissipation. Assuming that the force of friction is a function of velocity, the equation of motion becomes

$$\ddot{q} = f(q) + F(\dot{q}), \quad \dot{q}F(\dot{q}) \leq 0, \qquad (6.2.42)$$

the last relation emphasizing that *the damping force* $F(\dot{q})$ is opposite to the velocity; in fact, the equation (6.2.42) characterizes the *damped non-linear free oscillations*, in the unidimensional case (with a single degree of freedom). In the case of *forced oscillations*, this equation is completed in the form

$$\ddot{q} = f(q) + F(\dot{q}) + \mathscr{F}(t), \quad \dot{q}F(\dot{q}) \leq 0, \qquad (6.2.42')$$

where we have introduced also the perturbing force $\mathscr{F}(t)$. In the case in which the product $\dot{q}F(\dot{q})$ is positive for small values of the velocity modulus $|\dot{q}|$ and negative for great values of the same modulus, one obtains *self-sustained oscillations*.

As in the case of undamped oscillations, the equation (6.2.42) leads to

$$\frac{\mathrm{d}p}{\mathrm{d}q} = \frac{f(q) + F(p)}{p} \qquad (6.2.43)$$

in the phase space, a field of vectors being thus defined, excepting the positions of equilibrium (singular points). In the case in which the calling force $f(q)$ is linear ($f(q) = q$) one can build up a polygonal line, which approximates the integral curve; this procedure, due to Liénard, has been extended by J.L. Brown for the case in which $f(q)$ is a non-linear function (the case of non-linear free oscillations).

Chapter 7

PROBLEMS OF DYNAMICS OF THE PARTICLE

One of the most important cause (input) which leads to the motion of a mechanical system is the gravitational field, in particular the terrestrial one; some results concerning the motion of a free particle in vacuum or in a resistent medium are also given, and some considerations concerning the pendulary motion of a particle subjected to constraints are made. Other important classical problems of dynamics of the particle are then presented and notions on the stability of its equilibrium are given.

1. Motion of the particle in a gravitational field

First of all, we deal with some particular cases of motion of a free particle, i.e., the rectilinear and the plane motion, as well as the motion of a heavy particle; another case of motion of a particle subjected to constraints, that is the plane and the three-dimensional pendulary motion, is then studied.

1.1 Rectilinear and plane motion

Of a particular interest in the study of the three-dimensional motion of a free or constraint particle is the motion of its projection on a fixed straight line or plane; we are thus led to consider some particular cases of rectilinear and plane motions of a particle.

1.1.1 Rectilinear motion of a particle

We choose as *rectilinear trajectory* of a free particle P a straight line parallel to the Ox_1-axis, of equations $x_2 = x_2^0 = \text{const}$, $x_3 = x_3^0 = \text{const}$; the equations (6.1.22') show that the resultant **F** of the given forces must verify the conditions $F_2 = F_3 = 0$, hence it must have a fixed direction (the same as that of the Ox_1-axis); but the latter conditions are not sufficient to obtain the mentioned trajectory. Indeed, the above conditions lead to $\ddot{x}_2 = \ddot{x}_3 = 0$, hence to $x_k = \dot{x}_k^0(t - t_0) + x_k^0$, corresponding to the initial conditions $x_k(t_0) = x_k^0$, $\dot{x}_k(t_0) = \dot{x}_k^0$, $k = 2, 3$; to obtain the searched trajectory, one must have $\dot{x}_2^0 = \dot{x}_3^0 = 0$ too and we may state

Theorem 7.1.1. *The trajectory of a free particle is rectilinear if and only if the resultant of the given forces acting upon it has a fixed direction and its initial velocity has the same direction.*

In this case, the support of the force **F** coincides with the rectilinear trajectory, while ther equation of motion is reduced to ($x_2 = x_2^0, x_3 = x_3^0$)

$$m\ddot{x} = F,\qquad(7.1.1)$$

where $x = x(t)$, $F = F(x,\dot{x};t)$ and where, for the sake of simplicity, we have omitted the index 1; the initial conditions are of the form

$$x(t_0) = x_0,\quad v(t_0) = v_0.\qquad(7.1.1')$$

If the resultant of the given forces depends only on time ($F = F(t)$), then – by successive integrations – we get

$$mv(t) = m\dot{x}(t) = mv_0 + \int_{t_0}^{t} F(\tau)\mathrm{d}\tau,\qquad(7.1.2)$$

$$mx(t) = mx_0 + mv_0(t - t_0) + \int_{t_0}^{t} \mathrm{d}\tau \int_{t_0}^{\tau} F(\overline{\tau})\mathrm{d}\overline{\tau}.\qquad(7.1.2')$$

In particular, if $F = ma_0 = \text{const}$, we obtain

$$v(t) = v_0 + a_0(t - t_0),\quad x(t) = x_0 + v_0(t - t_0) + \frac{a_0}{2}(t - t_0)^2.\qquad(7.1.3)$$

Assuming that $F = F(x)$ (the resultant of the given forces depends only on the position), we may use the results obtained in Chap. 6, Subsec. 2.2.1 concerning the dynamics of a particle constrained to move on a given fixed curve; the formula (6.2.18') allows to write

$$t - t_0 = \pm \int_{x_0}^{x} \frac{\mathrm{d}\xi}{\sqrt{\varphi(\xi)}},\quad \varphi(x) = v^2 + \frac{2}{m}\int_{x_0}^{x} F(\xi)\mathrm{d}\xi,\quad \varphi(x_0) = v_0^2 \geq 0.\qquad(7.1.4)$$

We notice that the sign in the above formula corresponds to the sign of the initial velocity v_0. If $v_0 = 0$, then $\varphi(x_0) = 0$ and we must take the sign for which $\varphi(x) \geq 0$. If $\varphi(x)$ is a monotone increasing or decreasing function, then the sign + or –, respectively, is taken; noting that $\varphi'(x) = (2/m)F(x)$, it results that, in the case of a vanishing initial velocity, the particle moves on the trajectory, departing from the initial position, in the same direction as that of the force. In the above considerations, we assumed tacitly that, if $v_0 = 0$, then $\varphi'(x) \neq 0$, because $F(x_0) \neq 0$; hence, x_0 is only a simple zero for $\varphi(x)$.

Let x_1 and x_2 be two consecutive simple zeros of $\varphi(x)$, so that $x_1 < x_0 < x_2$; in this case

$$\varphi(x) = (x - x_1)(x_2 - x)\psi(x),\quad \psi(x) > 0,\quad x \in [x_1, x_2].\qquad(7.1.5)$$

Problems of dynamics of the particle

The particle $P(x)$ departs from the point $P_0 \equiv P(x_0)$ with the velocity $v_0 \neq 0$ (let be $v_0 > 0$) and oscillates between the positions $P_1 \equiv P(x_1)$ and $P_2 \equiv P(x_2)$ (Fig.7.1); assuming that the particle reaches the position P_2 at the moment t' with the velocity $v(t'-0) > 0$, then the position P_1 at the moment t'' with the velocity $v(t''-0) < 0$, and returns to P_0 at the moment t''' with the velocity $v(t'''-0) > 0$, we obtain

$$t' - t_0 = \int_{x_0}^{x_2} \frac{\mathrm{d}x}{\sqrt{(x-x_1)(x_2-x)\varphi(x)}}, \quad t'' - t' = -\int_{x_2}^{x_1} \frac{\mathrm{d}x}{\sqrt{(x-x_1)(x_2-x)\psi(x)}},$$

$$t''' - t'' = \int_{x_1}^{x_0} \frac{\mathrm{d}x}{\sqrt{(x-x_1)(x_2-x)\psi(x)}},$$

so that $t'' - t' = t''' - t'' + t' - t_0 = T/2$. After an interval of time

$$T = 2\int_{x_1}^{x_2} \frac{\mathrm{d}x}{\sqrt{(x-x_1)(x_2-x)\psi(x)}}, \tag{7.1.5'}$$

called *the period of the motion*, the particle returns to the same position P_0 with the same velocity; because the position P_0 is arbitrary, while T does not depend on this position, it follows that *the motion is periodic*. In this case, the function $x = x(t)$ given by the relation (7.1.4) is a periodic function.

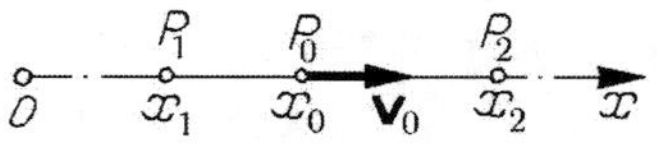

Figure 7.1. Rectilinear motion of a particle.

If the resultant of the given forces depends only on the velocity ($F = F(v)$), then we use the equation of motion written in the form (1.1.89''), obtaining thus

$$t - t_0 = \int_{v_0}^{v} \frac{m\mathrm{d}\eta}{F(\eta)}; \tag{7.1.6}$$

hence, if it is possible, we calculate $v = \dot{x} = \chi(t - t_0)$, wherefrom

$$x - x_0 = \int_{t_0}^{t} \chi(\tau - t_0)\mathrm{d}\tau, \tag{7.1.6'}$$

the initial conditions (which imply two constants of integration) being also verified, so that the problem is solved. If the function χ cannot be determined, then we may write

$$x - x_0 = \int_{v_0}^{v} \frac{m\eta\mathrm{d}\eta}{F(\eta)}, \tag{7.1.7}$$

being thus led to the parametric representation $x = x(v)$, $t = t(v)$, where v is *the hodographic variable*; thus, for any v with $F(v) \neq 0$, we may associate to the moment t a position x of the particle (the velocity of the particle at that moment is v). Assuming that from the relation (7.1.7) we obtain the function $v = \dot{x} = \varkappa(x - x_0)$, it results

$$t - t_0 = \int_{x_0}^{x} \frac{d\xi}{\varkappa(\xi - x_0)}, \qquad (7.1.7')$$

so that we may determine $x = x(t; t_0, x_0)$.

If the force F depends on two or three of the variables x, v, t, the problem becomes more intricate, and may be solved from case to case.

If $x = f(t; x_0, v_0)$ is given, one can enunciate the inverse problem: to determine the force F which leads to such a motion. We obtain thus $v = \dot{f}(t; x_0, v_0)$ and $a = \ddot{f}(t; x_0, v_0)$; eliminating the constants x_0 and v_0, we find the solution of the problem $F = m\ddot{f}(t; x_0(t; x, v), v_0(t; x, v)) = F(t; x, v)$, which is univocally determinate. If only a particular motion $x = f(t)$, which leads to $v = \dot{f}(t)$, $a = \ddot{f}(t)$, is known, then the solution of the problem is indeterminate; it becomes determinate if we introduce supplementary conditions, e.g., imposing that the force F do depend only on one of the variables t, x or v.

1.1.2 Plane motion of a particle

Let us suppose that the trajectory of a particle P *is contained in the fixed plane* $x_3 = x_3^0 = \text{const}$, parallel to the plane of co-ordinates Ox_1x_2; in this case, the resultant of the given forces $\mathbf{F}$ must verify the condition $F_3 = 0$, hence it must be parallel to the same fixed plane (it must be normal to a fixed direction, the same as that of the Ox_3-axis). But the latter condition is not sufficient to obtain the mentioned trajectory; indeed, the condition $F_3 = 0$ leads to $\ddot{x}_3 = 0$, wherefrom $x_3 = \dot{x}_3^0(t - t_0) + x_3^0$, corresponding to the initial conditions $x_3(t_0) = x_3^0$, $\dot{x}_3(t_0) = \dot{x}_3^0$. To can obtain the trajectory mentioned above, we must have $\dot{x}_3^0 = 0$ and may state

Theorem 7.1.2. *The trajectory of a free particle is a plane curve if and only if the resultant of the given forces acting upon it is parallel to a fixed plane (is normal to a fixed direction) and its initial velocity has the same property.*

Obviously, the plane which contains the trajectory (which contains the force $\mathbf{F}$ and the initial velocity $\mathbf{v}_0$ too) passes through the initial position of the particle, and the equations of motion are reduced to ($x_3 = x_3^0$)

$$m\ddot{x}_\alpha = F_\alpha(x_1, x_2, \dot{x}_1, \dot{x}_2; t), \quad \alpha = 1, 2; \qquad (7.1.8)$$

Problems of dynamics of the particle

the initial conditions are of the form ($x_\alpha = x_\alpha(t)$)

$$x_\alpha(t_0) = x_\alpha^0, \quad \dot{x}_\alpha(t_0) = \dot{x}_\alpha^0, \quad \alpha = 1,2. \tag{7.1.8'}$$

If the functions F_α fulfil certain conditions, the integration of the system of equations (7.1.8) may become easier. Such a case is that in which the motions of the particles associated to the projections of the particle P on the two axes in the plane may be studied independently, being governed by equations of the form

$$m\ddot{x}_1 = F_1(x_1, \dot{x}_1; t), \quad m\ddot{x}_2 = F_2(x_2, \dot{x}_2; t). \tag{7.1.9}$$

A particular case of a given force parallel to a fixed plane is the force of constant direction; the trajectory of the particle is thus in the plane determined by the initial velocity and by the direction of the force. Let us suppose that this plane is parallel to the Ox_1x_2-plane ($x_3 = x_3^0 = \text{const}$). The equation of motion can be written in the form $m\ddot{x}_1 = 0$, $m\ddot{x}_2 = F_2$, wherefrom

$$x_1 = v_1^0(t - t_0) + x_1^0; \tag{7.1.10}$$

in this case, one can eliminate x_1 and $\dot{x}_1$ from $F_2 = F_2(x_1, x_2, \dot{x}_1, \dot{x}_2; t)$, so that the second equation of motion will be given by

$$m\ddot{x}_2 = F_2(x_2, \dot{x}_2; t). \tag{7.1.10'}$$

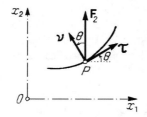

Figure 7.2. Plane motion of a particle.

The equations in intrinsic co-ordinates (6.2.19) are of the form (Fig.7.2)

$$m\dot{v} = F_2 \sin\theta, \quad m\frac{v^2}{\rho} = F_2 \cos\theta, \tag{7.1.11}$$

where ρ is the curvature radius of the trajectory; we notice also that

$$v\cos\theta = v_1^0. \tag{7.1.11'}$$

Eliminating the velocity v we get

$$F_2 \rho \cos^3\theta = m(v_1^0)^2 = \text{const}; \tag{7.1.12}$$

in the particular case in which $F_2 = \text{const}$, it results

$$\rho \cos^3 \theta = \text{const}, \qquad (7.1.12')$$

hence the intrinsic equation of a parabola.

Assuming that the functions $x_\alpha = f_\alpha(t; x_1^0, x_2^0, v_1^0, v_2^0)$ are given, one may enunciate the inverse problem: to determine the force **F** of components F_α, $\alpha = 1,2$, which leads to this motion. We can calculate $v_\alpha = \dot{f}_\alpha(t; x_1^0, x_2^0, v_1^0, v_2^0)$, $a_\alpha = \ddot{f}_\alpha(t; x_1^0, x_2^0, v_1^0, v_2^0)$; eliminating the constants of integration $x_1^0, x_2^0, v_1^0, v_2^0$, we find the solution $F_\alpha = F_\alpha(t; x_1, x_2, v_1, v_2)$, $\alpha = 1,2$, of the problem, which is univocally determinate. As in the previous subsection, if only one particular motion is known, then the solution of the problem is indeterminate; one must introduce supplementary conditions to obtain a determinate solution.

For instance, assuming that the trajectory of the particle is expressed in the form $x_2 = f(x_1)$, the force **F** which acts upon this particle being parallel to the Ox_2-axis, we may write

$$F_2 = m(v_1^0)^2 f''(x_1) + G(x_1, x_2, \dot{x}_1, \dot{x}_2; t)[x_2 - f(x_1)], \qquad (7.1.13)$$

where G is an arbitrary function of class C^0 (as in Chap. 6, Subsec. 1.1.4); the indetermination of the solution of the second basic problem is thus obvious.

More general, we may consider the motion of a particle P the projection of which (we assume that to each projection we associate a particle of the same mass as the particle P, but – for the sake of simplicity – we mention it no more) on a plane and on an axis non-parallel to that one or on three non-coplanar axes may be studied independently (as plane or rectilinear motions, respectively).

1.2 Motion of a heavy particle

In what follows, we consider first of all the general motion of a heavy particle in vacuum or in a resistent medium; the results thus obtained will be then particularized for the case of a rectilinear trajectory. Analogously, we present the study of the motion of a heavy particle constrained to stay on a curve or on a plane.

1.2.1 Motion of a heavy particle in vacuum

In the preceding subsection we have considered the motion of a particle acted upon by a force of constant direction. In particular, it is interesting to assume that the magnitude of the force is constant in time too; we will thus study the motion of a heavy particle P (the motion of a particle in the gravitational field of the Earth), of mass m, in vacuum. The equation of motion reads $\ddot{\mathbf{r}} = \mathbf{g}$, where $\mathbf{g}$ is the gravity acceleration, so that

$$\mathbf{r} = \frac{1}{2}\mathbf{g}(t - t_0)^2 + \mathbf{v}_0(t - t_0) + \mathbf{r}_0, \quad \mathbf{v} = \mathbf{g}(t - t_0) + \mathbf{v}_0, \qquad (7.1.14)$$

Problems of dynamics of the particle

where we took into account the initial conditions

$$\mathbf{r}(t_0) = \mathbf{r}_0, \quad \mathbf{v}(t_0) = \mathbf{v}_0; \qquad (7.1.14')$$

these results correspond to the formulae (5.1.35), which specify the motion of a particle of constant acceleration. As we know, the trajectory is a plane curve; indeed, the vector $\mathbf{r} - \mathbf{r}_0$ is a linear combination of the constant vectors $\mathbf{g}$ and $\mathbf{v}_0$. Without any loss of generality, we may suppose that $t_0 = 0$, $\mathbf{r}_0 = \mathbf{0}$, so that

$$\mathbf{r} = \frac{1}{2}\mathbf{g}t^2 + \mathbf{v}_0 t, \quad \mathbf{v} = \mathbf{g}t + \mathbf{v}_0; \qquad (7.1.14'')$$

hence, we obtain the remarkable relations

$$\mathbf{r} = -\frac{1}{2}\mathbf{g}t^2 + \mathbf{v}t = \frac{1}{2}(\mathbf{v} + \mathbf{v}_0)t. \qquad (7.1.14''')$$

We assume that $\mathbf{v}_0 \neq \mathbf{0}$ and that it has not the same support as $\mathbf{g}$; in this case, the velocity $\mathbf{v}$ cannot vanish, while the second relation (7.1.14''') allows to determine the velocity of the particle P by a simple drawing if its position is known or allows to determine graphically its position if its velocity is known.

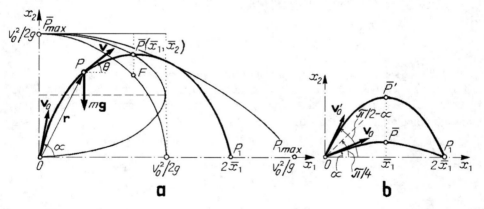

Figure 7.3. Motion of a heavy particle in vacuum: Cauchy problem (a); bilocal problem (b).

Projecting the equations (7.1.14'') on two orthogonal axes in the plane Ox_1x_2 (we take $\mathbf{g} = -g\mathbf{i}_2$, $\mathbf{v}_0 = v_0(\cos\alpha\,\mathbf{i}_1 + \sin\alpha\,\mathbf{i}_2)$, $-\pi/2 < \alpha < \pi/2$), we obtain the parametric equations of the trajectory (Fig.7.3,a)

$$x_1 = v_0 t \cos\alpha, \quad x_2 = -\frac{1}{2}gt^2 + v_0 t \sin\alpha, \qquad (7.1.15)$$

wherefrom, eliminating the time t, we may write

$$x_2 = -\frac{g}{2v_0^2 \cos^2\alpha}x_1^2 + x_1 \tan\alpha; \qquad (7.1.15')$$

hence, the trajectory of the particle P is a parabola, as it was shown in Subsec. 1.1.2 and in Chap. 5, Subsec. 1.3.1. The velocity of the particle at a given moment is given by

$$v_1 = v_0 \cos\alpha, \quad v_2 = -gt + v_0 \sin\alpha, \qquad (7.1.16)$$

so that

$$v = \sqrt{v_0^2 - 2gx_2} = v_0 \frac{\cos\alpha}{\cos\theta}, \qquad (7.1.16')$$

where we took into account the formula (7.1.11') too. Choosing the angle θ as generalized co-ordinate, taking into account (7.1.15), (7.1.16') and eliminating the time t, we may write

$$x_1 = -\frac{v_0^2 \cos^2\alpha}{g}(\tan\theta - \tan\theta_0),$$
$$x_2 = -\frac{v_0^2 \cos^2\alpha}{g}(\sec^2\theta - \sec^2\alpha). \qquad (7.1.15'')$$

The potential of the conservative force $m\mathbf{g}$ is $U = -mgx_2$, so that we may write the conservation theorem of mechanical energy in the form

$$m\frac{v^2}{2} = -mgx_2 + h, \quad h = \text{const}; \qquad (7.1.15''')$$

we find thus again the expression (7.1.16') of the magnitude of the velocity.

If $\alpha > 0$, then we obtain the basic problem of ballistics in the case in which the friction with the air is neglected. The particle (eventually, a projectile) reaches the highest point $\overline{P}$ of the trajectory for $v_2 = 0$, hence at the moment $\overline{t} = (v_0/g)\sin\alpha = v_2^0/g$; the co-ordinates of this point are

$$\overline{x}_1 = \frac{v_2^0}{2g}\sin^2\alpha = \frac{v_1^0 v_2^0}{g}, \quad \overline{x}_2 = \frac{v_0^2}{2g}\sin^2\alpha = \frac{(v_2^0)^2}{2g},$$

and we find again Torricelli's formula (5.1.37) in the form

$$v_2^0 = \sqrt{2gh}, \quad h = \overline{x}_2. \qquad (7.1.17)$$

We obtain thus the component v_2^0 of the velocity by which we may launch the projectile to reach the height h; because the angle α is not involved in the formula, one states that it remains valid for the motion on a vertical line ($\alpha = \pi/2$) too. The formula (7.1.16') may be written also in the form $v^2 = 2g(v_0^2/2g - x_2)$; we may thus state that the magnitude of the velocity at a given moment is equal to that of a falling particle, without initial velocity, from the height $v_0^2/2g$.

If $\alpha < 0$, then the particle departs from a point on the decreasing branch of the parabola.

The point P_1 of abscissa $2\overline{x}_1 = (v_0^2/g)\sin 2\alpha$ is the most distant point reached by the projectile on a horizontal plane, at the moment $2\overline{t}$, the magnitude of the velocity being the same as that at the initial moment; *the range of the throw* is maximal for $\alpha = \pi/4$, i.e., $2\overline{x}_{1\max} = v_0^2/g$. If we wish to reach a point P_1 of abscissa $2\overline{x}_1$, then the initial conditions must verify the relation $v_0^2 \sin 2\alpha = 2g\overline{x}_1$ (a bilocal problem). To the same magnitude v_0 of the initial velocity there correspond two angles: $\alpha < \pi/4$ and $\pi/2 - \alpha$ (symmetric with respect to the angle $\pi/4$, because $\pi/4 - \alpha = (\pi/2 - \alpha) - \pi/4$) under which one may reach the same point P_1 (Fig.7.3,b); in particular, if $v_0 = \sqrt{2g\overline{x}_1}$, then we have $\alpha = \pi/4$. To the two shooting angles there correspond the shooting heights $h = (v_0^2/2g)\sin^2 \alpha$ and $h = (v_0^2/2g)\cos^2 \alpha$.

The projectile passes through the point $P(\xi_1, \xi_2)$ if the condition

$$\frac{g\xi_1^2}{2v_0^2}\tan^2\alpha - \xi_1 \tan\alpha + \xi_2 + \frac{g\xi_1^2}{2v_0^2} = 0$$

is fulfilled; as in the particular case considered above, one may reach the point P shooting a projectile under two angles specified by

$$\tan\alpha = \frac{v_0^2}{g\xi_1}\left[1 \pm \sqrt{1 - \frac{2g}{v_0^2}\left(\xi_2 + \frac{g\xi_1^2}{2v_0^2}\right)}\right]. \tag{7.1.18}$$

To reach a point P by a projectile, that one must be in the interior of *the safety parabola* (Fig.7.3,a)

$$x_2 = -\frac{g}{2v_0^2}x_1^2 + \frac{v_0^2}{2g}, \tag{7.1.18'}$$

which passes through the points $\overline{P}_{\max}(0, v_0^2/2g)$ and $P_{1\max}(v_0^2/g, 0)$; no point in the exterior of this parabola may be reached by an initial velocity of magnitude v_0. This parabola is the envelope of the family of trajectories (7.1.15') for $v_0 = \text{const}$ and α variable.

The semilatus rectum of the parabola (7.1.15') is $p = (v_0^2/g)\cos^2\alpha$, so that the locus of the focus $F((v_0^2/2g)\sin 2\alpha, -(v_0^2/2g)\cos 2\alpha)$ is the quarter of circle (Fig.7.3,a)

$$x_1^2 + x_2^2 = \frac{v_0^4}{4g^2}, \tag{7.1.19}$$

the centre of which is the origin and which passes through the point $\overline{P}_{\max}$; all these parabolas have as directrix a parallel to the Ox_1-axis of equation $x_2 = v_0^2/2g$, which passes through the vertex of the safety parabola. The locus of the vertices of the trajectories (7.1.15') is the ellipse (Fig.7.3,a)

$$x_1^2 + 4x_2^2 = \frac{2v_0^2}{g}x_2,\qquad(7.1.20)$$

the minor axis of which is $O\overline{P}_{\max}$, the major axis being parallel to the Ox_1-axis (the half of it is equal to $v_0^2/2g$).

1.2.2 Motion of a heavy particle in a resistent medium

In the case of a particle P which is moving in a *resistent medium*, besides the given force (eventually, the gravity force) intervenes also a force **R**, called *resistance*, corresponding to the resistance of the medium; under the action of the forces **F** and **R**, the particle moves as a free one. As a matter of fact, the force **R** is the resultant of superficial actions (pressure and friction) on the body, which may be modelled as a particle. Such a body is, for instance, a projectile in motion, which has a spherical form and is not subjected to rotations; from the point of view of the mathematical modelling, the projectile is reduced to its centre of gravity. The resistent medium may be the air, the influence of which cannot be – in general – neglected; the respective problems form the so-called *external ballistics*. It is assumed that the force **R** has the same support as the velocity **v**, but is of opposite direction. The magnitude of this force depends – especially – on the magnitude v of the velocity; one may conceive also a dependence on the density or on the pressure of the air, on the absolute temperature, on the form of the projectile etc. (e.g., after Langevin, $R = apf(v/\sqrt{T})$, $a = \text{const}$, where p is the pressure of the air, while T is the absolute temperature). We will suppose that

$$\mathbf{R} = -mg\varphi(v)\,\text{vers}\,\mathbf{v},\quad \varphi(0) = 0,\quad \lim_{v\to\infty}\varphi(v) = \infty,\qquad(7.1.21)$$

where $\varphi(v)$ is a strictly increasing function (the resistance of the air increases together with the velocity v); there exists – obviously – a value v^* and only one for which $\varphi(v^*) = 1$.

Returning to the demonstration in Subsec. 1.1.1, we notice that the trajectory of the particle is rectilinear, e.g., parallel to the Ox_1-axis, of equations $x_2 = x_2^0 = \text{const}$, $x_3 = x_3^0 = \text{const}$, if $F_2 = F_3 = 0$, as it results from the equations

$$m\ddot{x}_i = F_i - mg\overline{\varphi}(v)\dot{x}_i,\quad i = 1,2,3,\quad \overline{\varphi}(v) = \varphi(v)/v.$$

But these conditions imposed to the force **F** lead to $m\ddot{x}_k + g\overline{\varphi}(v)\dot{x}_k = 0$, wherefrom

$$\dot{x}_k = \dot{x}_k^0 e^{-g\int_{t_0}^{t}\overline{\varphi}(v(\tau))\,d\tau},\quad k = 2,3;$$

Problems of dynamics of the particle 411

the trajectory coincides with that imposed if and only if $\dot{x}_k^0 = 0$, so that we may state

Theorem 7.1.1'. *The trajectory of a free particle in a resistent medium, modelled by the relation* (7.1.21), *is rectilinear if and only if the resultant of the given forces acting upon it has a fixed direction, and its initial velocity has the same direction.*

Analogously, proceeding as in Subsec. 1.1.2, we obtain

Theorem 7.1.2'. *The trajectory of a free particle in a resistent medium, modelled by the relation* (7.1.21), *is a plane curve if and only if the resultant of the given forces acting upon it is parallel to a fixed plane* (*is normal to a fixed direction*), *and its initial velocity has the same property.*

In the case of a heavy particle P in a resistent medium, the equation of motion is

$$\ddot{\mathbf{r}} = \mathbf{g} - g\overline{\varphi}(v)\dot{\mathbf{r}}, \qquad (7.1.22)$$

where we assume that – in general – the initial velocity $\mathbf{v}_0$ is not directed along the vertical of the position of launching ($\mathbf{v}_0$ has not the same direction as $\mathbf{g}$); corresponding to the Theorem 7.1.2', the trajectory is a plane curve (contained in a vertical plane). Using Frenet's trihedron, we may write

$$\dot{v} = -g[\sin\theta + \varphi(v)], \quad \frac{v^2}{\rho} = g\cos\theta, \qquad (7.1.22')$$

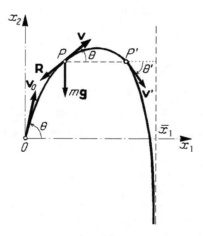

Figure 7.4. Motion of a heavy particle in a resistent medium.

where θ is the angle made by the velocity $\mathbf{v}$ with the Ox_1-axis. We notice that $\cos\theta \geq 0$, hence $-\pi/2 \leq \theta \leq \pi/2$; the concavity of the trajectory is directed towards the negative ordinates (Fig.7.4), so that to $ds > 0$ corresponds $d\theta < 0$ (the angle θ is decreasing). It follows $\rho = -ds/d\theta = -vdt/d\theta$, so that the second equation (7.1.22') takes the form

$$v\dot{\theta} = -g\cos\theta. \qquad (7.1.22'')$$

We have thus obtained a system of two differential equations (7.1.22'), (7.1.22") for the unknown functions $v = v(t)$ and $\theta = \theta(t)$, with the initial conditions $v(t_0) = v_0$, $\theta(t_0) = \theta_0$. Eliminating the time t, we may write the equation

$$\frac{dv}{d\theta} = v\left[\tan\theta + \frac{\varphi(v)}{\cos\theta}\right], \qquad (7.1.23)$$

which defines the function $v(\theta)$ with the initial condition $v(\theta_0) = v_0$. This equation of the hodograph of motion, which can be written also in the form

$$\frac{d(v\cos\theta)}{d\theta} = v\varphi(v), \qquad (7.1.23')$$

is *the basic equation of the external ballistics*. The equation (7.1.22") allows then to determine (usually, one takes $t_0 = 0$)

$$t = t_0 - \frac{1}{g}\int_{\theta_0}^{\theta} \frac{v\vartheta}{\cos\vartheta}d\vartheta, \qquad (7.1.23")$$

wherefrom – afterwards – we may obtain $\theta = \theta(t)$. Noting that $dx_1 = v\cos\theta dt$, $dx_2 = v\sin\theta dt$, there result the parametric equations of the trajectory in the form

$$x_1 = x_1^0 - \frac{1}{g}\int_{\theta_0}^{\theta} v^2(\vartheta)d\vartheta, \quad x_2 = x_2^0 - \frac{1}{g}\int_{\theta_0}^{\theta} v^2(\vartheta)\tan\vartheta d\vartheta, \qquad (7.1.23''')$$

where we take $x_1^0 = x_2^0 = 0$ if the particle (the projectile) is launched from the origin O. In the case of an object launched from an airplane at the height h we take $x_1^0 = 0$, $x_2^0 = h$; the initial velocity v_0 is the velocity of the airplane at the moment of launching the object.

From the second equation (7.1.22') one observes that (θ is only decreasing and greater than $-\pi/2$ for t finite, hence $\cos\theta > 0$) the velocity v is finite and non-zero. An extreme value of v is given by $dv/dt = 0$; we obtain thus $\varphi(v) = -\sin\theta$. Because the velocity v is finite, from (7.1.22") it follows that θ has an extreme value for $d\theta/dt = 0$, hence for $\cos\theta = 0$; but the angle θ is decreasing, so that we have $\lim_{t\to\infty}\theta = -\pi/2$. We notice that for $v > v^*$, $\varphi(v^*) = 1$, we have $\dot{v} < 0$, the function $\varphi(v)$ being monotone decreasing. Hence, the velocity v has an inferior limit ($v > 0$) and a superior one ($v \leq v^*$). The trajectory has a vertical asymptote $x_1 = \bar{x}_1$, with

$$\bar{x}_1 = \lim_{\theta \to -\frac{\pi}{2}+0} x_1 = \frac{1}{g}\int_{-\frac{\pi}{2}}^{\theta_0} v^2(\vartheta)d\vartheta, \qquad (7.1.24)$$

Problems of dynamics of the particle 413

and the corresponding velocity is given by $\lim_{\theta \to -\pi/2+0} v(\theta) = v^*$. Because of the resistance of the air, we notice that the range of throw of the projectile is smaller. Besides, for two points P and P' of the trajectory, which have the same ordinate x_2, it results $|\theta| < |\theta'|$; hence, the two branches (increasing and decreasing) of the trajectory are not symmetric. Applying the theorem of kinetic energy, we may write $\mathrm{d}(mv^2/2) = -mg\mathrm{d}x_2 - mg\varphi(v)v\mathrm{d}t$, so that, integrating between the points $P(t)$ and $P'(t')$, we obtain

$$\frac{1}{2}(v'^2 - v^2) = -g\int_t^{t'} \varphi[v(\tau)]v(\tau)\mathrm{d}\tau < 0,$$

wherefrom $v > v' > 0$.

Modelling the projectile as a rigid solid, one can take into account also its rotation, being led to a deviation from the vertical plane of the trajectory.

In particular, d'Alembert has considered *the law of resistance* $\varphi(v) = \lambda v^n$, $n > 0$, λ being a positive constant with dimension. The equation (7.1.23') leads to

$$\frac{\mathrm{d}}{\mathrm{d}\theta}(v\cos\theta) = \frac{\lambda(v\cos\theta)^{n+1}}{\cos^{n+1}\theta};$$

integrating, we get

$$v\cos\theta = \frac{v_0\cos\theta_0}{\{1 - n\lambda[\varepsilon_n(\theta) - \varepsilon_n(\theta_0)](v_0\cos\theta_0)^n\}^{1/n}}, \quad (7.1.25)$$

where we have introduced the integral

$$\varepsilon_n(\theta) = \int_0^\theta \frac{\mathrm{d}\vartheta}{\cos^{n+1}\vartheta}. \quad (7.1.25')$$

For small velocities, one can use *Stokes' law* ($n = 1$); thus, we obtain $\varepsilon_1(\theta) = \tan\theta$, so that

$$v(\theta) = \frac{v_0\cos\theta_0}{\cos\theta - \lambda v_0\sin(\theta - \theta_0)}. \quad (7.1.25'')$$

For velocities till 250 m/s one may take $n = 2$, obtaining *Euler's law*; we notice that

$$\varepsilon_2(\theta) = \frac{1}{2}\left[\frac{\tan\theta}{\cos\theta} + \ln\tan\left(\frac{\pi}{4} + \frac{\theta}{2}\right)\right]. \quad (7.1.26)$$

Let us consider now $n \in \mathbb{N}$; for n odd ($n = 2p - 1$), we have

$$\varepsilon_n(\theta) = \frac{\sin\theta}{2p-1}\left[\sec^{2p-1}\theta + \sum_{k=1}^{p-1}\frac{2^k(p-1)(p-2)\ldots(p-k)}{(2p-3)(2p-5)\ldots(2p-2k-1)}\sec^{2p-2k-1}\theta\right], \quad (7.1.26')$$

while for n even ($n = 2p$) we may write

$$\varepsilon_n(\theta) = \frac{\sin\theta}{2p}\left[\sec^{2p}\theta + \sum_{k=1}^{p-1}\frac{(2p-1)(2p-2)\ldots(2p-2k+1)}{2^k(p-1)(p-2)\ldots(p-k)}\sec^{2p-2k}\theta\right]$$
$$+ \frac{(2p-1)!!}{2^p\, p!}\ln\tan\left(\frac{\pi}{4} + \frac{\theta}{2}\right). \quad (7.1.26'')$$

The velocity $v(\theta)$ is then easily given by the formula (7.1.25), obtaining the time t and the parametric equations of the trajectory from the formulae (7.1.23"), (7.1.23''').

Legendre considers the resistance law $\varphi(v) = \lambda v^n + \mu$, where $\lambda, \mu, n > 0$, while $\mu < 1$ (otherwise, the particle – without initial velocity – comes against a resistance μmg and can no more fall); also in this case, the problem may be solved by quadratures.

We observe that, by the substitution $v[\sin\theta + \varphi(v)] = 1/y$, the equation (7.1.23) reads

$$\frac{dy}{dv} = v\left[\varphi^2(v) - 1\right]y^3 - \left[2\varphi(v) + v\frac{d\varphi(v)}{dv}\right]y^2; \quad (7.1.27)$$

Drach has determined all the forms of the function $\varphi(v)$ for which the solution of this equation may be obtained by quadratures.

1.2.3 Rectilinear motion of a heavy particle

If, in the motion of the heavy particle P considered above, the initial velocity vanishes ($\mathbf{v}_0 = \mathbf{0}$) or is collinear with $\mathbf{g}$, then the trajectory is rectilinear (along the local vertical). Taking the Ox-axis along the direction of the gravity acceleration $\mathbf{g}$, the equations (7.1.14) become

$$x = \frac{1}{2}g(t-t_0)^2 + v_0(t-t_0) + x_0, \quad v = g(t-t_0) + v_0, \quad (7.1.28)$$

for the motion in vacuum, with the initial conditions $x(t_0) = x_0$, $v(t_0) = v_0$; taking $t_0 = 0$ and $x_0 = 0$, we may write

$$x = \frac{1}{2}gt^2 + v_0 t, \quad v = gt + v_0, \quad (7.1.28')$$

without loosing anything from the generality.

Problems of dynamics of the particle 415

Eliminating the time t between the equations (7.1.28), we obtain the relation between the velocity v and the co-ordinate x in the form

$$v = \sqrt{v_0^2 + 2g(x - x_0)}, \qquad (7.1.28'')$$

corresponding to Torricelli's formula (7.1.17). A particle which is falling with the initial velocity v_0 from a height h (in this case $x - x_0 > 0$) comes down with the velocity $v = \sqrt{v_0^2 + 2gh}$ after an interval of time $(v_0/g)(\sqrt{1 + 2gh/v_0^2} - 1)$. A particle which is thrown up with the initial velocity v_0 (in this case $x - x_0 < 0$) attains the height $h = v_0^2/2g$ after an interval of time v_0/g; after another interval of time v_0/g, the particle comes back to the initial position with the same velocity v_0.

In the case of a resistent medium, it is convenient to distinguish between the descendent and the ascendent motion along the local vertical. In the first case, if we assume an initial velocity $v_0 > 0$, with the same direction as the Ox-axis, then the equation (7.1.22) leads to

$$\dot{v} = g[1 - \varphi(v)] = g[\varphi(v^*) - \varphi(v)], \qquad (7.1.29)$$

the velocity v^* being introduced in the previous subsection. It results that

$$t = t_0 + \frac{1}{g}\int_{v_0}^{v} \frac{\mathrm{d}\eta}{\varphi(v^*) - \varphi(\eta)}, \qquad (7.1.29')$$

and we may obtain $v = v(t)$; noting that $\mathrm{d}x = v\mathrm{d}t$, we get also

$$x = x_0 + \frac{1}{g}\int_{v_0}^{v} \frac{\eta\mathrm{d}\eta}{\varphi(v^*) - \varphi(\eta)}. \qquad (7.1.29'')$$

If $v_0 < v^*$, then $\varphi(v^*) - \varphi(v)$ is positive at the beginning, while the equation (7.1.29) shows that the velocity v increases. From (7.1.29') we notice that for $v \to v^*$ we have $t \to \infty$, so that v^* is a superior limit for the velocities; hence, the velocity v increases till this limit. Analogously, if $v_0 > v^*$, then the velocity v decreases till the limit value v^* (in this case $\dot{v} < 0$). We may thus state that, for any initial velocity v_0, the particle falls with a velocity which tends to become uniform (tends to v^* for $t \to \infty$); if $v_0 = v^*$, then the motion of the particle is uniform.

If a heavy body, which may be modelled by a heavy particle, is launched by a parachute, then the velocity is – at the beginning – increasing (a motion approximately uniform accelerated, as in vacuum, the resistance of the air being negligible); when the parachute is opening, the resistance of the air increases very much and the falling velocity tends to v^* (a motion approximately uniform). As well, let us consider two equal bodies (e.g., two whole spheres, of the same radius, but of different matter), modelled by two particles of masses m_1 and m_2; at equal velocities, these bodies come against the same resistance of the air. Hence, $m_1 g \varphi_1(v) = m_2 g \varphi_2(v)$, where $\varphi_1(v)$ and $\varphi_2(v)$ are the functions corresponding to the forces $\mathbf{R}_1 = \mathbf{R}_2$. Observing that $\varphi_1(v_1^*) = \varphi_2(v_2^*) = 1$ and making $v = v_1^*$, we may write $\varphi_1(v_1^*) = (m_2/m_1) \varphi_2(v_1^*) = 1$, so that $\varphi_2(v_1^*) = m_1/m_2$; hence, if $m_1 > m_2$, then it results $\varphi_2(v_1^*) > \varphi_2(v_2^*) = 1$, so that $v_1^* > v_2^*$. We may thus state that the heaviest particle has a limit velocity of falling in the air greater than that of the lighter one.

If the motion is ascendent and if we assume an initial velocity $v_0 > 0$ directed in the same direction as the Ox-axis (taken in the opposite direction of the gravity acceleration $\mathbf{g}$), then the equation (7.1.22) reads

$$\dot{v} = -g[1 + \varphi(v)]; \qquad (7.1.30)$$

we find thus

$$t = t_0 - \frac{1}{g} \int_{v_0}^{v} \frac{d\eta}{1 + \varphi(\eta)}, \quad x = x_0 - \frac{1}{g} \int_{v_0}^{v} \frac{\eta d\eta}{1 + \varphi(\eta)}. \qquad (7.1.30')$$

The particle attains a height

$$h = \frac{1}{g} \int_0^{v_0} \frac{v dv}{1 + \varphi(v)}, \qquad (7.1.30'')$$

after an interval of time

$$\overline{t} = \frac{1}{g} \int_0^{v_0} \frac{dv}{1 + \varphi(v)}. \qquad (7.1.30''')$$

If $\varphi(v) \equiv 0$, then we notice that one obtains greater values for h and t (the integrand will be greater); hence, *a heavy particle launched up along the vertical, with an initial velocity v_0, reaches in the air a smaller height in a shorter time as in the vacuum.* After an interval $\overline{t}$ of time, the particle stops and then comes down, as in the previous considerations, and – by coming down – has an initial zero velocity (hence, smaller than v^*). The particle reaches the initial position with a velocity v^0, specified by the relation

$$h = \frac{1}{g} \int_0^{v^0} \frac{v dv}{1 - \varphi(v)},$$

as it results form (7.1.29"); comparing with the relation (7.1.30"), in which the integrand is smaller, we notice that $v^0 < v_0$, so that *in the air the falling velocity is smaller than the launching one.* One returns to that position after an interval of time

$$t^0 = \frac{1}{g} \int_0^{v^0} \frac{dv}{1 - \varphi(v)}.$$

Problems of dynamics of the particle

To compare with the relation (7.1.30'''), we observe that

$$d\bar{t} = \frac{1}{g}\frac{dv_0}{1+\varphi(v_0)}, \quad dt^0 = \frac{1}{g}\frac{dv^0}{1-\varphi(v^0)};$$

as well

$$dh = \frac{1}{g}\frac{v_0 dv_0}{1+\varphi(v_0)} = \frac{1}{g}\frac{v^0 dv^0}{1-\varphi(v^0)}.$$

We get

$$d(\bar{t}^0 - \bar{t}) = \frac{1}{g}\left(\frac{v_0}{v^0} - 1\right)\frac{dv_0}{1+\varphi(v_0)} \geq 0,$$

so that $t^0 > \bar{t}$ (because for $v_0 = 0$ we have $\bar{t} = t^0 = 0$); hence, *in the air, the falling time is greater than that of rising* (for the same height h).

The integrals may be easily calculated for $\varphi(v) = \lambda v^n$, $n \in \mathbb{Q}$, $n > 0$, λ positive constant.

1.2.4 Motion of a heavy particle constrained to stay on a curve

Let P be a heavy particle constrained to move on a fixed curve C, neglecting the resistance of the air; assuming that the Ox-axis is directed along the local vertical, opposite to the gravity acceleration, we may write the conservation theorem of mechanical energy in the form

$$m\frac{v^2}{2} = -mgx + h, \quad h = \text{const},$$

wherefrom

$$v^2 = 2g(a - x), \quad a = \frac{h}{mg}. \tag{7.1.31}$$

Let be the plane Π of equation $x = a$ and P' the projection of the particle P on this plane; the velocity of the particle P is, in this case, given by

$$v = \sqrt{2gh}, \quad h = \overline{PP'}, \tag{7.1.31'}$$

and is equal to that of a heavy particle which falls from P' in P, without initial velocity.

Let us suppose that the curve C is closed. If the curve does not pierce the plane Π, that one being above it (we may assume to have an initial velocity v_0 for any initial position P_0 of applicate x_0, so that $a = x_0 + v_0^2/2g$ be as great as we wish), then the

418   MECHANICAL SYSTEMS, CLASSICAL MODELS

formula (7.1.31') shows that the velocity does never vanish; the motion is periodic, the point of maximal applicate having the minimal velocity, while that of minimal applicate has the maximal one.

If the velocity v_0 is not sufficient great, the plane pierces the curve C at the points $\overline{P}$ and $\overline{P}'$ (Fig.7.5). We suppose that the particle is launched from the point P_0 of minimal applicate x_0 with the initial velocity v_0; it reaches the point $\overline{P}$ after an interval of time

$$T = \frac{1}{\sqrt{2g}} \int_{x_0}^{a} \frac{ds}{\sqrt{a-x}}, \qquad (7.1.32)$$

where $ds = v dt$ is the element of arc on the arc $\widehat{P_0 P}$. If the tangent at $\overline{P}$ is not horizontal, then the particle returns at P_0 in a time T with the velocity v_0 and then reaches $\overline{P}'$ in a time T', which is calculated by the same formula (7.1.32), ds being an arc element on the arc $\widehat{P_0 \overline{P}'}$. Hence, the motion of the particle is oscillatory between the points $\overline{P}$ and $\overline{P}'$, each simple oscillation being of duration $T + T'$.

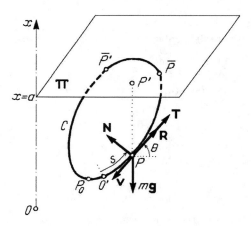

Figure 7.5. Motion of a heavy particle constrained to stay on a curve.

Let us consider now that the motion of the particle is with friction in a resistent medium; for the sake of simplicity we assume that the curve C is situated in a vertical plane. We choose an origin O' for the curvilinear co-ordinate s. As well, we introduce the normal constraint force $\mathbf{N}$, the tangential constraint force $\mathbf{T} = -fN\,\mathrm{vers}\,\mathbf{v}$, where f is the sliding friction coefficient, and the resistance $\mathbf{R} = -mg\varphi(v)\,\mathrm{vers}\,\mathbf{v}$. The equations of motion in intrinsic co-ordinates are written in the form

$$m\dot{v} = -mg[\sin\theta - \varphi(v)] + fN, \quad m\frac{v^2}{\rho} = N - mg\cos\theta, \qquad (7.1.33)$$

Problems of dynamics of the particle 419

where ρ is the curvature radius, while θ is the angle formed by the tangent to the curve C with the horizontal line. Eliminating the constraint force N between the two equations and noting that $2\dot{v} = \mathrm{d}(v^2)/\mathrm{d}s$, we obtain the equation

$$\frac{\mathrm{d}(v^2)}{\mathrm{d}s} = -2g[\sin\theta - f\cos\theta - \varphi(v)] + 2f\frac{v^2}{\rho}, \qquad (7.1.33')$$

which determines the unknown function $v = v(s)$; then one may calculate s and t by quadratures. The points on the curve for which

$$\sin\theta - f\cos\theta = \sin(\theta - \varphi)/\cos\varphi = 0$$

hence for which $\theta = \varphi$, where φ is the angle of sliding friction, represent limit positions of equilibrium for the particle.

1.2.5 Motion of a heavy particle constrained to stay on a plane

Let be a heavy particle P situated on a plane Π which makes the angle $0 < \alpha \le \pi/2$ with the horizontal plane (Fig.7.6). Assuming a sliding friction of coefficient $f = \tan\varphi$, the force of friction is Φ, $\Phi \le fN$, where N is the constraint force, normal to the plane; taking into account the condition (4.1.50) and observing that the given force has the magnitude $F = mg$ and the normal component $F_n = -mg\cos\alpha$, it results the condition of equilibrium $\cos^2\alpha \ge \cos^2\varphi$. Hence, if the angle α is at the most equal to the angle of friction, then the particle is in equilibrium in any position.

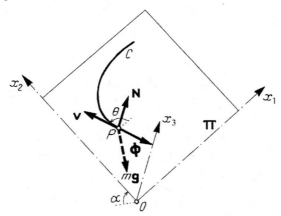

Figure 7.6. Motion of a heavy particle constrained to sliding friction on a plane.

If the particle is in motion and the mentioned condition is not fulfilled, then its position will be given by the equation

$$m\ddot{\mathbf{r}} = m\mathbf{g} - fN\frac{\mathbf{v}}{v} + \mathbf{N}, \quad N = mg\cos\alpha \qquad (7.1.34)$$

in the plane $x_3 = 0$; choosing the axes Ox_1 (the intersection of the plane Π with the horizontal one) and Ox_2 in this plane, we obtain

$$\ddot{x}_1 = \dot{v}\cos\theta - v\dot{\theta}\sin\theta = -fg\cos\alpha\cos\theta,$$
$$\ddot{x}_2 = \dot{v}\sin\theta + v\dot{\theta}\cos\theta = -g\sin\alpha - fg\cos\alpha\sin\theta,$$

where θ is the angle made by the velocity $\mathbf{v}$ with the Ox_1-axis. It results

$$\dot{v} = -fg\cos\alpha - g\sin\alpha\sin\theta, \quad v\dot{\theta} = -g\sin\alpha\cos\theta,$$

wherefrom

$$\frac{\mathrm{d}v}{\mathrm{d}\theta} = v\left(\tan\theta + \frac{f\cot\alpha}{\cos\theta}\right); \tag{7.1.34'}$$

this equation is of the form (7.1.23) and it may be analogously studied. Noting that $f\cot\alpha = \mathrm{const}$, we get

$$v\cos\theta = v_0\cos\theta_0\left[\tan\left(\frac{\pi}{4} - \frac{\theta_0}{2}\right)\cot\left(\frac{\pi}{4} - \frac{\theta}{2}\right)\right]^{f\cot\alpha} \tag{7.1.34''}$$

and then

$$x_1 = x_1^0 - \frac{1}{g\sin\alpha}\int_{\theta_0}^{\theta} v^2(\vartheta)\mathrm{d}\vartheta, \quad x_2 = x_2^0 - \frac{1}{g\sin\alpha}\int_{\theta_0}^{\theta} v^2(\vartheta)\tan\vartheta\,\mathrm{d}\vartheta, \tag{7.1.34'''}$$

with $-\pi/2 < \theta \leq \theta_0$. We notice that

$$\lim_{\theta \to -\pi/2+0} v = \begin{cases} \infty & \text{for } f\cot\alpha < 1, \\ 0 & \text{for } f\cot\alpha > 1, \end{cases}$$

assuming that $\theta_0 < \pi/2$.

In the case of a horizontal plane ($\alpha = 0$) it results $\dot{\theta} = 0$, hence $\theta = \theta_0$; *the motion is rectilinear and uniformly delayed.* The equation (7.1.34) can be decomposed in two equations

$$m\dot{\mathbf{v}} = -fN\frac{\mathbf{v}}{v}, \quad m\mathbf{g} + \mathbf{N} = \mathbf{0}, \tag{7.1.35}$$

as the vectors are contained in the Π-plane or are normal to this one. There results $N = mg$ and the velocity

$$\mathbf{v} = \mathbf{v}_0 e^{-\int_{t_0}^{t}\frac{fg}{v(\tau)}\mathrm{d}\tau}, \tag{7.1.35'}$$

Problems of dynamics of the particle 421

a confirmation that the trajectory is rectilinear. Choosing this trajectory as Ox-axis, we obtain

$$v = -fg(t - t_0) + v_0, \quad x = -\frac{fg}{2}(t - t_0)^2 + v_0(t - t_0) + x_0,$$
$$t_0 \leq t \leq t^* = t_0 + \frac{v_0}{fg}, \qquad (7.1.35'')$$

where the time t^* is given by the condition $v(t^*) = 0$; timing the time t^*, one may obtain – experimentally – the coefficient of sliding friction f. After a time $t^* - t_0$, the particle travels through the distance $l = v_0^2/2fg$; inversely, measuring l one determines the initial velocity

$$v_0 = \sqrt{2fgl}. \qquad (7.1.35''')$$

We may thus estimate the velocity of a car if we know the braking distance l.

If the initial velocity vanishes ($v_0 = 0$), then it results $\theta = 3\pi/2$, so that the sliding of the particle on the inclined plane takes place along the line of greatest slope, the trajectory being rectilinear; using the above equations, the acceleration at a given moment is specified by (along the Ox_2-axis)

$$a_2 = -g\frac{\sin(\alpha - \varphi)}{\cos\varphi}.$$

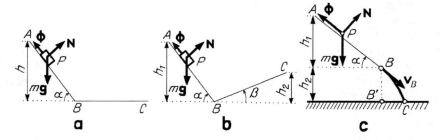

Figure 7.7. Motion of a heavy particle constrained to sliding friction on an inclined plane: motion continues on a horizontal plane (a); motion continues on an inclined plane (b); silo's problem (c).

Applying the theorem of kinetic energy in a finite form (6.1.48''), we obtain the velocity at the point B, assuming that the body, modelled as a particle, departs from A without initial velocity (Fig.7.7,a); noting that

$$T_A = 0, \quad T_B = \frac{1}{2}mv_B^2, \quad W_{AB} = mg\frac{h}{\sin\alpha}(\sin\alpha - f\cos\alpha) = mgh\frac{\sin(\alpha - \varphi)}{\sin\alpha\cos\varphi},$$

it follows

$$v_B = \sqrt{2gh}\sqrt{\frac{\sin(\alpha - \varphi)}{\sin\alpha \cos\varphi}}. \tag{7.1.36}$$

In the absence of frictions, $v_B = \sqrt{2gh}$, as in the case of a free falling along the vertical (Torricelli's formula). Applying further the theorem of kinetic energy, we see that the body moves on the horizontal line till the point C, so that (Fig.7.7,a)

$$\overline{BC} = h\frac{\sin(\alpha - \varphi)}{\sin\alpha \sin\varphi}; \tag{7.1.36'}$$

in the case of vanishing friction, C tends to infinity. If the plane BC makes the angle β with the horizontal line, the body moves till the height h_2, given by (Fig.7.7,b)

$$h_2 = h_1\frac{\sin\beta \sin(\alpha - \varphi)}{\sin\alpha \sin(\beta + \varphi)} < h_1; \tag{7.1.36''}$$

in the case in which the friction is vanishing, we have $h_2 = h_1$. This is *the problem of the sledge*. If, from the point B, the particle is falling freely till the point C (*the problem of the silo*, Fig.7.7,c), then the motion takes place along a parabola, so that

$$h_2 = \frac{1}{2}\frac{g}{v_B^2 \cos^2\alpha}\overline{B'C}^2 + \overline{B'C}\tan\alpha. \tag{7.1.36'''}$$

1.3 Pendulary motion

As we have seen in Subsec. 1.2.4, a heavy particle constrained to stay on a fixed curve has an oscillatory motion; such a motion is called also a *pendulary motion*. In what follows, we consider the case of a curve in a vertical plane, in particular the case in which the curve is a circle, an ellipse or a cycloid; we study then the general case of motion, as well as the case of small displacements in the neighbourhood of a stable position of equilibrium. Starting from the motion of a heavy particle on a surface of rotation, we present the general problem of the spherical pendulum too.

1.3.1 Simple pendulum

A *simple pendulum* (or *mathematical pendulum*) is a heavy particle which moves without friction on a circle C of radius l, situated in a vertical plane. The constraint may be bilateral (e.g., a ball modelled as a particle constrained to move in the interior of a circular tube (Fig.7.8,a) or a ball linked to the centre O of the circle by an inextensible and incompressible bar OP, of negligible mass with respect to that of the particle (Fig.7.8,b)) or *unilateral* (e.g., a ball linked to the centre O by an inextensible and perfectly flexible thread (Fig.7.8,c) or a ball constrained to move on a whole cylinder, which has a horizontal axis (Fig.7.8,d)).

If we choose the Ox-axis in the same direction as that of the gravity acceleration **g** (Fig.7.9), then the theorem of kinetic energy in finite form, applied between the points P_0 and P, allows to write

$$v^2 = v_0^2 - 2g(x_0 - x) = v_0^2 - 2gl(\cos\theta_0 - \cos\theta) = -2g(a - x),$$

$$a = x_0 - \frac{v_0^2}{2g},$$
(7.1.37)

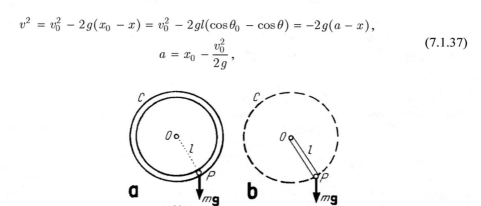

Figure 7.8. Motion of a heavy particle constrained to stay on a circle in a vertical plane: bilateral constraint (a,b); unilateral constraint (c,d).

where $v_0 = l\dot\theta_0$ is the modulus of the initial velocity (for the sake of simplicity, we will say the initial velocity) at the point P_0 (corresponding to the results in Subsec. 1.2.4). The equation $x = a$ is that of a straight line which may be reached by a particle launched up, along the local vertical, with the initial velocity v_0; the motion is characterized by the constant a in the case of a bilateral constraint. Indeed, if the straight line $x = a$ pierces the circle C ($-l < a < l$), then the motion is *oscillatory*, if this line is tangent to the circle ($a = -l$), then the motion is *asymptotic*, while if the line does not pierce the circle ($a < -l$), then the motion is *circular*. We notice that for $a = l$ we have $v_0 = 0$, corresponding a stable position of equilibrium; we cannot have $a > l$. From (7.1.37) it results that the velocity $v = l\dot\theta$ may vanish for an angle given by $\cos\theta = \cos\theta_0 - v_0^2/2gl$ or by $\sin^2(\theta/2) = \sin^2(\theta_0/2) + v_0^2/4gl$. This condition can never be satisfied if $v_0^2 > 4gl$ (or $\dot\theta_0^2 > 4\omega^2$, $\omega^2 = g/l$), the motion being circular. If $v_0^2 < 4gl$, then the condition may be fulfilled for some values of the angle θ_0, hence for some initial positions, e.g., for $\theta_0 = 0$; in this case, the motion is oscillatory. If $v_0^2 = 4gl$, then we must have $\theta_0 = 0$, the motion being asymptotic.

First of all, let us suppose that the motion is oscillatory; we denote $a = l\cos\alpha$, where $0 < \alpha < \pi$ is the angle corresponding to the limit position $\bar{P}$ (for which $v = 0$) of the particle P, specifying thus the amplitude of the motion. The relation (7.1.37) becomes

$$\dot{\theta}^2 = 2\omega^2(\cos\theta - \cos\alpha);\tag{7.1.38}$$

differentiating with respect to time, we may also write (we suppose $\dot{\theta} \neq 0$)

$$\ddot{\theta} + \omega^2 \sin\theta = 0.\tag{7.1.38'}$$

This equation (called *the equation of the mathematical pendulum*) is often encountered in problems of mechanics in one of the two equivalent forms mentioned above; in fact, the relation (7.1.38) corresponds to a first integral of the equation of motion (7.1.38').

The particle departs from the initial position P_0 with the velocity v_0 and travels up on the circle with a velocity which diminishes in intensity; at the extreme position $\overline{P}$, the velocity vanishes. Returning on the travelled arc of circle, the velocity increases; the particle passes through the initial position P_0 and reaches the lowest point P' of the trajectory, where it has the maximal velocity; then the velocity decreases till the particle reaches the point $\overline{P}'$ for which $\theta = -\alpha$. The particle returns then to P' and $\overline{P}$ a.s.o. Hence, the motion is oscillatory. From the relation (7.1.38), we also notice that the velocity $v(t)$ depends only on the position of the particle, being a periodic function of this position (of angle θ); integrating this equation with separate variables, we may write (during the motion we have $\cos\theta > \cos\alpha$)

$$t = t^0 + \frac{1}{\omega\sqrt{2}}\int_{\theta^0}^{\theta}\frac{\mathrm{d}\vartheta}{\sqrt{\cos\vartheta - \cos\alpha}},\tag{7.1.39}$$

where θ^0 corresponds to the position at the arbitrary moment t^0 (which may be different from the initial moment t_0). Hence, one can see that the interval of time $t - t^0$ depends only on the positions corresponding to the two moments; it results that *the oscillatory motion is periodical*, of *period T*. We notice too that changing the direction of motion on the arc of circle the sign of the velocity changes; its modulus remains the same when passing through the same point, so that the arc $\widehat{PP'}$ is travelled through in an interval of time $T/2$. Because the relation (7.1.38) is even with respect to θ, it results that, at points symmetric with respect to the Ox-axis, we have the same velocity (travelling up or down); hence, the arc $\widehat{P'P}$ is travelled through in a quarter of period. In this case, the period T is given by the relation

$$T = \frac{2\sqrt{2}}{\omega}\int_0^{\alpha}\frac{\mathrm{d}\vartheta}{\sqrt{\cos\vartheta - \cos\alpha}}.\tag{7.1.39'}$$

Observing that $\cos\theta - \cos\alpha = 2[\sin^2(\alpha/2) - \sin^2(\theta/2)]$ and denoting $\sin(\theta/2) = k\sin\varphi$, $k = \sin(\alpha/2)$, we may write

$$t = t^0 + \frac{1}{\omega}\int_{\varphi^0}^{\varphi}\frac{\mathrm{d}\psi}{\sqrt{1 - k^2\sin^2\psi}},\tag{7.1.40}$$

where φ^0 is specified by the relation $\sin(\theta^0/2) = k\sin\varphi^0$; denoting $\sin\varphi = z$, we may also write

$$t = t^0 + \frac{1}{\omega}\int_{z^0}^{z}\frac{\mathrm{d}\zeta}{\sqrt{(1-\zeta^2)(1-k^2\zeta^2)}}, \tag{7.1.40'}$$

where z^0 is specified by the relation $\sin\varphi^0 = z^0$. Introducing, after Legendre, *the elliptic integral of the first kind*

$$F(\varphi,k) = \int_0^{\varphi}\frac{\mathrm{d}\psi}{\sqrt{1-k^2\sin^2\psi}} = \int_0^{\sin\varphi}\frac{\mathrm{d}z}{\sqrt{(1-z^2)(1-k^2z^2)}}, \tag{7.1.41}$$

where φ is *the amplitude*, while k is *the modulus of the integral*, we obtain

$$t = t^0 + \frac{1}{\omega}\left[F(\varphi,k) - F(\varphi^0,k)\right]. \tag{7.1.40''}$$

Denoting $u = \omega t$, we may write

$$u - u^0 = F(\varphi,k) - F(\varphi^0,k), \tag{7.1.40'''}$$

where $u^0 = \omega t^0$. Taking $t^0 = 0$, with no loss of generality, and assuming that $\theta^0 = 0$, there results $\varphi^0 = z^0 = u^0 = F(\varphi^0,k) = 0$, so that

$$u = F(\varphi,k). \tag{7.1.42}$$

As it was noticed by Abel, it is easier to express the angle φ as a function of the variable u, in the form

$$\sin\varphi = \operatorname{sn} u, \tag{7.1.42'}$$

where sn is the symbol of *the elliptic sine* (*the amplitude sine*), one of *the Jacobi's elliptic functions*; analogously, one may use *the elliptic cosine* (*the amplitude cosine*), denoted by the symbol cn ($\cos\varphi = \operatorname{cn} u$).

Starting from the formula (7.1.39'), the period of the motion is given by

$$T = \frac{4}{\omega}K(k) = \frac{4}{\omega}\int_0^{\pi/2}\frac{\mathrm{d}\varphi}{\sqrt{1-k^2\sin^2\varphi}} = \frac{4}{\omega}\int_0^{1}\frac{\mathrm{d}z}{\sqrt{(1-z^2)(1-k^2z^2)}},$$
$$\omega = \sqrt{\frac{g}{l}}, \tag{7.1.43}$$

where $K(k) = F(\pi/2,k)$ is the complete elliptic integral of the first kind. Observing that $k^2 < 1$, one obtains the development into series (we use Newton's binomial series)

$$(1 - k^2 \sin^2 \varphi)^{-1/2} = 1 + \sum_{n=1}^{\infty} \frac{(2n)!}{2^{2n}(n!)^2} k^{2n} \sin^{2n} \varphi.$$

This series is absolute and uniform convergent on the interval $[0, 2\pi]$, and we may integrate it term by term; taking into account Wallis' formula

$$\int_0^{\pi/2} \sin^{2n} \varphi \, d\varphi = \frac{(2n)!}{2^{2n}(n!)^2} \frac{\pi}{2}, \qquad (7.1.44)$$

one obtains the period

$$T = 2\pi \sqrt{\frac{l}{g}} \left\{ 1 + \sum_{n=1}^{\infty} \frac{[(2n)!]^2}{2^{4n}(n!)^4} \sin^{2n} \frac{\alpha}{2} \right\}. \qquad (7.1.43')$$

Because we may develop $\sin(\alpha/2)$ into an absolute convergent series with respect to α too, we obtain also for T such a development, of the form

$$T = 2\pi \sqrt{\frac{l}{g}} \left(1 + \frac{\alpha^2}{16} + \frac{11}{12} \frac{\alpha^4}{16^2} + \ldots \right). \qquad (7.1.43'')$$

We observe that the ratio between the second and the first term of the series is equal to $\alpha^2/16$; as well, the ratio between the third and the second term is given by $(11/12)\alpha^2/16$ a.s.o. Hence, the series is rapidly convergent; practically, we may take

$$T = 2\pi \sqrt{\frac{l}{g}} \left(1 + \frac{\alpha^2}{16} \right). \qquad (7.1.43''')$$

If $\alpha = 0.4$ (it corresponds to the angle $22°55'06''$), then the correction brought by the second term of the development is not greater than 1%. The astronomic watches have amplitudes of $1°30'$, corresponding a correction of approximately $0.05‰$. In general, the period T depends on the angle α, but is independent of the mass m of the particle. In the case of small oscillations around a stable position of equilibrium (in Chap. 4, Subsec. 1.1.7 we have seen that the point P' represents a stable position of equilibrium), the equation (7.1.38') has the form (we approximate $\sin \theta$ by θ)

$$\ddot{\theta} + \omega^2 \theta = 0, \qquad (7.1.45)$$

wherefrom

$$\theta(t) = \alpha \cos(\omega t - \varphi), \qquad (7.1.45')$$

the angle φ being specified by the initial conditions, while the period is given by *Galileo's formula*

$$T = \frac{2\pi}{\omega} = 2\pi\sqrt{\frac{l}{g}};\qquad(7.1.45")$$

we notice that this result (intuited by considerations of homogeneity in Chap. 1, Subsec. 2.2.4) approximates the development into series (7.1.43"). The period T thus obtained depends on the length l of the pendulum and on the gravity acceleration g at the respective place on Earth's surface. Because this period does not depend on the amplitude α, we say that the respective motion is *isochronic* (the small oscillations around a stable position of equilibrium take place in the same interval of time). A particle P left free from $\bar{P}$ without initial velocity reaches the lowest position P' in a time equal to $T/4$, which does not depend on the initial position (angle α); in this case, this motion is called *tautochronous*.

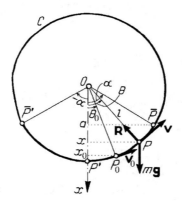

Figure 7.9. Simple pendulum.

The equation of motion along the principal normal to the trajectory is written in the form (Fig.7.9)

$$\frac{mv^2}{l} = R - mg\cos\theta,$$

where R is the constraint force directed towards the centre O; taking into account (7.1.37), we get

$$R = \frac{mg}{l}(3x - 2a) = mg(3\cos\theta - 2\cos\alpha).\qquad(7.1.38")$$

One obtains $R_{\max} = mg(3 - 2\cos\alpha)$. The constraint force diminishes if the particle P becomes closer to the extreme positions $\bar{P}$ and $\bar{P}'$; it vanishes for $x = 2a/3$, then it changes of sign (in case of a bilateral constraint). From (7.1.37) too, we notice that $x > a$ during the motion; the constraint force vanishes if $2a/3 > a$, hence for $a < 0$. On the other hand, $2a/3 > -l$, so that $a > -3l/2$; it results that the constraint force vanishes only for $-3l/2 < a < 0$ (hence, $\cos\alpha < 0$ and $\alpha > \pi/2$). Replacing the expression of a, we find the condition

$$\sqrt{2gx_0} < v_0 < \sqrt{g(3l + 2x_0)}, \qquad (7.1.46)$$

which must be fulfilled by the initial velocity, at the initial position, so that the particle be capable to reach the position (in any case, above the horizontal diameter) at which the constraint force may vanish. In particular, if $x_0 = l$, then we have

$$\sqrt{2gl} < v_0 < \sqrt{5gl}. \qquad (7.1.46')$$

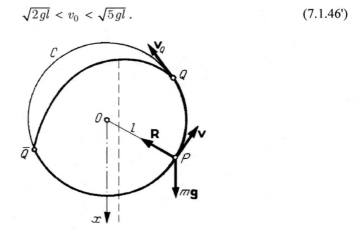

Figure 7.10. Simple pendulum; unilateral constraint.

If the constraint is unilateral, being obtained with the aid of a perfect flexible, torsionable and inextensible thread, then that one is tensioned if R is positive. At the moment at which $R = 0$ (at the point Q, Fig.7.10), the particle leaves the circle and moves as a free particle acted upon by its own weight. It will describe thus an arc of parabola of vertical axis, which is connected to the circle at the point in which the constraint force vanishes; the particle moves further till it meets again the circle at the point $\overline{Q}$. If $0 \leq v_0 \leq \sqrt{2gx_0}$, then the motion is oscillatory even in the case of a unilateral constraint.

1.3.2 Circular and asymptotic motion

If $v_0^2 > 4gl$, hence if $a < -l$, then the equation (7.1.38) takes the form

$$l^2\dot\theta^2 = 2g(l\cos\theta - a) = 2g\left(l - a - 2l\sin^2\frac{\theta}{2}\right)$$
$$= 2g(l - a)\left(1 - k^2\sin^2\frac{\theta}{2}\right), \qquad (7.1.47)$$

Problems of dynamics of the particle

where $k^2 = 2l/(l-a)$, $0 < k^2 < 1$; hence, the angular velocity $\dot\theta$ (the velocity v too) never vanishes, and maintains its sign. The motion of the particle becomes *circular* and periodic (the velocity depends only on the position).

Denoting $\tau = 2l/\sqrt{2g(l-a)}$, we get

$$t = t^0 + \frac{\tau}{2}\int_{\theta^0}^{\theta}\frac{d\vartheta}{\sqrt{1-k^2\sin^2\frac{\vartheta}{2}}} = t^0 + \tau\int_{z^0}^{z}\frac{d\zeta}{\sqrt{(1-\zeta^2)(1-k^2\zeta^2)}}, \qquad (7.1.47')$$

where θ^0 corresponds to an arbitrary moment t^0 (in general, distinct from the initial moment) and where we have made a change of variable $z = \sin(\theta/2)$, denoting also $z^0 = \sin(\theta^0/2)$. Assuming that $t^0 = \theta^0 = 0$, it results $z^0 = 0$, so that we may use the equations

$$\sin\frac{\theta}{2} = \operatorname{sn} u, \quad \cos\frac{\theta}{2} = \operatorname{cn} u, \qquad (7.1.47'')$$

where we have introduced the notation $u = t/\tau$. The period T in which the whole circumference is described is equal to twice the time necessary to arrive from the lowest point ($x = l$) to the highest one ($x = -l$); hence ($t^0 = \theta^0 = 0$)

$$T = \tau\int_0^\pi \frac{d\theta}{\sqrt{1-k^2\sin^2\frac{\theta}{2}}} = 2\tau\int_0^1 \frac{dz}{\sqrt{(1-z^2)(1-k^2z^2)}}$$

$$= \pi\tau\left[1 + \left(\frac{1}{2}\right)^2 k^2 + \left(\frac{1\cdot 3}{2\cdot 4}\right)^2 k^4 + \ldots\right]. \qquad (7.1.47''')$$

The corresponding constraint force is given by the same relation (7.1.38''); if $v_0 > \sqrt{g(3l+2x_0)}$, then the constraint force does not vanish and remains with its positive sign, and if the relation is an equality, then the constraint force vanishes at the highest point ($x = -l$). We notice that

$$R_{\max} = mg(3 - 2\cos\alpha), \quad R_{\min} = -mg(3 + 2\cos\alpha),$$
$$R_{\max} = -R_{\min} = 5mg. \qquad (7.1.48)$$

If $a = -l$, hence if $v_0^2 = 4gl$, then the equation of motion becomes

$$\dot\theta^2 = 2\omega^2(1 + \cos\theta) = 4\omega^2\cos^2\frac{\theta}{2}; \qquad (7.1.49)$$

by integration, one obtains

$$\tan\frac{1}{4}(\theta + \pi) = \tan\frac{1}{4}(\theta_0 + \pi)e^{\omega(t-t_0)}. \qquad (7.1.49')$$

For $t \to \infty$ we have $\theta \to \pi$; the particle reaches the highest point on the trajectory (a labile position of equilibrium) in an infinite time; the respective motion is called an *asymptotic motion*. The constraint force is given by

$$R = \frac{mg}{l}(3x + 2l) = mg(2 + 3\cos\theta); \qquad (7.1.49'')$$

in this case $R_{\max} = 5mg$, for $\theta = 0$. The minimal constraint force $R_{\min} = -mg$, for $\theta = \pi$, can be obtained only in the case of a bilateral constraint; in the case of a unilateral one, beginning with the point Q, determined by $x = -2l/3$ (hence, by $\cos\theta = -2/3$, corresponding to $\theta = 131°48'37''$), the motion is on an arc of parabola.

Analogously, one may study *the problem of the swing*, which may be modelled as a simple pendulum with a thread of length variable in time ($l = l(t)$).

1.3.3 Motion of a simple pendulum in a resistent medium

Introducing the resistance **R** of the medium, tangent to the trajectory and of direction opposite to that of the velocity, and writing the equation of motion along the tangent, one obtains

$$ml\ddot{\theta} = -mg\sin\theta - R. \qquad (7.1.50)$$

Considering a resistance proportional to the velocity (*viscous damping*), of the form $R = 2\lambda m l\dot{\theta}$, $\lambda > 0$, in the case of small oscillations ($\sin\theta \cong \theta$), the equation (7.1.50) becomes

$$\ddot{\theta} + 2\lambda\dot{\theta} + \omega^2\theta = 0; \qquad (7.1.51)$$

assuming that $\omega^2 > \lambda^2$ and denoting $\mu^2 = \omega^2 - \lambda^2$, we obtain the general integral

$$\theta(t) = e^{-\lambda t}(A\cos\mu t + B\sin\mu t), \qquad (7.1.51')$$

where the constants A and B may be determined by the initial conditions $\theta(t_0) = \theta_0$, $\dot{\theta}(t_0) = \dot{\theta}_0$. We may thus write (for the sake of simplicity, we assume that $t_0 = 0$)

$$\theta(t) = e^{-\lambda t}\left[\theta_0\cos\mu t + \frac{1}{\mu}(\lambda\theta_0 + \dot{\theta}_0)\sin\mu t\right], \qquad (7.1.52)$$

$$\dot{\theta}(t) = e^{-\lambda t}\left[\dot{\theta}_0\cos\mu t - \frac{1}{\mu}(\omega^2\theta_0 + \lambda\dot{\theta}_0)\sin\mu t\right]. \qquad (7.1.52')$$

If, in particular, we have $\dot{\theta}_0 = 0$, then the particle departs without initial velocity from the point P_0 and reaches the point P_1, where the velocity

$$\dot\theta = -\frac{1}{\mu}\omega^2\theta_0 e^{-\lambda t}\sin\mu t$$

vanishes at the moment $t_1 = \pi/\mu$ (Fig.7.11); then the motion follows the same law, the particle returning till the point P_2 at a time $t_2 = 2\pi/\mu$ a.s.o. The oscillations are isochronic, the period $T = 2\pi/\mu = 2\pi/\sqrt{\omega^2 - \lambda^2}$ (greater than that of the motion in vacuum) not depending on the amplitudes $\theta_0 > |\theta_1| > \theta_2 > |\theta_3| > \ldots$; we notice also that

$$|\theta_1|/\theta_0 = \theta_2/|\theta_1| = |\theta_3|/\theta_2 = \ldots = e^{-\pi\lambda/\mu},$$

so that the absolute values of the amplitudes form a geometric series of ratio $e^{-\pi\lambda/\mu}$. Hence, the motion is damped in an infinite time, the particle reaching the lowest position (stable position of equilibrium).

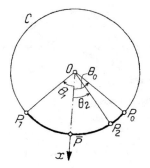

Figure 7.11. Simple pendulum in a resistent medium.

If, in the case of oscillations of finite amplitude, we consider a resistance proportional to the square of the velocity (*aerodynamic damping*), then $R = mlk^2\dot\theta^2$, and the equation (7.1.50) becomes

$$\ddot\theta + k^2\dot\theta^2 + \omega^2\sin\theta = 0 \qquad (7.1.53)$$

for an ascendent motion; in the case of a descendent motion, we replace k^2 by $-k^2$. Noting that $\ddot\theta = \dot\theta\, d\dot\theta/d\theta = d(\dot\theta^2)/2d\theta$, we may write the equation (7.1.53) in the form

$$\frac{1}{2}\frac{d(\dot\theta^2)}{d\theta} \pm k^2\dot\theta^2 = -\omega^2\sin\theta, \qquad (7.1.53')$$

the general integral of which is given by

$$\dot\theta^2 = Ce^{\mp 2k^2\theta} + \frac{2\omega^2}{4k^4 + 1}(\cos\theta \mp 2k^2\sin\theta), \qquad (7.1.53'')$$

where C is a constant which must be determined; the relation (7.1.53") represents, in fact, an equation with separate variables, and the quadrature may be calculated for small amplitudes.

1.3.4 Elliptic pendulum

Let P be a heavy particle constrained to move on an ellipse of semiaxes a and b of equation

$$\frac{x_1^2}{a^2} + \frac{x_2^2}{b^2} = 1, \qquad (7.1.54)$$

situated in a vertical plane, the Ox_1-axis being the descendent vertical. Using the parametric representation

$$x_1 = a\cos q, \quad x_2 = b\sin q, \quad 0 \le q < 2\pi,$$

the theorem of kinetic energy $\mathrm{d}\left[m(\dot{x}_1^2 + \dot{x}_2^2)/2\right] = mg\dot{x}_1$ leads to

$$(a^2\sin^2 q + b^2\cos^2 q)\ddot{q} + (a^2 - b^2)\sin q\cos q\,\dot{q}^2 = -ag\sin q. \qquad (7.1.55)$$

In the case of small motions around the point $P_0(a,0)$, which is a stable position of equilibrium, we have $\sin q \cong q$, $\cos q \cong 1$, $\dot{q}^2 \cong 0$, so that the equation (7.1.55) becomes

$$\ddot{q} + \omega^2 q = 0, \quad \omega^2 = \frac{ag}{b^2}; \qquad (7.1.55')$$

it results $q = q_0\cos\omega(t - t_0)$, the motion being periodic, isochronic, of period

$$T = \frac{2\pi b}{\sqrt{ag}}. \qquad (7.1.55'')$$

If $a = b = l$, then one obtains the results in Subsec. 1.3.1, for instance Galileo's formula (7.1.45").

1.3.5 Cycloidal pendulum

Let us consider the motion of a heavy particle P on a cycloid $\mathscr{C}$ with a horizontal basis, situated in a vertical plane and having the concavity towards the positive direction of the Ox_2-axis. The axis Ox_1 is tangent to the cycloid as its lowest point, while the Ox_2-axis (an ascendent one) is the symmetry axis of the cycloid (Fig.7.12). Starting from the definition of the cycloid as a locus (see Chap. 5, Subsec. 1.3.4), we obtain its parametric equations in the form

$$x_1 = a(\theta + \sin\theta), \quad x_2 = a(1 - \cos\theta), \quad \theta \in [-\pi,\pi],$$

where $x_2 = 2a$ is the straight line on which the generating circle C (of centre O' and radius a) of the cycloid is rolling without sliding. Departing from $\mathrm{d}s^2 = \mathrm{d}x_1^2 + \mathrm{d}x_2^2$, we find $\mathrm{d}s = 2a\cos(\theta/2)\mathrm{d}\theta = \sqrt{2a/x_2}\,\mathrm{d}x_2$; by integration, we obtain $s = 2\sqrt{2ax_2} = 4a\sin(\theta/2)$, so that $\mathrm{d}x_2/\mathrm{d}s = s/4a$. Euler's equations of motion are written in the form

$$m\dot{v} = m\ddot{s} = F_\tau = -mg\frac{\mathrm{d}x_2}{\mathrm{d}s}, \quad \frac{mv^2}{\rho} = F_\nu + R,$$

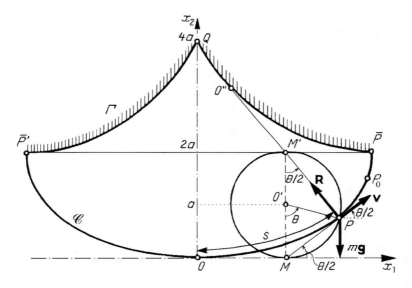

Figure 7.12. Cycloidal pendulum.

wherefrom it results

$$\ddot{s} + \omega^2 s = 0, \quad \omega^2 = \frac{g}{4a}; \tag{7.1.56}$$

hence,

$$s = s_0 \cos \omega t, \tag{7.1.56'}$$

assuming that the particle is launched without initial velocity from the point P_0, of curvilinear co-ordinate s_0, at the initial moment t_0. The period of the motion is

$$T = \frac{2\pi}{\omega} = 4\pi\sqrt{\frac{a}{g}} = 2\pi\sqrt{\frac{4a}{g}}; \tag{7.1.56''}$$

this period does not depend on the amplitude s_0, so that the oscillations are isochronic (independent of their magnitude). On the other hand, a particle in a free fall from the point P_0 reaches the point O (the lowest point of the cycloid) in the time $T/4$, which is independent of s_0, hence of the initial position; this is the property of *tautochronism of the cycloid*. We say that *the motion is tautochronous*, independent of the magnitude of the oscillations, the cycloid being thus a *tautochronous curve*. This property has been put in evidence by Huygens, which realized a cycloidal pendulum with the aid of the evolute of a cycloid, that one being a cycloid too. The thread by which is connected the particle P (unilateral constraint) is fixed at the point Q (cuspidal point of a cycloid concretely built up); but the resistances which intervene modify considerable the motion.

Taking into account the theorem of the kinetic energy, we may write $v^2 = 2g(x_2^0 - x_2)$, where the ordinate x_2^0 corresponds to the initial position P_0. Noting that

$$\mathrm{d}x_1 = a(1+\cos\theta)\mathrm{d}\theta = 2a\cos^2\frac{\theta}{2}\mathrm{d}\theta = \cos\frac{\theta}{2}\mathrm{d}s,$$

we have $F_\nu = -mg\mathrm{d}x_1/\mathrm{d}s = -mg\cos(\theta/2)$; because $\rho = \overline{PO''} = 2\overline{PM'} = 4a\cos(\theta/2)$, the second equation of Euler gives the constraint force

$$R = mg\left(\cos\frac{\theta}{2} + \frac{\cos\theta - \cos\theta_0}{2\cos\frac{\theta}{2}}\right). \tag{7.1.57}$$

If, in particular, $\theta_0 = \pm\pi$, hence if the particle travels through the cycloid without initial velocity, from one of the cuspidal points $\overline{P}$ or $\overline{P}'$, then it results

$$R = 2mg\cos\frac{\theta}{2} = -2F_\nu; \tag{7.1.57'}$$

in this case, we can state, after Euler, that *the modulus of the constraint force is the double of the modulus of the normal component of the own weight of the particle*.

1.3.6 Motion of a heavy particle on a surface of rotation

Let us consider the motion of a heavy particle P on a surface of rotation, the axis of rotation of which is vertical (Fig.7.13). The own weight mg of the particle and the constraint force **R** (the support of which pierces the Ox_3-axis) act in the meridian plane, their moments with respect to the symmetry axis vanishing; hence, we may write the first integral of areas for the projection P' of the particle P on the plane Ox_1x_2, hence also for the particle P, in the form (we use cylindrical co-ordinates)

$$r^2\dot\theta = r_0^2\dot\theta_0 = C, \tag{7.1.58}$$

where $r_0 = r(t_0)$, $\dot\theta_0 = \dot\theta(t_0)$. Because the constraint is scleronomic and the given force is conservative, we may use also the first integral of energy

Problems of dynamics of the particle 435

$$v^2 = \dot{r}^2 + r^2\dot{\theta}^2 + \dot{z}^2 = v_0^2 + 2g(z_0 - z), \tag{7.1.58'}$$

where $z_0 = z(t_0)$, $v_0 = v(t_0)$.

If the surface of rotation is given by $r = f(z)$ (the equation of the meridian curve $\mathscr{C}$), then we may eliminate the functions $r = r(t)$ and $\theta = \theta(t)$ between the equations (7.1.58), (7.1.58'), obtaining the equation with separable variables

$$\dot{z}^2\left(1 + f'^2\right) = v_0^2 + 2g(z_0 - z) - \frac{C^2}{f^2}, \quad f' = \frac{\mathrm{d}f}{\mathrm{d}z}, \tag{7.1.59}$$

which determines the applicate $z = z(t)$ by a quadrature; returning to the equation of the surface of rotation and to the first integral of areas, we obtain the other co-ordinates of the point P. In the case of a circular cylinder of radius l, the equation (7.1.59) becomes ($f = l$)

$$\dot{z}_0^2 = v_0^2 + 2g(z_0 - z) - \frac{C^2}{l^2}, \quad C = \text{const}, \tag{7.1.59'}$$

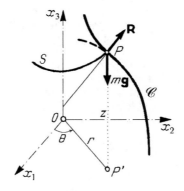

Figure 7.13. Motion of a heavy particle on a surface of rotation.

in the case of a circular cone of equation $r = kz$, $k = \text{const}$, we may write

$$\left(1 + k^2\right)\dot{z}^2 = v_0^2 + 2g(z_0 - z) - \frac{C^2}{k^2 z^2}, \quad C = \text{const}, \tag{7.1.59''}$$

while in the case of a sphere of radius l we obtain ($r^2 + z^2 = l^2$)

$$l^2\dot{z}^2 = \left[v_0^2 + 2g(z_0 - z)\right](l^2 - z^2) - C^2, \quad C = \text{const}. \tag{7.1.59'''}$$

If we represent the surface of rotation by the equation $z = \varphi(r)$, we may eliminate the functions $z = z(t)$ and $\theta = \theta(t)$; it results the equation

$$\dot{r}^2 \left(1 + \varphi'^2\right) = v_0^2 + 2g(z_0 - \varphi) - \frac{C^2}{r^2}, \quad \varphi' = \frac{d\varphi}{dr}, \quad C = \text{const}, \tag{7.1.60}$$

which specifies the radius $r = r(t)$ by a quadrature too.

Eliminating the time, we obtain the equation of the trajectory of the point P' in the form

$$\theta = \theta_0 + C \int_{r_0}^{r} \frac{d\rho}{\rho} \sqrt{\frac{1 + [\varphi'(\rho)]^2}{\{v_0^2 + 2g[z_0 - \varphi(\rho)]\}\rho^2 - C^2}}, \tag{7.1.61}$$

where $\theta_0 = \theta(t_0)$; assuming that the surface is algebraic, Kobb has put in evidence the cases in which the function $\theta = \theta(r)$ is expressed by means of elliptic functions.

In the case of a conservative force, the potential of which depends only on r, the problem may be solved only by quadratures too.

1.3.7 Spherical pendulum

A heavy particle which moves frictionless on a sphere of radius l is called *spherical pendulum*. The constraint may be bilateral or unilateral; in what follows we consider the case of a bilateral constraint. We choose the equatorial plane of the sphere as plane Ox_1x_2, the axis Ox_3 being along the descendent vertical; further it is convenient to use cylindrical co-ordinates (Fig.7.14). If the constant C of the first integral of areas (7.1.58) vanishes, then $\dot{\theta} = 0$, hence $\theta = \text{const}$; the trajectory of the particle is contained in a meridian plane of the sphere, being thus a great circle of it. The spherical pendulum is, in this case, a simple pendulum. If the constant C is non-zero, then we have to do with a non-degenerate spherical pendulum.

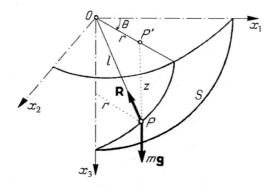

Figure 7.14. Spherical pendulum.

The equation (7.1.59''') becomes

$$l^2 \dot{z}^2 = P(z), \quad P(z) = \left[v_0^2 + 2g(z - z_0)\right](l^2 - z^2) - C^2, \tag{7.1.62}$$

so that

Problems of dynamics of the particle

$$t = t_0 \pm l \int_{z_0}^{z} \frac{d\zeta}{\sqrt{P(\zeta)}} ; \qquad (7.1.63)$$

the first integral (7.1.58) allows to determine the angle θ in the form

$$\theta = \theta_0 \pm Cl \int_{z_0}^{z} \frac{d\zeta}{(l^2 - \zeta^2)\sqrt{P(\zeta)}} . \qquad (7.1.63')$$

In the two above formulae one takes the sign of $\dot{z}_0 = \dot{z}(t_0)$, assuming that $\dot{z} \neq 0$. If $\dot{z}_0 = 0$, then one takes into account the increasing or decreasing of z, starting from the initial value z_0.

First of all, we assume that $\dot{z}_0 \neq 0$; in this case $P(z_0) > 0$ (from (7.1.58), (7.1.58'), (7.1.62) it results $P(z_0) = r_0^2(\dot{r}_0^2 + \dot{z}_0^2)$). However, during the motion one must have $P(z) \geq 0$ so that the integrals (7.1.63), (7.1.63') be real. Noting that $|z_0| < l$ (for $|z_0| = l$ we have the simple pendulum) and that $P(-\infty) = \infty$, $P(\pm l) = -C^2$, it results that the polynomial $P(z)$ is of the form

$$P(z) = -2g(z - z_1)(z - z_2)(z - z_3), \quad -\infty < z_3 < -l < z_2 < z_0 < z_1 < l. \qquad (7.1.64)$$

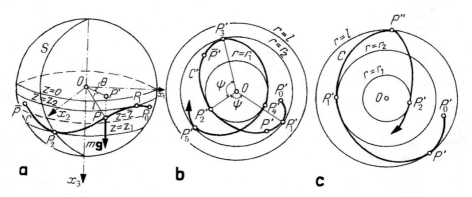

Figure 7.15. Spherical pendulum: zone of oscillation on the sphere (a); projection of the motion on the equatorial plane: case $z_2 > 0$ (b); case $z_2 < 0$ (c).

Hence, the particle P oscillates on the spherical zone contained between the parallel circles specified by $z = z_1$ and $z = z_2$ (to have $P(z) \geq 0$) (Fig.7.15,a). Viète's formula allows to write $z_3(z_1 + z_2) = -(l^2 + z_1 z_2)$; noting that $|z_1 z_2| < l^2$ and $z_3 < 0$, it results that $z_1 + z_2 > 0$. Hence, the parallel $z = \overline{z} = (z_1 + z_2)/2$, equidistant to the parallels $z = z_1$ and $z = z_2$, is always situated under the equatorial circle (in the austral hemisphere). Departing from P_0, let us suppose that z decreases (we have the sign – before the radical); the particle reaches P_1 on the parallel $z = z_2$, where the trajectory has a horizontal tangent (at this point $\dot z = 0$, but $\dot\theta \neq 0$). Then, the particle rotates about the vertical axis Ox_3 and reaches the point P_2 on the parallel $z = z_1$, where the tangent to the trajectory is horizontal too; further, the particle reaches the point P_3 on $z = z_2$ a.s.o. The particle P travels through the arc of trajectory $\overset{\frown}{P_1P_2}$ in the interval of time

$$\frac{T}{2} = l\int_{z_2}^{z_1} \frac{\mathrm{d}z}{\sqrt{P(z)}}; \qquad (7.1.65)$$

the same interval of time is necessary to travel through the arcs $\overset{\frown}{P_2P_3}$, $\overset{\frown}{P_3P_4}$ a.s.o. We notice that the meridian planes of the points of contact of the trajectory with the extreme parallels are planes of symmetry of this trajectory; indeed, for two points P and $\bar P$ of the same parallel z we have

$$\bar\theta - \theta_2 = \theta_2 - \theta = Cl\int_z^{z_1} \frac{\mathrm{d}\zeta}{(l^2 - \zeta^2)\sqrt{P(\zeta)}}.$$

The necessary intervals of time to travel through the arcs $\overset{\frown}{PP_2}$ and $\overset{\frown}{P_2\bar P}$ are equal too, being given by

$$l\int_z^{z_1} \frac{\mathrm{d}\zeta}{\sqrt{P(\zeta)}}.$$

We observe that after a time T we find again the same values for z and $\dot z$.

If $\dot z_0 = 0$, then we have $P(z_0) = 0$ too. Assuming that $P'(z_0)/2 = gr_0^2 - z_0 v_0^2 \neq 0$, the particle is launched from one of the extreme parallels (we have $z_1 = z_0$ or $z_2 = z_0$) with a horizontal initial velocity (hence, tangent to the respective parallel); in the case of the parallel $z = z_2$ one takes the sign + before the radical, while if one departs from the parallel $z = z_1$ the sign − is used. If $P'(z_0) = 0$, then the equation $P(z) = 0$ has a double root $z_1 = z_2 = z_0 = gr_0^2/v_0^2$ and one may write $l^2\dot z^2 = -2g(z - z_0)^2(z - z_3)$; noting that $z > z_3$, one can have the latter relation only for $z = z_0 = \mathrm{const}$. The trajectory of the particle is the parallel $z = z_0$ situated in the austral hemisphere, because $z_0 > 0$. The spherical pendulum is reduced – in this case – to a *circular conical pendulum*.

From (7.1.58), we notice that $\dot\theta$ has a constant sign; it results that the point P' (the projection of the particle P on the equatorial plane) rotates permanently in the same direction around the centre O, that direction being specified by the sign of the angular velocity $\dot\theta$, hence by the sign of C (if $C > 0$, then $\theta > 0$ too). The projection of the trajectory C of the particle P on the equatorial plane is the trajectory C' of the projection P'. If $z_2 \geq 0$, then the trajectory of this projection is contained between the concentric circles $r = r_1 = \sqrt{l^2 - z_1^2}$, $r = r_2 = \sqrt{l^2 - z_2^2}$, $r_1 < r_2$, without any point of inflection (Fig.7.15,b); the projection P' gives the impression that it describes an oval which is rotating in the direction of the motion (always in the same direction). One can show that $\widehat{P_1'OP_2'} = \widehat{P_2'OP_3'} = \ldots$; Puiseux proved that these angles are always greater than $\pi/2$, while Halphen and then Saint-Germain showed that they are at the most equal to π. The angle $\widehat{P_1'OP_5'} = 4\widehat{P_1'OP_2'}$ is the angle of precession and emphasizes the "delay" of the point P_5' with respect to the point P_1', so that the trajectory of the point P' cannot be closed. If $z_2 < 0$, then the corresponding parallel is above the equatorial circle, but we have still $r_1 < r_2$ (because $z_1 + z_2 > 0$); the trajectory of the projection P' is tangent to the equatorial circle, and we may have also inflection points. The properties which have been mentioned before are maintained (Fig.7.15,c).

With the aid of the substitution $z = z_1 - (z_1 - z_2)u^2$ and of the notation $k^2 = (z_1 - z_2)/(z_1 - z_3) < 1$, $\tau = l\sqrt{2}/\sqrt{g(z_1 - z_3)}$, we may write the formula (7.1.63) in the form

$$t = t_0 \pm \tau \int_0^u \frac{d\overline{u}}{\sqrt{(1-\overline{u}^2)(1-k^2\overline{u}^2)}}, \quad (7.1.66)$$

wherefrom (we take $t_0 = 0$ and the sign $+$ before the integral)

$$u = \operatorname{sn}\frac{t}{\tau}, \quad z = z_1 - (z_1 - z_2)\operatorname{sn}^2\frac{t}{\tau}; \quad (7.1.66')$$

hence, z is a doubly periodic function of t. The angular velocity $\dot{\theta}(t)$ is obtained as a rational function of $\operatorname{sn}(t/\tau)$; one can integrate by decomposing in simple elements, using Hermite's method, obtaining thus the function $\theta(t)$, which is not uniform. But Tissot and then Hermite have state how to obtain the co-ordinates z_1 and z_2 as uniform functions of t. Greenhill showed that, in the case in which the point P_0 is in the plane $z = 0$, the initial velocity being tangent to the equatorial circle, then θ and t are given by pseudoelliptic integrals, which may be expressed by elementary functions.

The constraint force R (Fig.7.14) is given by $R = -F_n + mv^2/\rho_n$; noting that $\rho_n = l$ and $F_n = -mz/l$, we get

$$R = \frac{m}{l}\left[v_0^2 + g(3z - 2z_0)\right]. \quad (7.1.67)$$

The force $\mathbf{R}$ is directed towards the centre O ($R > 0$ for $z > 0$) if the particle is in the austral hemisphere; if P is in the boreal hemisphere ($z_2 < 0$), then it is possible to have $R = 0$ for a parallel z. In the case of a unilateral constraint (the particle is at the extremity of a perfectly flexible and inextensible thread), after passing the parallel

$z = \overline{z}$, the particle may move further on an osculating parabola to the previous trajectory on the sphere, in a vertical plane, till it meets again the sphere; then, the motion is continued in conformity with the laws established before. In the case of a bilateral constraint, the constraint force may change its sign, being directed towards the interior of the sphere. We notice that, in general, the resultant of the forces $\mathbf{R}$ and $m\mathbf{g}$ is tangent to the trajectory at the point P, hence it is contained in the osculating plane to the curve at that point. If $\mathbf{R} = \mathbf{0}$, then the osculating plane is vertical; in this case, the trajectory of the projection P' on the equatorial plane has an inflection point at this point. In the case of the circular conical pendulum, the constraint force becomes $R = mgl/z_0 = mlv_0^2/r_0^2$; the projection of this force on the equatorial plane is $Rr_0/l = mv_0^2/r_0 = mr_0\dot{\theta}_0^2$, hence a centripetal force.

Projecting the equation of motion on the Cartesian co-ordinate axes, we obtain

$$m\ddot{x}_1 = -R\frac{x_1}{l}, \quad m\ddot{x}_2 = -R\frac{x_2}{l}, \quad m\ddot{x}_3 = mg - R\frac{x_3}{l};$$

in the case of small oscillations around the position $x_1 = x_2 = 0$, $x_3 = l$, which is a stable position of equilibrium, we notice that

$$x_3 = \sqrt{l^2 - (x_1^2 + x_2^2)} = l\left(1 - \frac{x_1^2 + x_2^2}{2l^2} + \ldots\right).$$

In a first approximation $x_3 = l$ (we neglect the second term with respect to unity), and we may assume that the motion takes place in the plane tangent at the lowest point of the sphere. In this case, the third equation of motion leads to $R = mg$; the first two equations may be written vectorially in the form

$$\ddot{\overline{\mathbf{r}}} + \frac{g}{l}\overline{\mathbf{r}} = \mathbf{0}, \tag{7.1.68}$$

where $\overline{\mathbf{r}}$ is the vector radius in the equatorial plane. One obtains thus an elliptic oscillator, which will be studied in Chap. 8, Subsec. 2.1.1. The motion is *periodic* and the trajectory is an ellipse, which may be travelled through in an interval of time given by $T = 2\pi\sqrt{l/g}$ (the period corresponding to a simple pendulum). An approximation of second order has been considered by Tisserand; other methods of approximation have been used by Resal and Sparre.

2. Other problems of dynamics of the particle

After considering the problems of Abel and Puiseux and of tautochronous motions, one deals with the motion on a brachistochrone or on a geodesic curve, as well as with other cases of motion. Some results concerning the stability of the equilibrium of a particle are given too.

2.1 Tautochronous motions. Motions on a brachistochrone and on a geodesic curve

In what follows, after presenting the problems of Abel and Puiseux, one considers – in particular – the study of tautochronous motions. A particular attention is paid to the motion on a brachistochrone curve or on a geodesic one, the study of which needs some notions of variational calculus.

2.1.1 Abel's problem

N.H. Abel tried to determine a curve C, contained in a vertical plane and passing through a given fixed point O, so that a heavy particle P, which is moving without friction on this curve, starting from the point P_0, situated at an applicate h above the point O, reaches that point, without initial velocity, in an interval of time $\tau = \tau(h)$, $h \in [0,a]$, assuming that the continuous function τ is given. We choose the horizontal as Ox_1-axis, the Ox_2-axis being the local ascendent vertical (Fig.7.16); the curve C is specified by the equation $s = \varphi(x_2)$, $\varphi(0) = 0$, $\varphi \in C^1[0,a]$, where s is the curvilinear abscissa. The conservation theorem of mechanical energy gives $v^2 = (\mathrm{d}s/\mathrm{d}t)^2 = 2g(h - x_2)$. Noting that $s(t)$ is a decreasing function, we may write $\mathrm{d}s = -\sqrt{2g(h - x_2)}\mathrm{d}t = \varphi'(x_2)\mathrm{d}x_2$; by integration ($x_2 = h$ for $t = 0$ and $x_2 = 0$ for $t = \tau$), we obtain an integral equation of the first kind, with a variable superior limit

$$\sqrt{2g}\tau(h) = \int_0^h \frac{\varphi'(x_2)\mathrm{d}x_2}{\sqrt{h - x_2}}. \qquad (7.2.1)$$

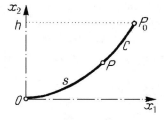

Figure 7.16. Abel's problem.

To solve this integral equation, which is the first one appeared in mathematical analysis, we use a particular ingenious procedure due to Abel; we multiply both members by $\mathrm{d}h/\sqrt{u - h}$, $u \in [0,a]$ and integrate with respect to h between the limits 0 and u, so that

$$\sqrt{2g}\int_0^u \frac{\tau(h)\mathrm{d}h}{\sqrt{u - h}} = \int_0^u \frac{\mathrm{d}h}{\sqrt{u - h}}\int_0^h \frac{\varphi'(x_2)\mathrm{d}x_2}{\sqrt{h - x_2}} = \int_0^u \varphi'(x_2)\mathrm{d}x_2 \int_{x_2}^u \frac{\mathrm{d}h}{\sqrt{(u - h)(h - x_2)}},$$

where the order of integration has been inverted (in fact, a double integral on a triangular domain bounded by the lines $x_2 = 0$, $x_2 = h$ and $h = u$ is calculated). Noting that the latter integral is equal to π, one obtains, finally,

$$\varphi(x_2) = \frac{\sqrt{2g}}{\pi} \int_0^{x_2} \frac{\tau(h)\,dh}{\sqrt{x_2 - h}}, \quad x_2 \in [0,a]. \tag{7.2.1'}$$

Following these ideas, Euler and Saladini deal with the determination of a curve C in a vertical plane, so that a heavy particle P, which is moving without friction on this curve, departing from the point P_0, without initial velocity, reaches an arbitrary point P in the same interval of time as that necessary if the particle would slide along the bisecant $P_0 P$, finding thus a lemniscate. O. Bonnet showed that this lemniscate has further the above mentioned property if we replace the gravitational field by a field of central forces of attraction, proportional to the distance to the pole P_0. As well, Fouret considers a particle which is subjected to the action of a field of conservative forces in a plane and which departs from the position O without initial velocity; a problem to determinate a family of homothetic curves C passing through O, so that the particle which departs from this pole describes an arc of curve C till the point P in the same interval of time in which the corresponding bisecant OP would be travelled through is put. The problem has a solution if the potential function is of the form $U(r,\theta) = \psi(r/\varphi(\theta))\varphi^2(\theta)$ in polar co-ordinates, φ and ψ being arbitrary functions of class C^1; the equation of the curve C is, in this case, of the form

$$r^2 = k^2 \varphi(\theta) e^{-\int \frac{\varphi(\theta)}{\varphi'(\theta)} d\theta}, \quad k = \text{const}. \tag{7.2.2}$$

Analogously, one may put the problem to determine the field of conservative forces which act upon a particle the trajectory of which is a curve from the family of curves C previously considered; if the equation of the curve C is of the form $r = k\omega(\theta)$, $k = \text{const}$, then the potential function is of the above mentioned form, where

$$\varphi(\theta) = \omega(\theta) e^{-\int \sqrt{[\omega'(\theta)/\omega(\theta)]^2 + 1}\, d\theta}. \tag{7.2.2'}$$

2.1.2 Puiseux's problem

In connection with tautochronous motions we consider – first of all – a particular problem, put and solved by Puiseux. It is thus asked to determine a law of force $F = F(x)$ for which a rectilinear and frictionless motion is tautochronous with respect to the pole O of the trajectory (Fig.7.17). The conservation theorem of mechanical energy leads to $m\dot{x}^2 = 2[\psi(x_0) - \psi(x)]$ where we denoted by

$$\psi(x) = -\int_0^x F(\xi)\,d\xi, \quad \psi(0) = 0,$$

Problems of dynamics of the particle

an increasing positive function, because $F(x)$ is a function obviously negative for $x > 0$, if we assume that $x_0 > 0$ (the force must be directed towards the pole O, because the particle must move towards that point). The particle reaches the point O in an interval of time equal to

$$\tau_0 = \sqrt{\frac{m}{2}} \int_0^{x_0} \frac{\mathrm{d}x}{\sqrt{\psi(x_0) - \psi(x)}}. \qquad (7.2.3)$$

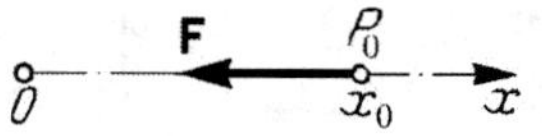

Figure 7.17. Puiseux's problem.

If we denote $\psi(x) = z$, $\psi(x_0) = z_0$, $z = z_0 u$ and $x = \chi(z)$, where χ is the inverse function, it results

$$\tau_0 = \sqrt{\frac{m}{2}} \int_0^{z_0} \frac{\chi'(z)\mathrm{d}z}{\sqrt{z_0 - z}} = \sqrt{\frac{m}{2}} \int_0^1 \frac{\chi'(z_0 u)\sqrt{z_0}\mathrm{d}u}{\sqrt{1 - u}};$$

the condition that τ_0 be not dependent on x_0, hence on z_0, is put in the form

$$\frac{\mathrm{d}\tau_0}{\mathrm{d}z_0} = \sqrt{\frac{m}{2}} \int_0^1 \frac{\chi''(z_0 u)z_0 u + \frac{1}{2}\chi'(z_0 u)}{\sqrt{z_0 - z_0 u}} \mathrm{d}u = \sqrt{\frac{m}{2}} \int_0^{z_0} \frac{z\chi''(z) + \frac{1}{2}\chi'(z)}{z_0\sqrt{z_0 - z}} \mathrm{d}z = 0$$

for any z_0, so that we must have $2z\chi''(z) + \chi'(z) = 0$ (otherwise we may choose a z_0 sufficiently small so that the integrand be of a constant sign, so that the condition would no more be fulfilled). We get $\chi(z) = 2c\sqrt{z}$, $c = \mathrm{const}$, an additive constant being equal to zero, because z and $x = \chi(z)$ vanish simultaneously; it results $z = \psi(x) = x^2 / 4c^2$, so that $F = -\psi'(x) = -x / 2c^2$. We see thus that the only force $F = F(x)$ which leads to a tautochronous rectilinear motion is a force in direct proportion to the distance (which will be studied in Chap. 8, Subsec. 2.2.1).

If the resultant of the given forces depends on the position as well as on the velocity, the problem is more intricate. Lagrange gave a law for the force for which the tautochronism takes place and from which – if the velocity is no more involved – one obtains the previous results. But this law does not contain all possible cases; e.g., Brioschi gave a more general formula.

2.1.3 Tautochronous motions

If we make $\tau(h) = \tau_0 = \mathrm{const}$ in the formula (7.2.1'), then we obtain

$$s = \varphi(x_2) = \frac{2\tau_0\sqrt{2g}}{\pi}\sqrt{x_2}, \qquad (7.2.4)$$

denoting $\tau_0^2 g / \pi^2 = a$, we find again the equation of the cycloid (Fig.7.12), considered in Subsec. 1.3.5. We have seen that the cycloid is a tautochronous curve, the respective motion being a tautochronous one with respect to the point of tautochronism O, which is reached by the particle, acted upon by its own weight, in the same interval of time τ_0, independent of the initial position, if the initial velocity vanishes. Besides, the cycloid is the only tautochronous curve with respect to a gravitational field.

In general, we say that a motion (hence, a curve C) is tautochronous if there exists a point O' (called point of tautochronism) on this curve, so that a particle which is acted upon by given forces of resultant $\mathbf{F}$ and which, departing from the position P_0, frictionless and without initial velocity, reaches the position O' in an interval of time independent of P_0. Projecting the equation of motion on the tangent to the curve, we may write

$$m\frac{d^2 s}{dt^2} = F_t = F_i \frac{dx_i}{ds} = f(s), \qquad (7.2.5)$$

where the second member is function only on the curvilinear co-ordinate s, if we assume that the resultant of the given forces depends only on the position ($\mathbf{F} = \mathbf{F}(\mathbf{r})$). The equation (7.2.5) is identical with the equation of the rectilinear motion on the Os-axis, if the particle is subjected to the action of a tangential force F_t.

Taking into account the results obtained in the preceding subsection, we must have $f(s) = -k^2 s$, so that the trajectory C be tautochronous, that one being a necessary and sufficient condition; the point of tautochronism $s = 0$ is – obviously – a stable position of equilibrium. The solution of the problem is indeterminate, to determine it being necessary a supplementary condition. For instance, one may put the condition that the curve C lays on a given fixed surface $\mathscr{F}(x_1, x_2, x_3) = 0$, adding the obvious relation $dx_i dx_i = ds^2$; one may obtain thus the parametric equations of the trajectory $x_i = x_i(s)$, $i = 1, 2, 3$, introducing two new constants of integration besides k^2. In the case of a conservative force, we obtain easily

$$U(x_1, x_2, x_3) = -\frac{k^2}{2}s^2 + K, \quad K = \text{const}. \qquad (7.2.6)$$

As well, we may put other conditions, e.g., that the curve C be tautochronous, with the same point of tautochronism for other given forces of resultant $\mathbf{F}'$; we introduce thus a new condition of the form

$$F_i' \frac{dx_i}{ds} = -k'^2 s, \quad k' = \text{const}. \qquad (7.2.5')$$

Hence, the curve obtained is tautochronous for a force $\lambda \mathbf{F} + \lambda' \mathbf{F}'$, λ, λ' constant positive scalars. If the second force is conservative too, of potential

Problems of dynamics of the particle

$$U'(x_1, x_2, x_3) = -\frac{k'^2}{2}s^2 + K', \quad K' = \text{const}, \qquad (7.2.6')$$

the tautochronous curve will stay on the surface

$$k'^2 [U(x_1, x_2, x_3) - K] = k^2 [U'(x_1, x_2, x_3) - K']. \qquad (7.2.7)$$

In the general case of a resistent medium of resistance $R = \varkappa(v)$, the resultant of the given forces depending also on the velocity ($\mathbf{F} = \mathbf{F}(\mathbf{r}, \mathbf{v})$), the equation of motion reads

$$m\frac{d^2 s}{dt^2} = F_i \frac{dx_i}{ds} + \varkappa\left(\frac{ds}{dt}\right); \qquad (7.2.5'')$$

we come thus back to the previous case of rectilinear motion for which a necessary and sufficient condition of tautochronism is not known, being possible to use – in particular – the law of force given by Lagrange.

2.1.4 Considerations on variational calculus

To study the motion on a brachistochrone, there are necessary some results concerning variational calculus. Let thus be an integral of the form

$$I(y) \equiv \int_{x_0}^{x_1} F(x; y, y') dx, \qquad (7.2.8)$$

where $F(x; y, y')$ is a known real function of arguments x, y and $y' \equiv dy/dx$, of class C^2 with respect to these arguments; the value of this integral depends on how is chosen the function $y = y(x)$, wherefrom the notation used, as well as the denomination of functional. We assume that the admissible arguments $y(x)$ are of class C^2 and that, at the extremities of the interval $[x_0, x_1]$, they take the given values y_0, y_1; in this case, the set $\{y(x)\}$ of the admissible arguments $y(x)$ may be seen as a family of smooth curves, passing through the points (x_0, y_0) and (x_1, y_1) of which we must choose one, which minimizes the functional $I(y)$. A necessary condition to determine this curve is *the Euler-Poisson equation*

$$F_y - \frac{dF_{y'}}{dx} = 0, \qquad (7.2.8')$$

associated to the variational problem $I(y) = \min$, where F_y and $F_{y'}$, represent the partial derivatives with respect to the corresponding arguments; developing, one obtains

$$F_{y'y'}\frac{d^2 y}{dx^2} + F_{y'y}\frac{dy}{dx} + F_{y'x} - F_y = 0. \qquad (7.2.8'')$$

If the condition (7.2.8') or (7.2.8") holds, we say that the functional is *stationary* on the curve $y(x)$. Because this condition is only necessary, one must then verify if the solution of the respective differential equation minimizes effectively the functional $I(y)$.

Let be now a functional of the form

$$I(u) \equiv \iint_D F(x_1, x_2; u, u_1, u_2) \mathrm{d}x_1 \mathrm{d}x_2 \qquad (7.2.9)$$

on a set $\{u(x_1, x_2)\}$ of functions of class C^2, which take continuously given values on the frontier of the domain D; F is a given function of class C^2 in the arguments x_1, x_2, u, $u_1 \equiv \partial u / \partial x_1$, $u_2 \equiv \partial u / \partial x_2$ on the domain of definition of those arguments. *The Euler-Ostrogradskiĭ equation* corresponding to the problem of minimum reads

$$F_u - (F_{u_1})_{,1} - (F_{u_2})_{,2} = 0, \qquad (7.2.9')$$

being a necessary condition too. In particular, in the case of the functional

$$I(u) \equiv \iint_D \left[u_1^2 + u_2^2 + 2f(x_1, x_2)u \right] \mathrm{d}x_1 \mathrm{d}x_2 \qquad (7.2.10)$$

we find an effective minimum given by *Poisson's equation*

$$\Delta u(x_1, x_2) = f(x_1, x_2) \qquad (7.2.10')$$

on the domain D.

Analogously, the functional

$$I(u) \equiv \iint_D F(x_1, x_2; u, u_1, u_2, u_{11}, u_{12}, u_{22}) \mathrm{d}x_1 \mathrm{d}x_2 \qquad (7.2.11)$$

leads to *the Euler-Ostrogradskiĭ equation*

$$F_u - (F_{u_1})_{,1} - (F_{u_2})_{,2} + (F_{u_{11}})_{,11} + (F_{u_{12}})_{,12} + (F_{u_{22}})_{,22} = 0. \qquad (7.2.11')$$

As well, for the functional

$$I(u) \equiv \iiint_D F(x_1, x_2, x_3; u, u_1, u_2, u_3) \mathrm{d}x_1 \mathrm{d}x_2 \mathrm{d}x_3, \qquad (7.2.12)$$

defined on a three-dimensional domain, we obtain *the Euler-Ostrogradskiĭ equation*

$$F_u - (F_{u_1})_{,1} - (F_{u_2})_{,2} - (F_{u_3})_{,3} = 0. \qquad (7.2.12')$$

In the case of several functions $y_k(x)$, $k = 1, 2, \ldots, n$, of the same independent variable x, the functional

Problems of dynamics of the particle

$$I(y_1, y_2, \ldots, y_n) \equiv \int_{x_0}^{x_1} F(x; y_1, y_2, \ldots, y_n, y_1', y_2', \ldots, y_n') \, dx \qquad (7.2.13)$$

leads to *the Euler-Lagrange system of equations*

$$F_{y_k} - \frac{d}{dx}(F_{y_k'}) = 0, \quad k = 1, 2, \ldots, n, \qquad (7.2.13')$$

which represent necessary conditions of stationarity.

2.1.5 Motion on a brachistochrone

The problem to determine a curve C passing through the points P^0 and P^1 is put, so that a particle P, which departs from P^0 with an initial velocity v_0 and is subjected to the action of a field of conservative forces $F = \operatorname{grad} U$, does slide frictionless along the curve from P^0 to P^1 in a minimal interval of time (Fig.7.18); a curve C which has this property is a *brachistochrone curve* for the field of given forces. The conservation theorem of mechanical energy leads to

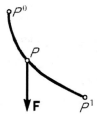

Figure 7.18. Motion of a particle on a brachistochrone.

$$\frac{m}{2}v^2 = \frac{m}{2}\left(\frac{ds}{dt}\right)^2 = U(x_1, x_2, x_3) + h, \quad h = \frac{m}{2}v_0^2 - U(x_1^0, x_2^0, x_3^0), \qquad (7.2.14)$$

wherefrom

$$t = \sqrt{m} \int_{P^0}^{P^1} \varphi(x_1, x_2, x_3) \, ds, \quad \varphi(x_1, x_2, x_3) = \frac{1}{\sqrt{2[U(x_1, x_2, x_3) + h]}}; \qquad (7.2.14')$$

noting that $\varphi = \varphi(s)$, by the agency of the functions $x_i = x_i(s)$, $i = 1, 2, 3$, and that $ds = \sqrt{x_i' x_i'} ds$, $x_i' = dx_i / ds$, we may write the Euler-Lagrange stationarity condition (7.2.13') in the form

$$\varphi_{,i} - \frac{d}{ds}(\varphi x_i') = 0, \quad i = 1, 2, 3. \qquad (7.2.15)$$

Vectorially, we have

$$\operatorname{grad} \varphi - \frac{d}{ds}(\varphi \boldsymbol{\tau}) = \mathbf{0}, \qquad (7.2.15')$$

where $\boldsymbol{\tau} = \mathrm{d}\mathbf{r}/\mathrm{d}s$ is the unit vector of the tangent to the curve. It results

$$\mathrm{grad}\,\varphi - \frac{\mathrm{d}\varphi}{\mathrm{d}s}\boldsymbol{\tau} - \frac{\varphi}{\rho}\boldsymbol{\nu} = \mathbf{0}, \tag{7.2.15''}$$

where we took into account the first Frenet formula, $1/\rho$ being the curvature of the curve, while $\boldsymbol{\nu}$ is the principal normal; a scalar multiplication by $\boldsymbol{\tau}$ leads to the identity $\mathrm{grad}\,\varphi \cdot \mathrm{d}\mathbf{r}/\mathrm{d}s - \mathrm{d}\varphi/\mathrm{d}s = 0$, so that the equations (7.2.15)-(7.2.15'') are reduced to only two equations.

We notice that the equation (7.2.15') is of the same form as the equation of equilibrium (4.2.57) of a perfectly flexible, torsionable and inextensible thread, where $\mathbf{T} = \varphi\boldsymbol{\tau}$, $T = \varphi$ being the tension in the thread, while $\mathbf{p}(s) = -\mathrm{grad}\,\varphi$ is the external conservative load on the unit length, which acts upon the thread, corresponding to the formula (4.2.58'); as well, the equations (7.2.15) may be put in correspondence with the equations (4.2.59). Taking into account (7.2.14'), it results $\varphi_{,i} = -\varphi^3 U_{,i}$, $i = 1,2,3$, so that $\mathbf{p}(s) = -\mathrm{grad}\,\varphi = \varphi^3 \mathrm{grad}\,U = \varphi^3 \mathbf{F}$, where $\mathbf{F}$ is the given force which acts upon the particle in motion on a brachistochrone. Projecting on the principal normal to the curve C, one obtains $p_\nu = \varphi^3 F_\nu = -T/\rho = -\varphi/\rho$, wherefrom $F_\nu = -1/\varphi^2 \rho$, relation which gives the curvature of the brachistochrone as a function of the given force which acts upon the particle. From (7.2.14), one may write $mv^2 = 1/\varphi^2$; on the other hand, the equation of motion in intrinsic co-ordinates reads $mv^2/\rho = F_\nu + R_\nu$, where $\mathbf{R}$ is the constraint force which acts upon the particle constrained to stay on the curve C, which must be determined. There results $F_\nu + R_\nu = 1/\rho\varphi^2$, so that $N_\nu = 2/\rho\varphi^2$ and

$$N_\nu = -2F_\nu; \tag{7.2.16}$$

we may state

Theorem 7.2.1 (*Euler*). *In the motion on a brachistochrone, the modulus of the normal constraint force is twice greater than the modulus of the normal component of the resultant of the given forces.*

If $T = \varphi = \mathrm{const}$ along the thread, we may suppose that one stays on a smooth surface $\varphi = \mathrm{const}$ (equipotential surface), $\mathbf{p}(s)$ being a constraint force $\mathbf{R}(s)$; in this case, the unit vector $\boldsymbol{\nu}$ coincides with the unit vector $\mathbf{n}$ of the normal to the surface and the brachistochrone is a geodesic curve of the surface. Noting that, in this case, $\mathbf{R}(s) = \varphi^3 \mathbf{F}(s)$, it results that such a situation is obtained only if the field of given forces is normal to the searched brachistochrone at each point of it. From this point of view, we may search such curves, situated on a surface and corresponding to a field of given conservative forces. The equation (7.2.15') reads

$$\mathrm{grad}\,\varphi + \lambda\,\mathrm{grad}\,f - \frac{\mathrm{d}}{\mathrm{d}s}(\varphi\boldsymbol{\tau}) = \mathbf{0}, \tag{7.2.17}$$

Problems of dynamics of the particle 449

where $f(x_1, x_2, x_3) = 0$ is the equation of the surface; to this equation there corresponds the equation (4.2.70) of the threads, and the study can be made analogously.

Taking $\mathbf{v}_0 = \mathbf{0}$ and choosing $U = mgx_3$, where Ox_3 is along the local descendent vertical, we get $h = 0$, so that $\varphi = 1/\sqrt{2mgx_3}$; the equations (7.2.15) lead to

$$\frac{1}{\sqrt{x_3}} \frac{dx_1}{ds} = C_1, \quad \frac{1}{\sqrt{x_3}} \frac{dx_2}{ds} = C_2, \quad C_1, C_2 = \text{const},$$

wherefrom $C_1 x_2 = C_2 x_1 + \text{const}$, the trajectory being in a vertical plane. Taking this plane as $x_2 = 0$ (hence $C_2 = 0$), it results ($x_1 = 0$ for $x_3 = 0$ and $1/C_1^2 = 2a$)

$$dx_1 = dx_3 \sqrt{\frac{x_3}{2a - x_3}};$$

we denote $x_3 = a(1 - \cos\theta)$ and get, by integration, $x_1 = a(\theta - \sin\theta)$, so that, in case of a gravitational field, the brachistochrone is a cycloid, the concavity of which is opposite to the direction of these forces. Euler's theorem has been verified in Subsec. 1.3.5 for this particular case.

2.1.6 Motion on a geodesic curve

The motion of a particle on a geodesic curve of a smooth fixed surface S has been emphasized in Chap. 6, Subsec. 2.2.2. To have such a trajectory, it is necessary and sufficient that $F_g = 0$, hence that the force $\mathbf{F}$ be in the osculating plane of the trajectory; on the other hand, the motion is uniform only if $F_\tau = 0$, hence if the given force is normal to the surface S and has an influence only upon the constraint force $\mathbf{R}$. Noting that along a geodesic line we have $\mathbf{n} = \mathbf{v}$, the normal $\mathbf{n}$ to the surface $f(x_1, x_2, x_3) = 0$ being collinear with $\text{grad}\, f$, we may write the equations of the geodesic lines in the form

$$\frac{d^2 \mathbf{r}}{ds^2} = \lambda \,\text{grad}\, f, \quad \frac{d^2 x_i}{ds^2} = \lambda f_{,i}, \quad i = 1, 2, 3, \tag{7.2.18}$$

where λ is an arbitrary scalar and where we took into account the first Frenet formula given by $\mathbf{v} = \rho d^2 \mathbf{r}/ds^2$. If the motion of the particle is uniform, then $ds = v_0 dt$, and the equations of motion along the geodesic curves are of the form

$$\ddot{\mathbf{r}} = \overline{\lambda}\,\text{grad}\, f, \quad \ddot{x}_i = \overline{\lambda} f_{,i}, \quad i = 1, 2, 3, \quad \overline{\lambda} = \lambda v_0^2. \tag{7.2.18'}$$

If P^0 and P^1 are two points on a geodesic line, then the distance l between these two points is

$$l = \int_{P^0}^{P^1} ds = \int_{P^0}^{P^1} \sqrt{x_i' x_i'}\, ds, \quad x_i' = \frac{dx_i}{ds}; \tag{7.2.19}$$

we put the problem to determine the functions $x_i = x_i(s)$ so that the functional l be stationary (in fact, minimal). Because these functions are linked by the relation $f(x_1, x_2, x_3) = 0$, we use the method of Lagrange's multiplier, searching functions for which the functional

$$\int_{P^0}^{P^1} \left(\sqrt{x_i' x_i'} + \lambda f \right) ds$$

is stationary, λ being an indeterminate parameter; the stationarity conditions (7.2.13') give $f_{,i} - dx_i'/ds = 0$, $i = 1,2,3$, thus finding again the equations (7.2.18), which coincide with the equations (7.2.17) for $\varphi = 1$. It results that the shortest way on a surface between two points of it is along the geodesic line which joints them and is unique. This property is characteristic for the geodesic lines, which are "the most straight lines" on the surface. In particular, if there are straight lines which stay on a surface (e.g., the case of a ruled surface), these ones are – obviously – geodesic lines; a particle which is acted upon by not one force and is launched along such a straight line travels through it corresponding to the principle of inertia, the constraint force vanishing.

If the surface is given by the parametric equations $x_i = x_i(q_1, q_2)$, $i = 1,2,3$, then the distance is written in the form

$$l = \int_{P^0}^{P^1} \sqrt{g_{\alpha\beta} q_\alpha' q_\beta'} \, ds, \quad q_\alpha' = \frac{dq_\alpha}{ds}, \quad g_{\alpha\beta} = g_{\beta\alpha} = \frac{dx_i}{dq_\alpha} \frac{dx_i}{dq_\beta}, \qquad (7.2.19')$$

where the Greek dummy indices correspond to the summation with respect to 1 and 2. The Euler-Lagrange equations lead to

$$2 \frac{d}{ds}\left(g_{\beta\gamma} q_\beta' \right) - \frac{\partial g_{\alpha\beta}}{\partial q_\gamma} q_\alpha' q_\beta' = 0, \quad \gamma = 1,2;$$

differentiating and noting that

$$\frac{\partial g_{\beta\gamma}}{\partial q_\alpha} q_\alpha' q_\beta' = \frac{1}{2}\left(\frac{\partial g_{\alpha\gamma}}{\partial q_\beta} + \frac{\partial g_{\beta\gamma}}{\partial q_\alpha} \right) q_\alpha' q_\beta',$$

we get the equations

$$g_{\beta\gamma} q_\beta'' + [\beta\delta, \gamma] q_\beta' q_\delta' = 0, \quad \gamma = 1,2, \qquad (7.2.19'')$$

where we have introduced Christoffel's symbol of the first kind. Multiplying by the normalized algebraic complement $g^{\alpha\gamma}$, summing and using Christoffel's symbol of second kind, there result the equations of the geodesic lines in the normal form

$$q_\alpha'' + \left\{ \begin{matrix} \alpha \\ \beta \ \gamma \end{matrix} \right\} q_\beta' q_\gamma' = 0, \quad \gamma = 1,2; \qquad (7.2.19''')$$

Problems of dynamics of the particle

we find thus again the equations (6.2.29), the latter ones being deduced in the hypothesis of a uniform motion. In this last case (we have $T = mv^2/2 = \text{const}$ too), we may write

$$v^2 = \left(\frac{ds}{dt}\right)^2 = g_{\alpha\beta}q'_\alpha q'_\beta \frac{d^2s}{dt^2} = g_{\alpha\beta}q'_\alpha q'_\beta v_0^2 = v_0^2,$$

what was to be expected, because $g_{\alpha\beta}q'_\alpha q'_\beta = 1$ represents a first integral of the system of equations (7.2.19'''), where one of these equations may be replaced by the respective first integral.

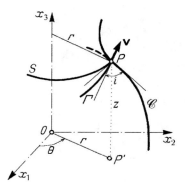

Figure 7.19. Motion of a particle on a geodesic curve of a surface of rotation.

Let us consider, in particular, the case in which the surface S is a surface of rotation of parametric equations $x_1 = r\cos\theta$, $x_2 = r\sin\theta$, $x_3 = z = \varphi(r)$, where r, θ, z are cylindrical co-ordinates (Fig.7.19); in this case, $ds^2 = (1+\varphi'^2)dr^2 + r^2 d\theta^2$, $\varphi' = d\varphi/dr$. In the case of a uniform motion $ds^2 = v_0^2 dt^2$, obtaining thus the first integral

$$(1+\varphi'^2)\dot{r}^2 + r^2\dot{\theta}^2 = v_0^2 = \text{const};\qquad(7.2.20)$$

we notice that the support of the constraint force passes through the rotation axis, so that a first integral of areas for the projection P' of the particle P on the plane Ox_1x_2 reads

$$r^2\dot{\theta} = r_0^2\dot{\theta}_0 = C = \text{const}.\qquad(7.2.20')$$

The solution of the problem is thus reduced to quadratures. Eliminating the time t between the two first integrals, we obtain $(1+\varphi'^2)dr^2 + r^2 d\theta^2 = r^4 d\theta^2/k^2$, $k^2 = C^2/v_0^2$, so that the equation of the geodesic curves passing through the point $(r_0, \theta_0, z_0 = \varphi(r_0))$ is given by

$$\theta = \theta_0 + k\int_{r_0}^{r}\sqrt{\frac{1+[\varphi'(\rho)]^2}{\rho^2 - k^2}}\frac{d\rho}{\rho};\qquad(7.2.21)$$

the constant k of the family of geodesic lines is determined imposing the condition that the respective line passes through a second given point. If the geodesic curve Γ pierces the meridian $\mathscr{C}$ of the point P by an incidence angle i, then the components of the velocity $\mathbf{v}$ along the meridian and the parallel of the point are $v\cos i$ and $v\sin i$, respectively. The moment of the velocity $\mathbf{v}$ with respect to the axis of rotation Ox_3 will be equal to C (corresponding to the first integral of areas) and is given only by the component $v\sin i$ with the arm level r (Fig.7.19); it results that $rv\sin i = C$. Because $v = v_0$, one obtains *Clairaut's formula*

$$r\sin i = k, \qquad (7.2.22)$$

where k is the constant of the family of geodesic lines (the ratio of the constants of the two first integrals). Reciprocally, if the relation (7.2.22) takes place for all the points of a curve on a surface, then that one is a geodesic line or a parallel of the surface of rotation.

2.2 Other applications

In what follows, some particular cases of motion of a particle will be presented, e.g., the motion of a particle on a circular helix; as well, some particular methods to solve problems of dynamics of the particle are considered, for instance the method of transformation of motions.

2.2.1 Motion of a particle on a circular helix

Let be a heavy particle, constrained to move on a circular helix of parametric equations $x_1 = R\cos\theta$, $x_2 = R\sin\theta$, $x_3 = p\theta/2\pi = R\theta\tan\alpha$, where p is the pitch of the helix, while α is the angle made by the helix with a horizontal parallel of the circular cylinder, of radius R, on which the helix is enveloped (see Chap. 5, Subsec. 1.3.3, Fig.5.9); in general, we assume that the motion takes place in a resistent medium, characterized by a viscous resistance λv, $\lambda > 0$. Euler's equations of motion read

$$m\dot{v} = F_\tau - \lambda v, \quad \frac{mv^2}{\rho} = F_\nu + R_\nu, \quad 0 = F_\beta + R_\beta.$$

If the initial velocity $\mathbf{v}_0$ is directed so that $\dot\theta < 0$ (the particle "moves down" on the helix), then we have

$$\boldsymbol{\tau} = \cos\alpha(\sin\theta\mathbf{i}_1 - \cos\theta\mathbf{i}_2) - \sin\alpha\mathbf{i}_3,$$
$$\boldsymbol{\nu} = -\cos\theta\mathbf{i}_1 - \sin\theta\mathbf{i}_2,$$
$$\boldsymbol{\beta} = -\sin\alpha(\sin\theta\mathbf{i}_1 - \cos\theta\mathbf{i}_2) - \cos\alpha\mathbf{i}_3.$$

so that, for the given force $\mathbf{F} = m\mathbf{g}$ we get the components

$$F_\tau = mg\sin\alpha, \quad F_\nu = 0, \quad F_\beta = -mg\cos\alpha.$$

Problems of dynamics of the particle

The equation of projection along the tangent leads to

$$v = \frac{g}{k}\sin\alpha\left(1 - e^{-kt}\right), \quad k = \frac{\lambda}{m}, \quad (7.2.23)$$

where we have put the initial condition $v(0) = 0$. But $v = -R\dot\theta/\cos\alpha$, so that we obtain ($R\theta(0)\tan\alpha = h$)

$$\theta = \frac{h}{R\tan\alpha} + \frac{g\sin 2\alpha}{2k^2 R}\left(1 - kt - e^{-kt}\right). \quad (7.2.23')$$

In particular, in the case of falling in vacuum ($\lambda = 0$, hence $k = 0$), we develop into series the exponential function, obtaining

$$v = gt\sin\alpha, \quad \theta = \frac{h}{R\tan\alpha} - \frac{gt^2\sin 2\alpha}{4R}. \quad (7.2.23'')$$

The other two intrinsic equations give the components of the constraint force.

2.2.2 The simple pendulum in a motion of rotation

Let us suppose that the vertical circle on which moves a heavy particle (in particular, the simple pendulum considered in Subsec. 1.3.1) is rotating with a constant angular velocity ω around its vertical diameter. The co-ordinates of the particle are

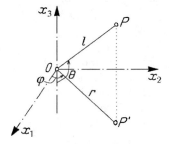

Figure 7.20. Simple pendulum in a motion of rotation.

$$x_1 = l\sin\theta\cos\omega t, \quad x_2 = l\cos\theta\sin\omega t, \quad x_3 = -l\cos\theta,$$

where the applicate x_3 is directed towards the local ascendent vertical (Fig.7.20). Because the constraint is rheonomic, we use Lagrange's equation (6.2.21), where

$$T = \frac{1}{2}ml^2\left(\dot\theta^2 + \omega^2\sin^2\theta\right), \quad Q = m\mathbf{g}\cdot\frac{d\mathbf{r}}{d\theta} = -mgl\sin\theta;$$

we obtain thus

$$\ddot{\theta} - \omega^2 \sin\theta \cos\theta + \frac{g}{l}\sin\theta = 0. \qquad (7.2.24)$$

Introducing the non-dimensional variable $\varphi = \omega t$, we also may write this equation in the form ($\mathrm{d}\varphi = \omega \mathrm{d}t$, $\theta' = \mathrm{d}\theta/\mathrm{d}\varphi$)

$$\theta'' = (\cos\theta - \lambda)\sin\theta, \quad \lambda = \frac{g}{l\omega^2}; \qquad (7.2.24')$$

multiplying by $2\dot{\theta}$ and integrating, one obtains the first integral

$$\dot{\theta}^2 - \left(\sin^2\theta + 2\lambda\cos\theta\right) = \mathrm{const}, \qquad (7.2.24'')$$

the equation $\theta = \theta(\varphi)$ of the trajectory being thus obtained by a quadrature.

The above considerations are valid for pendulary motions as well as for circular ones.

2.2.3 Transformation of motions

We – often – have seen that in the study of a motion one may use the results obtained for other simpler motions. For instance, we have seen that the study of arbitrary tautochronous motions is reduced to the study of rectilinear tautochronous motions; that is, in fact, a transformation of motions.

In Subsec. 2.1.5, an interesting analogy between the equilibrium of a perfectly flexible, torsionable and inextensible thread and a motion of a particle on a brachistochrone has been stated; that one may be considered as a transformation of motion too.

The motion of a particle on a smooth surface may be studied by means of the equations (6.2.24) with respect to Darboux's trihedron. Let us deform the surface, assuming that this is possible, so that the lengths of the lines drawn on the surface do remain invariable; in such a transformation, also the geodesic curvatures remain invariant. Let us modify the force **F** so that its projection on the plane tangent to the surface (hence, the components F_τ and F_g) do remain unchanged; in this case, the first and the third equation (6.2.24) which specify the motion remain – further – valid, and the motion is the same as in the first case (the study of this new motion may be much more simple); obviously, the constraint force given by the second equation (6.2.24) is another one. In the case of a developable surface, the problem may be – eventually – reduced to a plane problem. For instance, the trajectory of a heavy particle constrained to move on a vertical circular cylinder is obtained by enveloping on it a parabola of vertical axis (obtained by the motion of a heavy particle in a vertical plane). We notice that, in Subsec. 2.2.1, we have considered the case of a heavy particle constrained to move on a circular helix, which is a geodesic line of the circular cylinder on which lays the curve; but that problem is different from that mentioned above, which may be solved also by means of the equation (7.1.59'). If the initial velocity vanishes or is directed towards the local vertical, then the trajectory is just this vertical, which is a geodesic line of the circular cylinder too (in this case, $F_g = 0$), and the equation

Problems of dynamics of the particle

(7.1.59') leads to Torricelli's formula (7.1.28"). Analogously, the trajectory of a heavy particle constrained to move on a cone of rotation of vertical axis is obtained enveloping on this cone the plane trajectory of a particle acted upon by a central force, the magnitude of which is constant.

In the plane Ox_1x_2, let be a particle P of mass m, acted upon by a given force which depends only on the position ($\mathbf{F} = \mathbf{F}(\mathbf{r})$); the equations of motion are of the form

$$m\ddot{x}_\alpha = F_\alpha(x_1, x_2), \quad \alpha = 1, 2. \tag{7.2.25}$$

The space *homographic transformation*

$$x'_\alpha = \frac{a_{\alpha\beta}x_\beta + b_\alpha}{a_\gamma x_\gamma + c}, \quad a_{\alpha\beta}, b_\alpha, a_\gamma, c = \text{const}, \quad \alpha, \beta, \gamma = 1, 2, \tag{7.2.26}$$

and the time transformation for which

$$k\,dt' = \frac{dt}{(a_\gamma x_\gamma + c)^2} \tag{7.2.26'}$$

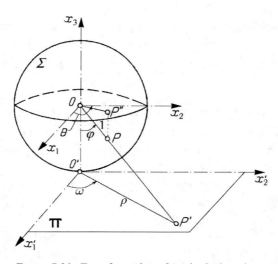

Figure 7.21. Transformation of a spherical motion.

lead to the motion of a particle P' acted upon by a force which depends only on the position ($\mathbf{F}' = \mathbf{F}'(\mathbf{r}')$) too; P. Appell showed that the trajectory of the particle P' is the homographic transformation of the trajectory of the particle P, the direction of the force $\mathbf{F}'$ being also the homographic transformation of the direction of the force $\mathbf{F}$. In particular, if $\mathbf{F}$ is a central force, then the force $\mathbf{F}'$ is a central one too or has a fixed direction. One may also prove that the most general transformation of the form $x'_\alpha = x'_\alpha(\mathbf{r})$, $\alpha = 1, 2$, $dt' = \lambda(\mathbf{r})dt$, so that for any $\mathbf{F} = \mathbf{F}(\mathbf{r})$ to have $\mathbf{F}' = \mathbf{F}'(\mathbf{r}')$, is a homographic transformation.

One may imagine also other transformations of interest, e.g., the transformation of a *spherical motion* (on a sphere) in a *plane motion*. Let thus be a sphere Σ of radius equal to unity (for the sake of simplicity) and of centre O, and a plane Π tangent to the sphere at the point O' (Fig.7.21). By a *central transformation*, used in the geographic map technique, to a point P of longitude θ and colatitude φ on the sphere we make to correspond a point P', of polar co-ordinates ρ, ω, at which the radius OP pierces the plane Π; thus, to the straight lines of the plane correspond great circles of the sphere and inversely. We notice that $\rho = \tan\varphi$, $\omega = \theta$. If the particle P is of mass equal to unity, then the corresponding equations of motion on the sphere read

$$\frac{d}{dt}\left(\sin^2\varphi\frac{d\theta}{dt}\right) = \Theta(\theta,\varphi), \quad \frac{d^2\varphi}{dt^2} - \sin\varphi\cos\varphi\left(\frac{d\theta}{dt}\right)^2 = \Phi(\theta,\varphi); \qquad (7.2.27)$$

as well, the equations of motion of a particle P' of mass equal to unity in the plane Π are written in the form

$$\frac{d^2\rho}{dt'^2} - \rho\left(\frac{d\omega}{dt'}\right)^2 = R(\rho,\omega), \quad \frac{d}{dt'}\left(\rho^2\frac{d\omega}{dt'}\right) = \Omega(\rho,\omega). \qquad (7.2.27')$$

P. Appell showed that, in the case of a central transformation for which $dt = \cos^2\varphi \, dt'$, the equations (7.2.27) become (7.2.27'), where $R = \Phi\cos^2\varphi$, $\Omega = \Theta\cos^2\varphi$. Obviously, the plane motion may be studied easier than the spherical one.

2.2.4 Motion on synchronous curves and surfaces

Let be – in a fixed plane – a family of curves $\{C\}$, which depend on a parameter and pass through the fixed point O. Let us launch from O, at the initial moment $t = t_0$, with initial velocities of equal modulus v_0, identical particles P, acted upon by forces which derive from a given potential; the locus of the positions of the particles P at the same moment t is the curve Γ. The curves $\{\Gamma\}$ form a family of curves which depend on the parameter t and are called *synchronous curves* of the curves $\{C\}$. If the family of curves $\{C\}$ belongs to the three-dimensional space, then the locus of the positions of the particles P at the same moment t is a surface Σ; the family of surfaces $\{\Sigma\}$ depends on the parameter t and forms *the synchronous surfaces* of the curves $\{C\}$.

Euler showed that, if $v_0 = 0$, the given force corresponding to a gravitational field, the lines $\{C\}$ being straight lines in a vertical plane and passing through O, then the synchronous curves $\{\Gamma\}$ are circles.

Let be the family of lines $\{C\}$, called *trajectories*, which pass through the point O, and the family of lines $\{C'\}$, called *synodal lines*, which pass through the same point. One may prove that there exists an infinity of given conservative forces **F** so that a particle departing from O with a given initial velocity, along one of the lines $\{C\}$, reaches a point P in the same interval of time in which it would reach it along that curve $\{C'\}$ which would pass through the same point. Obviously, the families of curves $\{C\}$ and $\{C'\}$ may invert their rôles. We mention that, if the synchronous lines are orthogonal to the trajectories, in a plane motion, then the latter ones coincide with the synodal lines, being brachistochrones for the considered forces. These problems have been studied by Fouret, Saint-Germain and Vâlcovici.

2.3 Stability of equilibrium of a particle

In what follows, we make some considerations concerning the stability of equilibrium of a free or constraint particle, completing thus the results in Chap. 4, Subsec. 1.1.5; the results thus obtained will be then used for representations in the phase plane.

2.3.1 Stability of equilibrium of a free particle

In Chap. 4, Subsec. 1.1.7, we have presented the Lagrange-Dirichlet theorem, which states that *a position P^0 of a free particle P, acted upon by a field of conservative forces is a position of stable equilibrium if the simple potential U has an isolated maximum at that point*. The demonstration which has been given has rather an intuitive character; we will use the conservation theorem of mechanical energy for a rigorous demonstration.

We notice that the potential $U(\mathbf{r})$ is determined making abstraction of an additive constant; choosing the point P^0 as origin O, we may take $U(0) = 0$. Let be a closed convex surface S which contains the point O (e.g., a sphere of centre O), of arbitrary small dimensions, so that in the interior and on the surface the function $U(\mathbf{r})$ be negative, vanishing only at O. We may assume that there exists $p > 0$ sufficiently small so that to have $-U > p$, hence $U + P < 0$ on the surface S. Let P_0 be an initial position of the particle P in the interior of the surface S, the corresponding velocity being v_0; taking into account the conservation theorem of mechanical energy, we may write $mv^2/2 = U + \left(mv_0^2/2 - U_0\right)$, $U_0 < 0$. We may determine the position and the magnitude of the velocity at the initial moment by the condition $mv_0^2/2 - U_0 < p$; to do this, it is sufficient – for instance – to take $mv^2/2 < p/2$, $-U_0 < p/2$. The first relation shows that $v_0 < \eta' = \sqrt{p/m}$. As well, the function U being continuous and vanishing at the origin, it results that there exists $\eta > 0$ so that $\overline{OP_0} < \eta$, corresponding $-U_0 < p/2$. Thus, if – in the interior of the surface S – we give to the particle an initial position at a distance of O less than η, with an initial velocity less than η', then the conservation theorem of mechanical energy leads to the inequality $mv^2/2 < U + p$, which proves that the particle cannot come out from the interior of the surface S; indeed, if the particle P would reach the surface S, then the sum $U + p$ would become negative, situation impossible if we take into account the previous relation. One may thus state that it corresponds an $\varepsilon > 0$, so that $\overline{OP} < \varepsilon$,

$P = P(t)$. As well, $mv^2/2 < p$, because $U < 0$; it results $v(t) < \sqrt{2p/m}$ $= \varepsilon' > 0$. The conditions that the point $P^0 \equiv O$ be a stable position of equilibrium are thus fulfilled and the Lagrange-Dirichlet theorem is proved for a free particle. For instance, for a free particle in rectilinear motion, acted upon by a force $F(x) = -kx$, $k > 0$, which derives from the simple potential $U(x) = -kx^2/2$, the origin of the co-ordinate axis represents a stable position of equilibrium.

This demonstration remains valid in the case of a generalized potential U, where the rôle of the function which has an isolated maximum is played by the scalar potential U_0.

In what concerns the reciprocal of this theorem, the problem is not sufficiently clarified. One may show that, in certain particular cases, a point may represent a stable position of equilibrium, the potential U having not an isolated maximum at that point. Analogously, it is proved that *a position of equilibrium P^0 is a position of labile equilibrium if the potential U has an isolated minimum at that point*. The conditions imposed by the Lagrange-Dirichlet theorem are – obviously – sufficient conditions. A more profound study of this problem will be made latter, after Lyapunov, in connection with the study of the stability of motion in the frame of Lagrangian or Hamiltonian mechanics. We mention the affirmation of T. Levi-Civita who said that "the instability is the rule, while the stability is rather an exception".

In the case of a non-conservative field of forces, the positions of equilibrium are recognized – in general – using the property of definition, that is perturbing arbitrarily such a position and studying the returning modality of the particle.

2.3.2 Stability of equilibrium of a particle subjected to constraints

In the case of a particle constrained to stay on a fixed smooth surface S, one may introduce the generalized forces $Q_\alpha(u,v)$, $\alpha = 1,2$, given by (4.1.47), as we have seen in Chap. 4, Subsec. 1.1.7; if $Q_1 du + Q_2 dv = d\overline{U}(u,v)$, that is a total differential, then we are led to the study of the extrema of the function $U = U(u,v)$, where u and v are generalized co-ordinates, the holonomic, scleronomic constraints being eliminated. We may always make so as to have a maximum equal to zero for $\overline{U}$ at the point P^0 coinciding with the origin ($\overline{U}(0,0) = 0$). We draw a closed curve C around the point P^0, on the surface S, so that to have $\overline{U} < 0$ on that curve; hence, there exists $p > 0$ so that $\overline{U} + p < 0$ on C. Displacing the particle from P^0 at a neighbouring point, in the interior of the curve C, we may follow the demonstration given in the previous subsection, so that the Lagrange-Dirichlet theorem is applicable in this case too.

Analogously, if the particle is constrained to stay on a fixed smooth curve C, we have seen in Chap. 4, Subsec. 1.1.7 too that the generalized force $Q(q)$, given by (4.1.48) may be introduced, hence the function $\overline{U}(q) = \int Q(q) dq$, being thus led to the study of the isolated maxima of that function. We assume that $\overline{U}(0) = 0$ at the point

P^0; then we follow the previous demonstration, displacing the particle in the interior of an arbitrary interval $[-\eta,\eta]$, which contains the point P^0.

Using the potential energy V given by (6.1.14), (6.1.14'} and starting from the Theorem 4.1.3, we may, finally, state

Theorem 7.2.2 (*Lagrange-Dirichlet*). *The position of equilibrium P^0 of a particle P subjected to holonomic, scleronomic constraints, in the presence of a field of conservative forces, the potential energy having an isolated minimum at that point, is a stable position of equilibrium.*

This statement contains the case of the free particle as well the case of a constraint one (in this case the potential energy must be expressed by means of generalized co-ordinates, eliminating the constraint relations), where the conservative field may derive from a simple or a generalized potential. *If the potential energy has an isolated maximum at the point P^0, then that point represents a position of labile equilibrium.*

In particular, in the case of a gravitational field $V(\mathbf{r}) = mgx_3$, one obtains the Theorem 4.1.2 of Torricelli for a particle constrained to stay on a fixed smooth curve or on a surface having the same properties.

2.3.3 Small oscillations of a heavy particle around the lowest point of a surface

Let be a surface S which passes through the origin O, so that the tangent plane at this point is horizontal, and the surface is over that plane in the vicinity of the respective point. Corresponding to Torricelli's theorem, the point O is a stable position of equilibrium for a heavy particle P. Taking the Ox_3-axis along the local ascendent vertical, the surface S may be represented in the neighbourhood of the point O by a Maclaurin series of the form

$$x_3 = \frac{1}{2}\left(\frac{x_1^2}{R_1} + \frac{x_2^2}{R_2}\right) + \varphi(x_1, x_2), \qquad (7.2.28)$$

where R_1 and R_2 are the principal radii of curvature (the extreme values of the radii ρ_n) of the surface at O, while $\varphi(x_1, x_2)$ corresponds to terms of at least the third degree with respect to the co-ordinates x_1, x_2. Let us consider the small motions of the particle P around the point O. The simple potential which corresponds to the gravitational field is $U(x_3) = -mgx_3$; eliminating the constraint relation (7.2.28) and neglecting terms of higher order, we obtain

$$\overline{U}(x_1, x_2) = -\frac{mg}{2}\left(\frac{x_1^2}{R_1} + \frac{x_2^2}{R_2}\right)$$

and the force which acts upon the particle is given by

$$\mathbf{F} = \operatorname{grad}\overline{U} = -mg\left(\frac{x_1}{R_1}\mathbf{i}_1 + \frac{x_2}{R_2}\mathbf{i}_2\right).$$

We obtain thus the equations of motion

$$\ddot{x}_1 = -\omega_1^2 x_1, \quad \ddot{x}_2 = -\omega_2^2 x_2, \quad \omega_1^2 = \frac{g}{R_1}, \quad \omega_2^2 = \frac{g}{R_2}. \tag{7.2.29}$$

Integrating, it results

$$x_1 = a_1 \cos(\omega_1 t - \varphi_1), \quad x_2 = a_2 \cos(\omega_2 t - \varphi_2), \tag{7.2.29'}$$

where the amplitudes a_1, a_2 and the phase shifts φ_1, φ_2 are determined by the initial conditions. In particular, if $R_1 = R_2 = R$, then one obtains the small motions corresponding to the spherical pendulum (see Subsec. 1.3.7); the general case will be studied in Chap. 8, Subsec. 2.2.5.

2.3.4 Representations in the phase plane. Topological methods

In Chap. 6, Subsec. 2.2.4 we have seen how one may study the motion of a particle with a single degree of freedom in the conservative case of equation $\ddot{q} = f(q)$ in the phase space of co-ordinates q, p; the results thus obtained may be applied for linear systems as well as for non-linear ones. As an example, we consider the simple pendulum (see Subsec. 1.3.1) for which ($q = \theta$)

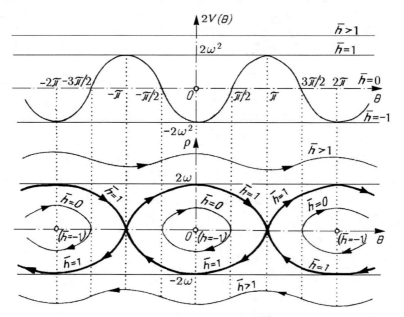

Figure 7.22. Simple pendulum; representation of the motion in the phase plane.

$$p^2 = 2\omega^2 \cos\theta + h, \quad p = \dot{\theta}, \quad V(\theta) = -\omega^2 \cos\theta, \quad \omega^2 = \frac{g}{l},$$

$$h = 2\omega^2 \bar{h}, \quad \bar{h} = -\cos\alpha,$$

$\bar{h}$ being a non-dimensional constant, while α is the amplitude. Representing $2V(\theta)$ as a function of θ (Fig.7.22), we see that the motion takes place only for $\bar{h} \in [-1,1]$; we may have also $\bar{h} > 1$, but it does not correspond to a real angle α, the motion being – in this case – circular. The condition $2V(\theta) \leq h$ allows to draw the curves $p = p(\theta)$, symmetric with respect to the $O\theta$-axis, function of various values of $\bar{h}$, in the phase space. For $\bar{h} \in (-1,1)$, e.g., for $\bar{h} = 0$, the motion is oscillatory (we have a simple pendulum). If $\bar{h} = 1$, then the motion is asymptotic, obtaining *the lines of separation* (drawn with a thicker line) in the phase space; for $\alpha = \pi$ it corresponds a labile position of equilibrium. For $\bar{h} = -1$ it results a stable position of equilibrium (a point in the phase space), corresponding $\alpha = 0$. Noting that, for $p > 0$, q increases at the same time as t, we have indicated by an arrow the direction of the motion in the phase space.

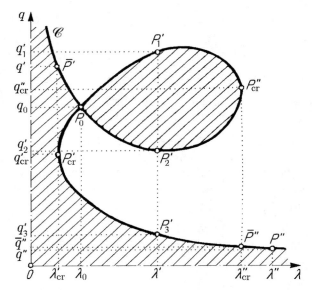

Figure 7.23. Topological structure of a phase trajectory depending on a parameter.

We notice that the separation lines are phase trajectories of the representative point in the phase space; they do not allow to pass from a type of motion to another one. We have seen that a *singular point* is specified by the equations $f(q) = 0$, $p = 0$, all the other points being *ordinary points*; it results that an ordinary point is characterized by a well defined direction of the tangent to the phase trajectory passing through this point. We may thus state

Theorem 7.2.3 (*Cauchy*). *Through each ordinary point of the phase plane passes a phase trajectory and only one.*

We notice that the equation (6.2.39') defines a field of vectors of components q, p, hence a field of velocities in the phase plane; hence, the singular point represents the point at which the velocity in the phase plane vanishes. *The topological methods* allow to study the general topological properties of the phase trajectories defined by the equation (6.2.39'). Taking into account the form of the phase trajectories in the neighbourhood of the points of stable equilibrium ($\bar{h} = -1$), such a singular point is called *centre*; analogous considerations lead to the denomination of *saddle point* for a singular point of labile equilibrium ($\bar{h} = 1$).

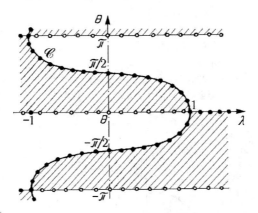

Figure 7.24. Topological structure of the phase trajectory of a simple pendulum in motion of rotation.

The topological structure of the phase trajectories may vary for some particular values of the parameter which appears in a first integral. After H. Poincaré, we introduce the parameter in the differential equation in the form $f(q,\lambda) = -\partial V(q,\lambda)/\partial q$, the positions of equilibrium being situated along the curve $\mathscr{C}$ of equation $f(q,\lambda) = 0$ (Fig.7.23). For various values of the parameter λ one obtains three positions of equilibrium (for $\lambda = \lambda'$ there correspond the points P_1', P_2', P_3' of ordinates q_1', q_2', q_3') or one position of equilibrium (for $\lambda = \lambda''$ there corresponds the point P'' of ordinate q''); one passes from three positions to only one position by critical values of the parameter λ ($\lambda = \lambda_{cr}', \lambda_{cr}''$), to which correspond the points P_{cr}', P_{cr}'', of ordinates q_{cr}', q_{cr}'', and the points P', P'' of ordinates q', q'', respectively. Noting that $dq/d\lambda = -f_\lambda'(q,\lambda)/f_q'(q,\lambda)$, it follows that the critical points correspond to the solutions of the equation $f_q'(q,\lambda) = 0$ (for which the tangent to the curve $f(q,\lambda) = 0$ is parallel to the axis Oq), assuming that $f_\lambda'(q,\lambda) \neq 0$. One may thus state that the points of equilibrium appear and disappear two by two. We suppose that the curve $\mathscr{C}$ is a Jordan one, which divides the plane in two regions. We notice that a straight line $\lambda = \lambda'$ pierces the curve $\mathscr{C}$, for instance at the point P_3'; if $f(q,\lambda') > 0$, hence if $V_q'(q,\lambda') < 0$, below the curve $\mathscr{C}$, then, for q increasing, $V_q'(q_3', \lambda') = 0$ on $\mathscr{C}$ and $V_q'(q,\lambda') > 0$, over the curve $\mathscr{C}$. It follows that $V(q_3', \lambda')$ represents an isolated minimum of the potential energy, and the Lagrange-Dirichlet theorem allows to state

Problems of dynamics of the particle 463

Theorem 7.2.4 (*Poincaré*). *The positions of equilibrium of a particle which moves after the law* $\ddot{q} = f(q,\lambda)$ *in a conservative field are stable if the domain* $f(q,\lambda) > 0$ *is under the curve* $f(q,\lambda) = 0$, $q > 0$, $\lambda > 0$, *and labile if this domain is over that one* (in Fig.7.23 the hatched domain corresponds to $f(q,\lambda) > 0$).

Let us apply these results to the case of the simple pendulum in a motion of rotation (see Subsec. 2.2.2), governed by the equation (7.2.24') of the form $\ddot{\theta} = f(\theta,\lambda) = (\cos\theta - \lambda)\sin\theta$. The curves $\mathscr{C}$ are given by the straight lines $\theta = 0$ and $\theta = \pm\pi$ and by the curve $\theta = \arccos\lambda$. Applying the Theorem 7.2.4, we find *stable branches* of the curve $\mathscr{C}$ (the points of equilibrium of centre type are denoted by whole little circles, i.e., $\theta = \arccos\lambda$ and $\theta = 0$, $\lambda > 1$, and $\theta = \pm\pi$, $\lambda < -1$), as well as *labile branches* (the points of equilibrium of saddle type are denoted by empty little circles, i.e., $\theta = 0$, $\lambda > 1$ and $\theta = \pm\pi$, $\lambda < -1$ (Fig.7.24). The points $\theta = 0$, $\lambda = 1$ and $\theta = \pm\pi$, $\lambda = -1$ are *points of ramification* of the equilibrium, while the values $\lambda_{cr} = \pm 1$ are *critical values (of bifurcation)* of the parameter λ, corresponding to those points. Taking into account (7.2.24'), it results that $\lambda > 0$, the domains of the figure being thus restraint; as well, to have $\lambda < 1$ the angular velocity ω must be sufficiently great. If a separation line passes through the singular point $\dot{\theta} = 0$, $\theta = 0$, then the first integral (7.2.24'') becomes

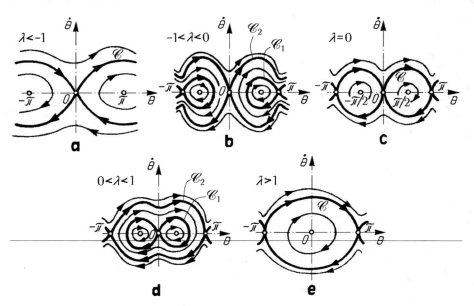

Figure 7.25. Phase space representation of the motion of a simple pendulum in rotation for various values of the parameter λ: $\lambda < -1$ (a); $-1 < \lambda < 0$ (b); $\lambda = 0$ (c); $0 < \lambda < 1$ (d); $\lambda > 1$ (e).

$$\dot{\theta}^2 = \sin^2\theta - 2\lambda(1 - \cos\theta), \quad \dot{\theta} = \pm 2\sin\frac{\theta}{2}\sqrt{\cos^2\frac{\theta}{2} - \lambda}; \qquad (7.2.30)$$

if such a line passes through the singular points $\dot{\theta} = 0$, $\theta = \pm \pi$, then the respective first integral has the form

$$\dot{\theta}^2 = \sin^2 \theta + 2\lambda(1 + \cos\theta), \quad \dot{\theta} = \pm 2\cos\frac{\theta}{2}\sqrt{\sin^2\frac{\theta}{2} + \lambda}. \quad (7.2.30')$$

For $\lambda < -1$, the singular points of saddle type $\dot{\theta} = 0$, $\theta = \pm\pi$ become singular points of centre type (Fig.7.25,a), passing through $\lambda'_{cr} = -1$; for $-1 < \lambda < 0$ appear two separation lines $\mathscr{C}_1$ and $\mathscr{C}_2$, the first of those ones surrounding two centres, while the point O becomes a singular point of saddle type (Fig.7.25,b). If $\lambda = 0$, hence if $\omega \to \infty$, then the curves $\mathscr{C}_1$ and $\mathscr{C}_2$ coincide with the curve $\mathscr{C}$ and form only one line of separation; in this case, the centres are of abscissae $\theta = \pm\pi/2$ (Fig.7.25,c). For $0 < \lambda < 1$ one obtains two separation lines $\mathscr{C}_1$ and $\mathscr{C}_2$, corresponding the equations (7.2.30) and (7.2.30'), respectively, which pass through the singular points of saddle type $\dot{\theta} = 0$, $\theta = 0$, and $\dot{\theta} = 0$, $\theta = \pm\pi$, respectively; in the interior of the loops of the curve $\mathscr{C}_1$ there exist two other singular points of centre type, having the abscissae $\theta = \pm 2\arccos\sqrt{\lambda}$ (Fig.7.25,d). If $\lambda = \lambda''_{cr} = 1$, then the curve $\mathscr{C}_1$ coincides with the singular point O, which becomes a point of centre type; for $\lambda > 1$ remains only one separation line $\mathscr{C}$ (Fig.7.25,e). We observe thus that the separation lines correspond to phase trajectories with different topological aspects.

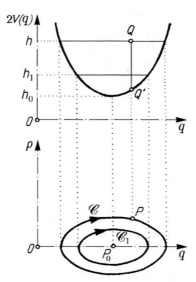

Figure 7.26. Potential function versus generalized co-ordinate diagram.

The above considerations allow to state, without demonstration,

Theorem 7.2.5 (*Poincaré*). *The closed phase trajectories of a particle which is moving after the law* $\ddot{q} = f(q,\lambda)$ *in a conservative field may surround only an odd number of*

singular points, the number of centres being greater than the number of singular points of saddle type.

Sometimes, it is difficult to build up phase trajectories by *analytical methods*, so that *approximate methods*, i.e., *graphical* or *grapho-analytical methods* are necessary. Thus, we may give a graphic representation of $2V(q)$ as function of q (Fig.7.26); drawing a parallel of applicate h_1 to Oq, we may measure the difference $\overline{Q'Q} = 2V(q) - h$. By the formula (6.2.40), the radical of this difference allows to specify the representative point P in the phase plane, setting up the phase trajectory $\mathscr{C}$ by points. Starting from $h = h_1$, we obtain the curve $\mathscr{C}_1$ a.s.o.; to $h = h_0$ corresponds the singular point of centre type P_0.

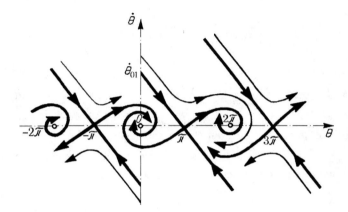

Figure 7.27. Phase space representation of the motion of a simple pendulum in a resistent medium.

In the case of a dissipative mechanical system with a single degree of freedom, of equation $\ddot{q} = f(q) + F(p)$, $pF(p) \leq 0$, we start from the considerations made in Chap. 6, Subsec. 2.2.5. Multiplying this equation, which corresponds to linear or non-linear damped free oscillations, by dq and integrating along a closed phase trajectory, it results

$$\oint p\,dp = \oint f(q)\,dq + \oint F(p)\,dq = 0;$$

noting that $\oint f(q)\,dq = 0$ too, we get

$$\oint F(p)\,dq = \int_T F(p)p\,dt = 0,$$

where we have integrated on a time interval equal to a period T. Because the product $pF(p)$ maintains a constant sign, it results that we cannot have such a relation, and the motion cannot be periodical.

We have seen in Subsec. 1.3.3 that, in the case of the motion of the simple pendulum in a resistent medium, the angular velocity is given by (7.1.53"); the condition $\dot\theta = \dot\theta_0 \geqslant 0$ for $\theta = 0$ leads to

$$\dot\theta^2 = \left(\dot\theta_0^2 - \frac{2\omega^2}{4k^4 + 1}\right)e^{\mp 2k^2\theta} + \frac{2\omega^2}{4k^4 + 1}(\cos\theta \mp 2k^2 \sin\theta), \qquad (7.2.31)$$

where the signs $\pm$ correspond to $\dot\theta_0 \gtrless 0$, respectively. The points $\dot\theta = 0$, $\theta = n\pi$, $n \in \mathbb{Z}$, correspond to positions of equilibrium; the equilibrium is stable for n even (the corresponding *singular points* are *of focus type*), while for n odd the equilibrium is labile (there correspond *singular points of saddle type* (Fig. 7.27)). If

$$\dot\theta_{0n}^2 = \frac{2\omega^2}{4k^4 + 1}\left(1 + e^{2k^2 n\pi}\right), \quad n \text{ odd}, \qquad (7.2.32)$$

then we notice that for $\dot\theta_0 < \dot\theta_{01}$ the particle oscillates, the motion being damped around the stable position of equilibrium $\dot\theta = 0$, $\theta = 0$; if $\dot\theta_0 = \dot\theta_{01}$, then one obtains the asymptotic motion of a particle. For $\dot\theta_{01} < \dot\theta_0 < \dot\theta_{03}$ the particle effects a complete rotation and then its oscillatory motion is damped; in general, if $\dot\theta_{0n} < \dot\theta_0 < \dot\theta_{0,n+2}$, n odd, then the particle effects $(n+1)/2$ complete rotations, passing then in a regime of damped oscillations around a stable position of equilibrium.

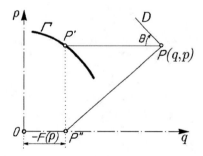

Figure 7.28. Application of Liénard's method in a phase space representation.

As in the case of conservative systems, in the case of non-conservative (dissipative) systems we may use approximate methods of phase trajectories in the phase plane. For example, in the case in which $f(q) = q$ one may use *Liénard's method*. Thus, one draws first the curve Γ of equation $q = -F(p)$ (Fig.7.28); starting from the representative point $P(q,p)$, one draws PP', $P' \in \Gamma$, parallel to Oq, and then $P'P''$, parallel to Op, $P'' \in Oq$. A perpendicular D at P to $P''P$ is inclined with respect to Oq by the angle θ given by $\tan\theta = [q + F(p)]/p$; taking into account (6.2.43), it results that the straight line D defines the slope of the phase trajectory at

P. Step by step, starting from an initial position, we obtain a polygonal line which approximates the searched phase trajectory. We mention also other approximate methods of computation, e.g., Drobov's method, Pell's method, the "delta" method etc.

In the above conditions, we have supposed that $pF(p) \leq 0$, hence that this product has a constant sign. But if this product is positive for small values of $|p|$, then the state of equilibrium is not yet known, a motion being developed starting from this state; if the product becomes then negative, the damping force being opposite to the velocity for great values of $|p|$, then the amplitude begins to be damped. The respective motion is automaintained and will be considered in Chap. 8, Subsecs 2.1.4 and 2.2.7.

Chapter 8

DYNAMICS OF THE PARTICLE IN A FIELD OF ELASTIC FORCES

In the problems studied till now, we considered – especially – the action of a gravitational field (a field of parallel forces, the supports of which pass through a fixed point situated in a plane at infinity) on the motion of a particle. We consider now the case in which the particle is acted upon by central forces (the supports of which pass through a fixed point at a finite distance). First of all, the general case of arbitrary central forces is presented, then the cases in which – after Bertrand's theorem – the orbit is a closed curve (the case of elastic forces and the case of forces of Newtonian attraction are considered). In this chapter we study, in detail, the motion of a particle in a field of elastic forces.

1. The motion of a particle acted upon by a central force

In what follows, we firstly give some general results concerning the equations of motion of the particle subjected to the action of a central force; we make also a qualitative study of the trajectory. Then it is shown that the problem of two particles leads to such a case of motion.

1.1 General results

The study of the motion of a particle subjected to the action of a central force leads to Binet's equation and formula, for which one obtains interesting qualitative results; Bertrand's theorem puts in evidence the two cases in which the trajectory is a closed curve.

1.1.1 Central forces. Binet's equation

Let us consider a particle P acted upon by a central force $\mathbf{F}$; the equation of motion is of the form

$$m\ddot{\mathbf{r}} = F\frac{\mathbf{r}}{r}. \tag{8.1.1}$$

As it was shown in Chap. 6, Subsec. 1.2.6, in this case the trajectory is a plane curve $\mathscr{C}$; taking the plane of the curve as Ox_1x_2-plane and using the results in Chap. 5, Subsec. 1.2.4, we may write the equations of motion in polar co-ordinates in the form

$$m(\ddot{r} - r\dot{\theta}^2) = F, \quad m(2\dot{r}\dot{\theta} + r\ddot{\theta}) = 0. \tag{8.1.1'}$$

The second equation leads to the first integral (corresponding to the formula (6.1.59))

$$2\Omega = r^2\dot{\theta} = rv_\theta = C, \quad C = r_0^2\dot{\theta}_0 = r_0 v_\theta^0 = r_0 v_0 \sin\alpha_0 = \text{const}, \tag{8.1.2}$$

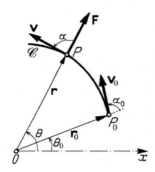

Figure 8.1. Particle acted upon by a central force.

where Ω is the areal velocity of the particle P, the constant C being specified by the initial conditions (Fig.8.1)

$$r_0 = r(t_0), \quad \theta_0 = \theta(t_0), \quad v_0 = v(t_0), \quad \dot{\theta}_0 = \dot{\theta}(t_0), \quad \alpha_0 = \sphericalangle(\mathbf{r}_0, \mathbf{v}_0). \tag{8.1.1''}$$

Taking into account (8.1.2), the first equation (8.1.1') may be written in the form

$$m\ddot{r} = \overline{F}, \tag{8.1.3}$$
$$\overline{F}(r,\theta,\dot{r},\dot{\theta};t) = F(r,\theta,\dot{r},\dot{\theta};t) + \frac{mC^2}{r^3} = F(r,\theta,\dot{r},\dot{\theta};t) + \frac{mv_\theta^2}{r}$$

too, where we have introduced *the apparent force* $\overline{F}$ (we notice that the supplementary force mv_θ^2/r is of the nature of a centrifugal force); the system of differential equations (8.1.2), (8.1.3) determines the functions $r = r(t)$, $\theta = \theta(t)$, the three integration constants which appear being specified by initial conditions. If $F = F(r,\dot{r};t)$, then the motion along the radius vector is given by the unidimensional equation of Newton, where the apparent force $\overline{F}$ is used, the angle θ being then obtained from the integral of areas.

Successively, we have

$$\dot{r} = \frac{\mathrm{d}r}{\mathrm{d}\theta}\dot{\theta} = \frac{C}{r^2}\frac{\mathrm{d}r}{\mathrm{d}\theta} = -C\frac{\mathrm{d}}{\mathrm{d}\theta}\left(\frac{1}{r}\right), \quad \ddot{r} = -C\frac{\mathrm{d}^2}{\mathrm{d}\theta^2}\left(\frac{1}{r}\right)\dot{\theta} = -\frac{C^2}{r^2}\frac{\mathrm{d}^2}{\mathrm{d}\theta^2}\left(\frac{1}{r}\right)$$

and, replacing in the equation (8.1.3), we obtain *Binet's equation* (we suppose that $\dot{F} \equiv \partial F/\partial t = 0$)

Dynamics of the particle in a field of elastic forces

$$\frac{d^2}{d\theta^2}\left(\frac{1}{r}\right) + \frac{1}{r} = -\frac{Fr^2}{mC^2}, \quad F = F(r,\theta,\dot{r},\dot{\theta}); \tag{8.1.4}$$

analogously, by eliminating $\dot{r}$ and $\dot{\theta}$ from the expression of the force F, we get a differential equation of second order, which determines the trajectory of the motion in the form

$$\frac{1}{r} = f(\theta; C_1, C_2). \tag{8.1.4'}$$

The initial conditions

$$f(\theta_0; C_1, C_2) = \frac{1}{r_0}, \quad f'(\theta_0; C_1, C_2) = -\frac{\dot{r}_0}{C} = -\frac{\cot\alpha_0}{r_0}, \tag{8.1.4''}$$

where we have put $\dot{r}_0 = v_0 \cos\alpha_0$ and $f' \equiv \partial f / \partial\theta$, allow to determine the integration constants C_1 and C_2. The integral of areas specifies the motion on the trajectory in the form

$$t = t_0 + \frac{1}{C}\int_{\theta_0}^{\theta} \frac{d\vartheta}{f^2(\vartheta; C_1, C_2)}. \tag{8.1.5}$$

Noting that $\mathbf{F}\cdot d\mathbf{r} = F(\mathbf{r}/r)\cdot d[r(\mathbf{r}/r)] = Fdr$, the theorem of kinetic energy leads to

$$\frac{mv^2}{2} - \frac{mv_0^2}{2} = \int_{r_0}^{r} F(\rho,\theta,\dot{\rho},\dot{\theta};t)d\rho.$$

If $F = F(r,\dot{\theta})$, hence if $F = F(r)$, then we may write a first integral of energy, hence a first integral of Binet's equation in the form

$$\left[\frac{d}{d\theta}\left(\frac{1}{r}\right)\right]^2 + \frac{1}{r^2} = \frac{1}{C^2}\left[v_0^2 + \frac{2}{m}\int_{r_0}^{r} F(\rho)d\rho\right], \tag{8.1.6}$$

noting that $v^2 = \dot{r}^2 + r^2\dot{\theta}^2 = \dot{r}^2 + C^2/r^2$; we may obtain this result multiplying both members of Binet's equation by $d(1/r)/d\theta$ and integrating. The given force is, in this case, conservative and we may introduce the simple potential $U = U(r)$, so that $F(r) = U'(r) = dU/dr$. The first integral (8.1.6) becomes

$$\frac{m\dot{r}^2}{2} = \frac{mC^2}{2}\left[\frac{d}{d\theta}\left(\frac{1}{r}\right)\right]^2 = \frac{mC^2}{2r^2}\left(\frac{dr}{d\theta}\right)^2 = \overline{U}(r) + h,$$

$$\overline{U}(r) = U(r) - \frac{mC^2}{2r^2}, \quad h = \frac{mv_0^2}{2} - U(r_0), \tag{8.1.6'}$$

where we have introduced *the apparent potential* $\bar{U}(r)$ and the constant of energy h; hence

$$\theta = \theta_0 \pm C \int_{r_0}^{r} \frac{\mathrm{d}(1/\rho)}{\sqrt{\varphi(\rho)}} = \theta_0 \mp C \int_{r_0}^{r} \frac{\mathrm{d}\rho}{\rho^2 \sqrt{\varphi(\rho)}}, \quad \varphi(r) = \frac{2}{m}[\bar{U}(r) + h], \qquad (8.1.6'')$$

the trajectory in polar co-ordinates being thus specified. The two first integrals used above allow also to put in evidence the motion of the particle along the trajectory, establishing its parametric equations in the form

$$t = t_0 + \frac{1}{C}\int_{\theta_0}^{\theta} r^2(\vartheta)\mathrm{d}\vartheta, \quad t = t_0 \pm \int_{r_0}^{r} \frac{\mathrm{d}\rho}{\sqrt{\varphi(\rho)}}. \qquad (8.1.6''')$$

If the potential is of the form $U(r) = k/r^s$, $k = \mathrm{const}$, $s \in \mathbb{Z}$, then the above integrals may be expressed by elementary functions only if $s = -2$ (harmonic oscillator), $s = -1$, $s = 1$ (Keplerian motion) and $s = 2$; if $s = -6, -4, 3, 4, 6$, then these integrals may be expressed by means of elliptic functions.

The sign before the radical is determined by the sign of the initial velocity $\dot{r}_0 = \dot{r}(t_0)$, as long as $\varphi(r) > 0$. If $\varphi(r) = 0$, it results that $v_r^0 = \dot{r}_0 = 0$, so that the velocity is normal to the radius vector at the initial moment; the motion along the radius vector takes place as this radius would be fixed, the force which acts upon the particle being $\bar{F}$. If this apparent force is positive (repulsive force), then r is increasing and takes the sign +; otherwise, one takes the sign –. In particular, let us suppose that $F = 0$ at the initial moment; in this case, the particle remains immovable for an observer on the radius vector, because the particle moves along this radius as if it would be fixed, the particle being launched without initial velocity at a point in which the apparent force vanishes. Hence, the trajectory is a circle of radius r_0, the motion being uniform (because the areal velocity is constant).

To have a circular trajectory, it is necessary that $\alpha_0 = \pm \pi/2$ (the velocity be normal to the radius vector at the initial moment, so that $C = \pm r_0 v_0$) and $F(r_0) + mC^2/r_0^3 = 0$. If $r = r_0$ (circular motion) and $\dot{\theta} = \dot{\theta}_0$ (uniform motion), then – during the motion – the equation (8.1.4) is identically verified; the initial conditions being fulfilled, the theorem of uniqueness ensures the searched solution. The velocity at the initial moment has thus the modulus

$$v_0 = \sqrt{\frac{-F(r_0)r_0}{m}}; \qquad (8.1.7)$$

hence, at the initial moment, the force F must be of attraction ($F(r_0) < 0$).

The relation (8.1.4) may be written also in the form

$$F = -\frac{mC^2}{r^2}\left[\frac{\mathrm{d}^2}{\mathrm{d}\theta^2}\left(\frac{1}{r}\right) + \frac{1}{r}\right]; \qquad (8.1.8)$$

Dynamics of the particle in a field of elastic forces 473

we obtain thus *Binet's formula*, which allows to solve the inverse problem: to determine the central force which, applied upon a given particle, imparts a plane trajectory to it, after the law of areas with respect to a fixed pole. Taking into account the equation (8.1.4') of the trajectory, we can write

$$F = -\frac{mC^2}{r^2}[f''(\theta) + f(\theta)] \qquad (8.1.8')$$

too, where $f'' \equiv \partial^2 f / \partial \theta^2$. If beforehand a form of F is not imposed, that one has a certain non-determination, taking into account the equation of the trajectory (the equation which links r to θ); eliminating θ, one obtains $F = F(r)$, form used the most times.

For instance, in case of trajectories to which corresponds the equation

$$r^k = a\cos k\theta + b, \quad a,b,k = \text{const}, \qquad (8.1.9)$$

choosing the fixed point as origin, we get

$$F(r) = -\frac{C^2}{r^{k+3}}\left[\frac{(k+1)(a^2 - b^2)}{r^k} + (k+2)b\right]; \qquad (8.1.9')$$

in particular, these trajectories may be conics with the pole at the focus ($k = -1$) or at the centre ($k = 1$), Pascal limaçons ($k = 2$, $b = 0$), lemniscates etc.

1.1.2 Qualitative study of orbits. Bertrand's theorem

Usually, the trajectory of a particle in a central field of forces is called *orbit* (even if it is not a closed curve). The relations (8.1.6')-(8.1.6''') determine the orbit and the motion on it only if $\dot{r}$, θ and t are real quantities, hence if $\varphi(r) \geq 0$; the apparent potential must verify the condition $\bar{U}(r) + h \geq 0$, which determines the domain of variation of r, corresponding to the motion of the particle; the solutions of the equation

$$\bar{U}(r) + h = 0 \qquad (8.1.10)$$

specify the frontier of the domain. From (8.1.6') one may see that the radial velocity vanishes on the frontier ($\dot{r} = 0$), the angular velocity being non-zero ($\dot{\theta} \neq 0$); if we would have $\dot{\theta} = 0$ at a point other than the origin, then the first integral of areas would lead to $C = 0$, that is to a rectilinear trajectory; hence, at the respective points the velocity is normal to the radius vector. On the frontier, $r(t)$ changes of sign, the respective point corresponding to a relative extremum for $r(t)$. The relation (8.1.2) shows that $\dot{\theta}(t)$ has a constant sign, so that $\theta(t)$ is a monotone function; the integrals (8.1.6''), (8.1.6''') must be calculated on intervals of monotony, the sign being chosen correspondingly. Let $r_{\min}$ and $r_{\max}$ be the extreme values which may be taken by r; the corresponding points on the orbit are called *apsides*. In this case,

$0 \leq r_{\min} \leq r \leq r_{\max}$. The roots of the equation (8.1.10) may be graphically determined, taking into consideration the intersection of the apparent potential $\bar{U} = \bar{U}(r)$ with the straight line $\bar{U} = -h$; the domain of variation of r corresponds to $\bar{U}(r) \geq -h$ (Fig.8.2).

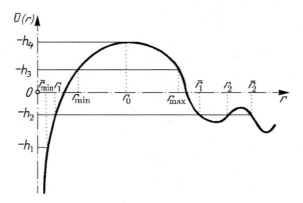

Figure 8.2. Diagram $\bar{U}(r)$ vs r.

Let us consider first of all the case of a motion to which corresponds the constant h_3 of energy. The radius $r_{\max}$ is finite, the orbit is bounded, while the trajectory is contained in the annulus determined by the circles $r = r_{\min}$ and $r = r_{\max}$ (we assume that $r_{\min} > 0$); the radii $r_{\min}$ and $r_{\max}$ are called *apsidal distances*. The points for which $r = r_{\min}$ are called *pericentres*, while those for which $r = r_{\max}$ are called *apocentres*. Noting that at an apsis the velocity is normal to the radius vector, which is the radius of a circle, it results that *the trajectory is tangent to the concentric circles at the corresponding apsides*. Choosing the origin of the angles after an apsis radius ($\theta_0 = 0$), called *apsidal line*, we may use the relation (8.1.6") for two points of the same radius vector r of the trajectory, on both sides of that line, r_0 being $r_{\min}$ or $r_{\max}$; it results that *the trajectory of the particle is symmetric with respect to an apsidal line*. The angle at the centre χ between two consecutive apsidal lines is constant; it is called *apsidal angle* and is given by

$$\chi = C \int_{r_{\min}}^{r_{\max}} \frac{\mathrm{d}r}{r^2 \sqrt{\varphi(r)}}. \tag{8.1.11}$$

It results that the angle at the centre between two consecutive pericentres (apocentres) is equal to 2χ. From the above mentioned properties, it results that *if one knows the arc of trajectory between two consecutive apsides, then one may set up geometrically the whole trajectory* (Fig.8.3). From (8.1.2) it results that $\dot{\theta}$ has a constant sign, so that the particle is rotating always in the same direction around the point O. A bounded orbit is closed if, after a finite number of such rotations, the particle returns to a previous position; thus, the condition $2\chi = 2\pi q$, $q \in \mathbb{Q}$ must be satisfied. Otherwise, the orbit is open and covers the annulus $r \in [r_{\min}, r_{\max}]$. We also notice that the apparent potential $\overline{U}(r)$ has a maximum at a point in the interior of the annulus, corresponding to $\overline{F}(r) = \mathrm{d}\overline{U}(r)/\mathrm{d}r = 0$. We mention that the equation $\varphi(r) = 0$ can have more than two roots (the case of the constant h_2 of energy, Fig.8.2). In this case, one obtains two possible annular domains (contained between the circles of radii r_1 and $\overline{r}_1$ or r_2 and $\overline{r}_2$, respectively); the motion takes place in that domain in

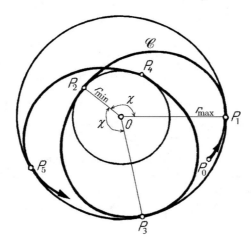

Figure 8.3. Orbit of a particle acted upon by a central force.

which is the initial position given by $r_0 = r(t_0)$. If $r_{\min} = 0$, then the particle passes through the pole O or stops at that point. Assuming that $C \neq 0$ (otherwise, the trajectory is rectilinear), the term $-mC^2/2r^2$ allows to have $\lim_{r \to 0} \overline{U}(r) = -\infty$, the "falling" towards O being thus hindered. The condition of "falling" towards O is given by the condition $\overline{U} \geq -h$, written in the form $r^2 U(r) - mC^2/2 \geq -hr^2$. To have $r_{\min} = 0$ it is necessary that

$$\lim_{r \to 0+0} \left[r^2 U(r) \right] \geq \frac{mC^2}{2},$$

hence that $U(r)$ tends to zero at least as A/r^2, $A > mC^2/2$, or as A/r^n, $A > 0$, $n > 2$.

If $r_{\min} = r_{\max} = r_0$, then the trajectory is a circle of radius r_0, corresponding $\overline{F}(r) = 0$ and the constant of energy $h_4 = -\overline{U}_{\max}$ (Fig.8.2); the considerations made at the previous subsection remain valid.

One may prove

Theorem 8.1.1 (*J. Bertrand*). *The only closed orbits corresponding to central forces are those for which* $s = -2$, $k < 0$, *in any initial conditions, or* $s = 1$, $k > 0$, *in certain initial conditions, assuming a potential of the form* $U(r) = k/r^s$, $k = \mathrm{const}$, $s \in \mathbb{Z}$.

1.1.3 Case of a force the modulus of which is in inverse proportion to the square of the distance to a fixed point

Jacobi has considered the case in which the central force is of the form $F = \gamma(\theta)/r^2$, hence it is in inverse proportion to the square of the distance to the pole O. Binet's equation (8.1.4) takes the form

$$\frac{\mathrm{d}^2}{\mathrm{d}\theta^2}\left(\frac{1}{r}\right) + \frac{1}{r} = -\frac{\gamma(\theta)}{mC^2}; \qquad (8.1.12)$$

by integration, one obtains

$$\frac{1}{r} = C_1 \cos\theta + C_2 \sin\theta + \overline{\gamma}(\theta), \qquad (8.1.12')$$

where $\overline{\gamma}(\theta)$ is a particular integral, which may always be calculated by quadratures. The integration constants are easily obtained by initial conditions of Cauchy type.

Analogously, we may consider central forces of the form k/r^3, $k = \mathrm{const}$, which lead to the equation

$$\frac{\mathrm{d}^2}{\mathrm{d}\theta^2}\left(\frac{1}{r}\right) + \left(1 + \frac{k}{mC^2}\right)\frac{1}{r} = 0, \qquad (8.1.13)$$

wherefrom it results the general integral

$$\frac{1}{r} = C_1 \cos\beta\theta + C_2 \sin\beta\theta, \quad \beta = \sqrt{1 + k/mC^2}. \qquad (8.1.13')$$

1.2 Other problems

We consider now two other problems, i.e.: the problem of two particles, which leads to the classical case of action of central forces and the problem of motion of a particle subjected to the action of a central force in a resistent medium. We study the phenomena of capture and diffraction too.

1.2.1 The problem of two particles. Capture. Diffraction

Let be two bodies which may be modelled by two particles P_1 and P_2 of position vectors $\mathbf{r}_1$ and $\mathbf{r}_2$ and of masses m_1 and m_2, respectively (Fig.8.4). We suppose that this system of particles is acted upon only by the internal forces $\mathbf{F}_{21} = -\mathbf{F}_{12} = F\,\mathrm{vers}\,\mathbf{r}_{12} = -F\,\mathrm{vers}\,\mathbf{r}_{21}$, $\mathbf{r}_{12} = \mathbf{r}_2 - \mathbf{r}_1 = -\mathbf{r}_{21}$, where $F > 0$ in case

Dynamics of the particle in a field of elastic forces

of *repulsive forces* and $F < 0$ in case of *forces of attraction*. From the equations of motion $m_1\ddot{\mathbf{r}}_1 = -\mathbf{F}_{21}$, $m_2\ddot{\mathbf{r}}_2 = \mathbf{F}_{21}$, amplifying by m_2 and m_1, respectively, and subtracting, one obtains

$$m\ddot{\mathbf{r}}_{12} = \mathbf{F}_{21}, \quad m\ddot{\mathbf{r}}_{21} = \mathbf{F}_{12}, \quad \frac{1}{m} = \frac{1}{m_1} + \frac{1}{m_2}, \tag{8.1.14}$$

where m is *the reduced mass*. These equations describe the motion of the particle P_2 with respect to the particle P_1 (as if the latter one would be fixed), as well as the motion of the particle P_1 with respect to the particle P_2; in such conditions, the force $\mathbf{F}_{21}$ and the force $\mathbf{F}_{12}$ are central forces, respectively, and one may use the results previously obtained.

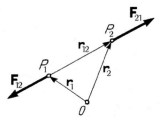

Figure 8.4. Problem of the two particles.

These considerations allow to study also the case in which $r_{\max}$ is not finite, the orbit being unbounded. We must have $-h \leq \lim_{r \to \infty} \overline{U}(r) = \lim_{r \to \infty} U(r) = U_\infty$. To fix the ideas, we assume that $U_\infty = 0$; because the integral $U_\infty = \int_r^\infty F(r) \mathrm{d}r$ must have a sense, we suppose that F tends to zero for $r \to \infty$ at least as $1/r^{1+\varepsilon}$, $\varepsilon > 0$. The condition of "escape" at infinity becomes $h \geq 0$. Indeed, if we assume that the particle P_1 is situated at the centre O, while the particle P_2 tends to infinity, the interaction ceases and one has $U_\infty = 0$; the particle P_2 is endowed, in this case, only with kinetic energy T, so that $T = h \geq 0$. The symmetry of the trajectory with respect to the line of the pericentre remains valid (the same as in the case of bounded orbits); we notice that the unbounded orbits have only one pericentre and pass only once through it.

If the particle P_2 comes from infinity and enters in interaction with the particle P_1, then may appear the phenomenon of *capture* ($r = r_{12}$ remains finite or vanishes for $t \to \infty$) or the phenomenon of *diffraction* ($r \to \infty$ for $t \to \infty$). We notice that

$$C = b\overline{v}, \quad h = \frac{m\overline{v}^2}{2}, \tag{8.1.15}$$

where $\overline{\mathbf{v}} = \mathbf{v}_2^0 - \mathbf{v}_1^0$ is the relative velocity at the initial moment, while b is the distance from P_1 to the support of the velocity. The formula (8.1.6''') where we take

$t_0 = 0$ gives the time in which the particle coming from infinity reaches a position at finite distance in the form

$$t = -\int_{-\infty}^{r} \frac{d\rho}{\sqrt{\varphi(\rho)}};\qquad(8.1.16)$$

we noticed that the radial velocity is directed towards the particle P_1. One has a real motion only if $\varphi(r) > 0$; we observe that $\varphi(-\infty) = 2h/m = \bar{v}^2 > 0$. Thus, if the time t increases, then the radius vector r decreases.

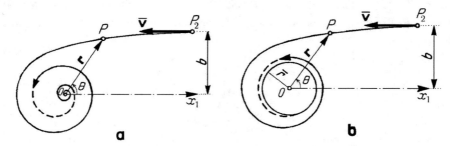

Figure 8.5. Capture: at a pole (a); on a circle (b).

For $t \to \infty$ we may have $r \to 0$, obtaining the phenomenon of *capture*; the trajectory of the particle P_2 is thus a spiral which tends to the pole O (Fig.8.5,a). But if there exists a radius $\bar{r}$ for which $\varphi(\bar{r}) = 0$, then we may write $\varphi(r) = (r - \bar{r})^\nu \psi(r)$, $\psi(\bar{r}) \neq 0$. The integral (8.1.16) is divergent if $\nu \geq 2$, while $r \to \bar{r}$ for $t \to \infty$; we obtain a phenomenon of capture too in which the trajectory of the particle P_2 is also a spiral wrapped up a circle of radius $r = \bar{r}$ (Fig.8.5,b). The phenomenon of capture may take place only in case of forces of attraction between the two particles.

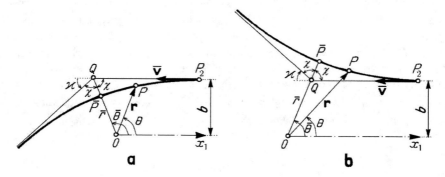

Figure 8.6. Diffraction: force of attraction (a); repulsive force (b).

If $\nu < 2$, then the integral (8.1.16) is convergent so that there exists a finite moment t for which $r = \bar{r}$, $\varphi(\bar{r}) = 0$ ($\bar{r} = \bar{r}_{\min}$ is the minimal value of r corresponding to a constant of energy h_1, Fig.8.2); the respective point $\overline{P}$ is a pericentre, while the apsidal line $O\overline{P}$ is a symmetry axis of the trajectory, which has the form in Fig.8.6,a, in case of forces of attraction, or the form in Fig.8.6,b, in case of repulsive forces. One obtains thus the phenomenon of *diffraction*. The apsidal angle is expressed in the form $\chi = \pi - \overline{\theta}$, in case of a force of attraction, and by $\chi = \overline{\theta}$, in case of a repulsive force, where the angle $\overline{\theta}$ is given by the formula (8.1.6")

$$\overline{\theta} = C \int_{-\infty}^{\overline{r}} \frac{\mathrm{d}r}{r^2 \sqrt{\varphi(r)}}, \quad \frac{mv^2}{2}\left(1 - \frac{b^2}{\overline{r}^2}\right) + U(\overline{r}) = 0. \qquad (8.1.17)$$

The angle $\varkappa$ formed by the asymptotic direction of the trajectory with the support of the velocity $\overline{\mathbf{v}}$ (in fact, the angle of the two asymptotes to the trajectory) is called *diffraction angle* and is given by $\varkappa = \mp(\pi - 2\overline{\theta})$, where one takes the sign $-$ or the sign $+$ as the diffraction is of attraction or is repulsive, respectively. If we denote by "prime" the quantities which intervene after the interaction (for $t \to \infty$), the relations (8.1.15) show that $b' = b$, $\overline{v}' = \overline{v}$; hence, the magnitude of the relative velocity is conserved, the velocity vector changing only with the diffraction angle.

1.2.2 Motion of a particle acted upon by a central force in a resistent medium

This case of motion generalizes the case previously considered; obviously, it leads to a plane trajectory $\mathscr{C}$ too, the resistance $\mathbf{R}$ of the medium being tangent to the trajectory and opposed to the motion (Fig.8.7). Hence, the equation of motion is of the form

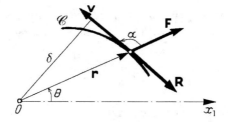

Figure 8.7. Particle acted upon by a central force in a resistent medium.

$$m\ddot{\mathbf{r}} = F\frac{\mathbf{r}}{r} - R\frac{\dot{\mathbf{r}}}{v}, \quad F = F(r,\theta,\dot{r},\dot{\theta};t), \quad R = R(r,\theta,\dot{r},\dot{\theta};t); \qquad (8.1.18)$$

in polar co-ordinates, we have

$$m(\ddot{r} - r\dot{\theta}^2) = F - R\frac{\dot{r}}{v}, \quad \frac{m}{r}\frac{\mathrm{d}}{\mathrm{d}t}(r^2\dot{\theta}) = -R\frac{r\dot{\theta}}{v}. \qquad (8.1.18')$$

Eliminating the resistance R between these equations, it results

$$F = m\left[\ddot{r} - r\dot{\theta}^2 - \frac{\dot{r}}{r^2\dot{\theta}}\frac{\mathrm{d}}{\mathrm{d}t}(r^2\dot{\theta})\right].$$

We return to the calculus in Subsec. 1.1.1 and assume that $\dot{F} = 0$; hence, we get

$$F = -\frac{m(r^2\dot\theta)^2}{r^2}\left[\frac{d^2}{d\theta^2}\left(\frac{1}{r}\right) + \frac{1}{r}\right] = -mv_\theta^2\left[\frac{d^2}{d\theta^2}\left(\frac{1}{r}\right) + \frac{1}{r}\right], \qquad (8.1.19)$$

generalizing thus Binet's formula (8.1.8).

In this case, the areal velocity is no more constant (we can no more write a first integral of areas). Noting that $\Omega = r^2\dot\theta/2$, the second equation (8.1.18') leads to

$$\frac{\dot\Omega}{\Omega} = \frac{d}{dt}(\ln\Omega) = -\frac{R}{mv}. \qquad (8.1.20)$$

We observe that $\Omega = \delta v/2$, where δ is the distance from the pole O to the support of the velocity $\mathbf{v}$ (Fig.8.7); we may write

$$R = -\frac{2m}{\delta}\dot\Omega \qquad (8.1.20')$$

too, so that the resistance of the medium is in direct proportion to the areal acceleration; as well, we obtain the relation

$$R = -\frac{4m\Omega}{\delta^2}\frac{d\Omega}{ds} = -\frac{2m}{\delta^2}\frac{d(\Omega^2)}{ds}. \qquad (8.1.20'')$$

In particular, in case of a resistance of the form $R = kmv^2$, $k = \text{const}$, $k > 0$, we obtain $d(\ln\Omega) = -kv dt = -k ds$, where s is the curvilinear co-ordinate along the trajectory; it results

$$\Omega = \Omega_0 e^{-ks}, \qquad (8.1.21)$$

so that the areal velocity decreases exponentially.

2. Motion of a particle subjected to the action of an elastic force

In case of a conservative central force for which $U(r) = -kr^2/2$ or $U(x) = -kx^2/2$, $k > 0$ (one of the two cases considered in Bertrand's theorem), one obtains an elliptic oscillator (with two degrees of freedom) or a linear oscillator (with only one degree of freedom), respectively; the force which derives from this potential is an *elastic force*. We consider also the corresponding damped and sustained (including self-sustained) oscillations. A particular attention is given to repulsive elastic forces ($k < 0$), as well as to a study of non-linear oscillations.

Dynamics of the particle in a field of elastic forces 481

2.1 Mechanical systems with two degrees of freedom

In what follows, we study the non-damped and damped elliptic oscillator, in case of elastic forces of attraction as well as in case of repulsive elastic forces; the case of self-sustained oscillations is also taken into consideration.

2.1.1 Elliptic oscillator

In case of central forces of the form $\mathbf{F} = F(r)\,\text{vers}\,\mathbf{r}$, we may assume for $F(r)$ a development into a Maclaurin series

$$F(r) = F(0) + \frac{F'(0)}{1!}r + \frac{F''(0)}{2!}r^2 + \ldots. \tag{8.2.1}$$

We take $F(0) = 0$, assuming that the pole O is a position of equilibrium. In case of small motions (for which $r \ll 1$, the unit having the dimension of a length) we neglect the powers of higher order. We denote $F'(0) = -k$, $k > 0$; the force is thus directed towards the centre O, considering that one as a stable position of equilibrium. A force of the form

$$\mathbf{F} = -kr\,\text{vers}\,\mathbf{r} = -k\mathbf{r} \tag{8.2.2}$$

is called *elastic force* (in fact, in the following, by elastic force we mean an *elastic force of attraction*).

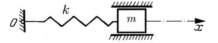

Figure 8.8. Model of an elastic force.

In the case in which the force $\mathbf{F}$ has a fixed support (along the Ox-axis) we may write $F(x) = -kx$, modelling thus the force by an elastic spring of *elastic constant* (*coefficient of elasticity*) k, fixed at one end and acted upon by a force $\mathbf{F}$ due to a mass m at the other end; the particle is modelled by a rigid body the motion of which is guided (Fig.8.8). The denomination given to this force is thus justified.

The corresponding apparent potential is

$$\bar{U}(r) = -\frac{k}{2}r^2 - \frac{mC^2}{2r^2} = -\frac{m}{2}\left(\omega^2 r^2 + \frac{C^2}{r^2}\right) = -\frac{m}{2}(\omega^2 + \dot{\theta}^2)r^2, \quad \omega = \sqrt{\frac{k}{m}}, \tag{8.2.3}$$

where we have taken into consideration the constant of areas (8.1.2); the graphic of this function is given in Fig.8.9, where the graphics of the two component terms are put into evidence. In case of a constant of energy h for which $\bar{U}(r) \geq -h$ we obtain a bounded orbit for which $r_{\min} \leq r \leq r_{\max}$. To $r_{\min} = r_{\max} = \bar{r}$ corresponds $\bar{U}_{\max}$; noting that $mC^2/2\bar{r}^2 = k\bar{r}^2/2$, because $(mC^2/2r^2)(kr^2/2) = mC^2 k/4 = $ const, it results

$$\bar{r} = \sqrt{\frac{|C|}{\omega}} = \sqrt{\frac{\bar{h}}{k}}, \quad \bar{h} = m\omega|C| = |C|\sqrt{km} = k\bar{r}^2, \qquad (8.2.3')$$

where we took into account (8.2.3) and we assumed that the motion takes place in the positive sense of the angle θ, so that $C > 0$. The extreme values of r are the roots of the equation $\varphi(r) = 2[\bar{U}(r) + h]/m = 0$ and are given by

$$r^2_{\substack{\max \\ \min}} = \frac{h}{m\omega^2}\left(1 \pm \sqrt{1 - \frac{C^2 m^2 \omega^2}{h^2}}\right) = \frac{h}{k}\left(1 \pm \sqrt{1 - \frac{mkC^2}{h^2}}\right) = \frac{h}{k}\left(1 \pm \sqrt{1 - \frac{\bar{h}^2}{h^2}}\right),$$

so that ($r_{\max} r_{\min} = \bar{r}^2$)

$$r_{\substack{\max \\ \min}} = \sqrt{\frac{h}{k}}\sqrt{1 \pm \sqrt{1 - \left(\frac{\bar{h}}{h}\right)^2}} = \sqrt{\frac{h}{2k}}\left(\sqrt{1 + \frac{\bar{h}}{h}} \pm \sqrt{1 - \frac{\bar{h}}{h}}\right)$$

$$= \frac{\bar{r}}{\sqrt{2}}\left(\sqrt{\frac{h}{\bar{h}} + 1} \pm \sqrt{\frac{h}{\bar{h}} - 1}\right), \qquad (8.2.3'')$$

the annulus which contains the orbit being thus specified.

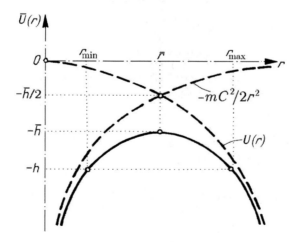

Figure 8.9. Diagram $\bar{U}(r)$ vs r, in case of an elastic force of attraction.

The trajectory is given by the relation (8.1.6"); choosing as Ox_1-axis the line of one of the apocentres ($\theta_0 = 0$), we have

Dynamics of the particle in a field of elastic forces

$$2\theta = -C \int_{1/r_{\max}^2}^{1/r^2} \frac{d(1/\rho^2)}{\sqrt{\frac{h^2}{m^2C^2} - \omega^2 - \left(\frac{C}{\rho^2} - \frac{h}{mC}\right)^2}} = \arccos \frac{\frac{C}{r^2} - \frac{\overline{h}}{h}\omega}{\frac{h}{\overline{h}}\omega\sqrt{1 - \left(\frac{\overline{h}}{h}\right)^2}} - \pi,$$

where we took into account (8.2.3'), (8.2.3"). Hence, it results

$$\frac{h}{\overline{h}}\omega\sqrt{1 - \left(\frac{\overline{h}}{h}\right)^2}\cos(2\theta + \pi) = \frac{h}{\overline{h}}\left\{\left[1 + \sqrt{1 - \left(\frac{\overline{h}}{h}\right)^2}\right]\sin^2\theta\right.$$
$$\left. + \left[1 - \sqrt{1 - \left(\frac{\overline{h}}{h}\right)^2}\right]\cos^2\theta - 1\right\} = \frac{C}{r^2} - \frac{\overline{h}}{h}\omega,$$

so that the orbit is an ellipse (Bertrand's theorem is verified for this case) of equation

$$\frac{1}{r^2} = \frac{\cos^2\theta}{a^2} + \frac{\sin^2\theta}{b^2} \qquad (8.2.3''')$$

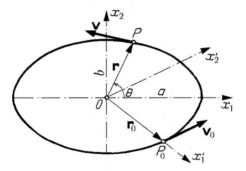

Figure 8.10. Elliptic oscillator.

in polar co-ordinates; the semiaxes are $a = r_{\max}$ and $b = r_{\min}$ (Fig.8.10). The mechanical system formed by a particle acted upon by an elastic force is called *elliptic oscillator*.

The motion on the trajectory is specified by the first formula (8.1.6'''), wherefrom ($t_0 = 0$)

$$t = \frac{1}{C}\int_0^\theta \frac{d\vartheta}{\frac{\cos^2\vartheta}{a^2} + \frac{\sin^2\vartheta}{b^2}} = \frac{1}{\omega}\arctan\left(\frac{a}{b}\tan\theta\right),$$

where we noticed that $C/ab = (\overline{h}/m\omega)/(\overline{h}/k) = \omega$. Taking into account (8.2.3'''), we get the parametric equations of the elliptic orbit in the form

$$\theta = \arctan\left(\frac{b}{a}\tan\omega t\right), \quad r = \sqrt{a^2\cos^2\omega t + b^2\sin^2\omega t}. \tag{8.2.4}$$

Using the first integral of areas and of mechanical energy, the components of the velocity are given by

$$\dot\theta = \frac{C}{r^2}\left(\frac{\overline{r}}{r}\right)^2\omega, \quad v_\theta = \frac{\overline{r}^2\omega}{r},$$

$$v_r = \dot r = \sqrt{\frac{2}{m}[\overline{U}(r) + h]} = \sqrt{v_0^2 + \omega^2(r_0^2 - r^2) - v_\theta^2}, \tag{8.2.4'}$$

where we took into account (8.1.6') and (8.2.3). We observe that the areal velocity is expressed in the form $\Omega = v\delta/2$, where δ is the distance from the pole O to the tangent to the trajectory at the considered point. Because $\Omega = \mathrm{const}$, it results that $v = v_{\max}$ for $\delta = r_{\min} = b$ (at the extremities of the minor diameter), while $v = v_{\min}$ for $\delta = r_{\max} = a$ (at the extremities of the major diameter).

In case of *the circular oscillator* we have $r = r_{\min} = r_{\max} = \overline{r}$, $\dot\theta = \omega$, $v_r = 0$, $v_0 = v_\theta = \omega\overline{r}$; it results that ω is the angular velocity and the motion is *uniform*.

Starting from the expression (8.1.2) of the elastic force, we may write the equation of motion (8.1.1) in the form

$$\ddot{\mathbf{r}} + \omega^2\mathbf{r} = \mathbf{0}, \tag{8.2.5}$$

where the pulsation $\omega > 0$ is given by (8.2.3). The initial conditions ($t_0 = 0$)

$$\mathbf{r}(0) = \mathbf{r}_0, \quad \mathbf{v}(0) = \mathbf{v}_0 \tag{8.2.5'}$$

lead to the solution of the problem in the form

$$\mathbf{r}(t) = \mathbf{r}_0\cos\omega t + \frac{\mathbf{v}_0}{\omega}\sin\omega t, \tag{8.2.5''}$$

$$\mathbf{v}(t) = \mathbf{v}_0\cos\omega t - \omega\mathbf{r}_0\sin\omega t; \tag{8.2.5'''}$$

we notice that $\mathbf{r}\times\mathbf{v} = \mathbf{r}_0\times\mathbf{v}_0$, corresponding to the first integral of areas. The vector $\mathbf{r}$ is a linear combination of the vectors $\mathbf{r}_0$ and $\mathbf{v}_0$; hence, the trajectory is a plane curve, excepting the case in which the cross product mentioned above vanishes (the vectors $\mathbf{r}_0$ and $\mathbf{v}_0$ are collinear). The trajectory does not pass through the origin because $\mathbf{r} \neq \mathbf{0}$, $\forall t$. We notice that $|\mathbf{r}| \leq |\mathbf{r}_0| + |\mathbf{v}_0/\omega|$, $\forall t$, so that all the points of the trajectory are at a finite distance. Hence, the trajectory is a closed curve, which surrounds the centre O, that one being a *stable position of equilibrium* (the orbit can be contained in the interior of a circle arbitrarily small, the velocity being arbitrarily small too); the motion is periodic because the particle returns to the same position ($\mathbf{r}(t+T) = \mathbf{r}(t)$) with the same velocity ($\mathbf{v}(t+T) = \mathbf{v}(t)$), after the same *period of time*

Dynamics of the particle in a field of elastic forces

$$T = \frac{2\pi}{\omega} = 2\pi\sqrt{\frac{m}{k}}. \tag{8.2.6}$$

The pole O is a centre of symmetry of the trajectory and of the motion, because $\mathbf{r}(t + T/2) = -\mathbf{r}(t)$, $\mathbf{v}(t + T/2) = -\mathbf{v}(t)$. The velocity vector is also finite; it is a continuous function, which is non-zero for all values of t, so that the motion takes place always in the same direction. With respect to a system of oblique co-ordinates $Ox_1'x_2'$, specified by the conjugate diameters corresponding to the vector $\mathbf{r}_0$ of the initial position and to the initial velocity $\mathbf{v}_0$ (Fig.8.10), one obtains the equation of the ellipse in the form

$$\frac{x_1'^2}{r_0^2} + \frac{x_2'^2}{(v_0/\omega)^2} = 1. \tag{8.2.7}$$

We notice that

$$\mathbf{r} \cdot \mathbf{v} = \mathbf{r}_0 \cdot \mathbf{v}_0 \cos 2\omega t + \frac{1}{2\omega}(v_0^2 - r_0^2\omega^2)\sin 2\omega t.$$

Hence, to obtain a circular oscillator ($\mathbf{r} \cdot \mathbf{v} = 0, \forall t$) it is necessary and sufficient that the initial conditions verify the relations $\mathbf{r}_0 \cdot \mathbf{v}_0 = 0$ and $v_0 = r_0\omega$.

The number which shows how many times the particle travels through the whole trajectory in a unit time is called *the frequency of the motion* and is given by

$$\nu = \frac{1}{T} = \frac{\omega}{2\pi} = \frac{1}{2\pi}\sqrt{\frac{k}{m}}; \tag{8.2.6'}$$

we notice that *the pulsation* $\omega = 2\pi\nu$ represents the number of periods in 2π units of time, the denomination of *circular frequency* used too being thus justified.

These results may be easily correlate with those previously obtained starting from the general theory of motion of a particle subjected to the action of a central force. We observe that one obtains the same results as in the case of small motions of a spherical pendulum around a stable position of equilibrium.

2.1.2 Case of a repulsive elastic force

In case of the motion around a labile position of equilibrium O, we consider a repulsive elastic force of the form

$$\mathbf{F} = kr \text{ vers } \mathbf{r} = k\mathbf{r}, \quad k > 0. \tag{8.2.8}$$

The corresponding apparent potential is

$$\bar{U}(r) = \frac{k}{2}r^2 - \frac{mC^2}{2r^2} = \frac{m}{2}\left(\omega^2 r^2 - \frac{C^2}{r^2}\right) = \frac{m}{2}(\omega^2 - \dot{\theta}^2)r^2, \quad \omega = \sqrt{\frac{k}{m}}; \tag{8.2.9}$$

the graphic of this function is represented in Fig.8.11, the radius r and the corresponding constant $\bar{h}$, given by the relations (8.2.3'), being put into evidence. In case of a constant of energy h for which $\bar{U}(r) \geq -h$ we obtain a unbounded orbit, which has a pericentre given by (the positive root of the equation $\varphi(r) = 0$)

$$r_{\min}^2 = \frac{h}{k}\left[\sqrt{1+\left(\frac{\bar{h}}{h}\right)^2}-1\right]. \qquad (8.2.9')$$

We are in the case of Fig.8.6,b, the integrals (8.1.6''), (8.1.6''') being convergent (we have $\varphi(r) = (r - r_{\min})\psi(r)$). Choosing as Ox_1-axis the line of the pericentre ($\theta_0 = 0$), the formula (8.1.6'') leads to

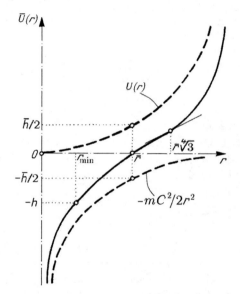

Figure 8.11. Diagram $\bar{U}(r)$ vs r, in case of a repulsive elastic force.

$$2\theta = -C\int_{1/r_{\min}^2}^{1/r^2}\frac{\mathrm{d}(1/\rho^2)}{\sqrt{\frac{h^2}{m^2C^2}+\omega^2-\left(\frac{C}{\rho^2}-\frac{h}{mC}\right)^2}} = \arccos\frac{\frac{C}{r^2}-\frac{h}{\bar{h}}\omega}{\frac{h}{\bar{h}}\omega\sqrt{1+\left(\frac{\bar{h}}{h}\right)^2}},$$

where we used the relations (8.2.3'), (8.2.9'). Hence, it results

$$\frac{h}{\bar{h}}\omega\sqrt{1+\left(\frac{\bar{h}}{h}\right)^2}\cos 2\theta = \frac{h}{\bar{h}}\omega\left\{\left[\sqrt{1+\left(\frac{\bar{h}}{h}\right)^2}+1\right]\cos^2\theta\right.$$
$$\left.-\left[\sqrt{1+\left(\frac{\bar{h}}{h}\right)^2}-1\right]\sin^2\theta - 1\right\} = \frac{C}{r^2} - \frac{h}{\bar{h}}\omega,$$

so that the orbit is an arc of hyperbola (Fig.8.12) of equation

$$\frac{1}{r^2} = \frac{\cos^2\theta}{a^2} - \frac{\sin^2\theta}{b^2} \qquad (8.2.10)$$

in polar co-ordinates; the semiaxes are given by

$$a^2 = r_{\min}^2, \quad b^2 = \frac{h}{k}\left[\sqrt{1+\left(\frac{\overline{h}}{h}\right)^2}+1\right], \quad ab = \overline{r}^2. \qquad (8.2.10')$$

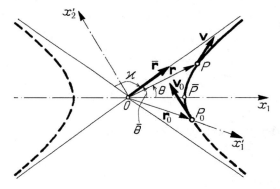

Figure 8.12. Trajectory of a particle acted upon by a repulsive elastic force.

The formulae (8.1.6''') allow then to study the motion of the particle along the trajectory. We notice that we have to do with a phenomenon of diffraction, where the angle of diffraction is given by

$$\varkappa = \pi - 2\overline{\theta} = \pi - C\int_{1/r_{\min}^2}^{0}\frac{\mathrm{d}(1/\rho^2)}{\sqrt{\frac{h^2}{m^2C^2}+\omega^2-\left(\frac{C}{\rho^2}-\frac{h}{mC}\right)^2}},$$

so that

$$\varkappa = \pi - \arccos\frac{-h}{\sqrt{h^2+\overline{h}^2}} = \arccos\frac{h}{\sqrt{h^2+\overline{h}^2}} = \arctan\frac{\overline{h}}{h}. \qquad (8.2.10'')$$

One observes that for $h < 0$ (hence $-h > 0$, we are above the Or-axis) we have $b < a$ and $\varkappa > \pi/2$, while for $h > 0$ (hence $-h < 0$, we are under the Or-axis) we have $b > a$ and $\varkappa < \pi/2$. For $h = 0$ one obtains $a = b = \sqrt{\overline{h}/k} = \overline{r}$ and $\varkappa = \pi/2$, the hyperbola being *equilateral* (*rectangular*).

In case of the force (8.2.8), the equation of motion (8.1.1) has the form

$$\ddot{\mathbf{r}} - \omega^2 \mathbf{r} = \mathbf{0}. \tag{8.2.11}$$

If we put the initial conditions (8.2.5'), the solution of the boundary value problem is given by

$$\mathbf{r}(t) = \mathbf{r}_0 \cosh \omega t + \frac{\mathbf{v}_0}{\omega} \sinh \omega t, \tag{8.2.11'}$$

$$\mathbf{v}(t) = \mathbf{v}_0 \cosh \omega t + \omega \mathbf{r}_0 \sinh \omega t. \tag{8.2.11''}$$

With respect to the system of oblique co-ordinates $Ox'_1 x'_2$, determined by the conjugate diameters which correspond to the vector $\mathbf{r}_0$ of the initial position and to the initial velocity $\mathbf{v}_0$ (Fig.8.12), it results that the trajectory is an arc of hyperbola of equation

$$\frac{x'^2_1}{r_0^2} - \frac{x'^2_2}{(v_0/\omega)^2} = 1, \tag{8.2.12}$$

the centre O being a *labile position of equilibrium* (the orbit cannot be contained in the interior of a circle arbitrarily small and the velocity of the particle may increase without any limit). The particle travels through the trajectory only once without returning to the initial position. We may write

$$\mathbf{r}(t) = \left(\mathbf{r}_0 + \frac{\mathbf{v}_0}{\omega} \tanh \omega t\right) \cosh \omega t, \quad \mathbf{v}(t) = \omega\left(\mathbf{r}_0 \tanh \omega t + \frac{\mathbf{v}_0}{\omega}\right) \cosh \omega t;$$

noting that $\lim_{t \to \infty} \tanh \omega t = 1$, it results that the asymptote to which tends the trajectory of the particle is specified by the vector

$$\overline{\mathbf{r}} = \mathbf{r}_0 + \frac{\mathbf{v}_0}{\omega}. \tag{8.2.12'}$$

As in the case of an elastic force of attraction, these results may be easily correlate to those previously obtained in the frame of the general theory of motion of a particle subjected to a repulsive central force.

2.1.3 Motion of a particle subjected to the action of an elastic and of a damping force

Let us suppose that in the motion of a particle subjected to the action of an elastic force of attraction (8.2.2) intervenes a *damping force* $\mathbf{\Phi} = -\Phi \operatorname{vers} \mathbf{v}$, tangent to the trajectory and having a direction opposite to that of the motion. If the magnitude of the damping force is proportional to the velocity, hence if $\mathbf{\Phi} = -k'\dot{\mathbf{r}}$, $k' > 0$ being a *damping coefficient*, then the equation of motion becomes

$$\ddot{\mathbf{r}} + 2\lambda \dot{\mathbf{r}} + \omega^2 \mathbf{r} = \mathbf{0}, \tag{8.2.13}$$

Dynamics of the particle in a field of elastic forces

with the constant $\lambda = k'/2m > 0$. The damping coefficient corresponding to the relation $\omega = \lambda$ is a *critical damping coefficient* k'_c; we notice that, in this case (k'_c does not depend on k'),

$$k'_c = 2m\omega = 2\sqrt{km}. \qquad (8.2.14)$$

We introduce also the non-dimensional *damping factor* (*the damping ratio, the critical damping fraction*)

$$\chi = \frac{k'}{k'_c} = \frac{\lambda}{\omega}. \qquad (8.2.14')$$

With the initial conditions (8.1.5'), we get

$$\mathbf{r}(t) = e^{-\lambda t}\left[\mathbf{r}_0 \cos\omega't + \frac{1}{\omega'}(\mathbf{v}_0 + \lambda\mathbf{r}_0)\sin\omega't\right], \qquad (8.2.15)$$

$$\mathbf{v}(t) = e^{-\lambda t}\left[\mathbf{v}_0 \cos\omega't - \frac{1}{\omega'}(\omega^2\mathbf{r}_0 + \lambda\mathbf{v}_0)\sin\omega't\right], \qquad (8.2.15')$$

where we have introduced the *pseudopulsation*

$$\omega' = \sqrt{\omega^2 - \lambda^2} = \omega\sqrt{1-\chi^2}, \qquad (8.2.15'')$$

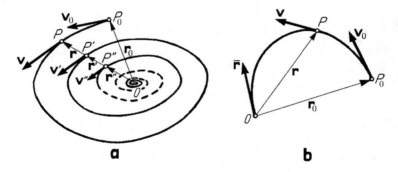

Figure 8.13. Damped pseudoelliptic oscillator (a); aperiodic damped motion (b).

assuming that $\chi < 1$, hence $\omega > \lambda$ (*subcritical damping*). The damping factor $e^{-\lambda t}$ transforms the trajectory which, in the absence of this factor, would be an ellipse in a spiral (the radius vector diminishes continuously); the particle tends, in an infinite time, to the origin O, with a velocity which tends to zero too (Fig.8.13,a). This mechanical system is called also *damped pseudoelliptic oscillator*, the respective motion of the particle being a *pseudoperiodic damped motion*. After equal intervals of time of *pseudoperiod*

$$T = \frac{2\pi}{\omega'} = \frac{2\pi}{\omega\sqrt{1-\chi^2}} = \frac{2\pi}{\omega}\left(1 + \frac{1}{2}\chi^2 + \frac{3}{8}\chi^4 + ...\right),$$

the particle reaches the points P', P'',..., which are situated on the common support of the position vectors $\mathbf{r}'$, $\mathbf{r}''$,...; the corresponding velocities $\mathbf{v}'$, $\mathbf{v}''$,... have the same direction. We observe that $r'/r = r''/r' = ... = e^{-\lambda T}$, $v'/v = v''/v' = ...$ $= e^{-\lambda T}$, obtaining a decrease in geometric progression of ratio $e^{-\lambda T}$ of the radius vector and of the velocity; the number

$$\delta = -\lambda T = -\frac{2\pi\lambda}{\omega'} = -\frac{2\pi\chi}{\sqrt{1-\chi^2}} = -2\pi\left(\chi + \frac{1}{2}\chi^3 + \frac{3}{8}\chi^5 + ...\right)$$

is called *logarithmic decrement* (for $\chi \ll 1$ we may take $-\lambda T \cong -2\pi\chi$), being equal to $\ln(r'/r) = \ln(v'/v) = ...$.

If $\chi = 1$, hence if $\omega = \lambda$ (*critical damping*), then we may write

$$\mathbf{r}(t) = e^{-\lambda T}[\mathbf{r}_0 + (\mathbf{v}_0 + \lambda\mathbf{r}_0)t], \quad \mathbf{v}(t) = e^{-\lambda T}[\mathbf{v}_0 - \lambda(\mathbf{v}_0 + \lambda\mathbf{r}_0)t]. \qquad (8.2.16)$$

The corresponding motion is damped; the trajectory starts from the point P_0 and tends, in an infinite interval of time, with a velocity which tends to zero, towards the centre O, which is an *asymptotic point* (Fig.8.13,b). Noting that we may write

$$\mathbf{r}(t) = te^{-\lambda T}\left[\frac{\mathbf{r}_0}{t} + \mathbf{v}_0 + \lambda\mathbf{r}_0\right], \quad \mathbf{v}(t) = te^{-\lambda T}\left[\frac{\mathbf{v}_0}{t} - \lambda(\mathbf{v}_0 + \lambda\mathbf{r}_0)\right]$$

and that we have $\lim_{t\to\infty} te^{-\lambda t} = 0$, it results that the tangent at O to the trajectory is specified by the vector

$$\overline{\mathbf{r}} = \mathbf{r}_0 + \frac{\mathbf{v}_0}{\lambda}. \qquad (8.2.16')$$

If $\chi > 1$, hence if $\omega < \lambda$ (*supercritical damping*), then we use the notation

$$\omega'' = \sqrt{\lambda^2 - \omega^2} = \omega\sqrt{\chi^2 - 1} \qquad (8.2.17)$$

and obtain

$$\mathbf{r}(t) = e^{-\lambda t}\left[\mathbf{r}_0 \cosh\omega''t + \frac{1}{\omega''}(\mathbf{v}_0 + \lambda\mathbf{r}_0)\sinh\omega''t\right], \qquad (8.2.17')$$

$$\mathbf{v}(t) = e^{-\lambda t}\left[\mathbf{v}_0 \cosh\omega''t - \frac{1}{\omega''}(\omega^2\mathbf{r}_0 + \lambda\mathbf{v}_0)\sinh\omega''t\right]. \qquad (8.2.17'')$$

Noting that we can write

$$\mathbf{r}(t) = e^{-\lambda t}\cosh\omega''t\left[\mathbf{r}_0 + \frac{1}{\omega''}(\mathbf{v}_0 + \lambda\mathbf{r}_0)\tanh\omega''t\right],$$

Dynamics of the particle in a field of elastic forces

$$\mathbf{v}(t) = e^{-\lambda t} \cosh \omega'' t \left[\mathbf{v}_0 - \frac{1}{\omega''}(\omega^2 \mathbf{r}_0 + \lambda \mathbf{v}_0) \tanh \omega'' t \right]$$

and that

$$\lim_{t \to \infty} e^{-\lambda t} \cosh \omega'' t = \frac{1}{2} \lim_{t \to \infty} e^{-(\lambda - \omega'')t} \left(1 + e^{-2\omega'' t} \right) = 0, \quad \lim_{t \to \infty} \tanh \omega'' t = 1,$$

it results that the trajectory of the particle has the same form as in the previous case (Fig.8.13,b); the tangent at the asymptotic point O will be specified by the vector

$$\overline{\mathbf{r}} = \mathbf{r}_0 + \frac{1}{\omega''}(\mathbf{v}_0 + \lambda \mathbf{r}_0). \qquad (8.2.17''')$$

Hence, the corresponding motion is strongly damped. We can say that both last considered cases are *aperiodic damped motions*.

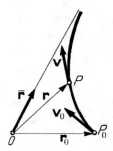

Figure 8.14. Aperiodic damped motion of a particle acted upon by a repulsive elastic force.

If the elastic force is repulsive, of the form (8.2.8), then the apparition of a damping force the modulus of which is proportional to the velocity leads to the equation of motion

$$\ddot{\mathbf{r}} + 2\lambda \dot{\mathbf{r}} - \omega^2 \mathbf{r} = \mathbf{0}; \qquad (8.2.18)$$

the initial conditions (8.2.5') lead to

$$\mathbf{r}(t) = e^{-\lambda t} \left[\mathbf{r}_0 \cosh \overline{\omega} t + \frac{1}{\overline{\omega}} (\mathbf{v}_0 + \lambda \mathbf{r}_0) \sinh \overline{\omega} t \right], \qquad (8.2.18')$$

$$\mathbf{v}(t) = e^{-\lambda t} \left[\mathbf{v}_0 \cosh \overline{\omega} t + \frac{1}{\overline{\omega}} (\omega^2 \mathbf{r}_0 - \lambda \mathbf{v}_0) \sinh \overline{\omega} t \right], \qquad (8.2.18'')$$

with the notation

$$\overline{\omega} = \sqrt{\lambda^2 + \omega^2} = \omega \sqrt{1 + \chi^2}. \qquad (8.2.18''')$$

We observe that

$$\lim_{t \to \infty} e^{-\lambda t} \cosh \overline{\omega} t = \frac{1}{2} \lim_{t \to \infty} e^{-(\overline{\omega} - \lambda)t} \left(1 + e^{-2\overline{\omega} t}\right) = \infty;$$

hence, the trajectory is similar to that in Fig.8.12 (the arc of hyperbola is deformed, the radii being smaller), but it is travelled through with a smaller velocity. The asymptote to which tends the trajectory is specified by the vector (Fig.8.14)

$$\overline{\mathbf{r}} = \mathbf{r}_0 + \frac{1}{\overline{\omega}}(\mathbf{v}_0 + \lambda \mathbf{r}_0). \tag{8.2.18^{iv}}$$

2.1.4 Self-sustained motions of a particle

Various types of motion considered in Subsecs 2.1.1 and 2.1.2 may be due to a given *perturbing force*, the respective motion being a *forced (sustained) motion*. If the force which maintains the motion is due to the motion itself, being of the form $\mathbf{\Phi} = k'\dot{\mathbf{r}}$, $k' > 0$, then the motion is called a *self-sustained motion*; in fact, the force $\mathbf{\Phi}$ is of the form $\mathbf{\Phi}' + \mathbf{\Phi}''$, where $\mathbf{\Phi}' = -\overline{k}'\dot{\mathbf{r}}$, $\overline{k}' > 0$, is a damped force, while $\mathbf{\Phi}'' = \overline{k}''\dot{\mathbf{r}}$, $\overline{k}'' > 0$, is a perturbing force, with $k' = \overline{k}'' - \overline{k}' > 0$ (if $\overline{k}'' < \overline{k}'$, then the motion is damped, while if $\overline{k}'' = \overline{k}'$, then the motion is non-damped). If the motion ceases, then the corresponding force disappears. We observe that in a forced motion that force exists independent of the motion and persists after its suppression.

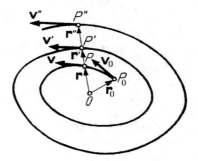

Figure 8.15. Non-damped pseudoelliptic oscillator.

In case of an elastic force of attraction, the equation of motion is of the form

$$\ddot{\mathbf{r}} - 2\lambda \dot{\mathbf{r}} + \omega^2 \mathbf{r} = \mathbf{0}, \tag{8.2.19}$$

where we use the notations of the preceding subsection. With the initial conditions (8.2.5'), we obtain

$$\mathbf{r}(t) = e^{\lambda t} \left[\mathbf{r}_0 \cos \omega' t + \frac{1}{\omega'}(\mathbf{v}_0 - \lambda \mathbf{r}_0) \sin \omega' t \right], \tag{8.2.19'}$$

$$\mathbf{v}(t) = e^{\lambda t} \left[\mathbf{v}_0 \cos \omega' t - \frac{1}{\omega'}(\omega^2 \mathbf{r}_0 - \lambda \mathbf{v}_0) \sin \omega' t \right], \tag{8.2.19''}$$

where we have introduced the pseudopulsation (8.2.15''), assuming that $\omega > \lambda$. The trajectory is a spiral and the particle is rotating around the centre O in the same

Dynamics of the particle in a field of elastic forces 493

direction till infinity (Fig.8.15). After a pseudoperiod $T = 2\pi/\omega'$, the particle returns on the same half-straight line with a velocity which has always the same direction for that line; the radii and the velocities are increasing in geometric progression of ratio $e^{\lambda t}$ for the same half-straight line, the number $\delta = \lambda T = 2\pi\lambda/\omega'$ being a *logarithmic increment*. The corresponding mechanical system may be called *non-damped pseudoelliptic oscillator*; the respective motion of the particle is a *pseudoperiodic non-damped motion*.

If $\omega = \lambda$, we may write

$$\mathbf{r}(t) = e^{\lambda t}[\mathbf{r}_0 + (\mathbf{v}_0 - \lambda \mathbf{r}_0)t], \quad \mathbf{v}(t) = e^{\lambda t}[\mathbf{v}_0 + \lambda(\mathbf{v}_0 - \lambda \mathbf{r}_0)t], \tag{8.2.20}$$

while if $\omega < \lambda$ we get

$$\mathbf{r}(t) = e^{\lambda t}\left[\mathbf{r}_0 \cosh \omega''t + \frac{1}{\omega''}(\mathbf{v}_0 - \lambda \mathbf{r}_0)\sinh \omega''t\right], \tag{8.2.21}$$

$$\mathbf{v}(t) = e^{\lambda t}\left[\mathbf{v}_0 \cosh \omega''t - \frac{1}{\omega''}(\omega^2 \mathbf{r}_0 - \lambda \mathbf{v}_0)\sinh \omega''t\right], \tag{8.2.21'}$$

where we used the notation (8.2.17). In both cases, one obtains an *aperiodic non-damped motion* (Fig.8.14), the asymptote to which tends the trajectory being specified by the vector

$$\overline{\mathbf{r}} = \mathbf{r}_0 - \frac{\mathbf{v}_0}{\lambda}, \tag{8.2.20'}$$

in the first case, and by the vector

$$\overline{\mathbf{r}} = \mathbf{r}_0 + \frac{1}{\omega''}(\mathbf{v}_0 - \lambda \mathbf{r}_0), \tag{8.2.21''}$$

in the second case, respectively.

In case of a repulsive elastic force, it results the equation of motion

$$\ddot{\mathbf{r}} - 2\lambda \dot{\mathbf{r}} - \omega^2 \mathbf{r} = \mathbf{0}, \tag{8.2.22}$$

where we used the same notations as above. We obtain

$$\mathbf{r}(t) = e^{\lambda t}\left[\mathbf{r}_0 \cosh \overline{\omega}t + \frac{1}{\overline{\omega}}(\mathbf{v}_0 - \lambda \mathbf{r}_0)\sinh \overline{\omega}t\right], \tag{8.2.22'}$$

$$\mathbf{v}(t) = e^{\lambda t}\left[\mathbf{v}_0 \cosh \overline{\omega}t + \frac{1}{\overline{\omega}}(\omega^2 \mathbf{r}_0 + \lambda \mathbf{v}_0)\sinh \overline{\omega}t\right], \tag{8.2.22''}$$

where we have put the initial conditions (8.2.5') and have introduced the notation (8.2.18'''). The motion of the particle is an aperiodic non-damped motion too (Fig.8.14); the trajectory tends to an asymptote specified by the vector

494   MECHANICAL SYSTEMS, CLASSICAL MODELS

$$\overline{r} = r_0 + \frac{1}{\omega}(v_0 - \lambda r_0). \quad\quad (8.2.22''')$$

2.2 Mechanical systems with a single degree of freedom

Projecting the motion of an elliptic oscillator on an axis in the plane of the corresponding trajectory, we obtain a linear oscillator, the most simple mechanical system with a single degree of freedom. We pass then to the case of a damped oscillator and to that of a sustained one. The phenomena of interference and of beats will be taken into consideration too, as well as the superposition of effects, which leads to small oscillations of a particle around a stable position of equilibrium.

2.2.1 Linear oscillator

We consider a particle acted upon by an *elastic force with a fixed support*, chosen as Ox-axis (of the form $F(x) = -kx$, $k > 0$), modelled as in Fig.8.8. With the notations in Subsec. 2.1.1, the equation of motion has the form

$$\ddot{x} + \omega^2 x = 0; \quad\quad (8.2.23)$$

with the initial conditions

$$x(0) = x_0, \quad v(0) = v_0, \quad\quad (8.2.23')$$

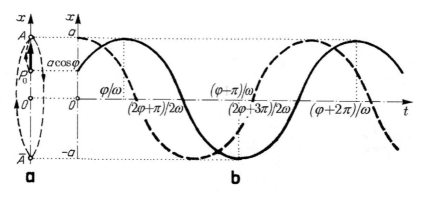

Figure 8.16. Linear oscillator: trajectory (a); diagram $x(t)$ vs $t(b)$.

we may write

$$x = x_0 \cos \omega t + \frac{v_0}{\omega} \sin \omega t = a \cos(\omega t - \varphi), \quad\quad (8.2.24)$$

$$v = v_0 \cos \omega t - \omega x_0 \sin \omega t = -a\omega \sin(\omega t - \varphi), \quad\quad (8.2.24')$$

where

$$a = \sqrt{x_0^2 + \frac{v_0^2}{\omega^2}}, \quad \varphi = \arctan \frac{v_0}{\omega x_0} \quad\quad (8.2.24'')$$

Dynamics of the particle in a field of elastic forces

are *the amplitude of the oscillation* (maximal elongation, the *elongation* $|x|$ being the distance from the centre of oscillation O to the position of the particle at a given moment) and *the phase shift* (the argument $\omega t - \varphi$ represents *the phase* at the moment t, the phase shift being calculated with respect to the phase ωt), respectively. The trajectory is the segment of line $\overline{A}A$, which is travelled through back and forth in the period of time (8.2.6), beginning with the initial position P_0 (Fig.8.16,a). We have thus to do with *oscillations* around *the oscillation centre* O, which is a *stable position of equilibrium*. Because *the period* T (and *the frequency* $\nu = 1/T$) is independent on the amplitude, it results that *the free linear oscillations with a single degree of freedom are isochronic*; on the other hand, the interval of time $T/4$ in which the segment of line AO is travelled through does not depend on the initial position A (does not depend on a), the velocity at that point vanishing, so that *the motion is tautochronous* too.

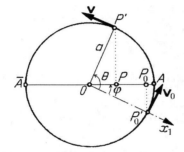

Figure 8.17. Linear oscillator as projection of a circular oscillator.

The mechanical system formed by a particle which describes a segment of a line, subjected to the action of an elastic force is called *linear oscillator*; that one may be also considered as a limit case of an elliptic oscillator, namely that in which one of the semiaxes of the ellipse tends to zero. We notice that a linear oscillator may be obtained too by projecting the motion of a circular oscillator (hence, of a particle P' with a velocity $\mathbf{v}$ of constant modulus $|\mathbf{v}| = a\omega$, which is in uniform motion on a circle) on a diameter $\overline{A}A$ of it (Fig.8.17); if the position of the diameter $\overline{A}A$ is specified by the angle φ with respect to the Ox_1-axis and if the angle $\theta = \omega t$, where ω is the angular velocity, gives the position of the radius OP', then we obtain the equation (8.2.24) of the linear oscillator. Any mechanical system with only one degree of freedom subjected to small oscillations around a stable position of equilibrium, e.g., the simple pendulum subjected to small oscillations, may be modelled by a linear oscillator.

Multiplying the equation (8.2.23) by $m\dot{x}$, we obtain

$$m\dot{x}\ddot{x} + kx\dot{x} = \frac{\mathrm{d}}{\mathrm{d}t}\left(\frac{m}{2}\dot{x}^2 + \frac{k}{2}x^2\right),$$

so that, taking into account (8.2.6'), (8.2.24) and (8.2.24'),

$$E = T - U = T + V = \frac{k}{2}a^2 = 2\pi^2\nu^2 ma^2; \quad (8.2.23'')$$

it results that, during the non-damped free linear oscillations with a single degree of freedom, a process of conservation of the mechanical energy takes place.

In general, a motion with a single degree of freedom of a mechanical system represents a *harmonic vibration* (*harmonic oscillation*, the term of oscillation being usually used if the particle returns always on the same trajectory) if it is of the form (8.2.24); we mention also that the term of oscillation may be used also for non-mechanical oscillations, but the term of vibration is used only for mechanical ones. The diagram of the considered motion is represented in Fig.8.16,b by a unbroken line (comparing with the dash line which corresponds to the vibration $x = a\cos\omega t$, the influence of the phase shift being thus put into evidence).

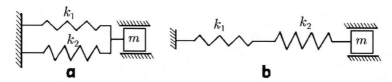

Figure 8.18. Model of a system of linear oscillators: parallel linkage (a); linkage in series (b).

Some vibrating mechanical systems may be physically modelled by a system formed of several elastic elements, of negligible masses, linked between them; such a system can be replaced, in general, by a single equivalent elastic element. In case of two springs of elastic constants k_1 and k_2, respectively, linked in parallel (Fig.8.18,a), the condition that the total elastic force for a displacement x be equal to the sum of the forces corresponding to each spring ($kx = k_1 x + k_2 x$) is put, wherefrom $k = k_1 + k_2$. In case of n springs *linked in parallel*, we may write (k/n is an *arithmetic mean*)

$$k = \sum_{i=1}^{n} k_i . \quad (8.2.25)$$

If the two springs are linked in series (Fig.8.18,b), then the condition that the total elongation x of the spring be equal to the sum of the elongations x_1 and x_2, of the component springs, respectively, is put ($x = x_1 + x_2$), the force in the spring being the same along it ($kx = k_1 x_1 + k_2 x_2$); it results $1/k = 1/k_1 + 1/k_2$. In case of n springs linked in series, we obtain (nk is a harmonic mean)

$$k = \frac{1}{\sum_{i=1}^{n} \frac{1}{k_i}} . \quad (8.2.25')$$

2.2.2 Modulated vibrations

In case of the motion of a mechanical system with a single degree of freedom, of the form

$$x(t) = a(t)\cos(\omega t - \varphi), \tag{8.2.26}$$

we say that we have to do with a *vibration modulated in amplitude*; in general, we assume that $a(t)$ varies little in *the quasi-period* $T = 2\pi/\omega$. The diagram of motion (the trajectory of which is also a segment of a line, as in the preceding case) has the aspect of a cosinusoid which is contained between the curves $x = \pm a(t)$ (Fig.8.19,a). We notice that the function (8.2.26) and its derivative $\dot{x}(t) = \dot{a}(t)\cos(\omega t - \varphi) - \omega a(t)\sin(\omega t - \varphi)$ have values equal to those of the functions $x(t) = \pm a(t)$ and of the respective derivatives $\dot{x}(t) = \pm \dot{a}(t)$ at the points of abscissae $t = \varphi/\omega$, $t = \varphi/\omega + T$,... and $t = \varphi/\omega + T/2$, $t = \varphi/\omega + 3T/2$,..., respectively; hence, the diagram of motion is tangent to the curves $x = \pm a(t)$. The intervals between two successive points of tangency are equal to T, as well as the intervals between two points in which the Ot-axis is pierced in the same direction. The respective motion is a quasi-periodic motion. Eventually, even the function $x = a(t)$ may be periodic.

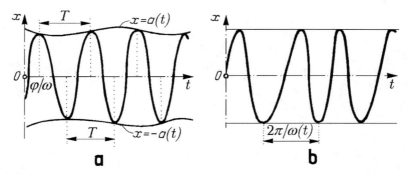

Figure 8.19. Modulated vibration: in amplitude (a); in frequency (b).

If the motion is definite in the form

$$x(t) = a\cos[\omega(t)t - \varphi], \tag{8.2.26'}$$

then we say that it is a *vibration modulated in frequency* (or *modulated in phase*); in general, one assumes that $\omega(t)$ varies little in a *pseudoperiod* $T = 2\pi/\omega(t)$. We may have, for instance, $\omega(t) = \omega_0 + \varepsilon \sin \omega t$, $\varepsilon/\omega_0 \ll 1$. The diagram of motion has the aspect in Fig.8.19,b.

2.2.3 Representations of harmonic vibrations

We have seen in Subsec. 2.2.1 that a harmonic vibration may be obtained by projecting a circular oscillator on one of its diameters; this observation has led Fresnel to elaborate a *vector method of representation of harmonic vibrations*; this method is particularly useful for the composition of these vibrations.

Thus, the harmonic vibration (8.2.24) may be represented by the vector $\overrightarrow{OQ}$ of modulus $\left|\overrightarrow{OQ}\right| = a$, the direction of which is given by the angle $\theta = \omega t - \varphi$, taken counterclockwise from the Ox_1-axis, which represents the phase of motion; this vector of constant modulus has a uniform rotation around the pole O, in positive or negative sense, as the angular velocity $\omega \gtrless 0$. If the point Q is projected at P on the Ox_1-axis, then $\overrightarrow{OP} = x$ (Fig.8.20). Observing that

$$\dot{x} = -a\omega \sin(\omega t - \varphi), \quad \ddot{x} = -a\omega^2 \cos(\omega t - \varphi),$$

it results the velocity $\overrightarrow{OP'} = \dot{x}$ and the acceleration $\overrightarrow{OP''} = \ddot{x}$, which correspond to a vector $\overrightarrow{OQ'}$ of modulus $a\omega$, the direction of which is obtained by rotating of $\pi/2$ the vector $\overrightarrow{OQ}$ counterclockwise or clockwise as $\omega \gtrless 0$ or to a vector $\overrightarrow{OQ''}$ of modulus $a\omega^2$, opposite to the vector $\overrightarrow{OQ}$, respectively.

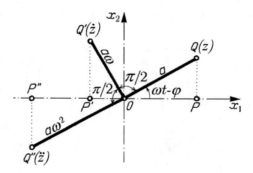

Figure 8.20. Vector representation of harmonic vibrations.

If we assume that the above vector *representation* is made *in the complex variables plane*, then to the point Q, hence to the vector $\overrightarrow{OQ}$, there corresponds the variable $z = x_1 + ix_2$; we may thus write (Fig.8.20)

$$z = a[\cos(\omega t - \varphi) + i\sin(\omega t - \varphi)] = ae^{i(\omega t - \varphi)}, \tag{8.2.27}$$

where $ae^{-i\varphi}$ is *the complex amplitude* of the oscillation. Differentiating successively, we obtain

$$\dot{z} = a\omega[-\sin(\omega t - \varphi) + i\cos(\omega t - \varphi)] = a\omega i e^{i(\omega t - \varphi)} = a\omega e^{i(\omega t - \varphi + \pi/2)}, \tag{8.2.27'}$$

$$\ddot{z} = -a\omega^2[\cos(\omega t - \varphi) + i\sin(\omega t - \varphi)] = -a\omega^2 e^{i(\omega t - \varphi)}$$
$$= a\omega^2 e^{i(\omega t - \varphi + \pi)}, \tag{8.2.27''}$$

Dynamics of the particle in a field of elastic forces 499

finding again the points Q' and Q'', respectively.

2.2.4 Composition of harmonic vibrations of the same direction. Interference. Beats. Harmonic analysis

We consider first of all two harmonic vibrations

$$x_1 = a_1 \cos(\omega t - \varphi_1), \quad x_2 = a_2 \cos(\omega t - \varphi_2),$$

which have the same direction and the same pulsation; their amplitudes and their phase shifts may be different. By the composition of these vibrations (in case of acoustic or light waves, the phenomenon is called interference too) we obtain also a harmonic vibration $x = x_1 + x_2 = a \cos(\omega t - \varphi)$, where

$$a = \sqrt{a_1^2 + a_2^2 + 2a_1 a_2 \cos(\varphi_2 - \varphi_1)}, \quad \varphi = \arctan \frac{a_1 \sin \varphi_1 + a_2 \sin \varphi_2}{a_1 \cos \varphi_1 + a_2 \cos \varphi_2}. \quad (8.2.28)$$

The term $2a_1 a_2 \cos(\varphi_2 - \varphi_1)$ is called *the term of interference* and leads to an *effect of interference stripes*. If $\varphi_2 - \varphi_1 = 2n\pi$, $n \in \mathbb{Z}$, then we obtain $a = a_1 + a_2$, while if $\varphi_2 - \varphi_1 = (2n+1)\pi$, $n \in \mathbb{Z}$, then we have $a = |a_1 - a_2|$; in the first case the interference is *constructive*, while in the second one it is *destructive*. Finally, if $a_1 = a_2$, then the destructive interference leads to *extinction* (zones in which the sound disappears, in case of acoustic waves, or zones of darkness, in case of light waves). If $\varphi_2 - \varphi_1 = n\pi/2, n \in \mathbb{Z}$, then $a = \sqrt{a_1^2 + a_2^2}, \varphi = \arctan(a_1/a_2)$. By composition of a certain number of harmonic vibrations one may effect an analogous computation.

If the two harmonic vibrations have not the same pulsation, being of the form

$$x_1 = a_1 \cos(\omega_1 t - \varphi_1), \quad x_2 = a_2 \cos(\omega_2 t - \varphi_2),$$

then their composition (by extension, the phenomenon bears the denomination of interference too) leads to an expression of the same form, modulated both in amplitude

$$a(t) = \sqrt{a_1^2 + a_2^2 + 2a_1 a_2 \cos[(\omega_1 - \omega_2)t - (\varphi_1 - \varphi_2)]} \quad (8.2.29)$$

and in phase

$$\varphi(t) = \arctan \frac{-a_1 \sin\left(\frac{\omega_1 - \omega_2}{2}t - \varphi_1\right) + a_2 \sin\left(\frac{\omega_1 - \omega_2}{2}t + \varphi_2\right)}{a_1 \cos\left(\frac{\omega_1 - \omega_2}{2}t - \varphi_1\right) + a_2 \cos\left(\frac{\omega_1 - \omega_2}{2}t + \varphi_2\right)}, \quad (8.2.29')$$

where $\omega = (\omega_1 + \omega_2)/2$. The motion thus obtained is no more harmonic, its form depending on the amplitudes, on the frequencies ratio, and on the phase shifts; it is a periodic motion only if the periods of the two component motions have a common multiple, hence only if $2\pi n_1/\omega_1 = 2\pi n_2/\omega_2$, $n_1, n_2 \in \mathbb{N}$ or $\omega_1/\omega_2 = n$, $n \in \mathbb{Q}$ (Figs.8.21, 8.22). The amplitude $a(t)$ has a variation between $a_{\min} = |a_1 - a_2|$ and

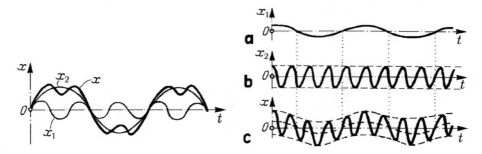

Figure 8.21-8.22. Composition of two harmonic vibrations for which $\omega_1/\omega_2 = n$, $n \in \mathbb{Q}$.

$a_{\max} = a_1 + a_2$, having maximal values (called *beats* in case of acoustic waves) at intervals of time given by the period $T_b = 2\pi/|\omega_1 - \omega_2|$ (Fig.8.23,a); the corresponding frequency is

$$\nu_b = \frac{1}{2\pi}|\omega_1 - \omega_2| = |\nu_1 - \nu_2|, \qquad (8.2.29'')$$

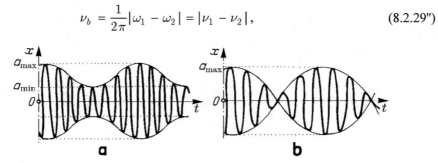

Figure 8.23. Beats: general case (a); simple beats (b).

hence it is equal to the absolute value of the difference of the frequencies of the component motions. One may thus syntonize two musical instruments (the period of the beats tends to infinity if the frequencies of the two instruments tend to be equal). The phenomenon is as much perceptible as the two amplitudes are closer. If $a_1 = a_2 = a$, then the formulae (8.2.29), (8.2.29') lead to

$$x = 2a\cos\left(\frac{\omega_1 - \omega_2}{2}t - \frac{\varphi_1 - \varphi_2}{2}\right)\cos\left(\frac{\omega_1 + \omega_2}{2}t - \frac{\varphi_1 + \varphi_2}{2}\right), \qquad (8.2.30)$$

hence to a product of two harmonic functions. In this case, $a_{\max} = 2a$, while $a_{\min} = 0$ (one obtains *the node of the beat*), the diagram of the motion being given in Fig.8.23,b; the beats are *simple beats*.

Using *Fresnel's method of representation*, we can compose the vectors $\overrightarrow{OQ_1}$ and $\overrightarrow{OQ_2}$, obtaining the vector $\overrightarrow{OQ}$, to which corresponds the motion $x = \overline{OP} = x_1 + x_2$, $x_1 = \overline{OP_1}$, $x_2 = \overline{OP_2}$ (Fig.8.24). Analogously, one may compose an arbitrary number of harmonic vibrations $x_i = a_i \cos(\omega_i t - \varphi_i)$, $i = 1, 2, ..., n$. The resultant motion is not – in general – harmonic, neither periodic; the motion is periodic if and only if the ratio between any two pulsations is a rational number ($\omega_i / \omega_j = n_{ij}$, $n_{ij} \in \mathbb{Q}$, $\forall i, j = 1, 2, ..., n$).

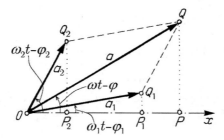

Figure 8.24. Fresnel's representation method.

The inverse problem, which consists in the determination of the harmonic components of a given periodic motion $x(t + T) = x(t)$ forms the object of *harmonic analysis*; this problem is of particular importance in physics, in engineering technology etc. Assuming that *Lejeune-Dirichlet's sufficient conditions* (the function $x(t)$ is piecewise continuous, having a finite number of points of discontinuity of the first kind and a finite number of maxima and minima on the time interval T) are fulfilled, we may decompose the motion in the form

$$x(t) = a_0 + a_1 \cos(\omega t - \varphi_1) + a_2 \cos(2\omega t - \varphi_2) + ...$$
$$... + a_n \cos(n\omega t - \varphi_n) + ... \qquad (8.2.31)$$

where $\omega = 2\pi / T$. We obtain thus *a finite Fourier representation* (an example of decomposition of a periodic motion in a sum of two harmonic vibrations is given in Fig.8.21) or a *development into a Fourier series*; the amplitudes $a_n = \sqrt{(a'_n)^2 + (a''_n)^2}$ and the phase shifts $\varphi_n = \arctan(a''_n / a'_n)$ are expressed by means of *the Fourier coefficients*

$$a_0 = \frac{1}{T} \int_T x(t) \, dt, \quad a'_n = \frac{2}{T} \int_T x(t) \cos n\omega t \, dt, \quad a''_n = \frac{2}{T} \int_T x(t) \sin n\omega t \, dt,$$
$$(8.2.31')$$

where the integration is effected along a period, beginning from an arbitrarily chosen point; the coefficient a_0 represents the mean value of the displacement $x(t)$. The oscillation $a_1 \cos(\omega t - \varphi_1)$ is of minimal frequency $\nu_{\min} = \omega / 2\pi$ (of maximal period

$T = 2\pi/\omega$, multiple of the periods of all the other oscillations, hence the period of the motion $x(t)$), being called *fundamental* (or *harmonic*) *oscillation*. The oscillation expressed by the term $a_n \cos(n\omega t - \varphi_n)$ has the frequency $\nu_n = n\nu_{\min}$ and is called *harmonic of nth order*. The set of amplitudes and the set of phases of the harmonic oscillations which compose a periodic motion form *the amplitudes spectrum* and *the phases spectrum*, respectively.

2.2.5 Composition of orthogonal harmonic vibrations. Small oscillations around a stable position of equilibrium

Let be the harmonic vibrations

$$x_1 = a_1 \cos(\omega_1 t - \varphi_1), \quad x_2 = a_2 \cos(\omega_2 t - \varphi_2)$$

along two orthogonal directions (corresponding to the Ox_1-axis and to the Ox_2-axis, respectively); by their composition, one obtains a motion the trajectory of which is contained in the rectangle $-a_1 \leq x_1 \leq a_1$, $-a_2 \leq x_2 \leq a_2$.

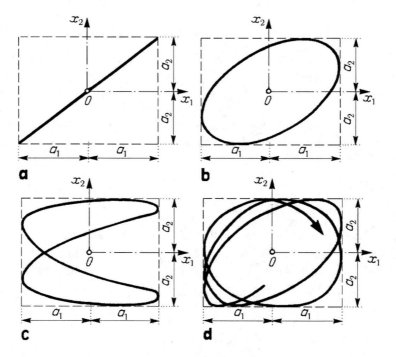

Figure 8.25. Composition of orthogonal harmonic vibrations: $\omega_1 = \omega_2$, $\varphi_1 = \varphi_2$ (a); $\omega_1 = \omega_2$, $\varphi_1 \neq \varphi_2$ (b); Lissajous' curves (c); $\omega_1/\omega_2 \neq n$, $n \in \mathbb{Q}$ (d).

If $\omega_1 = \omega_2$ and $\varphi_1 = \varphi_2$, then, eliminating the time t, it results $x_2 = a_2 x_1 / a_1$, hence the trajectory of the linear oscillator thus obtained is the diagonal of the mentioned rectangle (Fig.8.25,a). If $\omega_1 = \omega_2$ and $\varphi_1 \neq \varphi_2$, then it results an elliptic oscillator (Fig.8.25,b) of equation

$$a_2^2 x_1^2 - 2a_1 a_2 \cos(\varphi_1 - \varphi_2) x_1 x_2 + a_1^2 x_2^2 = a_1^2 a_2^2 \sin^2(\varphi_1 - \varphi_2),$$

obtained by elimination of the time t.

If $\omega_1 \neq \omega_2$ and $\omega_1 / \omega_2 = n$, $n \in \mathbb{Q}$ (the pulsations ω_1 and ω_2 are commensurable), then it exists a period T, the smallest common multiple of the periods of the two component motions, so that the motion is periodic, while the trajectory of the particle P is a closed curve, called *Lissajous curve*; the trajectory corresponding to the case $\omega_1 = 2\omega_2$, $\varphi_1 = \pi/4$, $\varphi_2 = 0$ is given in Fig.8.25,c. If the pulsations are not commensurable, then the motion can no more be periodic and the trajectory is an open curve which covers the whole rectangle mentioned above (Fig.8.25,d).

Analogously, one may compose three orthogonal harmonic vibrations

$$x_i = a_i \cos(\omega_i t - \varphi_i), \quad i = 1, 2, 3, \tag{8.2.32}$$

corresponding to the axes Ox_1, Ox_2 and Ox_3, respectively; one obtains thus a trajectory contained in the parallelepipedon $-a_i \leq x_i \leq a_i$, $i = 1, 2, 3$. However, one may thus study the small oscillations of a particle around the pole O, considered as a stable position of equilibrium. We assume that the force which acts upon the particle is conservative ($\mathbf{F} = \mathrm{grad}\, U$), the potential U being developable into series ($U = U_0 + U_1 + U_2 + ...$, U_i, $i = 0, 1, 2, ...$, homogeneous polynomial of ith degree in the co-ordinates x_1, x_2, x_3). Because the components of the force $\mathbf{F}$ are obtained as derivatives of the potential U, we may take $U_0 = 0$; as well, we must have $U_1 = 0$, because the origin is a position of equilibrium. Assuming that we have to do with small motions, we take $U = U_2$, neglecting the polynomials of higher order; we notice also that one must have $U_2 < 0$, the pole O being a stable position of equilibrium (the trajectory must be entirely at a finite distance). One may take

$$U(x_1, x_2, x_3) = -\frac{m}{2}\left(\omega_1^2 x_1^2 + \omega_2^2 x_2^2 + \omega_3^2 x_3^2\right) \tag{8.2.32'}$$

in this case; we are thus led to the equations of motion

$$\ddot{x}_1 + \omega_1^2 x_1 = 0, \quad \ddot{x}_2 + \omega_2^2 x_2 = 0, \quad \ddot{x}_3 + \omega_3^2 x_3 = 0, \tag{8.2.32''}$$

the solutions of which are the harmonic vibrations (8.2.32). The trajectory of the particle P is a closed curve (a Lissajous curve), the motion being periodic if and only if the pulsations (hence the periods and the frequencies too) are commensurable (are proportional to integer numbers $\omega_1 / n_1 = \omega_2 / n_2 = \omega_3 / n_3$, $n_1, n_2, n_3 \in \mathbb{Z}$). Otherwise, the trajectory is an open curve which covers entirely the above mentioned parallelepipedon; the particle does not pass twice through the same position, but may pass as close as possible to it in a sufficiently long time.

2.2.6 Damped linear oscillator

If, besides a linear elastic spring (Fig.8.8), we introduce, in parallel, a viscous damper with a damping coefficient $k' > 0$, then we obtain the physical model of a *damped linear oscillator* (Fig.8.26). Using the notations in Subsec. 2.1.3, it results the equation of motion (along the Ox -axis)

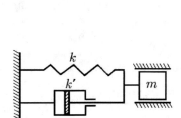

Figure 8.26. Model of a viscous damped linear oscillator.

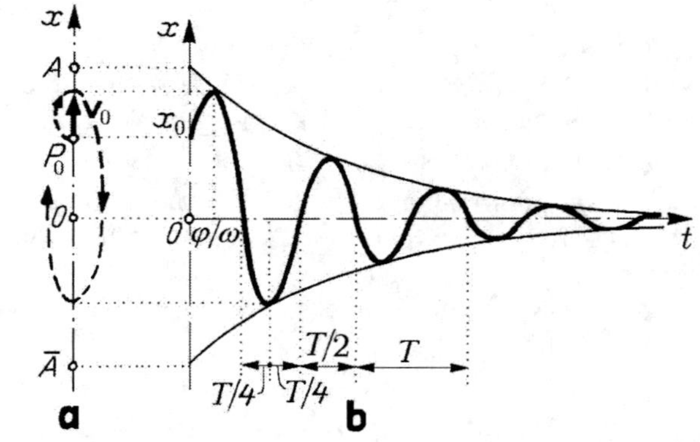

Figure 8.27. Damped linear oscillator: trajectory (a); diagram $x(t)$ vs t (b).

$$\ddot{x} + 2\lambda\dot{x} + \omega^2 x = 0; \qquad (8.2.33)$$

with the initial conditions (8.2.23'), we obtain the solution

$$x(t) = e^{-\lambda t}\left[x_0 \cos\omega' t + \frac{1}{\omega'}(v_0 + \lambda x_0)\sin\omega' t\right] = ae^{-\lambda t}\cos(\omega' t - \varphi), \qquad (8.2.33')$$

corresponding to a *subcritical damping* ($\chi < 1$). The motion is a *pseudoperiodic damped motion*, of *pseudoperiod* $T = 2\pi/\omega'$, the trajectory of which starts from the point P_0, being contained in the segment of a line $\overline{A}A$ and tending to the asymptotic point O after an infinity of oscillations around this pole (Fig.8.27,a). This motion constitutes a *vibration modulated in amplitude*, being strongly damped; the diagram of the motion has the aspect of a cosinusoid contained between the curves $x = \pm ae^{-\lambda t}$ and tangent to these ones at the points $t = \varphi/\omega'$, $t = \varphi/\omega' + T$,... and $t = \varphi/\omega' + T/2$, $t = \varphi/\omega' + 3T/2$,..., respectively (Fig.8.27,b).

In case of a *critical damping* ($\chi = 1$), we obtain an *aperiodic damped motion* given by

$$x(t) = e^{-\lambda t}\left[x_0 + (v_0 + \lambda x_0)t\right]. \qquad (8.2.33'')$$

If $v_0 > 0$, then the particle starts from the point P_0, reaches A at the moment $t' = v_0/\lambda(v_0 + \lambda x_0)$, and then changes of direction tending asymptotically to the centre O (Fig.8.28,a); the diagram of motion has a point of maximum for $t = t'$, tending asymptotically to zero (Fig.8.28,b). If $-\lambda x_0 \leq v_0 \leq 0$, then the particle starts from P_0, tending asymptotically to zero (Fig.8.29,a). The corresponding diagram has neither zeros, nor points of extremum (Fig.8.29,b); in the case in which $-\lambda x_0 /2 < v_0 < 0$ appears a point of inflection. If $v_0 < -\lambda x_0$, then the particle starts

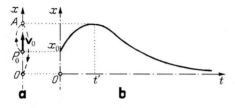

Figure 8.28. Linear critical damping; case $v_0 > 0$: trajectory (a); diagram $x(t)$ vs t (b).

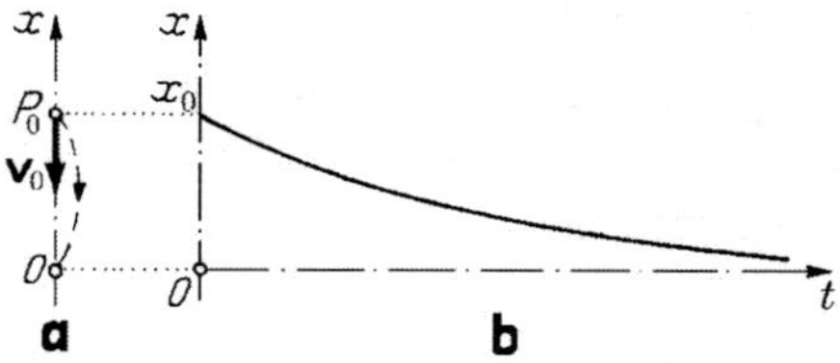

Figure 8.29. Linear critical damping; case $-\lambda x_0 \leq v_0 \leq 0$ (a); diagram $x(t)$ vs t (b).

from P_0, passes through the centre O at the moment $t'' = -x_0 /(v_0 + \lambda x_0)$ and reaches $\overline{A}$ at the moment t' and returns asymptotically towards the centre O (Fig.8.30,a); the diagram of motion pierces the Ot-axis at the point $t = t''$, has a minimum for $t = t'$, tending then asymptotically to zero with negative values (Fig.8.30,b). If the point P_0 is at the left of the pole O, hence if $x_0 < 0$, then one obtains, by symmetry, analogous results.

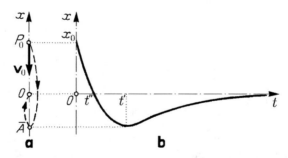

Figure 8.30. Linear critical damping; case $v_0 < -\lambda x_0$ (a); diagram $x(t)$ vs t (b).

A *supercritical damping* ($\chi > 1$) leads to an *aperiodic damped motion* of the form

$$x(t) = e^{-\lambda t}\left[x_0 \cosh \omega'' t + \frac{1}{\omega''}(v_0 + \lambda x_0)\sinh \omega'' t \right]. \qquad (8.2.33''')$$

In what concerns the trajectory and the diagram of motion, one obtains the same qualitative results as above as $v_0 > 0$, $-(\lambda + \omega'')x_0 \leq v_0 \leq 0$ or $v_0 < -(\lambda + \omega'')x_0$ (Figs.8.28-8.30); we observe that

$$t' = \frac{1}{\omega''}\operatorname{arg\,tanh}\frac{\omega''v_0}{\omega^2 x_0 + \lambda v_0}, \quad t'' = \frac{1}{\omega''}\operatorname{arg\,tanh}\frac{-\omega'' x_0}{v_0 + \lambda x_0}.$$

Multiplying the equation (8.2.33) by $m\dot{x}$, we obtain

$$\frac{\mathrm{d}}{\mathrm{d}t}(T+V) = \frac{\mathrm{d}}{\mathrm{d}t}(T-U) = -k'v^2, \tag{8.2.34}$$

so that the sum between the kinetic energy and the potential energy decreases (is *dissipated*) in time. In case of a subcritical damping, we observe that (abstraction of an additive constant)

$$E(t) = T(t) - U(t) = \frac{ma^2}{2}\mathrm{e}^{-2\lambda t}\left[\omega^2 + \lambda^2 \cos 2(\omega' t - \varphi) + \lambda\omega' \sin 2(\omega' t - \varphi)\right],$$

so that

$$E(t+T) = \mathrm{e}^{-2\lambda T} E(t), \tag{8.2.34'}$$

where $\delta = -\lambda T$ is the logarithmic decrement of motion; hence, *the mechanical energy of the pseudoperiodic damped linear oscillator decreases in geometric progression*. The relative energy dissipated in an interval of time equal to a pseudoperiod is given by

$$\varkappa = \frac{E(t) - E(t+T)}{E(t)} = 1 - \mathrm{e}^{2\delta}, \tag{8.2.34''}$$

being constant in time.

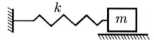

Figure 8.31. Model of a Coulombian damped linear oscillator.

The case considered above corresponds to a *viscous damping force* (the magnitude of which is proportional to the velocity). In the case of a *Coulombian dry damped force* (of constant magnitude during the motion), modelled physically as in Fig.8.31, the equation of motion is of the form

$$\ddot{x} + \omega^2 x + \frac{\Phi}{m}\operatorname{sign}\dot{x} = 0 \tag{8.2.35}$$

and leads to

$$x(t) = (x_0 \pm \delta_a)\cos\omega t + \frac{v_0}{\omega}\sin\omega t \mp \delta_a; \tag{8.2.35'}$$

we take $\operatorname{sign}\dot{x} = \pm 1$, denote by $\delta_a = \Phi/k$ the displacement along the spring of elastic constant k due to the force of dry friction Φ and put the initial conditions (8.2.23').

We must study the motion piecewise after the direction of the velocity $\dot{x}$ (in fact, on *semi-pseudoperiods*). Without any loss of generality, we may assume that the particle P starts from the point P_0 of abscissa $x_0 > 0$ without initial velocity ($v_0 = 0$); in these conditions, the motion can take place if and only if the damping force is less than the elastic force at the initial moment, hence if $\phi < kx_0$ or $\delta_a < x_0$. The particle begins to move with a negative velocity, so that its position is specified by $x(t) = (x_0 - \delta_a)\cos\omega t + \delta_a$, $0 \le t \le T/2$, till it reaches the point P_1 of abscissa $x_1 = -(x_0 - 2\delta_a)$, after a semi-pseudoperiod $T/2 = \pi/\omega$ (when the velocity $v(t) = -\omega(x_0 - \delta_a)\sin\omega t$ vanishes). If $x_1 > 0$, then $-x_0 + 2\delta_a < \delta_a$, the particle remaining further in permanent rest; hence, if the stop point is at the same part as the point of start (in particular, the initial position) with respect to the centre O, then the stopping is final. But if the point of stopping is situated on the other part of the centre O, then the particle moves further as the condition $x_0 - 2\delta_a > \delta_a$, hence the condition $x_0 > 3\delta_a$ is verified or not (Fig.8.32). If this condition is fulfilled, then the particle continues to move with a positive velocity, in an interval of time equal to a new semi-pseudoperiod, hence after the law $x(t) = (x_0 - 3\delta_a)\cos\omega t - \delta_a$, $T/2 \le t \le T$, which verifies the new conditions at the point P_1, at the moment $t = T/2$, till the point P_2 of abscissa $x_2 = x_0 - 4\delta_a$. An analogous reasoning is then made. Supposing

Figure 8.32. Coulombian damped linear oscillator. Trajectory.

that the conditions of motion are fulfilled, the particle reaches the point P_n of abscissa $x_n = (-1)^n (x_0 - 2n\delta_a)$ after n semi-periods; the abscissae of this oscillatory motion decrease in an arithmetic progression of ratio $-2\delta_a$. The motion ceases always after a finite number of semi-pseudoperiods, let be n semi-pseudoperiods. The particle passes over the point P_{n-1} and stops at the point P_n if $(2n-1)\delta_a < x_0 < (2n+1)\delta_a$. If $(2n-1)\delta_a < x_0 < 2n\delta_a$, hence if $n < x_0/2\delta_a + 1/2 < n + 1/2$, then the point P_n is at the same part of the centre O as the point P_{n-1}, if $2n\delta_a < x_0 < (2n+1)\delta_a$, hence if $n < x_0/2\delta_a < n + 1/2$, then the point P_n is at the other part of the centre O as the point P_{n-1}, while if $x_0 = 2n\delta_a$, hence if $n = x_0/2\delta_a$, then the point P_n at which the particle ceases to move coincides with the centre O. We denote $n_1 = \mathrm{E}[x_0/2\delta_a]$ and $n_2 = \mathrm{E}[x_0/2\delta_a + 1/2]$, where $\mathrm{E}[q]$ represents the greatest natural number contained in the number q; if $n_1 = n_2 = n$, then the particle stops at P_n, after n semi-pseudoperiods, the centre O being contained in the interior of the segment of a line $P_{n-1}P_n$, if $n_2 = n_1 + 1$, then $n_2 = n$, and the particle stops after n semi-pseudoperiods at the point P_n, on the same part as the point P_{n-1} with respect to the centre O, while if $n_1 = x_0 / 2\delta_a = n$, then the particle stops at the centre O, after an interval of time equal to the respective number of semi-pseudoperiods. The diagram of motion may be represented by a succession of arcs of cosinusoid, with amplitudes which decrease in arithmetic progression and of axes which differ by $\mp \delta_a$ with respect to the Ot-axis, the motion ceasing when the amplitude is less than δ_a (in Fig.8.33,b is taken $n = 5$); the points Q_0, Q_2, Q_4 and Q_1, Q_3, Q_5 of matching the arcs of cosinusoid are collinear, respectively, but the straight lines $x = \pm(x_0 - 4\delta_a t / T)$ are not tangent to those arcs at the respective points. We notice that the motion during each semi-pseudoperiod may be obtained as a projection of a uniform circular motion on a semicircle with the centre at Q of abscissa δ_a or at $\overline{Q}$, of abscissa $-\delta_a$, and of radii r_1, r_3, r_5 or r_2, r_4, respectively (Fig.8.33,a).

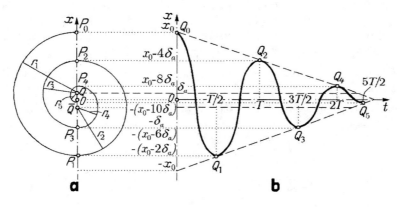

Figure 8.33. Coulombian damped linear oscillator. Trajectory as projection of circular uniform motions on semicircles (a); diagram $x(t)$ vs t (b).

The damping by dry friction forces is, in fact, a particular case of non-linear damping. Another important such case is that of a *hydraulic damping force*, in direct proportion to the square of the velocity, which leads to an equation of the form

$$\ddot{x} + 2\lambda \dot{x}^2 \operatorname{sign} \dot{x} + \omega^2 x = 0; \qquad (8.2.36)$$

proceeding as in Chap. 7, Subsec. 1.3.3, we may write the equivalent equation

$$\frac{1}{2} \frac{\mathrm{d}(\dot{x}^2)}{\mathrm{d}x} + 2\lambda \dot{x}^2 \operatorname{sign} \dot{x} + \omega^2 x = 0,$$

wherefrom we obtain the equation with separate variables

$$\dot{x}^2 = \overline{C} e^{-4\lambda \operatorname{sign} \dot{x}} + \frac{\omega^2}{8\lambda^2}(1 - 4\lambda x \operatorname{sign} \dot{x}), \qquad (8.2.36')$$

Dynamics of the particle in a field of elastic forces

the constant $\bar{C}$ being determined by the condition of vanishing the velocity at the points where that one changes of direction. We notice that, replacing x by $-x$ and $\dot{x}$ by $-\dot{x}$, one obtains the same result; hence, the representation of the motion in the phase space is symmetric with respect to the pole O, the representation in the upper semiplane being thus sufficient. Denoting $a = \omega^2/8\lambda^2$, $\bar{C} = aC$, we may write for the upper semiplane

$$v^2 = a\left(1 - 4\lambda x + Ce^{-4\lambda x}\right) = ae^{-4\lambda x}\left[C - (4\lambda x - 1)e^{4\lambda x}\right]. \qquad (8.2.36'')$$

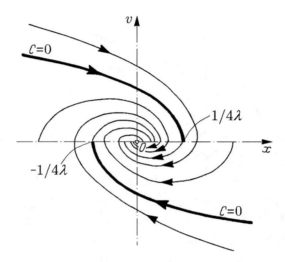

Figure 8.34. Hydraulic damped linear motion. Phase trajectory.

The points at which the phase trajectory pierces the Ox-axis are given by the equation $(4\lambda x - 1)e^{4\lambda x} = C$ (for $C < -1$ there is not one point, for $C = -1$ there exists the crunode $x = 0$, for $-1 < C < 0$ there exist the points $x_1 < 0$ and $x_2 > 0$, for $C \to 0$ there correspond $x_1 \to \infty$ and $x_2 \to 1/4\lambda$, while for $C \geq 0$ there exists only one point $x_2 \geq 1/4\lambda$). It results that for $-1 < C < 0$ one obtains closed phase trajectories, those ones having branches at infinity for $C > 0$. The separation curve is the parabola $v^2 = a(1 - 4\lambda x)$, corresponding to $C = 0$ (Fig.8.34).

2.2.7 Repulsive elastic forces. Self-sustained motions

In case of a *repulsive elastic force* of the form $F(x) = kx$, $k > 0$, the equation of motion is given by (we use the notations in Subsec. 2.1.2)

$$\ddot{x} - \omega^2 x = 0; \qquad (8.2.37)$$

with the initial conditions (8.2.23'), we obtain

$$x = x_0 \cosh \omega t + \frac{v_0}{\omega} \sinh \omega t, \quad v = v_0 \cosh \omega t + \omega x_0 \sinh \omega t. \quad (8.2.37')$$

To fix the ideas, we assume that $x_0 > 0$. If $v_0 > 0$, then the particle travels through a half straight line, starting from P_0, in the positive direction of the Ox-axis (Fig.8.35,a); the respective diagram of motion is given in Fig.8.35,b, the curve being asymptotic to a hyperbolic cosine. If $-\omega x_0 < v_0 < 0$, then the particle starts from P_0, reaches $\overline{A}$ and

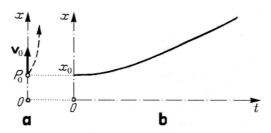

Figure 8.35. Particle acted upon by a repulsive elastic force; case $v_0 \geq 0$: linear trajectory (a); diagram $x(t)$ vs t (b).

then changes the direction of motion and tends to infinity in the positive sense of the Ox-axis (Fig.8.36,a); the diagram of motion has a minimum $x_{\min} = \sqrt{x_0^2 - (v_0/\omega)^2}$ for $t' = \arg\tanh(-v_0/\omega x_0)$, tending asymptotically to a hyperbolic cosine (Fig.8.36,b). If $v_0 < -\omega x_0$, then the particle starts from P_0, reaches O at the moment

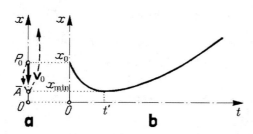
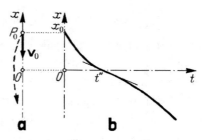

Figure 8.36. Particle acted upon by a repulsive elastic force; case $-\omega x_0 < v_0 < 0$: trajectory (a); diagram $x(t)$ vs t (b).

Figure 8.37. Particle acted upon by a repulsive elastic force; case $v_0 < -\omega x_0$: trajectory (a); diagram $x(t)$ vs t (b).

$t'' = \arg\tanh(-\omega x_0/v_0)$, tending then to infinity in the negative sense of the Ox-axis (Fig.8.37,a); the diagram of motion has a point of inflection at t'' on the Ot-axis and tends asymptotically to a hyperbolic cosine (Fig.8.37,b). All these motions are aperiodic and non-damped. If $v_0 = -\omega x_0$, then we obtain $x = x_0(\cosh \omega t - \sinh \omega t) = x_0 e^{-\omega t}$, so that the particle starts from P_0 and tends asymptotically towards the centre O, the motion being damped (it is an interesting case of damping due to the initial conditions) and aperiodic (Fig.8.29,a); the diagram of motion is of the same form as that in Fig.8.29,b.

If a *viscous damping force* intervenes too, then the equation of motion takes the form

$$\ddot{x} + 2\lambda\dot{x} - \omega^2 x = 0, \qquad (8.2.38)$$

with the notations in Subsec. 2.1.3; with the same initial conditions, it results

$$x(t) = e^{-\lambda t}\left[x_0 \cosh\overline{\omega}t + \frac{1}{\overline{\omega}}(v_0 + \lambda x_0)\sinh\overline{\omega}t\right], \quad \overline{\omega} = \sqrt{\lambda^2 + \omega^2}, \qquad (8.2.38')$$

$$v(t) = e^{-\lambda t}\left[v_0 \cosh\overline{\omega}t + \frac{1}{\overline{\omega}}(\omega^2 x_0 - \lambda v_0)\sinh\overline{\omega}t\right]. \qquad (8.2.38'')$$

One obtains the same qualitative results for the trajectory and for the diagram of motion, as $v_0 \geq 0$, as $-(\lambda + \overline{\omega})x_0 < v_0 < 0$, as $v_0 < -(\lambda + \overline{\omega})x_0$ or as $v_0 = -(\lambda + \overline{\omega})x_0$, respectively (Figs 8.35-8.37, 8.29); we observe that

$$t' = \frac{1}{\overline{\omega}}\arg\tanh\frac{\overline{\omega}v_0}{\lambda v_0 - \omega^2 x_0}, \quad t'' = \frac{1}{\overline{\omega}}\arg\tanh\frac{-\overline{\omega}x_0}{v_0 + \lambda x_0}.$$

In what concerns the diagram in Fig.8.37,b, the point of inflection is no more on the Ot-axis, but corresponds to

$$t''' = \frac{1}{\overline{\omega}}\arg\tanh\frac{\overline{\omega}(2\lambda v_0 - \omega^2 x_0)}{(\lambda^2 + \overline{\omega}^2)v_0 - \lambda\omega^2 x_0} > t''.$$

In case of *self-sustained motions* of the particle, we use the considerations in Subsec. 2.1.4. For an elastic force of attraction, the equation of motion is of the form

$$\ddot{x} - 2\lambda\dot{x} + \omega^2 x = 0. \qquad (8.2.39)$$

We obtain

$$x(t) = e^{\lambda t}\left[x_0 \cos\omega't + \frac{1}{\omega'}(v_0 - \lambda x_0)\sin\omega't\right] = ae^{\lambda t}\cos(\omega't - \varphi), \qquad (8.2.39')$$

hence a subcritical damping (we assume that $\omega > \lambda$, hence $\chi < 1$) for which the pseudopulsation ω' is given by (8.2.15''), the pseudoperiod being $T = 2\pi/\omega'$. The trajectory, which starts from the point P_0, has amplitudes more and more greater (Fig.8.38,a), being thus modulated in amplitude; the diagram of motion has the aspect of a cosinusoid contained between the curves $x = \pm ae^{\lambda t}$ and tangent to them (Fig.8.38,b). If $\omega = \lambda$ (critical damping, $\chi = 1$), we may write

$$x(t) = e^{\lambda t}[x_0 + (v_0 - \lambda x_0)t]. \qquad (8.2.39'')$$

If $0 < v_0 < \lambda x_0$, then the particle starts from P_0, reaches A at the moment $t' = v_0/\lambda(\lambda x_0 - v_0)$ and then changes of direction and tends to infinity in the negative direction of the Ox-axis (Fig.8.39,a); the diagram of motion has a maximum for $t = t'$ and pierces the Ot-axis at $t'' = x_0/(\lambda x_0 - v_0)$ (Fig.8.39,b). If $v_0 \geq \lambda x_0$,

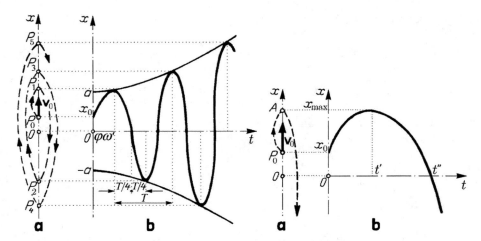

Figure 8.38. Self-sustained motion of a particle; subcritical damping. Trajectory (a); diagram $x(t)$ vs t (b).

Figure 8.39. Self-sustained motion of a particle; critical damping, case $0 < v_0 < \lambda x_0$: trajectory (a); diagram $x(t)$ vs t (b).

then the particle starts from P_0 and tends to infinity in the positive direction of the Ox-axis (Fig.8.35,a), the diagram of motion being that of Fig.8.35,b. If $v_0 \leq 0$, then the particle tends to infinity in the negative direction of the Ox-axis (Fig.8.40,a), while the diagram of motion pierces the Ot-axis at $t = t''$ (Fig.8.40,b). Analogously, for $\omega < \lambda$ (supercritical damping, $\chi > 1$) we obtain

$$x(t) = e^{\lambda t}\left[x_0 \cosh \omega'' t + \frac{1}{\omega''}(v_0 - \lambda x_0)\sinh \omega'' t\right], \qquad (8.2.39''')$$

the pseudopulsation ω'' being given by (8.2.17). In what concerns the trajectory and the diagram of motion, we obtain the same qualitative results as $0 < v_0 < (\lambda - \omega'')x_0$, as $v_0 \geq (\lambda - \omega'')x_0$ or as $v_0 \leq 0$, respectively (Figs 8.39, 8.35, 8.40); in this case

$$t' = \frac{1}{\omega''}\operatorname{arg tanh}\frac{\omega'' v_0}{\omega''^2 x_0 - \lambda v_0}, \quad t'' = \frac{1}{\omega''}\operatorname{arg tanh}\frac{\omega'' x_0}{\lambda x_0 - v_0}.$$

If the elastic force is *repulsive*, then the equation of motion has the form

$$\ddot{x} - 2\lambda \dot{x} - \omega^2 x = 0, \qquad (8.2.40)$$

and we are led to

$$x(t) = e^{\lambda t}\left[x_0 \cosh \overline{\omega}t + \frac{1}{\overline{\omega}}(v_0 - \lambda x_0)\sinh \overline{\omega}t\right], \qquad (8.2.40')$$

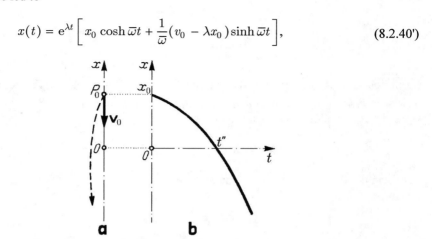

Figure 8.40. Self-sustained motion of a particle; critical damping, case $v_0 \leq 0$: trajectory (a); diagram $x(t)$ vs t (b).

with the same initial conditions and the notation (8.2.18'''). The trajectory and the diagram of motion are qualitatively given for $v_0 \geq 0$ and for $v_0 = -(\overline{\omega} - \lambda)x_0$ in Fig.8.35, for $-(\overline{\omega} - \lambda)x_0 < v_0 < 0$ in Fig.8.36 and for $v_0 < -(\overline{\omega} - \lambda)x_0$ in Fig.8.40; we notice that

$$t' = \frac{1}{\overline{\omega}}\operatorname{arg tanh}\frac{-\overline{\omega}v_0}{\lambda v_0 + \omega^2 x_0}, \quad t'' = \frac{1}{\overline{\omega}}\operatorname{arg tanh}\frac{\overline{\omega}x_0}{\lambda x_0 - v_0}.$$

2.2.8 Influence of perturbing forces. Resonance

The intervention of a perturbing force **F** leads to a *forced motion* (a *constraint* or *forced oscillation*) of a particle P, unlike the *free motion* (*free oscillation*) considered till now. The most simple physical model of a mechanical system with a simple degree of freedom acted upon by such a force is that of a *non-damped forced linear oscillator*, represented in Fig.8.41.

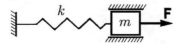

Figure 8.41. Model of a non-damped forced linear oscillator.

Let us consider firstly the case of a periodic force $F = F(t)$, which fulfils the Lejeune-Dirichlet sufficient conditions, that is which is developable into a Fourier series

$$F(t) = \alpha_0 + \alpha_1' \cos pt + \alpha_1'' \sin pt + \ldots + \alpha_n' \cos npt + \alpha_n'' \sin npt + \ldots, \qquad (8.2.41)$$

where

$$\alpha_0 = \frac{1}{T}\int_T F(t)\,\mathrm{d}t, \quad \alpha'_n = \frac{2}{T}\int_T F(t)\cos npt\,\mathrm{d}t, \quad \alpha''_n = \frac{2}{T}\int_T F(t)\sin npt\,\mathrm{d}t,$$
(8.2.41')

the period $T = 2\pi/p$ on which is effected the integration may begin from a point arbitrarily chosen. In case of a *non-damped linear oscillator subjected to a perturbing force*, the equation of motion is of the form

$$\ddot{x} + \omega^2 x = f(t),$$
(8.2.42)

where we use the previous notation and where

$$f(t) = \frac{1}{m}F(t).$$
(8.2.43)

We notice that the term α_0 in the expansion into a series leads only to a change of origin for $x(t)$, so that it may be neglected. For each term of the Fourier series one obtains a particular integral of the same type, leading to the corresponding motion; it is thus sufficient to consider the influence of only one term of the form

$$\alpha\cos(pt - \varphi), \quad \alpha = \alpha_1 = \frac{1}{m}\sqrt{\alpha_1'^2 + \alpha_1''^2}, \quad \varphi = \arctan\frac{\alpha''}{\alpha'}.$$
(8.2.44)

For $f(t) = \alpha\cos(pt - \varphi)$, we find

$$x(t) = x_0\cos\omega t + \frac{v_0}{\omega}\sin\omega t - \frac{\alpha}{\omega^2 - p^2}\left[\cos\varphi\cos\omega t + \frac{p}{\omega}\sin\varphi\sin\omega t - \cos(pt - \varphi)\right],$$
(8.2.42')

with the initial conditions (8.2.23'); we may also write

$$x(t) = a\cos(\omega t - \psi) + \frac{\alpha}{\omega^2 - p^2}\cos(pt - \varphi),$$
(8.2.42'')

where

$$a = \sqrt{\left(x_0 - \frac{\alpha\cos\varphi}{\omega^2 - p^2}\right)^2 + \frac{1}{\omega^2}\left(v_0 - \frac{\alpha p\sin\varphi}{\omega^2 - p^2}\right)^2}, \quad \psi = \arctan\frac{v_0 - \dfrac{\alpha p\sin\varphi}{\omega^2 - p^2}}{\omega\left(x_0 - \dfrac{\alpha\cos\varphi}{\omega^2 - p^2}\right)}.$$
(8.2.42''')

One observes thus that the motion of the particle may be obtained as an interference of two harmonic vibrations: *the proper vibration* (*the proper oscillation*) of pulsation ω

Dynamics of the particle in a field of elastic forces

and *the forced vibration* (*the forced oscillation*) of pulsation p, as it was shown in Subsec. 2.2.4. If, in particular, we assume homogeneous initial conditions ($x_0 = v_0 = 0$) and if the phase shift of the perturbing force vanishes ($\varphi = 0$), then it results

$$x(t) = \frac{\alpha}{\omega^2 - p^2}(\cos pt - \cos \omega t). \qquad (8.2.45)$$

If the pulsation p differs much from the pulsation ω ($p \ll \omega$ or $p \gg \omega$), then the diagram of motion is that in Fig.8.22 (the case $p \ll \omega$, hence a proper vibration of great pulsation "carried" by a forced vibration of small pulsation); we notice that the maximal elongation of the resultant motion is practically equal to the double of the amplitude of one of the motions ($x_{\max} \cong 2\alpha/(\omega^2 - p^2)$). If the two pulsations are close in magnitude, then one obtains the phenomenon of "beats" (Fig.8.23).

If $p = \omega$, then it results a non-determination in (8.2.45), as well as in (8.2.42'). If $p \to \omega$, then one obtains at the limit (we use L'Hospital's theorem)

$$x(t) = \frac{\alpha}{2\omega} t \sin \omega t, \qquad (8.2.46)$$

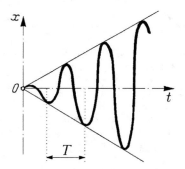

Figure 8.42. Phenomenon of resonance. Diagram $x(t)$ vs t.

for the law of motion (8.2.45). In case of the equation of motion (8.2.42') we get an analogous result (supplementary harmonic vibrations are added). The diagram of motion (8.2.46) is a sinusoid of amplitude modulated along the straight lines $x = \pm \alpha t/2\omega$ and of pseudoperiod $T = 2\pi/\omega$ (Fig.8.42). The amplitude increases very much, in arithmetic progression, and the phenomenon is called *resonance*, being extremely dangerous for civil and industrial constructions or for engine building; *the increasing velocity of the amplitude* is given by the slope

$$\frac{\alpha}{2\omega} = \frac{F_1/m}{2\sqrt{k/m}} = \frac{F_1}{2\sqrt{km}} = \frac{F_1}{k'_c}, \qquad (8.2.47)$$

hence it is in direct proportion to the amplitude $F_1 = m\alpha_1 = m\alpha$ of the perturbing force and in inverse proportion to the critical coefficient of damping (8.2.14). If

$x_{st} = F_1/k = F_1/m\omega^2$ is *the static displacement* produced by the force F_1 (corresponding to the relation of proportionality between the elastic force and the displacement), then we may express the amplitude A of the forced vibration in the form $A = x_{st}\mathscr{A}$, $\mathscr{A}$ being an *amplification factor of the forced vibration*, given by

$$\mathscr{A} = \frac{1}{1 - p^2/\omega^2} = \frac{1}{1 - \eta^2}, \qquad (8.2.48)$$

where we have introduced *the relative pulsation*

$$\eta = \frac{p}{\omega}, \qquad (8.2.48')$$

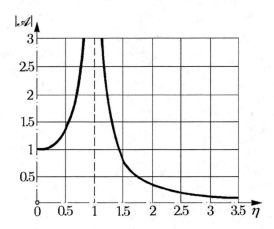

Figure 8.43. Phenomenon of resonance. Diagram $|\mathscr{A}|$ vs η.

which is a non-dimensional ratio. The diagram of the absolute value $|\mathscr{A}|$ is given in Fig.8.43.

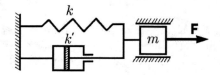

Figure 8.44. Model of a viscous damped forced linear motion.

If a viscous damping intervenes, then we use the physical model in Fig.8.44, being led to the equation of motion

$$\ddot{x} + 2\lambda\dot{x} + \omega^2 x = \alpha \cos pt, \qquad (8.2.49)$$

with the previously introduced notations (for the sake of simplicity, we assumed $\varphi = 0$). To fix the ideas, we assume to be in the case of a subcritical damping ($\chi < 1$); the motion of the particle is given by

$$x(t) = ae^{-\omega' t}\cos(\omega' t - \psi) + C_1 \cos pt + C_2 \sin pt, \qquad (8.2.49')$$

where

$$C_1 = \frac{(\omega^2 - p^2)\alpha}{(\omega^2 - p^2)^2 + 4\lambda^2 p^2}, \quad C_2 = \frac{2\lambda p\alpha}{(\omega^2 - p^2)^2 + 4\lambda^2 p^2}, \qquad (8.2.49'')$$

the last two terms corresponding to the forced motion. Taking into account the exponential term, the proper motion is rapidly damped, so that we may consider the forced motion in the form

$$x(t) = A\cos(pt - \varphi), \qquad (8.2.50)$$

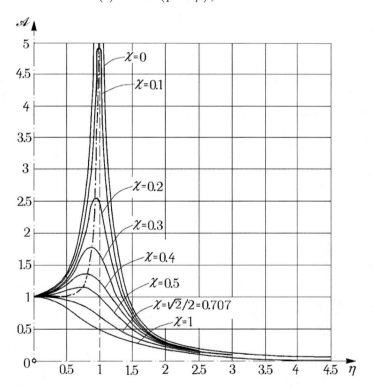

Figure 8.45. Viscous damped forced linear motion. Diagram $\mathscr{A}$ vs η.

with

$$A = \frac{\alpha}{\sqrt{(\omega^2 - p^2)^2 + 4\lambda^2 p^2}}, \quad \varphi = \arctan\frac{2\lambda p}{\omega^2 - p^2}. \qquad (8.2.50')$$

Using the notations introduced above and the damping factor (8.2.14'), we may also write

$$\mathcal{A} = \frac{1}{\sqrt{\left(1 - \eta^2\right)^2 + 4\chi^2\eta^2}}, \quad \varphi = \arctan\frac{2\chi\eta}{1 - \eta^2}. \tag{8.2.50''}$$

The diagram of the amplification factor $\mathcal{A} = \mathcal{A}(\eta)$ is given in Fig.8.45 for various values of the damping factor χ. We define a *resonance in amplitude* for the values

$$\eta = \eta_{\text{res}} = \sqrt{1 - 2\chi^2} < 1, \quad \chi \leq 1/\sqrt{2}, \tag{8.2.51}$$

for which the amplification factor has a maximum

$$\mathcal{A}_{\max} = \frac{1}{2\chi\sqrt{1 - \chi^2}} > \frac{1}{2\chi}. \tag{8.2.51'}$$

One observes that, in case of the phenomenon of resonance, the amplitude is as smaller as the damping is greater, the diagram of the function being planished for a great damping; the effect of damping appears especially in the vicinity of the resonance zone ($\eta \cong 1$). If the damping is very small ($\chi \ll 1$), then the resonance in amplitude appears for $\eta \cong 1$, the amplification factor being given by $\mathcal{A}_{\max} \cong 1/2\chi$. Eliminating χ between (8.2.51) and (8.2.51'), we get $\mathcal{A}_{\max} = 1/\sqrt{1 - \eta_{\text{res}}^4}$, that one being the locus of the points of maximum of the diagrams for various values of χ (represented by a dot-dash line). These points are at the left of the line $\eta = 1$; on this line one has $\mathcal{A} = 1/2\chi$.

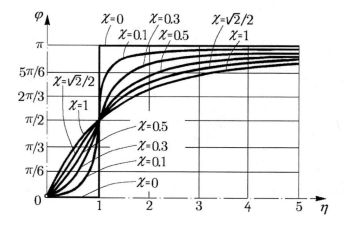

Figure 8.46. Viscous damped forced linear motion. Diagram φ vs η.

The diagram of the phase function $\varphi = \varphi(\eta)$ is given in Fig.8.46 for various values of the coefficient χ. We notice that, in case of a non-damped system, the phase is $\varphi = 0$, under the resonance ($\eta < 1$), the vibration being in phase with the perturbing

Dynamics of the particle in a field of elastic forces 519

force, or $\varphi = \pi$, over the resonance ($\eta > 1$), the vibration being in phase opposition with respect to the perturbing force; in case of a damped system there exists always a phase shift between the perturbing force and the vibration. For $\eta < 1$, as χ (hence, the damping) increases, so the phase shift between the motion and the perturbing force increases too, the motion remaining behind that force. For $\eta > 1$, as χ (hence, the damping) increases, so the phase shift decreases, the motion remaining behind the perturbing force too. For a very great η, the phase shift increases no matter the damping and the motion tends to be in opposition with the perturbing force. But the opposition is obtained rigorously only in the absence of the damping ($\eta = 0$). For $\eta = 1$ one obtains $\varphi = \pi/2$ for all the damping coefficients χ; one may define thus a *phase resonance* for which the vibration is in quadrature with the perturbing force.

Starting from the equation of motion

$$m\ddot{x} + k'\dot{x} + kx = F_1 \cos pt \qquad (8.2.52)$$

and multiplying by $\dot{x}$, we get

$$\frac{\mathrm{d}}{\mathrm{d}t}(T+V) = -k'\dot{x}^2 + F_1 \dot{x} \cos pt \,; \qquad (8.2.53)$$

it is thus seen that the mechanical energy does not remain constant, because the second member of this relation is – in general – non-zero. If we equate this member to zero, assuming that $\dot{x}$ N 0, we find

$$x = x_0 + \frac{F_1}{k'p} \sin pt \,; \qquad (8.2.53')$$

hence, the amplitude of the motion is $F_1/k'p$, corresponding to *the amplitude resonance* ($\mathscr{A} = 1/2\chi$ for $\eta = 1$).

2.2.9 Mechanical impedance. Transmissibility

Using the complex representation in Subsec. 2.2.3, we can write the equation of motion (8.2.52) in complex form

$$m\ddot{z} + k'\dot{z} + kz = F \,, \qquad (8.2.54)$$

where we denoted $z = z_0 \mathrm{e}^{\mathrm{i}pt}$ and $F = F_1 \mathrm{e}^{\mathrm{i}pt}$; replacing $\dot{z} = \mathrm{i}pz$, $\ddot{z} = -p^2 z$, we may write

$$F = Zz \,, \qquad (8.2.55)$$

where we have introduced *the mechanical impedance*

$$Z = k - mp^2 + \mathrm{i}k'p \,, \qquad (8.2.55')$$

which is a constant of proportionality, extending thus the notion of elastic constant. For an elastic spring, the impedance is $Z = k$, for a mass m we may write $Z = -mp^2$, while for a linear (viscous) damping it results $Z = ik'p$. We notice that the equation (8.2.54) may be obtained as a sum of vectors in the complex plane (Fig.8.47). Projecting on the direction of the force $\mathbf{F}$ and on a normal to it, one finds again the amplitude A and the phase shift φ given by (8.2.50'). Using the analogy with the constants corresponding to elastic springs (see Subsec. 2.2.1) and the formulae (8.2.25), (8.2.25'), we can express the impedance equivalent to n impedances Z_i, $i = 1, 2, ..., n$, linked in parallel, in the form

$$Z = \sum_{i=1}^{n} Z_i , \qquad (8.2.56)$$

while for the impedance equivalent to the same impedances linked in series we may write

$$Z = \frac{1}{\sum_{i=1}^{n} \frac{1}{Z_i}} . \qquad (8.2.56')$$

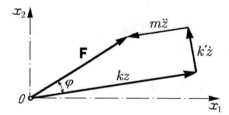

Figure 8.47. Viscous damped forced linear motion. Complex representation.

An examination of the physical model of a mechanical system constituted of an elastic spring and a linear damper which act in parallel on a mass (Fig.8.44) leads to the notion of *transmissibility* as the ratio C_T between the amplitude of the force $F(t) = kx + k'\dot{x}$ transmitted to the fixed element and the amplitude of the perturbing force $F(t)$ which acts upon the mass m. Introducing the impedance $\overline{Z} = k + ik'p$, $Z = k - mp^2 + ik'p$, it results

$$C_T = \frac{|\overline{Z}|}{|Z|} = \sqrt{\frac{k^2 + k'^2 p^2}{\left(k - mp^2\right)^2 + k'^2 p^2}} ;$$

with the aid of the previous notations, we may also write

$$C_T = \sqrt{\frac{1 + 4\chi^2 \eta^2}{\left(1 - \eta^2\right)^2 + 4\chi^2 \eta^2}} . \qquad (8.2.57)$$

Dynamics of the particle in a field of elastic forces

The transmissibility curves are given in Fig.8.48 for various values of the damping coefficient η. The maximal force transmitted to the fixed element is greater than the amplitude of the perturbing force if $C_T > 1$, hence for $0 < \eta < \sqrt{2}$, is less than this amplitude if $C_T < 1$, hence for $\eta > \sqrt{2}$, or is equal to the respective amplitude if $C_T = 1$, hence if $\eta = 0$ or $\eta = \sqrt{2}$. We notice also that for $\eta < \sqrt{2}$ the damping diminishes the transmissibility, while for $\eta > \sqrt{2}$ that one becomes smaller together with the damping.

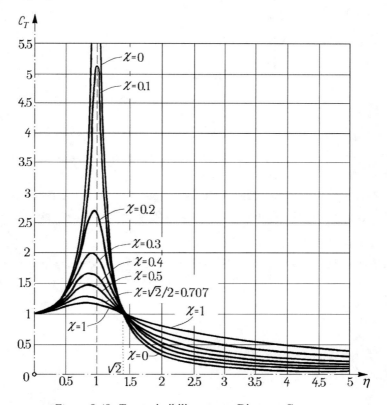

Figure 8.48. Transmissibility curves. Diagram C_T vs η.

2.2.10 Electro-mechanical analogy

Let be an R.L.C. *circuit*, constituted of an *ohmic resistance* R, a *loading inductance* L and a *condenser* of *capacity* C, connected in series with a *generator* having an *electromotive force* $E(t)$ (Fig.8.49). This force begins to act at the moment $t = 0$, when we close the circuit; for $t > 0$ a *current* of *intensity* $i(t) = \dot{q}(t)$, where $q(t)$ is *the charge*, is established.

Choosing as unknown function of the problem the charge $q(t)$, we may write the differential equation of second order with constant coefficients

$$L\ddot{q}(t) + R\dot{q}(t) + \frac{1}{C}q(t) = E(t), \quad t > 0, \tag{8.2.58}$$

corresponding to *Kirchhoff's second law*. Cauchy's problem for this circuit consists in the determination of the function $q(t)$ which satisfies the equation (8.2.58) for $t > 0$ and the initial conditions

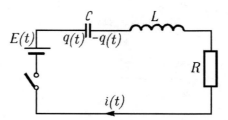

Figure 8.49. Electro-mechanical analogy.

$$q(0) = q_0, \quad \dot{q}(0) = i_0, \tag{8.2.58'}$$

where q_0 and i_0 are the charge and the intensity of the current at the moment $t = 0$, respectively. The correspondence $L \to m$, $R \to k'$, $1/C \to k$, $E(t) \to F(t)$, $q(t) \to x(t)$, $i(t) \to \dot{x}(t)$ leads to an interesting *electro-mechanical analogy*, which allows an experimental determination of the corresponding mechanical quantities.

2.2.11 Case of an arbitrary perturbing force

We consider first of all the non-damped motion with *arbitrary perturbing forces*, governed by an equation of the form (8.2.42), in case of elastic forces of attraction, or of the form

$$\ddot{x} - \omega^2 x = f(t) \tag{8.2.59}$$

in case of repulsive elastic forces. The fundamental solution of the operator $D_1 = d^2/dt^2 + \omega^2$ is

$$E_1(t) = \theta(t) \frac{\sin \omega t}{\omega}, \quad \omega > 0, \tag{8.2.60}$$

where $\theta(t)$ is Heaviside's distribution. As well, the distribution of function type

$$E_2(t) = -\frac{1}{2\omega} e^{-\omega|t|}, \quad \omega > 0, \tag{8.2.61}$$

is the fundamental solution of the operator $D_2 = d^2/dt^2 - \omega^2$; this distribution is generated by a continuous and everywhere differentiable function, excepting the origin. Using the considerations in App., Subsec. 3.3.1, we may write $D_i E_i(t) = \delta(t)$,

$i = 1,2$, where $\delta(t)$ is Dirac's distribution. If to the particular fundamental solution $E_2(t)$, the support of which is the real axis, we add the solution $e^{\omega t}/2\omega$, $\omega > 0$, corresponding to the homogeneous equation, then we find the fundamental solution

$$\overline{E}_2(t) = \theta(t)\frac{\sinh \omega t}{\omega}, \quad \omega > 0, \qquad (8.2.61')$$

the support of which belongs to the interval $[0,\infty)$; hence $D_2\overline{E}_2(t) = \delta(t)$. Denoting generically by $E_+(t)$ the fundamental solution corresponding to one of the two cases, we may write – for an arbitrary load $f(t)$ – the solution

$$x(t) = E_+(t) * f(t), \qquad (8.2.62)$$

in distributions, where we have introduced the convolution product. If $f(t)$ is an integrable function, it results

$$x(t) = \int_{-\infty}^{+\infty} f(t)E_+(t-\tau)\mathrm{d}\tau = \int_{-\infty}^{+\infty} f(t)G(t,\tau)\mathrm{d}\tau, \quad t \geq 0, \qquad (8.2.62')$$

where $G(t,\tau)$ is Green's function corresponding to one of the equations (8.2.42), (8.2.59).

Let us put initial conditions of the form

$$\lim_{t \to 0+0} x(t) = x_0, \quad \lim_{t \to 0+0} \dot{x}(t) = v_0, \qquad (8.2.63)$$

in case of a problem of Cauchy type. By a prolongation with zero for $t < 0$, we introduce the functions

$$\overline{x}(t) = \theta(t)x(t), \quad \overline{f}(t) = \theta(t)f(t); \qquad (8.2.64)$$

in this case, the equation (8.2.42), which is written for $t \geq 0$, is of the form

$$\frac{\tilde{\mathrm{d}}^2}{\mathrm{d}t^2}\overline{x}(t) + \omega^2\overline{x}(t) = \overline{f}(t),$$

where the sign "tilde" corresponds to the differentiation in the usual sense. Using the formula (1.1.50), we notice that

$$\frac{\mathrm{d}}{\mathrm{d}t}\overline{x}(t) = \frac{\tilde{\mathrm{d}}}{\mathrm{d}t}\overline{x}(t) + x_0\delta(t), \quad \frac{\mathrm{d}^2}{\mathrm{d}t^2}\overline{x}(t) = \frac{\tilde{\mathrm{d}}^2}{\mathrm{d}t^2}\overline{x}(t) + v_0\delta(t) + x_0\dot{\delta}(t),$$

so that one obtains

$$\ddot{\overline{x}}(t) + \omega^2\overline{x}(t) = \overline{f}(t) + v_0\delta(t) + x_0\dot{\delta}(t) = q(t), \qquad (8.2.65)$$

in distributions, where the differentiation takes place in the sense of the theory of distributions. The formula (8.2.62) allows to write

$$\overline{x}(t) = \frac{1}{\omega}\theta(t)\sin\omega t * \left[\overline{f}(t) + v_0\delta(t) + x_0\dot\delta(t)\right],$$

where we took into account (8.2.60). Effecting the convolution products, it results

$$\overline{x}(t) = \frac{1}{\omega}\theta(t)\sin\omega t * \overline{f}(t) + x_0\theta(t)\cos\omega t + \frac{v_0}{\omega}\theta(t)\sin\omega t, \qquad (8.2.65')$$

while, if $f(t)$ is a locally integrable function, then we get

$$x(t) = x_0\cos\omega t + \frac{v_0}{\omega}\sin\omega t + \frac{1}{\omega}\int_0^t f(\tau)\sin\omega(t-\tau)\mathrm{d}\tau, \quad t \geq 0; \qquad (8.2.65'')$$

the integral which intervenes is known as *the Duhamel integral*. If the perturbing force is a shock at the initial moment, then we have

$$f(t) = f_0\delta(t), \qquad (8.2.66)$$

where f_0 is a quantity the dimension of which is that of a velocity (dimensionally $[\delta(t)] = \mathrm{T}^{-1}$); we obtain

$$\overline{x}(t) = x_0\theta(t)\cos\omega t + \frac{f_0 + v_0}{\omega}\theta(t)\sin\omega t. \qquad (8.2.66')$$

We thus notice that the apparition of a shock at the initial moment is equivalent to the introduction of a supplementary initial velocity.

Considering the same problem for the equation (8.2.59), written in the form

$$\ddot{\overline{x}}(t) - \omega^2\overline{x}(t) = q(t), \qquad (8.2.67)$$

it results, analogously,

$$\overline{x}(t) = \frac{1}{\omega}\theta(t)\sinh\omega t * \overline{f}(t) + x_0\theta(t)\cosh\omega t + \frac{v_0}{\omega}\theta(t)\sinh\omega t; \qquad (8.2.67')$$

if $f(t)$ is a locally integrable function, then we get

$$x(t) = x_0\cosh\omega t + \frac{v_0}{\omega}\sinh\omega t + \frac{1}{\omega}\int_0^t f(\tau)\sinh\omega(t-\tau)\mathrm{d}\tau, \quad t \geq 0. \qquad (8.2.67'')$$

As well, in case of a perturbing force of the form (8.2.66), we have

$$\overline{x}(t) = x_0\theta(t)\cosh\omega t + \frac{f_0 + v_0}{\omega}\theta(t)\sinh\omega t, \qquad (8.2.66'')$$

Dynamics of the particle in a field of elastic forces

and we obtain a result analogous to that above.

In case of a bilocal problem, we put conditions of the form

$$x(t_1) = x_1, \quad x(t_2) = x_2, \tag{8.2.68}$$

assuming that $t \in [t_1, t_2]$. In case of Cauchy type conditions for $t = t_1$, we may write

$$x(t) = x_1 \cos\omega(t - t_1) + \frac{v_1}{\omega}\sin\omega(t - t_1) + \frac{1}{\omega}\int_{t_1}^{t} f(\tau)\sin\omega(t - \tau)d\tau,$$

considering an elastic force of attraction and assuming that $f(t)$ is a locally integrable function. The second bilocal condition (8.2.68) leads to

$$x_2 = x_1 \cos\omega(t_2 - t_1) + \frac{v_1}{\omega}\sin\omega(t_2 - t_1) + \frac{1}{\omega}\int_{t_1}^{t_2} f(\tau)\sin\omega(t_2 - \tau)d\tau,$$

so that we can determine the velocity v_1 at the initial moment; eliminating this velocity between the last two relations, one obtains the solution of the bilocal problem in the form

$$x(t) = \frac{1}{\omega\sin\omega(t_2 - t_1)}\Big\{\sin\omega(t_2 - t_1)\int_{t_1}^{t} f(\tau)\sin\omega(t - \tau)d\tau$$
$$-\sin\omega(t - t_1)\int_{t_1}^{t_2} f(\tau)\sin\omega(t_2 - \tau)d\tau + \omega[x_1 \sin\omega(t_2 - t)$$
$$+x_2 \sin\omega(t - t_1)]\Big\}, \quad t_2 \neq t_1 + k\frac{\pi}{\omega}, \quad k \in \mathbb{N}. \tag{8.2.69}$$

In case of a repulsive elastic force, it results

$$x(t) = \frac{1}{\omega\sinh\omega(t_2 - t_1)}\Big\{\sinh\omega(t_2 - t_1)\int_{t_1}^{t} f(\tau)\sinh\omega(t - \tau)d\tau$$
$$-\sinh\omega(t - t_1)\int_{t_1}^{t_2} f(\tau)\sinh\omega(t_2 - \tau)d\tau + \omega[x_1 \sinh\omega(t_2 - t)$$
$$+x_2 \sinh\omega(t - t_1)]\Big\} \tag{8.2.69'}$$

for the same bilocal problem.

In the general case of a law of motion of the form

$$m\ddot{x} + k'\dot{x} + kx = F(t), \tag{8.2.70}$$

where $-\infty < k, k' < \infty$, which corresponds to all possibilities of motion previously considered, we put Cauchy's problem in the form (8.2.63). By a prolongation with zero for $t < 0$, of the form (8.2.64),

$$\overline{F}(t) = \theta(t)F(t) \tag{8.2.64'}$$

and taking into account the connection between the derivatives in the usual sense and in the sense of the theory of distributions for the regular distributions thus obtained, we may write, as above, the corresponding equation in distributions

$$m\ddot{\bar{x}}(t) + k'\dot{\bar{x}}(t) + k\bar{x}(t) = \bar{F}(t) + (mv_0 + k'x_0)\delta(t) + mx_0\dot{\delta}(t) = Q(t), \qquad (8.2.70')$$

which includes the initial conditions. The solution of this problem is given by

$$\bar{x}(t) = E(t) * Q(t), \qquad (8.2.71)$$

where $E(t)$ is the fundamental solution in the sense of the theory of distributions, which verifies the equation

$$m\ddot{E}(t) + k'\dot{E}(t) + kE(t) = \delta(t). \qquad (8.2.71')$$

Applying the Laplace transform in distributions, we obtain

$$\left(mp^2 + k'p + k\right)\mathrm{L}[E(t)] = 1,$$

where p is the new variable in the space of transforms; it results

$$\mathrm{L}[E(t)] = \frac{1}{mp^2 + k'p + k} = \frac{1}{m(p - p_1)(p - p_2)},$$

$$p_1 = -\alpha + \beta, \quad p_2 = -\alpha - \beta, \quad \alpha = \frac{k'}{2m}, \quad \beta = \sqrt{\left(\frac{k'}{2m}\right)^2 - \frac{k}{m}}.$$

Three cases of integration are thus put in evidence. If

$$k'^2 > 4km, \qquad (8.2.72)$$

then we may write ($\beta^2 > 0$, $\beta \in \mathbb{R}$)

$$\mathrm{L}[E(t)] = \frac{1}{2m\beta}\left(\frac{1}{p - p_1} - \frac{1}{p - p_2}\right),$$

wherefrom

$$E(t) = \frac{\theta(t)}{2m\beta}\left(e^{p_1 t} - e^{p_2 t}\right) = \frac{1}{m\beta}\theta(t)e^{-\alpha t}\sinh\beta t. \qquad (8.2.72')$$

After effecting the convolution products, the solution (8.2.71) takes the form

$$\bar{x}(t) = \frac{1}{m\beta}\theta(t)e^{-\alpha t}\sinh\beta t * \bar{F}(t) + \theta(t)e^{-\alpha t}\left[x_0\cosh\beta t + \frac{1}{\beta}(v_0 + \alpha x_0)\sinh\beta t\right];$$

$$(8.2.73)$$

if $F(t)$ is a locally integrable function, then we can write

$$x(t) = e^{-\alpha t}\left[x_0 \cosh \beta t + \frac{1}{\beta}(v_0 + \alpha x_0)\sinh \beta t\right]$$
$$+ \frac{1}{m\beta}\int_0^t F(t-\tau)e^{-\alpha \tau}\sinh \beta\tau \mathrm{d}\tau, \quad t \geq 0. \tag{8.2.73'}$$

If $k' > 0$, then we denote $\alpha = \lambda > 0$ (viscous damping), while if $k' < 0$, then we denote $\alpha = -\lambda < 0$ (self-sustained oscillations). If $k > 0$ (elastic forces of attraction), then we denote $k/m = \omega^2$, $\beta = \omega''$ ($\lambda^2 > \omega^2$), while if $k < 0$ (repulsive elastic forces), then we denote $-k/m = \omega^2$, $\beta = \bar{\omega}$. We pass thus to the notation previously used.

If

$$k'^2 < 4km \tag{8.2.74}$$

and

$$\beta^2 < 0, \quad \beta \in \mathbb{C}, \quad \beta = i\omega', \quad \omega' = \sqrt{\frac{k}{m} - \left(\frac{k'}{2m}\right)^2} = \sqrt{\omega^2 - \lambda^2},$$

then we may write (one can have only $k > 0$, hence elastic forces of attraction)

$$E(t) = \frac{1}{m\omega'}\theta(t)e^{-\alpha t}\sin \omega' t; \tag{8.2.74'}$$

the solution of Cauchy's problem is

$$\bar{x}(t) = \frac{1}{m\omega'}\theta(t)e^{-\alpha t}\sin \omega' t * \bar{F}(t) + \theta(t)e^{-\alpha t}\left[x_0 \cos \omega' t + \frac{1}{\omega'}(v_0 + \alpha x_0)\sin \omega' t\right] \tag{8.2.75}$$

and, in the case in which $F(t)$ is a locally integrable function, we obtain

$$x(t) = e^{-\alpha t}\left[x_0 \cos \omega' t + \frac{1}{\omega'}(v_0 + \alpha x_0)\sin \omega' t\right]$$
$$+ \frac{1}{m\omega'}\int_0^t F(t-\tau)e^{-\alpha \tau}\sin \omega'\tau \mathrm{d}\tau, \quad t \geq 0. \tag{8.2.75'}$$

If

$$k'^2 = 4km, \tag{8.2.76}$$

then we have $\beta = 0$, so that $\mathrm{L}[E(t)] = 1/m(p+\alpha)^2$; the fundamental solution is

$$E(t) = \frac{1}{m}\theta(t)te^{-\alpha t}. \tag{8.2.76'}$$

The solution of the boundary value problem reads

$$\bar{x}(t) = \frac{1}{m}\theta(t)te^{-\alpha t} * \bar{F}(t) + \theta(t)e^{-\alpha t}[x_0 + (v_0 + \alpha x_0)t], \qquad (8.2.77)$$

while if $F(t)$ is a locally integrable function, then we get

$$x(t) = e^{-\alpha t}[x_0 + (v_0 + \alpha x_0)t] + \frac{1}{m}\int_0^t F(t-\tau)e^{-\alpha \tau}\tau d\tau, \quad t \geq 0. \qquad (8.2.77')$$

According to the above relations, for the last two cases we obtain a viscous damping or self-sustained oscillations for $\alpha = \pm \lambda$, respectively.

In the particular case of homogeneous initial conditions ($x_0 = v_0 = 0$) we remain only with the first terms of the formulae (8.2.73), (8.2.75) and (8.2.77).

The vibratory phenomena considered above, for which the parameters m, k, k' are constant, are – in general – in permanent régime. Such phenomena may be due, for instance, to a perturbing force, which is applied at the initial moment and remains constant in time, of the form

$$\bar{F}(t) = F_0\theta(t); \qquad (8.2.78)$$

we obtain thus

$$\bar{x}(t) = \begin{cases} -\dfrac{F_0}{k\beta}\left[e^{-\alpha t}(\alpha \sinh \beta t + \beta \cosh \beta t) - \beta\right] & \text{for } k'^2 > 4km, \\[6pt] \dfrac{F_0}{\alpha m}\left[1 - (1 + \alpha t)e^{-\alpha t}\right] & \text{for } k'^2 = 4km, \\[6pt] -\dfrac{F_0}{k\omega}\left[e^{-\alpha t}(\alpha \sin \omega t + \omega \cos \omega t) - \omega\right] & \text{for } k'^2 < 4km, \end{cases} \qquad (8.2.78')$$

for $t \geq 0$. Passing from a permanent régime to another one, a transitory phenomenon takes place; for instance, the transitory phenomena appear by the introduction or by the elimination of a perturbing force, as well as – in general – by any variation of the parameters of the considered phenomena. E.g., in case of a shock at the initial moment, given by

$$F(t) = F_0\delta(t), \qquad (8.2.79)$$

we obtain

$$\bar{x}(t) = \begin{cases} \dfrac{F_0}{m\beta}e^{-\alpha t}\sinh \beta t & \text{for } k'^2 > 4km, \\[6pt] \dfrac{F_0}{m}te^{-\alpha t} & \text{for } k'^2 = 4km, \\[6pt] \dfrac{F_0}{m\omega'}e^{-\alpha t}\sin \omega' t & \text{for } k'^2 < 4km, \end{cases} \qquad (8.2.79')$$

Dynamics of the particle in a field of elastic forces

for $t \geq 0$; this result corresponds to the fundamental solutions (8.2.72'), (8.2.76') and (8.2.74'), as well as to the velocity which appears in the previously considered case (it is obtained by differentiating the formula (8.2.78') with respect to time).

2.2.12 Oscillations with variable characteristics

The *oscillations with variable characteristics* (*parametric oscillations*) are the oscillations of a mechanical system the parameters of which (mass, frequency, dimensions, elastic coefficient, damping coefficient etc.) are functions of time. In the absence of a perturbing force, the mathematical model of such a mechanical phenomenon leads to a differential equation of the form

$$\ddot{x} + \alpha(t)\dot{x} + \beta(t)x = 0, \qquad (8.2.80)$$

where $\alpha(t)$ and $\beta(t)$ are periodic functions. By the substitution

$$x(t) = u(t)\mathrm{e}^{-(1/2)\int \alpha(t)\mathrm{d}t},$$

we obtain the equation

$$\ddot{u} + \gamma(t)u = 0, \quad \gamma(t) = \beta(t) - \frac{1}{2}\dot{\alpha}(t) - \frac{1}{4}\alpha^2(t). \qquad (8.2.81)$$

To deduce the aspect of the solution or to find an approximate solution of this equation may be a different task from case to case. The general solution can be written in the form

$$u(t) = \mathrm{e}^{\lambda t}\varphi(t) + \mathrm{e}^{\mu t}\psi(t), \qquad (8.2.81')$$

where the coefficients λ and μ are *characteristic coefficients* of the equation, while $\varphi(t)$ and $\psi(t)$ are periodic functions of period T (as well as the function $\gamma(t)$), which have to be determined. One may put the condition

$$u(t+T) = ku(t), \qquad (8.2.81'')$$

where $k = k(\lambda,\mu)$. Thus, if we know the motion during a period, then the motion in the following period is obtained by multiplying the elongation of the first period by the factor k a.s.o. If $|k| < 1$, then the elongations tend to zero and the oscillations are damped; if $|k| > 1$, then the elongations increase in time and the motion becomes instable.

For instance, in case of a mathematical pendulum of variable length $l = l(t)$, the equation of motion which specifies the generalized co-ordinate $\theta = \theta(t)$ is

$$\frac{\mathrm{d}}{\mathrm{d}t}(l^2\dot{\theta}) + gl\sin\theta = 0; \qquad (8.2.82)$$

in case of small oscillations, we obtain the equation

$$\ddot{\theta} + 2\frac{\dot{l}(t)}{l(t)}\dot{\theta} + \frac{g}{l(t)}\theta = 0. \tag{8.2.82'}$$

2.2.13 Non-linear oscillations

We say that a mechanical system represents a *non-linear oscillatory system* if, in the differential equation of motion, one or several characteristic parameters (mass, elastic coefficient, frequency etc.) depend on the displacement x. In general, the differential equation of these oscillations is of the form

$$\ddot{x} + f(x, \dot{x}; t) = 0, \tag{8.2.83}$$

resulting from Newton's equation (which puts in evidence the linkage between the elastic force and the displacement) or, in particular, of the form

$$\ddot{x} + f(x, \dot{x}) = 0, \tag{8.2.83'}$$

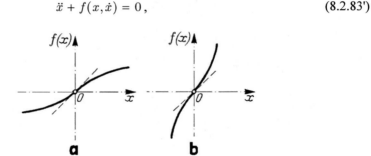

Figure 8.50. Non-linear vibrations determined by the equation $\ddot{x} + f(x) = 0$; diagrams $f(x)$ vs x for $x > 0$: case $f''(x) < 0$ (a); case $f''(x) > 0$ (b).

as the non-linear vibration is *non-autonomous* or *autonomous*, respectively. If the non-linear term depends only on x, then the function $f(x)$ which intervenes is called *arc characteristic*; the most times, in practice, the graphic of the function $f(x)$ is symmetric with respect to the origin ($f(x)$ is an odd function, that is $f(-x) = -f(x)$). If the graphic of the function $f(x)$ has the concavity towards down in the vicinity of the origin for $x > 0$, hence if $f''(x) < 0$ (Fig.8.50,a), then *the arc characteristic is weak*, while if, in the same vicinity, the graphic of the function $f(x)$ has the concavity towards up for $x > 0$, hence if $f''(x) > 0$ (Fig.8.50,b), then *the arc characteristic is strong*.

In case of great oscillations of the simple pendulum, the non-linear character of the phenomenon is put into evidence in the equation (7.1.38'). Developing $\sin\theta$ into a power series, we obtain, in a first approximation, the linear differential equation (7.1.45). In a second approximation (non-linear approximation, in which we take two terms in the series development), we may write

$$\ddot{\theta} + \omega^2\left(\theta - \frac{\theta^3}{6}\right) = 0; \qquad (8.2.84)$$

in this case, $f(\theta) = \omega^2\left(\theta - \theta^3/6\right)$ and $f''(\theta) = -\omega^2\theta < 0$ for $\theta > 0$, the arc characteristic being thus weak.

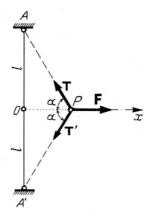

Figure 8.51. Motion of a particle situated at the middle of an elastic thread.

Let us consider also the case of a particle P of mass m, situated at the middle of an elastic thread fixed at the points A and A', $\overline{AA'} = 2l$ (e.g., the case of a training ball for boxing); in the position O of stable equilibrium, the tension in the thread is $\mathbf{T}_0$ (we suppose that the thread is tensioned). Perturbing the position of equilibrium by a displacement normal to AA', the particle reaches the position P (Fig.8.51), being acted by the tensions $\mathbf{T}$ and $\mathbf{T}'$, $T = T' = T_0 + k\left(\sqrt{l^2 + x^2} - l\right)$, and by the perturbing force $\mathbf{F}$, $F = 2T\cos\alpha = 2Tx/\sqrt{l^2 + x^2}$. Observing that

$$\frac{1}{\sqrt{l^2 + x^2}} = \frac{1}{l}\left(1 + \frac{x^2}{l^2}\right)^{-1/2} \cong \frac{1}{l}\left(1 - \frac{x^2}{2l^2}\right),$$

we obtain

$$F = 2kx + \frac{2(T_0 - kl)x}{\sqrt{l^2 + x^2}} \cong 2T_0\frac{x}{l} + (kl - T_0)\left(\frac{x}{l}\right)^3;$$

if $|x| \ll l$, then we remain only with the first term (the case of a linear elastic force). In the non-linear case considered (the second approximation), the equation of motion is

$$\ddot{x} + \alpha x + 2\beta x^3 = 0, \quad \alpha = \frac{2T_0}{l} > 0, \quad \beta = \frac{kl - T_0}{2l^3} > 0, \qquad (8.2.85)$$

where we take into account that the point O is a stable position of equilibrium, the force which acts upon the particle being directed towards this point (hence, it is equal to $-F$); we notice that the tension T_0 is of the form $k(l - l_0)$, $l_0 < l$, assuming that the thread is tensioned beginning with a length $2l_0$, for which the initial tension is zero, so that $kl > T_0$. Because $f(x) = \alpha x + 2\beta x^3$, we have $f''(x) = 12\beta x > 0$ for $x > 0$; hence, the arc characteristic is strong.

In the considered case, the force F is conservative ($2V(x) = \alpha x^2 + \beta x^4$), so that we may use the results in Chap. 6, Subsec. 2.2.4 and Chap. 7, Subsec. 2.3.4; we obtain thus the first integral

$$v^2 + \alpha x^2 + \beta x^4 = h, \quad h = \alpha a^2 + \beta a^4, \qquad (8.2.85')$$

where a is the maximal elongation, for which $v = 0$. The period is given by

$$T = 4\int_0^a \frac{dx}{\sqrt{h - (\alpha x^2 + \beta x^4)}} = 4\int_0^{\pi/2} \frac{d\theta}{\sqrt{\alpha + \beta a^2 (1 + \sin^2 \theta)}},$$

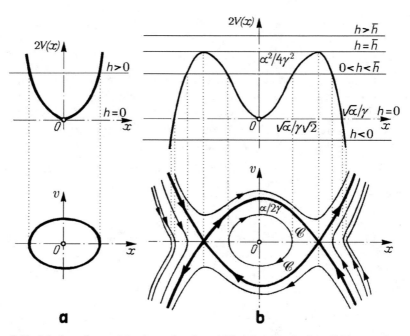

Figure 8.52. Motion of a particle situated at the middle of an elastic thread. Phase trajectories. Diagrams $2V(x)$ vs x and v vs x: case $\beta > 0$ (a); case $\beta < 0$ (b).

where we have made the substitution $x = a\sin\theta$. If $\beta > 0$ (strong spring, case considered above), then the frequency increases (the period decreases) together with the amplitude; the motion is periodic, the phase trajectories being closed for any initial conditions for which $h \geq 0$ (Fig.8.52,a).

Dynamics of the particle in a field of elastic forces

If $\beta = -\gamma^2 < 0$ (soft spring, weak characteristic), then the period increases together with the amplitude; the potential energy has two maxima and a minimum (Fig.8.52,b). For $h = \overline{h} = \alpha^2/4\gamma^2$ one obtains the separation curve $\mathscr{C}$, formed by arcs of parabola

$$P_1(x,v) \equiv 2\gamma v - 2\gamma^2 x^2 + \alpha = 0, \quad P_2(x,v) \equiv 2\gamma v + 2\gamma^2 x^2 - \alpha = 0.$$

For $h > \overline{h}$ the phase trajectories cover the whole phase plane. For $0 < h < \overline{h}$ ($\alpha > \gamma^2 a^2$) one obtains closed phase trajectories, interior to the curve $\mathscr{C}$, and open curves, exterior to that one, while for $h < 0$ ($\alpha < \gamma^2 a^2$) there are only open phase trajectories; the motion is periodic only for the closed curves, in initial conditions for which $P_1(x_0, v_0) > 0$, $P_2(x_0, v_0) < 0$. The point O is a singular point of centre type (stable equilibrium), while the points $(\pm\sqrt{\alpha}/\gamma\sqrt{2}, 0)$ are singular points of saddle type (labile equilibrium). Making $a^2 = \alpha/2\gamma^2$, one obtains

$$T = \frac{4\sqrt{2}}{\sqrt{\alpha}} \int_0^{\pi/2} \sec\theta\, d\theta$$

on the separation curve; we notice that

$$\int \sec\theta\, d\theta = \ln\tan\left(\frac{\pi}{4} + \frac{\theta}{2}\right),$$

so that the particle P reaches a labile position of equilibrium after an infinite time.

The cases considered above have led to equations of the form (8.2.83') and, precisely, of the form

$$\ddot{x} + f(x) = 0, \tag{8.2.83''}$$

corresponding to *the non-damped free vibrations of mechanical systems with non-linear elastic characteristics*. Let us assume that $f(x)$ is an odd function ($f(-x) = -f(x)$), symmetric with respect to the origin, and let us suppose that at the initial moment $t = 0$ the abscissa is x_0 and the velocity vanishes ($v_0 = 0$). In this case, multiplying the equation (8.2.83'') by $\dot{x}$ and integrating, one obtains

$$\frac{\dot{x}^2}{2} + \int_0^x f(\xi)\,d\xi = \int_0^{x_0} f(x)\,dx,$$

a relation which corresponds, in fact, to a conservation law of the mechanical energy. Taking into account that $\dot{x} = dx/dt$ and integrating between the limits $x = 0$ and $x = x_0$, we obtain the period (assuming that $x_0 > 0$, we take the sign minus before the radical, because between the initial position and that of equilibrium the motion takes place in the negative direction of the Ox-axis)

$$T = 4 \int_0^{x_0} \frac{\mathrm{d}x}{\sqrt{2 \int_x^{x_0} f(\xi) \mathrm{d}\xi}}. \tag{8.2.86}$$

In general,

$$t = \int_x^{x_0} \frac{\mathrm{d}\overline{x}}{\sqrt{2 \int_{\overline{x}}^{x_0} f(\xi) \mathrm{d}\xi}}. \tag{8.2.86'}$$

In particular, in the case of harmonic oscillations we have $f(x) = \omega^2 x$ and are led to

$$t = \int_x^{x_0} \frac{\mathrm{d}\overline{x}}{\omega \sqrt{x_0^2 - \overline{x}^2}} = \frac{1}{\omega} \arccos \frac{x}{x_0}.$$

As well, if $f(\theta) = \omega^2 \sin \theta$ (great oscillations of the mathematical pendulum), then the formula (8.2.86) allows to find again the period (7.1.39'). In general, for the equation (8.2.84) (and analogously for the equation (8.2.85)) the formula (8.2.86') leads to elliptic integrals of the first kind, as it was to be expected.

The damped free vibrations of mechanical systems with non-linear elastic characteristics lead to an equation of the form

$$\ddot{x} + f(x) + g(x, \dot{x}) = 0, \tag{8.2.87}$$

while *the parametric vibrations of non-damped mechanical systems* are modelled by an equation of the form

$$\ddot{x} + \omega^2(t) x = 0. \tag{8.2.87'}$$

Sometimes, in case of small non-linearities, we may introduce a small parameter ε, obtaining quasi-linear equations of the form

$$\ddot{x} + \omega_0^2 x + \varepsilon f(x) = 0, \tag{8.2.88}$$
$$\ddot{x} + \omega_0^2 x + \varepsilon [f(x) + g(\dot{x})] = 0, \tag{8.2.88'}$$
$$\ddot{x} + \omega_0^2 x + \varepsilon f(x, \dot{x}) = 0. \tag{8.2.88''}$$

We mention thus *the Van der Pol equation* (for self-excited vibrations)

$$\ddot{x} + x - \lambda(1 - x^2) \dot{x} = 0, \quad \lambda > 0, \tag{8.2.89}$$

Rayleigh's equation (for small non-linear dampings)

$$\ddot{x} + x + \varepsilon(m\dot{x}^3 - n\dot{x}) = 0, \quad m, n > 0, \tag{8.2.90}$$

Froude's equation (for dampings in a turbulent motion of fluids)

Dynamics of the particle in a field of elastic forces

$$\ddot{x} + \gamma x + p\dot{x}^2 + q\dot{x} = 0, \tag{8.2.90'}$$

Duffing's equation (for the response to a harmonic excitation of mechanical systems with non-linear elastic characteristics)

$$\ddot{x} + \alpha x + \beta x^3 + c\dot{x} = q\cos\omega t, \quad c \geq 0, \tag{8.2.91}$$

Mathieu's equation (for a harmonic variation in time of the rigidity)

$$\ddot{x} + \left(\omega_0^2 + \alpha_0^2 \cos\omega t\right)x = 0, \tag{8.2.92}$$

and *Hill's equation* (for the periodic variation in time of the rigidity)

$$\ddot{x} + \left(\omega_0^2 + \sum_{n=1}^{\infty} \alpha_n \cos n\omega t\right)x = 0; \tag{8.2.92'}$$

the study of these equations has put in evidence the most important non-linear phenomena and the corresponding basic methods of solution.

In general, for a *free vibratory motion* is searched a *periodic solution* (if there exists such a solution for the given initial conditions) and a study of its *stability* is made. In case of *forced vibrations*, the form of *the response curves of the mechanical system* (the relation between the amplitude and the frequency of the motion and the corresponding characteristics of the perturbing force) is searched too. The non-linear vibrations are non-isochronous, because the period T depends on the amplitude. As well, besides the problem of *static stability*, the problem of *dynamic stability* must be also considered. We have seen that the response of a damped linear system on which acts a sinusoidal perturbing force of pulsation ω is a harmonic vibration, having the same pulsation; in case of a non-linear system appear pulsations $n\omega$, $n \in \mathbb{N}$, too, called also *superharmonics* (multiples of the excitation pulsation) or even pulsations ω/n, $n \in \mathbb{N}$, called *subharmonics* (submultiples of the excitation pulsation). In case of the action of two independent perturbing forces upon a non-linear system, one can no more use the principle of superposition of effects; by the superposition of two distinct excitations, appears an interaction between the oscillations which arise, leading to the phenomenon of asynchronous suppression (if each of the independent vibrations is stable, then one of the motions destroys the stability conditions of the other one), to the phenomenon of *asynchronous excitation* (one of the independent vibrations is labile, the other one being stable and creating the conditions that the first one become stable too) or to the phenomenon of *carrying the pulsations* (if the independent vibrations have close pulsations which, in a certain zone of values, synchronize).

2.2.14 Computation methods

The problems which arise in the study of non-linear oscillations are difficult, so that their solution needs specific methods of computation, especially approximate ones. We have thus presented in Chap. 7, Subsec. 2.3.4 topological methods of computation in the phase space, which have been used in the previous subsection. In what follows we pass in review some approximate methods of computation.

Let be the quasi-linear equation (which intervenes in case of the mathematical pendulum)

$$\ddot{x} + \omega_0^2 x + \varepsilon x^3 = 0, \qquad (8.2.93)$$

a particular case of the equation (8.2.88). In *S.P. Timoshenko's method* we assume that, in a first approximation,

$$x(t) = a \cos \omega t \qquad (8.2.93')$$

represents a harmonic motion of pulsation ω, which differs not much from ω_0 and for which the initial conditions

$$x(0) = a, \quad \dot{x}(0) = 0 \qquad (8.2.93'')$$

are verified. We notice that $\omega_0^2 = \omega^2 + \Delta\omega^2$, $\Delta\omega^2 = \omega_0^2 - \omega^2$, and put the condition that the solution (8.2.93') verifies the equation (8.2.93), obtaining thus

$$\ddot{x} + \omega^2 x = -a\Delta\omega^2 \cos\omega t - \varepsilon a^3 \cos^3 \omega t = -\left(a\Delta\omega^2 + \frac{3}{4}\varepsilon a^3\right)\cos\omega t - \frac{1}{4}\varepsilon a^3 \cos 3\omega t.$$

The first term of the second member represents a perturbing element with the same pulsation as the proper one of the terms in the first member; but this term must be equated to zero ($a\Delta\omega^2 + 3\varepsilon a^3/4 = 0$) to can eliminate the phenomenon of resonance. With the same initial conditions, we obtain, in a second approximation, the solution (to the general solution of the homogeneous equation we add a particular solution of the complete equation)

$$x(t) = \left(a - \frac{\varepsilon a^3}{32\omega^2}\right)\cos\omega t + \frac{\varepsilon a^3}{32\omega^2}\cos 3\omega t, \qquad (8.2.93''')$$

with

$$\omega^2 = \omega_0^2 + \frac{3}{4}\varepsilon a^2. \qquad (8.2.94)$$

Analogously, we may use *the Ostrogradskiĭ-Lyapunov method*, based on successive approximations too. We choose thus the solution in the form of a polynomial in ε (we consider only the first three approximations)

$$x = x_0 + \varepsilon x_1 + \varepsilon^2 x_2, \qquad (8.2.95)$$

where $x_0 = x_0(t)$, $x_1 = x_1(t)$, $x_2 = x_2(t)$ are functions of class C^2; as well, we take (we retain only three terms)

$$\omega_0^2 = \omega^2 + C_1\varepsilon + C_2\varepsilon^2, \qquad (8.2.95')$$

where ω^2, C_1, C_2 are non-determinate constants. Replacing in (8.2.93) and equating to zero the coefficients of the powers of ε, we obtain

$$\ddot{x} + \omega^2 x_0 = 0, \quad \ddot{x}_1 + \omega^2 x_1 + C_1 x_0 + x_0^3 = 0,$$
$$\ddot{x}_2 + \omega^2 x_2 + C_1 x_1 + C_2 x_0 + 3 x_0^2 x_1 = 0.$$

The first of these equations leads (for x_0) to the first approximation (8.2.93') with the same initial conditions (8.2.93''); replacing x_0 in the second equation and determining C_1 so that the phenomenon of resonance be eliminate, we find again for $x = x_0 + \varepsilon x_1$ the same approximation (8.2.93'''). Analogously, we may determine x_2, thus the third approximation a.s.o. We notice that the two methods of successive approximations are – as a matter of fact – equivalent, differing only by the modality of approaching the computation. The non-linearity of the considered phenomenon has introduced superharmonics in the equation (8.2.93) (harmonics of third order in the approximation of second order).

In case of the equation (8.2.83'') we introduce the *square deviation* $\Delta = \left[f(x) - \omega^2 x \right]^2$ between the non-linear term and the linear approximation; using the solution in the linear case $x = a\cos\omega t$, we obtain *the mean square deviation* on the duration of a period

$$\mathsf{A}^2 = \frac{1}{T}\int_0^T \left[f(a\cos\omega t) - \omega^2 a\cos\omega t \right]^2 \mathrm{d}t = \frac{1}{2\pi}\int_0^{2\pi} \left[f(a\cos\theta) - \omega^2 a\cos\theta \right]^2 \mathrm{d}\theta,$$

where a change of variable $\omega t = \theta$ has been made. Imposing the condition that A^2 be minimal (*the least squares method*), hence the condition $\partial \mathsf{A}^2 / \partial \omega = 0$, we are led to

$$\omega^2 = \frac{1}{\pi a}\int_0^{2\pi} f(a\cos\theta)\cos\theta \, \mathrm{d}\theta, \qquad (8.2.96)$$

wherefrom we deduce $T = 2\pi/\omega$. In particular, in case of the equation (8.2.93), the formula (8.2.96) leads to the pulsation (8.2.94), so that

$$T = \frac{2\pi}{\sqrt{\omega_0^2 + \frac{3}{4}\varepsilon a^2}} \cong \frac{2\pi}{\omega_0}\left(1 - \frac{3}{8}\varepsilon \frac{a^2}{\omega_0^2}\right), \qquad (8.2.94')$$

in case of a small parameter ε. We notice that for $\varepsilon > 0$ (non-linear vibrations with strong characteristic) the period decreases with the amplitude, while for $\varepsilon < 0$ (non-linear vibrations with weak characteristic) the period increases with the amplitude; if $\varepsilon = 0$, then the vibrations are isochronous. We may thus easily determine, experimentally, the nature of these vibrations.

In *the harmonic balance method*, the condition that the term in $\cos\omega t$ of the development into series of the function $f(a\cos\omega t)$ be identical to the corresponding term $\omega^2 a\cos\omega t$ in the associated linear differential equation is put; thus, the formula (8.2.96) is found again.

Other approximate methods of computation are based on the so-called *variation of constants*, differing after the constants which are varied. We mention, e.g., *Van der Pol's averaging method*. In *the Krylov-Bogolyubov method*, the basis of which is *the first approximation theory*, the constants which vary are *the amplitude* and *the phase of the motion* of the non-linear system. However, the method has numerous variants, e.g. *the Bogolyubov-Mitropol'skiĭ variant*.

We mention *the equivalent linearization method* (*Krylov-Bogolyubov*) too, as well as *Galerkin's method*.

Thus, in case of the quasi-linear equation (8.2.88") we start from the harmonic solution

$$x = a\sin(\omega_0 t - \varphi), \quad \dot{x} = a\omega_0\cos(\omega_0 t - \varphi),$$

and take the amplitude $a = a(t)$ and the phase shift $\varphi = \varphi(t)$ as new unknowns, functions of time, which must be determined. Differentiating with respect to time, we obtain

$$\dot{x} = \dot{a}\sin(\omega_0 t - \varphi) - a\dot{\varphi}\cos(\omega_0 t - \varphi) + a\omega_0\cos(\omega_0 t - \varphi);$$

taking into account the expression of the velocity which corresponds to the linear case, there results the condition

$$\dot{a}\sin(\omega_0 t - \varphi) - a\dot{\varphi}\cos(\omega_0 t - \varphi) = 0.$$

Starting from the same expression of the velocity, we may write

$$\ddot{x} = \omega_0\dot{a}\cos(\omega_0 t - \varphi) + \omega_0 a\dot{\varphi}\sin(\omega_0 t - \varphi) - \omega_0^2 a\sin(\omega_0 t - \varphi),$$

and taking into account the equation (8.2.88"), we get

$$\omega_0\dot{a}\cos(\omega_0 t - \varphi) + \omega_0 a\dot{\varphi}\sin(\omega_0 t - \varphi) = -\varepsilon f(a\sin(\omega_0 t - \varphi), a\omega_0\cos(\omega_0 t - \varphi));$$

using also the condition previously obtained, we get the differential equations of first order (instead the differential equation of second order)

$$\begin{aligned}\dot{a} &= -\frac{\varepsilon}{\omega_0}f(a\sin(\omega_0 t - \varphi), a\omega_0\cos(\omega_0 t - \varphi))\cos(\omega_0 t - \varphi),\\ \dot{\varphi} &= -\frac{\varepsilon}{a\omega_0}f(a\sin(\omega_0 t - \varphi), a\omega_0\cos(\omega_0 t - \varphi))\sin(\omega_0 t - \varphi).\end{aligned} \quad (8.2.97)$$

We notice that the function f has the period $T = 2\pi/\omega_0$ and that $\dot{a}$ and $\dot{\varphi}$ vary slowly in time; these derivatives may be taken equal to their mean values $\dot{a}_{\mathrm{mean}}$ and

Dynamics of the particle in a field of elastic forces

$\dot\varphi_{\text{mean}}$ on a period (considered to be approximately equal to T). In a period, the angle $\psi = \omega_0 t - \varphi$ varies by 2π, so that

$$\dot a_{\text{mean}} = \frac{1}{2\pi}\int_0^{2\pi}\dot a\,d\psi = -\frac{\varepsilon}{2\pi\omega_0}\int_0^{2\pi} f(a\sin\psi, a\omega\cos\psi)\cos\psi\,d\psi,$$
$$\dot\varphi_{\text{mean}} = \frac{1}{2\pi}\int_0^{2\pi}\dot\varphi\,d\psi = -\frac{\varepsilon}{2\pi a\omega_0}\int_0^{2\pi} f(a\sin\psi, a\omega\cos\psi)\sin\psi\,d\psi,$$

(8.2.97')

corresponding to the first terms in the respective development into Fourier series. For a better approximation, one may take terms of higher rank in these expansions.

If $f = f(x)$, $f(-x) = -f(x)$, then we may write

$$\dot a = -\frac{\varepsilon}{2\pi\omega_0}\int_0^{2\pi} f(a\sin\psi)\cos\psi\,d\psi = -\frac{\varepsilon}{2\pi a\omega_0}\int_0^{2\pi} f(x)\,dx = 0,$$
$$a = a_0 = \text{const}.$$

(8.2.97'')

We obtain thus the approximate solution

$$x(t) = a\sin[\omega(a)t - \psi_0],$$

(8.2.97''')

the pulsation depending on the amplitude, as an effect of the non-linearity. Squaring the second relation (8.2.97') and neglecting ε^2, we find the remarkable relation

$$\omega^2(a) = \omega_0^2 + \frac{\varepsilon}{\pi a}\int_0^{2\pi} f(a\sin\psi)\sin\psi\,d\psi.$$

(8.2.97^{iv})

For instance, in case of the equation (8.2.84), we obtain

$$\omega^2 = \frac{g}{l} - \frac{g}{l}\frac{1}{\pi\alpha}\int_0^{2\pi}\frac{1}{6}\alpha^3\sin^4\psi\,d\psi = \frac{g}{l}\left(1 - \frac{\alpha^2}{8}\right),$$

where $\alpha = \theta_{\max}$ is the amplitude of the mathematical pendulum; it results the period

$$T = \frac{2\pi}{\omega} = 2\pi\sqrt{\frac{l}{g}}\frac{1}{\sqrt{1-\frac{\alpha^2}{8}}} \cong 2\pi\sqrt{\frac{l}{g}}\frac{1}{1-\frac{\alpha^2}{16}} \cong 2\pi\sqrt{\frac{l}{g}}\left(1 + \frac{\alpha^2}{16}\right),$$

finding again the approximate formula (7.1.43''') (if we limit ourselves to the first two terms in the considered developments).

Let be also the Van der Pol equation (8.2.89) with the initial conditions $x(0) = a_0$, $\dot x(0) = 0$. Observing that $\omega_0 = 1$, $\varepsilon = -\lambda$, $f(x,\dot x) = (1-x^2)\dot x$ (the unit having the necessary dimension), we start from the harmonic solution $x = a\sin(t-\varphi)$; the initial conditions lead to $x(0) = -a(0)\sin\varphi(0) = a_0$, $\dot x(0) = a(0)\cos\varphi(0) = 0$, so that $a(0) = -a_0$, $\varphi(0) = \pi/2$. The formulae (8.2.97') give the mean values

$$\dot{a}_{\mathrm{mean}} = \frac{\lambda}{2\pi}\int_0^{2\pi}(1-a^2\sin^2\psi)(a\cos\psi)\cos\psi\,\mathrm{d}\psi = \frac{1}{8}\lambda a(4-a^2),$$

$$\dot{\varphi}_{\mathrm{mean}} = \frac{\lambda}{2\pi a}\int_0^{2\pi}(1-a^2\sin^2\psi)(a\cos\psi)\sin\psi\,\mathrm{d}\psi = 0,$$

the system of differential equations of first order (8.2.97) becoming thus

$$\dot{a} = \frac{1}{8}\lambda a(4-a^2), \quad \dot{\varphi} = 0.$$

Separating the variables in the first equation, integrating by decomposition in simple fractions and taking into account the initial conditions for the new variables, we may write

$$x(t) = \frac{a_0}{\sqrt{(a_0/2)^2 + [1-(a_0/2)^2]\mathrm{e}^{-\lambda t}}}\cos t\,; \qquad (8.2.98)$$

this result corresponds to an aperiodic oscillatory motion, hence to a vibration modulated in amplitude, which tends asymptotically to 2 (in Fig.8.53 we assume that $a_0 < 2$).

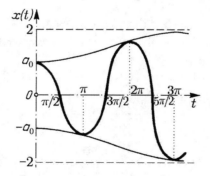

Figure 8.53. Van der Pol equation. Diagram $x(t)$ vs t.

The perturbations method, initiated by Poincaré, allows also an approximate study of the differential equations with a small parameter. Let thus be an autonomous mechanical system, the motion of which is modelled by a non-linear differential equation of the form (8.2.88″) for which $f(0,0) = 0$, with the initial conditions (8.2.93″). By the change of variable $\omega t = \tau$, where ω is a unknown pulsation, one obtains an equation of the form

$$\omega^2\frac{\mathrm{d}^2 x}{\mathrm{d}\tau^2} + \omega_0^2 x + \varepsilon\varphi\left(x, \frac{\mathrm{d}x}{\mathrm{d}\tau}, \omega^2\right) = 0, \qquad (8.2.99)$$

where φ is a given non-linear function; the initial conditions become $x = 0$, $\mathrm{d}x/\mathrm{d}\tau = 0$ for $\tau = 0$. In the perturbations method one considers both for the solution $x(\tau)$ and for the square of the unknown pulsation ω, developments into power series after the small parameter ε, i.e.

$$x(\tau) = x_0(\tau) + \varepsilon x_1(\tau) + \varepsilon^2 x_2(\tau) + \ldots, \qquad (8.2.100)$$

$$\omega^2 = \omega_0^2 + \varepsilon \omega_1^2 + \varepsilon^2 \omega_2^2 + \ldots. \qquad (8.2.100')$$

The initial conditions are verified if $x_0 = a$, $x_1 = x_2 = \ldots = 0$, $\mathrm{d}x_0/\mathrm{d}\tau = \mathrm{d}x_1/\mathrm{d}\tau = \mathrm{d}x_2/\mathrm{d}\tau = \ldots = 0$. Replacing in the equation (8.2.99), developing the function φ (which we assume to be analytical) into a Taylor series, arranging the terms after the powers of the small parameter ε, and equating to zero the coefficients of those powers (assuming successively, that one may neglect the higher powers of ε with respect to the lower ones), we get the equations

$$\omega_0^2 \frac{\mathrm{d}^2 x_0}{\mathrm{d}\tau^2} + \omega_0^2 x_0 = 0,$$

$$\omega_0^2 \frac{\mathrm{d}^2 x_1}{\mathrm{d}\tau^2} + \omega_0^2 x_1 = -\omega_1^2 \frac{\mathrm{d}^2 x_0}{\mathrm{d}\tau^2} - \varphi\left(x_0, \frac{\mathrm{d}x_0}{\mathrm{d}\tau}, \omega_0^2\right),$$

$$\omega_0^2 \frac{\mathrm{d}^2 x_2}{\mathrm{d}\tau^2} + \omega_0^2 x_2 = -\omega_1^2 \frac{\mathrm{d}^2 x_1}{\mathrm{d}\tau^2} - \omega_2^2 \frac{\mathrm{d}^2 x_0}{\mathrm{d}\tau^2} \qquad (8.2.100'')$$

$$- x_1 \left(\frac{\partial \varphi}{\partial x}\right)_0 - \frac{\mathrm{d}x_1}{\mathrm{d}\tau}\left(\frac{\partial \varphi}{\partial \dot{x}}\right)_0 - \omega_1^2 \left(\frac{\partial \varphi}{\partial \omega^2}\right)_0,$$

. . . .

In the first approximation, the small parameter ε is neglected; the first equation (8.2.100'') leads to the solution

$$x_0 = a \cos \tau = a \cos \omega t = a \cos \omega_0 t, \qquad (8.2.101)$$

which satisfies the initial conditions. In the second approximation, $\varepsilon^2, \varepsilon^3, \ldots$ are neglected with respect to ε, hence one takes $x(\tau) = x_0(\tau) + \varepsilon x_1(\tau)$, $\omega^2 = \omega_0^2 + \varepsilon \omega_1^2$. Replacing the solution (8.2.101) in the second equation (8.2.100''), it results

$$\omega_0^2 \left(\frac{\mathrm{d}^2 x_1}{\mathrm{d}\tau^2} + x_1\right) = a\omega_1^2 \cos\tau - \varphi\left(a \cos\tau, -a \sin\tau, \omega_0^2\right),$$

with the initial conditions $x_1 = 0$, $\mathrm{d}x_1/\mathrm{d}\tau = 0$ for $\tau = 0$. In the second member of the differential equation appear terms in $\cos\tau$ and $\sin\tau$ (secular terms), which lead to non-periodic particular solutions of the form $\tau\cos\tau$ and $\tau\sin\tau$. Because we search periodic solutions, we equate to zero the coefficients of those terms, obtaining supplementary conditions; we get thus ω_1 and may then integrate the differential equation. The procedure can be applied further, obtaining thus approximations of higher order (a solution $x(\tau)$ in the form of an expansion into a Fourier series).

In the particular case of the equation (8.2.93), the approximation of second order is given by

$$\frac{d^2 x_1}{d\tau^2} + x_1 = a \frac{\omega_1^2}{\omega_0^2} \cos \tau - \frac{a^3}{\omega_0^2} \cos^3 \tau = a \left[\left(\frac{\omega_1^2}{\omega_0^2} - \frac{3}{4} \frac{a^2}{\omega_0^2} \right) \cos \tau - \frac{a^2}{4\omega_0^2} \cos 3\tau \right],$$

wherefrom, taking into account the initial conditions, it results

$$x_1(\tau) = \frac{a^3}{32\omega_0^2} (\cos 3\tau - \cos \tau) + a \left(\frac{\omega_1^2}{\omega_0^2} - \frac{3}{4} \frac{a^2}{\omega_0^2} \right) \frac{\pi}{2} \sin \tau.$$

Equating to zero the secular term, we obtain $\omega_1^2 = 3a^2/4$, so that, in the second approximation, we have

$$x(t) = a \left[\left(1 - \varepsilon \frac{a^2}{32\omega_0^2} \right) \cos \omega t + \varepsilon \frac{a^2}{32\omega_0^2} \cos 3\omega t \right], \quad (8.2.102)$$

with the pulsation (8.2.94); we find again the period (8.2.94'). The dependence of the pulsation on the amplitude is thus put into evidence.

The perturbations method may be applied also in case of non-autonomous systems, the motion of which is modelled by differential equations of the form

$$\ddot{x} + \omega_0^2 x + \varepsilon f(x, \dot{x}; t) = 0, \quad (8.2.103)$$

where the function f is periodic with respect to t and has a known pulsation. By a phase shifting of the periodic solution with respect to the perturbing force, we may introduce the phase shift

$$\delta = \delta_0 + \varepsilon \delta_1 + \varepsilon^2 \delta_2 + \dots, \quad (8.2.104)$$

the computation procedure being, further, similar to that above.

Among the computation methods which may be applied, we mention also *the graphic methods (the isoclinic lines method, the delta method, the graphic methods for non-autonomous systems* etc.).

Chapter 9

NEWTONIAN THEORY OF UNIVERSAL ATTRACTION

In the preceding chapter, we have considered the action of an elastic force of attraction upon a particle, case in which, after Bertrand's theorem, the orbit is a closed curve. We will now study the action of forces of Newtonian attraction which, on the basis of the same theorem, lead to analogous trajectories; we pay a particular attention to those forces, because they are the most important ones of mechanical nature which are exerted by a body upon another one. The mathematical modelling of the universal forces of attraction corresponds to the classical model of mechanics (as it has been conceived by Newton) and to its gorgeous verification by astronomical observations (Kepler's laws). The results thus obtained will be applied to the study of planets' motions, to the problems of the artificial Earth satellites and of the interplanetary vehicles, to the motion at the atomic level etc.

1. Newtonian model of universal attraction

After considerations concerning the classical model of universal attraction, a study of the Newtonian potential is made; the modelling as particles of celestial bodies is just justified.

1.1 Principle of universal attraction

Starting from Kepler's laws, considered as laws of experimental nature (obtained by astronomical observations), we deduce – in what follows – the law of Newtonian attraction; we can make thus the connection with the gravitational fields.

1.1.1 Law of Newtonian attraction

Starting from the astronomical observations of his predecessors (especially those of the Dane Tycho Brahe at his observatory on the Ven isle, between Denmark and Sweden, and then as astronomer at the Imperial Court in Prague), Johann Kepler enounced three laws which are modelling the motion of planets in the solar system. Thus, the planets (modelled as particles) describe ellipses with respect to the Sun (considered to be situated at one of the foci), the motion being governed by the law of areas; the ratio of the cube of the semi-major axis to the square of the revolution time T is the same for all planets. However, Kepler tried to obtain a synthesis of these laws, without obtaining a final result (in an Aristotelian conception, he thought that the force is directed along the tangent to the trajectory). But Newton, using Kepler's laws, succeeded to set up the mathematical model of classical mechanics, deducing the expression of the force of universal attraction too; these results (taking into account the modality to obtain them, they may be considered as deriving from Kepler's laws, the latter ones being a mathematical model of mechanics) represent, in fact, the most important contributions of Newton to the development of mechanics. In contradistinction to Kepler's model, the Newtonian one has a general character and may be applied to all bodies of the real world.

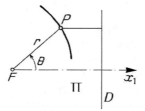

Figure 9.1. Conics.

To can express mathematically Kepler's laws, it is useful to introduce the equation of an ellipse and – in general – the equation of a conic in polar co-ordinates. To do this, we remember that a *conic* represents the locus of the points P of a plane Π for which the ratio of the distances to a fixed point $P \in \Pi$ (called *focus*) and to a fixed straight line $D \in \Pi$ (called *directrix*), respectively, is constant (the respective constant, denoted by e, is called *eccentricity*) (Fig.9.1); if $0 \le e \le 1$, then the conic is an *ellipse* ($e = 0$ corresponds to a circle), if $e = 1$, then it is a *parabola*, while if $e > 1$, then it is a *hyperbola*. The equation of the conic with respect to the focus F is written in the form

$$r = \frac{p}{1 + e\cos\theta}, \qquad (9.1.1)$$

where $p > 0$ is *the conic parameter* (*the semilatus rectum*).

Observing that the Sun is at the focus F, while the planet of mass m is at the point P, the forces which represent the action of one of the bodies upon the other one are internal forces in the system formed by the two bodies, modelled as particles. Assuming that the point F is fixed (we consider the motion of the planet relative to this point), the force **F** which acts upon the planet is a central force. Because the trajectory of the planet is an ellipse, it results that the point F represents a position of stable equilibrium, the force **F** being directed towards this point. The law of areas leads also to the conclusion that the force **F** is a central one. Replacing r given by (9.1.1) in Binet's formula (8.1.8), we get $F = -mC^2/pr^2$; hence, the magnitude F of the force is inverse-square to the distance between the Sun and the planet. As we have seen in Chap. 5, Subsec. 1.1.4, $\Omega = \dot{A}$, where the area A is the measure of the surface described by the radius vector; if that radius describes the whole ellipse, then, by integrating on the interval $[0,T]$, where T is *the revolution time* of the planet (the time in which the whole ellipse is described), we obtain $2\pi ab = CT$, where (6.1.57″) has been taken into account (the area of the ellipse of semiaxes a and b is πab).

Newtonian theory of universal attraction

From (9.1.1), one obtains (for $\theta = 0$ and $\theta = \pi$, respectively)

$$r_{\min} = \frac{p}{1+e}, \quad r_{\max} = \frac{p}{|1-e|}, \tag{9.1.1'}$$

corresponding to the apsidal points, called *pericentres* and *apocentres*, respectively. It results $r_{\max} \pm r_{\min} = 2a$, as we have to do with an ellipse or a hyperbola, respectively, $a > 0$ being *the semi-major axis*; in the first case, we have $r > 0$, so that $r_{\min} = \overline{FA}$, $r_{\max} = \overline{FA'}$ (Fig.9.2,a), while in the second case we have $r \geqslant 0$ as it corresponds to the branch of the hyperbola the focus of which is taken as origin or to the other branch, so that $r_{\min} = \overline{F'A'}$ and $r_{\max} = \overline{F'A}$ (Fig.9.2,b). We also notice that

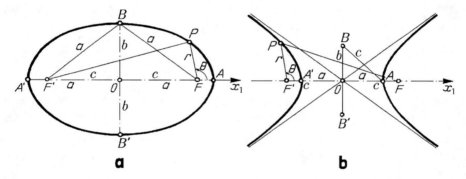

Figure 9.2. Conics. Ellipse (a). Hyperbola (b).

the ellipse is the locus of the points for which $\overline{PF} + \overline{PF'} = 2a = \text{const}$ (so that $\overline{BF} = \overline{BF'} = a$), while the hyperbola is the locus of the points for which $\overline{PF} - \overline{PF'} = 2a$. If $\overline{FO} = \overline{F'O} = c > 0$ is *the focal distance* to the centre, it results $p/(1+e) = |a-c|$, $p/|1-e| = a+c$, wherefrom $pe/|1-e^2| = c$, $p/|1-e^2| = a$; for both conics we get thus the eccentricity

$$e = \frac{c}{a} \tag{9.1.2}$$

and the parameter $p = a|1-e^2| = |a^2 - c^2|/a$. Observing that

$$a^2 = c^2 \pm b^2, \tag{9.1.2'}$$

where $b > 0$ is *the semi-minor* axis (the semi-major axis of *the conjugate hyperbola*, in case of a hyperbola), we may write

$$p = \frac{b^2}{a}. \tag{9.1.2''}$$

In this case, $C^2/p = 4\pi^2 a^2 b^2 / pT^2 = 4\pi^2 a^3 / T^2$, so that

$$F = -4\pi^2 \frac{ma^3}{r^2 T^2};\qquad (9.1.3)$$

taking into account the third law of Kepler (see Subsec. 2.1.4), there results

$$F = -\mu \frac{m}{r^2},\qquad (9.1.3')$$

where $\mu = \mu(M) > 0$, M being the Sun mass (μ depends only on the Sun, because it is independent of the particular planet considered). Newton assumed that this independence is linear, so that $\mu(M) = fM$, the force exerted by the Sun upon the planet of mass m, situated at the distance r, being thus given by

$$F = -f\frac{mM}{r^2}\qquad (9.1.3'')$$

and corresponding to the formula (1.1.84'). Indeed, if the action of the Sun upon a planet P_j of mass m_j is a force of modulus $\mu m_j / r^2$, $\mu = \mu(M)$, and if the planet P_j acts upon the Sun with a force of modulus $\mu_j M / r^2$, $\mu_j = \mu_j(m_j)$, then, corresponding to the principle of action and reaction, we may write $\mu m_j / r^2 = \mu_j M / r^2$, $j = 1,2,...$, wherefrom $\mu/M = \mu_1/m_1 = \mu_2/m_2 = ...$; the common ratio is just *the constant $f > 0$ of universal attraction*, which leads to the expression of *the universal (Newtonian) law of attraction*. This model of the force of attraction of the bodies in Universe has been extended by Newton for all the bodies, immaterial of the magnitudes of their masses, in the form (corresponding to the formula (1.1.84))

$$F = -f\frac{m_1 m_2}{r^2},\qquad (9.1.4)$$

where m_1 and m_2 are the masses of two bodies situated at the distance r one of the other (the distance between their centres of mass, the bodies being modelled as particles). As we have seen in Chap. 1, Subsec. 1.1.12, these forces are conservative. One obtains thus the universal (*Newtonian*) *law of attraction*.

The coefficient f may be obtained experimentally in a particular case and is given, in the CGS – system, by $f = 6.6732 \cdot 10^{-8} \cong 1/3871^2$ cm/g·s². The first determination in a laboratory has been made a century later by Cavendish, in 1798, with the aid of a *balance of torsion*. The apparatus is formed by a thread AB suspended at A and supporting at the end B a horizontal bar at the ends of which are two small spheres P_1 and P_2 ($\overline{BP_1} = \overline{BP_2}$) of masses m_1 and m_2, respectively. If we put in the vicinity of these spheres a fork with two spheres P_1' and P_2' of masses $m_1' \gg m_1$ and

Newtonian theory of universal attraction 547

$m_2' \gg m_2$, respectively (usually, $m_1 = m_2$ and $m_1' = m_2'$), then, due to the forces of Newtonian attraction which arise, the sphere P_1 is attracted by the sphere P_1' and the sphere P_2 is attracted by the sphere P_2'; thus, a circular motion of angle α takes place (Fig.9.3). The thread AB is subjected to torsion, wherefrom the denomination of the apparatus. Because the angle α is proportional to the couple formed by the Newtonian forces of attraction applied upon the particles P_1 and P_2, one can calculate the coefficient f.

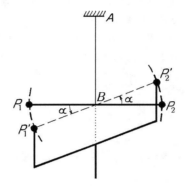

Figure 9.3. Balance of torsion.

Building up a mathematical model of universal attraction, in the frame of a classical model of mechanics, Newton opened a large outlook to the development of science; among others, this model is the basis of celestial mechanics, where the fundamental problem is that of n particles (e.g., The Sun, the Earth and the Moon – the problem of the three bodies, modelled as particles). Starting from the Newtonian theory of universal attraction, one could show, only by computation, that the Earth is an oblate *spheroid* (it is oblate at the poles), one could discover, only by computation too, new planets (thus, Leverrier discovered, in 1846, the planet Neptune, and – recently – a tenth planet has been intuited), one could predict the trajectories of the interplanetary vehicles (stating thus the basis of the theory of cosmic motions) etc.

In the following section we show that all celestial bodies, considered to be of quasi-spherical form, may be modelled as particles situated at their centres of mass and having a mass equal to that of the respective body.

1.1.2 Law of universal gravitation

We have seen in Chap. 1, Subsec. 1.1.12 that the gravity forces, due to the presence of a gravitational field, may be considered as particular cases of forces of Newtonian attraction (which, by extension, are called *forces of universal gravitation* too); one obtains thus the *law of universal gravitation*. Hence, it results the remarkable relation (1.1.85), which links the constant f to the mass M of the Earth, considered as a sphere of radius R, and to the gravity acceleration g.

The potential fmM/R (m is the mass of a body situated at the distance H from the Earth surface, along the local vertical) must be replaced by

$$f\frac{mM}{R+H} = f\frac{mM}{R}\left(1+\frac{H}{R}\right)^{-1} \cong f\frac{mM}{R}\left(1-\frac{H}{R}\right)$$
$$= f\frac{mM}{R} - f\frac{mM}{R^2}H \cong f\frac{mM}{R} - mgH,$$

where we took into account (1.1.85). This potential differs from that known (see formula (1.1.83)) by the additive constant term fmM/R. We notice that in case of the potential $fmM/(R+H)$ the zero level is at infinity, while in case of the potential $-mgH$ the zero level is at the Earth surface.

The motion of a particle in a *gravitational field* leads to the equation $m_i \mathbf{a} = m_g \mathbf{g}$, where, due to the equality between the inertial mass and the gravitational one (the relation (1.1.24)), these ones disappear from computation. In case of a motion in an *electric field* $\mathbf{E}$, which acts upon the particle with the force $e\mathbf{E}$, where e is *the electric charge*, the equation of motion is of the form

$$m_i \mathbf{a} = e\mathbf{E} ; \qquad (9.1.5)$$

in this case, the inertial mass does no more disappear, being different from the electric charge e.

Let us consider the motion of the Moon around the Earth; a force of attraction of the form (9.1.3), (9.1.3'), for which $\mu = 4\pi^2 a^3/T^2$, arises. A particle of mass equal to unity at the Earth surface is attracted to that one by a force equal to $1 \cdot g = 1 \cdot \mu/R^2$, being a constant which depends on the mass of the Earth of radius R. We obtain thus the relation

$$g = 4\pi^2 \frac{a^3}{R^2 T^2}, \qquad (9.1.6)$$

which allows to calculate the gravity acceleration. The trajectory of the Moon is quasi-circular, with $a \cong 60R \cong 384\,000$ km; it results $g \cong 4\pi^2 60^3 R/T^2$. Because $R \cong 6370$ km and $2\pi R \cong 40\,000$ km, the time of revolution of the Moon being 27 days 7 hours 43 minutes $= 39\,343 \cdot 60$ s, Newton has obtained $g \cong 9.8$ m/s^2, in a good concordance with the result previously obtained by Galileo. Taking into account what was shown above, it results that the ratio of the acceleration γ of a free fall of the Moon on the Earth to the gravity acceleration g is equal to the ratio $(R/a)^2 \cong 1/60^2$; taking $g = 9.81$ m/s^2, we get $\gamma = 2.72 \cdot 10^{-3}$ m/s^2, result which corresponds to that obtained in the study of Moon's trajectory. Thus, we get a brilliant confirmation for the coincidence of the two forces (the gravitational attraction and the weight at the Earth surface).

The relation (1.1.85) allows to determine the mass of the Earth in the form $M \cong gR^2/f$; observing that $M = 4\pi\mu R^3/3$, where μ is the mean unit mass, it results

$$\mu = \frac{3}{4\pi} \frac{g}{fR}. \tag{9.1.6'}$$

We get thus $\mu = 5.51$ g/cm^3. This density is much greater than that of the superior spherical strata of the Earth; we may thus conclude that the density is much greater towards the centre of the Earth.

As we have seen in Chap. 1, Subsec. 2.1.3, the mass can be taken as derived unit, which is expressed by means of units of length and of time as basic units, so that the universal constant f becomes equal to unity. We introduce thus *the natural hour* (which is approx. equal to 3871 seconds of mean time); this is the unit of time which must be adopted so that, for an arbitrary unit of length and a unit of mass corresponding to a cube of di1stillated water at $4°C$, equal to unity, to obtain $f = 1$.

1.2 Theory of Newtonian potential

To justify the modelling of celestial bodies as particles, we introduce – in what follows – the Newtonian potential and put into evidence some of its properties; we consider especially the surface and the volume potentials of the homogeneous sphere, as well as the potential of the spherical stratum. We give some results concerning the potential of the terrestrial spheroid too.

1.2.1 Newtonian potential

Corresponding to the universal gravitation law, a particle Q of position vector $\boldsymbol{\xi}$ and mass m acts upon a particle P of position vector $\mathbf{r}$ and mass equal to unity with a force of attraction given by

$$\mathbf{F} = -\frac{fm}{R^2} \operatorname{vers} \overrightarrow{QP} = \operatorname{grad} \frac{fm}{R} = f \operatorname{grad} U, \quad R = |\mathbf{r} - \boldsymbol{\xi}|, \tag{9.1.7}$$

where the potential (see the formula (1.1.84'))

$$U = \frac{m}{R} \tag{9.1.7'}$$

has been introduced (we have neglected an additive constant); thus, the projection of the force $\mathbf{F}$ on a direction of unit vector $\mathbf{n}$ is given by $f \operatorname{grad} U \cdot \mathbf{n} = f \partial U / \partial n$.

In case of several centres of attraction Q_j of position vectors $\boldsymbol{\xi}_j$ and masses m_j, $j = 1, 2, ..., n$, we obtain (in conformity with the principle of the parallelogram of forces)

$$\mathbf{F} = f \operatorname{grad} U, \quad U = \sum_{j=1}^{n} \frac{m_j}{R_j}, \quad R_j = |\mathbf{r} - \boldsymbol{\xi}_j|. \tag{9.1.8}$$

Let be, in general, a mechanical system $\mathscr{S}$ of geometric support Ω, at a finite distance; the potential of the force of attraction $\mathbf{F}$ exerted by $\mathscr{S}$ on the particle P is given by the Stieltjes integral

$$U(\mathbf{r}) = \int_\Omega \frac{\mathrm{d}m}{R}, \quad R = |\mathbf{r} - \boldsymbol{\xi}|, \tag{9.1.9}$$

where $m = m(\boldsymbol{\xi})$ is a distribution. Introducing the unit mass (1.1.71), (1.1.71"), we obtain the relation (9.1.8) for a discrete mechanical system of n particles Q_j of position vectors $\boldsymbol{\xi}_j$ and masses m_j, $j = 1, 2, \ldots, n$, or

$$U(\mathbf{r}) = \iiint_V \frac{\mu(\boldsymbol{\xi})}{R} \mathrm{d}V, \quad R = |\mathbf{r} - \boldsymbol{\xi}|, \tag{9.1.8'}$$

in case of a continuous mechanical system of unit mass $\mu(\boldsymbol{\xi}) \in C^1(V + S)$ (S is the frontier of the domain of volume V and verifies conditions of Lyapunov type).

If the continuum is homogeneous, then it results

$$U(\mathbf{r}) = \mu \iiint_V \frac{\mathrm{d}V}{R}, \quad R = |\mathbf{r} - \boldsymbol{\xi}|. \tag{9.1.8"}$$

We obtain thus the general form of *the Newtonian potential*, that is *the volume potential* given by the formula (A.2.86). In case of a two-dimensional mechanical system $\mathscr{S}$, we replace the volume integral by a surface one (it results a simple stratum potential of the form (A.2.87)), using a superficial unit mass; if the mechanical system is plane, then the integral is a double one. As well, in case of a one-dimensional mechanical system $\mathscr{S}$ we replace the volume integral by a curvilinear one, introducing a linear unit mass.

One can put in evidence following properties:
i) $U(\mathbf{r})$ is a continuous function in the whole space and vanishes at infinity.
ii) The derivatives of first order of $U(\mathbf{r})$ (hence, the vector field $\operatorname{grad} U(\mathbf{r})$ too) are continuous functions in the whole space and vanish at infinity; they are calculated by differentiation under the integral sign.
iii) The derivatives of second order of $U(\mathbf{r})$ have jumps by crossing the surface S; in particular, (one can make the connection with the formula (A.2.85))

$$\Delta U(\mathbf{r}) = \begin{cases} -4\pi\mu(\boldsymbol{\xi}) & \text{in the interior of } V, \\ 0 & \text{in the exterior of } S. \end{cases} \tag{9.1.10}$$

iv) The behaviour at infinity of the potential is given by

$$U(\mathbf{r}) = \frac{m}{r} + \frac{m}{r^2} \boldsymbol{\rho} \cdot \operatorname{vers} \mathbf{r} + \mathcal{O}\left(\frac{1}{r^3}\right). \tag{9.1.11}$$

For the first two properties we assume that $\mu(\boldsymbol{\xi})$ is a bounded and integrable on V function, while for the property iii) we consider that $\mu(\boldsymbol{\xi})$ is a function differentiable on V, their derivatives being bounded and differentiable functions. The first three properties may easily be proved for particles P in the exterior of the system $\mathscr{S}$. If the particle P belongs to the interior of the system $\mathscr{S}$, then that one becomes a singular point ($\mathbf{r} = \boldsymbol{\xi}$), and we isolate it by a sphere containing it; the volume integral must be calculated for the domain of volume V from which one subtracts the interior of that sphere, while the surface integral (which appears by using a formula of Gauss-Ostrogradskiĭ type) must be calculated for the sphere too. The integrands have no singularities in this case, no matter how small is the radius of the sphere; then, this radius is equated to zero.

Concerning the property iv), we denote $\xi = |\boldsymbol{\xi}| = |\overrightarrow{OQ}|$, $\varphi = \sphericalangle(\mathbf{r},\boldsymbol{\xi})$. From the triangle OPQ, it results (assuming that r is sufficiently great, ξ being bounded, we may use a binomial expansion or a Maclaurin series)

$$\frac{1}{|\overrightarrow{PQ}|} = \frac{1}{R} = \frac{1}{|\mathbf{r}-\boldsymbol{\xi}|} = \frac{1}{\sqrt{r^2 + \xi^2 - 2r\xi\cos\psi}} = \frac{1}{r}f\left(\frac{\xi}{r}\right)$$

$$= \frac{1}{r}\left[1 + \left(\frac{\xi}{r}\right)^2 - 2\frac{\xi}{r}\cos\psi\right]^{-1/2} = \frac{1}{r}\sum_{n=0}^{\infty}\frac{1}{n!}f^{(n)}(0) = \frac{1}{r}\sum_{n=0}^{\infty}\left(\frac{\xi}{r}\right)^n P_n(\cos\psi),$$

where $P_n(\cos\psi)$ are Legendre's polynomials; we have

$$P_n(\cos\psi) = \frac{1}{n!}f^{(n)}(0), \quad f(0) = 1, \quad f'(0) = \cos\psi, \quad f''(0) = 3\cos^2\psi - 1, \ldots$$

Replacing in (9.1.8'), we get

$$U(\mathbf{r}) = \frac{1}{r}\iiint_V \mu(\boldsymbol{\xi})\mathrm{d}V + \frac{1}{r^2}\iiint_V \xi\cos\psi\mu(\boldsymbol{\xi})\mathrm{d}V + \mathscr{O}\left(\frac{1}{r^3}\right),$$

so that $U(\mathbf{r})$ tends to zero as $1/r$ at infinity; observing that

$$m = \iiint_V \mu(\boldsymbol{\xi})\mathrm{d}V, \quad \boldsymbol{\rho} = \overrightarrow{OC} = \frac{1}{m}\iiint_V \boldsymbol{\xi}\mu(\boldsymbol{\xi})\mathrm{d}V,$$

where C is the centre of mass of the system $\mathscr{S}$, we obtain the formula (9.1.11).

If, in particular, we choose the origin at the point C, then we may write

$$U(\mathbf{r}) = U(r) = \frac{m}{r} + \mathscr{O}\left(\frac{1}{r^3}\right). \tag{9.1.12}$$

We may thus state that *one obtains a sufficient good approximation by replacing the mechanical system $\mathscr{S}$ of attraction by its centre of mass C, at which we assume to be concentrated the whole mass m of the system* (especially if the distance from the centre of attraction C to the particle P is sufficiently great). We show in the next subsection that this result is rigorous if the mechanical system $\mathscr{S}$ is with spherical symmetry.

From the formula (9.1.12) too, it results that the force obtained by applying the operator gradient *vanishes at infinity as* $1/r^2$. So, in the study of the solar system (the motion of the planets around the Sun) one can neglect the action of other celestial bodies, those ones being situated at distances practically infinite with respect to the distances of the planets to the Sun or between them.

The properties i)-iv) characterize the Newtonian potential. Indeed, if there exists another potential $U'(\mathbf{r})$ which has these properties, then $U(\mathbf{r}) - U'(\mathbf{r})$ is a harmonic function in the whole space, vanishing at infinity; from the theorem of maximum of harmonic functions it results that the two potentials differ by a constant, which – obviously – is equal to zero.

1.2.2 Potential of simple stratum of a homogeneous sphere

In case of a homogeneous sphere, *the potential of simple stratum* is given by the surface integral

$$U(\mathbf{r}) = \mu \iint_S \frac{\mathrm{d}S}{R}, \quad R = |\mathbf{r} - \boldsymbol{\xi}|, \qquad (9.1.13)$$

where the equation of the sphere S of radius ξ_0 reads $\xi = \xi_0$. Because of spherical symmetry, the potential depends only on the distance from the attracted particle to the centre of the sphere, so that $U = U(r)$, $r = |\mathbf{r}| = |\overrightarrow{OP}|$. Choosing the point P on the Ox_3-axis and using spherical co-ordinates (colatitude θ and longitude φ) for $Q \in \mathscr{S}$, we may write ($R^2 = r^2 + \xi_0^2 - 2r\xi_0 \cos\theta$)

$$U(\mathbf{r}) = \mu \iint_S \frac{\xi_0^2 \sin\theta \, \mathrm{d}\theta \, \mathrm{d}\varphi}{\sqrt{r^2 + \xi_0^2 - 2r\xi_0\cos\theta}} = \mu \int_0^{2\pi} \mathrm{d}\varphi \int_0^{\pi} \frac{\sin\theta \, \mathrm{d}\theta}{\sqrt{r^2 + \xi_0^2 - 2r\xi_0\cos\theta}}$$

$$= 2\pi\mu\xi_0^2 \left.\frac{\sqrt{r^2 + \xi_0^2 - 2r\xi_0\cos\theta}}{r\xi_0}\right|_0^{\pi} = \frac{2\pi\mu\xi_0}{r}(|r+\xi_0| - |r-\xi_0|)$$

$$= \begin{cases} -4\pi\mu\dfrac{\xi_0^2}{r} & \text{for } r \leq -\xi_0, \\ 4\pi\mu\xi_0 & \text{for } -\xi_0 \leq r \leq \xi_0, \\ 4\pi\mu\dfrac{\xi_0^2}{r} & \text{for } r \geq \xi_0; \end{cases}$$

observing that the total superficial mass is $m = 4\pi\mu\xi_0^2$, it results

$$U(\mathbf{r}) = U(r) = \begin{cases} \dfrac{m}{r} & \text{for } P \text{ exterior to the sphere } S, \\ \dfrac{m}{\xi_0} = \text{const} & \text{otherwise.} \end{cases} \qquad (9.1.14)$$

Newtonian theory of universal attraction

We can thus state that a particle situated outside the homogeneous sphere S is attracted towards its centre as if the whole superficial mass of attraction would be concentrated at this point (corresponding to the property iv) in the preceding subsection); if the particle is situated on the sphere or in its interior, no force of attraction is acting upon it. The potential $U(\mathbf{r})$ is harmonic ($\Delta U = 0$) in the whole space and vanishes at infinity.

We notice also that $\partial U / \partial n$ has a jump by passing through the surface S (the normal derivative is equal to $-4\pi\mu$ if P tends to S from the exterior and is equal to zero for P tending to S from the interior), although $U(\mathbf{r})$ is a continuous function in the whole space.

1.2.3 Volume potential of a homogeneous sphere

The volume potential of the homogeneous sphere $S(0, \xi_0)$ is given by (9.1.8"); with the same observations as at the preceding subsection, we may write (we use spherical co-ordinates ξ, θ, φ too; for P at the interior of the sphere S we make a separate integration, as $0 \leq \xi \leq r$ or as $r \leq \xi \leq \xi_0$)

$$U(\mathbf{r}) = \mu \iiint_V \frac{\xi^2 \sin\theta \, d\xi \, d\theta \, d\varphi}{\sqrt{r^2 + \xi^2 - 2r\xi\cos\theta}} = \mu \int_0^{2\pi} d\varphi \int_0^{\xi_0} \xi^2 d\xi \int_0^{\pi} \frac{\sin\theta \, d\theta}{\sqrt{r^2 + \xi^2 - 2r\xi\cos\theta}}$$

$$= \frac{2\pi\mu}{r} \int_0^{\xi_0} \xi(|r+\xi| - |r-\xi|) d\xi$$

$$= \begin{cases} 2\pi\mu\left(\xi_0^2 - \dfrac{r^2}{3}\right) & \text{for } P \text{ interior to the sphere } S, \\[4pt] \dfrac{4\pi\mu}{3} \dfrac{\xi_0^3}{r} & \text{otherwise.} \end{cases}$$

Observing that the total mass is $m = 4\pi\mu\xi_0^3/3$, it results

$$U(\mathbf{r}) = U(r) = \begin{cases} \dfrac{m}{2\xi_0}\left[3 - \left(\dfrac{r}{\xi_0}\right)^2\right] & \text{for } P \text{ interior to the sphere } S, \\[6pt] \dfrac{m}{r} & \text{otherwise.} \end{cases} \qquad (9.1.15)$$

As in case of the potential of simple stratum, a particle situated outside the homogeneous sphere S is attracted by its centre as if the whole mass would be concentrated at this point (Newton's *modelling of the celestial bodies as particles* is thus *justified*). The potential $U(\mathbf{r})$ and its derivatives of first order are continuous functions in the whole space; the derivatives of second order of this potential have discontinuities by crossing the surface S and ΔU is given by (9.1.10). For numerical computations we may take $\xi_0 = 6.37827 \cdot 10^6$ m.

In case of a homogeneous spherical stratum of mass m, contained between two spheres S_e and S_i of centre O and radii ξ_e and $\xi_i < \xi_e$, respectively, we can use the parallelogram principle (independence of the action of forces), obtaining thus

$$U(r) = \begin{cases} 2\pi\mu(\xi_e^2 - \xi_i^2) & \text{for } P \text{ interior to the sphere } S_i, \\ \dfrac{4\pi\mu}{3r}(\xi_e^3 - \xi_i^3) = \dfrac{m}{r} & \text{for } P \text{ exterior to the sphere } S_e, \end{cases} \quad (9.1.16)$$

where m is the mass of the spherical stratum. We are led to the same mechanical conclusions as above.

We notice that we can obtain the same results for the volume potential starting from the equation (9.1.10) written in spherical co-ordinates ($\Delta = (1/r^2)(d/dr)(r^2 d/dr)$) in case of spherical symmetry, corresponding to the formula (A.2.42')). For instance, in case of the volume potential of the sphere $S(0,\xi_0)$ we have

$$\frac{d}{dr}\left(r^2 \frac{dU(r)}{dr}\right) = 0,$$

wherefrom $U(r) = C_1/r + C_2$, $C_1, C_2 = \text{const}$; from (9.1.12) it results that $C_1 = m$, $C_2 = 0$.

1.2.4 Potential of the terrestrial spheroid

In a better approximation, the Earth must be considered as a *spheroid*. Taking into account the Maclaurin series in Subsec. 1.2.1, the volume potential is given by

$$U(\mathbf{r}) = \sum_{n=0}^{\infty} \frac{1}{r^{n+1}} \iiint_V \xi^n P_n(\cos\psi)\mu(\xi)dV, \quad (9.1.17)$$

for a particle P in the exterior of the Earth attractive mass. Corresponding to the first two Legendre polynomials, we notice that (the mass of the Earth and its static moment with respect to a plane normal to OP and passing through the pole O – centre of the Earth)

$$\iiint_V \mu(\xi)dV = m, \quad \iiint_V \xi\cos\psi\,\mu(\xi)dV = 0;$$

the third Legendre polynomial leads to

$$\iiint_V \xi^2(3\cos^2\psi - 1)\mu(\xi)dV = \iiint_V \xi^2(2 - 3\sin^2\psi)\mu(\xi)dV = 2I_O - 3I_\xi,$$

where I_O is the polar moment of inertia of the Earth with respect to its centre, while I_ξ is the axial moment of inertia with respect to OP. Choosing the principal axes of inertia $O2$ and $O3$ in the equatorial plane and the axis $O1$ along the axis of the Earth (so that $I_1 > I_2 > I_3$), we may write (we use the formulae (3.1.23) and (3.1.82''))

$2I_O = I_1 + I_2 + I_3$ and $I_\xi = I_1 n_1^2 + I_2 n_2^2 + I_3 n_3^2$, where n_i, $i = 1, 2, 3$, are the direction cosines of the straight line OP with respect to these axes; in spherical co-ordinates, we have $n_1 = \cos\theta$, $n_2 = \sin\theta \cos\varphi$, $n_3 = \sin\theta \sin\varphi$, so that

$$2I_O - 3I_\xi = I_1 \left(1 - 3\cos^2\theta\right) + I_2 \left(1 - 3\sin^2\theta \cos^2\varphi\right) + I_3 \left(1 - 3\sin^2\theta \sin^2\varphi\right)$$
$$= \left(I_1 - \frac{I_2 + I_3}{2}\right)\left(1 - 3\cos^2\theta\right) - \frac{3}{2}(I_2 - I_3)\sin^2\theta \cos 2\varphi.$$

Calculating the principal moments of inertia of the Earth, we get $(I_2 - I_3)/I_1 < 10^{-6}/3$; in this case, we may assume that $I_2 \cong I_3$ (which corresponds to the spheroidal model accepted for the Earth). Thus, we deduce

$$U(\mathbf{r}) = U(r) = \frac{m}{r} + \frac{3}{2r^3}(I_1 - I_2)\left(\frac{1}{3} - \cos^2\theta\right), \qquad (9.1.18)$$

where m is the mass of the Earth.

In case of the spherical model of the Earth, we find again the result in the preceding subsection (formula (9.1.15)).

For a more exact result, one may use the polynomial $P_3(\cos\psi)$ too.

These results are very important in the study of the motion of artificial satellites of the Earth.

2. Motion due to the action of Newtonian forces of attraction

We have seen that, in the classical model of mechanical systems, the celestial bodies are subjected only to the action of internal forces (forces of Newtonian attraction). This allows a study of the motion of planets, of artificial satellites of the Earth, of interplanetary vehicles, as well as of other types of motion at the atomic level etc. We notice that we will study, e.g., the motion of a celestial body (modelled as a particle) with respect to another celestial body, considered as fixed, hence the motion of a particle subjected to the action of a central force.

2.1 Motion of celestial bodies

After a general study of the motion of a particle acted upon by a central force of the nature of a Newtonian attraction force (elliptic, hyperbolic or parabolic trajectories), we will consider the motion of planets and comets; we put in evidence the deviation of the light ray too.

2.1.1 Rectilinear motion due to the action of a Newtonian force of attraction

If the initial velocity $\mathbf{v}_0$ is directed along $\overrightarrow{OP_0} = \mathbf{r}_0$, where P_0 is the initial position, then (as we have seen in Chap. 8, Subsec. 1.1.2) the trajectory of the particle P is *rectilinear*; it is assumed that there exists a centre of attraction at O (a particle of mass M), the particle P of mass m being subjected to a force of Newtonian attraction

$\mathbf{F} = -fmM\overrightarrow{OP}/x^3$ (we choose the corresponding trajectory as Ox-axis). The equation of motion

$$m\ddot{x} = m\dot{v} = m\frac{\mathrm{d}v}{\mathrm{d}x}\frac{\mathrm{d}x}{\mathrm{d}t} = mv\frac{\mathrm{d}v}{\mathrm{d}x} = -f\frac{mM}{x^2}$$

leads to $mv^2/2 = fmM/x + h$, hence to a conservation law of mechanical energy, h being the energy constant; the initial conditions ($x(0) = x_0$, $v(0) = v_0$) allow to write

$$v = \pm\sqrt{2fM\left(\frac{1}{x} - \frac{1}{x_0}\right) + v_0^2}, \quad h = \frac{m}{2}\left(v_0^2 - \frac{2fM}{x_0}\right). \tag{9.2.1}$$

We assume that $x_0 > 0$ (the positive direction of the Ox-axis is towards the initial position); one cannot have $x_0 = 0$, from the mechanical point of view. One takes the sign $\pm$ before the radical as $v_0 \gtrless 0$.

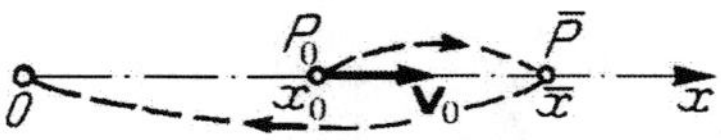

Figure 9.4. Rectilinear motion due to the action of a Newtonian force of attraction.

If $v_0 \leq 0$, then the particle comes near to the point O ($\mathrm{d}x/\mathrm{d}t < 0$) with a velocity increasing in absolute value, which tends to infinity for $x \to 0 + 0$. If $v_0 > 0$, then the particle moves away from the point O ($\mathrm{d}x/\mathrm{d}t > 0$); if $h \geq 0$, then the particle tends to infinity, while if $h < 0$ it stops at the point of abscissa $\bar{x} = -2fM/h$, where the velocity changes of sign and returns, as in the foregoing case (Fig.9.4).

Modelling the Earth as a particle of mass M and denoting $x_0 = R$, where R is the radius of the Earth, approximated as a sphere, the condition $h \geq 0$ leads to

$$v_0^2 \geq 2\frac{fM}{R}, \tag{9.2.2}$$

where we took into account (9.2.1). Introducing numerical values, we find that a particle from the Earth surface must be launched up along the local vertical with an initial velocity $v \geq v_{II}$ ($v_{II} \cong 11.2$ km/s being the second cosmic velocity, see Subsec. 2.2.2) so as not to return on the Earth; obviously, the resistance of the air has been neglected.

In case of a particle which falls on the Earth without initial velocity from the initial position $x = x_0 = \overline{OP_0}$, we obtain the velocity at the Earth surface (the falling velocity)

$$\bar{v} = -\sqrt{2fM\left(\frac{1}{R} - \frac{1}{x_0}\right)}; \tag{9.2.3}$$

if $x_0 = R + H$, $H \ll R$, and if we take into account Torricelli's formula (7.1.17), then we find again the formula (1.1.85). Also from (9.2.1), one obtains the falling time

$$\bar{t} = \int_{x_0}^{R} \frac{dx}{v} = -\int_{x_0}^{R} \frac{dx}{\sqrt{2fM\left(\frac{1}{x} - \frac{1}{x_0}\right)}} = \frac{1}{\sqrt{2fM}} \int_{R}^{x_0} \sqrt{\frac{xx_0}{x_0 - x}} dx. \qquad (9.2.3')$$

2.1.2 Curvilinear motion due to the action of a Newtonian force of attraction. Newton's problem. Runge-Lenz vector

Starting from the results obtained in Sec. 1.2, we may consider the motion of a particle in a field of central forces given by a potential of the form

$$U(r) = \frac{k}{r}; \qquad (9.2.4)$$

thus, the Newtonian gravitational field is of attraction ($k = fmM > 0$), while the Coulombian one (specified by (1.1.84")) may be an attractive or a repulsive field, depending on the relative signs of the charges in interaction ($k > 0$ or $k < 0$, respectively).

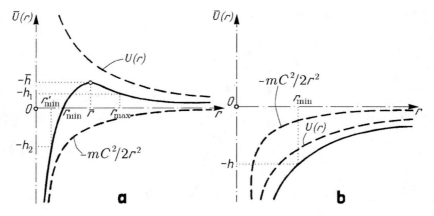

Figure 9.5. Apparent potential: attractive (a); repulsive (b).

We introduce the apparent potential (8.1.6') in the form

$$\bar{U}(r) = \frac{k}{r} - \frac{mC^2}{2r^2}.$$

In case of a potential of attraction ($k > 0$), we represent the apparent potential in Fig.9.5,a. At $\bar{U}(r) = -\bar{h}$ corresponds $\bar{U}_{\max} = \bar{U}(\bar{r})$, the orbit being circular; an elementary calculus shows that $\bar{h} = -k^2 / 2mC^2$, with an orbit radius $\bar{r} = mC^2 / k$. If $\bar{U}(r) \geq -h_1$, $0 < -h_1 < -\bar{h}$, then the orbit is contained in the circular annulus of radii $r_{\min}$ and $r_{\max}$, being a closed curve (in conformity to Bertrand's theorem). For $\bar{U}(r) \geq -h_2$, $-h_2 < 0$, we get a unbounded orbit with a pericentre at a distance $r'_{\min}$ from the centre of attraction. In case of a repulsive potential ($k < 0$), we represent the apparent potential in Fig.9.5,b; for $\bar{U}(r) \geq -h$, $-h < 0$, there result only unbounded orbits of pericentres at the distance $r_{\min}$ from the repulsive centre.

Let us consider now the case of a Newtonian potential of attraction. We choose the Ox_1-axis so as to be an apsidal line; the formulae (8.1.6'), (8.1.6") lead to the equation of the trajectory in polar co-ordinates in the form (we take $\theta_0 = 0$)

$$\theta = C \int_{r_{\min}}^{r} \frac{d\rho}{\rho^2 \sqrt{\frac{2}{m}\left(\frac{k}{r} - \frac{mC^2}{2r^2} + h\right)}} = \int_{1/r}^{1/r_{\min}} \frac{d(1/\rho)}{\sqrt{\frac{k^2}{m^2 C^4} + \frac{2h}{mC^2} - \left(\frac{1}{\rho} - \frac{k}{mC^2}\right)^2}}$$

$$= \arccos \frac{\frac{1}{r} - \frac{k}{mC^2}}{\frac{1}{mC}\sqrt{\frac{k^2}{C^2} + 2mh}},$$

where we took into account that $r = r_{\min}$ corresponds to $\theta = 0$. We obtain thus the equation (9.1.1) of a conic, with

$$p = \frac{mC^2}{k}, \quad e = \sqrt{1 + \frac{2mC^2 h}{k^2}}. \tag{9.2.5}$$

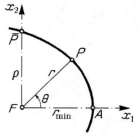

Figure 9.6. Curvilinear motion due to the action of a Newtonian force of attraction.

In Cartesian co-ordinates, there results (we notice that $x_1 = r \cos \theta$) (Fig.9.6)

$$x_1^2 + x_2^2 - (ex_1 - p)^2 = 0; \tag{9.2.5'}$$

the conic pierces the co-ordinate axes at the points $(r_{\min}, 0)$ and $(0, p)$, obtaining thus a geometric interpretation for the parameter of the conic too. Taking into account the eccentricity, it is seen that the trajectory is an ellipse, a parabola or a hyperbola as $h < 0$, $h = 0$ or $h > 0$, respectively; in particular, if $h = -k^2/2mC^2$, then we have $e = 0$, so that the ellipse is a circle.

Newtonian theory of universal attraction

We may obtain an equivalent form of this result starting from Binet's equation (8.1.4). Introducing the force of Newtonian attraction $F = -fmM/r^2$, where M is the mass of the attractive particle, we get the equation

$$\frac{d^2}{d\theta^2}\left(\frac{1}{r}\right) + \frac{1}{r} = \frac{fM}{C^2},$$

wherefrom

$$\frac{1}{r} = C_1 \cos(\theta - C_2) + \frac{fM}{C^2},$$

C_1, C_2 being two scalar integration constants; with the notations $C_1 = e/p$, $C_2 = \theta_1$, $p = C^2/fM$, we find again the equation of the conic with respect to the focus F and an axis inclined by θ_1 towards the apsidal line, in the form

$$r = \frac{p}{1 + e\cos(\theta - \theta_1)}. \tag{9.2.6}$$

If we put the initial conditions at the moment $t = t_0$ as in Chap. 8, Subsec. 1.1.1, then we may express the conic parameter in the form

$$p = \frac{C^2}{fM} = \frac{r_0^2 v_0^2 \sin^2 \alpha_0}{fM}. \tag{9.2.7}$$

We use the conditions (8.1.4") to determine the eccentricity e and the angle θ_1, obtaining thus

$$1 + e\cos(\theta_0 - \theta_1) = \frac{p}{r_0}, \quad e\sin(\theta_0 - \theta_1) = \frac{p}{r_0}\cot\alpha_0,$$

wherefrom

$$e^2 = \left(\frac{p}{r_0} - 1\right)^2 + \frac{p^2}{r_0^2}\cot^2\alpha_0 = 1 + \frac{p}{r_0}\left(\frac{p}{r_0 \sin^2\alpha_0} - 2\right)$$

$$= 1 + \frac{r_0 v_0^2 \sin^2 \alpha_0}{fM}\left(\frac{r_0 v_0^2}{fM} - 2\right), \tag{9.2.7'}$$

$$\tan(\theta_0 - \theta_1) = \frac{p\cot\alpha_0}{p - r_0} = \frac{r_0 v_0^2 \sin\alpha_0 \cos\alpha_0}{r_0 v_0^2 \sin^2\alpha_0 - fM}.$$

Hence, the trajectory is an ellipse, a parabola or a hyperbola as $r_0 v_0^2 < 2fM$, as $r_0 v_0^2 = 2fM$ or as $r_0 v_0^2 > 2fM$, respectively. The genus of the conic depends thus only on the initial distance to the centre of attraction (radius r_0), on the intensity of this centre (mass M) and on the magnitude of the initial velocity (velocity v_0), but does not depend on the direction of this velocity (angle α_0). For the condition $e = 0$, it results ($\alpha_0 = \pi/2$, so that $\sin \alpha_0 = 1$)

$$\frac{r_0 v_0^2}{fM}\left(\frac{r_0 v_0^2}{fM} - 2\right) = -1;$$

hence, the orbit is circular if $r_0 v_0^2 = fM$. These conditions are equivalent to those previously obtained, because the energy constant (the mechanical energy at the initial moment) is given by $h = mv_0^2/2 - fmM/r_0$.

The angle $\alpha = \sphericalangle(\mathbf{r}, \mathbf{v})$ (see Fig.8.1) is given by ($r^2\dot\theta = C$)

$$\tan\alpha = \frac{v_\theta}{v_r} = \frac{r\dot\theta}{\dot r} = \frac{C}{r\dot r} = \frac{r\mathrm{d}\theta}{\mathrm{d}r}; \qquad (9.2.8)$$

taking into account (9.2.6) we may also write

$$\tan\alpha = \frac{p}{re\sin(\theta - \theta_1)}. \qquad (9.2.8')$$

We denote $\theta_1 = \theta_\pi$ too, because for $\theta = \theta_\pi$ we obtain $r_{\min} = p/(1+e)$, hence the pericentre. The angle $\psi = \theta - \theta_\pi$ is called *true anomaly*, representing the angular distance of the particle with respect to the pericentre.

From the law of areas it results $C\mathrm{d}t = r^2\mathrm{d}\theta = r^2\mathrm{d}\psi$; taking into account the equation of the conic $r = p/(1 + e\cos\psi)$ and its parameter $p = C^2/fM$, we may also write

$$\frac{\mathrm{d}\psi}{(1 + e\cos\psi)^2} = \frac{C}{p^2}\mathrm{d}t = \frac{\sqrt{fM}}{p^{3/2}}\mathrm{d}t.$$

With the notation

$$\eta = \tan\frac{\psi}{2}, \qquad (9.2.9)$$

we obtain

$$\mathrm{d}\psi = \frac{2\mathrm{d}\eta}{1+\eta^2}, \quad \cos\psi = \frac{1-\eta^2}{1+\eta^2},$$

so that the law $\eta = \eta(t)$, hence the law $\psi = \psi(t)$ too, is given by the differential equation with separate variables

Newtonian theory of universal attraction

$$\frac{2}{(1+e)^2} \frac{(1+\eta^2)\,\mathrm{d}\eta}{(1+\gamma\eta^2)^2} = \frac{\sqrt{fM}}{p^{3/2}}, \quad \gamma = \frac{1-e}{1+e}, \qquad (9.2.9')$$

which will be integrated taking into account the nature of the conic.

The formula (8.1.6''') allows to find the law of motion of the particle along the trajectory in the form

$$t = t_0 + \int_{r_{\min}}^{r} \frac{\mathrm{d}\rho}{\sqrt{\dfrac{2k}{m\rho} - \dfrac{C^2}{\rho^2} + \dfrac{2h}{m}}}. \qquad (9.2.10)$$

In the study of Newton's problem we used till now two first integrals, corresponding to the conservation of the moment of momentum (a vector first integral, equivalent to three scalar first integrals) and to the conservation of the mechanical energy (a scalar first integral), respectively; hence, it results that the trajectory is a plane curve and that one can determine the motion on it (from the first integral of areas, which is a component of the first integral of moment of momentum, or from the first integral of mechanical energy). The formula (8.1.6'') or Binet's equation may be replaced by a third first integral specific for a field of Newtonian attraction. In case of a central force $\mathbf{F} = F\,\mathrm{vers}\,\mathbf{r}$, $F = F(\mathbf{r},\dot{\mathbf{r}};t)$, we may write, starting from Newton's equation

$$m\ddot{\mathbf{r}} \times \mathbf{K}_O = m\frac{\mathrm{d}}{\mathrm{d}t}(\dot{\mathbf{r}} \times \mathbf{K}_O) = F\frac{\mathbf{r}}{r} \times \mathbf{K}_O = \frac{mF}{r}\mathbf{r} \times (\mathbf{r} \times \dot{\mathbf{r}}),$$

where we took into account the conservation theorem of moment of momentum ($\dot{\mathbf{K}}_O = \mathbf{0}$); it results (we notice that $\mathbf{r} \cdot \dot{\mathbf{r}} = r\dot{r}$, in conformity to the formula (A.1.12))

$$\frac{\mathrm{d}}{\mathrm{d}t}(\dot{\mathbf{r}} \times \mathbf{K}_O) = \frac{F}{r}\mathbf{r} \times (\mathbf{r} \times \dot{\mathbf{r}}) = \frac{F}{r}\left[(\mathbf{r} \cdot \dot{\mathbf{r}})\mathbf{r} - r^2\dot{\mathbf{r}}\right] = F(\dot{r}\mathbf{r} - r\dot{\mathbf{r}}) = -r^2 F\left(\frac{\dot{\mathbf{r}}}{r} - \frac{\dot{r}}{r^2}\mathbf{r}\right),$$

so that

$$\frac{\mathrm{d}}{\mathrm{d}t}(\dot{\mathbf{r}} \times \mathbf{K}_O) = -r^2 F\frac{\mathrm{d}}{\mathrm{d}t}\left(\frac{\mathbf{r}}{r}\right).$$

Hence, if $r^2 F = \mathrm{const}$ (in the above considered case $F = -fmM/r^2$), then we may introduce the vector

$$\mathbf{R} = \dot{\mathbf{r}} \times \mathbf{K}_O - fmM\frac{\mathbf{r}}{r}, \qquad (9.2.11)$$

called *the Runge-Lenz vector*, which is conserved in time along the trajectory of the particle ($\mathrm{d}\mathbf{R}/\mathrm{d}t = \mathbf{0}$, hence $\mathbf{R} = \overline{\mathrm{const}}$), being a vector first integral of the motion, equivalent to three scalar first integrals. One obtains thus seven scalar first integrals which are not independent (we may have at the most six independent first integrals for a single particle). Observing the scalar product $\mathbf{R} \cdot \mathbf{K}_O = 0$ (two zero mixed products), it results that the Runge-Lenz vector, applied at the pole O of attraction, is contained in the plane of the motion (because $\mathbf{R} \perp \mathbf{K}_O$, $\mathbf{K}_O = \overrightarrow{\text{const}}$). Using the two previous first integrals, we can write ($\dot{\mathbf{r}} \perp \mathbf{K}_O$)

$$(\dot{\mathbf{r}} \times \mathbf{K}_O)^2 = |\dot{\mathbf{r}}|^2 \mathbf{K}_O^2 = 2mC^2\left(h + \frac{fmM}{r}\right);$$

it results then ($m(\dot{\mathbf{r}} \times \mathbf{K}_O) \cdot \mathbf{r} = (m\mathbf{r} \times \dot{\mathbf{r}}) \cdot \mathbf{K}_O = K_O^2$)

$$\left(\dot{\mathbf{r}} \times \mathbf{K}_O - fmM\frac{\mathbf{r}}{r}\right)^2 = (\dot{\mathbf{r}} \times \mathbf{K}_O)^2 - 2fmM(\dot{\mathbf{r}} \times \mathbf{K}_O) \cdot \frac{\mathbf{r}}{r} + f^2m^2M^2$$
$$= 2mC^2\left(h + \frac{fmM}{r}\right) - 2f\frac{M}{r}m^2C^2 + f^2m^2M^2 = 2mC^2h + f^2m^2M^2,$$

so that

$$|\mathbf{R}| = fmMe, \qquad (9.2.11')$$

where we have introduced the eccentricity (9.2.5) ($k = fmM$). Choosing the Ox_1-axis along the Runge-Lenz vector and denoting $\theta = \sphericalangle(\mathbf{R},\mathbf{r})$, we may write $\mathbf{R} \cdot \mathbf{r} = |\mathbf{R}|r\cos\theta$; as well, $\mathbf{R} \cdot \mathbf{r} = (\dot{\mathbf{r}} \times \mathbf{K}_O) \cdot \mathbf{r} - fmMr = K_O^2/m - fmMr = mC^2 - fmMr$. Equating the two expressions of the scalar product and taking into account (9.2.7), (9.2.11'), we find again the equation (9.1.1) of the conic. The Runge-Lenz vector allows thus to determine the equation of the trajectory on an algebraically way, its direction being from the centre of attraction to the pericentre.

2.1.3 Elliptic motion. Kepler's equation

If $h < 0$, then the trajectory of the particle acted upon by a force of Newtonian attraction (we can have only $k > 0$, e.g., $k = fmM$) is an ellipse; in *the elliptic motion* we write the equation (9.2.5') in the form $(1 - e^2)x_1^2 + x_2^2 + 2pex_1 = p^2$ (we have $0 \le e < 1$). We notice that we may write this equation also in the form

$$\frac{(x_1 + ae)^2}{a^2} + \frac{x_2^2}{b^2} = 1; \qquad (9.2.12)$$

the semiaxes (Fig.9.7)

$$a = \frac{p}{1-e^2} = -\frac{k}{2h}, \quad b = \frac{p}{\sqrt{1-e^2}} = C\sqrt{-\frac{m}{2h}} \qquad (9.2.12')$$

and the focal distance

$$c = ae = \frac{pe}{1-e^2} = -\frac{ke}{2h} = \sqrt{a^2 - b^2} \qquad (9.2.12'')$$

are thus put in evidence. We mention that, for a given potential (k is given), the semi-major axis of the ellipse depends only on the mechanical energy constant h. We may express the semiaxes of the ellipse by means of the initial conditions in the form

$$a = \frac{1}{2\left(\frac{1}{r_0} - \frac{mv_0^2}{2k}\right)}, \quad b = \frac{r_0 v_0 \sin\alpha_0}{\sqrt{\frac{2k}{mr_0} - v_0^2}}; \qquad (9.2.12''')$$

thus, we notice that a does not depend on the direction of the initial velocity (Fig.9.7). From (9.2.12') it results that the conservation theorem of mechanical energy may be written in the remarkable form

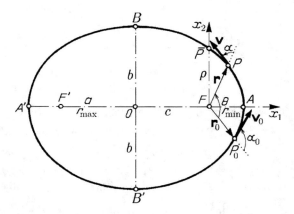

Figure 9.7. Elliptic motion due to the potential $U(r) = k/r$.

$$\frac{mv^2}{2} - \frac{fmM}{r} = -\frac{fmM}{2a}. \qquad (9.2.12^{iv})$$

To can determine the law of motion along the ellipse, we use the equation (9.2.10). From (9.1.1') and (9.2.12'), it results

$$k = -2ah, \quad C^2 = -\frac{2h}{m}\frac{p^2}{1-e^2} = -\frac{2h}{m}a^2(1-e^2), \quad r_{\min} = a(1-e),$$

so that

$$t = t_0 + \sqrt{-\frac{m}{2h}} \int_{(1-e)a}^{r} \frac{\rho d\rho}{\sqrt{a^2 e^2 - (a-\rho)^2}};$$

by a change of variable $\rho = a(1 - e\cos u)$, we may write

$$t = t_0 + a\sqrt{-\frac{m}{2h}} \int_0^u (1 - e\cos\bar{u})\mathrm{d}\bar{u},$$

wherefrom we get *Kepler's equation*

$$t = t_0 + a\sqrt{\frac{ma}{k}}(u - e\sin u). \qquad (9.2.13)$$

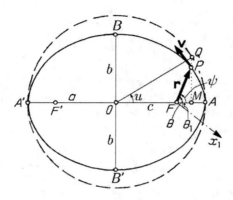

Figure 9.8. Elliptic motion in Kepler's representation.

We assume now that, in general, the Fx_1-axis does not coincide with the apsidal line ($\theta_1 \neq 0$, Fig.9.8), $\psi = \theta - \theta_1$ being *the true anomaly*. The equation of the conic takes the form $r(1 + e\cos\psi) = p = b^2/a = (a^2 - c^2)/a = a - ce$, where we used the notations (9.1.2), (9.1.2"); if we take into account the above change of variable, then we have $c + r\cos\psi = (a - r)/e = a\cos u$. The ordinate of the point P meets the director circle of the ellipse at Q; if $u = \widehat{QOF}$, then we notice that $\overline{OQ}\cos u = a\cos u$ is given by $\overline{OF} + \overline{FM} = c + r\cos\psi$, hence just the expression obtained above. The angle u has thus a simple geometric interpretation and is called *eccentric anomaly*. The Cartesian co-ordinates of the ellipse are easily obtained in the form (with respect to the axes in Fig.9.7)

$$x_1 = a(\cos u - e), \quad x_2 = a\sqrt{1 - e^2}\sin u, \qquad (9.2.14)$$

where we took into account the relations

$$r = a(1 - e\cos u), \quad a\cos u = c + r\cos\psi. \qquad (9.2.14')$$

One gets also

$$\cos\psi = \frac{a\cos u - c}{a(1 - e\cos u)}; \qquad (9.2.14'')$$

observing that

$$\frac{1-\cos\psi}{1+\cos\psi} = \frac{a+c}{a-c}\frac{1-\cos u}{1+\cos u} = \frac{1+e}{1-e}\frac{1-\cos u}{1+\cos u},$$

we obtain, finally, the relation which links the eccentric anomaly to the true anomaly in the form

$$\tan\frac{\psi}{2} = \sqrt{\frac{1+e}{1-e}}\tan\frac{u}{2}. \qquad (9.2.15)$$

Considering the equation (9.2.9'), we may write ($\gamma \neq 0$)

$$\int\frac{(1+\eta^2)\,d\eta}{(1+\gamma\eta^2)^2} = \frac{1}{\gamma}\int\frac{d\eta}{1+\gamma\eta^2} + \left(1-\frac{1}{\gamma}\right)\int\frac{d\eta}{(1+\gamma\eta^2)^2};$$

an integration by parts leads to

$$\int\frac{d\eta}{1+\gamma\eta^2} = \frac{\eta}{1+\gamma\eta^2} + 2\gamma\int\frac{\eta^2\,d\eta}{(1+\gamma\eta^2)^2} = \frac{\eta}{1+\gamma\eta^2} + 2\int\frac{d\eta}{1+\gamma\eta^2} - 2\int\frac{d\eta}{(1+\gamma\eta^2)^2},$$

so that, finally, we have

$$\int\frac{(1+\eta^2)\,d\eta}{(1+\gamma\eta^2)^2} = \frac{1}{2}\left(1-\frac{1}{\gamma}\right)\frac{\eta}{1+\gamma\eta^2} + \frac{1}{2}\left(1+\frac{1}{\gamma}\right)\int\frac{d\eta}{1+\gamma\eta^2}.$$

Thus, the equation (9.2.9') can be integrated in the form

$$\sqrt{\frac{fM}{p^3}}(t-t_0) = \frac{1}{(1+e)^2}\left[\left(1-\frac{1}{\gamma}\right)\frac{\eta}{1+\gamma\eta^2} + \left(1+\frac{1}{\gamma}\right)\int_0^\eta\frac{d\overline{\eta}}{1+\gamma\overline{\eta}^2}\right]$$

$$= \frac{2}{(1+e)(1-e^2)}\left(\int_0^\eta\frac{d\overline{\eta}}{1+\gamma\overline{\eta}^2} - e\frac{\eta}{1+\gamma\eta^2}\right). \qquad (9.2.16)$$

We notice that $0 \leq e < 1$, so that $0 < \gamma \leq 1$; it results

$$\int\frac{d\eta}{1+\gamma\eta^2} = \frac{1}{\sqrt{\gamma}}\arctan(\sqrt{\gamma}\eta).$$

By the change of variable $\sqrt{\gamma}\eta = \tan(u/2)$, we find the equation

$$\sqrt{\frac{fM}{p^3}}(t-t_0) = \frac{1}{(1+e)(1-e^2)\sqrt{\gamma}}(u-e\sin u) = \frac{1}{(1-e^2)^{3/2}}(1-e\sin u),$$

equivalent to Kepler's equation (9.2.13).

The particular case of circular motion ($e = 0$) has been considered in the preceding subsection; Kepler's equation (9.2.13) puts thus in evidence a uniform motion.

2.1.4 Keplerian motion. Kepler's laws

In the case in which the centre of attraction S of mass M, considered to be fixed, is the Sun, the particle in motion (with respect to the centre S) being a planet P, we have to do with the solar system. Analogously, one may consider the motion of a satellite of a planet with respect to that one, e.g., the motion of the Moon with respect to the Earth. We are in the case of the elliptic motion, considered above. Kepler's laws, enounced in Subsec. 1.1.1 as a synthesis of astronomical observations, are obtained now as a mathematical consequence of the Newtonian model of the considered mechanical system. The results thus obtained allow us to state

Theorem 9.2.1 (*Kepler*, I). *The motion of a planet around the Sun is an elliptic one, the Sun being at one of the foci.*

Theorem 9.2.2 (*Kepler*, II; *the law of areas*). *In the motion of a planet around the Sun, the radius vector of it describes equal areas in equal times.*

In celestial mechanics, the notation

$$n = \sqrt{\frac{k}{ma^3}} = \sqrt{\frac{fM}{a^3}} \qquad (9.2.17)$$

is used, so that Kepler's equation may be written in the form

$$u - e \sin u = n(t - t_0). \qquad (9.2.13')$$

We notice that to a variation 2π of the true anomaly there corresponds the same variation of the eccentric anomaly u. Kepler's equation (9.2.13') leads to the period T in which the planet P travels through the whole ellipse, hence effects a motion of revolution (the radius vector describes the whole area of the ellipse) in the form

$$T = \frac{2\pi}{n} = 2\pi a \sqrt{\frac{a}{fM}}; \qquad (9.2.17')$$

it results that n represents the circular frequency (called the mean motion too). We may also write

$$\frac{T^2}{a^3} = \frac{4\pi^2}{fM} \qquad (9.2.17'')$$

stating thus (the ratio $4\pi^2 / fM$ depends only on the mass of the Sun)

Theorem 9.2.3 (*Kepler*, III). *In the motion of planets around the Sun, the ratio of the square of the revolution time to the cube of the semi-major axis is the same for all planets.*

The eccentricity e of the orbits of the nine planets of the solar system (recently, a tenth planet has been discovered) is, in general, very small as it results from the

Table 9.1, in which the planets are put in the order of the distances to the Sun. We notice thus that, excepting Mercury and Pluto, the planets at the smallest and at the greatest distance to the Sun, respectively, the other planets (the Earth included) have an almost circular orbit; the closest to a circular is the orbit of Venus. The pericentre of the orbit is called *perihelion* (denoted by π), while the apocentre is called *aphelion* (denoted by α). Corresponding to the Theorem 9.2.2, a planet has the maximal velocity at the perihelion and the minimal one at the aphelion; between those points the velocity has a monotone variation. For instance, the Earth has the velocity $v_\pi = 30.27$ km/s at the perihelion and the velocity $v_\alpha = 29.27$ km/s at the aphelion (we take $v_{\text{mean}} \cong 30$ km/s). For the Moon, satellite of the Earth, we can make analogous observations, the corresponding eccentricity being $e = 0.066$; the point of the orbit which is the closest to the Earth is called *perigee*, while the point which is the most distant is called *apogee*.

Table 9.1

Planet	e	Planet	e	Planet	e
Mercury	0.2056	Mars	0.0933	Uranus	0.0471
Venus	0.0068	Jupiter	0.0484	Neptune	0.00855
Earth	0.0167	Saturn	0.0558	Pluto	0.2486

During the time, it was stated, by astronomical observations, with perfectionate instruments, that the theorems enounced above correspond to the reality only with a certain approximation. For instance, by the above calculation, one has obtained a immobile perihelion, while the observations put into evidence a displacement of this point of the planetary orbit, displacement which is more sensible for a planet close to the Sun, which has a great eccentricity; especially, Mercury has a *secular displacement* of 43.5" of its perihelion, which has been put in evidence in 1859 by U.-J.-J. Leverrier. To eliminate the non-correspondence between the theory and the observation, an improvement of the mathematical model of Newtonian attraction law has been attempted, by introducing an additional term or an exponent other than 2 at the denominator; as well, it has been admitted the existence of a planet not discovered yet. In this order of ideas, Weber modelled the phenomenon by introducing a non-conservative force of universal attraction of the form

$$\mathbf{F}_W = \left[1 - \frac{1}{c^2}\left(\dot{r}^2 + 2r\ddot{r}\right)\right]\mathbf{F}, \quad c = \text{const}, \tag{9.2.18}$$

but did not obtain satisfactory results concerning the deviation of Mercury's perihelion. Analogously, G. Armellini proposed a law of the form

$$\mathbf{F}_A = [1 + \varepsilon \dot{r}]\mathbf{F}, \quad \varepsilon = \text{const}, \tag{9.2.18'}$$

to may explain the existence of circular orbits. In both cases, $\mathbf{F}$ is a force of Newtonian attraction given by (9.1.7), (9.1.7'). But these non-concordances could be eliminated only in the frame of a non-classical model of mechanics: the relativistic model.

The elliptic motion of a planet is defined in space by six parameters, called *the elements of the elliptic motion*. Using an ecliptic heliocentric frame, we take as reference plane *the ecliptic plane* Π_e (which contains the orbit of the Earth), denoted as the Sxy-plane ($S \equiv F$). The Sx-axis passes through *the first point of Aries* (*the vernal point*) γ (on the celestial sphere of radius equal to unity, at the intersection of a plane parallel to the equatorial plane of the Earth, passing through the centre of mass of the Sun, and the ecliptic plane), corresponding to *the spring equinox*, while the Sy-axis passes, e.g., through the point corresponding to *the summer solstice*; the Sz-axis is directed towards the boreal pole of the ecliptic. The plane Π_p of the planet P orbit intersects the ecliptic plane at the line of nodes NN' (N and N' are on the celestial

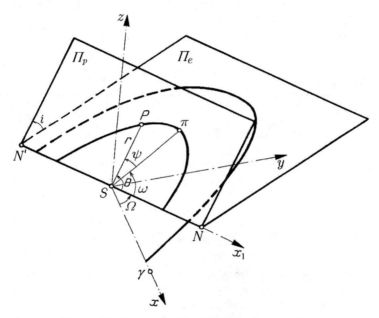

Figure 9.9. Elements of the elliptic motion of a planet.

sphere; N is *the ascending node* from which z passes from negative values to positive ones, while N' is *the descending node* from which z passes from positive values to negative ones); this plane is specified by the angle $\widehat{\gamma SN} = \Omega$, called *the longitude of ascending node*, and by its *inclination* i with respect to the ecliptic plane (Fig.9.9). Choosing SN as Sx_1-axis, we may specify the point π by $\widehat{NS\pi} = \omega = \theta_1$ (*the argument of perihelion*); the angle $\Omega + \omega$ (obtained by summing two angles in different planes) is called *the longitude of perihelion*. The position of the ellipse is thus given in its plane. The magnitude of the orbit is then put in evidence by the semi-major axis a and the eccentricity e. Finally, the motion along the ellipse is characterized by the period $T = 2\pi/n$ (*the period of the revolution motion*) and by the moment $t_\pi = t_0$ (the moment at which the planet P passes through the perihelion). Hence, the elements of the elliptic motion are Ω, i, ω, a, e and t_π; these six elements may be determined by measuring three directions from the Earth to the planet P, from three successive locations on the ecliptic.

We return now to Kepler's equation (9.2.13'), written in the form

$$u - e \sin u = \tau, \quad \tau = n(t - t_0), \tag{9.2.13''}$$

where τ is *the mean anomaly*. The solution of this equation is of the form $u = u(\tau)$; let us consider the function $\tau = \tau(u)$ too. For $u_1, u_2 \in \mathbb{R}$, $u_1 < u_2$, we may write $\tau(u_2) - \tau(u_1) = u_2 - u_1 - e(\sin u_2 - \sin u_1)$; but

$$|\sin u_2 - \sin u_1| \leq 2|\sin[(u_2 - u_1)/2]| \leq 2|(u_2 - u_1)/2| = |u_2 - u_1| = u_2 - u_1$$

and $0 \leq e < 1$, so that $\tau(u_2) > \tau(u_1)$. Hence, $\tau(u)$ is a continuous function on $\mathbb{R}$; the inverse function $u(\tau)$ exists and is uniform and continuous. Hence, there exists only one continuous function $u = u(\tau)$ which verifies Kepler's equation; in particular, for $\tau = m\pi$ it is seen that $u(m\pi) = m\pi$, $m = 0,1,2,\ldots$

Let us consider the series expansions

$$\sin u(\tau) = \sum_{m=1}^{\infty} a_m \sin m\tau, \quad \cos u(\tau) = b_0 + \sum_{m=1}^{\infty} b_m \cos m\tau,$$

with the Fourier coefficients

$$a_m = \frac{2}{\pi} \int_0^\pi \sin u(\tau) \sin m\tau \, d\tau, \quad b_0 = \frac{1}{\pi} \int_0^\pi \cos u(\tau) \, d\tau,$$

$$b_m = \frac{2}{\pi} \int_0^\pi \cos u(\tau) \cos m\tau \, d\tau,$$

b_0 being the mean value of $\cos u(\tau)$ on the interval $[0, \pi]$. We notice that

$$\int_0^\pi \sin u(\tau) \sin m\tau \, d\tau = -\frac{1}{m} \sin u(\tau) \cos m\tau \Big|_0^\pi + \frac{1}{m} \int_0^\pi \cos u(\tau) \cos m\tau \, du$$

$$= \frac{1}{2m} \int_0^\pi [\cos(u + m\tau) + \cos(u - m\tau)] \, du$$

$$= \frac{1}{2m} \int_0^\pi \{\cos[(m+1)u - me \sin u] + \cos[(m-1)u - me \sin u]\} \, du$$

$$= \frac{\pi}{2m}[J_{m+1}(me) + J_{m-1}(me)],$$

$$\int_0^\pi \cos u(\tau) \cos m\tau \, d\tau = \frac{1}{m} \cos u(\tau) \sin m\tau \Big|_0^\pi + \frac{1}{m} \int_0^\pi \sin u(\tau) \sin m\tau \, du$$

$$= \frac{\pi}{2m}[J_{m-1}(me) - J_{m+1}(me)],$$

$$\int_0^\pi \cos u(\tau)\mathrm{d}\tau = \int_0^\pi \cos u(\tau)[1 - e\cos u(\tau)]\mathrm{d}u$$
$$= \sin u\big|_0^\pi - \frac{e}{2}\left(u + \frac{1}{2}\sin 2u\right)\bigg|_0^\pi = -\frac{\pi}{2}e,$$

where we took into account Kepler's equation (9.2.13") and have introduced Bessel's functions of order m defined in the form

$$J_m(x) = \frac{1}{\pi}\int_0^\pi \cos(mu - x\sin u)\mathrm{d}u; \qquad (9.2.19)$$

using the recurrence relations

$$2mJ_m(x) = x[J_{m-1}(x) + J_{m+1}(x)], \quad 2\frac{\mathrm{d}}{\mathrm{d}x}J_m(x) = J_{m-1}(x) - J_{m+1}(x), \qquad (9.2.19')$$

we may, finally, write

$$\sin u(\tau) = \frac{2}{e}\sum_{m=1}^\infty \frac{1}{m}J_m(me)\sin m\tau,$$
$$\cos u(\tau) = -\frac{e}{2} + 2\sum_{m=1}^\infty \frac{1}{m^2}\frac{\mathrm{d}}{\mathrm{d}e}J_m(me)\cos m\tau. \qquad (9.2.20)$$

The relations (9.2.14') and (9.2.13") lead to

$$\frac{1}{a}r(\tau) = 1 + \frac{e^2}{2} - 2e\sum_{m=1}^\infty \frac{1}{m^2}\frac{\mathrm{d}}{\mathrm{d}e}J_m(me)\cos m\tau, \qquad (9.2.21)$$

$$u(\tau) = \tau + 2\sum_{m=1}^\infty \frac{1}{m}J_m(me)\sin m\tau. \qquad (9.2.21')$$

Introducing the notation

$$\lambda = \frac{1 - \sqrt{1-e^2}}{e} = \frac{e}{1 + \sqrt{1-e^2}} < 1 \qquad (9.2.22)$$

and observing that Euler's formula $e^{\pm i\varphi} = \cos\varphi \pm i\sin\varphi$ leads to $\tan\varphi = i(1 - e^{2i\varphi})/(1 + e^{2i\varphi})$, we may write the relation (9.2.15) in the form

$$\tan\frac{\psi}{2} = \frac{1+\lambda}{1-\lambda}\tan u, \qquad (9.2.15')$$

wherefrom

$$e^{i\psi} = \frac{1 - \lambda e^{-iu}}{1 - \lambda e^{iu}}e^{iu}; \qquad (9.2.15'')$$

Newtonian theory of universal attraction

applying now the logarithmic operator and taking into account the series expansion

$$\ln(1-x) = -\sum_{k=1}^{\infty} \frac{x^k}{k}, \quad 0 \le x < 1,$$

and Euler's formula, we get, finally,

$$\psi(\tau) = u(\tau) = 2 \sum_{m=1}^{\infty} \frac{\lambda^m}{m} \sin mu(\tau). \qquad (9.2.22')$$

We have thus put in evidence the polar co-ordinates $r = r(t)$ and $\psi = \psi(t)$ of the planet P in the plane of the orbit, as well as the eccentric anomaly $u = u(t)$ (solution of Kepler's equation) as function of the time $t = t_0 + \tau/n$.

2.1.5 Hyperbolic motion

The trajectory of the particle acted upon by a Newtonian force of attraction is a hyperbola if $h > 0$ (we may have $k > 0$, e.g., $k = fmM$, or $k < 0$, when we have only $h > 0$); in *the hyperbolic motion*, the equation (9.2.5') takes the form (we have $e > 1$) $(e^2 - 1)x_1^2 - x_2^2 - 2pex_1 = -p^2$. The equation reads

$$\frac{(x_1 - ae)^2}{a^2} - \frac{x_2^2}{b^2} = 1 \qquad (9.2.23)$$

too, where we have introduced the semiaxes (Fig.9.10)

$$a = \frac{p}{e^2 - 1} = \frac{|k|}{2h}, \quad b = \frac{p}{\sqrt{e^2 - 1}} = C\sqrt{\frac{m}{2h}} \qquad (9.2.23')$$

and the focal distance

$$c = ea = \frac{pe}{e^2 - 1} = \frac{|k|e}{2h} = \sqrt{a^2 + b^2}; \qquad (9.2.23'')$$

we notice that, for a given potential (k is given), the semi-major axis of the hyperbola depends only on the mechanical energy constant h. The semiaxes of the hyperbola may be expressed with the aid of the initial conditions in the form

$$a = \frac{|k|}{2\left(\frac{mv_0^2}{2} - \frac{k}{r_0}\right)}, \quad b = \frac{r_0 v_0 \sin \alpha}{\sqrt{v_0^2 - \frac{2k}{mr_0}}}; \qquad (9.2.23''')$$

as in case of the elliptic motion, a does not depend on the direction of the initial velocity. The inclinations of the asymptotes are $\pm b/a = \pm\sqrt{e^2 - 1} = \pm C\sqrt{2mh}/k$. In the case in which $k > 0$ the particle moves along the hyperbola branch which surrounds the centre of attraction F' (Fig.9.10,a), while, if $k < 0$, then the particle moves along the other hyperbola branch (Fig.9.10,b); in each of the mentioned cases, the other branch of the hyperbola leads to $r < 0$ (as in the case in Fig.9.2,b). We notice that $r_{\min} = p/(1+e) = a(e-1) < c$ for $k > 0$ and $r_{\min} = p/(e-1) = a(1+e) > c$ if $k < 0$; the apsidal point is reached only once by the particle during the motion along the trajectory. Taking into account (9.2.23'), it results the remarkable form (for $k > 0$)

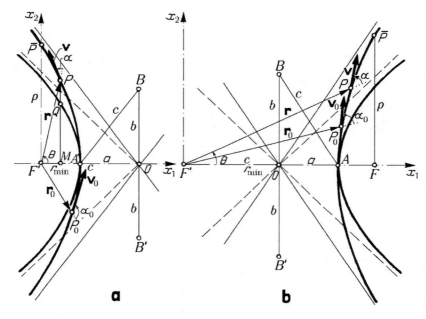

Figure 9.10. Hyperbolic motion due to the potential $U(r) = k/r$.
Case $k > 0$ (a); case $k < 0$ (b).

$$\frac{mv^2}{2} - \frac{fmM}{r} = \frac{fmM}{2a} \qquad (9.2.23^{\mathrm{iv}})$$

for the conservation theorem of mechanical energy.

The law of motion along a hyperbola branch is given by the equation (9.2.10). The relations (9.1.1') and (9.2.23') allow to write (for $k > 0$)

$$k = 2ah, \quad C^2 = \frac{2h}{m}\frac{p^2}{e^2-1} = \frac{2h}{m}a^2(e^2-1), \quad r_{\min} = a(e-1),$$

wherefrom

$$t = t_0 + \sqrt{\frac{m}{2h}} \int_{(e-1)a}^{r} \frac{\rho\,d\rho}{\sqrt{(a+\rho)^2 - a^2 e^2}};$$

the change of variable $\rho = a(e\cosh u - 1)$ leads to

$$t = t_0 + a\sqrt{\frac{m}{2h}} \int_0^u (e\cosh \bar{u} - 1)\,d\bar{u}.$$

Finally, we obtain the equation

$$e\sinh u - u = n(t - t_0), \qquad (9.2.24)$$

analogous to Kepler's one for the elliptic motion, n being given by (9.2.17); in this case, the parameter $u \in (-\infty, \infty)$ corresponds to the parametric equations $x_1 = \overline{OM} = -a\cosh u$, $x_2 = \overline{MQ} = a\sinh u$ of the rectangular hyperbola with the same centre O and the same semiaxis a (Fig.9.10,a). Taking into account the relations

$$r = a(e\cosh u - 1), \quad a\cosh u = c - r\cos\theta \qquad (9.2.25)$$

and the equation of the hyperbola, we get the Cartesian co-ordinates (Fig.9.10,a)

$$x_1 = a(e - \cosh u), \quad x_2 = a\sqrt{e^2 - 1}\sinh u. \qquad (9.2.25')$$

It results

$$\cos\theta = \frac{c - a\cosh u}{a(e\cosh u - 1)}, \qquad (9.2.25'')$$

wherefrom we obtain the relation between the angle θ and the parameter u in the form

$$\tan\frac{\theta}{2} = \sqrt{\frac{e+1}{e-1}}\tanh\frac{u}{2}. \qquad (9.2.26)$$

One can use also the equation (9.2.16) to study the motion of the particle along the trajectory; we have $\gamma < 0$, because $e > 1$. In this case

$$\int \frac{d\eta}{1 + \gamma\eta^2} = \frac{1}{\sqrt{-\gamma}}\arg\tanh(\sqrt{-\gamma}\eta);$$

by the change of variable $\sqrt{-\gamma}\eta = \tanh(u/2)$, we may write

$$\sqrt{\frac{fM}{p^3}}(t - t_0) = \frac{1}{(e+1)(e^2-1)\sqrt{-\gamma}}(e\sinh u - u) = \frac{1}{(e^2-1)^{3/2}}(e\sinh u - u),$$

obtaining thus an equation equivalent to the equation (9.2.24).

574   MECHANICAL SYSTEMS, CLASSICAL MODELS

In the particular motion on a rectangular hyperbola we have $e = \sqrt{2}$, so that $p = a$, while $\tan(\theta/2) = (1 + \sqrt{2})\tanh(u/2)$.

2.1.6 Deviation of the light ray

Starting from observations, it was stated that a light ray coming from a star $\mathscr{S}$ far off, its path passing in the vicinity of the Sun S, is deviated; its final direction (for the observer $\mathscr{O}$) makes an angle φ with its first direction. We consider thus the trajectory of a photon, assuming that this one, modelled as a particle, has a mass. The photon describes a conic in its motion in the gravitational field of the Sun. Assuming that the photon passes very close to the surface of the Sun (Fig.9.11), we may take $r_0 = 6.96 \cdot 10^8$ m, corresponding to Sun's radius (any point of the trajectory may be taken as initial position, in particular the apsidal point); as well, $v_0 = 3 \cdot 10^8$ m/s (the velocity of light in vacuum), while $M = 2 \cdot 10^{33}$ g (Sun's mass). The inequality $r_0 v_0^2 > 2fM$ is satisfied, so that the trajectory of the photon is an arc of hyperbola.

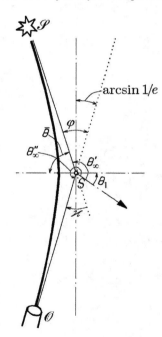

Figure 9.11. Deviation of the light ray.

From the equation (9.2.6) it results that the two asymptotes are specified by the equation $1 + e\cos(\theta - \theta_1) = 0$, which has always solutions, because $e > 1$. The angles θ_∞, corresponding to the two asymptotes, are given by

$$\theta_\infty = \theta_1 \pm \left(\frac{\pi}{2} + \arcsin\frac{1}{e}\right); \qquad (9.2.27)$$

we may write (Fig.9.11)

$$\theta'_\infty = \theta_1 + \frac{\pi}{2} + \arcsin\frac{1}{e}, \quad \theta''_\infty = \theta_1 + \frac{3\pi}{2} - \arcsin\frac{1}{e} \qquad (9.2.27')$$

too. We notice that each asymptote makes an angle equal to $\arcsin(1/e) \cong 1/e$ with the imaginary axis of the hyperbola; hence, the angle made by the two asymptotes, equal to the angle φ, corresponding to *the deviation of the light ray*, is given by

$$\varphi = 2\arcsin\frac{1}{e} \cong \frac{2}{e}. \qquad (9.2.27'')$$

However, this formula could be obtained also by using the results in Chap. 8, Subsec. 1.2.1 concerning the phenomenon of diffraction. In our case, the diffraction angle (Fig.9.11) is given by $\varkappa = -\pi + 2\bar{\theta} = -\pi + 2[\pi - (\theta'_\infty - \theta_1)] = \pi - 2(\theta'_\infty - \theta_1)$; from $\sin(\varkappa/2) = \cos(\theta'_\infty - \theta_1) = -1/e$ it results $\varkappa \cong -2/e$, the angle of diffraction being negative, because the centre S (the Sun) is of attraction. We notice that $|\varkappa| = \varphi$.

Because at the apsis the velocity is normal to the radius vector (which starts from S), we have $\alpha_0 = \pi/2$; in this case, from (9.2.7') it results (e is of an order of magnitude 10^6)

$$e = \frac{r_0 v_0^2}{fM} - 1 \cong \frac{r_0 v_0^2}{fM}. \qquad (9.2.28)$$

We may write

$$\varphi \cong \frac{2fM}{r_0 v_0^2}; \qquad (9.2.28')$$

numerically, we obtain $\varphi \cong 0.87$ s.

Astronomical observations of great precision made during the total solar eclipses (for the first time in May 1919), to may "catch" the light ray coming from the star $\mathscr{S}$, have put in evidence a double angle ($\varphi \cong 1.74$ s), obtaining thus a new non-concordance of the Newtonian model with the physical reality (besides the secular displacement of Mercury's perihelion, see Subsec. 2.1.4). However, one cannot be sure if these non-concordances are due to Newton's laws or to the Newtonian theory of gravitation (or to both mathematical models); some direct improvements of those models did not lead to convenient results, excepting the invariantive model built up by O. Onicescu (see Chap. 21, Sec. 3.2). These contradictions disappear in the frame of the general theory of relativity elaborated by A. Einstein.

2.1.7 Parabolic motion

If $h = 0$ (we can have only $k > 0$, e.g., $k = fmM$), then the trajectory of a particle acted upon by a Newtonian force of attraction is a parabola. Observing that $e = 1$, the equation (9.2.5') takes the form (Fig.9.12)

$$x_2^2 = -2p\left(x_1 - \frac{p}{2}\right) \qquad (9.2.29)$$

in *the parabolic motion*; the distance to the pericentre is given by

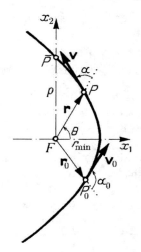

Figure 9.12. Parabolic motion due to the potential $U(r) = k/r$.

$$r_{\min} = \frac{p}{2} = \frac{C^2}{2fM} = \frac{r_0^2 v_0^2 \sin^2 \alpha_0}{2fM}. \qquad (9.2.29')$$

The conservation theorem of mechanical energy takes the form

$$\frac{mv^2}{2} = \frac{fmM}{r}. \qquad (9.2.30)$$

The formula (9.2.10) leads to

$$t = t_0 + \sqrt{\frac{m}{2k}} \int_{p/2}^{r} \frac{\rho \, d\rho}{\sqrt{\rho - p/2}},$$

wherefrom we get

$$t = t_0 + \frac{1}{3}\sqrt{\frac{2}{fM}}(r + p)\sqrt{r - p/2}. \qquad (9.2.31)$$

The equation (9.2.6) becomes (we consider the most general case in which $\theta_1 \neq 0$ and use the true anomaly $\psi = \theta - \theta_1$)

$$r = \frac{p}{1 + \cos\psi} = \frac{p}{2\cos^2\frac{\psi}{2}} = \frac{p}{2}\left(1 + \tan^2\frac{\psi}{2}\right), \qquad (9.2.29'')$$

so that

$$t = t_0 + \frac{p}{2}\sqrt{\frac{p}{fM}}\tan\frac{\psi}{2}\left(1 + \frac{1}{3}\tan^2\frac{\psi}{2}\right). \qquad (9.2.31')$$

Starting from (9.2.9'), we obtain $(1 + \eta^2)d\eta = 2\sqrt{fM/p^3}\,dt$; observing that $\eta = \tan(\psi/2)$, we find again the above results. Using the co-ordinates $x_1 = r\cos\psi$, $x_2 = r\sin\psi$, we get the Cartesian parametric equations of the trajectory

$$x_1 = \frac{p}{2}(1 - \eta^2), \quad x_2 = p\eta. \qquad (9.2.29''')$$

The equation (9.2.31') has only one real root $\tan(\psi/2)$ for a given t. The moment t_0 is obtained for $t = 0$; if $t_0 < 0$, then one cannot have $\psi = 0$, so that the particle starts from the initial position and describes an arc of parabola towards infinity, without passing through its vertex.

2.1.8 Motion of comets

Newton's research has been extended also on the comets known at his time, especially on the Halley comet, appeared in 1680; he assimilated thus the trajectories of the comets to very elongated ellipses, to which Kepler's laws may be applied. Based on Halley's observations, Newton concluded that in case of comets too we have $C^2/p = \text{const}$, being thus led to the same law of universal attraction. Hence, the law of motion is that obtained in Subsec. 2.1.2, and the motion of a comet may be also a parabolic one; corresponding to the results of the preceding subsection, the Sun is at one of the foci, as it was shown by Newton. The theorem of areas may be applied too.

The co-ordinates which specify the parabolic trajectory of a comet are Ω, i, ω, t_π and the perihelion distance $p/2$, hence only five independent parameters.

2.2 Problem of artificial satellites of the Earth and of interplanetary vehicles

Using the results obtained in the preceding subsection, we consider, in what follows, the problem of the artificial satellites of the Earth and of the interplanetary vehicles; we may thus put in evidence the cosmic velocities, the conditions of non-returning on the Earth, the conditions to become a satellite, and the conditions to escape in the cosmic space. As well, we make the study of the orbit of an artificial satellite.

2.2.1 Theory of artificial satellites of the Earth. Conditions to become a satellite

The general theory presented in Subsec. 2.1.2 may be used to study the launching and the motion of the artificial satellites of the Earth. To this goal, we assume that the Earth is a sphere of radius R, its mass being distributed with spherical geometry. If M is the mass of the Earth, in conformity with the formula (1.1.85) we may write $fM = gR^2$. In this case, the conservation theorem of mechanical energy takes the form

$$\frac{mv^2}{2} - \frac{mgR^2}{r} = h, \quad h = \frac{mv_0^2}{2} - \frac{mgR^2}{r_0}, \quad r_0 = R + H; \qquad (9.2.32)$$

we assume that the satellite P, modelled as a particle of mass m, is unbound from the launching rocket at the height H from Earth's surface, along the local vertical, at the point P_0 of position vector $\mathbf{r}_0$, the initial velocity being $\mathbf{v}_0$ (Fig.9.13). To obtain thus a satellite of the Earth, the trajectory of the particle P must be an ellipse; in this case $h < 0$, so that the initial velocity fulfils the condition

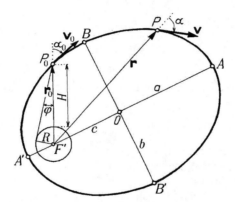

Figure 9.13. Launching of an artificial satellite of the Earth.

$$v_0^2 < \frac{2gR^2}{r_0} = \frac{2gR^2}{R + H}. \qquad (9.2.33)$$

The particle P becomes a satellite if, supplementary, the distance to the perigee is at least equal to the radius of the Earth ($r_{\min} \geq R$). Taking into account the results in Subsec. 2.1.3, we must have $a(1 - e) \geq R$, wherefrom $e \leq 1 - R/a$. Because $e \geq 0$, one must have $a \geq R$ too; associating the inequality (9.2.33) to (9.2.12'''), we have finally,

$$\frac{2R}{r_0} - 1 \leq \frac{v_0^2}{gR} < \frac{2R}{r_0}, \qquad (9.2.33')$$

obtaining thus inferior and superior limits for the initial velocity.

Taking into account the first relation (9.2.12'), we may express the condition $a(1-e) \geq R$ in the form $p/(1+e) \geq R$ or $p/R - 1 \geq e$ too; the first inequality (9.2.33') ensures that $p \geq R$ (because, in this case $e \geq 0$), so that we may also write $(p/R - 1)^2 \geq e^2$. The first relation (9.2.7') leads to

$$\left(\frac{p}{R} - 1\right)^2 \geq \left(\frac{p}{r_0} - 1\right)^2 + \frac{p^2}{r_0^2}\cot^2\alpha_0 \, ;$$

using the relation (9.2.7), we get then

$$\frac{v_0^2}{2gR^2}\left(\frac{r_0^2 \sin^2\alpha_0}{R^2} - 1\right) \geq \frac{1}{R} - \frac{1}{r_0}. \tag{9.2.34}$$

Because $1/R \geq 1/r_0$, it results that we must have $r_0^2 \sin^2\alpha_0 \geq R^2$ too, wherefrom

$$r_0 \sin\alpha_0 \geq R \, ; \tag{9.2.34'}$$

this relation states that the support of the velocity v_0 cannot pierce the Earth's sphere (at the limit, it may be tangent to it (Fig.9.12)). We notice that the relation

$$\frac{2R^2(r_0 - R)}{r_0(r_0^2\sin^2\alpha_0 - R^2)} \geq \frac{2R}{r_0} - 1$$

takes place; indeed, bringing to the same denominator, we read

$$(R - r_0\sin\alpha_0)^2 + 2Rr_0\sin\alpha_0(1 - \sin\alpha_0) \geq 0.$$

The equality can take place only in case of a launching from the surface of the Earth, tangent to it. From (9.2.33')-(9.2.34') it results

$$\frac{R}{r_0}\frac{R(r_0 - R)}{r_0^2\sin^2\alpha_0 - R^2} \leq \frac{v_0^2}{2gR} < \frac{R}{r_0} \leq \sin\alpha_0, \tag{9.2.35}$$

in this case. These inequalities represent the conditions to become a satellite; the first and the last inequality correspond to *the non-returning on the Earth* (to the *escape from the Earth*), while the second inequality corresponds to *the transformation of the body launched from the surface of the Earth in a satellite of that one* (elliptic trajectory).

If the satellite enters on the orbit under an incidence angle $\alpha_0 = \pi/2$, then the conditions (9.2.35) become (the last inequality is verified because $R \leq r_0$)

$$\frac{R}{r_0}\frac{R}{R + r_0} \leq \frac{v_0^2}{2gR} < \frac{R}{r_0}, \tag{9.2.35'}$$

while in the case in which $H \cong 0$, hence $r_0 \cong R$ (the launching point is very close to the surface of the Earth), we get

$$\sqrt{gR} \leq v_0 < \sqrt{2gR}. \qquad (9.2.35'')$$

2.2.2 Cosmic velocities

The inequalities (9.2.35'') may be written in the form

$$v_I \leq v_0 < v_{II}, \qquad (9.2.36)$$

where we have introduced the first and the second cosmic velocities, respectively, given by

$$v_I = \sqrt{gR}, \quad v_{II} = \sqrt{2gR} = \sqrt{2}v_I; \qquad (9.2.37)$$

taking $g = 9.81$ m/s^2 and $R = 6.38 \cdot 10^6$ m, we get $v_I = 7.905$ km/s and $v_{II} = 11.179$ km/s.

In general, the velocities for which a terrestrial body becomes a celestial one are called *cosmic velocities*. The two velocities put here in evidence are called *special cosmic velocities*. The first special cosmic velocity (*the circular cosmic velocity*) represents the smallest velocity by which a body can be launched from the surface of the Earth (tangent to it) without returning on the Earth; *the second special cosmic velocity* (*the parabolic cosmic velocity*) is the greatest velocity by which a body may be launched from the surface of the Earth so that to remain a satellite of it. Any cosmic velocity contained between the two mentioned velocities leads to an elliptic trajectory. A cosmic velocity equal to v_{II} leads to a parabolic trajectory, while a cosmic velocity greater than v_{II} leads to a hyperbolic one. In the latter cases, the celestial body so created can no more be a satellite of the Earth, having a non-bounded trajectory in the interplanetary space; eventually, it can be captured by another celestial body in the proximity of which may pass its trajectory.

The numerical results thus obtained have a certain degree of approximation; indeed, we should take into account the whole system of particles which are involved in motion (see Chap. 11 too), as well as the resistance of the atmosphere (on a trajectory of 200-300 km, till the satellite enters in the interplanetary space, where the resistance of the air is negligible). In this order of ideas, if the entrance on the trajectory takes place at a height H (measured along the local vertical), then the conditions (9.2.35') lead to (9.2.36) too, in the form $v_I^H \leq v_0 < v_{II}^H$, with

$$v_I^H = \frac{R\sqrt{2}}{\sqrt{r_0(R+r_0)}} v_I = \frac{(R/r_0)\sqrt{2}}{\sqrt{1+R/r_0}} v_I = \frac{v_I}{\sqrt{(1+H/R)(1+H/2R)}}$$

$$= \frac{v_I}{\sqrt{1+\frac{3H}{2R}+\frac{1}{2}\left(\frac{H}{R}\right)^2}} \cong \left[1 - \frac{3H}{4R} + \frac{19}{32}\left(\frac{H}{R}\right)^2\right] v_I, \qquad (9.2.37')$$

$$v_{II}^H = \sqrt{\frac{R}{r_0}} v_{II} = \frac{v_{II}}{\sqrt{1+H/R}} \cong \left[1 - \frac{1}{2}\frac{H}{R} + \frac{3}{8}\left(\frac{H}{R}\right)^2\right] v_{II}. \qquad (9.2.37'')$$

Taking, e.g., $H = 2 \cdot 10^5$ m, we get $v_I^H = (1 - 0.02355 + 0.00059)v_I \cong 0.977 v_I$ = 7.723 km/s and $v_{II}^H = (1 - 0.01570 + 0.00037)v_{II} \cong 0.985 v_{II} = 11.011$ km/s, hence velocities somewhat smaller. In general, we should take

$$v_I^H = \frac{\sqrt{2}R\sqrt{r_0 - R}}{\sqrt{r_0(r_0^2 \sin^2 \alpha_0 - R^2)}} v_I = \frac{\sqrt{2}\sqrt{H/R}}{\sqrt{(1+H/R)[(1+H/R)^2 \sin^2 \alpha_0 - 1]}} v_I; \qquad (9.2.37''')$$

obviously, the angle α_0 cannot be chosen arbitrarily, because the quantity under the radical sign must be positive.

Returning to the conservation theorem of mechanical energy (9.2.32) and assuming that a body (modelled as a particle) is launched from the Earth surface ($r_0 = R$), we may write

$$\frac{mv^2}{2} - \frac{mgR^2}{r} = \frac{mv_0^2}{2} - mgR; \qquad (9.2.38)$$

it results

$$v^2 = v_0^2 - 2gR\left(1 - \frac{R}{r}\right) = v_0^2 - v_{II}^2\left(1 - \frac{R}{r}\right). \qquad (9.2.38')$$

A particle can reach the point P, at a distance r from the centre of the Earth (Fig.9.13), with a velocity $v \neq 0$, only if

$$v_0 \geq \sqrt{1 - \frac{R}{r}} v_{II} \cong \left[1 - \frac{1}{2}\frac{R}{r} + \frac{1}{8}\left(\frac{R}{r}\right)^2\right] v_{II}. \qquad (9.2.39)$$

For instance, if $r = 60R$ (the distance from the centre of the Earth to the centre of the Moon), then we obtain $v_0 \geq (1 - 0.00833 + 0.00003)v_{II} \cong 0.992 v_{II} = 11.090$ km/s.

In general, the cosmic velocities corresponding to an arbitrary celestial body, of mass m, are given by

$$v_I^{r_0} = \sqrt{\frac{fm}{r_0}}, \quad v_{II}^{r_0} = \sqrt{2} v_I^{r_0}, \qquad (9.2.40)$$

where r_0 is the distance from the interplanetary vehicle to the centre of the considered celestial body. These formulae are useful to establish the conditions in which a body which is launched from the Earth surface and reaches a certain distance from another celestial body (Moon, Sun etc.) becomes a satellite of that one or continues its trajectory not being affected by a decisive influence; thus, if the velocity of the interplanetary vehicle is less than the second cosmic velocity of a celestial body in the vicinity of which it passes at the distance r_0, then the vehicle is captured by the respective celestial body.

2.2.3 Study of conditions to become a satellite

One can obtain a graphical image of conditions to become a satellite by introducing the non-dimensional variables

$$\xi = \frac{r_0}{R} - 1 = \frac{H}{R} \geq 0, \quad \eta = \frac{v_0^2}{gR} = \left(\frac{v_0}{v_I}\right)^2 > 0; \qquad (9.2.41)$$

the conditions (9.2.35), to which we associate also the first condition (9.2.33'), take the form

$$\frac{1-\xi}{1+\xi} \leq \frac{2\xi}{(1+\xi)\left[(1+\xi)^2 \sin^2 \alpha_0 - 1\right]} \leq \eta < \frac{2}{1+\xi}, \qquad (9.2.42)$$

with

$$\xi \geq \frac{1}{\sin \alpha_0} - 1. \qquad (9.2.42')$$

In a system of co-ordinate axes $O\xi\eta$ we draw the arcs of rectangular hyperbola γ_0 and $\overline{\gamma}_0$ of equations (Fig.9.14)

$$\eta = \frac{2}{1+\xi}, \quad \eta = \frac{1-\xi}{1+\xi}; \qquad (9.2.43)$$

the domain D, the points (ξ, η) of which verify the conditions (9.2.42) to become a satellite, is contained between these curves and the axes $O\eta$ ($\xi \geq 0$) and $O\xi$ ($\eta > 0$). To specify this domain we draw also the curves $\gamma = \gamma(\alpha_0)$ of equation

$$\eta = \frac{2\xi}{(1+\xi)\left[(1+\xi)^2 \sin^2 \alpha_0 - 1\right]}. \qquad (9.2.43')$$

Because

$$\frac{d\eta}{d\alpha_0} = -\frac{2\xi(1+\xi)^2 \sin 2\alpha_0}{(1+\xi)\left[(1+\xi)^2 \sin^2 \alpha_0 - 1\right]^2},$$

it results that $d\eta/d\alpha_0 < 0$ for $\alpha_0 \in (0, \pi/2)$, $d\eta/d\alpha_0 = 0$ for $\alpha_0 = \pi/2$ and $d\eta/d\alpha_0 > 0$ for $\alpha_0 \in (\pi/2, \pi)$. Hence, if α_0 increases from $\alpha_0 = 0$ (without that value), then the curve $\gamma(\alpha_0)$ goes down till a position corresponding to $\alpha_0 = \pi/2$; then $\gamma(\alpha_0)$ goes up together with the increasing of α_0. We notice that $\gamma(\alpha_0) = \gamma(\pi - \alpha_0)$. The curves $\gamma(\alpha_0)$ are over the curve $\overline{\gamma}_0$, so as it was to be expected (only the curve $\gamma_4 = \gamma(\pi/2)$ has a common point with the curve $\overline{\gamma}_0$). *The domain* $D = D(\alpha_0)$ *for which exists the possibility to become a satellite* (a height H being given, there corresponds a lot of possible initial velocities v_0) is contained between the curves $\overline{\gamma}_0$ and $\gamma(\alpha_0)$ (without the frontier $\overline{\gamma}_0$), which are piercing at the

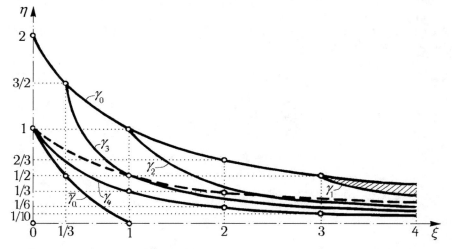

Figure 9.14. Graphic representation of the conditions to become a satellite in the $O\xi\eta$ - plane.

points of co-ordinates $\xi = \cot^2 \alpha_0$, $\eta = 2\sin^2 \alpha_0$. In Fig.9.14 are drawn the curves $\gamma_1 = \gamma(\pi/6)$, $\gamma_2 = \gamma(\pi/4)$, $\gamma_3 = \gamma(\pi/3)$ and $\gamma_4 = \gamma(\pi/2)$ (the domain $D(\pi/6)$ for which one gets a satellite is hatched). *The maximal such domain is* $D(\pi/2)$ *and is contained between the curves* $\overline{\gamma}_0$ *and* γ_4, *being bounded by the axis* $\xi = 0$. Indeed, only in this case we may have $H = 0$; if $\alpha_0 \neq \pi/2$, then a possibility to become a satellite takes place only for

$$\xi > \cot^2 \alpha_0. \tag{9.2.44}$$

The condition (9.2.42') is included in the condition (9.2.44) because $(1 - \sin \alpha_0)/\sin \alpha_0 < \cot^2 \alpha_0$.

The trajectory of the satellite is a circle if the eccentricity vanishes ($e = 0$); taking into account (9.2.7), (9.2.7') and observing that $fM = gR^2$, we must have simultaneously

$$\left(\frac{p}{r_0} - 1\right)^2 = 0, \quad \frac{p^2}{r_0^2}\cot^2 \alpha_0 = 0, \quad p = \frac{r_0^2 v_0^2 \sin^2 \alpha_0}{gR^2}.$$

It results that $\alpha_0 = \pi/2$, $v_0^2/gR = R/r_0$. With the notations introduced above, we may write

$$\eta = \frac{1}{1+\xi}; \qquad (9.2.45)$$

this arc of rectangular hyperbola (contained in the maximal domain $D(\pi/2)$), represented by a broken line in Fig.9.14, specifies the velocity v_0 by which a vehicle must be launched in the space from a given height H so that to become a satellite of the Earth, having a circular trajectory.

An analogous study has been effected by L. Dragoș, using the non-dimensional variables r_0/R and $v_0/\sqrt{gR}$.

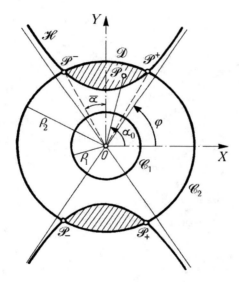

Figure 9.15. Graphic representation of the conditions to become a satellite in the OXY - plane.

A graphical interpretation of the conditions to become a satellite may be obtained in the hodographic plane too; following a study made by C. Iacob (who used the dimensional co-ordinates $v_0 \cos\alpha_0$ and $v_0 \sin\alpha_0$), we introduce the non-dimensional co-ordinates

$$\begin{aligned} X &= \frac{v_0}{v_I}\cos\alpha_0 = \frac{v_0}{\sqrt{gR}}\cos\alpha_0 = \sqrt{\eta}\cos\alpha_0, \\ Y &= \frac{v_0}{v_I}\sin\alpha_0 = \frac{v_0}{\sqrt{gR}}\sin\alpha_0 = \sqrt{\eta}\sin\alpha_0 \end{aligned} \qquad (9.2.46)$$

of the extremity of the position vector v_0/v_I. The conditions (9.2.35) to become a satellite take the form

$$\frac{1-\xi}{1+\xi} \le X^2 + Y^2 < \frac{2}{1+\xi}, \quad \xi(2+\xi)Y^2 - X^2 \ge \frac{2\xi}{1+\xi}, \tag{9.2.47}$$

to which we associate the condition (9.2.42'); we took into account the notation (9.2.41). The first inequalities show that the point $\mathscr{P}(X,Y)$ must be in the interior of the circular annulus, bounded by the circles $\mathscr{C}_1$ and $\mathscr{C}_2$ of radii $\rho_1 = \sqrt{(1-\xi)/(1+\xi)}$ and $\rho_2 = \sqrt{2/(1+\xi)}$, respectively, or on the internal frontier of it (Fig.9.15). As well, this point must be interior to the hyperbola $\mathscr{H}$ of equation (regions of the plane which do not contain the origin O)

$$\frac{Y^2}{\beta^2} - \frac{X^2}{\alpha^2} = 1 \tag{9.2.48}$$

and of semiaxes $\alpha = \sqrt{2\xi/(1+\xi)}$ and $\beta = \sqrt{2/(1+\xi)(2+\xi)}$; the foci $(0, \gamma)$ and $(0, -\gamma)$ are specified by $\gamma = \sqrt{2(1+\xi)/(2+\xi)}$. The asymptotes of the hyperbola make an angle φ given by $\cot\varphi = \sqrt{\xi(2+\xi)}$ with the OX-axis.

If we draw a tangent from the launching point P_0 to the Earth sphere, then we notice that the angle φ is just the angle made by this tangent with the straight line $P_0 F'$ (Fig.9.13). The condition (9.2.42') leads to $\cot\alpha_0 \le \xi(2+\xi)$, so that $\cot\alpha_0 \le \cot\varphi$; hence, $\alpha_0 \ge \varphi$ (the point $\mathscr{P}(X,Y)$ must be in the interior of the asymptotes' angle, which contains the hyperbola $\mathscr{H}$, condition which is fulfilled together with the second condition (9.2.47)). If $\alpha_0 < \varphi$ (the point $\mathscr{P}(X,Y)$ being in the interior of the asymptotes' angle, which does not contain the hyperbola $\mathscr{H}$), then the trajectory of the vehicle launched from the Earth surface pierces the Earth sphere immaterial of the launching velocity. These conditions are easily justified in Fig.9.15.

We have $\rho_1 < \beta < \rho_2$. Hence, one can launch satellites of the Earth from a given height H, that is for a given ξ, if the initial velocity v_0 and the direction of the launching α_0 are so that the point $\mathscr{P}(X,Y)$ be in the interior of the domain $\mathscr{D}$ bounded by arcs of hyperbola $\mathscr{H}$ and arcs of circle $\mathscr{C}_2$ (the hatched regions in Fig. 9.15). The points $\mathscr{P}^+, \mathscr{P}^-, \mathscr{P}_+$ and $\mathscr{P}_-$ of intersection of the hyperbola $\mathscr{H}$ with the circle $\mathscr{C}_2$ have the co-ordinates $\bar{X} = \pm\sqrt{2\xi}/(1+\xi)$, $\bar{Y} = \pm\sqrt{2}/(1+\xi)$. The angle $\bar{\alpha}$ formed by the OY-axis with the semi-lines $O\mathscr{P}^+$ and $O\mathscr{P}^-$ is given by $\bar{\alpha} = \arctan\sqrt{\xi}$; it results that the launching angle α_0 depends on ξ, hence on the launching height H, and that we must have $\alpha_0 \in [\pi/2 - \bar{\alpha}, \pi/2 + \bar{\alpha}]$.

The point $(0, 1/\sqrt{1+\xi})$ corresponds to a circular trajectory.

The discussion has been made for $Y > 0$; for $Y < 0$ one obtains analogous results and a trajectory symmetric to the first one.

In conclusion, if the point $\mathscr{P}(X,Y)$ belongs to the interior of the circle $\mathscr{C}_2$, then the trajectory is an ellipse (if $\mathscr{P} \in \mathscr{D}$, then the trajectory does not intersect the Earth's sphere), if the point $\mathscr{P}$ is on the circle $\mathscr{C}_2$, then the trajectory is a parabola, while if the point $\mathscr{P}$ belongs to the exterior of the circle $\mathscr{C}_2$, then the trajectory is an arc of hyperbola. A supplementary study is necessary to see if these arcs of parabola or of hyperbola intersect the Earth's sphere.

2.2.4 Study of the satellite orbit

As we have seen in Subsec. 2.2.2 too, in a more exact theory we must take into account the resistance of the atmosphere. As well, we must notice that the Earth is neither spherical, nor homogeneous, so that the potential created by it is only with a certain approximation equal to the potential of a particle situated at the mass centre of the Earth and at which would be concentrated its whole mass; we may use with a better approximation the formula (9.1.18) for the potential $U(r)$ of the terrestrial spheroid.

The equation of motion reads (m_s is the satellite mass)

$$m_i \mathbf{a} = f m_s \operatorname{grad} U - m_s \overline{\psi} \mathbf{v}, \qquad (9.2.49)$$

where we have introduced the resistance $\mathbf{R} = -m_s \overline{\psi}(t) \mathbf{v}$, $\overline{\psi}(t) = \psi(v(t))$. In spherical co-ordinates we can write (see Chap. 5, Subsecs. 1.1.3 and 1.2.4)

$$\ddot{r} - r\dot{\theta}^2 - r\sin^2\theta \dot{\varphi}^2 = f\frac{\partial U}{\partial r} - \overline{\psi}\dot{r},$$

$$\frac{\mathrm{d}}{\mathrm{d}t}(r^2\dot{\theta}) - \frac{1}{2}r^2 \sin 2\theta \dot{\varphi}^2 = f\frac{\partial U}{\partial \theta} - \overline{\psi}r^2\dot{\theta}, \qquad (9.2.49')$$

$$\frac{\mathrm{d}}{\mathrm{d}t}(r^2 \sin^2\theta \dot{\varphi}) = f\frac{\partial U}{\partial \varphi} - \overline{\psi}r^2 \sin^2\theta \dot{\varphi}.$$

The terrestrial spheroid has properties of axial symmetry, so that $\partial U / \partial \varphi = 0$; the third equation (9.2.49') leads thus to the first integral

$$r^2 \sin^2\theta \dot{\varphi} = r_0^2 \sin^2\theta_0 \dot{\varphi}_0 e^{-\int_{t_0}^{t} \overline{\psi}(\tau)\mathrm{d}\tau}, \qquad (9.2.49'')$$

with $r_0 = r(t_0)$, $\theta_0 = \theta(t_0)$, $\dot{\varphi}_0 = \dot{\varphi}(t_0)$. In particular, if the satellite is launched in a meridian plane ($v_\varphi(t_0) = r_0 \sin\theta_0 \dot{\varphi}_0 = 0$), then $v_\varphi = r \sin\theta \dot{\varphi} = 0$, hence $\varphi = \mathrm{const}$; thus, the satellite continues its motion in the meridian plane. In this case, the equations of motion are

$$\ddot{r} - r\dot{\theta}^2 = f\frac{\partial U}{\partial r} - \overline{\psi}\dot{r},$$

$$\frac{\mathrm{d}}{\mathrm{d}t}(r^2\dot{\theta}) = f\frac{\partial U}{\partial \theta} - \overline{\psi}r^2\dot{\theta}. \qquad (9.2.50)$$

If, in a first approximation, we take $\partial U / \partial \theta = 0$, then we get a second first integral ($\dot{\theta}_0 = \dot{\theta}(t_0)$)

$$r^2\dot{\theta} = r_0^2\dot{\theta}_0 e^{-\int_{t_0}^{t}\bar{\psi}(\tau)\mathrm{d}\tau}, \tag{9.2.50'}$$

which shows that the areal velocity $\Omega = r^2\dot{\theta}/2$ has an exponential variation.

The spherical co-ordinates $\theta = \theta(t)$ and $\varphi = \varphi(t)$ of the satellite S being determined, the problem to specify the orbit co-ordinates on the celestial sphere of radius equal to unity is put. We choose a geocentric frame of reference (which we assume to be inertial) with the origin at the Earth centre P; the principal plane is the equatorial one and the Px-axis is contained in this plane, being directed towards *the first point of Aries* γ (*the vernal equinoctial point*) (at the intersection of the equatorial plane with the ecliptic plane, which contains the trajectory described by the Earth); the Oz-axis is normal to this plane (it is the rotation axis of the Earth). The line of nodes PN is specified by the angle Ω in the plane Pxy; the plane of satellite's orbit, which passes through PN, is inclined on the principal plane by the angle i. The position S of the satellite is – in this case – given by $u = \mathrm{arc}\,NS$ (Fig.9.16). Applying the formulae of spherical trigonometry to the spherical triangle $NS'S$, rectangular in S', we get the relations

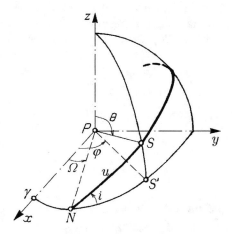

Figure 9.16. Trajectory of a satellite.

$$\cot i = \tan\theta\sin(\varphi - \Omega), \quad \cos u = \sin\theta\cos(\varphi - \Omega), \tag{9.2.51}$$

which give the angles i and u. We mention also the relation

$$\sin i \sin u = \cos\theta. \tag{9.2.51'}$$

2.3 Applications to the theory of motion at the atomic level

The motion of a particle electrically charged, e.g. of an electron in the vicinity of an atomic nucleus, must be studied in the frame of the quantic model of mechanics; however, one can obtain many interesting and useful results in the frame of the Newtonian model too. After presenting the classical model of the atom, based on

Hertz's oscillator, we pass to the study of Bohr's atom; by this occasion, we put in evidence Ritz's law too, which characterizes the lines spectrum of hydrogen.

2.3.1 Hertz's oscillator. Classical model of the atom

If an *electric charge* $e = e_0 \sqrt{\varkappa \varepsilon_0}$, where e_0 is *the rationalized electric charge*, ε_0 is *the permittivity of the vacuum*, while $\varkappa$ is a *coefficient of rationalization* ($\varkappa = 1$ in case of a non-rationalized system and $\varkappa = 4\pi$ in case of a rationalized one, e.g., the SI-system), passes from a condenser coat to another one (a rhythmical variation of the armatures' polarity by an alternative variation of the tension applied upon them), then one obtains an experimental device which emits electromagnetic waves, equivalent to a linear oscillator dipole of moment $p_0 = p/\sqrt{\varkappa \varepsilon_0} = e_0 x$, where x is the elongation, called *Hertz's oscillator*. This device emits (i.e. it loses) energy in the form of spontaneously radiated energy, expressed by the power

$$P = \frac{2}{3c^3} \overline{\dot{p}_0^2} = \frac{2}{3c^3} e_0^2 \overline{\ddot{x}^2}, \qquad (9.2.52)$$

where c is the velocity of light propagation in vacuum; the upper line indicates *the mean value*. Starting from the equation of motion (8.2.23), we may write $\overline{\ddot{x}^2} = \omega^4 \overline{x^2}$; taking into account the form (8.2.24) of the solution and observing that

$$\frac{1}{T} \int_0^T \cos^2(\omega t - \varphi) \mathrm{d}t = \frac{1}{2},$$

we obtain $\overline{\ddot{x}^2} = 8\pi^4 \nu^4 a^2$, where a is the amplitude, while ν is the frequency given by (8.2.6'). It results

$$P = \frac{16\pi^4}{3} \frac{e_0^2 \nu^2 a^4}{c^3}. \qquad (9.2.52')$$

Thus, by radiation, a quantity of mechanical energy $\mathrm{d}E/\mathrm{d}t = -P$ is lost from the mechanical energy E given by (8.2.23"). We can write $\mathrm{d}E/\mathrm{d}t = -\gamma E$, wherefrom

$$E = E_0 e^{-\gamma t}, \quad \gamma = \frac{8\pi^2}{3} \frac{e_0^2 \nu^2}{mc^3}, \qquad (9.2.53)$$

γ being *the radiation constant*; as it was shown by Max Planck, one obtains thus a damping effect.

The classical model of the atom is an oscillator of Hertz, that is an electron which oscillates around a position of equilibrium, where there is an atomic nucleus. But this mechanical system loses permanently energy, as we have seen above; because of this damped motion, after a sufficiently long (but finite) time, the electron falls towards the nucleus. Such a model of atom is labile and does not correspond to the reality; but it can be considered as satisfactory for a great lot of physical phenomena.

We notice that $E = Ka^2$, $K = \text{const}$; taking into account (9.2.53), it results also for the amplitude a variation of the form

$$a = a_0 e^{-\gamma t/2}, \qquad (9.2.54)$$

which characterizes a damped motion. Hence, the equation of motion of *the electron quasi-elastically linked* in the atom should be of a corresponding form.

If **F** is the force which arises under the influence of the atomic radiation, acting upon the oscillator itself and damping its motion, then we are led to the equation of motion $m\ddot{x} + m\omega^2 x = F$; multiplying by $\dot{x}$, we obtain $dE/dt = F\dot{x} = -(2e_0^2/3c^3)\ddot{x}^2$ (where we considered the non-averaged power), with $E = (mv^2 + m\omega^2 x^2)/2$. Taking into account (9.2.52) and observing that $\ddot{x}^2 = d(\dot{x}\ddot{x})/dt - \dot{x}\dddot{x}$, we may write

$$\overline{F\dot{x}} = \frac{2e_0^2}{3c^3}\overline{\dot{x}\dddot{x}} - \frac{2e_0^2}{3c^3}\dot{x}\ddot{x}\big|_0^T ;$$

assuming that the period T is sufficiently small, the second term may be neglected with respect to the first one and we can take

$$F = \frac{2e_0^2}{3c^3}\dddot{x}. \qquad (9.2.55)$$

We obtain thus the equation of motion of the electron quasi-elastically linked to the atom in the form

$$m\ddot{x} + m\omega^2 x - \frac{2e_0^2}{3c^3}\dddot{x} = 0. \qquad (9.2.56)$$

Introducing the radiation constant

$$\gamma = \frac{2e_0^2 \omega^2}{3mc^3}, \qquad (9.2.53')$$

the equation (9.2.56) becomes

$$\dddot{x} - \frac{\omega^2}{\gamma}\ddot{x} - \frac{\omega^4}{\gamma}x = 0. \qquad (9.2.56')$$

Choosing a solution of the form $x(t) = e^{\chi t}$, we find the condition $\gamma \chi^3 - \omega^2 \chi^2 - \omega^4 = 0$. Let us introduce the function $\varphi(\chi) = \gamma \chi^3 - \omega^2 \chi^2 - \omega^4$; we notice that $\varphi(-\infty) = -\infty$, $\varphi(0) = -\omega^4$, $\varphi(\infty) = \infty$. As well, $d\varphi/d\chi = 3\gamma\chi^2 - 2\omega^2 \chi$, $d^2\varphi/d\chi^2 = 6\gamma\chi - 2\omega^2$. We obtain the graphic representation in Fig.9.17.

We have thus only one real root of the form $\omega^2/\gamma + \overline{\gamma}$, $\overline{\gamma} \in (0,\gamma)$, because $\varphi(\omega^2/\gamma) = -\omega^4$, $\varphi(\omega^2/\gamma + \gamma) = \gamma^2(2\omega^2 + \gamma^2)$; the other two roots are $-\overline{\gamma}/2 \pm \overline{\omega}i$ (the sum of the roots is ω^2/γ). Hence, the general integral of the equation (9.2.56') is of the form (we use the other two Viète's relations between the roots and the coefficients)

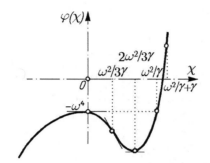

Figure 9.17. Graphic of the function $\varphi(\chi)$.

$$x(t) = e^{-\overline{\gamma}t/2}(A\cos\overline{\omega}t + B\sin\overline{\omega}t) + Ce^{\omega't}, \quad \overline{\gamma} \in (0,\gamma),$$

$$\overline{\omega}^2 = \frac{\overline{\gamma}}{\gamma}\omega^2 + \frac{3}{4}\overline{\gamma}^2 = \frac{\omega^2}{1 + \gamma\overline{\gamma}/\omega^2} - \frac{\overline{\gamma}^2}{4}, \quad \omega' = \frac{\omega^2}{\gamma} + \overline{\gamma}.$$
(9.2.57)

The phenomenon corresponds to damped oscillations, so that one can take $C = 0$.

2.3.2 Bohr's atom. Ritz's law

The classical model of the atom gives some useful information with a sufficient good approximation for some physical phenomena, but it is invalidated by other physical phenomena, e.g. the *spectral emissions* (in the form of *lines spectra*). In 1913, Niels Bohr, starting from Rutherford's conception, which considers a model similar to that of the solar system, develops a new model of the atom. Assuming the existence of a central nucleus, formed by positive charges, and of electrons with negative charges, attracted correspondingly to Coulomb's law (1.1.84"), one obtains a mechanical system in motion after Kepler's laws. As a consequence of the lose of energy, on a way analogous to that presented in the previous subsection, we are led to the motion of the axis of the elliptic trajectory, hence to the possibility of falling of the electron towards the nucleus (the so-called *atomic catastrophe*). To avoid such a phenomenon in the mathematical model, N. Bohr completed it with following principles: i) The electrons describe elliptic trajectories around the nucleus, after Kepler's laws, without lose of energy by radiation; from the set of all *possible trajectories*, these ones are *stationary*, characterized by *quantic conditions*. ii) In conformity to the quantification conditions, the integral of the generalized momentum corresponding to each generalized co-ordinate (to each degree of freedom) along the complete trajectory must be an integer multiple of *Planck's universal constant* $h = 6.626196 \cdot 10^{-27}$ erg·s for the stationarity trajectories. iii) The emission of radiations takes place only if the electron "jumps" from a stationary trajectory on another stationary trajectory, due to an external excitation.

In case of a single electron which moves around a nucleus, *Bohr's model* corresponds to the atom of hydrogen. We consider thus the motion of a particle of mass m, which has a negative electric charge and moves around a positive centre of attraction being acted upon by a Coulombian force which derives from the potential k/r, $k > 0$. In our case, $k = e_0^2 = e^2/\varkappa\varepsilon_0$, where $e = 1.6021917 \cdot 10^{-19}$ C (in coulombs) is the electric charge in the SI-system; the coefficient ε in (1.1.84") is, in this case, equal to unity.

Taking into account (9.2.6) and (9.2.12'), we may write the equation of the elliptic trajectory in the form

$$r = \frac{a(1-e^2)}{1+e\cos\psi} \tag{9.2.58}$$

$\psi = \theta - \theta_1$ being the real anomaly. Because the kinetic energy is given by $T = m(\dot r^2 + r^2\dot\theta^2)/2$, we define the generalized momenta corresponding to polar coordinates by

$$p_r = \frac{\partial T}{\partial \dot r} = m\dot r, \quad p_\theta = \frac{\partial T}{\partial \dot\theta} = mr^2\dot\theta = mr^2\dot\psi = mC, \tag{9.2.59}$$

where C is the constant of areas. The stationarity trajectories are thus specified by the quantification conditions

$$\oint p_r\,\mathrm{d}r = n'h, \quad \int_0^{2\pi} p_\theta\,\mathrm{d}\psi = nh, \quad n, n' \in \mathbb{N}, \tag{9.2.59'}$$

where h is Planck's constant; the integrals are calculated along the complete trajectory (for the first integral, r varies from $r_{\min}$ to $r_{\max}$ and again to $r_{\min}$). Observing that

$$m\dot r\,\mathrm{d}r = m\left(\frac{\mathrm{d}r}{\mathrm{d}\psi}\right)^2\dot\psi\,\mathrm{d}\psi = ma^2(1-e^2)^2\frac{e^2\sin^2\psi}{(1+e\cos\psi)^2}\dot\psi\,\mathrm{d}\psi$$

$$= mr^2\frac{e^2\sin^2\psi}{(1+e\cos\psi)^2}\dot\psi\,\mathrm{d}\psi = mC\frac{e^2\sin^2\psi}{(1+e\cos\psi)^2}\,\mathrm{d}\psi,$$

we may write

$$\oint p_r\,\mathrm{d}r = mCe^2\int_0^{2\pi}\frac{\sin^2\psi}{(1+e\cos\psi)^2}\,\mathrm{d}\psi = mCe\left(\left.\frac{\sin\psi}{1+e\cos\psi}\right|_0^{2\pi} - \int_0^{2\pi}\frac{\cos\psi\,\mathrm{d}\psi}{1+e\cos\psi}\right)$$

$$= mC\int_0^{2\pi}\left(\frac{1}{1+e\cos\psi} - 1\right)\mathrm{d}\psi = \left.\frac{2mC}{\sqrt{1-e^2}}\arctan\frac{\tan\frac{\psi}{2}}{\sqrt{\frac{1-e}{1+e}}}\right|_0^{2\pi} - 2\pi mC$$

$$= 2\pi mC \left(\frac{1}{\sqrt{1-e^2}} - 1 \right) = n'h \,;$$

as well

$$\int_0^{2\pi} p_\theta \, d\varphi = \int_0^{2\pi} mC \, d\psi = 2\pi mC = nh \,.$$

Finally, we get

$$mC = n\hbar, \quad e^2 = 1 - \frac{n^2}{(n+n')^2}, \quad n, n' \in \mathbb{N}, \tag{9.2.60}$$

where

$$\hbar = \frac{h}{2\pi} = 1.0545919 \cdot 10^{-34} \text{ J} \cdot \text{s} \tag{9.2.61}$$

is *the rationalized Planck's constant* (in the SI-system, in joule-seconds). The first relation (9.2.60) determines the magnitude of the ellipse, because (see the relations (9.2.7) and (9.2.12') too)

$$a = \frac{C^2}{(1-e^2)e_0^2/m} = \frac{mC^2}{(1-e^2)e_0^2} = \frac{n^2\hbar^2}{(1-e^2)me_0^2} = \frac{(n+n')^2\hbar^2}{me_0^2}, \tag{9.2.62}$$

while the second relation (9.2.60) specifies its form.

Taking into account (9.2.58) and (9.2.59), the kinetic energy is given by

$$T = \frac{m}{2}(\dot{r}^2 + r^2\dot{\theta}^2) = \frac{1}{2m}\left(p_r^2 + \frac{p_\theta^2}{r^2}\right) = \frac{mC^2}{2r^2}\left[\left(\frac{1}{r}\frac{dr}{d\psi}\right)^2 + 1\right]$$

$$= \frac{mC^2}{a^2(1-e^2)^2}\left(\frac{1+e^2}{2} + e\cos\psi\right),$$

while the potential energy reads

$$V = -\frac{e_0^2}{r} = -\frac{e_0^2}{a(1-e^2)}(1 + e\cos\psi)\,;$$

the mechanical energy results in the form

$$E = T + V = \frac{1}{a(1-e^2)}\left(\frac{mC^2}{2a}\frac{1+e^2}{1-e^2} - e_0^2\right) + \frac{e\cos\psi}{a(1-e^2)}\left[\frac{mC^2}{a(1-e^2)} - e_0^2\right],$$

so that, taking into account (9.2.62), we get

$$E = -\frac{e_0^2}{2a} = -\frac{me_0^4}{2\hbar} \frac{1}{(n+n')^2}. \tag{9.2.63}$$

From this basic formula in the theory of lines spectra, we see that *on a stationary trajectory with the quantic numbers n and n' the mechanical energy is constant.*

Considering two stationary trajectories of indices 1 and 2, the principle iii) shows that the emitted energy has a frequency given by $h\nu = E_2 - E_1$ (if $n_2 + n_2' > n_1 + n_1'$, then $E_2 > E_1$); we obtain *Ritz's law*

$$\nu = -R\left[\frac{1}{(n_1+n_1')^2} - \frac{1}{(n_2+n_2')^2}\right], \tag{9.2.64}$$

where

$$R = \frac{me_0^4}{4\pi\hbar^3} \tag{9.2.64'}$$

is *Rydberg's constant*.

We notice that the quantic numbers appear only in the combination $n + n'$, so that we may consider that we have only one quantic number, called *principal quantic number* (as only one degree of freedom would be quantified). The relation (9.2.64) emphasizes the lines spectrum of the hydrogen, with the various series obtained on an experimental way (Balmer, Fowler etc.).

Chapter 10

OTHER CONSIDERATIONS ON PARTICLE DYNAMICS

To complete the study of a mechanical system, which can be modelled by only one particle, we present some problems with a special character; we consider thus the motion with discontinuity, the motion of the particle with respect to a non-inertial frame of reference, as well as the motion of the particle of variable mass.

1. Motion with discontinuity

In the mathematical modelling of the motion of a particle we assumed, in general, that the vector $\mathbf{r}$ has derivatives of the first and the second order (its components are functions of class C^2); as well, we assumed that the force $\mathbf{F} = \mathbf{F}(\mathbf{r}, \dot{\mathbf{r}}; t)$ is continuous with respect to the position vector, the velocity and the time. In this case, we have seen that the equations of the problems that are put may be integrated with certain boundary conditions and in certain conditions of existence and uniqueness.

But, in case of many mechanical phenomena, the conditions of continuity mentioned above are not fulfilled; such phenomena cannot be easily integrated in the classical schemata based on usual functions. There appears thus the necessity to extend Newtonian mechanics and to complete its mathematical model with the aid of the methods of the theory of distributions. We establish thus the general equation of motion with discontinuity of a particle, stating then the general theorems corresponding to this motion.

1.1 Particle dynamics

After some general considerations concerning the trajectory, the velocity, the acceleration and the force, we establish, in what follows, the fundamental equation of motion for a free particle, as well as for a particle subjected to constraints. A special attention is paid for the motion of the heavy particle in vacuum.

1.1.1 General considerations

One of the mechanical phenomena which needs the introduction of methods of the theory of distributions is (as it was seen in Chap. 5, Subsec. 1.2.6) that in which the position vector is a continuous function, while the velocity and the acceleration have discontinuities of the first species at $t = t_0$, i.e. $\mathbf{v}(t_0 - 0) \neq \mathbf{v}(t_0 + 0)$ and $\dot{\mathbf{v}}(t_0 - 0) \neq \dot{\mathbf{v}}(t_0 + 0)$.

596   MECHANICAL SYSTEMS, CLASSICAL MODELS

A simple example of such a mechanical phenomenon is the motion of a heavy particle P on a trajectory ABC (a broken line, Fig.10.1,a), assuming, for the sake of simplicity, that it is frictionless; in fact, the character of the mechanical phenomenon does not change if the friction is taken into consideration. As it is easy to see, the trajectory of the particle P is represented by a continuous function, formed by two segments of a line AB and BC; we notice that the function which represents the trajectory has not derivatives of first and second order at B, so that we cannot determine the velocity, the acceleration and the constraint force at this point. We assume that at the moment $t = 0$ the particle is at A and that at the moment $t = t_0$ it reaches B; in this case, the acceleration modulus may be written in the form

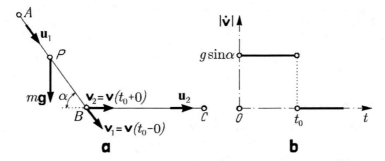

Figure 10.1. Particle in motion on a broken line. Trajectory (a); the acceleration modulus vs time (b).

$$|\dot{\mathbf{v}}| = g \sin \alpha [1 - \theta(t - t_0)] = \begin{cases} g \sin \alpha & \text{for } 0 \le t < t_0, \\ 0 & \text{for } t > t_0, \end{cases} \quad (10.1.1)$$

where $\mathbf{g}$ is the gravity acceleration (Fig.10.1,b), while θ is Heaviside's function. A discontinuity of the first species is thus put into evidence, because, at the moment $t = t_0$, we may write

$$\lim_{t \to t_0 - 0} |\dot{\mathbf{v}}| = g \sin \alpha, \quad \lim_{t \to t_0 + 0} |\dot{\mathbf{v}}| = 0.$$

We notice also that, besides the discontinuity of the modulus, the acceleration has also a discontinuity in direction; thus, the acceleration vector has a discontinuity of the first species at the moment $t = t_0$, given by

$$\lim_{t \to t_0 - 0} \dot{\mathbf{v}}(t) = \mathbf{a}_1 = g \sin \alpha \, \mathbf{u}_1, \quad \lim_{t \to t_0 + 0} \dot{\mathbf{v}}(t) = \mathbf{0}, \quad (10.1.1')$$

where the unit vector $\mathbf{u}_1$ specifies the line AB and is directed corresponding to the motion of the particle. The modulus of the velocity reads, in this case,

$$|\mathbf{v}| = g \sin \alpha [t - (t - kt_0)\theta(t - t_0)] = \begin{cases} gt \sin \alpha & \text{for } 0 \le t < t_0, \\ kgt_0 \sin \alpha & \text{for } t > t_0, \end{cases} \quad (10.1.2)$$

Other considerations on particle dynamics 597

where k, $0 \le k \le 1$, is a *coefficient of restitution*. Corresponding to the principle of inertia, the velocity of the particle on BC must be equal to the velocity at the right of B, hence equal to $\mathbf{v}(t_0 + 0)$. But this velocity is not known, so that it remains non-determinate on BC. To solve the problem, we assume that at the point of discontinuity takes place a *phenomenon of collision*; hence, we assume that the magnitude of the velocity at the right is proportional to its magnitude at the left, that is

$$|\mathbf{v}(t_0 + 0)| = k|\mathbf{v}(t_0 - 0)|. \qquad (10.1.3)$$

If the coefficient of restitution is equal to unity ($k = 1$), then the collision at B is *perfect elastic* and the magnitude of the velocity is a continuous function. Usually, this is the hypothesis that is assumed, although it is not the only possibility. Indeed, independent of the modulus of the velocity on the segment of a line BC, specified by the unit vector $\mathbf{u}_2$, directed corresponding to the motion, the velocity vector has a discontinuity of the first species at B, because

$$\lim_{t \to t_0 - 0} \mathbf{v}(t) = \mathbf{v}(t_0 - 0) = v(t_0 - 0)\mathbf{u}_1 = \mathbf{v}_1,$$
$$\lim_{t \to t_0 + 0} \mathbf{v}(t) = \mathbf{v}(t_0 + 0) = v(t_0 + 0)\mathbf{u}_2 = \mathbf{v}_2; \qquad (10.1.4)$$

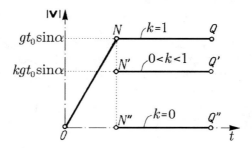

Figure 10.2. The collision model for a particle in motion on a broken line. The velocity modulus vs time.

$\mathbf{v}_1$ is the velocity of the particle P at B on AB, while $\mathbf{v}_2$ is the velocity of the same particle at B on BC. In Fig.10.2 is drawn the diagram of the modulus of the velocity as a function of time (the line segments ON and NQ for $k = 1$, the line segments ON and $N'Q'$ for $0 < k < 1$ and the segments ON and $N''Q''$ for $k = 0$, the case of a *plastic collision*).

The study of problems of the kind considered above needs the introduction of various mechanical quantities in the frame of the theory of distributions.

1.1.2 Trajectory. Velocity. Acceleration. Force. Principles of mechanics

Starting from the equation $\mathbf{r} = \mathbf{r}(t)$ of the trajectory, we introduce the velocity $\mathbf{v}(t) = \dot{\mathbf{r}}(t)$ and the acceleration $\mathbf{a}(t) = \dot{\mathbf{v}}(t) = \ddot{\mathbf{r}}(t)$ as continuous functions of t in the interval $[t', t'']$, excepting a finite number of discontinuities of the first species at the moments $t = t_i$, $t' < t_i < t''$, $i = 1,2,...,n$. In these conditions, the jump $\mathbf{V}_i$ of the velocity, given by (5.1.34'), allows to introduce *the complementary acceleration* $\mathbf{a}_c(t)$ in the form (5.1.34''); adding *the acceleration in the usual sense* $\tilde{\mathbf{a}}(t)$ to this acceleration, we obtain *the acceleration in the sense of the theory of distributions* $\mathbf{a}(t)$ (the formula (5.1.34''')), corresponding to the Theorem 5.1.2 of the generalized acceleration.

Taking into account Newton's law (1.1.89), we introduce the notations

$$\mathbf{F}(t) = m\mathbf{a}(t), \quad \tilde{\mathbf{F}}(t) = m\tilde{\mathbf{a}}(t), \quad \mathbf{F}_c(t) = m\mathbf{a}_c(t), \qquad (10.1.5)$$

where $\mathbf{F}(t)$ is *the generalized force* (in the sense of the theory of distributions), $\tilde{\mathbf{F}}(t)$ is *the force in the usual sense*, while $\mathbf{F}_c(t)$ is *the complementary force* due to the discontinuities; the relation (5.1.34''') becomes

$$\mathbf{F}(t) = \tilde{\mathbf{F}}(t) + \mathbf{F}_c(t). \qquad (10.1.5')$$

We thus state

Theorem 10.1.1 (*theorem of the generalized force*). *The generalized force* (*in the sense of the theory of distributions*), *which acts upon a particle, is equal to the sum of the force in the usual sense and the complementary force* (due to discontinuities), *which act upon the same particle.*

In this case, one can affirm that by introducing the generalized accelerations and forces, as well as by using the derivatives in the sense of the theory of distributions, *the second principle of mechanics*, given by the formula (1.1.89), may be used in the frame of this theory too.

In what concerns *the principle of inertia*, we assume that for the generalized force $\mathbf{F} = \mathbf{0}$ we have $\mathbf{a} = \mathbf{0}$, that is

$$\tilde{\mathbf{a}}(t) + \sum_{i=1}^{n} \mathbf{V}_i(t - t_i) = \mathbf{0}. \qquad (10.1.6)$$

Considering the arbitrary fundamental function $\varphi(t)$, the support of which does not contain the moments t_i, $i = 1,2,...,n$, the relation (10.1.6) leads to $(\tilde{\mathbf{a}}(t), \varphi(t)) = \mathbf{0}$, wherefrom we deduce that the function $\tilde{\mathbf{a}}(t)$ vanishes everywhere, excepting eventually at the moments t_i. Let be now a fundamental function $\varphi(t)$, the support of which contains only the moment t_i; we obtain $\mathbf{V}_i \varphi(t_i) = \mathbf{0}$, so that $\mathbf{V}_i = \mathbf{0}$. Proceeding in the same manner for $i = 1,2,...,n$, we may write $\mathbf{V}_i = \mathbf{0}$, $i = 1,2,...,n$, ; replacing in (10.1.6), it results $\tilde{\mathbf{a}}(t) = \mathbf{0}$, hence the particle has a rectilinear and uniform motion. In conclusion, the principle of inertia (which represents a criterion by which we may know if upon a particle is acting a force) can be enounced in the classical form.

The other three principles of mechanics may be applied as in the case of the classical model.

1.1.3 Basic equation of motion

The motion of a particle is modelled by the second principle of Newton. We consider thus the equation of motion (6.1.22), where $\mathbf{F}(t)$ is the generalized force (10.1.5') and the differentiation takes place in the sense of the theory of distributions; we assume that $\mathbf{r}(t)$ is defined for $t \in \mathbb{R}$. This represents one of the methods to express the equation of motion in distributions; the boundary conditions must be separately specified.

If we wish to include the boundary conditions in the equation of motion, then we use another procedure. We start from the same equation of motion (6.1.22), but the force is that in the usual sense $\widetilde{\mathbf{F}}$; differentiating in the usual sense too, we write

$$m \frac{\tilde{d}^2}{dt^2} \mathbf{r}(t) = \widetilde{\mathbf{F}}(t), \quad t > t_0. \tag{10.1.7}$$

We assume that, at the initial moment $t = t_0$, the position vector and the velocity are given by $\mathbf{r}_0$ and $\mathbf{v}_0$, respectively (initial conditions of Cauchy type). We introduce a new function $\bar{r}(t)$, defined for $t \in \mathbb{R}$ by the relation

$$\bar{\mathbf{r}}(t) = \theta(t - t_0)\mathbf{r}(t) = \begin{cases} \mathbf{0} & \text{for } t < t_0, \\ \mathbf{r}_0 & \text{for } t = t_0, \\ \mathbf{r}(t) & \text{for } t > t_0; \end{cases} \tag{10.1.8}$$

the velocity reads

$$\bar{\mathbf{v}}(t) = \theta(t - t_0)\mathbf{v}(t) = \begin{cases} \mathbf{0} & \text{for } t < t_0, \\ \mathbf{v}_0 & \text{for } t = t_0, \\ \mathbf{v}(t) & \text{for } t > t_0. \end{cases} \tag{10.1.8'}$$

We notice thus that $\mathbf{r}_0$ and $\mathbf{v}_0$ are the jumps of the position vector and of the velocity, respectively, at the initial moment $t = t_0$. Differentiating successively the distribution corresponding to the function (10.1.8), we get the velocity and the acceleration of the particle in the form

$$\bar{\mathbf{v}}(t) = \frac{d}{dt}\bar{\mathbf{r}}(t) = \frac{\tilde{d}}{dt}\bar{\mathbf{r}}(t) + \mathbf{r}_0 \delta(t - t_0),$$

$$\bar{\mathbf{a}}(t) = \frac{d^2}{dt^2}\bar{\mathbf{r}}(t) = \frac{\tilde{d}^2}{dt^2}\bar{\mathbf{r}}(t) + \mathbf{v}_0 \delta(t - t_0) + \mathbf{r}_0 \dot{\delta}(t - t_0).$$

The first of these relations corresponds to the distribution defined by (10.1.8'); as well, the velocity and the acceleration in the usual sense have been put into evidence. Replacing in the equation (10.1.7) and assuming that, besides the moment of discontinuity $t = t_0$, may appear also other moments of discontinuity, which introduce the complementary force $\mathbf{F}_c(t)$, we can write the basic equation of mechanics in the form

$$m \frac{\mathrm{d}^2}{\mathrm{d}t^2} \bar{\mathbf{r}}(t) = \mathbf{F}(t) + m\mathbf{v}_0 \delta(t - t_0) + m\mathbf{r}_0 \dot{\delta}(t - t_0), \qquad (10.1.9)$$

where the generalized force $\mathbf{F}(t)$ is given by (10.1.5), (10.1.5').

By means of the formula (A.3.33'), we may write a fundamental particular solution of the operator $\mathrm{d}^2/\mathrm{d}t^2$ in the form

$$E(t) = t\theta(t) = t_+; \qquad (10.1.10)$$

in this case, the solution of the equation (10.1.9) is given by

$$\bar{\mathbf{r}}(t) = t\theta(t) * \left[\frac{1}{m} \mathbf{F}(t) + \mathbf{v}_0 \delta(t - t_0) + \mathbf{r}_0 \dot{\delta}(t - t_0) \right],$$

wherefrom it results

$$\bar{\mathbf{r}}(t) = \frac{1}{m} t_+ * \mathbf{F}(t) + \mathbf{v}_0 (t - t_0)_+ + \mathbf{r}_0 \theta(t - t_0). \qquad (10.1.11)$$

If $\mathbf{F}(t)$ is a locally integrable function, then we may write

$$\mathbf{r}(t) = \frac{1}{m} \int_{t_0}^{t} (t - \tau) \mathbf{F}(\tau) \mathrm{d}\tau + (t - t_0) \mathbf{v}_0 + \mathbf{r}_0, \quad t \geq t_0. \qquad (10.1.11')$$

In particular, if $\tilde{\mathbf{F}}(t) = \mathbf{F}_0 = \overrightarrow{\mathrm{const}}$, $\mathbf{F}_c = \mathbf{0}$, then we get

$$\bar{\mathbf{r}}(t) = \frac{1}{2m}(t - t_0)^2_+ \mathbf{F}_0 + (t - t_0)_+ \mathbf{v}_0 + \theta(t - t_0)\mathbf{r}_0; \qquad (10.1.12)$$

this relation may be expressed also in the form

$$\mathbf{r}(t) = \frac{1}{2m}(t - t_0)^2 \mathbf{F}_0 + (t - t_0)\mathbf{v}_0 + \mathbf{r}_0, \quad t \geq t_0. \qquad (10.1.12')$$

In the case in which $\mathbf{F} = \mathbf{F}(\mathbf{r}, \dot{\mathbf{r}}; t)$, the basic equation remains, further, of the form (10.1.9), the force in the sense of the theory of distributions having, obviously, this general character. To solve this equation one can no more use the method given above, because one must take into account the new form of the differential equation of motion.

1.1.4 Case of a particle subjected to constraints

In case of a particle subjected to constraints, these ones must be eliminated by introducing constraint forces. Let us thus consider the case of a particle subjected to a single holonomic and rheonomic bilateral constraint (the case of two constraints may be studied analogously), expressed in the form $f(x_1, x_2, x_3; t) = 0$, where the function f is piecewise defined in the form

$$f(x_1, x_2, x_3; t) = \sum_{i=1}^{n} \overline{f}_i(x_1, x_2, x_3; t); \qquad (10.1.13)$$

we denoted

$$\overline{f}_i(x_1, x_2, x_3; t) = \begin{cases} f_i(x_1, x_2, x_3; t) & \text{for } t \in (t_{i-1}, t_i), \\ 0 & \text{for } t \notin (t_{i-1}, t_i), \end{cases} \qquad (10.1.13')$$

assuming that $t_0 < t_1 < t_2 < ... < t_{n-1} < t_n$ for the moments of discontinuity.

Eliminating the constraint and introducing *the generalized constraint force* $\mathbf{R}(t)$, the equation of motion of the particle, considered now as a free one, reads

$$m \frac{d^2 \overline{\mathbf{r}}(t)}{dt^2} = \mathbf{F}(t) + \mathbf{R}(t), \qquad (10.1.14)$$

where, for the sake of simplicity, in the generalized force $\mathbf{F}(t)$ we have introduced also the sum $m\mathbf{v}_0 \delta(t - t_0) + m\mathbf{r}_0 \dot{\delta}(t - t_0)$, which is of the nature of a complementary force.

The generalized constraint force $\mathbf{R}(t)$ may be expressed by the relation

$$\mathbf{R}(t) = \sum_{i=1}^{n} \mathbf{R}_i(t), \qquad (10.1.15)$$

where $\mathbf{R}_i(t)$ are the generalized constraint forces given by

$$\mathbf{R}_i(t) = \widetilde{\mathbf{R}}_i(t) + \mathbf{R}_{ic}(t). \qquad (10.1.15')$$

We denoted by $\widetilde{\mathbf{R}}_i(t)$ *the constraint forces in the usual sense*, which correspond to the moments $t \in (t_{i-1}, t_i)$, $i = 1, 2, ..., n$, and are expressed in the form

$$\widetilde{\mathbf{R}}_i(t) = \lambda_i(t) \nabla \overline{f}_i = \lambda_i(t) \operatorname{grad} \overline{f}_i, \qquad (10.1.15'')$$

where $\lambda_i(t)$ is a scalar, ∇ is Hamilton's operator and $\overline{f}_i \in C^1$; as well, $\mathbf{R}_{ic}(t)$ are *complementary constraint forces*, corresponding to the moments of discontinuity t_i,

$i = 0, 1, 2, ..., n$. The constraint forces in the usual sense are directed, in the intervals of definition, along the normals to the surfaces $f_i(x_1, x_2, x_3; t) = 0$, considered as rigid at the moment $t \in (t_{i-1}, t_i)$. On the intersection lines of these surfaces, that is on the lines corresponding to the moments t_i, these forces cannot be determined by means of the expressions (10.1.15").

To obtain a unitary expression of the generalized constraint force $\mathbf{R}_i(t)$, we use the equation of motion (10.1.14), written in the form

$$\mathsf{F}(t) \equiv \mathbf{F}(t) + \mathbf{R}(t) - m\frac{d^2\mathbf{\bar{r}}}{dt^2} = \mathbf{0}; \tag{10.1.14'}$$

this equation may be written everywhere, excepting the points which correspond to the moments t_i. We introduce also the moments of discontinuity if we replace the equation (10.1.14') by the equation

$$\prod_{i=1}^{n}(t - t_i)^{m_i + 1}\mathsf{F}(t) = \mathbf{0}, \quad m_i \in \mathbb{N}, \tag{10.1.14''}$$

which, obviously, may be written for $t \in \mathbb{R}$. In the frame of the usual functions, the solution of the equation (10.1.14") is (10.1.14'); to obtain a generalized solution, we assume that $\mathsf{F}(t)$ can be a distribution and we consider the equation (10.1.14") in the sense of the theory of distributions. Thus, the solution in distributions of the equation (10.1.14") includes, as a particular case, the solution (10.1.14') too.

To can make a study of the equation (10.1.14"), there are necessary some results concerning the structure of the distributions with punctual support. In this order of ideas, we state

Theorem 10.1.2. *A distribution $f(x)$ of a single variable satisfies the equation*

$$P(x)f(x) = 0, \tag{10.1.16}$$

where $P(x)$ is a polynomial, if and only if that one is expressed in the form

$$f(x) = \sum_{i=1}^{s} c_i \delta(x - x_i) + \sum_{k=1}^{r} \sum_{j=0}^{m_k - 1} c_j^k \delta^{(j)}(x - x_k'). \tag{10.1.16'}$$

Here, x_i, $i = 1, 2, ..., s$, are the simple roots of the polynomial $P(x)$, while x_k', $k = 1, 2, ..., r$, represent the multiple roots, the multiplicity order of which is m_k; the quantities c_i and c_j^k are constants, while δ is Dirac's distribution.

As well, we can state

Theorem 10.1.3. *If the support of the distribution $f(x)$ is formed by the points x_i, $i = 1, 2, ..., n$, then $f(x)$ is of the form*

$$f(x) = \sum_{i=1}^{n}\sum_{j=0}^{m_i} c_j^i \delta^{(j)}(x - x_i), \qquad (10.1.16'')$$

where c_j^i and m_i are constants, while δ is Dirac's distribution.

Taking into account the Theorem 10.1.2, we obtain the complementary constraint force in the form

$$\mathbf{R}_c = \sum_{i=1}^{n} \mathbf{R}_{ic} = \sum_{i=1}^{n}\sum_{j=0}^{m_i} \boldsymbol{\beta}_{ij} \delta^{(j)}(t - t_i); \qquad (10.1.17)$$

adding the solution corresponding to usual functions, it results the generalized constraint force

$$\mathbf{R}(t) = \sum_{i=1}^{n} \lambda_i(t) \nabla \overline{f}_i + \sum_{i=1}^{n}\sum_{j=0}^{m_i} \boldsymbol{\beta}_{ij} \delta^{(j)}(t - t_i). \qquad (10.1.17')$$

We may thus state (the result remains valid also in the case of two constraints)
Theorem 10.1.4 (*theorem of the generalized constraint force*). *The generalized constraint force which acts upon a particle subjected to holonomic and rheonomic bilateral constraints is equal to the sum of the constraint force in the usual sense at the moments in which that one is defined and the complementary constraint force due to the moments of discontinuity and expressed by means of Dirac's distribution and of its derivatives.*

This result shows that the operator ∇, applied in the sense of the theory of distributions, leads to a formula of the form

$$\nabla f = \widetilde{\nabla} f + \sum_{i=1}^{n}\sum_{j=0}^{m_i} \boldsymbol{\beta}_{ij} \delta^{(j)}(t - t_i), \qquad (10.1.18)$$

where $f(x_1, x_2, x_3; t)$ is a function of class C^1, excepting the points of discontinuity of the first species $t = t_i$, $i = 1, 2, ..., n$, and the symbol $\widetilde{\nabla}$ represents the operator ∇ in the usual sense.

We notice that the number m_i is indeterminate in the formulae (10.1.17), (10.1.17'); we make the same remark concerning the vectors $\boldsymbol{\beta}_{ij}$. The significance of those quantities is clear if we take into consideration the motion of the particle. Thus, comparing the equation (10.1.14), (10.1.17') with the equation (10.1.9), we see that $m_1 = 1$, while $\boldsymbol{\beta}_{10} = m\mathbf{v}_0$, $\boldsymbol{\beta}_{11} = m\mathbf{r}_0$; there are not other moments of discontinuity unlike the initial moment.

We must also notice that, in order to ensure the existence and the uniqueness of the solution, it is necessary to put supplementary conditions at the points of discontinuity. Thus, returning to the example in Subsec. 1.1.1, (Fig.10.1), it is necessary to give, besides the initial conditions at the point A (at the moment $t = 0$), the velocity to the right of the point B; otherwise, the motion on the line segment BC remains indeterminate. In conformity with the Cauchy-Lipschitz theorem, the force which acts upon the particle must be of class C^0; just this condition is not satisfied at the point B. To determine the motion on BC, the initial conditions at B must be given (at the moment $t = t_0$); hence, one has to specify $v(t_0 + 0)$ (the direction of the velocity is known). As we have seen, a relation of experimental nature between $\mathbf{v}(t_0 - 0)$ and $\mathbf{v}(t_0 + 0)$ is usually given.

1.1.5 Motion of a heavy particle in vacuum. Bilocal problems

Let us consider the motion of a heavy particle in vacuum, acted upon by the force $\mathbf{F} = m\mathbf{g}$, where $\mathbf{g}$ is the gravity acceleration. The corresponding *Cauchy problem* consists in the determination of the position vector $\mathbf{r} = \mathbf{r}(t)$ if the position vector $\mathbf{r}_0$ and the velocity $\mathbf{v}_0$ at the initial moment t_0 are given (see Chap. 7, Subsec. 1.2.1 too) (Fig.10.3,a). We are in the case $\mathbf{F} = \overrightarrow{\text{const}}$, so that the formula (10.1.12) allows to write

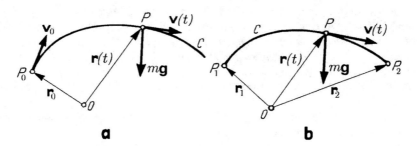

Figure 10.3. Motion of a heavy particle in vacuum. Cauchy's problem (a); bilocal problem (b).

$$\bar{\mathbf{r}}(t) = \frac{1}{2}(t - t_0)_+^2 \mathbf{g} + (t - t_0)_+ \mathbf{v}_0 + \theta(t - t_0)\mathbf{r}_0,$$
$$\bar{\mathbf{v}}(t) = (t - t_0)_+ \mathbf{g} + \theta(t - t_0)\mathbf{v}_0 + \delta(t - t_0)\mathbf{r}_0$$
(10.1.19)

or

$$\mathbf{r}(t) = \frac{1}{2}(t - t_0)^2 \mathbf{g} + (t - t_0)\mathbf{v}_0 + \mathbf{r}_0, \quad t \geq 0.$$
(10.1.19')

This result can be used to solve bilocal problems too; thus, if we have to determine the motion of a heavy particle in vacuum, for which the conditions (Fig.10.3,b)

$$\bar{\mathbf{r}}(t_1) = \mathbf{r}_1, \quad \bar{\mathbf{r}}(t_2) = \mathbf{r}_2$$
(10.1.20)

are fulfilled, then we may use as well a formula of the form (10.1.19), assuming that the velocity at the initial moment $t = t_1$ is known. It results

Other considerations on particle dynamics 605

$$\overline{\mathbf{r}}(t) = \frac{1}{2}(t - t_1)_+^2 \mathbf{g} + (t - t_1)_+ \mathbf{v}_1 + \theta(t - t_1)\mathbf{r}_1.$$

If we put also the second condition (10.1.20), then we can write

$$\mathbf{r}_2 = \frac{1}{2}(t_2 - t_1)_+^2 \mathbf{g} + (t_2 - t_1)_+ \mathbf{v}_1 + \theta(t_2 - t_1)\mathbf{r}_1 = \frac{1}{2}(t_2 - t_1)^2 \mathbf{g} + (t_2 - t_1)\mathbf{v}_1 + \mathbf{r}_1,$$

wherefrom

$$\mathbf{v}_1 = \frac{1}{t_2 - t_1}(\mathbf{r}_2 - \mathbf{r}_1) - \frac{1}{2}(t_2 - t_1)\mathbf{g};$$

we get thus the solution of the bilocal problem in the form

$$\overline{\mathbf{r}}(t) = \frac{1}{2}(t - t_1)_+ [(t - t_1)_+ - (t_2 - t_1)]\mathbf{g} + \frac{1}{t_2 - t_1}(t - t_1)_+(\mathbf{r}_2 - \mathbf{r}_1) + \theta(t - t_1)\mathbf{r}_1$$

(10.1.21)

or

$$\mathbf{r}(t) = -\frac{1}{2}(t - t_1)(t_2 - t)\mathbf{g} + \frac{1}{t_2 - t_1}[(t - t_1)\mathbf{r}_2 + (t_2 - t)\mathbf{r}_1], \quad t \in [t_1, t_2].$$

(10.1.21')

The velocity is given by

$$\mathbf{v}(t) = \left[t - \frac{1}{2}(t_1 + t_2)\right]\mathbf{g} + \frac{1}{t_2 - t_1}(\mathbf{r}_2 - \mathbf{r}_1), \quad t \in [t_1, t_2]. \qquad (10.1.21'')$$

Let us consider also the case of *the mixed bilocal problem* for which the boundary conditions

$$\overline{\mathbf{r}}(t_1) = \mathbf{r}_1, \quad \overline{\mathbf{v}}(t_2) = \mathbf{v}_2 \qquad (10.1.22)$$

are put. If the velocity at the initial moment $t = t_1$ would be known too, then we could write

$$\overline{\mathbf{v}}(t) = (t - t_1)_+ \mathbf{g} + \theta(t - t_1)\mathbf{v}_1 + \delta(t - t_1)\mathbf{r}_1;$$

if we put now the second bilocal condition (10.1.22), then it results $\mathbf{v}_2 = (t_2 - t_1)\mathbf{g} + \mathbf{v}_1$, wherefrom $\mathbf{v}_1 = \mathbf{v}_2 - (t_2 - t_1)\mathbf{g}$. The trajectory of the particle is thus given by the relation

$$\overline{\mathbf{r}}(t) = (t - t_1)_+ \left[\frac{1}{2}(t - t_1)_+ - (t_2 - t_1)\right]\mathbf{g} + (t - t_1)_+ \mathbf{v}_2 + \theta(t - t_1)\mathbf{r}_1 \qquad (10.1.23)$$

or by the relation

$$\mathbf{r}(t) = -\frac{1}{2}(t - t_1)(2t_2 - t_1 - t)\mathbf{g} + (t - t_1)\mathbf{v}_2 + \mathbf{r}_1, \quad t \in [t_1, t_2]. \quad (10.1.23')$$

The velocity reads

$$\mathbf{v}(t) = (t - t_2)\mathbf{g} + \mathbf{v}_2, \quad t \in [t_1, t_2]. \quad (10.1.23'')$$

1.2 General theorems

In the following, we introduce some mechanical quantities (moment, moment of momentum, work, kinetic energy, impulse of the force, impulse of the moment of the force) in the frame of the theory of distributions. This allows the statement of the general theorems corresponding to motions with discontinuous characteristics or to the case of elastic collisions.

1.2.1 Momentum. Moment of momentum. Work. Kinetic energy

The momentum

$$\mathbf{H}(t) = m\mathbf{v}(t) \quad (10.1.24)$$

of a particle is a continuous function in the considered interval of time $[t', t'']$, excepting the moments t_i, $i = 1, 2, \ldots, n$, where appear discontinuities of the first species. The jump of the momentum at the moment of discontinuity t_i is given by

$$(\Delta \mathbf{H})_i = m\mathbf{V}_i = m[\mathbf{v}(t_i + 0) - \mathbf{v}(t_i - 0)], \quad (10.1.24')$$

being expressed by means of the velocity at the same moment.

We introduce also the quantities

$$\int_{t'}^{t''} \mathbf{F}(t)\mathrm{d}t, \quad \int_{t'}^{t''} \tilde{\mathbf{F}}(t)\mathrm{d}t, \quad \int_{t'}^{t''} \mathbf{F}_c(t)\mathrm{d}t, \quad (10.1.25)$$

which represent *the impulse of the generalized force, of the force in the usual sense* and *of the complementary force*, respectively, in the time interval $[t', t'']$; we have adopted classical notations for the first and the third of these quantities, to have a uniform symbolism, although the respective integrals have not sense, in general, from the point of view of the theory of distributions. Observing that

$$\int_{t'}^{t''} \mathbf{F}_c \mathrm{d}t = \sum_{i=1}^{n} \int_{t'}^{t''} m\mathbf{V}_i \delta(t - t_i)\mathrm{d}t = \sum_{i=1}^{n} m\mathbf{V}_i, \quad t' < t < t'',$$

and taking into account (10.1.5'), we may write

$$\int_{t'}^{t''} \mathbf{F}(t)\mathrm{d}t = \int_{t'}^{t''} \tilde{\mathbf{F}}(t)\mathrm{d}t + \sum_{i=1}^{n} (\Delta \mathbf{H})_i, \quad (10.1.26)$$

wherefrom we state

Theorem 10.1.5 (*theorem of impulse of the generalized force*). *The impulse of the generalized force which acts upon a particle, in a certain time interval, is equal to the sum of the impulse of the force in the usual sense, which acts upon that particle, in the same time interval, and the sum of the jumps of the momentum of the particle, corresponding to the moments of discontinuity.*

In case of the problem considered in Subsec. 1.1.1 we may write ($t' = 0$, $t'' = t_0$)

$$\int_0^{t_0} \mathbf{F}(t)\mathrm{d}t = mgt_0 \sin\alpha \mathbf{u}_1 + m(\mathbf{v}_2 - \mathbf{v}_1).$$

The moment of momentum of a particle is introduced by means of the relation

$$\mathbf{K}_O(t) = \mathbf{r}(t) \times \mathbf{H}(t) = \mathbf{r}(t) \times [m\mathbf{v}(t)]; \tag{10.1.27}$$

its jump at the moment of discontinuity t_i is given by

$$(\Delta\mathbf{K}_O)_i = m\mathbf{r}_i(t) \times \mathbf{v}(t_i + 0) - m\mathbf{r}_i(t) \times \mathbf{v}(t_i - 0) = m\mathbf{r}(t_i) \times \mathbf{V}_i = \mathbf{r}(t_i) \times (\Delta\mathbf{H})_i \tag{10.1.27'}$$

and is expressed with the aid of the jump of the momentum (or of the jump of the velocity) at the same moment.

We introduce also the quantities

$$\int_{t'}^{t''} \mathbf{M}_O(\mathbf{F}(t))\mathrm{d}t, \quad \int_{t'}^{t''} \mathbf{M}_O(\widetilde{\mathbf{F}}(t))\mathrm{d}t, \quad \int_{t'}^{t''} \mathbf{M}_O(\mathbf{F}_c(t))\mathrm{d}t, \tag{10.1.28}$$

which represent *the impulse of the moment of the generalized force, of the moment of the force in the usual sense* and *of the moment of the complementary force*, respectively, in the time interval $[t', t'']$. In general, the first and the third of these integrals have no sense from the point of view of the theory of distributions; but we adopted classical notations, so as to have a uniform symbolism. Starting from the relation (10.1.5'), we can write

$$\int_{t'}^{t''} \mathbf{M}_O(\mathbf{F}(t))\mathrm{d}t = \int_{t'}^{t''} \mathbf{M}_O(\widetilde{\mathbf{F}}(t))\mathrm{d}t + \int_{t'}^{t''} \mathbf{M}_O(\mathbf{F}_c(t))\mathrm{d}t$$

and are led to

$$\int_{t'}^{t''} \mathbf{M}_O(\mathbf{F}(t))\mathrm{d}t = \int_{t'}^{t''} \mathbf{M}_O(\widetilde{\mathbf{F}}(t))\mathrm{d}t + \sum_{i=1}^{n}(\Delta\mathbf{K}_O)_i; \tag{10.1.29}$$

thus, we may state

Theorem 10.1.6 (*theorem of impulse of the moment of the generalized force*). *The impulse of the moment with respect to a given pole of the generalized force, which acts upon a particle, in a certain time interval, is equal to the sum of the impulse of the moment with respect to the same pole of the force in the usual sense, which acts upon* *that particle, in the same time interval, and the sum of the jumps of the moment of momentum of the particle, corresponding to the moments of discontinuity.*

In case of the problem in Subsec. 1.1.1, we can write

$$\int_0^{t_0} \mathbf{r}(t) \times \mathbf{F}(t) \mathrm{d}t = \int_0^{t_0} \mathbf{r}(t) \times \widetilde{\mathbf{F}}(t) \mathrm{d}t + \mathbf{r}(t_0) \times [m(\mathbf{v}_2 - \mathbf{v}_1)]$$
$$= \int_0^{t_0} \mathbf{r}(t) \times \widetilde{\mathbf{F}}(t) \mathrm{d}t + \overrightarrow{AB} \times (m\mathbf{v}_2),$$

assuming that the point A is chosen as pole of the position vectors.

The work effected by the force $\mathbf{F}(t)$ in the interval of time $[t',t'']$ is given by

$$W = \int_{t'}^{t''} \mathbf{F}(t) \cdot \mathrm{d}\mathbf{r}(t) ; \qquad (10.1.30)$$

we introduce the notations

$$W_F = \int_{t'}^{t''} \mathbf{F}(t) \cdot \mathrm{d}\mathbf{r}(t), \quad W_{\widetilde{F}} = \int_{t'}^{t''} \widetilde{\mathbf{F}}(t) \cdot \mathrm{d}\mathbf{r}(t), \quad W_{F_c} = \int_{t'}^{t''} \mathbf{F}_c(t) \cdot \mathrm{d}\mathbf{r}(t), \quad (10.1.31)$$

where W_F is *the work of the generalized force* $\mathbf{F}(t)$, $W_{\widetilde{F}}$ is *the work of the force in the usual sense* $\widetilde{\mathbf{F}}(t)$ and W_{F_c} is *the work of the complementary force* $\mathbf{F}_c(t)$, in the time interval $[t',t'']$. The first and the third of these integrals have no sense the point of view of the theory of distributions; but for the uniformity of the symbolism we adopt these classical notations. Taking into account (10.1.5'), we may write

$$W_F = W_{\widetilde{F}} + W_{F_c}, \qquad (10.1.31')$$

obtaining thus

Theorem 10.1.7 (*theorem of work*). *The work of the generalized force which acts upon a particle in a certain time interval is the sum of the work of the force in the usual sense, which acts upon that particle, in the same time interval, and the work of the complementary force, which acts upon the same particle in the considered time interval.*

The kinetic energy of the particle is given by the relation

$$T = \frac{1}{2} m v^2(t). \qquad (10.1.32)$$

If t_i is a moment of discontinuity, then we may write the relations

$$\lim_{t \to t_i \pm 0} T(t) = \frac{1}{2} m \left[\lim_{t \to t_i \pm 0} v(t) \right]^2 = \frac{1}{2} m v^2(t_i \pm 0) \qquad (10.1.32')$$

hence, there results that the moments of discontinuity of the velocity are the moments of discontinuity of the kinetic energy too. We obtain, as well,

Other considerations on particle dynamics   609

$$(\Delta T)_i = \frac{1}{2} m \left[v^2(t_i + 0) - v^2(t_i - 0) \right]. \tag{10.1.32''}$$

1.2.2 General theorems in case of motions with discontinuous characteristics

The second principle of mechanics takes the form

$$\mathbf{F}(t) = \frac{\mathrm{d}}{\mathrm{d}t}[m\mathbf{v}(t)] = \frac{\mathrm{d}}{\mathrm{d}t}\mathbf{H}(t) \tag{10.1.33}$$

in the frame of the theory of distributions; if discontinuities (in the usual sense) do not appear, then it becomes

$$\tilde{\mathbf{F}}(t) = \frac{\tilde{\mathrm{d}}}{\mathrm{d}t}[m\mathbf{v}(t)] = \frac{\tilde{\mathrm{d}}}{\mathrm{d}t}\mathbf{H}(t). \tag{10.1.33'}$$

We notice that

$$\frac{\mathrm{d}}{\mathrm{d}t}\mathbf{H}(t) = \frac{\tilde{\mathrm{d}}}{\mathrm{d}t}\mathbf{H}(t) + \sum_{i=1}^{n} (\Delta \mathbf{H})_i \, \delta(t - t_i),$$

where the jump of the momentum is given by (10.1.24'); taking into account also the relation

$$\mathbf{F}_c = \sum_{i=1}^{n} (\Delta \mathbf{H})_i \, \delta(t - t_i), \tag{10.1.34}$$

which gives the complementary force, we find the relation

$$\frac{\mathrm{d}}{\mathrm{d}t}\mathbf{H}(t) = \mathbf{F}(t) = \tilde{\mathbf{F}}(t) + \mathbf{F}_c(t), \tag{10.1.35}$$

being thus led to

Theorem 10.1.8 (*theorem of momentum*). *The derivative with respect to time, in the sense of the theory of distributions, of the momentum of a free particle is equal to the generalized force which acts upon that particle.*

As we have seen in Subsec. 1.1.4, if the particle is subjected to bilateral constraints, then we must introduce *the generalized constraint force* (in the sense of the theory of distributions) $\mathbf{R}(t)$, given by

$$\mathbf{R}(t) = \tilde{\mathbf{R}}(t) + \mathbf{R}_c(t), \tag{10.1.36}$$

where $\tilde{\mathbf{R}}(t)$ is *the constraint force in the usual sense*, while $\mathbf{R}_c(t)$ is *the complementary constraint force* due to the discontinuities. In this case, the theorem of momentum takes the form

$$\frac{\mathrm{d}}{\mathrm{d}t}\mathbf{H}(t) = \mathbf{F}(t) + \mathbf{R}(t).\tag{10.1.35'}$$

The complementary constraint force is expressed in the form

$$\mathbf{R}_c(t) = \sum_{i=1}^{n}\mathbf{R}_{ic}(t) = \sum_{i=1}^{n}m\mathbf{V}_i\delta(t - t_i).$$

Returning to the problem in Subsec. 1.1.1, we notice that the constraint force at the point B is given by

$$\mathbf{R} = m\mathbf{V}\delta(t - t_0) = m(\mathbf{v}_2 - \mathbf{v}_1)\delta(t - t_0).$$

If we have $k = 1$ in the relation (10.1.2) (perfect elastic collision), then it results $|\mathbf{v}_2| = |\mathbf{v}_1|$; the jump of the constraint force will have as direction the internal bisectrix of $\widehat{ABC}$, while the trajectory will be tangent to the external bisectrix. If $k = 0$ (perfect plastic collision), then we have $\mathbf{v}_2 = \mathbf{0}$ (the second limit case).

If the velocity and the acceleration are continuous functions (in the usual sense), then we can write the theorem of moment of momentum in the form

$$\frac{\widetilde{\mathrm{d}}}{\mathrm{d}t}\mathbf{K}_O(t) = \mathbf{r}(t) \times \widetilde{\mathbf{F}}(t) = \mathbf{M}_O(\widetilde{\mathbf{F}}(t)).$$

But if the velocity and the acceleration have discontinuities of the first species, then we notice that the moment of momentum $\mathbf{K}_O(t)$ and its derivative in the sense of the theory of distributions have the same moments of discontinuity; we may write

$$\frac{\mathrm{d}}{\mathrm{d}t}\mathbf{K}_O(t) = \frac{\widetilde{\mathrm{d}}}{\mathrm{d}t}\mathbf{K}_O(t) + \sum_{i=1}^{n}(\Delta\mathbf{K}_O)_i\,\delta(t - t_i),$$

where the jump of the moment of momentum is given by (10.1.27'). In this case, we get

$$\frac{\mathrm{d}}{\mathrm{d}t}\mathbf{K}_O(t) = \mathbf{M}_O(\mathbf{F}(t)) = \mathbf{M}_O(\widetilde{\mathbf{F}}(t)) + \mathbf{M}_O(\mathbf{F}_c(t)),\tag{10.1.37}$$

the moment of the complementary force being given by

$$\mathbf{M}_O(\mathbf{F}_c(t)) = \sum_{i=1}^{n}\mathbf{r}(t_i) \times (\Delta\mathbf{H})_i\,\delta(t - t_i) = \sum_{i=1}^{n}(\Delta\mathbf{K}_O)_i\,\delta(t - t_i).\tag{10.1.38}$$

In this case, we can state

Theorem 10.1.9 (*theorem of moment of momentum*). *The derivative with respect to time, in the sense of the theory of distributions, of the moment of momentum with respect to a given pole of a free particle is equal to the moment, with respect to the same pole, of the generalized force which acts upon the respective particle.*

Other considerations on particle dynamics 611

Introducing the generalized force too, in case of a particle subjected to bilateral constraints, the theorem of moment of momentum takes the form

$$\frac{\mathrm{d}}{\mathrm{d}t}\mathbf{K}_O(t) = \mathbf{M}_O(\mathbf{F}(t)) + \mathbf{M}_O(\mathbf{R}(t)). \qquad (10.1.37')$$

In the classical case (in the usual sense), the theorem of kinetic energy has the finite form

$$W = T(t'') - T(t').$$

For the derivative with respect to time of the kinetic energy we may write

$$\frac{\mathrm{d}}{\mathrm{d}t}T(t) = \frac{\tilde{\mathrm{d}}}{\mathrm{d}t}T(t) + \sum_{i=1}^{n}(\Delta T)_i\,\delta(t-t_i),$$

where the jump of the kinetic energy, corresponding to the moment of discontinuity t_i, is given by (10.1.32''). Using the relation (10.1.31), we obtain

$$W_F = \int_{t'}^{t''}\frac{\tilde{\mathrm{d}}}{\mathrm{d}t}T(t)\,\mathrm{d}t + \sum_{i=1}^{n}\int_{t'}^{t''}(\Delta T)_i\,\delta(t-t_i)\,\mathrm{d}t,$$

wherefrom

$$W_F = T(t'') - T(t') + \sum_{i=1}^{n}(\Delta T)_i. \qquad (10.1.39)$$

We notice thus that the complementary work is just the jump of the kinetic energy at the respective moment of discontinuity. We may state

Theorem 10.1.10 (*theorem of kinetic energy in finite form*). *The work of the generalized force which acts upon a particle in a certain interval of time is equal to the sum of the difference between the kinetic energy at the final moment and the kinetic energy at the initial one and the sum of the jumps of the kinetic energy of that particle, corresponding to the moments of discontinuity.*

In case of the problem considered in Subsec. 1.1.1, we have

$$W_F = \int_0^{t_0} mg\sin\alpha\,\mathrm{d}r + (\Delta T)_0 = mg\overline{AB}\sin\alpha + \frac{1}{2}m\left(v_2^2 - v_1^2\right);$$

if $k = 1$ (perfect elastic collision), then the jump of the kinetic energy vanishes.

1.2.3 General theorems in case of elastic collisions

Starting from the notion of generalized force and of impulse of the generalized force, we introduce the notion of *percussion of a particle*, by means of the definition formula

$$\mathbf{P} = \lim_{t''-t' \to 0+0} \int_{t'}^{t''} \mathbf{F}(t) \mathrm{d}t, \qquad (10.1.40)$$

where the limit is considered in the sense of the theory of distributions. Obviously, neither in this case the integral written above has not sense from the point of view of the theory of distributions; but we use this symbolism, to be closer to the classical one. We assume, as well, that the time interval $[t',t'']$ contains only one moment of discontinuity t_0 and is thus that $|t'' - t'| < \varepsilon$, $\varepsilon > 0$ arbitrary. In this case, we have not to do with usual forces, but with generalized ones, more precisely, the distribution $\delta(t - t_0)$ appears; thus, the phenomenon of collision is no more introduced in mechanics as a special phenomenon, but as a usual one, where the principles of mechanics are applied in the conditions enounced in Subsec. 1.1.2.

Using a mean value theorem, we may write

$$\lim_{t''-t' \to 0+0} \int_{t'}^{t''} \widetilde{\mathbf{F}}(t) \mathrm{d}t = \mathbf{0}, \quad \lim_{t''-t' \to 0+0} \int_{t'}^{t''} \mathbf{M}_O\left(\widetilde{\mathbf{F}}(t)\right) \mathrm{d}t = \mathbf{0}.$$

Hence, the impulse of the force and the impulse of the moment of the force in the usual sense are quantities which can be neglected with respect to the impulse of the complementary force and of the moment of the complementary force due to the discontinuities, respectively; it results

$$\lim_{t''-t' \to 0+0} \int_{t'}^{t''} \mathbf{F}(t) \mathrm{d}t = m\mathbf{v}_0, \quad \lim_{t''-t' \to 0+0} \int_{t'}^{t''} \mathbf{M}_O(\mathbf{F}(t)) \mathrm{d}t = \mathbf{r}_0 \times (m\mathbf{v}_0),$$

where $\mathbf{r}_0$ is the position vector corresponding to the moment of discontinuity, while $\mathbf{v}_0$ is the jump of the velocity, corresponding to the same moment. The Theorem 10.1.8 takes the form

$$(\Delta \mathbf{H})_0 = \mathbf{P}, \qquad (10.1.41)$$

so that we may state

Theorem 10.1.11 (*theorem of momentum*). *The jump of the momentum of a free particle at a moment of discontinuity is equal to the percussion which acts upon that particle at the same moment.*

Using the Theorem 10.1.9 and the notation (10.1.40), we get

$$(\Delta \mathbf{K}_O)_0 = \mathbf{r}_0 \times (\Delta \mathbf{H})_0 = \mathbf{r}_0 \times \mathbf{P}; \qquad (10.1.42)$$

thus, we state

Theorem 10.1.12 (*theorem of moment of momentum*). *The jump of the moment of momentum with respect to a given pole of a free particle (which is equal to the moment with respect to that pole of the jump of the momentum of the particle) at a moment of discontinuity is equal to the moment with respect to the same pole of the percussion which acts upon that particle at the same moment.*

Other considerations on particle dynamics 613

Let $\mathbf{v}'$ and $\mathbf{v}''$ be the velocities of the particle before and after collision, respectively; we can write the theorem of momentum (10.1.41) in the form

$$m\mathbf{v}_0 = m(\mathbf{v}'' - \mathbf{v}') = \mathbf{P}. \tag{10.1.43}$$

A scalar product of this relation by $\mathbf{v}''$ leads to $m(v'')^2 - m\mathbf{v}' \cdot \mathbf{v}'' = \mathbf{P} \cdot \mathbf{v}''$ or $T'' - T' + T_0 = \mathbf{P} \cdot \mathbf{v}''$, where ($\mathbf{v}_0 = \mathbf{v}'' - \mathbf{v}'$)

$$T' = \frac{1}{2}m(v')^2, \quad T'' = \frac{1}{2}m(v'')^2, \quad T_0 = \frac{1}{2}m(\mathbf{v}'' - \mathbf{v}')^2 = \frac{1}{2}mv_0^2 \tag{10.1.44}$$

are *the kinetic energy before* and *after collision* and *the kinetic energy of the lost velocities*, respectively. The variation of the kinetic energy is given by

$$(\Delta T)_0 = T'' - T'; \tag{10.1.44'}$$

thus, we may write

$$(\Delta T)_0 + T_0 = \mathbf{P} \cdot \mathbf{v}'' \tag{10.1.45}$$

and we can state
Theorem 10.1.13 (*theorem of kinetic energy*). *The sum of the variation of the kinetic energy of a free particle at a moment of discontinuity and the kinetic energy of the lost velocity at the same moment is equal to the scalar product of the percussion which acts upon the particle by the velocity after that moment of discontinuity.*
  If the relation

$$\mathbf{P} \cdot \mathbf{v}'' = 0 \tag{10.1.46}$$

takes place, which can happen, e.g., if the velocity of the particle vanishes after collision, then we obtain the relation

$$(\Delta T)_0 + T_0 = 0, \tag{10.1.47}$$

so that we may state
Theorem 10.1.14 (*Carnot*). *If, in the motion of a free particle subjected to collision, the condition* (10.1.46) *is fulfilled, then the sum of the variation of the kinetic energy of that particle at a moment of discontinuity and the kinetic energy of the lost velocity at the same moment is equal to zero.*
  A scalar product of the relation (10.1.43) by $\mathbf{v}'$ leads to $m\mathbf{v}' \cdot \mathbf{v}'' - m(v')^2 = \mathbf{P} \cdot \mathbf{v}'$ or to $T'' - T' - T_0 = \mathbf{P} \cdot \mathbf{v}'$, wherefrom

$$(\Delta T)_0 - T_0 = \mathbf{P} \cdot \mathbf{v}'; \tag{10.1.45'}$$

we may thus state

Theorem 10.1.13' (*analogous to the theorem of kinetic energy*). *The difference between the variation of the kinetic energy of a free particle at a moment of discontinuity and the kinetic energy of the lost velocity at the same moment is equal to the scalar product of the percussion which acts upon this particle by its velocity before that moment of discontinuity.*

If, in particular,

$$\mathbf{P} \cdot \mathbf{v}' = 0, \qquad (10.1.46')$$

it results

$$(\Delta T)_0 = T_0. \qquad (10.1.47')$$

Thus, we state

Theorem 10.1.14' (*analogous to Carnot's theorem*). *If the condition* (10.1.46') *is fulfilled, in the motion of a free particle subjected to collision, then the variation of the kinetic energy of that particle, at a moment of discontinuity, is equal to the kinetic energy of the lost velocity at the same moment.*

Summing the relations (10.1.45) and (10.1.45'), we get

$$(\Delta T)_0 = \frac{1}{2} \mathbf{P} \cdot (\mathbf{v}' + \mathbf{v}''), \qquad (10.1.48)$$

so that it results

Theorem 10.1.15 (*Kelvin*). *The variation of the kinetic energy of a free particle at a moment of discontinuity is equal to the scalar product of the percussion which acts upon the particle by the semi-sum of both the velocities before and after the phenomenon of discontinuity.*

Subtracting (10.1.45') from (10.1.45), we get

$$T_0 = \frac{1}{2} \mathbf{P} \cdot (\mathbf{v}'' - \mathbf{v}') = \frac{1}{2} \mathbf{P} \cdot \mathbf{v}_0; \qquad (10.1.48')$$

we may thus write

Theorem 10.1.15' (*analogous of Kelvin's theorem*). *The kinetic energy of the lost velocity of a free particle at a moment of discontinuity is equal to half of the scalar product of the percussion which acts upon the particle by the jump of the velocity at the moment of discontinuity.*

The phenomenon of collision will be studied in detail in Chap. 13, §1, where the corresponding mathematical model is better put in evidence in case of a mechanical system which is not reduced to only one particle.

2. Motion of a particle with respect to a non-inertial frame of reference

The results obtained till now concerning the motion of a particle are using the fundamental law of Newton, written with respect to an inertial frame of reference. Taking into account that not all frames are inertial (or cannot be approximated to inertial ones), it is necessary to perform a study of the motion of a particle with respect to a non-inertial frame (starting, e.g., from a geocentric or heliocentric frame); we obtain thus important results concerning terrestrial mechanics too.

2.1 Relative motion. Relative equilibrium

Starting from the results obtained in Chap. 5, Sec. 3.1, concerning kinematics of the relative motion of a particle, we make, in what follows, a study of dynamics of the relative motion of that particle; in particular, we get results for the relative equilibrium of it. The application of general theorems leads to the law of motion with respect to a movable frame; one can thus specify the set of inertial frames which forms the Galileo-Newton group. As well, we put in evidence the general theorems and conservation theorems of mechanics and the principle of equivalence.

2.1.1 Dynamics of the relative motion of a particle

Let be an *inertial frame of reference* $O'x_1'x_2'x_3'$, which is considered to be *fixed*, and a *non-inertial frame* (*movable frame*) $Ox_1x_2x_3$ in motion with respect to the fixed one. Newton's law (1.1.89) for a particle P is written with respect to an inertial frame (in absolute motion); in Chap. 5, Sec. 3.1 we have seen that the absolute motion is obtained by the composition of the relative motion and the transportation one (a vector composition for the velocities, while for the accelerations one must add the Coriolis acceleration too). Starting from the formula (5.3.5) of composition of the accelerations and multiplying both members by the mass m, we can write $m\mathbf{a}_a = m\mathbf{a}_t + m\mathbf{a}_r + m\mathbf{a}_C$, where $\mathbf{a}_t, \mathbf{a}_r, \mathbf{a}_C$ are the transportation, relative and Coriolis accelerations, respectively, given by (5.3.4). Taking into account the equation of motion of a free particle ($m\mathbf{a}_a = \mathbf{F}$, where $\mathbf{F}$ is the resultant of the given forces), we get

$$m\mathbf{a}_r = \mathbf{F} + \mathbf{F}_t + \mathbf{F}_C, \qquad (10.2.1)$$

where

$$\begin{aligned} \mathbf{F}_t &= -m\mathbf{a}_t = -m[\mathbf{a}_O' + \dot{\boldsymbol{\omega}} \times \mathbf{r} + \boldsymbol{\omega} \times (\boldsymbol{\omega} \times \mathbf{r})], \\ \mathbf{F}_C &= -m\mathbf{a}_C = -2m\boldsymbol{\omega} \times \mathbf{v}_r \end{aligned} \qquad (10.2.1')$$

are *complementary forces* (*the transportation force* and *the Coriolis force*, respectively); these forces are added to the given force $\mathbf{F}$ and allow writing the equation of motion in a non-inertial frame of reference. We notice that the forces $\mathbf{F}_t$ and $\mathbf{F}_C$ depend on $\mathbf{a}_O'$ and $\boldsymbol{\omega}$, hence on the acceleration of the movable frame pole and on the rotation vector of this frame with respect to the fixed (inertial) one, respectively; hence, *the motion with respect to a non-inertial frame can be determined only starting from an inertial one.*

The complementary forces are called also *inertial forces* because their magnitude is proportional to the inertial mass; these forces are applied to the particle in motion (unlike the forces of inertia (6.2.11) which are applied to the agent which provokes the motion). The inertial forces are real ones with respect to an observer linked to a non-inertial frame of reference; e.g., *the centrifugal force*, which appears in a motion of rotation, is a transportation force.

Introducing *the relative force*

$$\mathbf{F}_r = \mathbf{F} + \mathbf{F}_t + \mathbf{F}_C, \qquad (10.2.2)$$

we may write the equation (10.2.1) in the form

$$m\mathbf{a}_r = \mathbf{F}_r, \quad \mathbf{F}_r = \mathbf{F}_r(\mathbf{r}, \mathbf{v}_r; t); \qquad (10.2.2')$$

we put initial conditions of Cauchy type

$$\mathbf{r}(t_0) = \mathbf{r}_0, \quad \mathbf{v}_r(t_0) = \mathbf{v}_r^0 \qquad (10.2.2'')$$

and notice that one can use all the results given in Chap. 6, stating thus

Theorem 10.2.1 (*theorem of the relative motion*). *The equation of motion of a particle with respect to an inertial frame of reference maintains its form with respect to a non-inertial one if the given force is replaced by the force relative to the latter frame.*

As well, we state

Theorem 10.2.2 (*theorem of the relative force*). *The relative force (with respect to a non-inertial frame of reference) is equal to the sum of the given force and the complementary forces (the force of transportation and the Coriolis force) with respect to an inertial one.*

In case of a particle subjected to bilateral constraints, we use the axiom of liberation of constraints, introducing *the constraint force* $\mathbf{R}$. The equation of motion (10.2.1) becomes

$$m\mathbf{a}_r = \mathbf{F} + \mathbf{F}_t + \mathbf{F}_C + \mathbf{R}; \qquad (10.2.1'')$$

taking into account (10.2.2), we may also write

$$m\mathbf{a}_r = \mathbf{F}_r + \mathbf{R} \qquad (10.2.2''')$$

too. We notice that the Theorems 10.2.1 and 10.2.2 remain still valid.

Thus, the problem of motion with respect to a non-inertial frame of reference may be reduced to a corresponding problem with respect to an inertial frame, chosen conveniently (with respect to which the Newtonian model of mechanics is verified with a sufficient good approximation).

2.1.2 Particular cases of non-inertial frames of reference

Let us consider first of all the case of a *non-inertial frame of reference in* a *motion of translation* with respect to an inertial one; hence, we assume that $\boldsymbol{\omega} = \mathbf{0}$. In this case, the equation of motion is ($\mathbf{F}_t = -m\mathbf{a}'_O$, $\mathbf{F}_C = \mathbf{0}$)

$$m\mathbf{a}_r = \mathbf{F} - m\mathbf{a}'_O. \tag{10.2.3}$$

We assume, in particular, that the particle P is acted upon by its own weight ($\mathbf{F} = m\mathbf{g}$), so that we can write

$$\mathbf{a}_r = \mathbf{g} - \mathbf{a}'_O. \tag{10.2.4}$$

In the case in which $\mathbf{a}'_O \parallel \mathbf{g}$ we have *the elevator problem*. If $\mathbf{a}'_O$ has the same direction as $\mathbf{g}$ (it is directed towards the centre of the Earth), then the particle P seems to be lighter ($\mathbf{a}_r = \mathbf{g} - \mathbf{a}'_O$), while if $\mathbf{a}'_O$ has a direction opposite to that of $\mathbf{g}$ ($\mathbf{a}_r = \mathbf{g} + \mathbf{a}'_O$), then the particle seems to be heavier. In particular, if the elevator is in a free falling ($\mathbf{a}'_O = \mathbf{g}$), then the apparent weight of the particle with respect to it vanishes; we are in *the case of imponderability*.

Moreover, let us suppose that $\mathbf{a}'_O = \overrightarrow{\text{const}}$. If, for instance, the particle is in free falling ($v_0 = 0$ for $t = 0$) from a height h with respect to the floor of the elevator ($x = 0$, the Ox-axis of unit vector $\mathbf{i}$ along the ascendent vertical), we get, in the movable non-inertial frame, $x = -(g + a'_O)t^2/2 + h$, so that the falling time is given by

$$T = \sqrt{\frac{2h}{g + a'_O}}, \tag{10.2.5}$$

assuming that $g + a'_O > 0$ ($\mathbf{a}'_O = a'_O \mathbf{i}$, with $a'_O > 0$, of an opposite direction to that of the gravitation $\mathbf{g} = -g\mathbf{i}$, or with $a'_O < 0$, of the same direction but with $|\mathbf{a}'_O| < |\mathbf{g}|$). If $g + a'_O = 0$, then the particle is immobile, while if $g + a'_O < 0$ ($a'_O < 0$ with $|\mathbf{a}'_O| > |\mathbf{g}|$), then the particle goes up along the vertical Ox with respect to an observer linked to the movable frame. If $a'_O = 0$, then the movable frame becomes an inertial one, finding again the usual laws of falling.

Another important particular case is that of a *non-inertial frame of reference in motion of rotation* with respect to a fixed Ox'_3-axis ($\mathbf{v}'_O = \mathbf{a}'_O = \mathbf{0}$, $\boldsymbol{\omega} = \omega \mathbf{i}_3$, $\omega = \omega(t)$, Fig.10.4). The transportation force is given by ($\mathbf{r} = \overrightarrow{OP'} + \overrightarrow{P'P} = \overrightarrow{OP'} + \mathbf{r}^0$)

$$\mathbf{F}_t = -m\dot{\boldsymbol{\omega}} \times \mathbf{r} - m\boldsymbol{\omega} \times (\boldsymbol{\omega} \times \mathbf{r}) = -m\dot{\boldsymbol{\omega}} \times \mathbf{r}^0 - m\boldsymbol{\omega} \times (\boldsymbol{\omega} \times \mathbf{r}^0)$$
$$= -m\dot{\omega}\mathbf{i}_3 \times \mathbf{r}^0 + m\omega^2 \mathbf{r}^0, \tag{10.2.6}$$

corresponding to Rival's formula (5.2.6'''), and the Coriolis force by

$$\mathbf{F}_C = -2m\boldsymbol{\omega} \times \mathbf{v}_r = -2m\omega \mathbf{i}_3 \times \mathbf{v}_r . \tag{10.2.6'}$$

In case of a uniform rotation ($\dot{\boldsymbol{\omega}} = \mathbf{0}$), we have

$$\mathbf{F}_t = m\omega^2 \mathbf{r}^0 . \tag{10.2.6''}$$

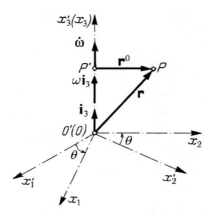

Figure 10.4. Non-inertial frame of reference in motion of rotation with respect to a fixed axis.

Assuming that the particle P is moving in the Ox_1x_3-plane (with respect to the non-inertial frame) and taking into account (10.2.6), (10.2.6'), we may write the equation of motion (10.2.1) in components, in the form

$$m\ddot{x}_1 = F_1 + m\omega^2 x_1, \quad 0 = F_2 - m\dot{\omega}x_1 - 2m\omega\dot{x}_1, \quad m\ddot{x}_3 = F_3;$$

observing then that $\omega = \dot{\theta}$ and replacing x_1 by r and x_3 by z, we find again the equations of motion of the particle in cylindrical co-ordinates (6.1.26), with respect to a fixed (inertial) frame.

By composition of the two particular cases considered above, one can obtain the case of a *non-inertial frame of reference in a finite motion of rototranslation* with respect to an inertial one. In the case of a non-inertial frame of reference in an arbitrary motion with respect to an inertial one, one may use the results obtained in Chap. 5, Subsec. 2.3.1 concerning the instantaneous helical motion.

If the forces do not intervene explicitly in the relative motion, then the composition of motions can be reduced to the composition of velocities. Let be thus the case of a non-inertial frame in motion of translation with respect to an inertial one, with the velocity $\mathbf{v}'_O$; if the particle has a relative velocity $\mathbf{v}_r$, then the composition of the velocities leads to $\mathbf{v}' = \mathbf{v}'_O + \mathbf{v}_r$. Assuming that the velocities $\mathbf{v}'_O$ and $\mathbf{v}_r$ are collinear (e.g., a boat which is moving with the velocity v_r along a river, the velocity of which with respect to the riversides is v'_O), we have $v' = v'_O + v_r$ (the boat is moving in the same direction as the river) or $v' = v_r - v'_0$ (the boat is moving in a direction opposite to that of the river).

As well, we mention *the meeting problem* of a particle P_1, which has a rectilinear and uniform motion of velocity $\mathbf{v}_1$, and a particle P_2, which has a curvilinear and uniform motion of velocity $\mathbf{v}_2$; the latter velocity will be collinear with the vector $\overrightarrow{P_2P_1}$, tangent to the trajectory of the particle P_2, which has to be determined (Fig.10.5). We choose an inertial frame of reference $P_1^0 x_1' x_2'$, linked to the initial position P_1^0 of the particle P_1, and a non-inertial frame $P_1 x_1 x_2$, linked to a location of the particle P_1 in uniform translation with respect to the first frame; in fact, this second frame is inertial too, its motion being specified by the equations (we assume that the two particles start from the points P_1^0 and P_2^0, respectively, at the initial moment $t = 0$) $x_1' = x_1$, $x_2' = v_1 t + x_2$. Observing that $\mathbf{v}_2 = v_2 \text{ vers} \overrightarrow{P_2P_1}$, we get the differential equations, which determine the trajectory of the particle P_2 in the movable frame, in the form ($\mathbf{v}_2$ has the components $\dot{x}_1$ and $\dot{x}_2$)

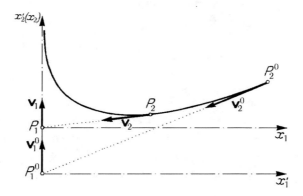

Figure 10.5. The meeting problem of two particles.

$$\dot{x}_1 = -\frac{v_2 x_1}{\sqrt{x_1^2 + x_2^2}}, \quad \dot{x}_2 = -v_1 - \frac{v_2 x_2}{\sqrt{x_1^2 + x_2^2}};$$

dividing member by member the two equations (eliminating the time t), it results

$$\frac{dx_2}{dx_1} = \frac{x_2}{x_1} + \frac{v_1}{v_2}\sqrt{1 + \left(\frac{x_2}{x_1}\right)^2}.$$

Observing that $dx_2 / dx_1 = x_2 / x_1 + (x_1 / dx_1) d(x_2 / x_1)$ and integrating, we can write the equation of the trajectory with respect to the movable frame in the form

$$x_2 = \frac{x_1}{2}\left[\left(\frac{x_1}{a}\right)^{v_1/v_2} - \left(\frac{x_1}{a}\right)^{-v_1/v_2}\right]. \tag{10.2.7}$$

Eliminating $\sqrt{x_1^2 + x_2^2}$ between the two differential equations, we may write $\dot{x}_1 = (v_1 + \dot{x}_2)(x_1/x_2)$; taking into account the previous observation, as well as (10.2.7), it results

$$dt = \frac{1}{v_1}\left(\frac{x_2}{x_1}dx_1 - dx_2\right) = -\frac{1}{v_1}x_1 d\left(\frac{x_2}{x_1}\right)$$

$$= -\frac{a}{2\sqrt{2}}\left[\left(\frac{x_1}{a}\right)^{v_1/v_2} + \left(\frac{x_1}{a}\right)^{-v_1/v_2}\right]d\left(\frac{x_1}{a}\right),$$

wherefrom, assuming that $v_1 \neq v_2$,

$$t = \bar{t} - \frac{x_1}{2}\left[\frac{(x_1/a)^{v_1/v_2}}{v_1 + v_2} - \frac{(x_1/a)^{-v_1/v_2}}{v_1 - v_2}\right]. \qquad (10.2.7')$$

Calculating $x_1 = x_1(t)$ and then $x_2 = x_2(t)$, we obtain the motion of the particle along the trajectory. The equation of the trajectory with respect to the fixed frame is of the form

$$x_2' = v_1\bar{t} + \frac{v_2 x_1'}{2}\left[\frac{(x_1'/a)^{v_1/v_2}}{v_1 + v_2} + \frac{(x_1'/a)^{-v_1/v_2}}{v_1 - v_2}\right]. \qquad (10.2.7'')$$

The constants a and $\bar{t}$ are specified by (10.2.7), (10.2.7') with the initial conditions $x_1 = x_1^0 = x_1'^0$, $x_2 = x_2^0 = x_2'^0$, for $t = 0$. For $x_1 \to 0$ we have $x_2 \to 0$, $t \to \bar{t}$ if $v_1 < v_2$ and $x_2 \to -\infty$, $t \to \infty$ if $v_1 > v_2$. As a consequence, if $v_1 < v_2$, then the particle P_2 meets the particle P_1 at the moment $t = \bar{t}$, at the point of co-ordinates $x_1' = 0$, $x_2' = v_1\bar{t}$ with respect to the inertial frame. But if $v_1 > v_2$, then the two particles do not meet. The distance between them is given by $\overline{P_2 P_1}^2 = x_1^2 + x_2^2$, so that

$$\overline{P_2 P_1} = \frac{x_1}{2}\left[\left(\frac{x_1}{a}\right)^{v_1/v_2} + \left(\frac{x_1}{a}\right)^{-v_1/v_2}\right] = \frac{a}{2}\left[\left(\frac{x_1}{a}\right)^{1+v_1/v_2} + \left(\frac{x_1}{a}\right)^{1-v_1/v_2}\right]; \qquad (10.2.8)$$

the minimum of this distance is obtained for

$$x_1 = a\left(\frac{v_1 - v_2}{v_1 + v_2}\right)^{v_2/2v_1}, \qquad (10.2.8')$$

so that

$$\overline{P_2 P_1}_{1\min} = \frac{a v_1}{\sqrt{v_1^2 - v_2^2}}\left(\frac{v_1 - v_2}{v_1 + v_2}\right)^{v_2/2v_1} \qquad (10.2.8'')$$

Other considerations on particle dynamics 621

at the moment

$$t = \overline{t} + \frac{2av_1v_2}{(v_1^2 - v_2^2)\sqrt{v_1^2 - v_2^2}}\left(\frac{v_1 - v_2}{v_1 + v_2}\right)^{v_2/2v_1}. \tag{10.2.8'''}$$

If $v_1 = v_2 = v$, then we get the equation of the trajectory with respect to the movable frame

$$x_2 = \frac{x_1^2 - a^2}{2a}, \tag{10.2.9}$$

the motion being specified by

$$t = \overline{t} - \frac{a}{2v}\left[\frac{1}{2}\left(\frac{x_1}{a}\right)^2 + \ln\frac{x_1}{a}\right]; \tag{10.2.9'}$$

as well,

$$\overline{P_2P_1} = \frac{x_1^2 + a^2}{2a}. \tag{10.2.9''}$$

The two particles do not meet; the minimal distance between them is obtained for $x_1 = 0$ at the moment $t \to \infty$ and is equal to $a/2$.

The particle P_1 can be a man (master) and the particle P_2 his dog.

2.1.3 General theorems in the relative motion

Let be a free particle P of mass m, the position of which is specified by the vector $\mathbf{r}'$ with respect to an inertial (fixed) frame of reference $O'x_1'x_2'x_3'$ and by the vector $\mathbf{r}$ with respect to a non-inertial (movable) frame $Ox_1x_2x_3$; starting from the relation $\mathbf{r}' = \mathbf{r}_O' + \mathbf{r}$, where $\mathbf{r}_O'$ is the position vector of the pole of the non-inertial frame with respect to the inertial one, and using the relations in Chap. 5, Sec. 3.1, we find the relation $\mathbf{v}' = \mathbf{v}_O' + \mathbf{v}_r + \boldsymbol{\omega} \times \mathbf{r}$, where $\boldsymbol{\omega}$ is the angular velocity vector (the rotation vector) of the movable frame. We obtain thus the relation between the momenta with respect to the two frames in the form

$$\mathbf{H}' = \mathbf{H} + m\mathbf{v}_O' + m\boldsymbol{\omega} \times \mathbf{r}, \tag{10.2.10}$$

with $\mathbf{H} = m\mathbf{v}_r$. Using the relation (A.2.37) between the absolute and the relative derivatives of a vector, we can write

$$\frac{d\mathbf{H}'}{dt} = \frac{\partial \mathbf{H}}{\partial t} + \boldsymbol{\omega} \times (m\mathbf{v}_r) + m\mathbf{a}_O' + m\dot{\boldsymbol{\omega}} \times \mathbf{r} + m\boldsymbol{\omega} \times (\mathbf{v}_r + \boldsymbol{\omega} \times \mathbf{r}),$$

wherefrom, taking into account (6.1.45) and (10.2.1'), (10.2.2), it results

$$\dot{\mathbf{H}} = \frac{\partial \mathbf{H}}{\partial t} = \mathbf{F}_r = \mathbf{F} + \mathbf{F}_t + \mathbf{F}_C. \qquad (10.2.11)$$

However, this result was to be expected, taking into account (10.2.2'); we may state
Theorem 10.2.3 (*theorem of momentum*). *The derivative with respect to time of the momentum of a free particle, in a non-inertial frame of reference, is equal to the relative force which acts upon the particle in that frame.*
Analogously, we obtain

$$\dot{\mathbf{K}}_O = \frac{\partial \mathbf{K}_O}{\partial t} = \frac{\partial}{\partial t}(\mathbf{r} \times \mathbf{H}) = \mathbf{r} \times \mathbf{F}_r = \mathbf{M}_O + \mathbf{r} \times (\mathbf{F}_t + \mathbf{F}_C), \qquad (10.2.12)$$

where, taking into account (10.2.1'), we notice that

$$\mathbf{r} \times \mathbf{F}_t = m\big[(\dot{\boldsymbol{\omega}} \cdot \mathbf{r})\mathbf{r} - r^2 \dot{\boldsymbol{\omega}} - \mathbf{a}'_O + (\boldsymbol{\omega} \cdot \mathbf{r})(\boldsymbol{\omega} \times \mathbf{r})\big]; \qquad (10.2.13)$$

thus, we state
Theorem 10.2.4 (*theorem of moment of momentum*). *The derivative with respect to time of the moment of momentum of a free particle, in a non-inertial frame of reference, with respect to its pole, is equal to the moment of the relative force which acts upon the particle in that frame (the sum of the moment of the resultant of the given forces which act upon the particle and the moment of the complementary forces), with respect to the same pole.*
The Theorems 10.2.3 and 10.2.4 lead to the relation

$$\dot{\tau}_O(\mathbf{H}) = \frac{\partial \tau_O(\mathbf{H})}{\partial t} = \tau_O(\mathbf{F}_r) = \tau_O(\mathbf{F}) + \tau_O(\mathbf{F}_t) + \tau_O(\mathbf{F}_C), \qquad (10.2.14)$$

obtaining thus
Theorem 10.2.5 (*theorem of torsor*). *The derivative with respect to time of the torsor of the momentum of a free particle, in a non-inertial frame of reference, with respect to its pole, is equal to the torsor of the relative force which acts upon the particle in that frame (the sum of the torsor of the resultant of the given forces which act upon the particle and the torsor of the complementary forces), with respect to the same pole.*
It results, as well,

$$\mathrm{d}T = \mathrm{d}\left(\frac{m}{2}v_r^2\right) = \mathrm{d}W_{F_r} = \mathrm{d}W + \mathrm{d}W_{F_t} = \mathbf{F}_r \cdot \mathrm{d}\mathbf{r} = \mathbf{F} \cdot \mathrm{d}\mathbf{r} + \mathbf{F}_t \cdot \mathrm{d}\mathbf{r} \qquad (10.2.15)$$

because, taking into account (10.2.1'), we have $\mathrm{d}W_{F_C} = \mathbf{F}_C \cdot \mathrm{d}\mathbf{r} = 0$; we may thus state
Theorem 10.2.6 (*theorem of kinetic energy*). *The differential of the kinetic energy of a free particle with respect to a non-inertial frame of reference is equal to the elementary work of the relative force which acts upon it in this frame (the sum of the elementary work of the resultant of the given forces which act upon the particle and the work of the transportation force).*

Other considerations on particle dynamics 623

Dividing the relation (10.2.15) by dt, we get

$$\dot{T} = \frac{\partial T}{\partial t} = P_r = \mathbf{F}_r \cdot \mathbf{v}_r = P + \mathbf{F}_t \cdot \mathbf{v}_r \qquad (10.2.16)$$

so that we may state

Theorem 10.2.6' (*theorem of kinetic energy; second form*). *The derivative with respect to time of the kinetic energy of a free particle in a non-inertial frame of reference is equal to the power of the relative force which acts upon it in this frame* (*the sum of the power of the resultant of the given forces which act upon the particle and the power of the transportation force*).

In case of a particle subjected to bilateral constraints, we apply the axiom of liberation of constraints and introduce the constraint force $\mathbf{R}$; the formulae (10.2.11), (10.2.12), (10.2.14) - (10.2.16) take the form

$$\dot{\mathbf{H}} = \frac{\partial \mathbf{H}}{\partial t} = \mathbf{F}_r + \mathbf{R} = \mathbf{F} + \mathbf{F}_t + \mathbf{F}_C + \mathbf{R}, \qquad (10.2.11')$$

$$\dot{\mathbf{K}}_O = \frac{\partial \mathbf{K}_O}{\partial t} = \mathbf{r} \times (\mathbf{F}_r + \mathbf{R}) = \mathbf{M}_O + \mathbf{r} \times (\mathbf{F}_t + \mathbf{F}_C) + \overline{\mathbf{M}}_O, \qquad (10.2.12')$$

$$\dot{\tau}_O(\mathbf{H}) = \frac{\partial \tau_O(\mathbf{H})}{\partial t} = \tau_O(\mathbf{F}_r) + \tau_O(\mathbf{R}) = \tau_O(\mathbf{F}) + \tau_O(\mathbf{F}_t) + \tau_O(\mathbf{F}_C) + \tau_O(\mathbf{R}),$$
$$(10.2.14')$$

$$dT = dW_{F_r} + dW_R = dW + dW_{F_t} + dW_R = \mathbf{F}_r \cdot d\mathbf{r} + \mathbf{R} \cdot d\mathbf{r}$$
$$= \mathbf{F} \cdot d\mathbf{r} + \mathbf{F}_t \cdot d\mathbf{r} + \mathbf{R} \cdot d\mathbf{r}, \qquad (10.2.15')$$

$$\dot{T} = \frac{\partial T}{\partial t} = \mathbf{F}_r \cdot \mathbf{v}_r + \mathbf{R} \cdot \mathbf{v}_r = P + \mathbf{F}_t \cdot \mathbf{v}_r + \mathbf{R} \cdot \mathbf{v}_r, \qquad (10.2.16')$$

so that we can state corresponding theorems. In case of holonomic and scleronomic constraints we have $dW_R = \mathbf{R} \cdot d\mathbf{r} = 0$, hence $\mathbf{R} \cdot \mathbf{v}_r = 0$.

Starting from the above results, we can find, in certain conditions, first integrals and may state conservation theorems with respect to a non-inertial frame of reference. Thus, using the theorems of Chap. 6, Subsec. 1.2.5, we obtain, for a free particle:

Theorem 10.2.7 (*conservation theorem of momentum*). *The momentum* (*and the velocity*) *of a free particle with respect to a non-inertial frame of reference is conserved in time if and only if the relative force which acts upon the particle in this frame* (*the sum of the resultant of the given forces which act upon the particle and the complementary forces*) *vanishes.*

Theorem 10.2.8 (*conservation theorem of moment of momentum*). *The moment of momentum of a free particle with respect to the pole of a non-inertial frame of reference is conserved in time if and only if the moment of the relative force which acts upon it in this frame* (*the sum of the moment of the resultant of the given forces which act upon the particle and the moment of the complementary forces*), *with respect to the same pole, vanishes.*

Theorem 10.2.9 (*conservation theorem of torsor*). *The torsor of the momentum of a free particle with respect to the pole of a non-inertial frame of reference is conserved in* time if and only if the torsor of the relative force which acts upon it in this frame (*the sum of the resultant of the given forces which act upon the particle and the complementary forces*), *with respect to the same pole, vanishes*.

Theorem 10.2.10 (*conservation theorem of the mechanical energy*). *The mechanical energy of a free particle with respect to a non-inertial frame of reference is conserved in time if and only if the sum of the given forces which act upon the particle and the transportation force is conservative.*

The conditions in the latter theorem (in which the mechanical energy contains also an energy due to the transportation force) are not so easy to fulfil. If the resultant **F** of the given forces is a conservative one, which derives from a simple or from a generalized potential, then we can write

$$\mathrm{d}E = \mathrm{d}(T + V) = \mathbf{F}_t \cdot \mathrm{d}\mathbf{r}. \tag{10.2.17}$$

In particular, if $\mathbf{a}'_O = \mathbf{0}$ (hence $\overrightarrow{\mathbf{v}'_O = \text{const}}$) and $\overrightarrow{\boldsymbol{\omega} = \text{const}}$, hence if the transportation motion is a finite motion of rototranslation of constant velocities of translation and rotation (the movable frame has a motion of rotation with a constant angular velocity around an axis which passes through its pole, that one having a rectilinear and uniform motion with respect to the fixed frame), we can write

$$\mathbf{F}_t \cdot \mathrm{d}\mathbf{r} = -m\mathbf{a}_t \cdot \mathrm{d}\mathbf{r} = -m\boldsymbol{\omega} \times (\boldsymbol{\omega} \times \mathbf{r}) \cdot \mathrm{d}\mathbf{r} = -m(\boldsymbol{\omega}, \boldsymbol{\omega} \times \mathbf{r}, \mathrm{d}\mathbf{r}) = m(\boldsymbol{\omega} \times \mathbf{r}, \boldsymbol{\omega}, \mathrm{d}\mathbf{r})$$
$$= m(\boldsymbol{\omega} \times \mathbf{r}) \cdot (\boldsymbol{\omega} \times \mathrm{d}\mathbf{r}) = \mathrm{d}\left[\frac{m}{2}(\boldsymbol{\omega} \times \mathbf{r})^2\right]$$

we obtain thus a first integral (called, sometimes, the generalized first integral of the energy, because it is reduced to the first integral of the energy if $\boldsymbol{\omega} = \mathbf{0}$) of the form

$$E = T + V = \frac{m}{2}(\boldsymbol{\omega} \times \mathbf{r})^2 + h, \quad h = \text{const}. \tag{10.2.17'}$$

In case of a particle subjected to constraints we get results analogous to those in Chap. 6, Subsec. 2.1.3. If the constraints are holonomic and scleronomic, then the relations (10.2.17), (10.2.17') maintain their form.

2.1.4 Inertial frames of reference. Galileo-Newton group

Starting from the equation (10.2.2), (10.2.2'), we will search the movable frames (specified by the acceleration $\mathbf{a}'_O$ of the pole of the frame and by its angular velocity $\boldsymbol{\omega}$) for which a free particle P is moving after the law

$$m\mathbf{a}_r = \mathbf{F}, \tag{10.2.18}$$

hence after the same law as in the case of the fixed (inertial) frame. We notice that, in this case, the sum of the complementary forces must vanish; hence, we must have

$$-m(\mathbf{a}_t + \mathbf{a}_C) = -m[\mathbf{a}'_O + \dot{\boldsymbol{\omega}} \times \mathbf{r} + \boldsymbol{\omega} \times (\boldsymbol{\omega} \times \mathbf{r}) + 2\boldsymbol{\omega} \times \mathbf{v}_r] = \mathbf{0} \tag{10.2.19}$$

for any point at which the particle may be (for any **r**) and for any relative velocity $\mathbf{v}_r$ of it.

We assume that, starting from the same position vector **r**, the particle P can have the relative velocity $\mathbf{v}'_r$ in one case of motion and the relative velocity $\mathbf{v}''_r$ in another case of motion. Imposing the condition (10.2.19) in both cases and subtracting a relation from the other, we find $\boldsymbol{\omega} \times (\mathbf{v}'_r - \mathbf{v}''_r) = \mathbf{0}$ for any $\mathbf{v}'_r$ and $\mathbf{v}''_r$; it results that we must have $\boldsymbol{\omega} = \mathbf{0}$ (hence, $\dot{\boldsymbol{\omega}} = \mathbf{0}$ too). Returning to the above condition, it follows that $\mathbf{a}'_O = \mathbf{0}$.

From the latter condition we see that one passes from the fixed frame to the movable one by a transformation of space co-ordinates (to which a time transformation may be added) of the form (6.1.42″), which forms a group with seven parameters. The condition $\boldsymbol{\omega} = \mathbf{0}$ shows that the movable frame can have only a finite geometric rotation of the form (6.1.43), concerning its relative position with respect to the fixed frame, so that the most general transformation which corresponds to the passing from a frame to another one, modelling the motion of a particle in the form (10.2.18), is given by (6.1.44) and forms *the Galileo-Newton group* with ten parameters, studied in Chap. 6, Subsec. 1.2.3. One can thus apply the Theorem 6.1.6 (of relativity) of Galileo, the movable frame being – in this case – an inertial frame too, with respect to which the law of motion maintains its form (acting only the given force and, eventually, the constraint one). If we write the relation of transformation (6.1.44) for two particles P_1 and P_2, then we have

$$\mathbf{r}'_1 = \boldsymbol{\alpha}\mathbf{r}_1 + \mathbf{v}_0 t + \mathbf{r}_0, \quad \mathbf{r}'_2 = \boldsymbol{\alpha}\mathbf{r}_2 + \mathbf{v}_0 t + \mathbf{r}_0, \quad t' = t + t_0,$$

the tensor $\boldsymbol{\alpha}$ corresponding to a finite rotation of the movable frame; we obtain thus $\mathbf{r}'_2 - \mathbf{r}'_1 = \boldsymbol{\alpha}(\mathbf{r}_2 - \mathbf{r}_1)$, wherefrom

$$(\mathbf{r}'_2 - \mathbf{r}'_1)^2 = [\boldsymbol{\alpha}(\mathbf{r}_2 - \mathbf{r}_1)]^2 = [\boldsymbol{\alpha}(\mathbf{r}_2 - \mathbf{r}_1)] \cdot [\boldsymbol{\alpha}(\mathbf{r}_2 - \mathbf{r}_1)] = [\boldsymbol{\alpha}\boldsymbol{\alpha}(\mathbf{r}_2 - \mathbf{r}_1)] \cdot (\mathbf{r}_2 - \mathbf{r}_1)$$
$$= [\boldsymbol{\delta}(\mathbf{r}_2 - \mathbf{r}_1)] \cdot (\mathbf{r}_2 - \mathbf{r}_1) = (\mathbf{r}_2 - \mathbf{r}_1)^2,$$

$\boldsymbol{\delta}$ being Kronecker's tensor. Hence, *the distance between two particles remains invariant in a transformation of the Galileo-Newton group*; the forces which depend only on distances (e.g., the forces of Newtonian attraction) have the same property of invariance. In this case, taking into account the invariance of the acceleration, it results that the mass of the particle is invariant too (constant, property iii) of the mass; see also Chap. 1, Subsec. 1.1.6).

If, in a non-inertial system of reference, we determine experimentally the sum $\mathbf{F}_t + \mathbf{F}_C$ of the complementary forces, then one calculate the quantities $\mathbf{a}'_O$ and $\boldsymbol{\omega}$ which specify the motion of the frame (neglecting a rectilinear and uniform motion of translation, which cannot be put in evidence by mechanical experiments). As a matter of fact, by no experiment of physical (not only mechanical) nature a preferential inertial (e.g., "fixed") frame cannot be put in evidence, all inertial systems being thus equivalent.

2.1.5 Principle of equivalence

Let us suppose that the particle P is subjected to the action of a uniform gravitational field (the terrestrial gravitational field) $\mathbf{F} = m\mathbf{g}$ and let us consider the motion with respect to a non-inertial frame of reference in rectilinear and uniform accelerated motion of translation (hence, for which $\mathbf{a}'_O = \mathbf{g}$, $\boldsymbol{\omega} = \mathbf{0}$); in this case, $\mathbf{F}_t = -m\mathbf{g}$ and $\mathbf{F}_C = \mathbf{0}$, so that $\mathbf{F}_r = \mathbf{0}$, the equation of motion (10.2.2') leading thus to $\mathbf{a}_r = \mathbf{0}$. We notice that an observer, situated in an elevator which is moving in the direction of the gravitation and is acted upon by a force equal to $m\mathbf{g}$, cannot perceive the gravitational field (particularly, if he could not be in contact with the universe of the exterior of the elevator); it is the case of imponderability, considered in Subsec. 2.1.2.

The mass plays two different rôles in the identity $m\mathbf{a}'_O = m\mathbf{g}$; in the left member is put in evidence the aspect of *inertial mass* (which leads to the inertial transportation force), while in the right member appears the aspect of *gravity mass* (called *heavy mass* too) (see also Chap. 1, Subsecs 1.1.6 and 1.2.1). This observation led A. Einstein to state *the principle of equivalence* between the gravity mass and the inertial one in the general relativistic model of mechanics; thus, the two properties of the mass represent two different aspects of the same material quantity. This result has been confirmed by the experiments performed by Eötvös and Zeeman, by Southern and Zeeman etc.

2.1.6 Relative equilibrium

The location $\mathbf{r} = \mathbf{r}_0$ is called position of relative equilibrium of a free particle P (which is in relative rest with respect to a non-inertial frame of reference) if the equation of motion (10.2.2), (10.2.2') with the initial conditions $\mathbf{r}(t_0) = \mathbf{r}_0$, $\mathbf{v}_r(t_0) = \mathbf{0}$ admits as solution $\mathbf{r}(t) = \mathbf{r}_0$, $\forall t \geq t_0$; in this case, we have $\mathbf{v}_r(t) = \mathbf{0}$, $\mathbf{a}(t) = \mathbf{0}$, $\forall t \geq t_0$. It results $\mathbf{F}_C = \mathbf{0}$, so that

$$\mathbf{F}_r = \mathbf{F} + \mathbf{F}_t = \mathbf{0}, \qquad (10.2.20)$$

and we can state (the condition (10.2.20) is sufficient too, because $\mathbf{F}_C = \mathbf{0}$ leads to $\mathbf{v}_r = \mathbf{0}$, the rectilinear and uniform motion with respect to the movable frame being thus excluded)

Theorem 10.2.11 (*theorem of relative equilibrium*). *A free particle is in relative equilibrium (with respect to a non-inertial frame of reference) if and only if the sum of the resultant of the given forces which act upon that particle and the transportation force vanishes.*

If $\mathbf{a}'_O = \mathbf{0}$, $\dot{\boldsymbol{\omega}} = \mathbf{0}$ (the non-inertial frame has a finite motion of rototranslation with constant velocities of translation and rotation), then the transportation force

$$\mathbf{F}_t = -m\boldsymbol{\omega} \times (\boldsymbol{\omega} \times \mathbf{r}) \qquad (10.2.21)$$

is a centrifugal one.

In case of a particle subjected to bilateral constraints, the necessary and sufficient condition of relative equilibrium (with respect to a non-inertial frame of reference) reads

$$\mathbf{F}_r + \mathbf{R} = \mathbf{F} + \mathbf{F}_t + \mathbf{R} = \mathbf{0}, \qquad (10.2.20')$$

where $\mathbf{R}$ is the constraint force.

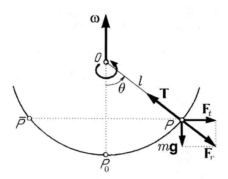

Figure 10.6. Heavy particle constrained to move on a circle in uniform rotation about a vertical diameter of it.

Let us consider the case of a heavy particle P of mass m, constrained to move on a circle in uniform rotation about a vertical diameter of it, with the angular velocity $\boldsymbol{\omega}$ (along one of the axes of the inertial frame of pole O); for instance, let us suppose that the particle P is a small heavy sphere, situated at the end of a perfect inextensible thread of length l (Fig.10.6). We choose the non-inertial frame with the pole at O too, the axis of which coincides with the rotation axis of the movable frame. The particle is acted upon by the gravity force $m\mathbf{g}$, by the centrifugal force $F_t = m|\boldsymbol{\omega} \times (\boldsymbol{\omega} \times \mathbf{r})|$ $= ml\omega^2 \sin\theta$ and by the constraint force $\mathbf{T}$ (the tension in the thread). Projecting the equation of equilibrium ($m\mathbf{g} + \mathbf{F}_t + \mathbf{T} = \mathbf{0}$) on the normal to the force $\mathbf{T}$ and along the direction of it, we get

$$F_t \cos\theta = mg\sin\theta, \quad T = F_t \sin\theta + mg\cos\theta.$$

The positions of relative equilibrium are given by

$$\sin\theta = 0, \quad \cos\theta = \frac{g}{l\omega^2}, \quad \forall \omega > \sqrt{\frac{g}{l}}; \qquad (10.2.22)$$

hence, excepting the point P_0 (which corresponds to $\theta = 0$), we obtain two symmetric positions of relative equilibrium P and $\overline{P}$. For P_0 it results the tension $T = mg$, while for P and $\overline{P}$ the tension is $T = ml\omega^2$.

Let be an ideal liquid in rest with respect to a vessel in uniform accelerated translation with respect to an inertial frame of pole O'; the non-inertial frame is linked to the vessel, the velocity of the pole O being $\mathbf{a}'_O = \overrightarrow{\text{const}}$ with respect to the pole O'. If $\mathbf{a}'_O = \mathbf{0}$, then the free surface of the liquid (in rest with respect to the fixed frame) is a horizontal plane. A particle P, of mass m, of the free surface of the fluid (Fig.10.7,a) is in equilibrium under the action of the gravity force $m\mathbf{g}$, of the transportation force $\mathbf{F}_t = -m\mathbf{a}'_O$ and of the constraint force $\mathbf{N}$ (normal to the separation surface). If the liquid is in relative equilibrium with respect to the vessel in motion of translation, then the free surface is a plane inclined with respect to the horizontal one by an angle α given by $\tan\alpha = a'_O / g$, while the constraint force is $N = m\sqrt{a'^2_O + g^2}$.

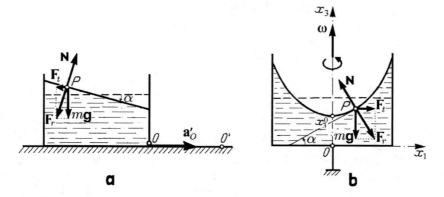

Figure 10.7. Relative equilibrium of a liquid with respect to a vessel in uniform accelerated translation (a) or in uniform motion of rotation about a vertical axis of symmetry (b).

If the vessel has a uniform motion of rotation about a vertical axis of symmetry, linked to the fixed frame, the movable frame being connected to the vessel, then a particle P, of mass m, of the separation surface of the fluid is acted upon by the gravity force $m\mathbf{g}$, by the centrifugal force $\mathbf{F}_t = -m\boldsymbol{\omega}\times(\boldsymbol{\omega}\times\mathbf{r})$, of magnitude $F_t = m\omega^2 x_1$, and by the constraint force $\mathbf{N}$ (Fig.10.7,b). Projecting the equation of relative equilibrium ($m\mathbf{g} + \mathbf{F}_t + \mathbf{N} = \mathbf{0}$) on the tangent at P to the meridian curve of the free surface, we get $m\omega^2 x_1 \cos\alpha = mg\sin\alpha$; observing that $\tan\alpha = \mathrm{d}x_3/\mathrm{d}x_1$, it results $\mathrm{d}x_3/\mathrm{d}x_1 = \omega^2 x_1/g$, the meridian curve being a parabola of equation $x_3 = (\omega^2/2g)x_1^2 + x_3^0$. Hence, the free surface is an axial-symmetrical paraboloid of equation

$$x_3 = \frac{\omega^2}{2g}(x_1^2 + x_2^2) + x_3^0. \qquad (10.2.23)$$

As well, the constraint force is given by $N = mg/\cos\alpha$.

2.2 Elements of terrestrial mechanics

After a study of the influence of the transportation force (inclusive of the centrifugal force) and of Coriolis' force on the motion of a particle on Earth's surface (using a geocentric or a heliocentric frame), some phenomena due to non-inertial frames on the Earth surface are considered. Thus, the phenomenon of tides is explained, the terrestrial acceleration is calculated, the deviation of the plummet from the local vertical, the deviation towards the east in free falling and Baer's law are determined and the idea of imponderability is explained; as well, Foucault's pendulum, very important for the knowledge of terrestrial motions, is presented. In particular, one obtains the case of relative equilibrium.

2.2.1 Geocentric and heliocentric frames

In the study of motion of a particle on the Earth surface, we considered till now that the local frame of reference is inertial (absolute); we obtained thus the motion of the particle with respect to this frame, where the motion of the Earth was not taken into consideration (an approximation of the physical reality). Assuming that the Earth is spherical, we choose the orthonormal frame with the pole at a point O on the boreal

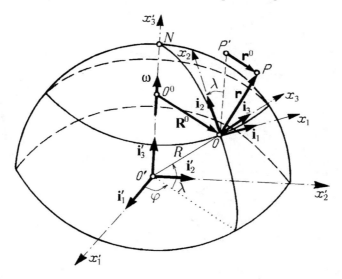

Figure 10.8. Motion of a particle with respect to a geocentric (inertial) frame of reference.

hemisphere; the Ox_1-axis is tangent to the parallel of the point O, being directed towards the east point, the Ox_2-axis is tangent to the local meridian, being directed towards the north, while the Ox_3-axis is directed towards the ascendent local vertical (the same direction as $\overrightarrow{O'O}$, O' being the centre of the Earth (Fig.10.8). If a heavy particle falls free from the height h (sufficient great, without other influences) and if we assume that the considered frame of reference is inertial, then the equations of the rectilinear trajectory are $x_1 = x_2 = 0$, $x_3 = h - gt^2/2$; but, experimentally, it is seen that a deviation towards the east point, along the parallel ($x_1 \neq 0$) takes place. This deviation is, obviously, due to Earth's motion of rotation (the local frame of reference is non-inertial). The most simple frame which can be chosen as an inertial one is *the geocentric frame of reference* (Ptolemy's frame), assuming that the pole is at the centre O' of the Earth; usually, the equatorial plane is taken as principal plane, the $O'x_1'$-axis being directed towards the vernal equinoctial point (on the celestial sphere of radius equal to unity, at the intersection of the equatorial plane with the ecliptic plane, which contains the orbit of the Earth), while the $O'x_3'$-axis is normal to that plane (hence, it is the rotation axis of the Earth) (Fig.10.8). Practically, a system of stars considered "fixed", which allows to specify the axes of the considered frame, is used.

The local frame of reference is rigidly linked to the Earth, hence it has a motion of rotation defined by the vector $\boldsymbol{\omega}$, directed along the axis of the poles, from the south to the north. We notice that a complete rotation of the Earth about its axis takes place in a *sidereal day*, which is smaller than a *mean solar day* by 3 min 56 s; hence, a sidereal day has $24 \cdot 60 \cdot 60 - (3 \cdot 60 + 56) = 86164$ s (seconds of mean solar time), so that the magnitude of the vector is given by

$$\omega = \frac{2\pi}{86164} \cong 7.292 \cdot 10^{-5} \text{ rad/s}, \quad (10.2.24)$$

the unit of time being the second. In the non-inertial frame of unit vectors $\mathbf{i}_j$, $j = 1,2,3$, the rotation vector is given by

$$\boldsymbol{\omega} = \omega \cos \lambda \mathbf{i}_2 + \omega \sin \lambda \mathbf{i}_1, \quad (10.2.25)$$

where λ, $\lambda \in [0, \pi/2]$ in the boreal hemisphere and $\lambda \in [-\pi/2, 0]$ in the austral hemisphere, is the *local latitude*. The unit vectors of the axes of the two frames are linked by the relations

$$\begin{aligned}
\mathbf{i}_1 &= -\sin\varphi \mathbf{i}_1' + \cos\varphi \mathbf{i}_2', \\
\mathbf{i}_2 &= -\sin\lambda \cos\varphi \mathbf{i}_1' - \sin\lambda \sin\varphi \mathbf{i}_2' + \cos\lambda \mathbf{i}_3', \\
\mathbf{i}_3 &= \cos\lambda \cos\varphi \mathbf{i}_1' + \cos\lambda \sin\varphi \mathbf{i}_2' + \sin\lambda \mathbf{i}_3',
\end{aligned} \quad (10.2.26)$$

where φ, $\varphi \in [0, 2\pi)$, is the *local longitude*. Taking into account (A.2.36) and the fact that the pole of the movable frame has a circular motion along the parallel of radius $R\cos\lambda$, where R is Earth's radius, we get

$$\omega_1 = 0, \quad \omega_2 = \dot\varphi \cos\lambda, \quad \omega_3 = \dot\varphi \sin\lambda; \quad (10.2.25')$$

observing that $\omega = \dot\varphi$, the formulae (10.2.25') correspond to the expression (10.2.25) previously obtained for the rotation vector, what was to be expected, taking into account the results in Chap. 5, Subsec. 2.2.2.

Other considerations on particle dynamics 631

But there are some phenomena for which the use of a geocentric frame of reference leads to results which are not in concordance with the physical reality; in this case, one must choose another frame as inertial one, i.e. a *heliocentric frame* (Copernicus' frame), with the pole O'' at the centre of mass of the Sun (very close to the centre of mass of the solar system, which – as it will be shown in Chap. 11, Subsec. 1.2.5 – has a rectilinear and uniform motion). The heliocentric frame may be ecliptic or equatorial, as it has been shown in Chap. 1, Subsec. 1.1.4, its axes $O''x_1''$, $O''x_2''$, $O''x_3''$ being specified correspondingly. The geocentric frame previously considered is, in this case, a non-inertial one, its axes having fixed directions (the corresponding rotation vector $\boldsymbol{\omega}'$ is equal to zero); indeed, corresponding to Kepler's first law, the Earth has also a motion of revolution which is – in fact – a motion of translation.

If neither the heliocentric frame of reference cannot be considered to be inertial, then one can choose a *galactocentric frame* a.s.o.

2.2.2 Motion of a particle at the Earth surface. Relative equilibrium

Let us consider, first of all, the case of a heliocentric frame (an inertial frame), the geocentric one being non-inertial. The equation of motion (10.2.2') with respect to the latter frame of reference is written in the form (neglecting the effect of the centrifugal force due to the motion of revolution of the Earth; Coriolis' force vanishes, because the motion is plane)

$$m\mathbf{a}_r' = \mathbf{F}' - m\mathbf{a}_{O'}'', \qquad (10.2.27)$$

where $\mathbf{r}_{O'}'' = \overrightarrow{O''O'}$, while $\mathbf{F}'$ is the force which acts upon the particle P of mass m.

Let ($\mathbf{f}_\mathscr{C}$ is, in fact, an acceleration)

$$m_Q \mathbf{f}_\mathscr{C}(Q) = -f \frac{m_\mathscr{C} m_Q}{\overline{CQ}^3} \overrightarrow{CQ} \qquad (10.2.28)$$

be the force of universal attraction by which a celestial body $\mathscr{C}$ of centre C and mass $m_\mathscr{C}$ acts upon a particle Q of mass m_Q. With respect to the heliocentric frame, the equation of motion of an element of mass $\mathrm{d}m_E = \mu(Q)\mathrm{d}V$ of the Earth, situated at the point of position vector $\mathbf{r}_Q$, reads

$$\mu(Q)\mathbf{a}_Q''\mathrm{d}V = \sum_j \mu(Q)\mathbf{f}_{\mathscr{C}_j}(Q)\mathrm{d}V,$$

where we have put into evidence the action of the celestial bodies $\mathscr{C}_j$, $j = 1, 2, \ldots$, upon that element; but (V is the volume of the Earth)

$$\int_V \mu(Q)\mathbf{a}_Q''\mathrm{d}V = \frac{\mathrm{d}^2}{\mathrm{d}t^2}\int_V \mu(Q)\mathbf{r}_Q''\mathrm{d}V = \frac{\mathrm{d}^2}{\mathrm{d}t^2}\left(m_E \mathbf{r}_Q''\right) = m_E \mathbf{a}_Q'',$$

where we took into account the formula (3.1.3), which gives the position of the centre of mass O' of the Earth, so that we get

$$m_E \mathbf{a}''_{O'} = \int_V \mu(Q) \sum_j \mathbf{f}_{\mathcal{C}_j}(Q) \mathrm{d}V, \quad m_E = \int_V \mu(Q) \mathrm{d}V.$$

Finally, the equation of motion (10.2.27) becomes

$$m\mathbf{a}'_r = \mathbf{F} + m[\mathbf{f}_E(P) + \mathbf{f}_M(P)] + m\sum_j \mathbf{f}_{P_j}(P)$$

$$- \frac{m}{m_E} \int_V \mu(Q) \left[\mathbf{f}_S(Q) + \mathbf{f}_M(Q) + \sum_j \mathbf{f}_{P_j}(Q) \right] \mathrm{d}V, \quad (10.2.29)$$

where $\mathbf{F}$ is the resultant of the other forces which act upon the particle P (forces of resistance of the medium, forces of friction, forces of electromagnetic nature etc.) and where we have put into evidence the action of the Sun, of the Moon and of the other planets P_j, $j = 1, 2, \ldots$. Neglecting the action of the planets in comparison with the action of the Sun and of the Moon, because of the great distances from these planets to the Earth, and applying a mean value theorem, we can write, with a good approximation,

$$m_E \mathbf{a}''_{O'} = [\mathbf{f}_S(O') + \mathbf{f}_M(O')] \int_V \mu(Q) \mathrm{d}V,$$

so that

$$\mathbf{a}''_{O'} = \mathbf{f}_S(O') + \mathbf{f}_M(O') = -f \frac{m_S}{\overline{O''O'}^3} \overrightarrow{O''O'} - f \frac{m_M}{\overline{O_M O'}^3} \overrightarrow{O_M O'}, \quad (10.2.28')$$

where O_M is the centre of the Moon. The equations (10.2.29) become

$$m\mathbf{a}'_r = \mathbf{F} + m\mathbf{f}_E(P) + m[\mathbf{f}_S(P) - \mathbf{f}_S(O')] + m[\mathbf{f}_M(P) - \mathbf{f}_M(O')]. \quad (10.2.29')$$

Observing that

$$|\mathbf{f}_E(P)| = f\frac{m_E}{R^2}, \quad |\mathbf{f}_S(O')| = f\frac{m_S}{\overline{O''O'}^2}, \quad |\mathbf{f}_M(O')| = f\frac{m_M}{\overline{O_M O'}^2},$$

we may write

$$\frac{|\mathbf{f}_S(O')|}{|\mathbf{f}_E(P)|} = \frac{m_S}{m_E}\left(\frac{R}{\overline{O''O'}}\right)^2 \cong 6.040 \cdot 10^{-4},$$

$$\frac{|\mathbf{f}_M(O')|}{|\mathbf{f}_E(P)|} = \frac{m_M}{m_E}\left(\frac{R}{\overline{O_M O'}}\right)^2 \cong 3.379 \cdot 10^{-6},$$

where we took into account the relations $m_S = 3.330 \cdot 10^5 m_E$, $m_E \cong 81.301 m_M$, $\overline{O''O'} \cong 2.348 \cdot 10^4 R$, $\overline{O_M O'} \cong 60.336 R$ (in fact, the distance $\overline{O''O'}$ put into

Other considerations on particle dynamics 633

evidence is the distance from the centre O'' of the Sun to the centre of mass of the mechanical system formed by the Earth and the Moon; but this latter centre is inside the terrestrial surface, at approximative $R/2 \ll \overline{O''O'}$ from the centre O', so that it may be identified, with a good approximation, with O'). Hence, the influence of the forces of attraction of the Sun and of the Moon can be neglected with respect to the force of attraction of the Earth. Let us consider the forces $\mathbf{f}_S(P)$ and $\mathbf{f}_M(P)$ too. We will have $|\mathbf{f}_S(P)|_{\min} = |\mathbf{f}_S(P_1)|$, $|\mathbf{f}_S(P)|_{\max} = |\mathbf{f}_S(P_2)|$ (Fig.10.9), as well as

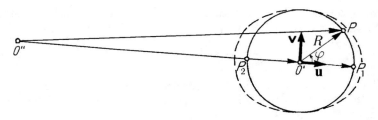

Figure 10.9. The influence of the forces of attraction of the Sun and of the Moon with respect to the force of attraction of the Earth.

$$\frac{|\mathbf{f}_S(P_1)|}{|\mathbf{f}_E(P)|} \cong \frac{|\mathbf{f}_S(P_2)|}{|\mathbf{f}_E(P)|} \cong \frac{|\mathbf{f}_S(O')|}{|\mathbf{f}_E(P)|};$$

analogously, $|\mathbf{f}_M(P)|_{\min} = |\mathbf{f}_M(P_1)|$ and $|\mathbf{f}_M(P)|_{\max} = |\mathbf{f}_M(P_2)|$, so that $(\overline{O_M P_1} \cong 61.336 R, \overline{O_M P_2} \cong 59.336 R)$

$$\frac{|\mathbf{f}_M(P_1)|}{|\mathbf{f}_E(P)|} = \frac{m_M}{m_E}\left(\frac{R}{\overline{O_M P_1}}\right)^2 \cong 3.269 \cdot 10^{-6},$$

$$\frac{|\mathbf{f}_M(P_2)|}{|\mathbf{f}_E(P)|} = \frac{m_M}{m_E}\left(\frac{R}{\overline{O_M P_2}}\right)^2 \cong 3.494 \cdot 10^{-6}.$$

In fact, we are interested in the differences (components along $\mathbf{u} = \text{vers}\,\overrightarrow{O'P_1}$)

$$-|\mathbf{f}_M(P_1)| + |\mathbf{f}_M(O')| = -f\frac{m_M}{\overline{O_M P_1}^2} + f\frac{m_M}{\overline{O_M O'}^2} = |\mathbf{f}_M(O')|\left[1 - \left(1 + \frac{R}{\overline{O_M O'}}\right)^{-2}\right]$$

$$\cong |\mathbf{f}_M(O')|\left[1 - \left(1 - \frac{R}{\overline{O_M O'}} + \ldots\right)\right] \cong 2\frac{R}{\overline{O_M O'}}|\mathbf{f}_M(O')|$$

$$-|\mathbf{f}_M(P_2)| + |\mathbf{f}_M(O')| = |\mathbf{f}_M(O')|\left[1 - \left(1 - \frac{R}{\overline{O_M O'}}\right)^{-2}\right] \cong -2\frac{R}{\overline{O_M O'}}|\mathbf{f}_M(O')|;$$

in this case ($\overline{OP_2} \leq \overline{OP} \leq \overline{OP_1}$)

$$\frac{|\mathbf{f}_M(P) - \mathbf{f}_M(O')|}{|\mathbf{f}_E(P)|} \leq 2\frac{m_M}{m_E}\left(\frac{R}{\overline{O_M O'}}\right)^3 \cong 1.120 \cdot 10^{-7}.$$

Analogously, we get

$$\frac{|\mathbf{f}_S(P) - \mathbf{f}_S(O')|}{|\mathbf{f}_E(P)|} \leq 2\frac{m_S}{m_E}\left(\frac{R}{\overline{O'' O'}}\right)^3 \cong 5.145 \cdot 10^{-8}.$$

In conclusion, to study phenomena connected to an isolated particle, situated on the Earth surface, we can use – with a very good approximation – the equation

$$m\mathbf{a}'_r = \mathbf{F} + m\mathbf{f}_E(P), \quad (10.2.29'')$$

written with respect to a geocentric frame of reference, which is non-inertial.

Equating to zero the relative motion, we obtain the condition of *relative equilibrium* (with respect to the Earth) in the form

$$\mathbf{F} + m\mathbf{f}_E(P) = \mathbf{0}. \quad (10.2.30)$$

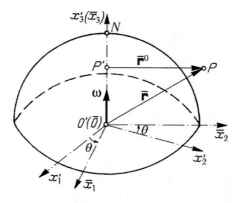

Figure 10.10. Relative equilibrium of a particle with respect to the Earth.

Let be a movable frame of reference $\overline{O}\overline{x}_1\overline{x}_2\overline{x}_3$ (non-inertial), rigidly linked to the Earth, so that $\overline{O} \equiv O'$ and $\overline{O}\overline{x}_3 \equiv O'x'_3$, its position being specified by the angle θ (Fig.10.10); we may write the equations of motion with respect to the new frame in the form

$$m\overline{\mathbf{a}}_r = \mathbf{F} + m\mathbf{f}_E(P) - m\dot{\boldsymbol{\omega}} \times \mathbf{r} - m\boldsymbol{\omega} \times (\boldsymbol{\omega} \times \overline{\mathbf{r}}) - 2m\boldsymbol{\omega} \times \overline{\mathbf{v}}_r, \quad (10.2.31)$$

where $\boldsymbol{\omega}$ is the rotation vector ($\omega = \dot{\theta}$), while $\overline{\mathbf{r}} = \overrightarrow{OP}$. We notice that $\boldsymbol{\omega} = \omega \mathbf{i}'_3$, where ω is given by (10.2.24); using Rivals' formula (5.2.6''') and neglecting $\dot{\boldsymbol{\omega}}$ (the angular velocity $\boldsymbol{\omega}$ is, with a good approximation, constant, the variation in time of the poles' axis being negligible), we may write

Other considerations on particle dynamics 635

$$m\overline{\mathbf{a}}_r = \mathbf{F} + m\mathbf{f}_E(P) + m\omega^2 \overline{\mathbf{r}}^0 - 2m\boldsymbol{\omega} \times \overline{\mathbf{v}}_r,\qquad(10.2.31')$$

where $\overline{\mathbf{r}}^0 = \overrightarrow{P'P}$ (Fig.10.10). The condition of relative equilibrium (with respect to the movable frame) is given by

$$\mathbf{F} + m\mathbf{f}_E(P) + m\omega^2 \overline{\mathbf{r}}^0 = \mathbf{0}.\qquad(10.2.32)$$

If we choose the movable frame rigidly linked to the Earth, with the pole at a point O on the Earth surface (Fig.10.8) (because the particle is in motion in the vicinity of this pole), the equation of motion with respect to a local frame (non-inertial) becomes

$$m\mathbf{a}_r = \mathbf{F} + m\mathbf{f}_E(P) + m\left[\omega^2 \mathbf{R}^0 - \boldsymbol{\omega} \times (\boldsymbol{\omega} \times \mathbf{r})\right] - 2m\boldsymbol{\omega} \times \mathbf{v}_r;\qquad(10.2.33)$$

we notice that the acceleration $\mathbf{a}'_O$ with respect to the frame $O'x'_1x'_2x'_3$, considered to be fixed, is given by $\mathbf{a}'_O = -\omega^2 \mathbf{R}^0$, with $\overrightarrow{O^0 O} = \mathbf{R}^0$ (the pole O has a uniform motion along a parallel of the Earth, of radius R^0). In components, we may write (we assume that the force $\mathbf{f}_E(P)$ is applied in the vicinity of the pole O, along the descendent vertical of that pole, the gravitational field being uniform)

$$\begin{aligned}
m\ddot{x}_1 &= F_1 - 2m\omega(\dot{x}_3 \cos\lambda - \dot{x}_2 \sin\lambda) + m\omega^2 x_1,\\
m\ddot{x}_2 &= F_2 - 2m\omega\dot{x}_1 \sin\lambda - m\omega^2 \sin\lambda[(x_3 + R)\cos\lambda - x_2 \sin\lambda],\qquad(10.2.33')\\
m\ddot{x}_3 &= F_3 - m|\mathbf{f}_E(P)| + 2m\omega\dot{x}_1 \cos\lambda\\
&\quad + m\omega^2 \cos\lambda[(x_3 + R)\cos\lambda - x_2 \sin\lambda].
\end{aligned}$$

In general, this system of equations cannot be exactly integrated; we must use approximate methods of calculation, e.g. the method of the small parameter (after Poincaré), developing the solution in a power series after that parameter (from (10.2.24) it results that $\varepsilon = \omega T$, where T is a time specific to the considered mechanical phenomenon, ε being a non-dimensional small parameter)

$$\mathbf{r} = \mathbf{r}_0 + \varepsilon \mathbf{r}_1 + \varepsilon^2 \mathbf{r}_2 + \ldots.\qquad(10.2.34)$$

If we replace in (10.2.33') and identify the powers of the same order as ε, we can write successively

$$m\ddot{x}_1^{(0)} = F_1,\quad m\ddot{x}_2^{(0)} = F_2,\quad m\ddot{x}_3^{(0)} = F_3 - m|\mathbf{f}_E(P)|,\qquad(10.2.35)$$

$$\ddot{x}_1^{(1)} = -2\left(\dot{x}_3^{(0)} \cos\lambda - \dot{x}_2^{(0)} \sin\lambda\right),\quad \ddot{x}_2^{(1)} = -2\dot{x}_1^{(0)} \sin\lambda,$$
$$\ddot{x}_3^{(1)} = 2\dot{x}_1^{(0)} \cos\lambda,\ldots\qquad(10.2.35')$$

The approximation of *n*th order is obtained by quadratures, starting from the approximation of (*n*-1)th order; obviously, the convergence of the solution must be also verified. The approximation of order zero (10.2.35) corresponds to the neglect of the motion of rotation of the Earth. For instance, in case of initial conditions of the form $\mathbf{r}(0) = \mathbf{r}^0$, $\mathbf{v}(0) = \mathbf{v}^0$, we have

$$\mathbf{r}_0(0) = \mathbf{r}^0, \quad \mathbf{v}_0(0) = \mathbf{v}^0, \qquad (10.2.36)$$
$$\mathbf{r}_1(0) = \mathbf{0}, \quad \mathbf{v}_1(0) = \mathbf{0}, \dots \qquad (10.2.36')$$

a.s.o. If the initial conditions depend on ω, then one uses systematically expansions in power series with respect to the small parameter ε and to each approximation are put initial conditions which multiply the same power of ε.

The condition of relative equilibrium with respect to the local frame $Ox_1x_2x_3$ is written in the form

$$\mathbf{F} + m\mathbf{f}_E(P) + m\left[\omega^2 \mathbf{R}^0 - \boldsymbol{\omega} \times (\boldsymbol{\omega} \times \mathbf{r})\right] = \mathbf{0}. \qquad (10.2.37)$$

We put in evidence, in what follows, the influence of each term which appears in this equation.

2.2.3 Acceleration of the centre of the Earth. Tide

The translation acceleration $\mathbf{a}''_{O'}$ of the centre O' of the Earth with respect to the heliocentric frame $O''x_1''x_2''x_3''$ is due, as we have seen, to the forces of attraction of the Sun, of the Moon and of other planets (the latter ones may be neglected because of the great distances from those planets to the Earth). The magnitude $a''_{O'}$ of the acceleration (due to the Sun and to the Moon) is given by $a''_{O'} = a''_{O'}(S) + a''_{O'}(M)$, where

$$a''_{O'}(S) = f\frac{m_S}{\overline{O''O'}^2}, \quad a''_{O'}(M) = f\frac{m_M}{\overline{O_MO'}^2} = \frac{m_M}{m_S}\left(\frac{\overline{O''O'}}{\overline{O_MO'}}\right)^2 a''_{O'}(S);$$

taking $f = 6.673 \cdot 10^{-8}$ cm^3/g·s^2, $m_S = 1.989 \cdot 10^{33}$ g, $m_M = 7.347 \cdot 10^{25}$ g, $\overline{O''O'} = 1.496 \cdot 10^{13}$ cm, $\overline{O_MO'} = 3.844 \cdot 10^{10}$ cm, we obtain $a''_{O'}(S) = 5.931 \cdot 10^{-1}$ cm/s^2 (which leads to a force of 0.5931 dyne per 1 g mass) and $a''_{O'}(M) = 0.00559 a''_{O'}(S) = 0.0033$ cm/s^2 (corresponding 0.0033 dyne per 1 g mass).

For a better approximation of the influence of that acceleration, let us return to the equation (10.2.29'); we evaluate thus the magnitude of the force

$$m[\mathbf{f}_S(P) - \mathbf{f}_S(O')] = -f\frac{mm_S}{\overline{O''P}^3}\overrightarrow{O''P} + f\frac{mm_S}{\overline{O''O'}^3}\overrightarrow{O''O'}. \qquad (10.2.38)$$

Assuming that the Sun is at an infinite distance, hence that $\overrightarrow{O''P} \parallel \overrightarrow{O''O'}$, the component of this force along the unit vector $\mathbf{u} = \text{vers}\,\overrightarrow{O''O'}$ (Fig.10.9) is given by

Other considerations on particle dynamics 637

$$m[\mathbf{f}_S(P) - \mathbf{f}_S(O')]_u = -f\frac{mm_S}{(\overline{O''O'} + R\cos\varphi)^2} + f\frac{mm_S}{\overline{O''O'}^2}$$

$$= f\frac{mm_S}{\overline{O''O'}^2}\left[1 - \left(1 + \frac{R}{\overline{O''O'}}\cos\varphi\right)^{-2}\right] = f\frac{mm_S}{\overline{O''O'}^2}\left[1 - \left(1 - 2\frac{R}{\overline{O''O'}}\cos\varphi + ...\right)\right],$$

wherefrom

$$m[\mathbf{f}_S(P) - \mathbf{f}_S(O')]_u \cong 2f\frac{mm_S}{\overline{O''O'}^3}R\cos\varphi = 2m|\mathbf{f}_S(O')|\frac{R}{\overline{O''O'}}\cos\varphi, \quad (10.2.39)$$

with $\varphi = \sphericalangle(\overrightarrow{O'P_1}, \overrightarrow{O'P})$. But $\overrightarrow{O''P} = \overrightarrow{O''O'} + \overrightarrow{O'P}$, so that the force (10.2.38) reads

$$m[\mathbf{f}_S(P) - \mathbf{f}_S(O')] = fmm_S\left[\left(\frac{1}{\overline{O''O'}^3} - \frac{1}{\overline{O''P}^3}\right)\overrightarrow{O''O'} - \frac{1}{\overline{O''P}^3}\overrightarrow{O'P}\right];$$

but $\overline{O''P}^2 = \overline{O''O'}^2 + R^2 + 2\overline{O''O'}R\cos\varphi$, so that

$$\left(\frac{\overline{O''O'}}{\overline{O''P}}\right)^3 = \left[1 + 2\frac{R}{\overline{O''O'}}\cos\varphi + \left(\frac{R}{\overline{O''O'}}\right)^2\right]^{-3/2}$$

$$= 1 - 3\frac{R}{\overline{O''O'}}\cos\varphi + \frac{3}{2}(5\cos^2\varphi - 1)\left(\frac{R}{\overline{O''O'}}\right)^2 + ...$$

Projecting on the unit vector **u** and neglecting the terms in $(R/\overline{O''O'})^2$ with respect to $R/\overline{O''O'}$, we find again the component (10.2.39); the projection on the unit vector **v** (normal to **u**, Fig.10.9) is given by

$$m[\mathbf{f}_S(P) - \mathbf{f}_S(O')]_v \cong f\frac{mm_S}{\overline{O''O'}^3}R\sin\varphi = -m|\mathbf{f}_S(O')|\frac{R}{\overline{O''O'}}\sin\varphi, \quad (10.2.39')$$

the corresponding component being directed towards the diameter P_2P_1.

Hence, the particle P is acted upon, besides the gravity force $m\mathbf{f}_E(P)$, by a force due to the attraction of the Sun. We have seen, at the preceding subsection, that this force can be neglected; but, in case of a continuous mechanical system of great dimensions, the contribution of the terms due to this force of attraction is added and becomes noticeable, so that it must be taken into consideration. If such a mechanical system is rigidly connected to the Earth, e.g. its solid crust, then the effect of the force of attraction is not observed, but if this system is not rigidly linked to the Earth (the case of seas and oceans or of fluid masses in the interior of the terrestrial sphere), then appears the phenomenon called *tide*. The component along **u** has extreme values for $\varphi = 0$ (the face opposite to the Sun, for which the water is repulsed) and for $\varphi = \pi$ (the face towards the Sun, when the water is attracted) and we have (Fig.10.9)

$$m[\mathbf{f}_S(P) - \mathbf{f}_S(O')]_{u_{\text{extr}}} \cong \pm 2m|\mathbf{f}_S(O')|\frac{R}{\overline{O''O'}}, \qquad (10.2.40)$$

the component along $\mathbf{v}$ vanishing in this case. The component along $\mathbf{u}$ is equal to zero for $\varphi = \pm\pi/2$, while the component along $\mathbf{v}$ has extreme values

$$m[\mathbf{f}_S(P) - \mathbf{f}_S(O')]_{v_{\text{extr}}} \cong \mp m|\mathbf{f}_S(O')|\frac{R}{\overline{O''O'}}. \qquad (10.2.40')$$

In this zone, the particles of fluid are attracted by a greater force towards the Earth. At a point on the Earth surface, the force of attraction is varying in time, because the angle φ varies together with the rotation of the Earth. The extreme values of the magnitude of this force (hence, the phenomenon of tide, *the flood* and *the ebb*, respectively) appear twice in 24 hours (once when the Sun passes at the local meridian and once when it passes at the opposite one). It is obvious that the level of the tides is greater at the equatorial zone and tends to zero near the polar circle.

One can study the influence of the Moon too, obtaining analogous formulae; thus

$$m[\mathbf{f}_M(P) - \mathbf{f}_M(O')]_{u_{\text{extr}}} \cong \pm 2m|\mathbf{f}_M(O')|\frac{R}{\overline{O_M O'}}, \qquad (10.2.41)$$

$$m[\mathbf{f}_M(P) - \mathbf{f}_M(O')]_{v_{\text{extr}}} \cong \mp m|\mathbf{f}_M(O')|\frac{R}{\overline{O_M O'}}. \qquad (10.2.41')$$

Introducing the numerical data mentioned above, we notice that

$$\frac{m[\mathbf{f}_M(P) - \mathbf{f}_M(O')]_{u_{\text{extr}}}}{m[\mathbf{f}_S(P) - \mathbf{f}_S(O')]_{u_{\text{extr}}}} = \frac{m[\mathbf{f}_M(P) - \mathbf{f}_M(O')]_{v_{\text{extr}}}}{m[\mathbf{f}_S(P) - \mathbf{f}_S(O')]_{v_{\text{extr}}}}$$

$$= \frac{|\mathbf{f}_M(O')|}{|\mathbf{f}_S(O')|}\frac{\overline{O''O'}}{\overline{O_M O'}} = \frac{m_M}{m_S}\left(\frac{\overline{O''O'}}{\overline{O_M O'}}\right)^3 = 2.177;$$

it results that the effect of the attraction force of the Moon on the tides is approximative twice greater than that of the Sun, although the mass of the Moon is smaller than that of the Sun, because it is closer to the Earth than the Sun.

2.2.4 State of imponderability

Let us consider the translation motion of a spatial vehicle, the mass centre O of which describes an ellipse around the Earth. With respect to a non-inertial frame of reference with the pole at O, we can write the equation of motion of a particle P of mass m, situated in the vehicle, in the form

$$m\mathbf{a}_r = \mathbf{F} + m[\mathbf{f}_E(P) - \mathbf{f}_E(O)] = \mathbf{F} + m\left[-f\frac{m_E}{\overline{O'P}^3}\overrightarrow{O'P} + f\frac{m_E}{\overline{O'O}^3}\overrightarrow{O'O}\right], \qquad (10.2.42)$$

Other considerations on particle dynamics 639

where O' is the centre of mass of the Earth, the chosen fixed geocentric frame being considered inertial; we have neglected the force of attraction of the spatial vehicle, as well as the attraction forces of the Sun, of the Moon and of other planets, the equation (10.2.42) being similar to the equation (10.2.28').

Observing that $\overrightarrow{O'P} = \overrightarrow{O'O} + \overrightarrow{OP}$, we may also write

$$m\mathbf{a}_r = \mathbf{F} + fmm_E\left[\left(\frac{1}{\overline{O'O}^3} - \frac{1}{\overline{O'P}^3}\right)\overrightarrow{O'O} - \frac{1}{\overline{O'P}^3}\overrightarrow{OP}\right].$$

If $\varphi = \sphericalangle\left(\overrightarrow{O'O}, \overrightarrow{OP}\right)$, then we have

$$\frac{\overline{O'P}}{\overline{O'O}} = \left[1 + 2\frac{\overline{OP}}{\overline{O'O}}\cos\varphi + \left(\frac{\overline{OP}}{\overline{O'O}}\right)^2\right]^{1/2} = 1 + \frac{\overline{OP}}{\overline{O'O}}\cos\varphi + \ldots,$$

so that $\overline{O'P} \cong \overline{O'O}$; on the other hand $\overline{OP} \ll \overline{O'P}$. The equation of motion is thus reduced to $m\mathbf{a}_r = \mathbf{F}$. Assuming that the particle P is not in contact with the walls of the cabin and that upon it there are not acting resistent forces of the medium, of an electromagnetic or of another nature, due to certain facilities existing in the vehicle, it results that $\mathbf{F} = \mathbf{0}$, hence $\mathbf{a}_r = \mathbf{0}$ too; thus, *the state of imponderability* in the interior of the spatial vehicle may be justified.

2.2.5 Centrifugal force

We have seen that the transportation force, in a motion with respect to the geocentric frame of reference, contains – besides the force due to the acceleration of the Earth centre and the force due to the angular acceleration vector $\dot{\boldsymbol{\omega}}$, which can be neglected, as we have mentioned before – also *the centrifugal force* (corresponding to *the centripetal acceleration*)

$$\mathbf{F}_c = -m\boldsymbol{\omega} \times (\boldsymbol{\omega} \times \overline{\mathbf{r}}) = m\omega^2 \overline{\mathbf{r}}^0, \tag{10.2.43}$$

in conformity to the results in Subsec. 2.2.2; this force is normal to the rotation vector $\boldsymbol{\omega}$, having the direction towards the exterior of the Earth. But, for particles at the Earth surface, it is more convenient to use a local frame $Ox_1x_2x_3$ (Fig.10.8), so that we obtain the centrifugal force

$$\mathbf{F}_c = m\left[\omega^2\mathbf{R}^0 - \boldsymbol{\omega} \times (\boldsymbol{\omega} \times \mathbf{r})\right], \tag{10.2.44}$$

of components

$$\begin{aligned}F_{c1} &= m\omega^2 x_1, \\ F_{c2} &= -m\omega^2 \sin\lambda[(x_3 + R)\cos\lambda - x_2\sin\lambda], \\ F_{c3} &= m\omega^2 \cos\lambda[(x_3 + R)\cos\lambda - x_2\sin\lambda].\end{aligned} \tag{10.2.44'}$$

For the sake of simplicity, we assume that the particle is situated at the origin of the local frame ($P \equiv O$); it results ($F_c = mR\omega^2 \cos\lambda$)

$$F_{c1} = 0, \quad F_{c2} = -\frac{1}{2}m\omega^2 R \sin 2\lambda, \quad F_{c3} = m\omega^2 R \cos^2 \lambda, \qquad (10.2.44'')$$

where λ is the local latitude (Fig.10.11).

As it can be noticed, the vector component $\mathbf{F}_{c2}$ (which vanishes at the equator and at the poles, having an extreme value of 1.693 dyne per 1 g mass at the latitude of $45°$, both in the boreal and in the austral hemispheres) is tangent to the meridian circle, being directed towards the south in the boreal hemisphere and towards the north in the austral one; hence, at any point on the Earth surface, this component has the tendency to deviate the vertical direction (of the plummet) towards the equator. The vector component $\mathbf{F}_{c3}$ (which vanishes at the poles and has, at the equator, a maximal value of 3.386 dyne per 1 g mass, which represents $3.386/980.6 \cong 0.003453$, hence approximate 0.35% from the weight of a mass of 1 g) is along the local vertical (along the radius of the Earth, considered to be spherical), being directed towards the exterior of the Earth; this component has the tendency to decrease the effect of attraction exerted by the Earth upon the considered particle.

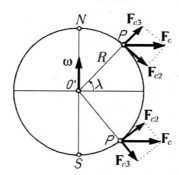

Figure 10.11. Action of the centrifugal force on the Earth surface.

The ratio of the centrifugal force due to the motion of revolution of the Earth to the centrifugal force due to its motion of rotation is given approximately by

$$\left(\frac{1}{365.25}\right)^2 \frac{\overline{O''O'}}{R} \cong 0.176;$$

the equation (10.2.27) and then the equation (10.2.33) are thus justified.

2.2.6 Deviation of the plummet from the local vertical

For a systematic study of the influence of the centrifugal force at the Earth surface, we consider a particle P, hanged up by a thread at the fixed point Q, rigidly connected to the Earth (a *plummet*). The particle P of gravity mass m_g is acted upon by the

Other considerations on particle dynamics

terrestrial force of attraction $m_g \mathbf{f}_E(P) = m_g \mathbf{g}'$, where $\mathbf{g}'$ is *the gravity acceleration* (*the theoretical acceleration*) and by the tension $\mathbf{T}$ in the thread. If the Earth would be immovable, perfect spheric and with a distribution of mass with spherical symmetry, then the particle P would be on the line QO', and the *theoretical* acceleration would be the *practical* one (which is determined experimentally); but, due to the rotation of the Earth, appears the centrifugal force $\mathbf{F}_c$ too, so that the equation of relative equilibrium (10.2.37) reads

$$m_g \mathbf{g}' + \mathbf{T} + \mathbf{F}_c = \mathbf{0}. \tag{10.2.45}$$

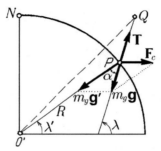

Figure 10.12. Deviation of the plummet from the local vertical.

Denoting $m_g \mathbf{g}' + \mathbf{F}_c = m_g \mathbf{g}$, where $\mathbf{g}$ is *the terrestrial acceleration* (*the practical acceleration*), we obtain (Fig.10.12)

$$m_g \mathbf{g} + \mathbf{T} = \mathbf{0}. \tag{10.2.45'}$$

The theoretical vertical PO' is thus replaced by *the practical vertical* QP (the local latitude being thus *the astronomical latitude* λ, instead of *the geographical latitude* λ'). In fact, a deviation of angle α (the angle between the theoretical vertical and the practical one) of the plumb line is put into evidence.

Projecting the relation (10.2.45) on a direction normal to the practical vertical, we get ($|\mathbf{F}_c| = m_i R \omega^2 \cos \lambda'$, where m_i is the inertial mass)

$$-m_g g' \sin \alpha + m_i R \omega^2 \cos \lambda' \sin(\lambda' + \alpha) = 0,$$

wherefrom ($k = m_i / m_g$)

$$\tan \alpha = \frac{k R \omega^2 \sin \lambda' \cos \lambda'}{g' - k R \omega^2 \cos^2 \lambda'}. \tag{10.2.46}$$

The denominator of this expression is always positive, because the term $k R \omega^2 \cos^2 \lambda'$ $= 3.386 k \cos^2 \lambda'$ cm/s² is, in any case, less than $g' \cong g \cong 981$ cm/s²; hence, $\tan \alpha \gtreqless 0$, so that $\alpha \gtreqless 0$, as $0 < \lambda' < \pi/2$ or $-\pi/2 < \lambda' < 0$. In conclusion, the plummet is deviated from the theoretical local vertical towards the equator, as it was shown in the preceding subsection. Differentiating the function (10.2.46), one can show that $\alpha_{\max}$ is obtained for

$$\cos 2\lambda' = \frac{kR\omega^2}{2g' - kR\omega^2}.$$

Taking into account the numerical data mentioned above, it results, with a very good approximation, $\cos 2\lambda' \cong 0$, hence $\lambda' \cong 45°$; in this case

$$\alpha_{\max} \cong (\tan \alpha)_{\max} \cong \frac{kR\omega^2}{2g' - kR\omega^2}. \qquad (10.2.47)$$

If we notice that $g'^2 \gg kR\omega^2 \cos^2 \lambda'(2g' - kR\omega^2)$, then we can also write

$$\cos \alpha \cong 1 - k\frac{R\omega^2}{g'}\cos^2 \lambda', \quad \sin \alpha \cong k\frac{R\omega^2}{2g'}\sin 2\lambda', \qquad (10.2.46')$$

the signs which are chosen corresponding to both hemispheres; the second of those formulae played a particularly important rôle in the theory of mechanical systems, being used to state experimentally the equality between the gravitational mass and the inertial one. To this scope, Eötvös and Zeeman used a balance of torsion (Eötvös's balance), stating that, at any point of the Earth surface (and for all points having the same latitude), one obtains the same angle α for particles of different masses (an approximation of 10^{-8} with respect to unity). On the basis of the mentioned formula, it results $k = \text{const}$. Hence, the gravity mass differs from the inertial one by a constant factor k, which can be taken equal to unity ($k = 1$). Recent experiments of Dicke have put in evidence this result with an approximation of 10^{-10} with respect to unity. As well, they have been extended to the intimate structure of the atom.

If we take $k = 1$ in (10.2.47) and put $g' \cong g \cong 980.6 \text{ cm/s}^2$, we find $\alpha_{\max} \cong 0.001730$ rad or $\alpha_{\max} \cong 5'56.92''$, hence the deviation of the plummet with respect to the local theoretical vertical is at the most of six minutes, so that it can be neglected.

2.2.7 Calculation of the terrestrial acceleration

From (10.2.45) and (10.2.45'), it results (we take $k = 1$)

$$\mathbf{g} = \mathbf{g}' + R\omega^2 \cos \lambda' \text{ vers } \mathbf{F}_c, \qquad (10.2.48)$$

so that

$$g^2 = g'^2 - 2g'R\omega^2 \cos^2 \lambda' + R^2\omega^4 \cos^2 \lambda'. \qquad (10.2.48')$$

Developing into series, we may write

$$\frac{g}{g'} = 1 - \frac{R\omega^2}{g'}\cos^2\lambda' + \frac{1}{8}\frac{R^2\omega^4}{g'^2}\sin^2 2\lambda' + \ldots, \tag{10.2.49}$$

obtaining thus the terrestrial acceleration g (the practical acceleration), as a function of the gravity acceleration g' (the theoretical acceleration), at any latitude on the Earth surface; replacing R by $R+h$, we get g as function of the altitude h from the sea level too. The terms which contain $\left(R\omega^2/g'\right)^2$ can be, in general, neglected. Thus, we obtain, with a good approximation, the relation which links the terrestrial acceleration to the gravity one

$$g = g'\left(1 - \frac{R\omega^2}{g'}\cos^2\lambda'\right). \tag{10.2.50}$$

At the poles ($\lambda' = \pm\pi/2$) we have $g = g'$; in reality, this equality is approximate, because the Earth is not a sphere, but a spheroid and the distribution of masses is only approximately with spherical symmetry. At the equator ($\lambda' = 0$), at the sea level, we get

$$g_e = g'\left(1 - \frac{R\omega^2}{g'}\right); \tag{10.2.50'}$$

we notice that this result is an exact one, corresponding also to the formula (10.2.48'). Besides, the maximal difference $|g - g'|$ is obtained at the equator. Numerically, we have $R\omega^2/g' \cong 3.4636 \cdot 10^{-3} \cong 1/288.717 \cong 1/289 = (1/17)^2$, where we have considered the Earth as a sphere of radius $R = 6.371 \cdot 10^8$ cm and we have assumed that $g' \cong g \cong 978.1$ cm/s^2. Hence, if the Earth would be rotating about its axis with an angular velocity of $17\omega \cong 1.240 \cdot 10^{-3}$ rad/s, the particles situated at the equator would be weightless.

In reality, the formulae obtained above have a certain degree of approximation, due to the model of sphere assumed for the Earth. If one takes into account that the Earth is a spheroid (e.g., in the classical theory of Clairaut), then one obtains the possibility to determine the gravity acceleration g' corresponding to a parallel on the Earth surface; the theoretical acceleration g may be thus calculated. As a matter of fact, even using the formulae (10.2.48)-(10.2.50) one cannot obtain g without knowing g'; but one can state that the difference $g' - g > 0$ is sufficiently small.

Using a model closer to the physical reality, one can establish the formula

$$g = 980.60509 - 2.5028\cos 2\lambda - 0.0003h, \tag{10.2.51}$$

which gives the terrestrial acceleration in cm/s^2, as a function of the geographical latitude λ (the practical latitude) and of the height h in cm, over the sea level. We find thus, at the sea level ($h = 0$), $g_e = 978.1$ cm/s^2 at the equator (result used before) and $g_p = 983.1$ cm/s^2 at the poles. At the latitude $\lambda = 45°$ (which corresponds with a good approximation for Bucharest too) one obtains $g_{45°} = 980.6$ cm/s^2 which represents the mean of the extreme values of the terrestrial acceleration. We notice that $(g_p - g_e)/g_{45°} = 0.0050989$, so that the theoretical acceleration has a variation of maximum 0.5%.

Projecting the relation (10.2.45) on the practical vertical QP (Fig.10.12), we get

$$m_g g = m_g g' \cos\alpha - m_i R\omega^2 \cos\lambda' \cos(\lambda' + \alpha);$$

if we put $m_g = m_i$, neglect the angle α with respect to λ', and take $\cos\alpha \cong 1$, then we find again the formula (10.2.50) for $h = 0$.

Observing that $\mathbf{g} \cdot \mathbf{g}' = gg' \cos\alpha$, the formula (10.2.48) leads to

$$g^2 + g'^2 - 2gg' \cos\alpha = R^2 \omega^4 \cos^2 \lambda';$$

taking into account (10.2.48'), we find

$$\cos\alpha = \frac{g'}{g}\left(1 - \frac{R\omega^2}{g'}\cos^2 \lambda'\right), \qquad (10.2.52)$$

relation which puts into evidence the degree of approximation of the formula (10.2.50) (approximation $\cos\alpha \cong 1$). We may then also calculate (we take further into consideration the relation (10.2.48'))

$$\sin\alpha = \frac{R\omega^2}{2g}\sin 2\lambda' = \frac{g'}{g}\frac{R\omega^2}{2g'}\sin 2\lambda'; \qquad (10.2.52')$$

as a matter of fact, this relation could be obtained as projection of the relation (10.2.45) on a normal to the theoretical vertical PO' (Fig.10.12). The relations (10.2.52), (10.2.52') are exact, in comparison with the relations (10.2.46') and correspond to the relation (10.2.46) (for $k = 1$). As well, the calculation of the angle $\alpha_{\max}$ for $\lambda' = 45°$ is justified too.

The formula (10.2.52) is an exact one in the frame of the spherical model used for the Earth. Eliminating $R\omega^2$ between (10.2.52) and (10.2.52'), we find the relation

$$g = g' \frac{\sin\lambda'}{\sin(\lambda' + \alpha)}, \qquad (10.2.53)$$

which is exact too. As a matter of fact, this result may be obtained by projecting the relation (10.2.45) on a normal to the centrifugal force $\mathbf{F}_c$ (Fig.10.12).

2.2.8 Deviation towards the east in the free falling of a heavy particle

Let us return to the equations of motion of a heavy particle at the Earth surface (10.2.33'), where we take $\mathbf{F} = \mathbf{0}$, and let us put $|\mathbf{f}_E(P)| = g' \cong g$, corresponding to the results of the preceding subsection. The first approximation (10.2.35) with the initial conditions (10.2.36) leads to

$$x_1^{(0)} = v_1^0 t + x_1^0, \quad x_2^{(0)} = v_2^0 t + x_2^0, \quad x_3^{(0)} = -\frac{1}{2}gt^2 + v_3^0 t + x_3^0. \quad (10.2.54)$$

Assuming that the particle is falling from a height h with the velocity $\mathbf{v}^0$ from a point of the local vertical (in what follows, by local vertical we mean only the theoretical vertical, directed towards the centre of the Earth, considered to be spheric), we take $x_1^0 = x_2^0 = 0$, $x_3^0 = h$; in case of a free falling ($\mathbf{v}^0 = \mathbf{0}$), the motion takes place as the Earth would be fixed (in a first approximation the motion of rotation of the Earth is neglected).

In a second approximation, the equations (10.2.35') with the initial conditions (10.2.36') allow to write

$$x_1^{(1)} = \frac{1}{3}gt^3 \cos\lambda + \left(v_2^0 \sin\lambda - v_3^0 \cos\lambda\right)t^2, \quad x_2^{(1)} = -v_1^0 t^2 \sin\lambda, \quad x_3^{(1)} = v_1^0 t^2 \cos\lambda.$$

Remaining at this approximation (hence, neglecting ω^2 as well as higher powers of ω), we get the solution

$$x_1(t) = \frac{1}{3}\omega g t^3 \cos\lambda + \left(v_2^0 \sin\lambda - v_3^0 \cos\lambda\right)\omega t^2 + v_1^0 t + x_1^0,$$

$$x_2(t) = -v_1^0 \omega t^2 \sin\lambda + v_2^0 t + x_2^0, \quad (10.2.54')$$

$$x_3(t) = \left(v_1^0 \omega \cos\lambda - \frac{1}{2}g\right)t^2 + v_3^0 t + x_3^0,$$

which contains also the influence of the motion of rotation of the Earth (in a first approximation).

In case of a free falling of a heavy particle along the local vertical ($\mathbf{v}^0 = \mathbf{0}$) from the height h, we get

$$x_1 = \frac{1}{3}\omega g t^3 \cos\lambda, \quad x_2 = 0, \quad x_3 = -\frac{1}{2}gt^2 + h; \quad (10.2.55)$$

the results in Subsec. 2.2.1 are thus completed with a component in the positive direction of the Ox_1-axis. Hence, the heavy particle in free falling does not describe the descendent vertical, but an arc of semicubic parabola

$$x_1^2 = \frac{8}{9}\frac{\omega^2}{g}(h - x_3)^3 \cos^2\lambda, \quad (10.2.55')$$

and reaches the Earth after an interval of time $t = \sqrt{2h/g}$ with a deviation towards the east (corresponding to $x_1 > 0$, Fig.10.8) equal to

$$\delta = x_{1\max} = \frac{2\omega}{3}\sqrt{\frac{2}{g}}h^{3/2}\cos\lambda. \tag{10.2.55''}$$

Considering the mean value $g = 980.6$ cm/s^2 and taking into account (10.2.24), we obtain

$$\delta = 2.195 \cdot 10^{-6} h^{3/2} \cos\lambda, \tag{10.2.55'''}$$

where h is taken in cm, the deviation being obtained in the same units. The experimental verifications have put in evidence a good concordance with the theoretical results. Thus, in 1831, Reich has made experiments in the mines of Freiburg (to avoid the influence of the wind and of the other influences at the Earth surface), at a latitude $\lambda = 51°$ and a falling height $h = 1.585 \cdot 10^4$ cm; the average value $\delta_{\exp} = 2.83$ cm of 106 experiments has been in good correspondence with the theoretical result $\delta = 2.76$ cm. We notice that for a height $h = 10^4$ cm one has $\delta = 2.195\cos\lambda$, and the deviation is obtained in cm; e.g., at Bucharest ($\lambda = 45°$) one has $\delta = 1.55$ cm. At the poles ($\lambda = \pm\pi/2$) we get $\delta = 0$. The maximal deviation takes place at the equator ($\lambda = 0$, $g = 978.1$ cm/s^2) and is given by $\delta = 2.198 \cdot 10^{-6} h^{3/2}$ (δ and h in cm); hence, $\delta_{\max} \cong 2.20$ cm in free falling from $h = 10^4$ cm $= 100$ m.

If we launch a heavy particle along the ascendent local vertical ($\mathbf{r}^0 = \mathbf{0}$, $v_1^0 = v_2^0 = 0$, $v_3^0 > 0$), we get

$$x_1 = \frac{1}{3}\omega g t^3 \cos\lambda - v_3^0 \omega t^2 \cos\lambda, \quad x_2 = 0, \quad x_3 = -\frac{1}{2}gt^2 + v_3^0 t; \tag{10.2.56}$$

the component $v_3 = -gt + v_3^0$ of the velocity along the vertical vanishes at the moment $t' = v_3^0/g$, for which we get

$$x_3(t') = \frac{1}{2g}(v_3^0)^2, \quad x_1(t') = -\frac{2\omega}{3g^2}(v_3^0)^3 \cos\lambda < 0,$$

$$v_1(t') = -\frac{\omega}{g}(v_3^0)^2 \cos\lambda < 0.$$

The rotation of the Earth about its axis leads thus to a deviation towards the west of the particle. Taking the position and the velocity at the moment t' as initial conditions and studying further the motion of the particle, we find that it takes place also in a plane normal to the local meridian; the particle returns on the Earth at the moment $t'' = 2t' = 2v_3^0/g$ at the point of co-ordinates $x_1(t'') = -(4/3)(\omega/g^2)(v_3^0)^3 \cos\lambda = 2x_1(t') < 0$, $x_3(t'') = 0$, tangent at the vertical of this location ($v_1(t'') = 0$,

$v_3(t'') = -v_3^0$); it results further a deviation towards the west, equal to that obtained in case of the ascendent motion (unlike the case of the free falling, the influence of the velocity $v_1(t') < 0$ takes place).

In case of an approximation of higher order, when terms in ω^2 intervene too, it is necessary to take into account the variation of g with the height, as well as the force of attraction of the Moon.

2.2.9 Coriolis' force. Baer's law

The deviation along a tangent to the local parallel, considered in the preceding subsection, is due to the action of *Coriolis' force* $\mathbf{F}_C = -2m\boldsymbol{\omega} \times \mathbf{v}_r$. The equations of motion of the heavy particle (10.2.33') read (with respect to the local frame of reference, Fig.10.8)

$$\ddot{x}_1 = 2\omega(\dot{x}_2 \sin\lambda - \dot{x}_3 \cos\lambda), \quad \ddot{x}_2 = -2\omega\dot{x}_1 \sin\lambda,$$
$$\ddot{x}_3 = -g + 2\omega\dot{x}_1 \cos\lambda, \tag{10.2.57}$$

in the absence of the centrifugal force; integrating, with the initial conditions $\mathbf{r}(0) = \mathbf{r}^0$, $\mathbf{v}(0) = \mathbf{v}^0$, we can write

$$\dot{x}_1 = 2\omega\left[(x_2 - x_2^0)\sin\lambda - (x_3 - x_3^0)\cos\lambda\right] + v_1^0,$$
$$\dot{x}_2 = -2\omega(x_1 - x_1^0)\sin\lambda + v_2^0, \tag{10.2.57'}$$
$$\dot{x}_3 = -gt + 2\omega(x_1 - x_1^0)\cos\lambda + v_3^0.$$

This system of linear differential equations can be easily integrated; as a matter of fact, eliminating x_2 and x_3 between the first equation (10.2.57) and the last two equations (10.2.57'), we get

$$\ddot{x}_1 + 4\omega^2 x_1 = 4\omega^2 x_1^0 + 2\omega g t \cos\lambda + 2\omega(v_2^0 \sin\lambda - v_3^0 \cos\lambda). \tag{10.2.57''}$$

By integration, it results

$$x_1(t) = -\frac{1}{2\omega}\left[(v_3^0 \cos\lambda - v_2^0 \sin\lambda)(1 - \cos 2\omega t) - \left(v_1^0 - \frac{g\cos\lambda}{2\omega}\right)\sin 2\omega t\right]$$
$$+ \frac{g\cos\lambda}{2\omega}t + x_1^0. \tag{10.2.58}$$

Replacing in the last two equations (10.2.57') and integrating, we obtain also the solutions

$$x_2(t) = -\frac{\sin\lambda}{2\omega}\left[(v_3^0 \cos\lambda - v_2^0 \sin\lambda)\sin 2\omega t + \left(v_1^0 - \frac{g\cos\lambda}{2\omega}\right)(1 - \cos 2\omega t)\right]$$
$$- \frac{g\sin 2\lambda}{4}t^2 + \cos\lambda(v_2^0 \cos\lambda + v_3^0 \sin\lambda)t + x_2^0. \tag{10.2.58'}$$

$$x_3(t) = \frac{\cos\lambda}{2\omega}\left[\left(v_3^0\cos\lambda - v_2^0\sin\lambda\right)\sin 2\omega t + \left(v_1^0 - \frac{g\cos\lambda}{2\omega}\right)(1 - \cos 2\omega t)\right]$$
$$-\frac{g\sin^2\lambda}{2}t^2 + \sin\lambda\left(v_2^0\cos\lambda + v_3^0\sin\lambda\right)t + x_3^0. \qquad (10.2.58'')$$

With the aid of the series expansions $\sin 2\omega t = 2\omega t - (2\omega)^3 t^3/6 + \ldots$, $\cos 2\omega t = 1 - (2\omega)^2 t^2/2 + \ldots$ and neglecting the terms in ω^2 or of higher degree, we find again the solution (10.2.54'), obtained in the preceding subsection; as a matter of fact, in the following considerations we will use this solution, which puts in evidence (from a qualitative point of view), in a simpler way, various mechanical phenomena.

If we assume that the heavy particle is launched in a meridian plane ($x_1 = 0$), then we take $\mathbf{r}^0 = \mathbf{0}$ and $v_1^0 = 0$; we obtain

$$x_1(t) = \frac{1}{3}\omega g t^3 \cos\lambda + \omega\left(v_2^0\sin\lambda - v_3^0\cos\lambda\right)t^2, \qquad (10.2.59)$$

so that the particle is deviated from the initial plane. As well, if the particle is launched in a plane normal to the local meridian ($\mathbf{r}^0 = \mathbf{0}$ and $v_2^0 = 0$), it results

$$x_2(t) = -v_1^0\omega t^2 \sin\lambda, \qquad (10.2.59')$$

hence a tendency of deviation from this plane.

Let us consider the motion in a plane tangent to the Earth sphere ($x_3 = 0$) at the considered location (we take $\mathbf{r}^0 = \mathbf{0}$ and $v_3^0 = 0$); we can write (with respect to the local frame of reference, Fig.10.8)

$$x_1(t) = \frac{1}{3}\omega g t^3 \cos\lambda + v_2^0\omega t^2 \sin\lambda + v_1^0 t,$$
$$x_2(t) = -v_1^0\omega t^2 \sin\lambda + v_2^0 t, \qquad (10.2.60)$$
$$x_3(t) = \left(v_1^0\omega\cos\lambda - \frac{1}{2}g\right)t^2.$$

We notice that $g \gg 2v_1^0\omega\cos\lambda$ (taking into account (10.2.24), we have $g > 2v_1^0\omega\cos\lambda$, even if the component v_1^0 of the initial velocity is equal to the second cosmic velocity); it results $x_3(t) < 0$. Hence, the particle remains on the Earth surface and we may assume that the motion (in the vicinity of the initial position with respect to the Earth sphere) takes place in the tangent plane $x_3 = 0$. A constraint force $\mathbf{N}$ (normal to the tangent plane, hence along the local vertical), which impedes the particle to leave the plane $x_3 = 0$ is also acting in this case. The theorem of kinetic energy (10.2.16') allows to write

Other considerations on particle dynamics 649

$$\frac{d}{dt}\left(\frac{1}{2}mv_r^2\right) = m\mathbf{g}\cdot\mathbf{v}_r + \mathbf{N}\cdot\mathbf{v}_r = 0,$$

so that $v_r = v_0 = $ const (result which can be accepted with a good approximation). We can thus express the components of the velocity in the form $v_1(t) = v_0 \cos\alpha$, $v_2(t) = v_0 \sin\alpha$, where $\alpha = \alpha(t)$ is the angle made by the velocity $\mathbf{v}_r$ with the Ox_1-axis. Replacing in one of the first two equations (10.2.57) (where we make $\dot{x}_3 = 0$), we find, after identification, $\dot\alpha = -2\omega\sin\lambda$; hence,

$$\alpha(t) = \alpha_0 - 2\omega t \sin\lambda. \tag{10.2.61}$$

Thus, we get

Theorem 10.2.12 (*Baer's law*). *In the boreal hemisphere, the angle α decreases, corresponding to a deviation towards the right of the heavy particle, while in the austral hemisphere the angle α increases, corresponding to a deviation towards the left of it.*

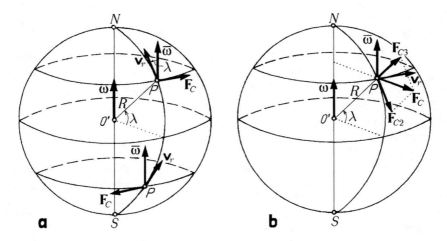

Figure 10.13. Action of the Coriolis force upon a particle on the Earth surface: launching along the tangent to the meridian (a); launching in the boreal hemisphere along the tangent to the parallel towards the east (b).

Taking into account (10.2.61), one can show, by integration, that

$$x_1(t) = -\frac{v_0}{2\omega\sin\lambda}\sin(\alpha_0 - 2\omega t \sin\lambda),$$

$$x_2(t) = \frac{v_0}{2\omega\sin\lambda}\cos(\alpha_0 - 2\omega t \sin\lambda),$$

so that the trajectory of the particle is circular; the radius of the circle is $v_0/2\omega|\sin\lambda|$ and is very great (we take into account (10.2.24)) and we can approximate it by its tangent at the initial moment.

We notice that, with respect to the gravity force, we have $|2m\boldsymbol{\omega} \times \mathbf{v}_r|/mg \leq 2\omega v_r/g \cong 1.487 \cdot 10^{-5} v_r$, where v_r is expressed in m/s; it results that Coriolis' force may be, in general, neglected in case of a particle subjected to relative small velocities.

Let us consider, in particular, the case in which the particle P is launched in the boreal hemisphere, along the tangent to the meridian, e.g., towards the north (Fig.10.13,a); in this case, Coriolis' force will be directed towards the east, tangent to the parallel which passes through P, having the magnitude $F_C = 2m\omega v_r \sin \lambda$. From (10.2.59) we obtain a deviation towards the east given by

$$x_1(t) = \frac{1}{3}\omega g t^3 \cos \lambda + \omega v_2^0 t^2 \sin \lambda > 0.$$

If the particle P would be moving along the meridian towards the south point, then the Coriolis' force would be directed towards the west. In both cases corresponds a deviation towards the right, in conformity to Baer's law. In the austral hemisphere, the phenomenon is symmetric to that of the boreal hemisphere, with respect to the equatorial plane; thus, if the relative velocity is directed towards the north point, then Coriolis' force will be directed towards the west. Coriolis' force vanishes at the equator and is maximal at the poles ($F_{C\max} = 2m\omega v_r$); starting from one of the poles, the deviation will always take place towards the west. Thus, after a time t, the distance vt will be travelled through, corresponding a linear deviation $\omega v_r t^2$ and an angular deviation equal to $\omega v_r t^2 / v_r t = \omega t$, hence equal to the angle by which the Earth is rotating in that interval of time. Hence, a projectile launched from a pole has a rectilinear trajectory; its apparent deviation is due to the fact that the Earth is rotating. Thus, after 1 min 30 s one obtains an angular deviation of 0.0065628 rad, hence of approximative 22′34″, which cannot be neglected.

The effect of Baer's law is considerable in case of great continuous mechanical systems. For instance, *the right bank of the rivers* which run from the south towards the north or from the north towards the south, in the boreal hemisphere, is caving more then the left bank; we mention thus the rivers at the north of Asia (which run from the south to the north), which have the tendency of displacement towards the east. Due to the Coriolis force too, at the railway lines which are directed approximately along a meridian, being travelled through in a unique direction, *the rail at the right* is subjected to a greater wear (the east rail if the direction of circulation is from the south to the north or the west one in case of a circulation in the opposite direction). As well, *the trade winds* are directed towards the equator; the colder air (hence, heavier) tends to replace the warmer one (hence, lighter) in the boreal hemisphere from the north towards the south, while in the austral hemisphere from the south towards the north. Indeed, the wind is a mass of air in motion; in the absence of the Coriolis force, the direction of the motion corresponds to the gradient of the atmosphere pressure (from a high pressure to a low one, normal to the isobar lines). The intervention of the Coriolis force leads to a deviation towards the west (Fig.10.14,a) for the boreal hemisphere. In the state of equilibrium, the configuration of the wind is stationary; Coriolis' force is in equilibrium with the forces due to the pressure, so that the wind becomes parallel to the isobar lines.

Other considerations on particle dynamics 651

A zone of low pressure surrounded by isobar lines forms a *cyclone*. Due to Coriolis' force, the wind is circulating around the cyclone, counterclockwise in the boreal hemisphere (Fig.10.14,b) and clockwise in the austral one.

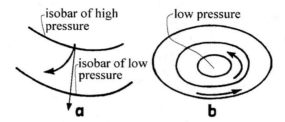

Figure 10.14. Formation of trade winds (a) and of cyclones (b) in the boreal hemisphere.

Let us study now the case in which the particle P is launched, in the boreal hemisphere, along the tangent to the parallel, e.g., towards the east (Fig.10.13,b); in this case, the Coriolis's force is contained in the meridian plane and is normal to the poles' line, being directed towards the exterior of the Earth (the same direction as the centrifugal force) and having the magnitude $F_C = 2m\omega v_r$. We decompose the force $\mathbf{F}_C$ in two components: the force $\mathbf{F}_{C3}$ along the ascendent local vertical, which has the tendency to make smaller the weight of the particle, and the force $\mathbf{F}_{C2}$, tangent to the meridian and directed towards the south, which has the tendency to deviate the particle in that direction (in conformity to the formula (10.2.59')); if the particle P is launched in the austral hemisphere, towards the east too, then to the component $\mathbf{F}_{C2}$ of Coriolis' force corresponds a deviation towards the north (hence, in both cases, towards the equator). In the case in which the particle is launched towards the west, in any of the two hemispheres, Coriolis' force and its components have an opposite direction; thus, the component $\mathbf{F}_{C3}$ has the tendency to augment the weight of the particle, while the component $\mathbf{F}_{C2}$ leads to a deviation towards the poles (towards the north in the boreal hemisphere and towards the south in the austral one).

The effects of the Coriolis' force can be put into evidence at the atomic level too. Thus, the polyatomic molecules have an aggregate motion of rotation, while the atoms oscillate around their positions of equilibrium; hence, the atoms are in relative motion with respect to a frame rigidly linked to the molecule. The Coriolis' forces are non-zero, leading to a displacement of the atoms in a direction normal to that of the original oscillations.

2.2.10 Foucault's pendulum

Let us consider the motion of a heavy particle on a fixed sphere at the Earth surface in the boreal hemisphere; the spherical pendulum thus obtained is called *the Foucault pendulum*, after Léon Foucault, who made a famous experiment with such a pendulum in 1815, at the Pantheon, in Paris. We assume that the particle P (in fact a spheric ball), of mass m, is hanged up at the end of a flexible and inextensible thread, of length l, fixed at the pole O of the local (non-inertial) frame of reference. The equation of motion is written in the form

$$m\mathbf{a}_r = m\mathbf{g} - 2m\boldsymbol{\omega} \times \mathbf{v}_r + \mathbf{T}, \tag{10.2.62}$$

where $\mathbf{T}$ is the tension in the thread (the constraint force arises because the particle is constrained to stay on the sphere of centre O and radius l) and where the influence of the centrifugal force has been neglected with the respect to Coriolis's force (ω^2 has been neglected with respect to ω). In components, we obtain (Fig.10.15)

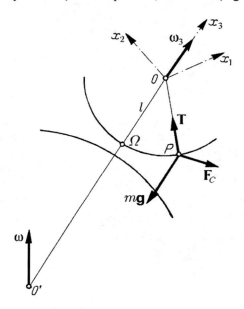

Figure 10.15. The Foucault pendulum.

$$\begin{aligned} m\ddot{x}_1 &= 2m\omega(\dot{x}_2 \sin\lambda - \dot{x}_3 \cos\lambda) - T\frac{x_1}{l}, \\ m\ddot{x}_2 &= -2m\omega\dot{x}_1 \sin\lambda - T\frac{x_2}{l}, \\ m\ddot{x}_3 &= -mg + 2m\omega\dot{x}_1 \cos\lambda - T\frac{x_3}{l}; \end{aligned} \tag{10.2.62'}$$

the relation of holonomic and scleronomic constraint

$$x_1^2 + x_2^2 + x_3^2 - l^2 = 0 \tag{10.2.62''}$$

is added to these equations. The elementary work of the constraint force is, in this case, equal to zero, so that the theorem of kinetic energy (10.2.15') leads to

$$d\left(\frac{1}{2}mv_r^2\right) = m\mathbf{g} \cdot d\mathbf{r} = -mg\,dx_3, \tag{10.2.63}$$

wherefrom we get the first integral of energy

$$v_r^2 = -2g(x_3 + h), \quad h = -x_3(0), \quad v_r(0) = 0. \tag{10.2.63'}$$

From (10.2.62") it results (one introduces the sign minus, taking into account the zone in which is the particle, Fig.10.15)

$$x_3 = -l\left[1 - \frac{1}{l^2}(x_1^2 + x_2^2)\right]^{1/2} = -l\left[1 - \frac{1}{2l^2}(x_1^2 + x_2^2) + ...\right].$$

We assume that the length l is very large and that the mass m is sufficiently great. The pendulum has a small displacement from its position of equilibrium Ω, oscillating without initial velocity with respect to the Earth; in this case, it effects small oscillations around the position of equilibrium, which is a stable one. We can thus state that the ratios x_1/l and x_2/l are very small (we say, generally, that they are of order of magnitude $\varepsilon > 0$, that is $x_1/l = \mathscr{O}(\varepsilon)$, $x_2/l = \mathscr{O}(\varepsilon)$); it results $x_3 = -l[1 + \mathscr{O}(\varepsilon^2)]$, $h = -l[1 + \mathscr{O}(\varepsilon^2)]$ and $h + x_3 = \mathscr{O}(\varepsilon^2)$. From (10.2.62"), (10.2.63'), we can write

$$x_1\dot{x}_1 + x_2\dot{x}_2 + x_3\dot{x}_3 = 0, \quad \dot{x}_1^2 + \dot{x}_2^2 + \dot{x}_3^2 = -2g(h + x_3) = \mathscr{O}(\varepsilon),$$

so that $\dot{x}_1 = \mathscr{O}(\varepsilon)$, $\dot{x}_2 = \mathscr{O}(\varepsilon)$, $\dot{x}_3 = \mathscr{O}(\varepsilon)$, $x_3\dot{x}_3/l = -x_1\dot{x}_1/l - x_2\dot{x}_2/l = \mathscr{O}(\varepsilon^2)$ and then $\dot{x}_3 = \mathscr{O}(\varepsilon^2)$. The tension T being of the order of unity, the first two equations (10.2.62') show that we have at least $\ddot{x}_1 = \mathscr{O}(\varepsilon)$, $\ddot{x}_2 = \mathscr{O}(\varepsilon)$. Differentiating the relation (10.2.62") once more with respect to time, we can write $(x_3/l)\ddot{x}_3 = -(\dot{x}_1^2 + \dot{x}_2^2 + \dot{x}_3^2)/l - (x_1/l)\ddot{x}_1 - (x_2/l)\ddot{x}_2 = \mathscr{O}(\varepsilon^2)$, so that $\ddot{x}_3 = \mathscr{O}(\varepsilon^2)$. Using the previous evaluations and observing that, in this case, $x_3 \cong -l$, the third equation (10.2.62') leads to

$$T \cong mg, \tag{10.2.64}$$

the tension in the thread being approximately equal to that corresponding to the position of equilibrium.

Returning to the first two equations (10.2.62'), we find the equations of motion in the plane $x_3 = -l$, tangent to the position of equilibrium Ω (which corresponds to small oscillations around this stable position of equilibrium), in the form (we consider the axes Ωx_1 and Ωx_2 parallel to the axes Ox_1 and Ox_2, respectively; for the sake of simplicity, we use the same notations for the co-ordinates)

$$\ddot{x}_1 = -\frac{g}{l}x_1 + 2\omega\dot{x}_2 \sin\lambda, \quad \ddot{x}_2 = -\frac{g}{l}x_2 - 2\omega\dot{x}_1 \sin\lambda; \tag{10.2.65}$$

in a vector form, we get

$$\ddot{\boldsymbol{\rho}} = -\frac{g}{l}\boldsymbol{\rho} - 2\boldsymbol{\omega}_3 \times \dot{\boldsymbol{\rho}}, \tag{10.2.65'}$$

where $\boldsymbol{\omega}_3 = \omega_3 \mathbf{i}_3 = \omega \sin \lambda \mathbf{i}_3$ is the vector component of the rotation vector $\boldsymbol{\omega}$ along the local vertical, while $\boldsymbol{\rho}$ is the position vector in the tangent plane. A vector product by $\boldsymbol{\rho}$ leads to (we have $\boldsymbol{\rho} \cdot \boldsymbol{\omega}_3 = 0$)

$$\boldsymbol{\rho} \times \ddot{\boldsymbol{\rho}} = \frac{d}{dt}(\boldsymbol{\rho} \times \dot{\boldsymbol{\rho}}) = -2\boldsymbol{\rho} \times (\boldsymbol{\omega}_3 \times \dot{\boldsymbol{\rho}}) = -2(\boldsymbol{\rho} \cdot \dot{\boldsymbol{\rho}})\boldsymbol{\omega}_3 = -\frac{d}{dt}(\rho^2)\boldsymbol{\omega}_3,$$

wherefrom

$$\boldsymbol{\rho} \times \dot{\boldsymbol{\rho}} + \rho^2 \boldsymbol{\omega}_3 = \mathbf{C}, \quad \mathbf{C} = \overrightarrow{\text{const}}, \tag{10.2.66}$$

corresponding to a first integral of moment of momentum; scalarly, we have

$$x_1 \dot{x}_2 - x_2 \dot{x}_1 + (x_1^2 + x_2^2)\omega \sin \lambda = C, \quad C = \text{const}. \tag{10.2.66'}$$

In polar co-ordinates ρ, θ, we may express the first integrals (10.2.63'), (10.2.66') in the form

$$\dot{\rho}^2 + \left(\frac{g}{l} + \dot{\theta}^2\right)\rho^2 = 2g(l - h), \quad \rho^2 \dot{\theta} + \rho^2 \omega \sin \lambda = C, \tag{10.2.67}$$

characterizing thus the motion in the plane $\Omega x_1 x_2$.

We introduce a system of axes $\Omega \xi_1 \xi_2$, movable with respect to the system $\Omega x_1 x_2$, which is rotating about the local vertical Ox_3 with the angular velocity $-\boldsymbol{\omega}_3 = -\omega \sin \lambda \mathbf{i}_3$ (in the sense north-east-south-west, hence clockwise). Taking into account (10.2.1), (10.2.1') and starting from the equation (10.2.65'), we can write the equation of motion of the particle P with respect to this last movable frame in the form

$$\frac{\partial^2 \boldsymbol{\rho}}{\partial t^2} = -\frac{g}{l}\boldsymbol{\rho} - 2\boldsymbol{\omega}_3 \times \dot{\boldsymbol{\rho}} - (-\dot{\boldsymbol{\omega}}_3) \times \boldsymbol{\rho} - (-\boldsymbol{\omega}_3) \times [(-\boldsymbol{\omega}_3) \times \boldsymbol{\rho}] - 2(-\boldsymbol{\omega}_3) \times \frac{\partial \boldsymbol{\rho}}{\partial t},$$

where $\partial \boldsymbol{\rho}/\partial t$ and $\partial^2 \boldsymbol{\rho}/\partial t^2$ are the relative velocity and acceleration, respectively. We notice that $\dot{\boldsymbol{\omega}}_3 = 0$ and $\dot{\boldsymbol{\rho}} = d\boldsymbol{\rho}/dt = \partial \boldsymbol{\rho}/\partial t + (-\boldsymbol{\omega}_3) \times \boldsymbol{\rho}$, so that (we have $\boldsymbol{\rho} \cdot \boldsymbol{\omega}_3 = 0$)

$$\frac{\partial^2 \boldsymbol{\rho}}{\partial t^2} = -\frac{g}{l}\boldsymbol{\rho} + \boldsymbol{\omega} \times (\boldsymbol{\omega} \times \boldsymbol{\rho}) = -\overline{\omega}^2 \boldsymbol{\rho}, \quad \overline{\omega}^2 = \frac{g}{l} + \omega_3^2 = \frac{g}{l} + \omega^2 \sin^2 \lambda. \tag{10.2.68}$$

This is the equation of motion of a particle P of mass m, attracted by the centre Ω with an elastic force

$$-m\overline{\omega}^2 \boldsymbol{\rho} = -m\left(\frac{g}{l} + \omega^2 \sin^2 \lambda\right)\boldsymbol{\rho}; \tag{10.2.68'}$$

Other considerations on particle dynamics 655

hence, the trajectory is an ellipse or a segment of a line (degenerate ellipse), fixed with respect to the frame $\Omega\xi_1\xi_2$ (Fig.10.16). This ellipse is rotating with respect to the frame $\Omega x_1 x_2$, linked to the Earth, about the local vertical, with the angular velocity $\omega_3 = \omega\sin\lambda$, in the sense east-south-west-north (clockwise); hence, if the particle pendulates in the plane $x_3\Omega P_0$, then this plane is rotating with the angular velocity ω_3 in the sense indicated above.

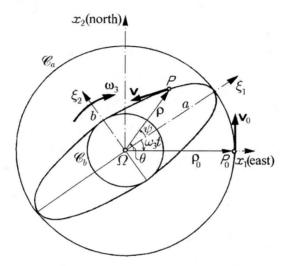

Figure 10.16. The Foucault pendulum. Elliptic trajectory in the tangent plane with respect to the non-inertial frame of reference $\Omega\xi_1\xi_2$.

If we change the variable $\theta = \psi - \omega_3 t$ and notice that (because $\boldsymbol{\rho}\times(\partial\boldsymbol{\rho}/\partial t)$ $= \rho^2\dot\psi\mathbf{i}_3$, Fig.10.16)

$$\left(\frac{d\boldsymbol{\rho}}{dt}\right)^2 = \left(\frac{\partial\boldsymbol{\rho}}{\partial t}\right)^2 + (\boldsymbol{\rho}\times\boldsymbol{\omega}_3)^2 + 2\left(\frac{\partial\boldsymbol{\rho}}{\partial t},\boldsymbol{\omega}_3,\boldsymbol{\rho}\right) = \dot\rho^2 + \rho^2\omega_3^2 + 2\omega_3\rho^2\dot\psi,$$

then we can express the first integrals (10.2.67) with respect to the frame $\Omega\xi_1\xi_2$ in the form

$$\dot\rho^2 + \rho^2\dot\psi^2 + \left(\frac{g}{l} + \omega^2\sin^2\lambda\right)\rho^2 = 2g(l-h), \quad \rho^2\dot\psi = C; \quad (10.2.67')$$

we took into account that

$$\frac{g}{l} + \omega^2\sin^2\lambda = \frac{g}{l}\left(1 + \frac{l}{g}\omega^2\sin^2\lambda\right) = \frac{g}{l}[1 + \mathcal{O}(10^{-8}\ldots 10^{-7})],$$

in the first integral, as it will be seen further. As a matter of fact, by calculation it has been obtained the term $(g/l + 2\omega^2\sin^2\lambda)\rho^2$; but by the mentioned approximation one has obtained a result corresponding to the elastic force (10.2.68'). Considering the elastic force of the form $-m(g/l)\mathbf{\rho}$, we write the first integral

$$\dot{\rho}^2 + \rho^2\dot{\psi}^2 + \frac{g}{l}\rho^2 = 2g(l-h), \qquad (10.2.67'')$$

correspondingly. We have thus put into evidence the first integral of energy (the potential energy corresponds to the conservative elastic force (10.2.68')) and the first integral of areas.

The particle is launched from a position of rest P_0, of position vector $\mathbf{\rho}_0$ (situated on the Ωx_1-axis, because for $t = 0$ we have $\theta = \psi$), with a zero initial velocity with respect to the frame $\Omega x_1 x_2$ linked to the Earth ($\mathrm{d}\mathbf{\rho}/\mathrm{d}t = \mathbf{0}$, hence $\mathrm{d}\theta/\mathrm{d}t = 0$ for $t = 0$); in case of a change of variable of the form $\theta = \psi - \omega_3 t - \varphi$, $\varphi = \mathrm{const}$, we start from an initial position P_0 non-situated on the Ωx_1-axis. If $P_0 \in \Omega x_1$, then we get $\dot{\psi}(0) = \omega_3$, so that the constant of areas is given by $C = \rho_0^2 \omega_3 = \rho_0^2 \omega \sin\lambda$. The initial velocity is $\mathbf{v}_0 = (\partial\mathbf{\rho}/\partial t)_0 = \mathbf{\omega}_3 \times \mathbf{\rho}_0$, normal to $\mathbf{\rho}_0$ and of magnitude $v_0 = \omega_3 \rho_0 = \rho_0 \omega \sin\lambda = (\rho\dot{\psi})_0$ (Fig.10.16); it results, as well, $\dot{\rho}(0) = 0$. Analogously, one obtains the energy constant $h = l - \rho_0^2/2l - \rho_0^2 \omega_3^2/g$. The first integrals (10.2.67') become thus

$$\dot{\rho}^2 + \rho^2\dot{\psi}^2 + \left(\frac{g}{l} + \omega_3^2\right)(\rho^2 - \rho_0^2) = \rho_0^2\omega_3^2, \quad \rho^2\dot{\psi} = \rho_0^2\omega_3. \qquad (10.2.67''')$$

We notice also that the initial position is one of the extremities of the major axis of the elliptic trajectory. The ellipse is travelled through by the particle in an opposite direction to that of the rotation of the frame $\Omega\xi_1\xi_2$ with respect to the frame $\Omega x_1 x_2$ (counterclockwise).

In contradistinction to Foucault's pendulum, in case of the spherical pendulum, (see Chap. 7, Subsec. 1.3.7), in the hypothesis of small oscillations, the elliptic trajectory is fixed; but, in general, the particle oscillates between two parallel circles, while the first integral of areas shows that the meridian plane of the pendulum is rotating about the vertical in the direction indicated by the initial velocity, which is the same with that in which the movable ellipse, which approximates the projection of the particle on the tangent plane $x_3 = -l$ is travelled through. This result is obtained because the spheric pendulum is considered with respect to a local (inertial) frame; but Foucault's pendulum is studied with respect to a local frame, considered to be non-inertial, taking thus into account the influence of the Coriolis force.

In case of Foucault's pendulum, the semiaxes of the ellipse are

$$a = \rho_0, \quad b = \frac{\rho_0\omega_3}{\omega} = \frac{\rho_0\omega_3}{\sqrt{g/l + \omega_3^2}} \cong \rho_0\omega_3\sqrt{\frac{l}{g}}; \qquad (10.2.69)$$

the ratio

Other considerations on particle dynamics

$$\frac{b}{a} = \frac{\omega_3}{\sqrt{g/l + \omega_3^2}} \cong \omega_3 \sqrt{\frac{l}{g}} \qquad (10.2.69')$$

is, in general, very small, and the eccentricity

$$e = \sqrt{1 - \frac{b^2}{a^2}} = \left(1 + \frac{l}{g}\omega_3^2\right)^{-1/2} = 1 - \frac{l\omega^2}{2g} + \ldots = 1 - \mathscr{O}\left(10^{-8}\ldots 10^{-7}\right) \qquad (10.2.69'')$$

is smaller than the unity, but very close to it.

The ellipse is travelled through in a period

$$\tau = \frac{2\pi}{\omega} = \frac{2\pi}{\sqrt{g/l + \omega_3^2}} \cong 2\pi \sqrt{\frac{l}{g}} \qquad (10.2.70)$$

and effects a complete rotation north-east-south-west (clockwise) in an interval of time equal to

$$T = \frac{2\pi}{\omega_3} = \frac{2\pi}{\omega \sin \lambda}. \qquad (10.2.70')$$

From (10.2.69')-(10.2.70') it results

$$\frac{b}{a} = \frac{\tau}{T} \qquad (10.2.71)$$

and we can state

Theorem 10.2.13 (*Chevilliet*). *In the motion of Foucault's pendulum, the ratio of the two semiaxes (minor and major) of the ellipse of projection is equal to the ratio of the period in which the ellipse is travelled through to the period of complete rotation of it.*

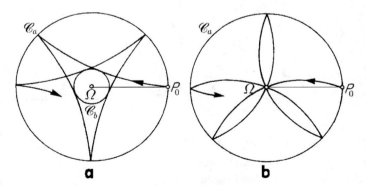

Figure 10.17. The Foucault pendulum. The trajectory with respect to the inertial frame of reference Ox_1x_2 if $\dot\psi \neq 0$ (a) and if $\dot\psi = 0$ (b), for $\dot\theta = 0$.

At the points for which $\dot\rho = 0$, the velocities are normal to the corresponding radii vectors, the trajectory being normal to those radii and tangent to the circles with Ω as centre, having as radii the above mentioned ones; if we make $\dot\rho = 0$ in (10.2.67''') and eliminate $\dot\psi$, then we find again the semiaxes a and b given by (10.2.69). Indeed, in its rotation, the ellipse is contained between the circles $\mathscr{C}_a$ and $\mathscr{C}_b$ of radii a and b, respectively. One observes that the semi-major axis a does not depend neither on the location on the Earth surface, nor on the initial conditions; but the semi-minor axis b depends on the latitude λ, as well as on the initial conditions (radius ρ_0). Moreover, the initial conditions play an important rôle, specifying the nature of the trajectory of the particle with respect to the frame $\Omega\xi_1\xi_2$ (segment of a line or ellipse) and with respect to the frame $\Omega x_1 x_2$. Thus, for $\rho = \rho_0$ we obtain $\dot\psi = \omega_3$, hence $\dot\theta = 0$, while $\rho = \rho_0 \omega_3 / \overline{\omega}$ leads to $\dot\psi = \overline{\omega}^2 / \omega_3 = (g/l + \omega_3^2)/\omega_3$, wherefrom $\dot\theta = g/l\omega_3$. The trajectory of the particle with respect to the frame $\Omega x_1 x_2$ can be represented as in Fig.10.17,a; the points of tangency to the circle $\rho = a$ are cuspidal points ($\dot\theta = 0$), while at the points for which $\rho = \rho_{\min}$ the trajectory is tangent to the circle $\rho = b$ ($\dot\theta \neq 0$). If $\dot\psi = 0$, then $C = 0$ and the trajectory is a segment of a line, with respect to the frame $\Omega\xi_1\xi_2$; hence, we get $\dot\theta = -\omega_3$, the trajectory with respect to the frame $\Omega x_1 x_2$ having the form of a multifolium (in particular, quadrifolium) (Fig.10.17,b). In case of other initial conditions (if $d\theta/dt \neq 0$ for $t = 0$) one obtains trajectories as in Figs 10.18,a,b.

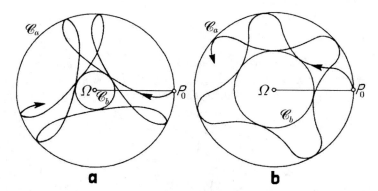

Figure 10.18. The Foucault pendulum. Trajectories with respect to the inertial frame of reference $Ox_1 x_2$ for $\dot\theta \neq 0$.

The numerical data in Foucault's experiment have been $m = 2.8 \cdot 10^4$ g and $l = 6.7 \cdot 10^3$ cm, Paris' latitude being $\lambda = 48°50'$ to which corresponds $g = 9.809 \cdot 10^2$ cm/s²; from the formulae (10.2.70), (10.2.70') it results $\tau = 16.42$ s, $T = 114458$ s $= 31$ hours 47 min 38 s $\cong 32$ hours. These theoretical data have been in very good concordance with the experiment; at the same time, the quantitative evaluations previously made are justified. A thorough study can be found in the dissertation of H. Kamerlingh Onnes (Groningen, 1879). In Cologne, at the latitude

$\lambda = 50°50'30''$, Garthe has obtained (with $l = 5 \cdot 10^3$ cm, $m = 1.7 \cdot 10^4$ g and $r_0 = 3 \cdot 10^2$ cm) a rotation of $11°38'51''$, the rotation observed experimentally being of $11°37'40.8''$ (hence, a very good concordance). In Bucharest, at the latitude $\lambda = 45°$, one obtains $T \cong 121854$ s = 33 hours 50 min 54 s $\cong 34$ hours.

In the austral hemisphere, the ellipse is rotating in the sense east-north-west-south (counterclockwise).

Because the theoretical results obtained starting from the hypothesis of rotation of the Earth about the poles' axis are in good concordance with the experimental ones, we can state that the Earth is indeed rotating about this axis; moreover, the universal attraction law which has been put in evidence – at the beginning – for cosmic bodies, extends its validity for the phenomena at the Earth surface too. We mention that an observer localized in an inertial frame (e.g., a heliocentric frame) would see the pendulum oscillating only in the same plane (assuming that the trajectory of the particle with respect to the frame $\Omega \xi_1 \xi_2$ is a segment of a line), the Earth being in rotation with respect to this plane.

These conclusions have a particular importance for the knowledge of our planet and put in evidence the interest presented by Foucault's experiment. We must mention also that the motion of rotation of the Earth has been stated by astronomical observations, before this famous experiment; but Foucault's study puts theoretically in evidence the motion of the Earth, astronomical observations being no more necessary (which cannot be made if, for instance, the Earth would be covered by a thick stratum of clouds, as Venus, the only planet which is rotating about its axis from west to east).

3. Dynamics of the particle of variable mass

There exist bodies the mass of which is variable in time; it can decrease (e.g., the mass of a rocket, which is acted upon by a propulsive force due to the ejection of an explosive material – fine particles, gas, internal liquid – emission phenomenon) or increase (e.g., a planet on which fine cosmic particles of a nebula encountered in its way are falling – capture phenomenon). We may consider also other examples, as: an aerostat which lifts by throwing down the ballast over the border, the splinting of a device processed at the lathe etc. In the case in which such a body can be modelled as a particles arises the problem to obtain the equation governing this motion and to integrate it in various particular cases.

3.1 Mathematical model of the motion. General theorems

To can set up a mathematical model of motion of a particle of variable mass, hence to find a law of motion which may be reduced to Newton's equation in case of a constant mass, we start from a classical mathematical model, corresponding to a discrete mechanical system, the general theorems (especially, the theorem of momentum) allowing then to establish the equation of motion which is governing the considered mechanical phenomenon. We may then state the corresponding general theorems, which extend those of the particle of constant mass.

3.1.1 Meshcherskiĭ's model. Levi-Civita's equations

We assume, in what follows, that the motion of a free particle P of variable mass $m = m(t)$ takes place by the detachment (emission) of some parts of it (it corresponds the decrease of the mass); this phenomenon puts in evidence the apparition of internal forces, which are called reactive forces. We assume thus that, at the moment of detachment of some parts of the particle (in fact, the particle P is a mechanical system

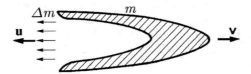

Figure 10.19. Meshcherskiĭ's mathematical model of a particle of variable mass.

which is emitting particles, e.g., a rocket which is emitting particles of gas, Fig.10.19), takes place a phenomenon analogous to that of collision. We consider the motion of the particle with respect to an inertial (fixed, absolute) frame of reference, so that – at the moment t – it has the velocity $\mathbf{v}(t)$ and the momentum $\mathbf{H}(t) = m\mathbf{v}(t)$. In the interval of time Δt, a part of mass $-\Delta m$, $\Delta m > 0$, is detached from the particle, with the absolute velocity $\mathbf{u}$ and with a relative velocity $\mathbf{w}$ with respect to a non-inertial frame of reference, attached to the particle in motion; on the basis of the principle of action and reaction, appears a reactive force $\mathbf{R}$ (corresponding to a collision force in the mentioned analogy). The momentum of the system formed by the particle without the detached part (after emission) and the detached (emitted) part (see Chap. 11, Subsec. 1.1.1) is given by $\mathbf{H}(t + \Delta t) = [m - (-\Delta m)](\mathbf{v} + \Delta'\mathbf{v}) - \mathbf{u}\Delta m$, where $\Delta'\mathbf{v}$ represents the variation of the velocity of the particle P of variable mass, in the interval of time Δt, due to the process of emission of a part of it. The considered mechanical system is a closed one, so that we can apply the conservation theorem of momentum (see Chap. 11, Subsec. 1.2.5), hence $(m + \Delta m)(\mathbf{v} + \Delta'\mathbf{v}) - \mathbf{u}\Delta m = m\mathbf{v}$. Neglecting the terms of higher order, we obtain $\Delta'\mathbf{v} = (\Delta m / m)(\mathbf{u} - \mathbf{v})$, determining thus the variation of the velocity of the particle of variable mass m, due to the emission of mass $-\Delta m$. Introducing the influence of the given forces of resultant $\mathbf{F}$ too, Newton's equation gives $\Delta''\mathbf{v} = (\mathbf{F}/m)\Delta t$. Finally, on the basis of parallelogram's principle, we may write $\Delta \mathbf{v} = \Delta'\mathbf{v} + \Delta''\mathbf{v}$, so that, dividing by Δt and passing to the limit for $\Delta t \to 0$, it results I.V. Meshcherskiĭ's equation (we assume that, in the interval of time Δt, the velocity $\mathbf{v}$ of the particle P has a continuous variation and is not influenced by the collision effect due to the emission of mass, taking into account the inertia of the mass of the particle and that, before the emission, the emitted mass had the same velocity as the particle P)

$$m\dot{\mathbf{v}} = \mathbf{F} + \dot{m}(\mathbf{u} - \mathbf{v}), \quad \dot{m} < 0, \qquad (10.3.1)$$

obtained by him in 1897; it was found again in 1898 by K.E. Tsiolkovskiĭ and applied to the study of the rockets with several steps. We can write this equation also in the form

Other considerations on particle dynamics 661

$$\frac{\mathrm{d}}{\mathrm{d}t}(m\mathbf{v}) = \dot{\mathbf{H}} = \mathbf{F} + \dot{m}\mathbf{u}, \quad \dot{m} < 0. \tag{10.3.1'}$$

Observing that the relative velocity of the emitted part with respect to a non-inertial frame of reference, attached to the particle in motion, is given by $\mathbf{w} = \mathbf{u} - \mathbf{v}$, we can introduce *the reactive force*

$$\mathbf{R} = \dot{m}(\mathbf{u} - \mathbf{v}) = \dot{m}\mathbf{w}, \quad \dot{m} < 0, \tag{10.3.2}$$

so that Meshcherskiĭ's equation takes the form

$$m\dot{\mathbf{v}} = \mathbf{F} + \mathbf{R}, \quad \dot{m} < 0; \tag{10.3.1''}$$

we state thus (with respect to an inertial frame of reference)

Theorem 10.3.1 (*Meshcherskiĭ*). *The product of the mass of a free particle of variable mass by its acceleration is equal to the sum of the resultant of the given forces and the reactive force which acts upon the particle.*

In the case in which the mass of the particle P is increasing, due to a phenomenon of capture of a mass Δm, $\Delta m > 0$, we can make an analogous study. Maintaining the previous notations, we may write $\mathbf{H}(t) = m\mathbf{v} + \mathbf{u}\Delta m$, corresponding to the system formed by the particle P and the mass Δm which is captured; as well, we have $\mathbf{H}(t + \Delta t) = (m + \Delta m)(\mathbf{v} + \Delta'\mathbf{v})$. Using once more the conservation theorem of momentum and the principle of the parallelogram and passing to the limit, it results

$$\dot{\mathbf{H}} = \mathbf{F} + \dot{m}\mathbf{u}, \quad \dot{m} > 0, \tag{10.3.3}$$

as well as

$$m\dot{\mathbf{v}} = \mathbf{F} + \dot{m}\mathbf{w} = \mathbf{F} + \mathbf{R}, \quad \dot{m} > 0, \tag{10.3.3'}$$

the Theorem 10.3.1 remaining still valid.

In the study of capture of meteorites by a planet, T. Levi-Civita assumed that the absolute velocity $\mathbf{u}$ of the captured masses vanishes (the absolute quadratic mean velocity of the captured cloud of particles is negligible with respect to the velocity of the planet); the equation (10.3.3) becomes

$$\frac{\mathrm{d}}{\mathrm{d}t}(m\mathbf{v}) = \dot{\mathbf{H}} = \mathbf{F}, \quad \dot{m} > 0, \tag{10.3.4}$$

obtaining thus Levi-Civita's equation. We may state

Theorem 10.3.2 (*Levi-Civita*). *If the absolute velocity of the masses captured by a free particle of variable mass vanishes, then we can express the theorem of momentum as in the case of a particle of constant mass.*

The theorem is valid also in the case of emission of mass.

Analogously, if the relative velocity $\mathbf{w}$ of the emitted masses vanishes (uniform emission of particles in all directions, from a celestial body), then the equation (10.3.1") becomes

$$m\dot{\mathbf{v}} = \mathbf{F}, \quad \dot{m} < 0, \qquad (10.3.5)$$

and we can state

Theorem 10.3.3 (*Levi-Civita*). *If the masses emitted by a free particle of variable mass have a vanishing relative velocity with respect to it, then Newton's equation of motion maintains its form.*

Projecting Meshcherskiĭ's equation on the co-ordinate axes, we get

$$m\ddot{x}_i = F_i + R_i, \quad i = 1,2,3. \qquad (10.3.1''')$$

In particular, if the relative velocity of the masses emitted by the particle P of variable mass is directed along the tangent of unit vector $\boldsymbol{\tau}$ to its trajectory, the magnitude of the velocity of the particle being constant in time, we can write (the velocity $\mathbf{w}$ is opposite to the velocity $\mathbf{v}$)

$$m\dot{\mathbf{v}} = \mathbf{F} - \dot{m}w\boldsymbol{\tau} = \mathbf{F} - \dot{m}\frac{w}{v}\mathbf{v}, \quad \dot{m} < 0. \qquad (10.3.6)$$

Often, by a convenient change of variable, we may obtain remarkable forms for the above equations. Let us suppose, for instance, that the absolute velocity $\mathbf{u}$ of the masses emitted by a particle of variable mass vanishes. Meshcherskiĭ's equation (10.3.1) takes the form

$$m\dot{\mathbf{v}} = \mathbf{F} - \dot{m}\mathbf{v}, \qquad (10.3.4')$$

corresponding to Levi-Civita's equation (10.3.4). By the change of variable

$$d\tau = \frac{dt}{m(t)}, \qquad (10.3.7)$$

we get, successively,

$$\mathbf{v} = \frac{d\mathbf{r}}{dt} = \frac{1}{m}\frac{d\mathbf{r}}{d\tau}, \quad \mathbf{a} = -\frac{\dot{m}}{m^2}\frac{d\mathbf{r}}{d\tau} + \frac{1}{m^2}\frac{d^2\mathbf{r}}{d\tau^2};$$

replacing in the equation (10.3.3'), we can write, finally,

$$M(t)\frac{d^2\mathbf{r}}{d\tau^2} = \mathbf{F}, \quad M(t) = \frac{1}{m(t)}, \qquad (10.3.7')$$

obtaining thus an equation of motion, which has a form analogous to that of the classical Newton equation.

3.1.2 Meshcherskiĭ's generalized equation

In some cases, the motion of the free particle P of variable mass takes place with *simultaneous emission and capture of mass*. We mention thus the turbo-jet airplanes

Other considerations on particle dynamics 663

(the captured particles of air are evacuated together with the products of the combustion in the motor), the jet-propelled ships, the captive balloons (the ballast is thrown and the connecting cable is lengthened) etc., which can be modelled as particles of variable mass. Assuming that, in the interval of time Δt, the particle P loses (emits) a mass $-\Delta m^-$, which has the absolute velocity $\mathbf{u}_-$, and captures a mass Δm^+, which has an absolute velocity $\mathbf{u}_+$, using the results previously obtained and the principle of the parallelogram, we can develop a unitary theory, writing *the Meshcherskiĭ generalized equation* in the form (Fig.10.20)

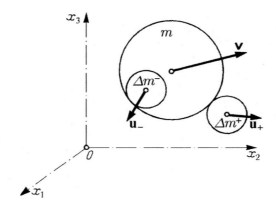

Figure 10.20. Mathematical model for Meshcherskiĭ's generalized equation.

$$m\dot{\mathbf{v}} = \mathbf{F} + \dot{m}^-(\mathbf{u}_- - \mathbf{v}) + \dot{m}^+(\mathbf{u}_+ - \mathbf{v}), \quad \dot{m}^- < 0, \quad \dot{m}^+ > 0. \qquad (10.3.8)$$

Introducing *the reactive force* (in general, it accelerates the motion of the particle, especially if its direction is close to that of the velocity $\mathbf{v}$)

$$\mathbf{R}_- = \dot{m}^-(\mathbf{u}_- - \mathbf{v}) = \dot{m}^-\mathbf{w}_-, \quad \dot{m}^- < 0, \qquad (10.3.9)$$

due to the process of emission, and *the braking force* (in general, it brakes the motion of the particle, especially if its direction is close to the direction of the velocity $\mathbf{v}$)

$$\mathbf{R}_+ = \dot{m}^+(\mathbf{u}_+ - \mathbf{v}) = \dot{m}^+\mathbf{w}_+, \quad \dot{m}^+ > 0, \qquad (10.3.9')$$

due to the process of capture, we may write the equation (10.3.8) in the form

$$m\dot{\mathbf{v}} = \mathbf{F} + \mathbf{R}_- + \mathbf{R}_+. \qquad (10.3.8')$$

We notice that, in general, the processes of emission and capture are independent one of the other and independent of the mass $m(t)$ of the particle P. In the case in which the particle is not free, being subjected to constraints, the corresponding constraint forces must intervene.

In particular, if only a *process of emission* takes place, then we obtain

$$m\dot{\mathbf{v}} = \mathbf{F} + \mathbf{R}_-, \qquad (10.3.10)$$

while in case of only a *process of capture*, we have

$$m\dot{\mathbf{v}} = \mathbf{F} + \mathbf{R}_+ ; \tag{10.3.10'}$$

in both equations intervenes an instantaneous variation of mass, which can be independent of the mass $m(t)$. If the instantaneous variation of the captured mass is equal to the instantaneous variation of the emitted mass ($-\dot{m}^- = \dot{m}^+$), it results

$$m\dot{\mathbf{v}} = \mathbf{F} + \dot{m}^-(\mathbf{u}_- - \mathbf{u}_+), \tag{10.3.10''}$$

where the difference of velocities between the parentheses corresponds, in fact, to a relative velocity. We notice that the mass m of the particle at the moment t is given by

$$m = m_0 + m^+ + m^-, \quad \dot{m}^+ > 0, \quad \dot{m}^- < 0 ; \tag{10.3.11}$$

in case of a process of emission, we have $m^+ = 0$, so that $m = m_0 + m^-$ and $\dot{m} = \dot{m}^-$, $\dot{m} < 0$, while in case of a process of capture $m^- = 0$, hence $m = m_0 + m^+$, $\dot{m} = \dot{m}^+$, $\dot{m} > 0$. Thus, we can write the equation (10.3.10'') in the form

$$m\dot{\mathbf{v}} = \mathbf{F} + \dot{m}(\mathbf{u}_- - \mathbf{u}_+), \tag{10.3.10'''}$$

obtaining an equation of the form (10.3.1).

The equations of motion established above, in the frame of the Newtonian model of mechanics, can be used if a law of variation of mass as function of time is given, in general of the form

$$m = m_0 f(t), \quad f(0) = 1, \tag{10.3.11'}$$

m_0 being the initial mass at the moment $t = 0$ (considered to be the beginning of the process of mass variation). We mention – especially – the case in which $f(t)$ is a *linear function* and the reactive force is constant

$$f(t) = 1 - \alpha t, \quad R = -\dot{m}w = \alpha m_0 w, \quad \alpha = \text{const}, \tag{10.3.12}$$

and the case in which $f(t)$ is an *exponential function*, the relative acceleration being constant and the reactive force variable

$$f(t) = e^{-\alpha t}, \quad \frac{1}{m}R = -\frac{\dot{m}}{m}w = \alpha w, \quad R = \alpha m_0 w e^{-\alpha t}, \quad \alpha = \text{const}. \tag{10.3.12'}$$

Reciprocally, if the relative acceleration R/m is constant, then the law of variation of mass is exponential and if, instead the sign – we take the sign + in the formulae (10.3.12), (10.3.12'), then one obtains the results corresponding to the phenomenon of capture.

Other considerations on particle dynamics 665

3.1.3 Inverse problem of dynamics of the particle of variable mass

In *the direct problem* (the first fundamental problem) of a free particle of variable mass, the given and the reactive force, as well as the braking force (hence, the law of mass variation) which act upon the particle are given, and one must determine its trajectory. In *the inverse problem* (the second fundamental problem), the motion of the particle is known and the law of mass variation is given and one must determine the forces which act upon the particle, or the forces which act upon the particle are given and the law of mass variation must be determined. Obviously, one can imagine various types of *mixed problems*.

Let us consider, e.g., the motion of a heavy particle of variable mass along the local vertical, in a resistent medium; choosing the ascendent vertical as axis, we can write Meshcherskiĭ's equation in the form

$$m\dot{v} = -mg - m_0 g \varphi(v) - \dot{m}w,$$

where $m_0 g \varphi(v)$ is *the resistance of the medium*. Taking into account (10.3.11'), we get

$$f\dot{v} = -gf - g\varphi - \dot{f}w.$$

Assuming now that the motion of the particle is known, being given by $x = x(t)$, we can calculate $v = v(t)$ and than $\varphi(v(t)) = \overline{\varphi}(t)$; observing that $w = w(t)$ and denoting

$$p(t) = \frac{\dot{v}(t) + g}{w}, \quad q(t) = \frac{g\overline{\varphi}(t)}{w}, \tag{10.3.13}$$

it results the linear differential equation of first order

$$\dot{f}(t) + p(t)f(t) + q(t) = 0, \tag{10.3.13'}$$

which, by integration, reads

$$f(t) = e^{-\int p(t) \mathrm{d}t} \left[C - \int q(t) e^{\int p(t) \mathrm{d}t} \mathrm{d}t \right], \tag{10.3.13''}$$

the law of mass variation being thus determined. First of all, let us suppose that the motion is in vacuum ($\varphi(v) = 0$). If $v = v_0 = \text{const}$, then it results $p(t) = g/w = \text{const}$ and $q(t) = 0$, while $f(t) = Ce^{-pt}$; taking into account (10.3.11'), we find an exponential law for the mass variation

$$m = m_0 e^{-\frac{g}{w}t}. \tag{10.3.14}$$

If $v = a_0 t + v_0$, $\dot{v} = a_0$, then we get, analogously,

$$m = m_0 e^{-\frac{a_0+g}{w}t}, \qquad (10.3.14')$$

result which is a generalization of the previous one. Let us suppose now that the motion takes place in a resistent medium, for which $\varphi(v) = kv^\gamma$, $\gamma = \text{const}$. If $v = v_0 = \text{const}$, then we obtain $\overline{\varphi}(t) = kv_0^\gamma = \text{const}$ and $p(t) = g/w = \text{const}$, $q(t) = kgv_0^\gamma/w = \text{const}$; it results, finally,

$$m = m_0 e^{-\frac{g}{w}t}\left[1 + kv_0^\gamma\left(1 - e^{\frac{g}{w}t}\right)\right] = m_0\left(1 + kv_0^\gamma\right)e^{-\frac{g}{w}t} - m_0 kv_0^\gamma. \qquad (10.3.14'')$$

If $v = a_0 t + v_0$, $\dot{v} = a_0$, we get $\varphi(t) = k(a_0 t + v_0)^\gamma$ and

$$f(t) = e^{-\frac{a_0+g}{w}t}\left[C - \frac{kg}{w}\int(a_0 t + v_0)^\gamma e^{\frac{a_0+g}{w}t}dt\right];$$

for $\gamma = 1$ (resistance proportional to the velocity) we may write

$$m = m_0 e^{-\frac{a_0+g}{w}t}\left\{1 + \frac{kg}{a_0+g}\left[\left(v_0 - \frac{a_0 w}{a_0+g}\right)\left(1 - e^{\frac{a_0+g}{w}t}\right) - a_0 t e^{\frac{a_0+g}{w}t}\right]\right\}$$

$$= m_0\left[1 + \frac{kg}{a_0+g}\left(v_0 - \frac{a_0 w}{a_0+g}\right)\right]e^{-\frac{a_0+g}{w}t} - \frac{m_0 kg}{a_0+g}\left(v_0 - \frac{a_0 w}{a_0+g} + a_0 t\right).$$

$$(10.3.14''')$$

3.1.4 Motion of a particle of variable mass in absence of given forces. Tsiolkovskiĭ's first problem

In absence of given forces, the equation of motion of a free particle P of variable mass becomes

$$m\dot{\mathbf{v}} = \dot{m}\mathbf{w} = \mathbf{R}. \qquad (10.3.15)$$

In *Tsiolkovskiĭ's first problem*, the relative velocity $\mathbf{w}$, of constant magnitude, is along the tangent to the trajectory and opposite to the velocity $\mathbf{v}$; we can thus write

$$m\dot{\mathbf{v}} = -\dot{m}\frac{w}{v}\mathbf{v}. \qquad (10.3.15')$$

Let us suppose that the motion is rectilinear, along the Ox-axis; in a scalar form, we have

$$m\dot{v} = -\dot{m}w. \qquad (10.3.15'')$$

Taking into account (10.3.11') and integrating with respect to time, we get $v = -w\ln f + C$, $C = \text{const}$; we obtain *Tsiolkovskiĭ's formula*

$$v = v_0 + w\ln\frac{m_0}{m}, \qquad (10.3.16)$$

Other considerations on particle dynamics 667

with the initial condition $v(0) = v_0$. Let us take a zero initial condition ($v_0 = 0$) and let $m = m_0$ be the total mass emitted till the moment t; introducing *Tsiolkovskiĭ's number* $Z = (m_0 - m)/m$, we may write the velocity at a given moment in the form

$$v = w\ln(1 + Z) \cong 2.3 w \log(1 + Z). \qquad (10.3.16')$$

This formula allows to obtain some conclusions particularly important in practice. Thus, the velocity of the particle of variable mass at the end of *the active segment* (the end of the process of emission, hence of the combustion in a rocket) is as greater as the relative velocity of emission is greater; this velocity increases logarithmically with Tsiolkovskiĭ's number (in fact, with the ratio of the mass at the initial moment to the mass at the final moment) and does not depend on the variation law of mass (it does not depend on the régime of work of rocket's motor). Hence, to obtain velocities as great as possible of the particle of variable mass at the end of the process of emission, it is more advantageous to increase the relative velocity of the evacuated masses than to increase Z (the reserve of fuel). The motor of the rocket must be improved and the fuel must be conveniently chosen.

If the relative velocity (of emission) changes of direction, remaining constant in modulus (reactive braking), when the particle P reaches the velocity v_1, then is put the problem to determine the supplementary reserve of mass, necessary to equate to zero its velocity v_2 (for the landing of the rocket). Let us suppose that at the velocity $v_1 = w\ln(m_0/m_1)$ the mass of the particle is m_1; we can write

$$v_2 = v_1 - w\ln\frac{m_1}{m_2} = 0,$$

where m_2 is the mass of the particle which corresponds to $v_2 = 0$. It results $m_0/m_1 = m_1/m_2$ or $m_0/m_2 = (m_1/m_2)^2$. If $Z' = (m_1 - m_2)/m_2$ is Tsiolkovskiĭ's number which defines the total reserve of mass, then we can write $1 + Z = (1 + Z')^2$, wherefrom

$$Z = Z'(Z' + 2); \qquad (10.3.17)$$

we obtain thus the total reserve of mass necessary to landing as function of Tsiolkovskiĭ's number at the end of *the driving line segment*.

Observing that $v = dx/dt$, starting from (10.3.16), and taking into account (10.3.11'), we can write

$$x = x_0 + v_0 t - w\int_0^t \ln f(\tau)d\tau. \qquad (10.3.18)$$

In the particular case in which the relative velocity of emission vanishes ($w = 0$), it results a uniform motion ($x = x_0 + v_0 t$). If the mass has a linear variation of the form (10.3.12), then we get

$$x = x_0 + v_0 t + \frac{w}{\alpha}[(1-\alpha t)\ln(1-\alpha t) + \alpha t], \tag{10.3.18'}$$

while if the mass has an exponential variation of the form (10.3.12'), then we obtain

$$x = x_0 + v_0 t + \frac{1}{2} w\alpha t^2; \tag{10.3.18''}$$

the parameter α is called *unit mass of consumption* and characterizes the consumption of mass with respect to the initial mass. In the first case, the reactive force is constant and is given by (10.3.12), while in the second case the relative acceleration of emission is constant and the reactive force has an exponential variation, being given by (10.3.12').

3.1.5 Theorem of momentum

Starting from Meshcherskiĭ's equation (10.3.1), we can extend the general theorems of mechanics to the dynamics of the free particle of variable mass; we mention that the generalized Meshcherskiĭ equation does no more lead to interesting results.

The equation of motion (10.3.1) leads to the formula (10.3.1') or to the formula (10.3.3) and we may state (with respect to an inertial frame of reference)

Theorem 10.3.4 (*theorem of momentum*). *The derivative with respect to time of the momentum of a free particle of variable mass* $m = m(t)$ *is equal to the sum of the resultant of the given forces which act upon the particle and the product of the derivative with respect to time* $\dot{m}$ *of the mass of the particle, which characterizes its variation, by the absolute velocity of the emitted or captured mass* ($\dot{m} < 0$ *corresponds to the phenomenon of emission, while* $\dot{m} > 0$ *corresponds to the phenomenon of capture*).

We notice that the product $\dot{m}\mathbf{u}$ is of the nature of a force (different from the reactive force definite by the relation (10.3.2)).

By integration with respect to time, for $t \in [t_1, t_2]$, we obtain

$$\Delta \mathbf{H} = \mathbf{H}(t_2) - \mathbf{H}(t_1) = m_2 \mathbf{v}_2 - m_1 \mathbf{v}_1 = \int_{t_1}^{t_2} \mathbf{F}(t) \mathrm{d}t + \int_{t_1}^{t_2} \dot{m}(t) \mathbf{u}(t) \mathrm{d}t$$

$$= \int_{t_1}^{t_2} \mathbf{F}(t) \mathrm{d}t + \int_{m_1}^{m_2} \mathbf{u}(t) \mathrm{d}m(t) \tag{10.3.19}$$

and may thus state

Theorem 10.3.5 (*theorem of variation of momentum*). *The variation of the momentum of a free particle of variable mass in a given interval of time is equal to the sum of the impulse of the resultant of the given forces which act upon it and the impulse of the mass emitted or captured in the same interval of time.*

If the absolute velocity of the emitted or captured masses vanishes ($\mathbf{u} = \mathbf{0}$), then we find again the Theorem 10.3.2, and the formula (10.3.19) is reduced to the formula (6.1.45'').

If the absolute velocity $\mathbf{u}$ is constant in time ($\mathbf{u} = \mathbf{u}_0$), then we get

Other considerations on particle dynamics 669

$$m_2\mathbf{w}_2 - m_1\mathbf{w}_1 = -\int_{t_1}^{t_2} \mathbf{F}(t)\mathrm{d}t, \quad \mathbf{w}_1 = \mathbf{u}_0 - \mathbf{v}_1, \quad \mathbf{w}_2 = \mathbf{u}_0 - \mathbf{v}_2, \qquad (10.3.19')$$

where we have introduced the relative velocities $\mathbf{w}$ at the initial and at the final moment, respectively.

If the relative velocity $\mathbf{w}$ vanishes ($\mathbf{u} = \mathbf{v}$), then we may write

$$\Delta \mathbf{H} = m_2\mathbf{v}_2 - m_1\mathbf{v}_1 = \int_{t_1}^{t_2} \mathbf{F}(t)\mathrm{d}t + \int_{m_1}^{m_2} \mathbf{v}(t)\mathrm{d}m(t). \qquad (10.3.19'')$$

In the case in which the resultant of the given forces vanishes ($\mathbf{F} = \mathbf{0}$) the formula (6.1.45") leads to a conservation theorem of momentum ($m_1\mathbf{v}_1 = m_2\mathbf{v}_2$); assuming that the initial moment is $t_1 = 0$, while the final moment is t, we may write

$$\mathbf{v}(t) = \frac{m_0}{m}\mathbf{v}_0 = \frac{1}{f(t)}\mathbf{v}_0 \qquad (10.3.20)$$

too, the velocity being of constant direction. In general, we can write a scalar first integral if the projection (component) of the sum $\mathbf{F} + \dot{m}\mathbf{u}$ along a fixed direction vanishes; if

$$\mathbf{F} + \dot{m}\mathbf{u} = \mathbf{0}, \qquad (10.3.21)$$

then we obtain a conservation theorem of momentum. Let us express the given force in the form $\mathbf{F} = \mathbf{a}$, where $\mathbf{a}$ is a vector component of the acceleration $\dot{\mathbf{v}}$; the above condition takes the form $m\mathbf{a} + \dot{m}\mathbf{u} = \mathbf{0}$. Hence, taking into account (10.3.11"), we can state

Theorem 10.3.6 (*conservation theorem of momentum*). *The momentum of a free particle of variable mass is conserved in time if and only if the masses emitted or captured are moving after a law given by the relation*

$$\mathbf{u} = \lambda(t)\mathbf{a}, \quad \lambda(t) = -\frac{1}{(\mathrm{d}/\mathrm{d}t)\ln f}. \qquad (10.3.21')$$

In fact, integrating once more, we get a new vector first integral, corresponding to the motion law of the particle.

From the formula (6.1.45"), we obtain

$$m\mathbf{v}_2 - m\mathbf{v}_1 = \mathbf{F}_0(t_2 - t_1) \qquad (10.3.22)$$

if $\mathbf{F} = \mathbf{F}_0 = \overrightarrow{\mathrm{const}}$.

3.1.6 Theorem of moment of momentum. Theorem of areas

Starting from the relation (10.3.1') or from the relation (10.3.3) and by means of a vector product at left by $\mathbf{r}$, we get

$$\frac{\mathrm{d}}{\mathrm{d}t}[\mathbf{r} \times (m\mathbf{v})] = \dot{\mathbf{K}}_O = \mathbf{M}_O + \mathbf{r} \times (\dot{m}\mathbf{u}), \quad \mathbf{M}_O = \mathbf{r} \times \mathbf{F} \qquad (10.3.23)$$

and we state

Theorem 10.3.7 (*theorem of moment of momentum*). *The derivative with respect to time of the moment of momentum of a free particle of variable mass, with respect to a given pole, is equal to the sum of the moment of the resultant of the given forces which act upon it and the moment of the product of the derivative with respect to time of the mass of the particle, which characterizes its variation, by the absolute velocity of the mass emitted or captured, with respect to the same pole.*

By integration with respect to time, for $t \in [t_1, t_2]$, we get

$$\Delta \mathbf{K}_O = \mathbf{K}_O(t_2) - \mathbf{K}_O(t_1) = \mathbf{r}_2 \times (m_2 \mathbf{v}_2) - \mathbf{r}_1 \times (m_1 \mathbf{v}_1)$$
$$= \int_{t_1}^{t_2} \mathbf{M}_O(t)\mathrm{d}t + \int_{m_1}^{m_2} \mathbf{r}(t) \times \mathbf{u}(t)\mathrm{d}m(t). \qquad (10.3.24)$$

If the absolute velocity of the masses emitted or captured vanishes ($\mathbf{u} = \mathbf{0}$), then this formula is reduced to the formula (6.1.46") and we can state

Theorem 10.3.8. *If the absolute velocity of the emitted or captured masses by a free particle of variable mass vanishes, then we may express the theorem of moment of momentum as in the case of a particle of constant mass.*

If the moment of the resultant of given forces vanishes ($\mathbf{M}_O = \mathbf{0}$) too, then we obtain a conservation theorem of the moment of momentum ($m_2 \mathbf{r}_2 \times \mathbf{v}_2 = m_1 \mathbf{r}_1 \times \mathbf{v}_1$). Assuming that the initial moment is $t_1 = 0$, while the final one is t, and introducing the areal velocity given by (5.1.16), we may also write

$$\mathbf{\Omega}(t) = \frac{m_0}{m}\mathbf{\Omega}_0 = \frac{1}{f(t)}\mathbf{\Omega}_0; \qquad (10.3.25)$$

hence, the areal velocity is of constant direction. In general, we obtain a scalar first integral if the projection (component) of the sum $\mathbf{M}_O + \mathbf{r} \times (\dot{m}\mathbf{u})$ on a fixed direction vanishes. If $\mathbf{r} \times (\mathbf{F} + \dot{m}\mathbf{u}) = \mathbf{0}$, then we obtain a conservation theorem of moment of momentum. We notice that, if a relation of the form (10.3.21) takes place, then we can write not only a conservation theorem of momentum, but a conservation theorem of moment of momentum ($\mathbf{K}_O = \mathbf{C} = \overrightarrow{\mathrm{const}}$) too. As in the case of the particle of constant mass, the trajectory of the particle P of variable mass is plane and it results

$$\mathbf{\Omega} = \frac{1}{2m}\mathbf{C}. \qquad (10.3.26)$$

Because $m = m(t)$, we can no more write a theorem of areas. We may state

Theorem 10.3.9. *To conserve the moment of momentum of a free particle of variable mass, it is sufficient that the emitted or captured masses move after a law given by the relation* (10.3.21').

Other considerations on particle dynamics 671

The relation (10.3.23) may be written also in the form

$$\mathbf{r} \times (m\dot{\mathbf{v}}) = m\frac{\mathrm{d}}{\mathrm{d}t}(\mathbf{r} \times \mathbf{v}) = 2m\dot{\boldsymbol{\Omega}} = \mathbf{r} \times \mathbf{F} + \mathbf{r} \times (\dot{m}\mathbf{w}). \qquad (10.3.26')$$

We have

$$\boldsymbol{\Omega} = \overrightarrow{\mathrm{const}} \Leftrightarrow \mathbf{r} \times \mathbf{F} + \mathbf{r} \times (\dot{m}\mathbf{w}) = \mathbf{r} \times (\mathbf{F} + \mathbf{R}) = \mathbf{0} \,;$$

hence, we can state

Theorem 10.3.10 (*conservation theorem of areal velocity*). *In case of a free particle of variable mass, the areal velocity with respect to a given pole is conserved only and only if the sum of the moment of the resultant of the given forces which act upon the particle and the moment of the product of the derivative with respect to time of the particle mass, which characterizes its variation by the relative velocity* (*with respect to the particle*) *of the emitted or captured mass, with respect to the same pole, vanishes.*

We can mention some particular cases in which a conservation theorem of areal velocity takes place, hence a *theorem of areas* too. Thus, if the resultant $\mathbf{F}$ of the given forces is a central one, passing through the pole O, and if the support of the relative velocity $\mathbf{w} = \mathbf{u} - \mathbf{v}$ passes through the same pole, then it results $\boldsymbol{\Omega} = \overrightarrow{\mathrm{const}}$; in particular, we can have $\mathbf{w} = \mathbf{0}$ (hence, $\mathbf{u} = \mathbf{v}$) or $\mathbf{F} = \mathbf{0}$. One obtains the same result if the resultant $\mathbf{F}$ of the given forces equilibrates the reactive force $\mathbf{R}$.

In the case in which the absolute velocity $\mathbf{u}$ of the emitted and captured masses is collinear with the velocity $\mathbf{v}$ of the particle P of variable mass ($\mathbf{u} = \lambda \mathbf{v}$), $\lambda = \lambda(t)$, the force $\mathbf{F}$ being a central one, we may write the theorem of moment of momentum in the form

$$\frac{\mathrm{d}}{\mathrm{d}t}(\mathbf{r} \times m\mathbf{v}) = \lambda \mathbf{r} \times (\dot{m}\mathbf{v}) \,;$$

introducing the areal velocity, we find

$$\dot{\boldsymbol{\Omega}} = (\lambda - 1)\frac{\dot{m}}{m}\boldsymbol{\Omega} \,.$$

Because $\boldsymbol{\Omega}$ and $\mathrm{d}\boldsymbol{\Omega}$ are collinear vectors, it results that $\boldsymbol{\Omega}(t)$ and $\boldsymbol{\Omega}(t + \mathrm{d}t)$ are collinear vectors at any moment t, the areal velocity $\boldsymbol{\Omega}$ being thus a vector of constant direction; hence, the trajectory of the particle P is a curve contained in a plane which passes through the fixed pole O. A scalar product of the relation obtained above by $\boldsymbol{\Omega}$ leads to (we notice that $\dot{\boldsymbol{\Omega}} \cdot \boldsymbol{\Omega} = \dot{\Omega}\Omega$)

$$\dot{\Omega} = (\lambda - 1)\frac{\dot{m}}{m}\Omega \,;$$

denoting $\boldsymbol{\Omega} = \Omega \mathbf{u}$, $\mathbf{u} = \mathrm{vers}\,\boldsymbol{\Omega}$, we can write $\dot{\boldsymbol{\Omega}} = \dot{\Omega}\mathbf{u} + \Omega\dot{\mathbf{u}}$, wherefrom, taking into account the above relations, we get $\dot{\mathbf{u}} = \mathbf{0}$ or $\mathbf{u} = \overrightarrow{\mathrm{const}}$, hence the same conclusion as above. Integrating and using polar co-ordinates in the plane of motion, we have

672   MECHANICAL SYSTEMS, CLASSICAL MODELS

$$2\Omega = r^2\dot{\theta} = Ce^{\int (\lambda - 1)\frac{dm}{m}}. \tag{10.3.27}$$

If $\lambda = \text{const}$, then it results

$$2\Omega = r^2\dot{\theta} = Cm^{\lambda - 1}; \tag{10.3.27'}$$

for $\lambda = 1$ we find again the previous result, in which the areal velocity is conserved. This case is encountered in the external ballistics of the particle of variable mass.

As in the particular case of constant mass, the Theorems 10.3.4 and 10.3.7 allow us to write

$$\dot{\tau}_O(\mathbf{H}) = \tau_O(\mathbf{F}) + \tau_O(\dot{m}\mathbf{u}) \tag{10.3.28}$$

and to state

Theorem 10.3.11 (*theorem of torsor*). *The derivative with respect to time of the torsor of a free particle of variable mass, with respect to a given pole, is equal to the sum of the torsor of the resultant of the given forces which act upon it and the torsor of the product of the derivative with respect to time of the particle mass, which characterizes its variation, by the absolute velocity of the emitted or captured mass, with respect to the same pole.*

Obviously, the condition (10.3.21) leads to a conservation theorem of torsor.

Starting from (10.3.1") and (10.3.26'), we may write

$$\tau_O(m\dot{\mathbf{v}}) = \tau_O(\mathbf{F}) + \tau_O(\mathbf{R}), \tag{10.3.28'}$$

where $\mathbf{R}$ is the reactive force.

3.1.7 Theorem of kinetic energy

Starting from Meshcherskiĭ's relation (10.3.1) or (10.3.3') and with the aid of a scalar product by $d\mathbf{r} = \mathbf{v}dt$, we can write

$$m\mathbf{v} \cdot d\mathbf{v} + v^2 dm = \mathbf{F} \cdot d\mathbf{r} + \mathbf{u} \cdot \mathbf{v} dm;$$

introducing the kinetic energy $T = mv^2/2$ and the elementary work of the given forces $dW = \mathbf{F} \cdot d\mathbf{r}$, it results

$$dT + \frac{1}{2}v^2 dm = dW + \dot{m}\mathbf{u} \cdot d\mathbf{r}, \tag{10.3.29}$$

so that we may state

Theorem 10.3.12 (*theorem of kinetic energy*). *The sum of the differential of the kinetic energy and the semiproduct of the differential of mass by the square of the velocity of a free particle of variable mass is equal to the sum of the elementary work of the resultant of the given forces which act upon the particle and the elementary work of the product of the derivative with respect to time of the particle mass, which characterizes its variation, by the absolute velocity of the emitted or captured mass.*

We notice that one may write the relation (10.3.29) also in the form

$$dT = dW + \frac{dm}{m}(\mathbf{H} \cdot \mathbf{u} - T); \qquad (10.3.29')$$

introducing the relative velocity $\mathbf{w} = \mathbf{u} - \mathbf{v}$ and the reactive force $\mathbf{R} = \dot{m}\mathbf{w}$, it results

$$md\left(\frac{v^2}{2}\right) = dW + dW_R, \qquad (10.3.29'')$$

too, where $dW_R = \mathbf{R} \cdot d\mathbf{r}$ is the elementary work of the reactive force. Thus, we state

Theorem 10.3.12' (*theorem of kinetic energy; second form*). *The product of the mass of a particle of variable mass by the differential of its kinetic energy, assuming that it has a mass equal to unity, is equal to the sum of the elementary work of the resultant of the given forces which act upon the particle and the elementary work of the reactive force.*

Dividing by dt, we may also write

$$m\frac{d}{dt}\left(\frac{v^2}{2}\right) = P + P_R, \qquad (10.3.29''')$$

so that we can state

Theorem 10.3.12'' (*theorem of kinetic energy; third form*). *The product of the mass of a free particle of variable mass by the derivative of its kinetic energy with respect to time, assuming a mass equal to unity, is equal to the sum of the power of the resultant of the given forces which act upon the particle and the power of the reactive force.*

In the case in which the absolute velocity of the masses emitted or captured is equal to zero ($\mathbf{u} = \mathbf{0}$), we obtain

$$dT + \frac{1}{2}v^2 dm = dW \qquad (10.3.30)$$

or

$$\frac{1}{m}d(mT) = dW; \qquad (10.3.30')$$

if the relative velocity of these masses vanishes ($\mathbf{w} = \mathbf{0}$, hence $\mathbf{u} = \mathbf{v}$), then it results

$$dT - \frac{1}{2}v^2 dm = dW, \qquad (10.3.31)$$

as well as

$$md\left(\frac{v^2}{2}\right) = dW \qquad (10.3.31')$$

or

$$m\frac{\mathrm{d}}{\mathrm{d}t}\left(\frac{v^2}{2}\right) = P. \qquad (10.3.31'')$$

One can thus state corresponding theorems for the two problems of Levi-Civita.

If the absolute velocity **u** of the emitted or captured masses is normal to the velocity **v** of the particle (**u** · **v** = 0), then we get the same results as in the case **u** = **0**.

3.2 Motion of a particle of variable mass in a gravitational field

We present, in what follows, the motion of a particle of variable mass in a gravitational field; we mention, especially, the motion along a vertical, in vacuum or in a resistent medium, in case of a linear or exponential variation of mass. As well, we consider the motion of a particle in a fixed plane (the external ballistics problem).

3.2.1 Tsiolkovskiĭ's second problem

Let be a free particle P of variable mass, which is moving along the local vertical; in *Tsiolkovskiĭ's second problem* one assumes that the particle, of weight $m\mathbf{g}$, is launched up with an initial velocity $\mathbf{v}_0$, at the moment $t = 0$, the relative velocity $\mathbf{w} = \overline{\mathrm{const}}$ of the emitted masses being descendent (in this case, the reactive force is directed towards up, because $\dot{m} < 0$, Fig.10.21). Taking the Ox-axis along the ascendent vertical, we may write Meshcherskiĭ's equation (10.3.1") in the form (neglecting the air friction)

Figure 10.21. Tsiolkovskiĭ's second problem.

$$m\dot{v} = -mg - \dot{m}w$$

or in the form

$$\dot{v} = -g - w\frac{\mathrm{d}}{\mathrm{d}t}\ln f(t);$$

Other considerations on particle dynamics 675

taking into account the conditions at the initial moment $t = 0$ ($m = m_0$, $\mathbf{v} = \mathbf{v}_0$), it results the velocity

$$v(t) = v_0 - gt + w \ln \frac{m_0}{m}, \qquad (10.3.32)$$

the position of the particle P being given by ($x(0) = 0$)

$$x(t) = -\frac{1}{2}gt^2 + v_0 t - w \int_0^t \ln f(\tau) \mathrm{d}\tau. \qquad (10.3.32')$$

In case of a linear variation of mass, given by (10.3.12), we obtain

$$x(t) = -\frac{1}{2}gt^2 + v_0 t + \frac{w}{\alpha}[(1 - \alpha t)\ln(1 - \alpha t) + \alpha t], \qquad (10.3.33)$$

while in case of an exponential variation of mass of the form (10.3.12'), we have

$$x(t) = -\frac{1}{2}gt^2 + v_0 t + \frac{1}{2}\alpha w t^2. \qquad (10.3.33')$$

The particle reaches a height for which $v(\bar{t}) = 0$; in case of the motion given by the law (10.3.33'), we get $\bar{t} = v_0 / (g - \alpha w)$, so that

$$h = x(\bar{t}) = \frac{v_0^2}{2(g - \alpha w)}. \qquad (10.3.34)$$

If $\alpha = g/w$, then one observes that $v = v_0$, the motion being uniform (relative equilibrium), while if $\alpha > g/w$, then the reactive force is greater than the weight. From the expression of the time $\bar{t}$ it results $\alpha < g/w$, so that the particle is moving in a field of acceleration $g - \alpha w > 0$; one observes thus that the formula (10.3.34) corresponds to Torricelli's formula (7.1.17).

3.2.2 Motion of a particle of variable mass along a vertical, in a field of Newtonian attraction

As above, we consider the motion in vacuum of a particle P of variable mass along the local vertical, with an initial velocity v_0, directed towards up, assuming an exponential variation of mass, of the form (10.3.12'); we suppose that the particle is acted upon by a Newtonian force of attraction, in inverse proportion to the square of the distance between the particle and the centre O of the Earth. Meshcherskiĭ's equation of motion (10.3.1") reads

$$m\dot{v} = -\frac{g_0 R^2}{x^2}m - \dot{m}w,$$

where g_0 is the gravity acceleration at the Earth surface (at the sea level), for $x = R$, the Ox-axis being along the ascendent vertical. Taking into account the variation of mass, we obtain ($v = \mathrm{d}x/\mathrm{d}t$)

$$v\frac{\mathrm{d}v}{\mathrm{d}x} = \alpha w - \frac{g_0 R^2}{x^2}$$

and, integrating with respect to x, with the condition $v = v_0$ for $x = R$, the velocity is given by

$$v^2 = v_0^2 + 2(x - R)\left(\alpha w - \frac{g_0 R}{x}\right). \tag{10.3.35}$$

By a change of variable $x = R(1 + z)$, we can write

$$\frac{R}{x} = \frac{1}{1+z} = 1 - z + z^2 + \mathcal{O}(z^3),$$

where we neglect z^3 and higher powers for heights x relative small. The relation (10.3.35) becomes

$$v^2 = v_0^2 + 2g_0 Rz\left(\frac{\alpha w}{g_0} - 1 + z\right) + \mathcal{O}(z^3),$$

wherefrom (we notice that $v = R\mathrm{d}z/\mathrm{d}t$)

$$\frac{\mathrm{d}z}{\sqrt{\alpha + \beta z + z^2}} = \lambda \mathrm{d}t, \quad \alpha = \frac{v_0^2}{2g_0 R}, \quad \beta = \frac{\alpha w}{g_0} - 1, \quad \lambda = \sqrt{\frac{2g_0}{R}}. \tag{10.3.36}$$

If we take $x = R$, hence $z = 0$ for $t = 0$, and if we assume that $v_0 = 0$, hence $\alpha = 0$, then we get, by integration, $z = \beta(1 - \cosh \lambda)/2 = \beta \sinh^2(\lambda t/2)$; the height $\bar{x} = x - R = Rz$ with respect to the Earth surface is given by

$$\bar{x} = R\beta \sinh^2 \frac{\lambda t}{2} = \frac{\alpha w - g_0}{g_0} R \sinh^2 \sqrt{\frac{g_0}{2R}} t. \tag{10.3.36'}$$

Denoting by t_1 the time in which the active line segment is travelled through, its length results in the form

$$h = \frac{\alpha w - g_0}{g_0} R \sinh^2 \sqrt{\frac{g_0}{2R}} t_1, \tag{10.3.37}$$

the velocity at its end being given by

$$v_1^2 = 2h\left(\alpha w - \frac{g_0 R}{R+h}\right). \tag{10.3.37'}$$

The motion of the particle P along the passive segment of a line is governed by the equation (with $m_1 = m(t_1)$)

$$m_1 \dot v = m_1 v \frac{\mathrm d v}{\mathrm d x} = -\frac{g_0 R^2}{x^2} m_1,$$

which leads to (with the condition $v = v_1$ for $x = R+h$, at the end of the active segment of a line)

$$v^2 = v_1^2 + 2g_0 R^2 \left(\frac{1}{x} - \frac{1}{R+h}\right);$$

for $v = 0$ we obtain the total height $H = x - R$, given by

$$H = \frac{1}{\dfrac{1}{R+h} - \dfrac{v_1^2}{2g_0 R^2}} - R. \tag{10.3.37''}$$

In case of an *instantaneous combustion* ($h = 0$), we have

$$H = \frac{v_1^2}{2g_0 - \dfrac{v_1^2}{R}}; \tag{10.3.37'''}$$

if $v_1 \ll \sqrt{2g_0 R}$, then we find again Torricelli's formula, corresponding to the case of the particle of constant mass.

The study may be performed for an arbitrary x too.

Let us consider now for the mass of the particle P a linear variation of the form

$$m = m_0(1 - \alpha t) = m_0\left(1 - \beta \frac{t}{t_1}\right),$$

where $\beta = (m_0 - m_1)/m_0$, m_1 being the mass at the end of the active segment (at the end of the combustion, $t = t_1$) while $m_0 - m_1$ is *the consumed mass*. The equation of motion reads

$$\dot v = \frac{\dfrac{\beta}{t_1} w}{1 - \beta \dfrac{t}{t_1}} - g,$$

where $g = g_0 R^2 / x^2$ is the gravity acceleration, $g = g_0$ taking place for $x = R$ (at the Earth surface – sea level). Assuming, at the beginning, that $g = g_0 = \text{const}$, we obtain

$$v = -w \ln\left(1 - \beta \frac{t}{t_1}\right) - g_0 t \,;$$

at the end of the active segment, we may write ($v(t_1) = \overline{v}_1$)

$$\overline{v}_1 = -w \ln(1 - \beta) - g_0 t_1, \qquad (10.3.38)$$

this formula being used in external ballistics.

If we take into account the variation of g with the altitude, then we have to integrate the equation

$$\ddot{x} + g_0 \frac{R^2}{x^2} - \frac{\dfrac{\beta}{t_1} w}{1 - \beta \dfrac{t}{t_1}} = 0,$$

observing that we can consider $m = m_0 f(t) = m_0 \overline{f}(x)$, it results (we have $\ddot{x} = v \mathrm{d}v / \mathrm{d}x$)

$$v_1^2 = 2 g_0 \left(\frac{R^2}{R+h} - R\right) + 2 \frac{\beta}{t_1} w \int_R^{R+h} \frac{\mathrm{d}x}{\overline{f}(x)}, \qquad (10.3.39)$$

where $v = v_1$ and $x = R + h$ for $t = t_1$. Integrating, analogously, the equation of motion for which one takes, with approximation, $g = g_0$, we read

$$\overline{v}_1^2 = -2 g_0 h + 2 \frac{\beta}{t_1} w \int_R^{R+h} \frac{\mathrm{d}x}{\overline{f}(x)} \,; \qquad (10.3.39')$$

subtracting one formula from the other, it results, finally,

$$v_1^2 = \overline{v}_1^2 + 2 g_0 \frac{h^2}{R+h} = [w \ln(1-\beta) + g_0 t_1]^2 + 2 g_0 \frac{h^2}{R+h}, \qquad (10.3.38')$$

where we took into account (10.3.38). We get thus the velocity at the end of the active segment, as a function of the altitude at the respective moment. We notice that in the formula (10.3.39') it has been taken the same function $x = x(t)$ to express the mass variation $m = m_0 f(t)$ as in the formula (10.3.39), which can be assumed only as a first approximation.

The acceleration at the end of the combustion (for $t = t_1$ we have $g = g_1$) reads

Other considerations on particle dynamics 679

$$\gamma = \frac{\beta}{1-\beta} \frac{w}{t_1} - g_1. \qquad (10.3.40)$$

3.2.3 Rectilinear motion of a particle of variable mass in a resistent medium

Let us, further, consider the motion of a particle P of variable mass along the local vertical, with an initial velocity v_0, directed towards up in a resistent medium; we assume that the mass is varying in time after the linear law (10.3.12). Meshcherskiĭ's equation of motion (10.3.1") is written in the form (the resistance $\mathbf{Q}$ and the relative velocity $\mathbf{w}$ are directed opposite to the motion, Fig.10.22)

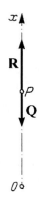

Figure 10.22. Rectilinear motion of a particle of variable mass in a resistent medium.

$$m\dot v = -Q(v) - \dot m w. \qquad (10.3.41)$$

If the magnitude of the resistance is proportional to the velocity ($Q = m_0 k v$, $k = \text{const}$) and if we take, in general, $m = m_0 f(t)$, we obtain

$$\dot v + \frac{k}{f} v + \frac{\dot f}{f} w = 0,$$

wherefrom

$$v(t) = e^{-k\int \frac{dt}{f(t)}} \left[C - w \int \frac{\dot f(t)}{f(t)} e^{k\int \frac{dt}{f(t)}} dt \right], \quad C = \text{const}. \qquad (10.3.42)$$

Observing that $f(t) = 1 - \alpha t$ and by the initial condition $v(0) = v_0$, we get

$$v(t) = \left(v_0 - \frac{\alpha w}{k} \right)(1 - \alpha t)^{k/\alpha} + \frac{\alpha w}{k}; \qquad (10.3.42')$$

integrating once more and taking $x(0) = 0$, we obtain

$$x(t) = \frac{\alpha w}{k}t + \frac{1}{k+\alpha}\left(v_0 - \frac{\alpha w}{k}\right)\left[1 - (1-\alpha t)^{(k+\alpha)/\alpha}\right]. \qquad (10.3.42'')$$

If the magnitude of the resistance is proportional to the square of the velocity ($Q = m_0 k v^2$), then one integrates the equation

$$(1 - \alpha t)\dot{v} = \alpha w - k v^2,$$

by separating the variables; we put the initial condition $v(0) = 0$ and get

$$v = \sqrt{\frac{\alpha w}{k}} \frac{1 - (1-\alpha t)^{2\sqrt{kw/\alpha}}}{1 + (1-\alpha t)^{2\sqrt{kw/\alpha}}}. \qquad (10.3.42''')$$

For kw/α sufficiently great, we can take, with a good approximation, $v \cong \sqrt{\alpha w/k}$. In aerodynamics one obtains $m_0 k = b\rho A/2$, where b is the aerodynamical coefficient, ρ is the air density, while A is a characteristic area of the body, modelled as a particle; one obtains thus an approximative expression of the velocity in the form

$$v = \sqrt{\frac{2\alpha G_0 w}{b\rho g_0 A}}, \qquad (10.3.43)$$

where G_0 and g_0 are the weight of the particle and the gravity acceleration at the initial moment $t = 0$, respectively.

Let us consider now the projection on a horizontal line of the motion of a particle of variable mass in a resistent medium (we take the Ox-axis in the horizontal plane). We assume that $Q = k_Q v^2$ and $P = k_P v^2$, where P is the lift (the ascensional force); if k_Q and k_P are proportional to the air density, then the ratio $\varkappa = k_P/k_Q$ is independent on this density. It must be, permanently, an equilibrium between the weight of the particle and the lift ($mg = P$), so that the equation of motion (10.3.41) reads

$$\dot{v} = -\frac{g}{\varkappa} + \frac{\alpha w}{1-\alpha t},$$

where we considered a linear variation of the mass too. In the case in which $x = \mathrm{const}$, we get, by integration (with the initial condition $v(0) = v_0$),

$$v(t) = v_0 - \frac{g}{\varkappa}t - w\ln(1-\alpha t); \qquad (10.3.44)$$

integrating once more (with $x(0) = 0$), it results

$$x(t) = (v_0 + w)t - \frac{g}{2\varkappa}t^2 + \frac{w}{\alpha}(1-\alpha t)\ln(1-\alpha t), \qquad (10.3.44')$$

Other considerations on particle dynamics 681

hence the equation of motion of the particle on the active segment of a line. These results allow to solve also some interesting problems of optimum.

3.2.4 The balloon problem

A problem in the frame of those studied above is that of a balloon of weight $m\mathbf{g}$, which rises along the vertical by a continuous throwing of the ballast over the border. We assume that the balloon is acted upon by an ascensional force $\mathbf{A}$, corresponding to Archimedes' theorem (hence, a force equal to the weight of a volume of air corresponding to that of the balloon), and by the resistance of the air, the magnitude of which is in direct proportion to the square of the velocity ($Q(v) = kv^2$, $k = \text{const}$). The equation of motion along the ascendent vertical reads

$$m\ddot{x} = -mg + A - k\dot{x}^2 + \dot{m}w, \qquad (10.3.45)$$

hence it is a differential equation of the first order of Riccati's type in $x(t)$, which may be integrated by two quadratures if a particular integral is known.

If, e.g., the condition that the motion of the balloon be uniform is put ($\dot{x} = v_0$, $v_0 = \text{const}$), then we get an equation with separate variables, which leads to (with $w = \text{const}$ and the initial condition $m(0) = m_0$)

$$m(t) = \frac{Q - kv_0^2}{g}\left(1 - e^{-gt/w}\right) + m_0 e^{-gt/w}. \qquad (10.3.45')$$

If the ballast $M(t)$ is thrown in a sufficiently long time, so that the factor $e^{-gt/w}$ be practically zero for $M(t) \to 0$ together with $t \to \infty$, then we have $m_0 - M_0 = (Q - kv_0^2)/g$, $M_0 = M(0)$, and the relation (10.3.45') leads to

$$M(t) = M_0 e^{-gt/w}; \qquad (10.3.45'')$$

hence, if the mass of the ballast varies after an exponential law, then the ascensional motion of the balloon is uniform.

3.2.5 External ballistic problem

Let us consider a more general case of motion of a heavy particle P of variable mass in the air, at the Earth surface. We assume that the trajectory is a plane curve (see Subsec. 3.1.6 too), the particle being acted upon by the weight $m\mathbf{g}$, by the reactive force $\mathbf{R} = R\boldsymbol{\tau} = \dot{m}\mathbf{w}$, $\dot{m} < 0$ ($\boldsymbol{\tau}$ is the unit vector of the tangent to the trajectory in the direction of the motion) and by the resistance $\mathbf{Q} = -Q\boldsymbol{\tau}$, with $Q = (b/2)\rho_0(\rho/\rho_0)Av^2$ (Fig.10.23), where b is the aerodynamical coefficient, ρ and ρ_0 represent the density of the air at a given height and at the Earth surface, respectively, while A is a characteristic area of the body modelled as a particle. We notice that we can take, in general, $b = \psi(v)\varphi(x_3)$, so that

$$Q = Kh(x_3)f(v), \quad K = \frac{1}{2}\rho_0 A, \quad h(x_3) = \frac{\rho}{\rho_0}\varphi(x_3), \quad f(v) = v^2\psi(v). \quad (10.3.46)$$

Denoting by θ the angle made by the tangent to the trajectory with the Ox_1-axis, we have $\cos\theta = \dot{x}_1/v$, $\sin\theta = \dot{x}_3/v$ so that we may write the equations of motion along the considered axes in the form

$$\ddot{x}_1 = -\left[\frac{K}{m}h(x_3)f(v) + \frac{\dot{m}}{m}w\right]\frac{\dot{x}_1}{v}, \quad \ddot{x}_3 = -g - \left[\frac{K}{m}h(x_3)f(v) + \frac{\dot{m}}{m}w\right]\frac{\dot{x}_3}{v}.$$
(10.3.47)

In Frenet's frame of reference, of unit vectors $\boldsymbol{\tau}, \boldsymbol{\nu}$, it results

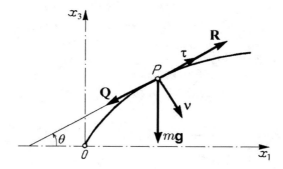

Figure 10.23. The external ballistic problem.

$$\dot{v} + g\sin\theta + \frac{K}{m}h(x_3)f(v) + \frac{\dot{m}}{m}w = 0, \quad -\frac{v^2}{R} + g\cos\theta = v\dot{\theta} + g\cos\theta = 0, (10.3.47')$$

where we took into account the definition relation of the curvature radius $1/R = -\mathrm{d}\theta/\mathrm{d}s = -\dot{\theta}/v$. These equations are used in the external ballistics of the projectiles of variable mass.

3.3 Mathematical pendulum. Motion of a particle of variable mass in a field of central forces

The classical problems of the dynamics of a particle can be taken again in the case of a particle of variable mass. In the following, we consider the mathematical pendulum of variable mass and the motion of a particle of variable mass in a field of central forces.

3.3.1 Mathematical pendulum of variable mass

Let be a particle P of variable mass and weight $m\mathbf{g}$, constrained to move on a circle of radius l and centre O, situated in a vertical plane (Fig.7.9). We assume that the motion takes place in vacuum, the absolute value of the emitted masses vanishing

Other considerations on particle dynamics 683

($\mathbf{u} = \mathbf{0}$). Observing that $v = l\dot\theta$, where $\theta = \theta(t)$ is the generalized co-ordinate which specifies the location of the particle, the theorem of kinetic energy, written in the form (10.3.30), leads to

$$\frac{1}{2}\mathrm{d}\left(ml^2\dot\theta^2\right) + \frac{1}{2}l^2\dot\theta^2\mathrm{d}m = -mgl\sin\theta\,\mathrm{d}\theta;$$

if we exclude the case of equilibrium ($\dot\theta \neq 0$), then it results

$$\ddot\theta + \frac{\dot m}{m}\dot\theta + \omega^2\sin\theta = 0, \quad \omega^2 = \frac{g}{l}, \tag{10.3.48}$$

obtaining an equation which generalizes the classical equation (7.1.38') of the mathematical pendulum.

Writing the equation of motion

$$m\dot{\mathbf{v}} = \mathbf{F} + \overline{\mathbf{R}} + \mathbf{R}, \quad \mathbf{R} = -\dot m \mathbf{v}, \tag{10.3.49}$$

in projection on the principal normal, we obtain the magnitude of the constraint force $\overline{\mathbf{R}}$ in the form (with the direction towards the pole O)

$$\overline{R} = m\frac{v^2}{l} + mg\cos\theta = mg\left(\frac{v^2}{gl} + \cos\theta\right) = mg\left[\left(\frac{\dot\theta}{\omega}\right)^2 + \cos\theta\right]. \tag{10.3.49'}$$

If the variation of mass is after an exponential law of the form $m = m_0 e^{\alpha t}$, $\alpha = \mathrm{const}$, then the equation (10.3.48) becomes

$$\ddot\theta + \alpha\dot\theta + \omega^2\sin\theta = 0; \tag{10.3.48'}$$

it has thus the same form as the equation of the pendulum of constant mass in a resistent medium, for which the resistance is proportional to the velocity (see Chap. 7, Subsec. 1.3.3).

In case of small oscillations around a stable position of equilibrium ($\theta = 0$) we can take $\sin\theta \cong \theta$, and the equation (10.3.48) becomes linear

$$\ddot\theta + \frac{\dot m}{m}\dot\theta + \omega^2\theta = 0. \tag{10.3.48''}$$

Meshcherskiĭ considered the case of a mass with linear variation. Thus, if the mass has a variation of the form $m = m_0(1 - \alpha t)$, $\alpha > 0$, the equation (10.3.48) reads

$$\ddot\theta - \frac{\alpha}{1-\alpha t}\dot\theta + \omega^2\theta = 0. \tag{10.3.50}$$

By a change of variable $\tau = (1 - \alpha t)\omega/\alpha$, one obtains Bessel's equation

$$\frac{d^2\theta}{d\tau^2} + \frac{1}{\tau}\frac{d\theta}{d\tau} + \theta = 0, \qquad (10.3.50')$$

the solution of which is written by means of Bessel's function of the first species and of order zero in the form

$$\theta(\tau) = CJ_0(\tau) = \sum_{n=0}^{\infty}(-1)^n \frac{\tau^{2n}}{2^{2n}(n!)^2}; \qquad (10.3.50'')$$

we get

$$\theta(t) = \theta_0 \frac{J_0\left[\frac{\omega}{\alpha}(1-\alpha t)\right]}{J_0\left(\frac{\omega}{\alpha}\right)}, \qquad (10.3.50''')$$

with the initial condition $\theta(\omega/\alpha) = \theta_0$ for $t = 0$.

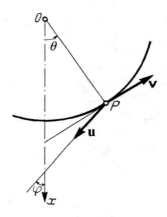

Figure 10.24. Mathematical pendulum of variable mass.

If the mass has a variation after the law $m = m_0(1 + \alpha t)$, $\alpha > 0$, then one obtains the equation

$$\ddot{\theta} + \frac{\alpha}{1+\alpha t}\dot{\theta} + \omega^2\theta = 0; \qquad (10.3.51)$$

in fact, Meshcherskiĭ has considered the equation ($\beta > 0$)

$$\ddot{\theta} + \frac{\beta}{1+\alpha t}\dot{\theta} + \omega^2\theta = 0, \qquad (10.3.51')$$

which has a somewhat more general character. By a change of variable $\tau = (1+\alpha t)\omega/\alpha$ and of function $\vartheta = \theta(1+\alpha t)^{(\beta-\alpha)/2\alpha}$, we obtain Bessel's equation

Other considerations on particle dynamics 685

$$\frac{d^2\vartheta}{d\tau^2} + \frac{1}{\tau}\frac{d\vartheta}{d\tau} + \left(1 - \frac{n^2}{\tau^2}\right) = 0, \quad n = \frac{\beta-\alpha}{2\alpha}, \qquad (10.3.51'')$$

which can be integrated with the aid of Bessel's functions $J_{\pm n}$, as n is an integer or a fraction; in particular, if $\beta = \alpha$, then we find again the preceding result (changing the sign, from $-\alpha$ to α).

In the general case in which $\mathbf{u} \neq \mathbf{0}$, the theorem of kinetic energy, written in the form (10.3.29) leads to

$$\ddot{\theta} + \frac{\dot{m}}{m}\dot{\theta} + \omega^2 \sin\theta = \frac{\dot{m}}{ml}u\sin(\theta - \varphi). \qquad (10.3.52)$$

where φ is the angle made by the absolute velocity $\mathbf{u}$ with the Ox-axis (Fig.10.24).

Analogously, one can study the motion of the mathematical pendulum in a resistent medium.

3.3.2 Motion of a particle of variable mass in a field of central forces

We consider the motion of a particle of variable mass $m = m(t)$, $\dot{m} < 0$, acted upon by a central force, for which the relative velocity of the emitted masses vanishes ($\mathbf{w} = \mathbf{0}$). It is assumed that the central force is of attraction, its magnitude being in inverse proportion to the square of the distance to the fixed point and in direct proportion to the square of the mass; the equation of motion corresponding to the Theorem 10.3.3 is written in the form

$$\ddot{\mathbf{r}} = -\frac{m}{r^3}\mathbf{r}; \qquad (10.3.53)$$

this situation may occur, for instance, in the study of the motion of a particle of variable mass with respect to another particle having the same mass, both particles being acted upon by forces of Newtonian attraction with a gravity constant equal to unity ($f = 1$). In conformity to the results in Subsec. 3.1.6, the trajectory is a plane curve.

Projecting on the co-ordinate axes Ox_k, $k = 1,2$, the equations of motion read

$$\ddot{x}_k + \frac{m}{r^3}x_k = 0, \quad k = 1,2, \quad r^2 = x_1^2 + x_2^2; \qquad (10.3.53')$$

by a change of function and of variable

$$x_k = m^p \xi_k, \quad k = 1,2, \quad d\tau = m^q dt, \qquad (10.3.54)$$

where $p, q \in \mathbb{N}$ must be determined, these equations become

$$m^{p+2q}\frac{d^2\xi_k}{d\tau^2} + (2p+q)m^{p+q-1}\dot{m}\frac{d\xi_k}{d\tau} + p(p-1)m^{p-2}\dot{m}^2\xi_k$$
$$+ pm^{p-1}\ddot{m}\xi_k + m^{-2p+1}\frac{\xi_k}{\rho^3} = 0, \quad k = 1,2, \quad \rho^2 = \xi_1^2 + \xi_2^2.$$

For the sake of simplicity, we put $2p + q = 0$, $p + 2q = -2p + 1$, obtaining $p = -1$, $q = 2$; the above equations take the form

$$\frac{d^2 \xi_k}{d\tau^2} + \left[\frac{1}{\rho^3} + \frac{1}{m^3}\frac{d^2}{dt^2}\left(\frac{1}{m}\right)\right]\xi_k = 0, \quad k = 1,2. \tag{10.3.54'}$$

We suppose, after A.S. Lapin, that $d^2(1/m)/dt^2 = 0$, hence that the law of mass variation is of the form

$$m = \frac{m_0}{1 - \alpha t}, \quad \alpha > 0; \tag{10.3.55}$$

the equations of motion read

$$\frac{d^2 \xi_k}{d\tau^2} + \frac{\xi_k}{\rho^3} = 0, \quad k = 1,2, \tag{10.3.55'}$$

in this case. If, after MacMillan, we put $d^2(1/m)/dt^2 = -\alpha^2 m^3/4m_0^4$, $\alpha = \text{const}$, the equations of motion become

$$\frac{d^2 \xi_k}{d\tau^2} + \left(\frac{1}{\rho^3} - \frac{\alpha^2}{4m_0^4}\right)\xi_k = 0, \quad k = 1,2, \tag{10.3.56}$$

corresponding a law of mass variation of the form

$$m = \frac{m_0}{\sqrt{1 - \alpha t}}, \quad \alpha > 0. \tag{10.3.56'}$$

Let us consider now the case of two particles P and P' of masses $m = m(t)$ and $m' = \text{const}$, respectively, acted upon by Newtonian forces of attraction, with a gravity constant equal to unity ($f = 1$). Assuming that the absolute velocity of the emitted masses vanishes ($\mathbf{u} = \mathbf{0}$) and using the law of mass variation (10.3.55), we can write the vector equation of motion of the particle P with respect to the particle P' (chosen as origin) in the form

$$m\dot{\mathbf{v}} = -\frac{mm'}{r^3}\mathbf{r} - \dot{m}\mathbf{v}; \tag{10.3.57}$$

the corresponding scalar equations read

$$\ddot{x}_k + \frac{m'}{r^3}x_k + \frac{\alpha}{1 - \alpha t}\dot{x}_k = 0, \quad k = 1,2, \quad r^2 = x_1^2 + x_2^2. \tag{10.3.57'}$$

By a change of function and variable

$$x_k = (1-\alpha t)^2 \xi_k, \quad k=1,2, \quad dt = (1-\alpha t)^3 d\tau, \tag{10.3.58}$$

we are led to the equations

$$\frac{d^2\xi_k}{d\tau^2} + \frac{m'}{\rho^3}\xi_k = 0, \quad k=1,2, \quad \rho^2 = \xi_1^2 + \xi_2^2 \tag{10.3.58'}$$

if we notice that

$$(1-\alpha t)^3 \frac{d}{dt}\left[(1-\alpha t)^3 \frac{d\xi_k}{dt}\right] = \frac{d^2\xi_k}{d\tau^2};$$

we see that the equations (10.3.58') are of the same form as the equations (10.3.53'). Consequently, we can replace the study of the particle P of variable mass $m = m(t)$, in the plane Ox_1x_2, by the study of a particle Π of constant mass $m' = \text{const}$ in the plane $O\xi_1\xi_2$; the particle Π is the image of the particle P. Multiplying the equation (10.3.58') by $d\xi_k/d\tau$, summing with respect to k and integrating, we get the first integral of energy for the image particle

$$\left(\frac{d\xi_1}{d\tau}\right)^2 + \left(\frac{d\xi_2}{d\tau}\right)^2 = 2\frac{m'}{\rho} + 2h, \tag{10.3.59}$$

where h is an integration constant (the energy constant).

We can make also a direct study of the system of equations (10.3.57'). Taking into account (10.3.58), we may write the first integral (10.3.59) in the form

$$(1-\alpha t)^2\left(\dot{x}_1^2 + \dot{x}_2^2\right) + 4\alpha(1-\alpha t)(x_1\dot{x}_1 + x_2\dot{x}_2) + 4\alpha^2 r^2$$
$$= \frac{2m'(1-\alpha t)^2}{r} + 2h. \tag{10.3.60}$$

Multiplying the equation (10.3.57') by x_j, making successively $k=1$, $j=2$ and $k=2$, $j=1$, and subtracting, it results

$$(1-\alpha t)\frac{d}{dt}(x_1\dot{x}_2 - \dot{x}_1 x_2) = -\alpha(x_1\dot{x}_2 - \dot{x}_1 x_2),$$

wherefrom (we choose conveniently the integration constant)

$$x_1\dot{x}_2 - \dot{x}_1 x_2 = \frac{C}{m}(1-\alpha t), \quad C = \text{const}; \tag{10.3.60'}$$

we obtain this result also if we make $\lambda = 0$ in the relation (10.3.27') and express the areal velocity in Cartesian co-ordinates (if $\mathbf{u} = \mathbf{0}$, then the condition $\mathbf{u} = \lambda\mathbf{v}$, $\mathbf{v} \neq \mathbf{0}$, leads to $\lambda = 0$). We obtain thus two first integrals, which make easier the integration of the system of equations (10.3.57').

Integrating the third relation (10.3.58) and using the condition $\tau = 0$ for $t = 0$ (the same origin on the time-axis both for the particle and its image), we get $\tau = \left[1/(1-\alpha t)^2 - 1\right]/2\alpha$, wherefrom $\tau \in (-1/2, \infty)$, corresponding ($t \in (-\infty, 1/\alpha)$)

$$t = \frac{1}{\alpha}\left(1 - \frac{1}{\sqrt{1+2\alpha\tau}}\right). \tag{10.3.61}$$

From (10.3.58), it results that, at the initial moment, the particle P coincides with its image Π; at a certain moment τ (to which corresponds t by the relation (10.3.61)), the straight line which connects the particle P to its image Π passes through the centre of attraction O. As well, we notice that

$$r = (1 - \alpha t)^2 \rho = \frac{\rho}{1 + 2\alpha\tau}. \tag{10.3.61'}$$

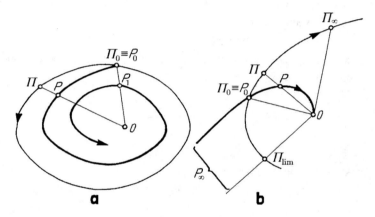

Figure 10.25. Motion of a particle P of variable mass in a field of central forces. Elliptical (a) and parabolical (b) trajectory of the image Π.

Using the results obtained in Chap. 9, Subsec. 2.1.2, we can state that the image Π describes an ellipse, a parabola or a hyperbola as $h < 0$, $h = 0$ or $h > 0$, respectively. Let us assume that the trajectory of the image Π is an ellipse, travelled through, beginning from the point $\Pi_0 \equiv P_0$ (at the moment $\tau = t = 0$), in a period equal to T; at the moment τ, the image of the particle is at Π, while the particle is at P, with $r < \rho$ (Fig.10.25,a). After a period T, the image of the particle will coincide with Π_0, while the particle reaches the location P_1 of radius vector $r_1 = r_0(1 - \alpha t_1)^2 = r_0/(1 + \alpha T)$, $r_0 = \overline{OP_0} = \rho_0$, at the moment $t_1 = (1 - 1/\sqrt{1 + 2\alpha T})/\alpha$. Further, in its motion along the elliptical trajectory, the particle starts from the position P_{n-1} and reaches the position P_n in an interval of time

$$t_n - t_{n-1} = \frac{1}{\alpha}\left[\frac{1}{\sqrt{1+2(n-1)\alpha T}} - \frac{1}{\sqrt{1+2n\alpha T}}\right], \tag{10.3.62}$$

Other considerations on particle dynamics 689

and the distance $\overline{P_{n-1}P_n}$ is given by

$$\overline{P_{n-1}P_n} = r_{n-1} - r_n = \frac{2\alpha T r_0}{(1 + 2n\alpha T)[1 + 2(n-1)\alpha T]}. \qquad (10.3.62')$$

It is easy to see that the particle P approaches the centre of attraction O along a spiral, the motion being periodical and asymptotically damped towards this centre; the distance between two successive turns becomes smaller in an interval of time which becomes smaller too.

If the trajectory of the image Π is an arc of parabola, then the motion starts from the point $P_0 \equiv \Pi_0$ for which $r_0 = \rho_0$ and, while the image Π describes the arc of parabola till the point Π_∞ ($\tau \to \infty$ and $t \to 1/\alpha - 0$), the particle P reaches the centre O, the tangent to the trajectory passing through Π_∞ (Fig.10.25,b). For $\tau \to -1/2\alpha + 0$ we have $t \to -\infty$; the image Π will tend to the position $\Pi_{\lim}$, while the particle P tends to P_∞ along a curve which meets the straight line $O\Pi_{\lim}$ at the very same point. In fact, we can assume that the motion starts at P and tends to O; thus, the motion of the particle is aperiodic and strongly damped.

The case in which the image Π of the particle describes a branch of hyperbola leads to an analogous result.

3.4 Applications of Meshcherskiĭ's generalized equation

In some important problems for technics, in which the variation of mass takes place both by emission and capture, one must use Meshcherskiĭ's generalized equation in the form (10.3.8')-(10.3.9'); in what follows, we consider the motion of the aircraft with jet propulsion as well as the motion of a propelled ship.

3.4.1 Motion of an aircraft with jet propulsion

The displacement of an *aircraft with jet propulsion* takes place by capture of the air and then by eliminating it. To study the motion of such an aircraft modelled as a particle of variable mass, we assume that: i) the change of location of the mass centre of the aircraft with respect to its case, due to the fuel consumption, is negligible; ii) one neglects the motion of the air masses in the interior of the aircraft; iii) the relative velocities of the captured and emitted masses are considered to be collinear with the velocity of the mass centre of the aircraft. In this case, the motion is rectilinear, along the Ox-axis, and the equation (10.3.8) reads

$$m\dot{v} = F - \dot{m}^- w_- - \dot{m}^+ v, \quad \dot{m}^- < 0, \quad \dot{m}^+ > 0, \qquad (10.3.63)$$

assuming that the relative velocity of the emitted masses is constant ($\mathbf{w}_- = \overrightarrow{\text{const}}$) and is directed opposite to the motion, and that the absolute velocity of the captured masses vanishes ($\mathbf{u}_+ = \mathbf{0}$). In technics, it is considered that the rates of flow of capture and emission are constant, verifying the relation $\dot{m}^- = -\gamma \dot{m}^+$, where $\gamma \geq 1$ characterizes the variation of mass due to the combustion of the mixture fuel-air; the equation (10.3.63) becomes (we denote $w_- = w$)

$$m\dot{v} = -m_0 k^2 v - \dot{m}^-\left(w - \frac{v}{\gamma}\right), \qquad (10.3.63')$$

where we assumed that the resistance of the air is proportional to the velocity, in a horizontal flight. Taking into account the above hypotheses, we may take $\dot{m}^- = -m_0\alpha$, $\dot{m}^+ = m_0\alpha/\gamma$, so that, taking into account (10.3.11) too, it results

$$m = m_0(1 - \lambda t), \quad \lambda = \alpha\left(1 - \frac{1}{\gamma}\right) > 0. \qquad (10.3.64)$$

The equation of motion reads

$$(1 - \lambda t)\dot{v} = \alpha w - \sigma v, \quad \sigma = k^2 + \frac{\alpha}{\lambda}, \qquad (10.3.63'')$$

wherefrom, with the initial condition $v(0) = v_0$, it results

$$v(t) = \frac{\alpha w}{\sigma} - \left(\frac{\alpha w}{\sigma} - v_0\right)(1 - \lambda t)^{\sigma/\lambda}; \qquad (10.3.65)$$

observing that $\sigma/\lambda \gg 1$, and $1 - \lambda t < 1$, we obtain the limit velocity

$$v_{\lim} = \frac{\alpha w}{\sigma} = \frac{\alpha\gamma w}{\alpha + k^2\gamma}. \qquad (10.3.65')$$

Integrating with respect to time and using the initial condition $x(0) = 0$, we may write

$$x(t) = \frac{\alpha w}{\sigma}t + \frac{\alpha w - \sigma v_0}{\lambda\sigma(\lambda + \sigma)}\left[(1 - \lambda t)^{(\lambda+\sigma)/\lambda} - 1\right]. \qquad (10.3.66)$$

The length of the active segment of a line is thus determined.

3.4.2 Motion of a propelled ship

Analogously, we can study the motion of a *propelled ship* which has a displacement by absorption of water at its prow with the aid of a pump P and by elimination of it with a great velocity at its poop (Fig.10.26). If we put the pump to work and if we neglect the mass of the consumed fuel, then we may assume that the mass of the ship remains practically equal to m_0, so that it can be modelled as a particle of variable mass, for which the instantaneous variation of the captured mass is equal to the instantaneous variation of the emitted mass. We consider that the pipe through which circulates the water is horizontal and that at the initial moment $t = 0$ the velocity of the ship is $v_0 < w_+$; we study the motion of the ship in the interval of time in which its velocity increases from v_0 to w_+. If the water is absorbed through a section of area

A^+ and is eliminated through a section of area A^-, then the flux is the same ($w_+ A^+ = w_- A^-$), wherefrom

$$w_- = \sigma w_+, \quad \sigma = \frac{A^+}{A^-}; \tag{10.3.67}$$

the mass of water absorbed in a unit of time is thus equal to the mass of water eliminated in the same interval of time ($\mu m_+ A^+ = \mu m_- A^-$). The equation (10.3.10") leads to

$$m_0 \dot{v} = -m_0 k v^2 + \mu w_- A^- (w_- - w_+),$$

wherefrom (we denote $w_+ = w$)

$$\dot{v} = -kv^2 + qw^2, \quad q = \frac{\mu}{m_0} A^+; \tag{10.3.68}$$

the resistance of the air is considered to be proportional to the square of the velocity. The velocity of the ship does not decrease ($\dot{v} > 0$) if $q > 0$, hence if $\sigma > 1$ (the area of the absorption section must be greater than the area of the exit section).

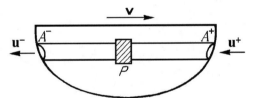

Figure 10.26. Motion of a propelled ship.

By integration, we get

$$v = \sqrt{\frac{q}{k}} \frac{1 - C e^{-2\sqrt{qk}wt}}{1 + C e^{-2\sqrt{qk}wt}} w, \quad C = \frac{\sqrt{q}w - \sqrt{k}v_0}{\sqrt{q}w + \sqrt{k}v_0}, \tag{10.3.68'}$$

with the initial condition $v(0) = v_0$; if $v_0 = 0$, then we have ($C = 1$)

$$v = \sqrt{\frac{q}{k}} \frac{1 - e^{-2\sqrt{qk}wt}}{1 + e^{-2\sqrt{qk}wt}} w = \sqrt{\frac{q}{k}} w \tanh(\sqrt{qk}wt). \tag{10.3.69}$$

In this case, the necessary time for the velocity v of the ship be equal to w is given by

$$T = \frac{1}{2\sqrt{qkw}} \ln \frac{\sqrt{q} + \sqrt{k}}{\sqrt{q} - \sqrt{k}}. \tag{10.3.69'}$$

APPENDIX

The appendix contains elements of vector calculus, as well as notions on the field theory and on the theory of distributions. These results are presented without demonstration or with a concise one, representing a review of known results or complements of such results.

1. Elements of vector calculus

In the following, we deal with vector analysis and with exterior differential calculus; the notions of vector calculus can be found in several chapters of the work and are linked – especially – to the systems of forces. For a better understanding of the principal properties of the vectors and taking into account that we apply techniques of vector calculus to the study of mechanical systems in an Euclidean three-dimensional space E_3, we consider the vectors in the vector three-dimensional space L_3, introduced in Chap. 1, Subsec. 1.1.2, using oriented segments of line as geometric representations of them; their tensor properties have been emphasized in Chap. 3, Subsecs 1.2.2 and 1.2.3. However, some results which will be given hold in a n-dimensional vector space L_n too.

1.1 Vector analysis

A free, bound or sliding vector is a *function of the independent variable* $t \in [t_0, t_1]$ if the parameters which determine it are functions of this variable. In general, we suppose that we have to do with free vectors; however, the results obtained are valid also for the other types of vectors, excepting special cases. Let be the vector

$$\mathbf{V} = \mathbf{V}(t), \quad V_i = V_i(t), \quad i = 1, 2, 3, \qquad (A.1.1)$$

with respect to an orthonormed frame of reference; thus, various operations which will be defined in connection with the vector $\mathbf{V}$ correspond to operations effected on its components in a system of orthogonal Cartesian co-ordinates or – eventually – in another system of co-ordinates. In the following, we deal with functions or vector mappings $t \to \mathbf{V}(t)$, in the mentioned case, in which a single variable is involved, as well as in the case in which they depend on several variables. Without many details, the results known for scalar functions may be adapted for vector ones.

1.1.1 Limits. Continuity

We say that the vector $\mathbf{V}(t)$ tends to a *limit* $\mathbf{V}^0$ for $t \to t^0$, $t^0 \in T \equiv [t_0, t_1]$, and we have

$$\lim_{t \to t^0} \mathbf{V}(t) = \mathbf{V}^0, \tag{A.1.2}$$

if we may write

$$\lim_{t \to t^0} V_i(t) = V_i^0, \quad i = 1, 2, 3, \tag{A.1.2'}$$

for its components; analogously, we may define *the limits at the right* and *at the left*.

We say that the vector $\mathbf{V}$ is a continuous function (of class C^0) if its components are continuous functions. Obviously, the domain of definition of the vector function is specified by the domain (eventually, domains) of definition of its components; in general, we assume that all its components have the same domain of definition. Similar properties may be obtained, in the same way, in the case of vectors depending on several independent variables.

1.1.2 Differentiation of vectors

We say that *the vector function* $t \to \mathbf{V}(t)$ is *differentiable* in $t \in T$ if the limit

$$\lim_{h \to 0} \frac{\mathbf{V}(t+h) - \mathbf{V}(t)}{h} = \mathbf{V}'(t) = \dot{\mathbf{V}}(t) \tag{A.1.3}$$

exists; the notation by a "point" for the derivative is used in the case in which *the independent variable t is the time*, as it will be assumed in what follows. *The differential* of the vector $\mathbf{V}(t)$ is

$$d\mathbf{V}(t) = \dot{\mathbf{V}}(t) dt, \tag{A.1.4}$$

so that its *derivative* may be written in the form

$$\dot{\mathbf{V}}(t) = \frac{d\mathbf{V}(t)}{dt} \tag{A.1.3'}$$

too. If the vector is given in the form $\mathbf{V}(t) = V_j(t)\mathbf{i}_j$, we obtain

$$\dot{\mathbf{V}}(t) = \dot{V}_j(t)\mathbf{i}_j, \tag{A.1.5}$$

which may be a definition relation of the derivative. *The derivatives of higher order* $\ddot{\mathbf{V}}(t)$, $\dddot{\mathbf{V}}(t), \ldots, \mathbf{V}^{(n)}(t)$ can be analogously defined. We say that the vector $\mathbf{V}(t)$ is of class $C^n(T)$ if its components in a system of co-ordinate axes (in particular, in a system of orthogonal Cartesian co-ordinates) are of class $C^n(T)$ (the derivative of nth order exists and is differentiable; n finite or infinite).

Appendix

The modulus of the derivative $\dot{\mathbf{V}}$ is given by

$$|\dot{\mathbf{V}}| = \sqrt{\dot{V}_i \dot{V}_i}, \qquad (A.1.6)$$

while the modulus of the differential $d\mathbf{V}$ reads

$$|d\mathbf{V}| = \sqrt{dV_i dV_i}. \qquad (A.1.6')$$

Following formulae of differentiation

$$\frac{d}{dt}(\mathbf{V}_1 + \mathbf{V}_2) = \dot{\mathbf{V}}_1 + \dot{\mathbf{V}}_2, \qquad (A.1.7)$$

$$\frac{d}{dt}(\lambda \mathbf{V}) = \dot{\lambda}\mathbf{V} + \lambda\dot{\mathbf{V}}, \quad \lambda = \lambda(t) \text{ scalar}, \qquad (A.1.7')$$

$$\frac{d}{dt}(\mathbf{V}_1 \cdot \mathbf{V}_2) = \dot{\mathbf{V}}_1 \cdot \mathbf{V}_2 + \mathbf{V}_1 \cdot \dot{\mathbf{V}}_2, \qquad (A.1.8)$$

$$\frac{d}{dt}(\mathbf{V}_1 \times \mathbf{V}_2) = \dot{\mathbf{V}}_1 \times \mathbf{V}_2 + \mathbf{V}_1 \times \dot{\mathbf{V}}_2, \qquad (A.1.8')$$

$$\frac{d}{dt}(\mathbf{V}_1, \mathbf{V}_2, \mathbf{V}_3) = (\dot{\mathbf{V}}_1, \mathbf{V}_2, \mathbf{V}_3) + (\mathbf{V}_1, \dot{\mathbf{V}}_2, \mathbf{V}_3) + (\mathbf{V}_1, \mathbf{V}_2, \dot{\mathbf{V}}_3), \qquad (A.1.8'')$$

$$\frac{d}{dt}\mathbf{V}[u(t)] = \frac{d\mathbf{V}}{du}\frac{du}{dt} = \mathbf{V}'_u \dot{u}, \quad u(t) \text{ scalar}, \qquad (A.1.9)$$

are easily obtained. From the relation $\mathbf{V}^2(t) = V^2(t)$ one has $\mathbf{V} \cdot d\mathbf{V} = V dV$, so that we may write

$$d|\mathbf{V}| \leq |d\mathbf{V}|. \qquad (A.1.10)$$

The equality takes place in the case of a vector of constant direction ($\mathbf{V} = V(t)\mathbf{u}$, $\mathbf{u} = \overrightarrow{\text{const}}$). In the case of a vector $\mathbf{V}(t)$ of constant modulus ($V^2 = \text{const}$) we have $\mathbf{V} \cdot \dot{\mathbf{V}} = 0$; hence, *the derivative of a vector of constant modulus is a vector normal to that one*. As well,

$$\mathbf{V} = \overrightarrow{\text{const}} \Leftrightarrow d\mathbf{V} = \mathbf{0}. \qquad (A.1.11)$$

If two vectors $\mathbf{V} = \mathbf{V}(t)$ and $\mathbf{W} = \mathbf{W}(t)$ have the same direction, hence the same unit vector $\mathbf{u} = \mathbf{u}(t)$, we may write $\mathbf{V} = V\mathbf{u}$, $\mathbf{W} = W\mathbf{u}$, so that $\mathbf{V} \cdot d\mathbf{W} = V\mathbf{u} \cdot (\mathbf{u}dW + Wd\mathbf{u})$; using the previous results, we get

$$\mathbf{V} \cdot d\mathbf{W} = V dW. \qquad (A.1.12)$$

Let be the plane Ox_1x_2 and a point P, specified by the position vector $\mathbf{r} = \mathbf{r}(t)$ (Fig.A.1); let us also consider the unit vector $\mathbf{i}_r(t) = \text{vers}\,\mathbf{r}(t)$ and the angle $\theta(t)$

formed by that vector with the Ox_1-axis. We have $\mathbf{i}_r = \cos\theta\mathbf{i}_1 + \sin\theta\mathbf{i}_2$; we introduce also the unit vector $\mathbf{i}_\theta = -\sin\theta\mathbf{i}_1 + \cos\theta\mathbf{i}_2$, obtained by a positive rotation of right angle of the unit vector $\mathbf{i}_r$.

We may write $\dot{\mathbf{i}}_r = -\sin\theta\dot{\theta}\mathbf{i}_1 + \cos\theta\dot{\theta}\mathbf{i}_2$, wherefrom

$$\dot{\mathbf{i}}_r = \dot{\theta}\mathbf{i}_\theta, \qquad (A.1.13)$$

formula which allows to calculate *the derivative of a unit vector*, which is *contained in a fixed plane* (or is parallel to a fixed plane), *passing through a fixed point*. It is obvious that, using the same formula, we may write also

$$\dot{\mathbf{i}}_\theta = -\dot{\theta}\mathbf{i}_r. \qquad (A.1.13')$$

The unit vectors $\mathbf{i}_r$ and $\mathbf{i}_\theta$ define a system of orthogonal co-ordinates, the point P having the polar co-ordinates r and θ; in this system, a vector $\mathbf{V}$ is written in the form

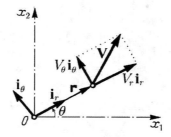

Figure A.1. Polar co-ordinates.

$$\mathbf{V} = V_r\mathbf{i}_r + V_\theta\mathbf{i}_\theta. \qquad (A.1.14)$$

We have $\dot{\mathbf{V}} = \dot{V}_r\mathbf{i}_r + V_r\dot{\mathbf{i}}_r + \dot{V}_\theta\mathbf{i}_\theta + V_\theta\dot{\mathbf{i}}_\theta$, wherefrom, taking into account (A.1.13), (A.1.13'), we get

$$\dot{\mathbf{V}} = \left(\dot{V}_r - V_\theta\dot{\theta}\right)\mathbf{i}_r + \left(\dot{V}_\theta + V_r\dot{\theta}\right)\mathbf{i}_\theta. \qquad (A.1.14')$$

In particular, if $V_\theta = 0$, $V_r = \text{const}$, then we may write *the derivative of a vector of constant modulus*, the support of which passes through a fixed point, in the form

$$\dot{\mathbf{V}} = V_r\dot{\theta}\mathbf{i}_\theta. \qquad (A.1.14'')$$

Let be an ordered system of several independent variables $q_1, q_2, \ldots, q_s$; if the point $(q_1, q_2, \ldots, q_s)$ describes a domain D in the corresponding s-dimensional space, we can define *the function* or *the vector mapping* $(q_1, q_2, \ldots, q_s) \to \mathbf{V} = \mathbf{V}(q_1, q_2, \ldots, q_s)$, which may be written also in the canonical form

Appendix

$$\mathbf{V} = V_j(q_1, q_2, \ldots, q_s)\mathbf{i}_j. \tag{A.1.15}$$

We define *the partial derivatives* of the first order by

$$\frac{\partial \mathbf{V}}{\partial q_h} = \frac{\partial V_j}{\partial q_h}\mathbf{i}_j, \quad h = 1, 2, \ldots, s; \tag{A.1.15'}$$

as in the case of vector functions of a single variable, we obtain *the differential*

$$d\mathbf{V} = \sum_{h=1}^{s} \frac{\partial \mathbf{V}}{\partial q_h} dq_h = \frac{\partial \mathbf{V}}{\partial q_h} dq_h, \tag{A.1.15''}$$

where we have introduced the summation convention of Einstein in the s-dimensional space. Analogously, *partial derivatives as well as differentials of higher order* may be defined.

Considering the mappings $t \to q_h(t)$, $h = 1, 2, \ldots, s$, and assuming that the vector $\mathbf{V}$ may depend also explicitly on the variable t, we can write *the total* (or *substantial*) *derivative* of the vector function in the form

$$\frac{d\mathbf{V}}{dt} = \frac{\partial \mathbf{V}}{\partial q_h}\frac{dq_h}{dt} + \frac{\partial \mathbf{V}}{\partial t}, \tag{A.1.16}$$

the total differential being

$$d\mathbf{V} = \frac{\partial \mathbf{V}}{\partial q_h} dq_h = \frac{\partial \mathbf{V}}{\partial t} dt; \tag{A.1.16'}$$

if $q_1, q_2, \ldots, q_s$ are the co-ordinates of a point in the considered s-dimensional space, t being the time variable, then we say that

$$\frac{\partial \mathbf{V}}{\partial q_h}\frac{dq_h}{dt} = \frac{\partial \mathbf{V}}{\partial q_h}\dot{q}_h \tag{A.1.16''}$$

represents *the space derivative* of the vector function, while

$$\frac{\partial \mathbf{V}}{\partial t} = \dot{\mathbf{V}} \tag{A.1.16'''}$$

represents *the time derivative* of this function (the partial derivative with respect to the time t).

We say that the vector $\mathbf{V}$ is of class $C^n(D)$ with respect to a variable or with respect to a set of variables if its components are of class $C^n(D)$ with respect to that variable or with respect to all variables, respectively. The computation of the mixed derivative of order $m \leq n$ of a vector does not depend on the order of differentiation if the components of the vector have this property and that one takes place if the vector is of class $C^n(D)$ with respect to all variables; in particular, we may write

$$\frac{\partial^2 \mathbf{V}}{\partial q_i \partial q_j} = \frac{\partial^2 \mathbf{V}}{\partial q_j \partial q_i}, \quad i,j = 1,2,...,s, \qquad (A.1.17)$$

in this case (Schwarz's theorem).

1.1.3 Sequences and series of vectors

Let be *the sequence of vectors* $\{\mathbf{V}_n, n \in \mathbb{N}\}$; we say that this sequence tends to the vector $\mathbf{V}$ for $n \to \infty$ and we write

$$\lim_{n \to \infty} \mathbf{V}_n = \mathbf{V} \qquad (A.1.18)$$

if

$$\lim_{n \to \infty} V_{ni} = V_i, \quad i = 1,2,3. \qquad (A.1.18')$$

As well, let be *the series* of general term

$$\mathbf{V}_n = V_{nj}\mathbf{i}_j; \qquad (A.1.19)$$

we say that this series is *convergent* if the series of general terms V_{nj}, $j = 1,2,3$, are convergent. Analogously, we may introduce series of vector functions.

Let be given the vector function $t \to \mathbf{V}(t)$, $t \in T$; we may write a development in the neighbourhood of the moment t in the form

$$\mathbf{V}(t+h) = \mathbf{V}(t) + \frac{h}{1!}\dot{\mathbf{V}}(t) + \frac{h^2}{2!}\ddot{\mathbf{V}}(t) + ... + \frac{h^n}{n!}\mathbf{V}^{(n)}(t) + \mathbf{R}_n, \qquad (A.1.20)$$

where the rest is given by

$$\mathbf{R}_n = \frac{h^{n+1}}{(n+1)!} V_j^{(n+1)}(t+\tau_j)\mathbf{i}_j, \quad \tau_j \in (0,h), \quad j = 1,2,3; \qquad (A.1.20')$$

obviously, this development is equivalent to three developments for the three components of the vector $\mathbf{V}(t)$. If

$$\lim_{n \to \infty} \mathbf{R}_n = \mathbf{0}, \qquad (A.1.20'')$$

then we obtain a development into a *Taylor series*. Obviously, we assume that $\mathbf{V}(t)$ is of class $C^{n+1}(D)$ or of class $C^\infty(D)$, respectively. In particular, for $t = 0$, assuming that this moment belongs to the interval of definition, we obtain a development into a *Maclaurin series*

Appendix 699

$$\mathbf{V}(h) = \mathbf{V}(0) + \frac{h}{1!}\dot{\mathbf{V}}(0) + \frac{h^2}{2!}\ddot{\mathbf{V}}(0) + \ldots + \frac{h^n}{n!}\mathbf{V}^{(n)}(0) + \ldots . \quad (A.1.21)$$

In the case of a vector function of several variables one can write analogous developments.

1.1.4 Integration of vectors

Let be the vector function $t \to \mathbf{V}(t)$, $t \in [t', t'']$, and let be $T \equiv [t_0, t_1] \subset [t', t'']$; we say that the function $\mathbf{V}$ is *integrable* on T if its components are integrable functions on T. We may write

$$\int_{t_0}^{t_1} \mathbf{V}(t)\mathrm{d}t = \mathbf{i}_j \int_{t_0}^{t_1} V_j(t)\mathrm{d}t \quad (A.1.22)$$

in this case. In what follows, we consider only *Riemann integrals*; obviously, one may take into consideration also other types of integrals of vector functions. Let now be the vector

$$\mathbf{W}(t) = \int_{t_0}^{t} \mathbf{V}(\tau)\mathrm{d}\tau, \quad t_0, t \in T; \quad (A.1.23)$$

it results

$$\frac{\mathrm{d}\mathbf{W}}{\mathrm{d}t} = \mathbf{V}. \quad (A.1.23')$$

The solution of this equation may be written in the form

$$\mathbf{W}(t) = \int \mathbf{V}(t)\mathrm{d}t + \mathbf{C}, \quad \mathbf{C} = \overrightarrow{\mathrm{const}}, \quad (A.1.23'')$$

where we have introduced *the primitive* of a vector function. We mention following properties:

$$\int_{t_0}^{t_1} \mathbf{V}(t)\mathrm{d}t = \int_{t_0}^{t_2} \mathbf{V}(t)\mathrm{d}t + \int_{t_2}^{t_1} \mathbf{V}(t)\mathrm{d}t, \quad t_2 \in T, \quad (A.1.24)$$

$$\int_{t_0}^{t_1} \mathbf{V}_1(t)\mathrm{d}t + \int_{t_0}^{t_1} \mathbf{V}_2(t)\mathrm{d}t = \int_{t_0}^{t_1} [\mathbf{V}_1(t) + \mathbf{V}_2(t)]\mathrm{d}t, \quad (A.1.24')$$

$$\int_{t_0}^{t_1} \lambda \mathbf{V}(t)\mathrm{d}t = \lambda \int_{t_0}^{t_1} \mathbf{V}(t)\mathrm{d}t, \quad \lambda = \mathrm{const}, \quad (A.1.25)$$

$$\int_{t_0}^{t_1} \mathbf{C}\lambda(t)\mathrm{d}t = \mathbf{C}\int_{t_0}^{t_1} \lambda(t)\mathrm{d}t, \quad \mathbf{C} = \overrightarrow{\mathrm{const}}, \quad \lambda(t) \text{ scalar}, \quad (A.1.25')$$

$$\int_{t_0}^{t_1} \mathbf{C}\cdot\mathbf{V}(t)\mathrm{d}t = \mathbf{C}\cdot\int_{t_0}^{t_1} \mathbf{V}(t)\mathrm{d}t, \quad \mathbf{C} = \overrightarrow{\mathrm{const}}, \quad (A.1.26)$$

$$\int_{t_0}^{t_1} \mathbf{C}\times\mathbf{V}(t)\mathrm{d}t = \mathbf{C}\times\int_{t_0}^{t_1} \mathbf{V}(t)\mathrm{d}t, \quad \mathbf{C} = \overrightarrow{\mathrm{const}}. \quad (A.1.26')$$

Let be a point P of position vector $\mathbf{r}$; the vector mapping $q \to \mathbf{r}(q)$, $q \in Q \equiv [q',q'']$, of class $C^1(Q)$ determines a curve C, locus of the point P. Let us consider the curvilinear abscissa defined by the function $q \to s(q)$, and two points P_0 and P_1 on the curve C, of curvilinear abscissae $s_0 = s(q_0)$ and $s_1 = s(q_1)$, respectively. We define a vector function $s \to \mathbf{V}(s)$ in any point of the curve C too (Fig.A.2). We introduce thus the *curvilinear vector integral*

$$\int_{\widehat{P_0 P_1}} \mathbf{V}(s)\mathrm{d}s = \int_{s_0}^{s_1} \mathbf{V}(s)\mathrm{d}s = \int_{q_0}^{q_1} \mathbf{V}(s(q))s'(q)\mathrm{d}q, \tag{A.1.27}$$

equivalent to three scalar curvilinear integrals, corresponding to the components of the vector $\mathbf{V}(s)$. Noting that the position vector $\mathbf{r}(q) = x_j(q)\mathbf{i}_j$ of the point P has the derivative $\mathbf{r}'(q) = x'_j(q)\mathbf{i}_j$, the latter vector has the same direction as the tangent to the curve C at the point P, and the differential $\mathrm{d}\mathbf{r} = \mathbf{i}_j \mathrm{d}x_j$ has the same property. We introduce the curvilinear integral

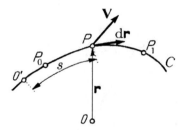

Figure A.2. Curvilinear vector integral.

$$W_{\widehat{P_0 P_1}}(\mathbf{V}) = \int_{\widehat{P_0 P_1}} \mathbf{V}(\mathbf{r}) \cdot \mathrm{d}\mathbf{r} = \int_{\widehat{P_0 P_1}} V_j \mathrm{d}x_j = \int_{q_0}^{q_1} V_j(q) x'_j(q) \mathrm{d}q, \tag{A.1.28}$$

which represents *the work* of the vector $\mathbf{V} = \mathbf{V}(\mathbf{r})$ along the curve C, between the points P_0 and P_1; obviously, the direction of travelling through that curve is from P_0 to P_1. The work of a vector is a scalar quantity. We denote by

$$\mathrm{d}W = \mathbf{V}(\mathbf{r}) \cdot \mathrm{d}\mathbf{r} \tag{A.1.28'}$$

the elementary work, which – in general – is not an exact differential. We notice that the work of the sum of n vectors applied at the same point is equal to the sum of the works of those vectors; this result is obvious, taking into account the property of distributivity of the scalar product with respect to the addition of vectors.

In the case of a closed curve C, we consider the curvilinear vector integral

$$\oint_C \mathbf{V}(s)\mathrm{d}s = \oint_C \mathbf{V}(s(q))s'(q)\mathrm{d}q \tag{A.1.29}$$

Appendix 701

too, the direction of travelling through being that indicated (the counterclockwise). Analogously, we may also consider the work of the vector **V** along the closed curve C, in the form

$$W_C(\mathbf{V}) = \oint_C \mathbf{V}(\mathbf{r}) \cdot \mathrm{d}\mathbf{r} = \oint_C V_j \mathrm{d}x_j ; \qquad (A.1.29')$$

this work is called *the circulation of the vector* **V** on the closed curve C. We mention that the curvilinear vector integrals along a closed curve do not depend on the point from which the travelling through of the curve begins.

Let be a surface Σ, which is represented in a parametric form by $\mathbf{r} = \mathbf{r}(u,v)$, $(u,v) \in D$, as well as the vector function $(u,v) \to \mathbf{V}(u,v)$, defined at the point P, of position vector **r**. If $S \subset \Sigma$ and if the vector function $\mathbf{V}(u,v)$ is integrable on S, then we may introduce *the surface vector integral* in the form

$$\iint_S \mathbf{V}(P)\mathrm{d}S = \mathbf{i}_j \iint_S V_j(P)\mathrm{d}S, \qquad (A.1.30)$$

where $\mathrm{d}S$ is the element of area; obviously, the vector function $\mathbf{V}(u,v)$ is integrable on S if its components have the same property. We may express the surface integral by means of the variables u and v too. As well, we can consider also the surface integrals for which S is a closed surface.

Let be a domain $D \subset \mathbb{R}^3$ and let be the vector mapping $\mathbf{r} \to \mathbf{V}(\mathbf{r})$, defined for $P \in D$, where **r** represents the position vector of the point P; we say that the vector function $\mathbf{V}(\mathbf{r})$ is integrable if its components are integrable functions. In this case, we may introduce *the volume vector integral*

$$\iiint_D \mathbf{V}(\mathbf{r})\mathrm{d}\tau = \mathbf{i}_j \iiint_D V_j(\mathbf{r})\mathrm{d}\tau, \qquad (A.1.31)$$

where $\mathrm{d}\tau = \mathrm{d}x_1 x_2 x_3$ is the volume element.

1.1.5 Curvilinear co-ordinates

Let us consider, in what follows, the vector mapping $(q_1,q_2,q_3) \to \mathbf{V}(q_1,q_2,q_3)$, $(q_1,q_2,q_3) \in D \subset \mathbb{R}^3$, and the point P of position vector **r**, defined by (Fig.A.3)

$$\mathbf{r}(q_1,q_2,q_3) = x_j(q_1,q_2,q_3)\mathbf{i}_j ; \qquad (A.1.32)$$

if the point (q_1,q_2,q_3) describes the domain D, then the point P describes a domain V. Through each point of the domain V may pass three co-ordinate lines, that is the curves $q_2,q_3 = \mathrm{const}$, $q_3,q_1 = \mathrm{const}$ and $q_1,q_2 = \mathrm{const}$; the co-ordinates on these co-ordinate lines are called *curvilinear co-ordinates*. The link between the Cartesian and the curvilinear co-ordinates will be expressed in the form

$$x_j = x_j(q_1,q_2,q_3), \quad j = 1,2,3, \qquad (A.1.33)$$

where $x_j \in C^1(D)$; the transformation (A.1.33) is locally reversible only if the functional determinant J does not vanish

$$J = \det\left[\frac{\partial(x_1, x_2, x_3)}{\partial(q_1, q_2, q_3)}\right] \neq 0. \qquad (A.1.34)$$

We assume that this transformation is one-to-one, that is to a curvilinear system of co-ordinates (q_1, q_2, q_3) corresponds a single point P and reciprocally.

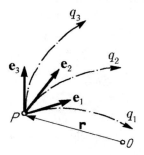

Figure A.3. Curvilinear co-ordinates.

If we consider the mappings $t \to q_i(t)$, $i = 1, 2, 3$, $t \in [t_0, t_1]$, then the point P describes a curve C, the tangent at that point being specified by

$$d\mathbf{r} = \frac{\partial \mathbf{r}}{\partial q_i} dq_i; \qquad (A.1.35)$$

the vectors

$$\mathbf{e}_i = \frac{\partial \mathbf{r}}{\partial q_i}, \quad i = 1, 2, 3, \qquad (A.1.36)$$

are tangent to the co-ordinate curves and form a *local basis*, because

$$(\mathbf{e}_1, \mathbf{e}_2, \mathbf{e}_3) = J \neq 0. \qquad (A.1.34')$$

The arc element $ds = |d\mathbf{r}|$ on the curve C is given by

$$ds^2 = d\mathbf{r}^2 = g_{ij} dq_i dq_j, \qquad (A.1.37)$$

where

$$g_{ij} = g_{ji} = \frac{\partial \mathbf{r}}{\partial q_i} \cdot \frac{\partial \mathbf{r}}{\partial q_j} = \mathbf{e}_i \cdot \mathbf{e}_j, \quad i, j = 1, 2, 3; \qquad (A.1.38)$$

Appendix

the metrics of the considered *Euclidean space* is thus defined. The volume element, that is the volume of the curvilinear parallelepipedon built up with the vectors $\mathbf{e}_1 dq_1$, $\mathbf{e}_2 dq_2$, $\mathbf{e}_3 dq_3$ is given by

$$dV = (\mathbf{e}_1, \mathbf{e}_2, \mathbf{e}_3) dq_1 dq_2 dq_3 = J dq_1 dq_2 dq_3, \qquad (A.1.39)$$

assuming that we have to do with a positive basis. Using Gramm's determinant (2.1.42'), we may write

$$g = \det[g_{ij}] = \det[\mathbf{e}_i \cdot \mathbf{e}_j] = (\mathbf{e}_1, \mathbf{e}_2, \mathbf{e}_3)^2 = J^2 \qquad (A.1.34'')$$

too, so that

$$dV = \sqrt{g}\, dq_1 dq_2 dq_3. \qquad (A.1.39')$$

In the case of a system of *orthogonal curvilinear co-ordinates* we have $g_{ij} = 0$, $i \neq j$, and

$$g_{11} = \mathbf{e}_1^2 = \left(\frac{\partial \mathbf{r}}{\partial q_1}\right)^2 = H_1^2 = \frac{1}{h_1^2}, \quad g_{22} = \mathbf{e}_2^2 = \left(\frac{\partial \mathbf{r}}{\partial q_2}\right)^2 = H_2^2 = \frac{1}{h_2^2},$$

$$g_{33} = \mathbf{e}_3^2 = \left(\frac{\partial \mathbf{r}}{\partial q_3}\right)^2 = H_3^2 = \frac{1}{h_3^2}, \qquad (A.1.40)$$

H_1, H_2, H_3 being *Lamé's coefficients*, while h_1, h_2, h_3 are *differential parameters of first order*; it results

$$g = g_{11} g_{22} g_{33} = (H_1 H_2 H_3)^2 = \frac{1}{(h_1 h_2 h_3)^2}. \qquad (A.1.40')$$

The element of arc is given by

$$ds^2 = ds_1^2 + ds_2^2 + ds_3^2 = (H_1 dq_1)^2 + (H_2 dq_2)^2 + (H_3 dq_3)^2$$

$$= \left(\frac{dq_1}{h_1}\right)^2 + \left(\frac{dq_2}{h_2}\right)^2 + \left(\frac{dq_3}{h_3}\right)^2. \qquad (A.1.40'')$$

and the element of volume reads

$$dV = H_1 H_2 H_3 dq_1 dq_2 dq_3 = \frac{dq_1 dq_2 dq_3}{h_1 h_2 h_3}. \qquad (A.1.40''')$$

A system of *spherical co-ordinates* (r, θ, φ) is linked to the orthogonal Cartesian co-ordinates (see Fig.1.5,c) by the relations

$$x_1 = r\sin\theta\cos\varphi, \quad x_2 = r\sin\theta\sin\varphi, \quad x_3 = r\cos\theta, \quad r \geq 0,$$
$$0 \leq \theta \leq \pi, \quad 0 \leq \varphi < 2\pi; \tag{A.1.41}$$

the element of arc is expressed in the form ($H_1 = 1$, $H_2 = r$, $H_3 = r\sin\theta$)

$$\mathrm{d}s^2 = \mathrm{d}r^2 + r^2\mathrm{d}\theta^2 + r^2\sin^2\theta\mathrm{d}\varphi^2 = \mathrm{d}r^2 + \mathrm{d}s_\theta^2 + \mathrm{d}s_\varphi^2, \tag{A.1.41'}$$

while the element of volume is given by ($g = r^4\sin^2\theta$)

$$\mathrm{d}V = r^2\sin\theta\,\mathrm{d}r\mathrm{d}\theta\mathrm{d}\varphi. \tag{A.1.41''}$$

The functional determinant fulfils the condition $J = r^2\sin\theta \neq 0$ if $r > 0$, $0 < \theta < \pi$.

The system of *cylindrical co-ordinates* (r,θ,z) is linked to the orthogonal Cartesian co-ordinates (see Fig.1.5,b) by the relations

$$x_1 = r\cos\theta, \quad x_2 = r\sin\theta, \quad x_3 = z, \quad r \geq 0, \quad 0 \leq \theta < 2\pi,$$
$$-\infty < z < \infty; \tag{A.1.42}$$

the element of arc is given by ($H_1 = H_3 = 1$, $H_2 = r$)

$$\mathrm{d}s^2 = \mathrm{d}r^2 + r^2\mathrm{d}\theta^2 + \mathrm{d}z^2 = \mathrm{d}r^2 + \mathrm{d}s_\theta^2 + \mathrm{d}z^2 \tag{A.1.42'}$$

and the element of volume is expressed in the form ($g = r^2$)

$$\mathrm{d}V = r\mathrm{d}r\mathrm{d}\theta\mathrm{d}z. \tag{A.1.42''}$$

As well, to have $J = r \neq 0$, it is necessary that $r > 0$.

Differentiating the formula (A.1.38) with respect of the variable q_k, we may write

$$\mathbf{e}_{i,k}\cdot\mathbf{e}_j + \mathbf{e}_i\cdot\mathbf{e}_{j,k} = g_{ij,k}, \quad i,j,k = 1,2,3, \tag{A.1.43}$$

where the index at the right to the comma indicates the differentiation with respect to the corresponding variable; we write again this relation by circular permutations

$$\mathbf{e}_{j,i}\cdot\mathbf{e}_k + \mathbf{e}_j\cdot\mathbf{e}_{k,i} = g_{jk,i}, \quad \mathbf{e}_{k,j}\cdot\mathbf{e}_i + \mathbf{e}_k\cdot\mathbf{e}_{i,j} = g_{ki,j}, \quad i,j,k = 1,2,3. \tag{A.1.43'}$$

Summing the relations (A.1.43') and subtracting the relation (A.1.43), we may express *Christoffel's symbols of first species* in the form

$$[ij,k] = \begin{bmatrix} i\ j \\ k \end{bmatrix} = \Gamma_{kij} = \mathbf{e}_{i,j}\cdot\mathbf{e}_k = \frac{1}{2}(-g_{ij,k} + g_{jk,i} + g_{ki,j}),$$
$$i,j,k = 1,2,3, \tag{A.1.44}$$

where we have introduced the most used notations; to obtain this result, we have taken into consideration that $\mathbf{e}_{i,j} = \mathbf{e}_{j,i}$, $i,j = 1,2,3$, due to the relation of definition (A.1.38) and to the property of the mixed derivatives of second order of the position vectors $\mathbf{r} \in C^2(D)$ of not depending on the order of differentiation. *The Christoffel symbols of second species* are defined in the form

$$\begin{Bmatrix} k \\ i\ j \end{Bmatrix} = \Gamma_{ij}^k = g^{kl}\,[ij,l]\,, \quad i,j,k = 1,2,3\,, \tag{A.1.45}$$

where g^{ij} is *the normalized algebraic complement* (the algebraic complement divided by g) of the element g_{ij} of $\det[g_{ij}]$; we have $g^{ij} = g^{ji}$ because $g_{ij} = g_{ji}$. We notice that the relations

$$g_{ik}g^{kj} = \delta_i^j, \quad g^{ik}g_{kj} = \delta_j^i, \quad i,j = 1,2,3, \quad \det|g^{ij}| = \frac{1}{g} \tag{A.1.43''}$$

take place. Christoffel's symbols are symmetric with respect to the indices i and j, so that

$$[ij,k] = [ji,k], \quad \begin{Bmatrix} k \\ i\ j \end{Bmatrix} = \begin{Bmatrix} k \\ j\ i \end{Bmatrix}, \quad i,j,k = 1,2,3\,; \tag{A.1.46}$$

hence, there are 18 distinct symbols of each species. Multiplying the relation (A.1.45) by g_{km} and taking into account (A.1.43''), we get

$$[ij,k] = g_{kl}\begin{Bmatrix} l \\ i\ j \end{Bmatrix}, \quad i,j,k = 1,2,3\,. \tag{A.1.45'}$$

Christoffel's symbols are defined by the relations (A.1.44), (A.1.45) in the case of a linear space L_n too.

1.2 Exterior differential calculus

In what follows, we introduce the external product of vectors as well as differential forms of various orders; in connection with these forms, we put then in evidence the operator of exterior differentiation.

1.2.1 External product of vectors

A (free) *n-dimensional vector* is a mathematical entity characterized by an ordered set of n numbers V^i, $i = 1,2,...,n$; using the way indicated in Chap. 1, Subsec. 1.1.2, we may set up an *n-dimensional vector space* (*the linear space L_n*, which has the same properties as the linear space L_3). *There exist*, in this space, *at the most n independent* *linear vectors*; an ordered set of n independent linear vectors $\{\mathbf{e}_i, i = 1, 2, \ldots, n\}$ forms a *basis*, and an arbitrary vector $\mathbf{V}$ may be written in the form

$$\mathbf{V} = \sum_{i=1}^{n} V^i \mathbf{e}_i, \qquad (\text{A.1.47})$$

where V^i are *the components of the vector* in this basis (the contravariant components, but – as till now – we do not use the notions of contravariance and covariance). *The external product* or *the bivector* $\mathbf{V}_1 \wedge \mathbf{V}_2$ (it coincides with the vector product for $n = 3$; we replace the sign "$\times$" by the sign "$\wedge$") of two vectors $\mathbf{V}_1, \mathbf{V}_2$ is defined by the properties ($\mathbf{V}, \mathbf{V}_1, \mathbf{V}_2$ are vectors; λ_1, λ_2 are scalars):

  i) $(\lambda_1 \mathbf{V}_1 + \lambda_2 \mathbf{V}_2) \wedge \mathbf{V} = \lambda_1 (\mathbf{V}_1 \wedge \mathbf{V}) + \lambda_2 (\mathbf{V}_2 \wedge \mathbf{V})$
  $\mathbf{V} \wedge (\lambda_1 \mathbf{V}_1 + \lambda_2 \mathbf{V}_2) = \lambda_1 (\mathbf{V} \wedge \mathbf{V}_1) + \lambda_2 (\mathbf{V} \wedge \mathbf{V}_2)$
  (distributivity with respect to addition of vectors);
  ii) $\mathbf{V} \wedge \mathbf{V} = \mathbf{0}$;
  iii) $\mathbf{V}_1 \wedge \mathbf{V}_2 + \mathbf{V}_2 \wedge \mathbf{V}_1 = \mathbf{0}$ (anticommutativity).

We may write

$$\mathbf{V}_1 \wedge \mathbf{V}_2 = \sum_{i=1}^{n} V_1^i \mathbf{e}_i \wedge \sum_{j=1}^{n} V_2^j \mathbf{e}_j = \sum_{i=1}^{n} \sum_{j=1}^{n} V_1^i V_2^j \mathbf{e}_i \wedge \mathbf{e}_j; \qquad (\text{A.1.48})$$

inverting i by j and taking into account the property iii), it results

$$\mathbf{V}_1 \wedge \mathbf{V}_2 = \frac{1}{2} \sum_{i=1}^{n} \sum_{j=1}^{n} \left(V_1^i V_2^j - V_1^j V_2^i \right) \mathbf{e}_i \wedge \mathbf{e}_j. \qquad (\text{A.1.48}')$$

The properties ii) and iii) lead to the relation

$$\mathbf{V}_1 \wedge \mathbf{V}_2 = \sum_{i=1}^{n} \sum_{j=1}^{n} \left(V_1^i V_2^j - V_1^j V_2^i \right) \mathbf{e}_i \wedge \mathbf{e}_j, \quad i < j. \qquad (\text{A.1.48}'')$$

We denote by $\wedge^2 L_n$ the vector space of the bivectors defined on L_n (corresponding to this notation $\wedge^1 L_n = L_n$); noting that the set $\{\mathbf{e}_i \wedge \mathbf{e}_j, 1 \leq i < j \leq n\}$ forms a basis in this space, its dimension is given by

$$\dim \wedge^2 L_n = 1 + 2 + \ldots + (n-1) = \frac{n(n-1)}{2} = C_n^2. \qquad (\text{A.1.49})$$

In general, a *p-vector* $\mathbf{V}_1 \wedge \mathbf{V}_2 \wedge \ldots \wedge \mathbf{V}_p$ in L_n is defined by the properties ($\mathbf{W}_1, \mathbf{W}_2, \mathbf{V}_1, \mathbf{V}_2, \ldots, \mathbf{V}_p$ are vectors; λ_1, λ_2 are scalars):

i) $(\lambda_1 \mathbf{W}_1 + \lambda_2 \mathbf{W}_2) \wedge \mathbf{V}_2 \wedge ... \wedge \mathbf{V}_p = \lambda_1 (\mathbf{W}_1 \wedge \mathbf{V}_2 \wedge ... \wedge \mathbf{V}_p)$
$+ \lambda_2 (\mathbf{W}_2 \wedge \mathbf{V}_2 \wedge ... \wedge \mathbf{V}_p)$
(and analogous relations, obtained by replacing the other vectors $\mathbf{V}_i$, $i = 2, 3, ..., p$, which form the p-vector by linear combinations);

ii) $\mathbf{V}_1 \wedge \mathbf{V}_2 \wedge ... \wedge \mathbf{V}_p = \mathbf{0}$ if and only if $\mathbf{V}_i = \mathbf{V}_j$, $i \neq j$;

iii) The external product $\mathbf{V}_1 \wedge \mathbf{V}_2 \wedge ... \wedge \mathbf{V}_p$ changes its sign if two factors of it permute.

Let be $\wedge^p L_n$ the vector space formed by p-vectors defined on L_n, $2 \leq p \leq n$. The set $\{\mathbf{e}_{i_1} \wedge \mathbf{e}_{i_2} \wedge ... \wedge \mathbf{e}_{i_p}, 1 \leq i_1 < i_2 < ... < i_p \leq n\}$ forms a basis in this space, so that

$$\dim \wedge^p L_n = C_n^p, \tag{A.1.49'}$$

where we have introduced the combination symbol of n things p at a time; in particular, $\dim \wedge^n L_n = C_n^n = 1$.

In the case of a *trivector* $\mathbf{V}_1 \wedge \mathbf{V}_2 \wedge \mathbf{V}_3$ we may write

$$\mathbf{V}_1 \wedge \mathbf{V}_2 \wedge \mathbf{V}_3 = \sum_{i=1}^{n} \sum_{j=1}^{n} \sum_{k=1}^{n} V_1^i V_2^j V_3^k \mathbf{e}_i \wedge \mathbf{e}_j \wedge \mathbf{e}_k; \tag{A.1.50}$$

inverting two upper indices and taking into account the property iii), we find $3! = 6$ different representations of the external product (obtained by permuting the indices i, j and k and by introducing the sign minus in the case of an odd permutation). Summing these six representations, we obtain a representation of the form

$$\mathbf{V}_1 \wedge \mathbf{V}_2 \wedge \mathbf{V}_3 = \frac{1}{3!} \sum_{i=1}^{n} \sum_{j=1}^{n} \sum_{k=1}^{n} V^{ijk} \mathbf{e}_i \wedge \mathbf{e}_j \wedge \mathbf{e}_k, \tag{A.1.50'}$$

where the scalar V^{ijk} is totally skew-symmetric with respect to the upper indices i, j, k. In general, a p-vector $\boldsymbol{\varphi}$ may be represented by

$$\boldsymbol{\varphi} \equiv \mathbf{V}_1 \wedge \mathbf{V}_2 \wedge ... \wedge \mathbf{V}_p = \frac{1}{p!} \sum_{i_k=1}^{n} V^{i_1 i_2 ... i_p} \mathbf{e}_{i_1} \wedge \mathbf{e}_{i_2} \wedge ... \wedge \mathbf{e}_{i_p}, \tag{A.1.51}$$

where the scalar $V^{i_1 i_2 ... i_p}$ is totally skew-symmetric with respect to the upper indices i_k, $k = 1, 2, ..., p$ (the summation takes place for all upper indices). For $p = 2$ one obtains the representation (A.1.48').

One can define a p-vector $\boldsymbol{\varphi} \equiv \mathbf{V}_1 \wedge \mathbf{V}_2 \wedge ... \wedge \mathbf{V}_p$ and a q-vector $\boldsymbol{\psi} \equiv \mathbf{W}_1 \wedge \mathbf{W}_2 \wedge ... \wedge \mathbf{W}_q$ on L_n; their external product is a $p+q$-vector, $p + q \leq n$, defined in the form

$$\varphi \wedge \psi \equiv (V_1 \wedge V_2 \wedge ... \wedge V_p) \wedge (W_1 \wedge W_2 \wedge ... \wedge W_q)$$
$$= V_1 \wedge ... \wedge V_p \wedge W_1 \wedge ... \wedge W_q. \quad (A.1.52)$$

We mention following properties:
i) $\varphi \wedge (\lambda_1 \psi_1 + \lambda_2 \psi_2) = \lambda_1 \varphi \wedge \psi_1 + \lambda_2 \varphi \wedge \psi_2$ (*distributivity with respect to addition*; ψ_1, ψ_2 are p-vectors, λ_1, λ_2 are scalars);
ii) $(\varphi \wedge \psi) \wedge \chi = \varphi \wedge (\psi \wedge \chi)$ (*associativity*; χ is r-vector; $p + q + r \leq n$);
iii) $\varphi \wedge \psi = (-1)^{pq} \psi \wedge \varphi$.

If one of the numbers p, q is even, then the external product is commutative, otherwise it is anticommutative.

1.2.2 Differential forms. Exterior derivative

We call *differential form of first degree* in $x \equiv (x_1, x_2, ..., x_n) \in E_n$ the expression

$$\omega = \sum_{i=1}^{n} a_i \, dx_i, \quad a_i = \text{const}. \quad (A.1.53)$$

If between the basis $\{e_i\}$ of the linear space L_n and $\{dx_i\}$, $i = 1, 2, ..., n$, is established an isomorphism, then ω is an element of the space $\wedge^1 L_n$, introduced in the preceding subsection. Analogously, a *form of pth degree* in x is an element of the space $\wedge^p L_n$, being expressed by

$$\omega = \sum a_{i_1 i_2 ... i_p} dx_{i_1} \wedge dx_{i_2} \wedge ... \wedge dx_{i_p}, \quad a_{i_1 i_2 ... i_p} = \text{const}. \quad (A.1.53')$$

In the case of a form of pth degree on $D \subset E_n$, $a_i = a_i(x_1, x_2, ..., x_n)$ may be smooth functions (differentiable as much as it is necessary) on D. We denote by $F^p(D)$ the set of the forms of pth degree on D; in this case, $F^0(D)$ represents the set of smooth functions.

Let be the forms of first degree

$$\omega_1 = a_i(x_1, x_2, x_3) dx_i, \quad \omega_2 = b_j(x_1, x_2, x_3) dx_j; \quad (A.1.54)$$

the *external product* introduced in the previous subsection is calculated in the form (dx_i play the rôle of vectors e_i)

$$\omega_1 \wedge \omega_2 = (a_2 b_3 - a_3 b_2) dx_2 \wedge dx_3 + (a_3 b_1 - a_1 b_3) dx_3 \wedge dx_1$$
$$+ (a_1 b_2 - a_2 b_1) dx_1 \wedge dx_2. \quad (A.1.54')$$

One obtains thus a *form of second degree*.

The operator $d: F^p(D) \to F^{p+1}(D)$, called (after Cartan) *operator of exterior differentiation*, exists and is unique, being defined by the properties (ω, ω_1 and ω_2 are p-forms):

i) $d(\omega_1 + \omega_2) = d\omega_1 + d\omega_2$;

ii) $d(\omega_1 \wedge \omega_2) = d\omega_1 \wedge \omega_2 + (-1)^{\mathrm{grad}\,\omega_1} \omega_1 \wedge d\omega_2$;

iii) $d(d\omega) = 0$ (Poincaré's lemma);

iv) $df = \sum_{j=1}^{n} \dfrac{\partial f}{\partial x_j} dx_j$ for any function f.

These properties are independent of the system of co-ordinates. One may show that the exterior derivative of the form (A.1.53') is expressed by

$$d\omega = \sum \frac{\partial a_{i_1 i_2 \ldots i_p}}{\partial x_j} dx_j \wedge dx_{i_1} \wedge dx_{i_2} \wedge \ldots \wedge dx_{i_p}. \qquad (A.1.55)$$

In the particular case $n = 3$ one may introduce *the gradient* operator in the form

$$df = \frac{\partial f}{\partial x_1} dx_1 + \frac{\partial f}{\partial x_2} dx_2 + \frac{\partial f}{\partial x_3} dx_3, \qquad (A.1.56)$$

where f is a function defined on $D \subset E_3$. For a form of first degree on D

$$\omega = a_1(x_1, x_2, x_3) dx_1 + a_2(x_1, x_2, x_3) dx_2 + a_3(x_1, x_2, x_3) dx_3 \qquad (A.1.57)$$

the exterior derivative

$$d\omega = \sum \frac{\partial a_i}{\partial x_j} dx_j \wedge dx_i = \left(\frac{\partial a_3}{\partial x_2} - \frac{\partial a_2}{\partial x_3}\right) dx_2 \wedge dx_3$$
$$+ \left(\frac{\partial a_1}{\partial x_3} - \frac{\partial a_3}{\partial x_1}\right) dx_3 \wedge dx_1 + \left(\frac{\partial a_2}{\partial x_1} - \frac{\partial a_1}{\partial x_2}\right) dx_1 \wedge dx_2 \qquad (A.1.57')$$

allows to introduce *the* operator *curl*. As well, the exterior derivative

$$d\omega = \left(\frac{\partial a_1}{\partial x_1} + \frac{\partial a_2}{\partial x_2} + \frac{\partial a_3}{\partial x_3}\right) dx_1 \wedge dx_2 \wedge dx_3, \qquad (A.1.58)$$

corresponding to the form of second degree on D

$$\omega = a_1 dx_2 \wedge dx_3 + a_2 dx_3 \wedge dx_1 + a_3 dx_1 \wedge dx_2, \qquad (A.1.58')$$

introduces *the* operator *divergence*.

2. Notions of field theory

In what follows, we deal with conservative vectors, with the operator gradient, as well as with the introduction of the curl and the divergence of a vector; differential operators of second order are considered too and some integral formulae are given. We mention also the absolute and the relative derivatives.

2.1 Conservative vectors. Gradient

Let be a point $P(x_1, x_2, x_3) \in D \subset \mathbb{R}^3$. In what follows, the vector mapping $(x_1, x_2, x_3) \to \mathbf{V}(x_1, x_2, x_3)$ defines a *vector field* ($\mathbf{V} : D \to L$); the respective vectors are bound vectors, the points P being their points of application. This field is a *steady* one; if $\mathbf{V} = \mathbf{V}(x_1, x_2, x_3; t)$, then the vector field defined on $D \times [t_0, t_1]$ is *non-stationary*. We express the vector $\mathbf{V} = \mathbf{V}(x_1, x_2, x_3) = \mathbf{V}(\mathbf{r})$ in the form

$$\mathbf{V}(x_1, x_2, x_3) = V_j(x_1, x_2, x_3)\mathbf{i}_j, \qquad (A.2.1)$$

where $V_j \in C^1(D)$, $j = 1, 2, 3$; let us introduce the vector fields

$$\frac{\partial \mathbf{V}}{\partial x_i} = \mathbf{V}_{,i} = \frac{\partial V_j}{\partial x_i}\mathbf{i}_j = V_{j,i}\mathbf{i}_j, \quad i = 1, 2, 3, \qquad (A.2.2)$$

and the differential

$$d\mathbf{V} = \mathbf{V}_{,i}dx_i, \qquad (A.2.3)$$

where the index at the right of the comma specifies the derivative with respect to the corresponding variable. The curves for which the tangents at each point P are directed along the vectors $\mathbf{V} = \mathbf{V}(\mathbf{r})$ of the field are called *vector lines* (or *field lines*); the lines form a congruence of curves. Because the differential $d\mathbf{r}$ is tangent to these lines, their vector equation is of the form

$$\mathbf{V}(\mathbf{r}) \times d\mathbf{r} = 0; \qquad (A.2.4)$$

scalarly, these lines are given by a system of differential equations of first order

$$\frac{dx_1}{V_1} = \frac{dx_2}{V_2} = \frac{dx_3}{V_3}. \qquad (A.2.4')$$

Let be C a curve which is not a field line; on the basis of the theorem of existence and uniqueness for a system of differential equations of the form (A.2.4'), through each point of the curve C passes a field line (the integral curve of the system (A.2.4')). The surface generated by these curves is called *field surface*.

In the case of a non-steady field, *the differential* is of the form

$$d\mathbf{V} = \mathbf{V}_{,i}dx_i + \dot{\mathbf{V}}dt, \qquad (A.2.3')$$

where $\dot{\mathbf{V}} = \partial \mathbf{V}/\partial t$, and *the total derivative* is given by (we consider the mapping $t \to \mathbf{r}(t)$)

$$\frac{d\mathbf{V}}{dt} = \mathbf{V}_{,i}\dot{x}_i + \dot{\mathbf{V}}, \qquad (A.2.3'')$$

Appendix 711

being the sum of the *space* and *time derivatives*; in the case of a steady field remains only the space derivative.

In what follows we introduce some particular fields of vectors.

2.1.1 Conservative vectors. The nabla operator

Let us consider a scalar function $(x_1, x_2, x_3) \to U(x_1, x_2, x_3)$, $(x_1, x_2, x_3) \in D \subset \mathbb{R}^3$; the function $U \in C^1(D)$ defines a scalar field ($U : D \to \mathbb{R}^3$), because to each point P, which is of position vector $\mathbf{r}(x_1, x_2, x_3)$ one can associate the scalar $U = U(P)$. This scalar field is *steady*; we may consider also *non-steady* fields of the form $U(x_1, x_2, x_3; t)$, defined on $D \times [t_0, t_1]$. If it is necessary, the function U may have also continuous derivatives of higher order. Let be a vector field $\mathbf{V}$, defined by the relations

$$V_i = U_{,i}, \quad i = 1, 2, 3. \tag{A.2.5}$$

Considering a unit vector $\mathbf{n}(n_i)$, we notice that

$$\mathbf{V} \cdot \mathbf{n} = U_{,i} n_i = \frac{\partial U}{\partial n}; \tag{A.2.6}$$

hence, the components of $\mathbf{V}$ with respect to a new three-orthogonal trihedron of reference $Ox_1' x_2' x_3'$ are $\partial U / \partial x_i'$, $i = 1, 2, 3$. Thus, the definition given to the vector field does not depend on the chosen co-ordinate system. Such a field is called a *conservative field* (which derives from the potential U); the corresponding vectors are called *conservative vectors* and the field U is called *potential*. Assuming that $U = U(\mathbf{r}; t)$, one obtains a vector field defined by the same formulae (A.2.5); this is a *quasi-conservative field* and the corresponding vectors are *quasi-conservative vectors*. Analogously, the function U is called *quasi-potential*.

We notice that one can formally write $\mathbf{V} = U_{,j} \mathbf{i}_j = (\mathbf{i}_j \partial / \partial x_j) U$; applying the vector differential operator

$$\mathbf{\nabla} = \mathbf{i}_j \frac{\partial}{\partial x_j} = \mathbf{i}_j \partial_j, \tag{A.2.7}$$

which is called *nabla* (or *del*) and was introduced by Hamilton, we get

$$\mathbf{V} = \mathbf{\nabla} U. \tag{A.2.8}$$

In the case of a quasi-potential, *the differential* is of the form

$$\mathrm{d}U = U_{,j} \mathrm{d}x_j + \dot{U} \mathrm{d}t = \mathbf{\nabla} U \cdot \mathrm{d}\mathbf{r} + \dot{U} \mathrm{d}t, \tag{A.2.9}$$

where $\dot{U} = \partial U / \partial t$, while *the total (substantial) derivative* is given by (we consider the mapping $t \to \mathbf{r}(t)$)

$$\frac{\mathrm{d}U}{\mathrm{d}t} = U_{,j}\dot{x}_j + \dot{U} = \nabla U \cdot \dot{\mathbf{r}} + \dot{U} \qquad (\text{A.2.9'})$$

and is the sum of the *space derivative* and the *time derivative*; in the case of a potential remains only the space derivative.

We observe also that, in the case of a conservative vector field, *the elementary work of a conservative vector* is given by

$$\mathrm{d}W = \nabla U \cdot \mathrm{d}\mathbf{r} = \mathrm{d}U, \qquad (\text{A.2.10})$$

so that it is a total differential; it results

$$W_{\widehat{P_0 P_1}} = U(P_1) - U(P_0), \qquad (\text{A.2.10'})$$

where we started from the formula (A.1.28). Hence, in the case of a conservative vector, the work between two points does not depend on the path, but only on the values of the potential at its extremities; analogously, using the formula (A.1.29'), we notice that the work of a conservative vector on a closed curve (*circulation of a conservative vector*) vanishes if the corresponding domain D is simply connected.

2.1.2 Equipotential surfaces. Gradient

Let be

$$U(x_1, x_2, x_3) = C, \quad C = \text{const} \qquad (\text{A.2.11})$$

the equation of a surface, which is the locus of the points for which the scalar potential is constant; we assume that $U \in C^1$. This surface is called an *equipotential surface*; if $U(\mathbf{r})$ defines an arbitrary scalar field, then the surface (A.2.11) is called also a *level surface*. If $U = U(x_1, x_2, x_3; t)$, then we get an *equiquasi-potential surface*

$$U(x_1, x_2, x_3; t) = C, \quad C = \text{const}, \qquad (\text{A.2.11'})$$

which is a variable surface in time. Let us suppose that through the point $\left(x_1^0, x_2^0, x_3^0\right)$ pass two equipotential surfaces, so that for that point we may write $U\left(x_1^0, x_2^0, x_3^0\right) = C_1$, $U\left(x_1^0, x_2^0, x_3^0\right) = C_2$, $C_1, C_2 = \text{const}$; it follows that $0 = C_1 - C_2$. Hence, the two equipotential surfaces coincide, admitting that the function U is uniform. If two equipotential surfaces do not coincide, then they have not common points.

Applying the operator ∇ to a scalar function, one obtains a vector function which is called *the gradient* of the scalar function. Hence,

$$\mathrm{grad}\, U = \nabla U, \qquad (\text{A.2.12})$$

so that the operator ∇ transforms the scalar field in a vector one. Thus, the gradient allows to appreciate the variation of a scalar function, obtaining also its derivatives of

first order. We observe thus that a field of conservative vectors is a field of gradients. We mention the properties:
  i) $\mathrm{grad}(U_1 + U_2) = \mathrm{grad}\,U_1 + \mathrm{grad}\,U_2$ (distributivity with respect to addition of scalars);
  ii) $\mathrm{grad}\,CU = C\,\mathrm{grad}\,U$, $C = \mathrm{const}$;
  iii) $\mathrm{grad}\,C = \mathbf{0}$, $C = \mathrm{const}$.

The relation (A.2.6) may be written in the form

$$\mathbf{n} \cdot \mathrm{grad}\,U = \frac{\partial U}{\partial n}, \qquad (\mathrm{A}.2.6')$$

so that the gradient forms a vector field, independent on the chosen system of co-ordinate axes; $\partial U / \partial n$ is the derivative of the scalar field U in the direction of the unit vector $\mathbf{n}$. Because

$$\mathrm{grad}\,U = U_{,j}\mathbf{i}_j = \partial_j U \mathbf{i}_j, \qquad (\mathrm{A}.2.12')$$

it results that the gradient of the function U is normal to the equipotential surface $U(x_1, x_2, x_3) - C = 0$. As well, if we take $\mathbf{n} = \mathrm{vers}\,\mathrm{grad}\,U$ in the relation (A.2.12'), then the gradient of the scalar function U is a vector directed in the same direction for which the value of the function is increasing. The congruence of gradient lines is thus normal to the family of corresponding equipotential surfaces, the direction of travelling through being that in which the value of the scalar function U is increasing. These properties hold also for a quasi-conservative scalar field. If $\mathrm{d}\mathbf{r}$ is along the tangent to a curve C, which is travelled through by a point $P(x_1, x_2, x_3)$, then it results – by equating to zero the expression (A.2.9) – that, for a quasi-potential scalar field, the curve C cannot stay on an equiquasi-potential surface; but – in exchange – for a potential scalar field, the curve C may belong to a corresponding equipotential surface. We can verify the above mentioned properties, e.g., in the case of the scalar potential

$$U(x_1, x_2, x_3) = x_i x_i = x_1^2 + x_2^2 + x_3^2 \qquad (\mathrm{A}.2.13)$$

and in the case of the quasi-potential scalar

$$U(x_1, x_2, x_3; t) = x_i x_i - \frac{1}{c^2} t^2 = x_1^2 + x_2^2 + x_3^2 - \frac{1}{c^2} t^2, \quad c = \mathrm{const}. \qquad (\mathrm{A}.2.13')$$

The necessary and sufficient conditions for the scalar functions

$$V_i = V_i(x_1, x_2, x_3; t), \quad i = 1, 2, 3 \qquad (\mathrm{A}.2.14)$$

to be the components of a quasi-conservative vector are of the form

$$\epsilon_{ijk}\,\partial_j V_k = \epsilon_{ijk}\,V_{k,j} = 0, \quad i = 1, 2, 3; \qquad (\mathrm{A}.2.15)$$

they result from the relation (A.2.5) and from the condition that the mixed derivatives of second order of the quasi-potential $U = U(x_1, x_2, x_3; t)$ be immaterial on the order of differentiation (we assume that $U \in C^2(D)$, $(x_1, x_2, x_3) \in D \subset \mathbb{R}^3$, with respect to the space variables).

To a determined potential $U = U(x_1, x_2, x_3)$ one may add an arbitrary constant, without any change of its properties. We mention also the properties:

iv) $\operatorname{grad}(U_1 U_2) = U_1 \operatorname{grad} U_2 + U_2 \operatorname{grad} U_1$;

v) $\operatorname{grad} f(U) = f'(U) \operatorname{grad} U$, $f \in C^1$;

vi) $\operatorname{grad} f(U_1, U_2) = f'_{U_1} \operatorname{grad} U_1 + f'_{U_2} \operatorname{grad} U_2$;

vii) $f(U) \operatorname{grad} U = \operatorname{grad} \int f(U) \mathrm{d}U$, f integrable.

These properties hold for a quasi-potential field too. Concerning the position vector $\mathbf{r}$, we may write

$$\operatorname{grad} r = \frac{1}{r} \mathbf{r}, \quad \operatorname{grad} f(r) = \frac{f'(r)}{r} \mathbf{r}, \quad f \in C^1, \tag{A.2.16}$$

$$\operatorname{grad}(\mathbf{C} \cdot \mathbf{r}) = \mathbf{C}, \quad \mathbf{C} = \overline{\text{const}}. \tag{A.2.16'}$$

In the case of a potential $U = U(x_1, x_2, x_3) = U(\mathbf{r})$, the formula (A.2.9) allows to write the differential

$$\mathrm{d}U = \operatorname{grad} U \cdot \mathrm{d}\mathbf{r}. \tag{A.2.17}$$

Introducing the mapping $s \to \mathbf{r}(s)$, which defines a curve C, it results that $U = U(s)$, so that

$$\frac{\partial U}{\partial s} = \frac{\mathrm{d}U}{\mathrm{d}s} = \operatorname{grad} U \cdot \frac{\mathrm{d}\mathbf{r}}{\mathrm{d}s} = \operatorname{grad} U \cdot \boldsymbol{\tau}, \tag{A.2.18}$$

obtaining thus the derivative on the direction of the unit vector $\boldsymbol{\tau}$ of the tangent to this curve; the partial derivative is, in this case, equal to the total derivative. By means of the above introduced operator ∇, one may conceive the symbol d of the total differential as an operator, in the form of a scalar product

$$\mathrm{d} = \mathrm{d}\mathbf{r} \cdot \nabla = \mathrm{d}\mathbf{r} \cdot \operatorname{grad}. \tag{A.2.19}$$

The formula (A.2.6') leads to the operator derivative in the direction of the unit vector $\mathbf{n}$, which may be expressed in the form

$$\frac{\partial}{\partial n} = \mathbf{n} \cdot \nabla = \mathbf{n} \cdot \operatorname{grad}; \tag{A.2.20}$$

if $\mathbf{r} = \mathbf{r}(s)$ and $\mathbf{n} = \boldsymbol{\tau}$, then we get the operator

$$\frac{\partial}{\partial s} = \frac{\mathrm{d}}{\mathrm{d}s} = \mathbf{r}'(s) \cdot \boldsymbol{\nabla} = \boldsymbol{\tau} \cdot \mathrm{grad} \, . \tag{A.2.20'}$$

Thus, one introduces the linear scalar differential operator

$$\mathbf{A} \cdot \mathrm{grad} = \mathbf{A} \cdot \boldsymbol{\nabla} = A_i \frac{\partial}{\partial x_i} = A_i \partial_i \, , \tag{A.2.21}$$

where $\mathbf{A}$ is a constant or variable given vector. Applying this operator to the scalar field U, one obtains the scalar

$$(\mathbf{A} \cdot \mathrm{grad})U = (A_i \partial_i)U = A_i U_{,i} \tag{A.2.21'}$$

as well, but applying the operator to the vector field $\mathbf{V}$, it results the vector

$$(\mathbf{A} \cdot \mathrm{grad})\mathbf{V} = (A_j \partial_j)(V_k \mathbf{i}_k) = A_j V_{k,j} \mathbf{i}_k \, . \tag{A.2.21''}$$

In particular, we get the total differential of a vector field $\mathbf{V} = \mathbf{V}(x_1, x_2, x_3)$ in the form

$$\mathrm{d}\mathbf{V} = (\mathrm{d}\mathbf{r} \cdot \mathrm{grad})\mathbf{V} \, , \tag{A.2.22}$$

and the derivative in the direction of the unit vector $\mathbf{n}$ is given by

$$\frac{\partial \mathbf{V}}{\partial n} = (\mathbf{n} \cdot \mathrm{grad})\mathbf{V} \, ; \tag{A.2.23}$$

for $\mathbf{r} = \mathbf{r}(s)$ and $\mathbf{n} = \boldsymbol{\tau}$ one obtains

$$\frac{\partial \mathbf{V}}{\partial s} = \frac{\mathrm{d}\mathbf{V}}{\mathrm{d}s} = (\boldsymbol{\tau} \cdot \mathrm{grad})\mathbf{V} \, . \tag{A.2.23'}$$

In the system of curvilinear co-ordinates (q_1, q_2, q_3), introduced in Subsec. 1.1.5, the moduli of the gradients of the surfaces of co-ordinates $q_i = \mathrm{const}$, $i = 1, 2, 3$, are given by the differential parameters of the first order

$$|\mathrm{grad}\, q_i| = \sqrt{g^{ii}} = h_i \, (!), \quad i = 1, 2, 3 \, . \tag{A.2.24}$$

In orthogonal curvilinear co-ordinates, the gradient operator has the form

$$\mathrm{grad} = \frac{1}{H_1} \mathbf{i}_1 \frac{\partial}{\partial q_1} + \frac{1}{H_2} \mathbf{i}_2 \frac{\partial}{\partial q_2} + \frac{1}{H_3} \mathbf{i}_3 \frac{\partial}{\partial q_3} \, . \tag{A.2.25}$$

In spherical co-ordinates, we obtain

$$\operatorname{grad} = \mathbf{i}_r \frac{\partial}{\partial s_r} + \mathbf{i}_\theta \frac{\partial}{\partial s_\theta} + \mathbf{i}_\varphi \frac{\partial}{\partial s_\varphi} = \mathbf{i}_r \frac{\partial}{\partial r} + \frac{1}{r} \mathbf{i}_\theta \frac{\partial}{\partial \theta} + \frac{1}{r \sin \theta} \mathbf{i}_\varphi \frac{\partial}{\partial \varphi}, \qquad (A.2.25')$$

while in cylindrical co-ordinates we may write

$$\operatorname{grad} = \mathbf{i}_r \frac{\partial}{\partial s_r} + \mathbf{i}_\theta \frac{\partial}{\partial s_\theta} + \mathbf{i}_z \frac{\partial}{\partial s_z} = \mathbf{i}_r \frac{\partial}{\partial r} + \frac{1}{r} \mathbf{i}_\theta \frac{\partial}{\partial \theta} + \mathbf{i}_z \frac{\partial}{\partial z}. \qquad (A.2.25'')$$

2.2 Differential operators of first and second order

One may set up vector differential operators of first order and scalar differential operators of second order by means of the vector differential operator ∇; such operators may appear in problems of mechanics.

2.2.1 Vector differential operators of first order

Besides the scalar differential operator $\mathbf{A} \cdot \nabla$, we introduce also *the vector differential operator*

$$\mathbf{A} \times \nabla = \epsilon_{jkl} A_j \mathbf{i}_l \partial_k = \epsilon_{jkl} A_j \mathbf{i}_l \frac{\partial}{\partial x_k}, \qquad (A.2.26)$$

where $\mathbf{A}$ is a constant or variable given vector; applying this operator to the scalar field U, one obtains the vector

$$(\mathbf{A} \times \nabla) U = \epsilon_{jkl} A_j U_{,k} \mathbf{i}_l. \qquad (A.2.26')$$

The differential operator "$\nabla \cdot$" applied to a vector $\mathbf{V}$ defines *the divergence* of the vector; we have thus

$$\operatorname{div} \mathbf{V} = \nabla \cdot \mathbf{V} = \frac{\partial V_i}{\partial x_i} = \partial_i V_i = V_{i,i}. \qquad (A.2.27)$$

This is a scalar quantity, hence invariant to a change of co-ordinate axes. We mention the properties:
 i) $\operatorname{div}(\mathbf{V}_1 + \mathbf{V}_2) = \operatorname{div} \mathbf{V}_1 + \operatorname{div} \mathbf{V}_2$;
 ii) $\operatorname{div}(\lambda \mathbf{V}) = \lambda \operatorname{div} \mathbf{V} + \mathbf{V} \cdot \operatorname{grad} \lambda$, $\lambda = \lambda(\mathbf{r})$ scalar.
In orthogonal curvilinear co-ordinates, we obtain

$$\operatorname{div} \mathbf{V} = \frac{1}{H_1 H_2 H_3} \left[\frac{\partial}{\partial q_1}(H_2 H_3 V_1) + \frac{\partial}{\partial q_2}(H_3 H_1 V_2) + \frac{\partial}{\partial q_3}(H_1 H_2 V_3) \right]; \qquad (A.2.28)$$

in particular, in spherical co-ordinates, we have

$$\operatorname{div} \mathbf{V} = \frac{1}{r^2} \frac{\partial}{\partial r}(r^2 V_r) + \frac{1}{r \sin \theta} \frac{\partial}{\partial \theta}(\sin \theta V_\theta) + \frac{1}{r \sin \theta} \frac{\partial V_\varphi}{\partial \varphi}, \qquad (A.2.28')$$

while in cylindrical co-ordinates we may write

$$\operatorname{div} \mathbf{V} = \frac{1}{r}\frac{\partial}{\partial r}(rV_r) + \frac{1}{r}\frac{\partial V_\theta}{\partial \theta} + \frac{\partial V_z}{\partial z}. \qquad (A.2.28'')$$

The differential operator "$\nabla \times$" applied to a vector $\mathbf{V}$ leads to the *curl* of this vector; we may write

$$\operatorname{curl} \mathbf{V} = \nabla \times \mathbf{V} = \epsilon_{jkl}\frac{\partial V_k}{\partial x_j}\mathbf{i}_l = \epsilon_{jkl}\,\partial_j V_k \mathbf{i}_l = \epsilon_{jkl}\,V_{k,j}\mathbf{i}_l, \qquad (A.2.29)$$

this definition being immaterial of the co-ordinate axes too. We mention the properties:
  i) $\operatorname{curl}(\mathbf{V}_1 + \mathbf{V}_2) = \operatorname{curl}\mathbf{V}_1 + \operatorname{curl}\mathbf{V}_2$;
  ii) $\operatorname{curl}(\lambda\mathbf{V}) = \lambda\operatorname{curl}\mathbf{V} + \mathbf{V}\times\operatorname{grad}\lambda$, $\lambda = \lambda(\mathbf{r})$ scalar.
In orthogonal curvilinear co-ordinates we have

$$\operatorname{curl}\mathbf{V} = \frac{1}{H_2 H_3}\left[\frac{\partial}{\partial q_2}(H_3 V_3) - \frac{\partial}{\partial q_3}(H_2 V_2)\right]\mathbf{i}_1 + \frac{1}{H_3 H_1}\left[\frac{\partial}{\partial q_3}(H_1 V_1) - \frac{\partial}{\partial q_1}(H_3 V_3)\right]\mathbf{i}_2$$

$$+ \frac{1}{H_1 H_2}\left[\frac{\partial}{\partial q_1}(H_2 V_2) - \frac{\partial}{\partial q_2}(H_1 V_1)\right]\mathbf{i}_3; \qquad (A.2.30)$$

in particular, in spherical co-ordinates we may write

$$\operatorname{curl}\mathbf{V} = \frac{1}{r\sin\theta}\left[\frac{\partial}{\partial\theta}(\sin\theta V_\varphi) - \frac{\partial V_\theta}{\partial\varphi}\right]\mathbf{i}_r + \frac{1}{r}\left[\frac{1}{\sin\theta}\frac{\partial V_r}{\partial\varphi} - \frac{\partial}{\partial r}(rV_\varphi)\right]\mathbf{i}_\theta$$

$$+ \frac{1}{r}\left[\frac{\partial}{\partial r}(rV_\theta) - \frac{\partial V_r}{\partial\theta}\right]\mathbf{i}_\varphi, \qquad (A.2.30')$$

while in cylindrical co-ordinates we get

$$\operatorname{curl}\mathbf{V} = \left(\frac{1}{r}\frac{\partial V_z}{\partial\theta} - \frac{\partial V_\theta}{\partial z}\right)\mathbf{i}_r + \left(\frac{\partial V_r}{\partial z} - \frac{\partial V_z}{\partial r}\right)\mathbf{i}_\theta + \frac{1}{r}\left[\frac{\partial}{\partial r}(rV_\theta) - \frac{\partial V_r}{\partial\theta}\right]\mathbf{i}_z. \qquad (A.2.30'')$$

Concerning the operators grad, div and curl we mention the formulae

$$\operatorname{grad}(\mathbf{V}_1\cdot\mathbf{V}_2) = (\mathbf{V}_2\cdot\nabla)\mathbf{V}_1 + (\mathbf{V}_1\cdot\nabla)\mathbf{V}_2 + \mathbf{V}_1\times\operatorname{curl}\mathbf{V}_2 + \mathbf{V}_2\times\operatorname{curl}\mathbf{V}_1, \qquad (A.2.31)$$

$$\operatorname{div}(\mathbf{V}_1\times\mathbf{V}_2) = \mathbf{V}_2\cdot\operatorname{curl}\mathbf{V}_1 - \mathbf{V}_1\cdot\operatorname{curl}\mathbf{V}_2, \qquad (A.2.31')$$

$$\operatorname{curl}(\mathbf{V}_1\times\mathbf{V}_2) = (\mathbf{V}_2\cdot\nabla)\mathbf{V}_1 - (\mathbf{V}_1\cdot\nabla)\mathbf{V}_2 + \mathbf{V}_1\operatorname{div}\mathbf{V}_2 - \mathbf{V}_2\operatorname{div}\mathbf{V}_1. \qquad (A.2.31'')$$

In particular, the formula (A.2.31) leads to

$$\frac{1}{2}\operatorname{grad}\mathbf{V}^2 = (\mathbf{V}\cdot\nabla)\mathbf{V} + \mathbf{V}\times\operatorname{curl}\mathbf{V}. \qquad (A.2.32)$$

These results have been obtained by the methods of vector algebra, effecting formal calculations by means of the vector operator ∇.

A vector field for which

$$\operatorname{curl} \mathbf{V} = \nabla \times \mathbf{V} = \mathbf{0} \tag{A.2.33}$$

is called *irrotational*. We notice that a field of gradients

$$\mathbf{V} = \operatorname{grad} U = \nabla U \tag{A.2.33'}$$

is irrotational ($\operatorname{curl} \operatorname{grad} U = \mathbf{0}$); hence, the fields of quasi-conservative vectors (in particular, conservative) are irrotational, being the only ones which have the mentioned property. A vector field for which

$$\operatorname{div} \mathbf{V} = \nabla \cdot \mathbf{V} = 0 \tag{A.2.34}$$

is called *solenoidal*. It is easy to see that a field of curls

$$\mathbf{V} = \operatorname{curl} \mathbf{W} = \nabla \times \mathbf{W} \tag{A.2.34'}$$

is solenoidal ($\operatorname{div} \operatorname{curl} \mathbf{W} = 0$); one may show that this is the only vector field which has the property (A.2.34).

2.2.2 Absolute and relative derivatives

Let $\mathbf{V}$ be a vector expressed in the form $V'_j \mathbf{i}'_j$ with respect to a fixed orthonormed frame of reference $O' x'_1 x'_2 x'_3$ and in the canonical form $V_j \mathbf{i}_j$ with respect to a movable orthonormed frame $O x_1 x_2 x_3$. The mobility of the latter frame is characterized by the mappings $t \to \mathbf{i}_j(t)$, $t \in [t_0, t_1]$, $i = 1, 2, 3$; as well the vector V defines a vector field by the mapping $t \to \mathbf{V}(t)$. The independent variable t may be – eventually – the time. Differentiating with respect to t, one obtains

$$\frac{\mathrm{d}\mathbf{V}}{\mathrm{d}t} = \frac{\mathrm{d}V_j}{\mathrm{d}t} \mathbf{i}_j + V_j \frac{\mathrm{d}\mathbf{i}_j}{\mathrm{d}t} = \dot{V}_j \mathbf{i}_j + V_j \dot{\mathbf{i}}_j .$$

We denote

$$\frac{\partial \mathbf{V}}{\partial t} = \dot{V}_j \mathbf{i}_j , \tag{A.2.35}$$

that is *the relative derivative* of the vector $\mathbf{V}$ with respect to the movable frame. Let be the vector

$$\boldsymbol{\omega} = \frac{1}{2} \in_{jkl} (\dot{\mathbf{i}}_j \cdot \mathbf{i}_k) \mathbf{i}_l ; \tag{A.2.36}$$

Appendix 719

we notice that

$$\boldsymbol{\omega} \times \mathbf{V} = \epsilon_{lmn} \, \omega_l V_m \mathbf{i}_n = \frac{1}{2} \epsilon_{ljk} \epsilon_{lmn} \, V_m \left(\mathbf{i}_j \cdot \mathbf{i}_k \right) \mathbf{i}_n = V_j \left(\mathbf{i}_j \cdot \mathbf{i}_k \right) \mathbf{i}_k = V_j \mathbf{i}_j,$$

where we took into account a formula of the form (2.1.46') and the relation $\mathbf{i}_j \cdot \mathbf{i}_k + \mathbf{i}_j \cdot \dot{\mathbf{i}}_k = 0$ (consequence of the relation (2.1.14)). Finally, we obtain the relation

$$\frac{d\mathbf{V}}{dt} = \frac{\partial \mathbf{V}}{\partial t} + \boldsymbol{\omega} \times \mathbf{V}, \qquad (A.2.37)$$

where $d\mathbf{V}/dt$ is *the absolute derivative* of the vector $\mathbf{V}$ with respect to the fixed frame of reference. In particular, we obtain *Poisson's formulae* (the unit vectors $\mathbf{i}_k$ are fixed with respect to the movable frame, so that $\partial \mathbf{i}_k / \partial t = \mathbf{0}$)

$$\dot{\mathbf{i}}_k = \boldsymbol{\omega} \times \mathbf{i}_k, \quad k = 1,2,3. \qquad (A.2.38)$$

From the formula (A.2.37), which links the absolute derivative of a vector to its relative one, one sees that these derivatives are – in general – not equal. The two derivatives are equal only in the case in which $\boldsymbol{\omega} \times \mathbf{V} = \mathbf{0}$, hence if the vectors $\mathbf{V}$ and $\boldsymbol{\omega}$ are collinear or if $\boldsymbol{\omega} = \mathbf{0}$; this happens if $\dot{\mathbf{i}}_j = \mathbf{0}$, $j = 1,2,3$, hence if the movable frame moves without any rotation (the unit vectors $\mathbf{i}_j$, $j = 1,2,3$, are of constant direction). We state also that $d\boldsymbol{\omega}/dt = \partial \boldsymbol{\omega}/\partial t = \dot{\boldsymbol{\omega}}$.

2.2.3 Scalar differential operators of second order

We introduce *the scalar differential operator of second order*

$$\Delta = \boldsymbol{\nabla}^2 = \operatorname{div} \operatorname{grad} = \partial_i \partial_i = \frac{\partial}{\partial x_i} \frac{\partial}{\partial x_i} = \frac{\partial^2}{\partial x_1^2} + \frac{\partial^2}{\partial x_2^2} + \frac{\partial^2}{\partial x_3^2} \qquad (A.2.39)$$

called *the operator of Laplace* (*Laplacian*). If it is applied to a scalar $U = U(x_1, x_2, x_3)$, then one obtains

$$\Delta U = \boldsymbol{\nabla}^2 U = \partial_i \partial_i U = U_{,ii}, \qquad (A.2.39')$$

while it is applied to a vector, then we get

$$\Delta \mathbf{V} = \Delta V_j \mathbf{i}_j = V_{j,kk} \mathbf{i}_j. \qquad (A.2.39'')$$

A scalar function $U \in C^2(D)$, which verifies the equation

$$\Delta U = 0 \qquad (A.2.40)$$

in the domain D, is called *harmonic function* in this domain; analogously, a scalar function $U \in C^4(D)$, which verifies the equation

$$\Delta^2 U = \nabla^4 U = 0 \tag{A.2.41}$$

in the domain D, is called *biharmonic function* in this domain. One may introduce – on this way – *polyharmonic functions* too. In orthogonal curvilinear co-ordinates it results

$$\Delta = \frac{1}{H_1 H_2 H_3} \left[\frac{\partial}{\partial q_1} \left(\frac{H_2 H_3}{H_1} \frac{\partial}{\partial q_1} \right) + \frac{\partial}{\partial q_2} \left(\frac{H_3 H_1}{H_2} \frac{\partial}{\partial q_2} \right) + \frac{\partial}{\partial q_3} \left(\frac{H_1 H_2}{H_3} \frac{\partial}{\partial q_3} \right) \right]; \tag{A.2.42}$$

in particular, in spherical co-ordinates, we get

$$\Delta = \frac{1}{r^2} \frac{\partial}{\partial r} \left(r^2 \frac{\partial}{\partial r} \right) + \frac{1}{r^2 \sin \theta} \frac{\partial}{\partial \theta} \left(\sin \theta \frac{\partial}{\partial \theta} \right) + \frac{1}{r^2 \sin^2 \theta} \frac{\partial^2}{\partial \varphi^2}, \tag{A.2.42'}$$

while, in cylindrical co-ordinates, one has

$$\Delta = \frac{1}{r} \frac{\partial}{\partial r} \left(r \frac{\partial}{\partial r} \right) + \frac{1}{r^2} \frac{\partial^2}{\partial \theta^2} + \frac{\partial^2}{\partial z^2}. \tag{A.2.42''}$$

The operator of Laplace is of *elliptic type*. Analogously, we introduce *the operator of d'Alembert of hyperbolic type* (d'Alembertian) in the form

$$\Box_c = \Delta - \frac{1}{c^2} \frac{\partial^2}{\partial t^2}, \quad c = \text{const}; \tag{A.2.43}$$

assuming that $U = U(x_1, x_2, x_3; t)$, the equation

$$\Box_c U = 0, \tag{A.2.44}$$

where $U \in C^2(D)$, is called *the wave equation*, while c is the propagation velocity of these waves. Analogously, the equation

$$\Box_c \mathbf{V} = \mathbf{0}, \tag{A.2.44'}$$

where $\mathbf{V} = \mathbf{V}(x_1, x_2, x_3; t)$, corresponds to three scalar equations of wave propagation. If $U \in C^4(D)$, then we may introduce *the double wave equation*

$$\Box_1 \Box_2 U = 0 \tag{A.2.45}$$

too, where the operators

Appendix

$$\Box_i = \Delta - \frac{1}{c_i^2} \frac{\partial^2}{\partial t^2}, \quad c_i = \text{const}, \quad i = 1, 2, \quad \text{(A.2.45')}$$

correspond to two simple wave equations. Analogously, one may introduce functions which verify a *poly-wave equation*.

We introduce also *Nicolescu's caloric operator of parabolic type*

$$\underline{\Box} = \Delta - \frac{1}{a} \frac{\partial}{\partial t}, \quad a = \text{const}, \quad \text{(A.2.46)}$$

where a is *the thermic diffusivity*; assuming that $U = U(x_1, x_2, x_3; t)$, the equation

$$\underline{\Box} U = 0, \quad \text{(A.2.47)}$$

where $U \in C^2(D)$, is called *the caloric equation*; analogously, one may introduce *polycaloric functions*.

Between the differential operators grad, div, curl and Δ takes place the relation

$$\text{curl curl } \mathbf{V} = \text{grad div } \mathbf{V} - \Delta \mathbf{V}. \quad \text{(A.2.48)}$$

We notice the relations

$$\Delta(x_i U) = 2U_{,i} + x_i \Delta U, \quad i = 1, 2, 3, \quad \text{(A.2.49)}$$

$$\Delta(r^2 U) = 6U + 4\mathbf{r} \cdot \text{grad } U + r^2 \Delta U, \quad \text{(A.2.49')}$$

which, if U is a harmonic function, become

$$\Delta(x_i U) = 2U_{,i}, \quad i = 1, 2, 3, \quad \text{(A.2.50)}$$

$$\Delta(r^2 U) = 6U + 4\mathbf{r} \cdot \text{grad } U; \quad \text{(A.2.50')}$$

we mention the formula

$$\Delta(\mathbf{r} \cdot \mathbf{V}) = 2 \text{ div } \mathbf{V} + \mathbf{r} \cdot \Delta \mathbf{V}, \quad \text{(A.2.51)}$$

which, if the vector $\mathbf{V}$ is harmonic, becomes

$$\Delta(\mathbf{r} \cdot \mathbf{V}) = 2 \text{ div } \mathbf{V}. \quad \text{(A.2.51')}$$

Also, one verifies the relations

$$\Box_i (\mathbf{r} \cdot \mathbf{V}) = 2 \text{ div } \mathbf{V} + \mathbf{r} \cdot \Box_i \mathbf{V}, \quad i = 1, 2, \quad \text{(A.2.52)}$$

analogous to the formula (A.2.51), which – in the case when the vector $\mathbf{V}$ satisfies a simple wave equation – becomes

$$\square_i(\mathbf{r}\cdot\mathbf{V}) = 2\operatorname{div}\mathbf{V}, \quad i = 1,2. \tag{A.2.52'}$$

If

$$\mathbf{V} = \operatorname{curl}\mathbf{W}, \tag{A.2.53}$$

then the formulae (A.2.51) and (A.2.52) take the form

$$\Delta(\mathbf{r}\cdot\operatorname{curl}\mathbf{W}) = \mathbf{r}\cdot\Delta\operatorname{curl}\mathbf{W} = \mathbf{r}\cdot\operatorname{curl}\Delta\mathbf{W}, \tag{A.2.54}$$

$$\square_i(\mathbf{r}\cdot\operatorname{curl}\mathbf{W}) = \mathbf{r}\cdot\square_i\operatorname{curl}\mathbf{W} = \mathbf{r}\cdot\operatorname{curl}\square_i\mathbf{W}, \quad i = 1,2. \tag{A.2.54'}$$

2.2.4 Theorems of Almansi and Boggio type

Sometimes, the study of certain partial derivative equations may be reduced to the study of several equations of the same type but of a smaller order. We can thus state
Theorem A.2.1 (*of Almansi type*). *If $\mathscr{D}$ is a differential operator with constant coefficients, of order m, in s variables $q_1, q_2, \ldots, q_s$, and if we assume that $U = U(q_1, q_2, \ldots, q_s)$, then the solution of the equation*

$$\mathscr{D}^n U = 0, \tag{A.2.55}$$

where $U \in C^{mn}(D)$, may be written in the form

$$U = U_0 + q_1 U_1 + q_1^2 U_2 + \ldots + q_1^{n-1} U_{n-1}, \tag{A.2.55'}$$

where $U_0, U_1, \ldots, U_{n-1}$ are functions of the same variables, which verify the equations

$$\mathscr{D} U_i = 0, \quad i = 0, 1, 2, \ldots, n-1; \tag{A.2.55''}$$

obviously, the variable q_1 may be replaced by anyone of the other $n-1$ independent variables and the functions U_i can be of the class C^p, $p < mn$.

In particular, let be the biharmonic equation (A.2.41) and two harmonic functions U_1 and U_2 in the domain D, which satisfy the equation (A.2.40). The biharmonic function U may be expressed univocally by means of Almansi's formula

$$U = U_1 + x_1 U_2; \tag{A.2.56}$$

hence, a biharmonic function is – in a certain manner – equivalent to two harmonic functions. We notice that one may replace the variable x_1 by one of the variables x_2 or x_3; as well, we may write

$$U = U_1 + r^2 U_2. \tag{A.2.56'}$$

Analogously, let $\mathscr{D}_i$, $i = 1, 2, \ldots, p$, be differential operators of order m_i in the variables $q_1, q_2, \ldots, q_s$ and let be a function $U = U(q_1, q_2, \ldots, q_s)$; we may state

Theorem A.2.2 (*of Boggio type*). *If* $\mathscr{D}_i$ *are permutable differential operators, prime two by two,*

$$\mathscr{D}_i\mathscr{D}_j = \mathscr{D}_j\mathscr{D}_i, \quad i,j = 1,2,\ldots,p, \tag{A.2.57}$$

then the solution of the equation

$$\mathscr{D}_1\mathscr{D}_2\ldots\mathscr{D}_p U = 0, \tag{A.2.57'}$$

where $U \in C^{m_1+m_2+\ldots+m_p}(D)$, *may be written in the form*

$$U = U_1 + U_2 + \ldots + U_p, \tag{A.2.57''}$$

$U_1, U_2, \ldots, U_p$ *being functions of the same variables and the same class, which satisfy the equations*

$$\mathscr{D}_i U_i = 0, \quad i = 1,2,\ldots,p. \tag{A.2.57'''}$$

The condition (A.2.57) is fulfilled, for instance, in the case of operators with constant coefficients. In particular, observing that Laplace's operator (A.2.39) can be written in the form $\Delta = (x_1 + \mathrm{i}x_2)(x_1 - \mathrm{i}x_2)$, $\mathrm{i} = \sqrt{-1}$, in the two-dimensional case, a harmonic function can be written in the form

$$U = U_1 + U_2, \ U_1 = U_1(x_1 + \mathrm{i}x_2), \ U_2 = U_2(x_1 - \mathrm{i}x_2), \tag{A.2.40'}$$

obtained by a change of variable of the form $z = x_1 + \mathrm{i}x_2$, $\bar{z} = x_1 - \mathrm{i}x_2$, so that one gets the equations $\partial U_1/\partial \bar{z} = 0$ and $\partial U_2/\partial z = 0$, respectively.

In the case of the double wave equation (A.2.45) we obtain

$$U = U_1 + U_2, \tag{A.2.58}$$

where U_1 and U_2 verify the equations

$$\square_1 U_1 = 0, \quad \square_2 U_2 = 0, \tag{A.2.58'}$$

d'Alembert's operators being given by (A.2.45').

2.3 Integral formulae

We have considered discrete systems of bound and sliding vectors in Chap. 2, Sec. 2.2; in what follows, we will deal with continuous systems of such vectors. Besides, a continuous system of bound vectors is a field of vectors. We show thus the form taken in this case by the results in Chap. 2, Subsec. 2.2; we introduce also some integral formulae useful in practice.

2.3.1 Continuous systems of vectors

Let be a *continuous system of vectors* {**V**}, which may be bound or sliding ones; assuming that at any point P, of position vector $\mathbf{r}$, belonging to a domain D, is applied a vector $\mathbf{V}d\tau$, we may consider the vector field $\mathbf{V} = \mathbf{V}(\mathbf{r})$, proportional to the volume element $d\tau$ at any point of the domain. We introduce *the resultant* $\mathbf{R}$ of the system of vectors in the form of a free vector, given by

$$\mathbf{R} = \iiint_D \mathbf{V}(\mathbf{r})d\tau \qquad (A.2.59)$$

and *the resultant moment*, as a bound vector, applied at the pole O and given by

$$\mathbf{M}_O = \iiint_D \mathbf{r} \times \mathbf{V}(\mathbf{r})d\tau ; \qquad (A.2.60)$$

thus, we obtain *the torsor* $\tau_O\{\mathbf{V}\} = \{\mathbf{R}, \mathbf{M}_O\}$ of the continuous system of vectors. The components of the resultant $\mathbf{R}$ are given by

$$R_i = \iiint_D V_i(\mathbf{r})d\tau , \quad i = 1,2,3 , \qquad (A.2.59')$$

while the components of the resultant moment $\mathbf{M}_O$ are written in the form

$$M_{Ox_i} = \epsilon_{ijk} \iiint_D x_j V_k(\mathbf{r})d\tau , \quad i = 1,2,3 . \qquad (A.2.60')$$

A continuous system of sliding vectors {**V**} is equivalent to zero if and only if its torsor with respect to an arbitrary pole is equal to zero (condition (2.2.25)); this condition is only necessary in the case of a continuous system of bound vectors. Taking into account (A.2.59') and (A.2.60'), it results that this condition is equivalent to six equations of projection on the three axes of co-ordinates. As well, two continuous systems of sliding vectors {**V**} and {**V′**} are equivalent if and only if their torsors with respect to the same arbitrary pole are equal (condition (2.2.26)); this condition is only necessary in the case of a continuous system of bound vectors.

Analogous considerations can be made, in particular, for a continuous system of parallel or coplanar vectors. Let be, for instance, a *continuous system of parallel vectors*, the direction of which is specified by the unit vector $\mathbf{u}$; we may write

$$\mathbf{V}(\mathbf{r}) = V(\mathbf{r})\mathbf{u} , \qquad (A.2.61)$$

where $V(\mathbf{r})$ represents the component of the vector $\mathbf{V}(\mathbf{r})$ along the given direction. If

$$V = \iiint_D V(\mathbf{r})d\tau , \qquad (A.2.62)$$

then we obtain the resultant

Appendix

$$\mathbf{R} = V\mathbf{u}; \quad (A.2.62')$$

if $V \neq 0$, then we put in evidence the centre C of the continuous system of parallel vectors, specified by the position vector

$$\boldsymbol{\rho} = \frac{1}{V} \iiint_D V(\mathbf{r}) \mathbf{r} d\tau. \quad (A.2.63)$$

The properties emphasized for the point C remain valid for a continuous system of vectors too.

2.3.2 Stokes' formula

Let be a closed curve C, situated on the sufficiently smooth surface Σ, limiting on it a simply connected domain S (reducible – by continuous deformation – to a point, without leaving the surface Σ) (Fig.A.4,a). The surface Σ is oriented by means of the unit vector $\mathbf{n}$ of the normal to it; as well, we assume a direction of travelling through the curve C. Let us consider the vector mapping $(x_1, x_2, x_3) \to \mathbf{V}(x_1, x_2, x_3)$, with $\mathbf{V} \in C^1(D)$, where D is a domain which includes $S + C$. One may prove *Stokes' formula*

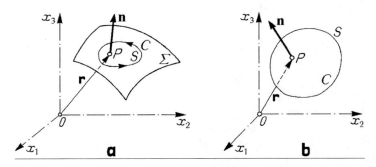

Figure A.4. The Stokes formula (a). The Gauss-Ostrogradskiĭ formula (b).

$$\oint_C \mathbf{V} \cdot d\mathbf{r} = \iint_S \mathbf{n} \cdot \operatorname{curl} \mathbf{V} dS; \quad (A.2.64)$$

because the left member of the formula depends only on the curve C, we may replace the surface S by any other surface $S_0 \subset D$, which satisfies analogous conditions.

As we have seen in Subsec. 1.1.4, the curvilinear integral is called *circulation* (it represents the work of a field of vectors), while the surface integral represents the *flux* of a field of curls; in the above mentioned conditions, it results that the circulation of a field of vectors along a closed curve is equal to the flux of the curl of the very same field of vectors through a sufficiently smooth arbitrary surface, bounded by the given curve. We also mention that the circulation of a field of irrotational vectors, hence of a field of conservative vectors, vanishes. With respect to the orthonormed frame of reference $Ox_1x_2x_3$, we may write

$$\oint_C V_i dx_i = \epsilon_{ijk} \iint_S n_i V_{k,j} dS; \qquad (A.2.64')$$

in particular, if $V_1 = F \in C^1(D)$, $V_2 = V_3 = 0$, then it results

$$\oint_C F dx_1 = \iint_S (n_2 F_{,3} - n_3 F_{,2}) dS. \qquad (A.2.65)$$

If in Stokes' formula (A.2.64) we concentrate all the surface at a point of position vector **r**, then we get

$$\mathrm{pr}_\mathbf{n}\, \mathrm{curl}\, \mathbf{V} = \mathbf{n} \cdot \mathrm{curl}\, \mathbf{V} = \lim_{S \to 0} \frac{\oint_C \mathbf{V} \cdot d\mathbf{r}}{S}. \qquad (A.2.66)$$

Hence, we may determine the projection of the vector $\mathrm{curl}\,\mathbf{V}$ on an arbitrary axis of unit vector **n** (hence the very same vector), without any reference to a frame; thus, the definition given to the curl in Subsec. 2.2.1 has an intrinsic value, being immaterial on the frame.

2.3.3 Gauss-Ostrogradskiĭ formula

Let be a sufficiently smooth closed surface S, limiting a domain D, and a field of vectors $\mathbf{V} = \mathbf{V}(\mathbf{r})$ (Fig.A.4,b); we assume that $\mathbf{V} \in C^1(D + S)$. We may write *the Gauss-Ostrogradskiĭ formula* in the form

$$\iint_S \mathbf{V} \cdot \mathbf{n} dS = \iiint_D \mathrm{div}\, \mathbf{V} d\tau, \qquad (A.2.67)$$

where **n** is the unit vector of the external normal to the surface; the surface integral represents the flux of the field of vectors through the surface S, so that the formula is called *the flux-divergence formula* too. The flux of a solenoidal vector vanishes. In the orthonormed frame $Ox_1 x_2 x_3$ we obtain

$$\iint_S V_j n_j dS = \iiint_D V_{j,j} d\tau. \qquad (A.2.67')$$

In particular, if we consider the component $V_j = F \in C^1(D + S)$, the other components vanishing, we may write

$$\iint_S F n_j dS = \iiint_D F_{,j} d\tau, \quad j = 1,2,3; \qquad (A.2.68)$$

multiplying by the unit vector $\mathbf{i}_j$ and summing, we get

$$\iint_S F \mathbf{n} dS = \iiint_D \mathrm{grad}\, F d\tau. \qquad (A.2.68')$$

If we take F of the form $\epsilon_{ijk} V_j$ in (A.2.68), we obtain

$$\iint_S \epsilon_{ijk}\, n_j V_k \mathrm{d}S = \iiint_D \epsilon_{ijk}\, V_{k,j}\mathrm{d}\tau,$$

so that

$$\iint_S \mathbf{n} \times \mathbf{V} \mathrm{d}S = \iiint_D \operatorname{curl} \mathbf{V} \mathrm{d}\tau. \tag{A.2.69}$$

Let be a domain $D' \subset D$, reducible by continuous deformation to a point of position vector $\mathbf{r}$ and S' the surface (sufficiently smooth) which bounds it. Starting from the Gauss-Ostrogradskiĭ formula (A.2.67), we may represent the divergence of the vector $\mathbf{V}$ in the form

$$\operatorname{div} \mathbf{V} = \lim_{D' \to 0} \frac{\iint_{S'} \mathbf{V} \cdot \mathbf{n} \mathrm{d}S}{D'}; \tag{A.2.70}$$

hence, the definition given in Subsec. 2.2.1 is immaterial of the frame and has an intrinsic value.

2.3.4 Green's formulae

The formula (A.2.67) leads to

$$\iint_S \frac{\partial U}{\partial n} \mathrm{d}S = \iiint_D \Delta U \mathrm{d}\tau \tag{A.2.71}$$

for a field of conservative vectors of the form (A.2.33'), where we took into account (A.2.6') and the definition of Laplace's operator. If $\mathbf{V} = \psi \mathbf{W}$, $\psi = \psi(x_1, x_2, x_3)$ scalar, then we may write

$$\iint_S \psi \mathbf{W} \cdot \mathbf{n} \mathrm{d}S = \iiint_D (\psi \operatorname{div} \mathbf{W} + \mathbf{W} \cdot \operatorname{grad} \psi) \mathrm{d}\tau, \tag{A.2.72}$$

where we used the property ii) of the divergence. If, after this, we take $\mathbf{W} = \operatorname{grad} \varphi$, $\varphi = \varphi(x_1, x_2, x_3)$ scalar, then we get

$$\iint_S \psi \frac{\partial \varphi}{\partial n} \mathrm{d}S = \iiint_D (\psi \Delta \varphi + \operatorname{grad} \varphi \cdot \operatorname{grad} \psi) \mathrm{d}\tau; \tag{A.2.73}$$

inverting φ and ψ and subtracting the relation thus obtained from (A.2.73), it results

$$\iint_S \left(\psi \frac{\partial \varphi}{\partial n} - \varphi \frac{\partial \psi}{\partial n}\right) \mathrm{d}S = \iiint_D (\psi \Delta \varphi - \varphi \Delta \psi) \mathrm{d}\tau. \tag{A.2.74}$$

The formulae (A.2.71), (A.2.73) and (A.2.74) are known as *Green's formulae*.

2.3.5 The differentiation formula for integrals which depend on a parameter

We consider, in what follows, integrals of functions which depend on a parameter t, very useful in applications. Let thus be the curvilinear integral on the curve C

$$I_C \equiv \int_C F_i(x_1, x_2, x_3; t) \mathrm{d}x_i ,\qquad (A.2.75)$$

where F_i, $i = 1, 2, 3$, are functions of class C^1 on the domain of definition, with respect both to the variables x_j, $j = 1, 2, 3$, and to the time t; we have

$$\frac{\mathrm{d}I_C}{\mathrm{d}t} = \int_C \frac{\mathrm{d}}{\mathrm{d}t} F_i(x_1, x_2, x_3; t) \mathrm{d}x_i .\qquad (A.2.75')$$

If $x_j = x_j(t)$, $j = 1, 2, 3$, then we may calculate the total derivative under the integral operator.

Analogously, for the surface integral

$$I_S \equiv \iint_S F(x_1, x_2, x_3; t) \mathrm{d}S \qquad (A.2.76)$$

we may write

$$\frac{\mathrm{d}I_S}{\mathrm{d}t} = \iint_S \frac{\mathrm{d}}{\mathrm{d}t} F(x_1, x_2, x_3; t) \mathrm{d}S .\qquad (A.2.76')$$

We notice that both the curve C and the surface S may be open or closed.

For the volume integral

$$I_D \equiv \iiint_D F(x_1, x_2, x_3; t) \mathrm{d}\tau \qquad (A.2.77)$$

we obtain, as well,

$$\frac{\mathrm{d}I_D}{\mathrm{d}t} = \iiint_D \frac{\mathrm{d}}{\mathrm{d}t} F(x_1, x_2, x_3; t) \mathrm{d}\tau .\qquad (A.2.77')$$

We obtain analogous results if $\mathbf{I}_S$ or $\mathbf{I}_D$ are vectors, the functions under the integral operator being vector functions.

If the curve C, the surface S or the volume D depend on the parameter t too, being variable quantities, then these formulae must be completed, taking into account the respective limits. Let thus be *the volume integral*

$$I_{D(t)} \equiv \iiint_{D(t)} F(x_1, x_2, x_3; t) \mathrm{d}\tau ,\qquad (A.2.78)$$

where $x_j = x_j(t)$, $j = 1, 2, 3$ and $\mathrm{d}\tau = \mathrm{d}x_1 \mathrm{d}x_2 \mathrm{d}x_3$. By a change of variable of the form $x_i = x_i(x_1^0, x_2^0, x_3^0; t)$, $i = 1, 2, 3$, including the condition that the Jacobian of the transformation be non-zero ($J \equiv \det[\mathrm{d}x_i / \mathrm{d}x_j^0] \neq 0$), we obtain

$$\frac{dI_{D(t)}}{dt} = \iiint_{D(t)} \left(\frac{dF}{dt} + F\operatorname{div}\mathbf{v}\right)d\tau = \iiint_{D(t)} \left[\frac{\partial F}{\partial t} + \operatorname{div}(F\mathbf{v})\right]d\tau, \quad (A.2.79)$$

where we have introduced the velocity $\mathbf{v} = d\mathbf{r}/dt$, we used the total derivative (A.2.9'), and we took into account the property ii) of the divergence. With the aid of the Gauss-Ostrogradskiĭ formula (A.2.67), we may also write

$$\frac{dI_{D(t)}}{dt} = \iiint_{D(t)} \frac{\partial F}{\partial t} d\tau + \iint_{S(t)} (\mathbf{v}\cdot\mathbf{n})F dS, \quad (A.2.79')$$

$\mathbf{n}$ being the external normal to the surface $S(t)$. In the case of a vector field

$$\mathbf{I}_{D(t)} \equiv \iiint_{D(t)} \mathbf{V}(\mathbf{r};t)d\tau, \quad (A.2.80)$$

we obtain, analogously,

$$\frac{d\mathbf{I}_{D(t)}}{dt} = \iiint_{D(t)} \left(\frac{d\mathbf{V}}{dt} + \mathbf{V}\operatorname{div}\mathbf{v}\right)d\tau = \iiint_{D(t)} \frac{\partial \mathbf{V}}{\partial t} d\tau + \iint_{S(t)} (\mathbf{v}\cdot\mathbf{n})\mathbf{V} dS. \quad (A.2.80')$$

For *the surface integral*

$$\mathbf{I}_{S(t)} \equiv \iint_{S(t)} \mathbf{V}(\mathbf{r};t)\cdot\mathbf{n} dS \quad (A.2.81)$$

one may show that

$$\frac{d\mathbf{I}_{S(t)}}{dt} = \iint_{S(t)} \left[\frac{\partial \mathbf{V}}{\partial t} + \mathbf{v}\operatorname{div}\mathbf{V} + \operatorname{curl}(\mathbf{V}\times\mathbf{v})\right]\cdot\mathbf{n} dS; \quad (A.2.81')$$

by means of Stokes' formula (A.2.64), we can write

$$\frac{d\mathbf{I}_{S(t)}}{dt} = \iint_{S(t)} \left(\frac{\partial \mathbf{V}}{\partial t} + \mathbf{v}\operatorname{div}\mathbf{V}\right)\cdot\mathbf{n} dS + \oint_{C(t)} (\mathbf{V}\times\mathbf{v})\cdot d\mathbf{r}. \quad (A.2.81'')$$

Let be *the curvilinear integral*

$$I_{C(t)} \equiv \iiint_{C(t)} F(\mathbf{r};t)ds \quad (A.2.82)$$

too, where $\mathbf{r} = \mathbf{r}(\lambda;t)$, $\lambda \in [\lambda_0,\lambda_1]$, is the parametric equation of the curve $C(t)$; we get

$$\frac{dI_{C(t)}}{dt} = \int_{C(t)} \left[\frac{\partial F}{\partial t} + \mathbf{v}\cdot\operatorname{grad} F + \frac{\frac{\partial \mathbf{r}}{\partial \lambda}\cdot\frac{\partial \mathbf{v}}{\partial \lambda}}{\left(\frac{\partial \mathbf{r}}{\partial \lambda}\right)^2}\right] ds. \quad (A.2.82')$$

2.3.6 Basic integral formula. Newtonian potentials

The third formula of Green (A.2.74) allows a study of Poisson's equation. To this goal, we write the respective formula for a function $u \in C^2(D)$ and for a function

$$v = \frac{1}{4\pi R}, \quad R = |\mathbf{r} - \boldsymbol{\xi}|, \qquad (A.2.83)$$

which satisfies the equation of Laplace (excepting the point $\mathbf{r} = \boldsymbol{\xi}$, which is a singular one) and represents the basic solution in the sense of the theory of distributions of Poisson's equation; the presence of the factor $1/4\pi$ is due to the fact that – in distributions – we have

$$\Delta v = -\delta(\mathbf{r}), \qquad (A.2.84)$$

where δ is Dirac's distribution. To calculate the integrals in the formula (A.2.74), we isolate the singular point $\mathbf{r} = \boldsymbol{\xi}$ by a sphere containing this point; the volume integral is calculated for the domain D without the interior of this sphere, while the surface integral is calculated for the sphere too. If the radius of the sphere, centred at the singular point, tends to zero, then we obtain *the basic integral formula*

$$u(\mathbf{r}) = -\frac{1}{4\pi}\iiint_D \frac{\Delta u}{R}\,\mathrm{d}\tau + \frac{1}{4\pi}\iint_S \left[\frac{1}{R}\frac{\partial u}{\partial n} - u\frac{\partial}{\partial n}\left(\frac{1}{R}\right)\right]\mathrm{d}S. \qquad (A.2.85)$$

This formula remains valid also for infinite domains if the function u is regular at infinity. It puts in evidence properties of Laplace's operator, to which it associates the identical operator and the normal derivative operator.

The formula (A.2.85) introduces three Newtonian potentials, linked by the basic formula, i.e.: *the volume potential*

$$F_1(\mathbf{r}) = \iiint_D \frac{f_1(\xi)}{R}\,\mathrm{d}\tau, \qquad (A.2.86)$$

the potential of simple stratum

$$F_2(\mathbf{r}) = \iint_S \frac{f_2(\xi)}{R}\,\mathrm{d}S \qquad (A.2.87)$$

and *the potential of double stratum*

$$F_3(\mathbf{r}) = -\iint_S f_3(\xi)\frac{\partial}{\partial n}\left(\frac{1}{R}\right)\mathrm{d}S, \qquad (A.2.87')$$

f_1, f_2, f_3 being the respective densities. The study of those potentials allows to find the solution of Poisson's equation if the value of the function or of its normal derivative is given on the frontier.

Appendix 731

3. Elements of theory of distributions

As it was shown in Chap. 1, Subsec. 1.1.7, in the study of discontinuous phenomena and for their representation in a unitary form, together with the continuous ones, it is necessary to use some notions of the theory of distributions. In what follows, we give some results concerning the composition of distributions and the integral transforms in distributions; as well, we introduce the notion of basic solution of a differential equation in the sense of the theory of distributions. These notions acquire thus a larger interest.

3.1 Composition of distributions

In general, the product of two distributions has no meaning; we have seen that the product by a function of class C^∞ has sense. That is why we will define products of a special type (composition of distributions). We introduce thus the direct (or tensor) product and the convolution product.

3.1.1 Direct product of two distributions

Let $x \equiv (x_1, x_2, ..., x_n)$ be a point of the *n*-dimensional Euclidean space X^n and $y \equiv (y_1, y_2, ..., y_m)$ a point of the *m*-dimensional Euclidean space Y^m; by direct Cartesian product $X^n \times Y^m$ of the two Euclidean spaces we mean a new *n+m*-dimensional Euclidean space, built up of the points $(x, y) \equiv (x_1, x_2, ..., x_n, y_1, y_2, ..., y_m)$, where – obviously – we have put in evidence the co-ordinates of a point of this space, in the order in which they have been written.

The direct product $f(x) \times g(y)$ of two distributions $f(x)$ and $g(y)$, defined on the basic spaces $K_x (x \in X^n)$ and $K_y (y \in Y^m)$, respectively, is given by the relation

$$(f(x) \times g(y), \varphi(x,y)) = (f(x), (g(y), \varphi(x,y))), \quad (A.3.1)$$

where $\varphi(x, y)$ is a basic function defined on $X^n \times Y^m$; this product is a distribution defined on the basic space $K_x \times K_y$. In the case of usual functions, this product coincides with their usual product. We mention the properties:
 i) $f(x) \times g(y) = g(y) \times f(x)$ *(commutativity)*;
 ii) $[f(x) \times g(y)] \times h(z) = f(x) \times [g(y) \times h(z)] = f(x) \times g(y) \times h(z)$
  (associativity).
The first of these properties allows to write the definition relation (A.3.1) also in the form

$$(f(x) \times g(y), \varphi(x,y)) = (g(y), (f(x), \varphi(x,y))). \quad (A.3.1')$$

The second property takes into account the fact that the direct product may be defined for an arbitrary finite number of distributions.

Let D_{x_1} and D_{x_2} be two differential operators with respect to the variables x_1 and x_2, respectively; we may write the relation

$$D_{x_1} D_{x_2} [f(x_1) \times g(x_2)] = D_{x_1} f(x_1) \times D_{x_2} g(x_2). \tag{A.3.2}$$

In particular, we get

$$\frac{\partial^2}{\partial x_1 \partial x_2}[\theta(x_1) \times \theta(x_2)] = \frac{d\theta(x_1)}{dx_1} \times \frac{d\theta(x_2)}{dx_2} = \delta(x_1) \times \delta(x_2) = \delta(x_1, x_2). \tag{A.3.3}$$

3.1.2 The convolution product of two distributions

Let $f(x)$ and $g(x)$ be locally integrable functions of x; their convolution product is the function defined by

$$f(x) * g(x) = \int_{-\infty}^{\infty} f(\xi) g(x - \xi) d\xi; \tag{A.3.4}$$

obviously, the definition remains valid for $x \in \mathbb{R}^n$. If the functions $f(x)$ and $g(x)$ are continuous, then their convolution product is a continuous function too. In order that the convolution product may exist, it is necessary that the functions $f(x)$ and $g(x)$ should satisfy certain conditions; thus, a sufficient condition in this respect is that the supports of the two functions $f(x)$ and $g(x)$ be compact.

If $f(x)$ and $g(x)$ are two distributions on $\mathbb{R}^n$, then their *convolution product* $f(x) * g(x)$ represents a new distribution on $\mathbb{R}^n$, defined by the formula

$$\begin{aligned}(f(x) * g(x), \varphi(x)) &= (f(x) \times g(y), \varphi(x + y)) \\ &= (f(x), (g(y), \varphi(x + y))) = (g(y), (f(x), \varphi(x + y)));\end{aligned} \tag{A.3.5}$$

this definition is reduced to the first one in case of usual functions. We may show that the convolution product has a meaning if one of the following conditions is satisfied:
  i) one of the distributions $f(x)$, $g(x)$ has a compact support;
  ii) the distributions $f(x)$ and $g(x)$ have the supports bounded on the same side.
Thus, if $f(x) = 0$ for $x < a$ and $g(x) = 0$ for $x < b$, then the supports of the two distributions are bounded at the left.

We remark that the convolution product may be defined for an arbitrary finite number of distributions.

Under the conditions required for the existence of the convolution product one may prove the properties:
  i) $f(x) * g(x) = g(x) * f(x)$ (*commutativity*);
  ii) $[f(x) * g(x)] * h(x) = f(x) * [g(x) * h(x)] = f(x) * g(x) * h(x)$
    (*associativity*).
We notice that

$$\delta(x) * f(x) = f(x) * \delta(x) = f(x); \quad (A.3.6)$$

hence, Dirac's distribution is a *unit element* for the product of convolution. We have, as well,

$$\delta(x - a) * f(x) = f(x) * \delta(x - a) = f(x - a). \quad (A.3.6')$$

If D is an arbitrary differential operator, then we may write

$$D[f(x) * g(x)] = Df(x) * g(x) = f(x) * Dg(x). \quad (A.3.7)$$

3.2 Integral transforms in distributions

A strong tool for the integration of differential equations is the method of integral transforms. We give, in the following, some general results concerning Fourier and Laplace transforms.

3.2.1 Fourier transform of a distribution

If $f(x)$ is a real or complex function of the real variable $x \in \mathbb{R}$, which satisfies Dirichlet's conditions (it is bounded, piecewise monotone and has at the most a finite number of points of discontinuity of the first species) and is absolutely integrable, then the function

$$F(u) = \int_{-\infty}^{\infty} f(x) e^{iux} dx, \quad i = \sqrt{-1}, \quad (A.3.8)$$

exists and is called *the Fourier transform of the function* $f(x)$; we shall write

$$F[f(x)] = F(u) = f(\tilde{u}), \quad (A.3.8')$$

noting that the variable u is real. In general, *the image function* $F(u)$ is complex, although the function $f(x)$ may be a real function. Assuming that the function $F(u)$ is given, the equality (A.3.8) may be considered as an integral equation with respect to the unknown function $f(x)$ under the integral symbol; the solution of this integral equation is written in the form

$$f(x) = \frac{1}{2\pi} \int_{-\infty}^{\infty} F(u) e^{-iux} du. \quad (A.3.9)$$

The function $f(x)$ is called *the inverse Fourier transform of the function* $F(u)$; we have

$$f(x) = F^{-1}[F[f(x)]] = F^{-1}[F(u)]. \quad (A.3.9')$$

Let $\varphi(x)$ be a basic complex function of a real variable x; hence, $\varphi(x) \in C^\infty$ and has a compact support (e.g., $|x| \leq a$). By the formula (A.3.8), the Fourier transform of the basic function is

$$\mathrm{F}[\varphi(x)] = \psi(u) = \int_{-\infty}^{\infty} \varphi(x) e^{iux} \mathrm{d}x. \tag{A.3.10}$$

The function $\psi(u)$ may be defined also for complex values $s = u + iv$, namely

$$\psi(s) = \int_{-\infty}^{\infty} \varphi(x) e^{isx} \mathrm{d}x = \int_{-\infty}^{\infty} \varphi(x) e^{-vx} e^{iux} \mathrm{d}x. \tag{A.3.10'}$$

The set of functions $\psi(s) = \mathrm{F}[\varphi(x)]$, where the support of the basic functions is included in the segment $[-a, a]$, forms the vector space $Z(a)$. We denote by

$$Z = \bigcup_a Z(a), \quad K = \bigcup_a K(a) \tag{A.3.11}$$

the new complex linear space; then, Z' is the set of linear and continuous functionals defined on Z (ultradistributions).

If $\mathrm{F}(s)$ is a distribution defined on Z and $f(x)$ is a distribution defined on K, then the functional $\mathrm{F}(s) \in Z'$, specified by the equality of the Parseval type

$$(\mathrm{F}(s), \psi(s)) = 2\pi(f(x), \varphi(x)), \tag{A.3.12}$$

is called *the Fourier transform of the distribution* $f(x)$ and is denoted by

$$\mathrm{F}(s) = \mathrm{F}[f(x)]. \tag{A.3.13}$$

We can also write

$$(\mathrm{F}[f(x)], \mathrm{F}[\varphi(x)]) = 2\pi(f(x), \varphi(x)). \tag{A.3.12'}$$

Analogously, one may introduce the Fourier transform of a distribution of several variables.

The classical properties of the Fourier transform are maintained in the form:

i) $P\left(\dfrac{\mathrm{d}}{\mathrm{d}s}\right) \mathrm{F}(s) = P\left(\dfrac{\mathrm{d}}{\mathrm{d}s}\right) \mathrm{F}[f(x)] = \mathrm{F}[P(ix)f(x)]$, P polynomial;

ii) $\mathrm{F}\left[P\left(\dfrac{\mathrm{d}}{\mathrm{d}x}\right) f(x)\right] = P(-is)\mathrm{F}[f(x)] = P(-is)\mathrm{F}(s)$;

iii) $\mathrm{F}^{-1}[\mathrm{F}[f(x)]] = f(x)$;

iv) $\mathrm{F}[\mathrm{F}[f(x)]] = 2\pi f(-x)$.

We denote by F^{-1} *the inverse operator* defined on Z'.

We can prove the relation

Appendix 735

$$F[f(x) \times g(y)] = F[f(x)] \times F[g(y)] \qquad (A.3.14)$$

for the direct product of two distributions. As well, in connection with the convolution product, one may show the relations

$$F[f(x) * g(x)] = F[f(x)]F[g(x)], \qquad (A.3.15)$$
$$F[f(x)g(x)] = F[f(x)] * F[g(x)]. \qquad (A.3.15')$$

The first relation is valid if $f(x) \in S'$ and $g(x)$ is a distribution with bounded support; the second relation is valid if $f(x) \in S'$ and the function $g(x) \in C^\infty$ is such that $f(x)g(x) \in S'$ and the support of its Fourier transform is bounded.

We mention the Fourier transforms:

$$F[\delta(x)] = 1, \qquad (A.3.16)$$
$$F[\delta(x_1, x_2, ..., x_n)] = 1, \qquad (A.3.16')$$
$$F[\delta(x - a)] = e^{iua}, \qquad (A.3.16'')$$
$$F[\theta(x)] = \pi\delta(u) + \frac{i}{u}, \qquad (A.3.17)$$
$$F[x_+] = -\frac{1}{u^2} - i\pi\delta'(u), \qquad (A.3.17')$$
$$F[1(x)] = 2\pi\delta(u). \qquad (A.3.17'')$$

3.2.2 Laplace transform of a distribution

Let $f(x)$ be a complex function of a real variable, which satisfies the conditions:
i) $f(x) = 0$ for $x < 0$;
ii) $f(x)$ is piecewise differentiable;
iii) $|f(x)| \leq Me^{ax}$, where M is a positive constant, while the non-negative constant a represents the incremental ratio of the function.

Then the function $L(p)$ of the complex variable $p = u + iv$, defined by the expression

$$L(p) = \int_0^\infty f(x)e^{-px}dx \qquad (A.3.18)$$

is called *the Laplace transform of the function* $f(x)$ and is denoted by

$$L[f(x)] = L(p). \qquad (A.3.18')$$

The function $f(x)$ is also called *the original function* and the function $L(p)$ *the image function*. To the Laplace transform thus defined there corresponds an inverse Laplace transform, given by

$$L^{-1}[L(p)] = f(x) = \frac{1}{2\pi i} \int_{u-i\infty}^{u+i\infty} L(p) e^{px} dp, \quad u > a. \tag{A.3.19}$$

If $f(x)$ is a distribution having its support on the half-line $x \geq 0$ and is such that the distribution $f(x)e^{-px}$ is a temperate distribution, then

$$L[f(x)] = (f(x), e^{-px}) \tag{A.3.20}$$

represents *the Laplace transform of that distribution*. It is obvious that the relation (A.3.20) generalizes the relation (A.3.18).

We mention the properties:

i) $L[f(x-a)] = e^{-pa} L[f(x)]$ *(the delay theorem)*;

ii) $L[f(kx)] = \frac{1}{k} L\left(\frac{p}{k}\right)$, $k > 0$ *(the theorem of similitude)*;

iii) $L[f(x)e^{qx}] = L(p-q)$ *(the theorem of translation; the damping theorem)*.

One may give analogous results for the distributions of several variables.

In the case of a derivative of a distribution one may write

$$L[f'(x)] = pL[f(x)]. \tag{A.3.21}$$

For a convolution product it results

$$L[f(x) * g(x)] = L[f(x)]L[g(x)]. \tag{A.3.22}$$

We may write the Laplace transforms:

$$L[\delta(x)] = 1, \tag{A.3.23}$$
$$L[\delta(x_1, x_2, \ldots, x_n)] = 1, \tag{A.3.23'}$$
$$L\left[\delta^{(m)}(x)\right] = p^m, \quad m = 0, 1, 2, \ldots. \tag{A.3.23''}$$

3.3 Applications to the study of differential equations. Basic solutions

The theory of distributions is particularly useful in the study of ordinary or partial differential equations, as well as in the case of various boundary value problems. We shall give first some general results concerning the basic solutions and then we shall deal with the problem of obtaining them for some particular differential equations.

3.3.1 Ordinary differential equations

Let be the *linear ordinary differential equation* with constant coefficients

$$Dy(x) \equiv y^{(n)}(x) + a_1 y^{(n-1)}(x) + \ldots + a_n y(x) = f(x), \tag{A.3.24}$$

where $f(x)$ is a distribution. The distribution $E(x)$ which satisfies the equation

Appendix 737

$$DE(x) = \delta(x) \tag{A.3.25}$$

is called *the basic solution* of the equation (A.3.24) and is of the form

$$E(x) = Y(x) + E_+(x), \tag{A.3.26}$$

where $Y(x)$ is *the general solution of the homogeneous equation*

$$DY(x) = 0, \tag{A.3.27}$$

while $E_+(x)$ is a *particular basic solution* (corresponding to the non-homogeneous equation (A.3.25)). We shall give a simple method for determining this solution; to this end, we determine first the solution $Y(x)$ which satisfies the initial conditions

$$Y(0) = Y'(0) = Y''(0) = \dots = Y^{(n-2)}(0) = 0, \quad Y^{(n-1)}(0) = 1; \tag{A.3.28}$$

one can prove that a basic particular solution is, in this case, given by

$$E_+(x) = Y(x)\theta(x). \tag{A.3.28'}$$

The basic solution of a differential equation is useful to determine its general solution; thus, *the general solution* of the equation (A.3.24) is given by

$$y(x) = E(x) * f(x). \tag{A.3.29}$$

Let now be again the equation (A.3.24) with $x > 0$, $f(x)$ being a continuous function having the support in $[0, \infty)$; in the case of *initial conditions of Cauchy type*

$$y^{(k)}(0) = y_k, \quad k = 0, 1, 2, \dots, n-1, \tag{A.3.30}$$

the solution of the equation (A.3.24) is expressed in the form

$$y(x) = E_+(x) * f(x)\theta(x) + \sum_{k=0}^{n-1} h_k \frac{\tilde{d}^k}{dx^k} E_+(x), \tag{A.3.30'}$$

where the coefficients h_k, $k = 0, 1, 2, \dots, n-1$, are given by

$$h_k = y_{n-k-1} + a_1 y_{n-k-2} + \dots + a_{n-k-1} y_0. \tag{A.3.30''}$$

The solution of the homogeneous equation

$$Dy(x) = 0 \tag{A.3.31}$$

for $x \geq 0$, with the initial conditions of Cauchy type (A.3.30), is given by

$$y(x) = \sum_{k=0}^{n-1} h_k \frac{\tilde{\mathrm{d}}^k}{\mathrm{d}x^k} E_+(x). \tag{A.3.31'}$$

In particular, in the case of the differential equation

$$y^{(n)}(x) = f(x), \tag{A.3.32}$$

with initial conditions of the form (A.3.30), we obtain the basic particular solution

$$E_+(x) = \theta(x) \frac{x^{n-1}}{(n-1)!} = \frac{1}{(n-1)!} x_+^{n-1}, \quad x \in \mathbb{R}; \tag{A.3.33}$$

the solution of the boundary value problem is given by

$$y(x) = y_0 + y_1 x + y_2 \frac{x^2}{2!} + \ldots + y_{n-1} \frac{x^{n-1}}{(n-1)!} + \frac{1}{(n-1)!} \int_0^x (x-\xi)^{n-1} f(\xi) \mathrm{d}\xi, \tag{A.3.32'}$$

where the latter integral, which represents the convolution product, is known as *the Cauchy formula*; for $n = 2$, we get

$$E_+(x) = x_+ . \tag{A.3.33'}$$

The above ideas may be extended to *systems of ordinary differential equations with constant coefficients*.

3.3.2 General considerations on partial differential equations

Problems similar to those in the preceding subsection may be put in the case of *partial differential equations*. Let thus be

$$P\left(\frac{\partial}{\partial x_1}, \frac{\partial}{\partial x_2}, \ldots, \frac{\partial}{\partial x_m}; \frac{\partial}{\partial t}\right) u(x_1, x_2, \ldots, x_m; t) = 0 \tag{A.3.34}$$

a homogeneous linear partial differential equation of *n*th order with respect to the variable t, with constant coefficients. For example, *the Cauchy problem* for this equation consists in the determination of the function $u(x_1, x_2, \ldots, x_m; t)$ which satisfies the equation (A.3.34) and the initial conditions

$$\begin{aligned} u(x_1, x_2, \ldots, x_m; t_0) &= u_0(x_1, x_2, \ldots, x_m), \\ \frac{\partial}{\partial t} u(x_1, x_2, \ldots, x_m; t_0) &= u_1(x_1, x_2, \ldots, x_m), \\ &\cdots\cdots\cdots\cdots\cdots\cdots\cdots\cdots\cdots\cdots \\ \frac{\partial^{n-1}}{\partial t^{n-1}} u(x_1, x_2, \ldots, x_m; t_0) &= u_{n-1}(x_1, x_2, \ldots, x_m). \end{aligned} \tag{A.3.34'}$$

To solve this problem, we consider the function

$$\bar{u}(x_1, x_2, \ldots, x_m; t) = u(x_1, x_2, \ldots, x_m; t)\theta(t - t_0), \qquad (A.3.35)$$

as well as the corresponding regular distribution; taking into account the formula which links the derivative in the sense of the theory of distributions to its derivative in the usual sense and using the initial conditions (A.3.34'), it results

$$\frac{\partial}{\partial t}\bar{u}(x_1, x_2, \ldots, x_m; t) = \frac{\tilde{\partial}}{\partial t}\bar{u}(x_1, x_2, \ldots, x_m; t) + u_0(x_1, x_2, \ldots, x_m)\delta(t - t_0),$$

$$\frac{\partial^2}{\partial t^2}\bar{u}(x_1, x_2, \ldots, x_m; t) = \frac{\tilde{\partial}^2}{\partial t^2}\bar{u}(x_1, x_2, \ldots, x_m; t) + u_1(x_1, x_2, \ldots, x_m)\delta(t - t_0)$$

$$+ u_0(x_1, x_2, \ldots, x_m)\dot{\delta}(t - t_0), \qquad (A.3.36)$$

. . . .

$$\frac{\partial^n}{\partial t^n}\bar{u}(x_1, x_2, \ldots, x_m; t) = \frac{\tilde{\partial}^n}{\partial t^n}\bar{u}(x_1, x_2, \ldots, x_m; t) + u_{n-1}(x_1, x_2, \ldots, x_m)\delta(t - t_0)$$

$$+ u_{n-2}(x_1, x_2, \ldots, x_m)\dot{\delta}(t - t_0) + \ldots + u_0(x_1, x_2, \ldots, x_m)\delta^{(n-1)}(t - t_0).$$

Noting that the derivatives in the usual sense with respect to the variable t of the function $\bar{u}(x_1, x_2, \ldots, x_m; t)$ are equal to the corresponding derivatives of the function $u(x_1, x_2, \ldots, x_m; t)$ for $t \geq t_0$, the equation (A.3.34) takes the form

$$P\left(\frac{\partial}{\partial x_1}, \frac{\partial}{\partial x_2}, \ldots, \frac{\partial}{\partial x_m}; \frac{\partial}{\partial t}\right)\bar{u}(x_1, x_2, \ldots, x_m; t) = f(x_1, x_2, \ldots, x_m; t) \qquad (A.3.37)$$

in distributions, where $f(x_1, x_2, \ldots, x_m; t)$ is a given distribution, which contains the initial conditions considered above.

Thus, we call basic solution of the equation (A.3.37) the distribution $E(x_1, x_2, \ldots, x_m; t)$ which satisfies the equation

$$P\left(\frac{\partial}{\partial x_1}, \frac{\partial}{\partial x_2}, \ldots, \frac{\partial}{\partial x_m}; \frac{\partial}{\partial t}\right)E(x_1, x_2, \ldots, x_m; t) = \delta(x_1, x_2, \ldots, x_m; t). \qquad (A.3.38)$$

The solution of the above Cauchy problem is given by (A.3.35), where

$$\bar{u}(x_1, x_2, \ldots, x_m; t) = E(x_1, x_2, \ldots, x_m; t) * f(x_1, x_2, \ldots, x_m; t), \qquad (A.3.37')$$

the convolution product corresponding to all $m + 1$ variables.

It should be noted that some equations of mathematical physics cannot be always deduced directly in the space of distribution, owing to the difficulties encountered in modelling physical phenomena. In general, the equations which describe such phenomena are obtained first by classical methods. Next, an extension is effected, where the unknown functions take zero values, so that they be defined on the whole space; the derivatives, considered in the usual sense, are expressed by relations which connect derivatives in the sense of the theory of distributions to the derivatives in the usual sense of a distribution corresponding to an almost everywhere continuous function, having a finite number of discontinuities of the first species. In this way, the unknowns of the problem will be regular distributions; then it will be assumed that these unknowns may be arbitrary distributions. Another possibility, which is frequently used is to suppose, from the very beginning, that the unknowns of the problem are arbitrary distributions, assuming the same form in distributions for the differential equation obtained by classical methods (obviously, these ones are no longer valid for the whole space). However, there are not general methods for passing to differential equations in distributions.

3.3.3 Equations of elliptic type

Let be *Poisson's equation*

$$\Delta u(x_1, x_2, x_3) = f(x_1, x_2, x_3),\qquad(\text{A.3.39})$$

where $f(x_1, x_2, x_3)$ is a given distribution; the basic solution is of the form

$$E(x_1, x_2, x_3) = -\frac{1}{4\pi r}, \quad r = \sqrt{x_1^2 + x_2^2 + x_3^2}.\qquad(\text{A.3.39'})$$

Analogously, for the equation

$$\Delta\Delta u(x_1, x_2, x_3) = f(x_1, x_2, x_3)\qquad(\text{A.3.40})$$

we have

$$E(x_1, x_2, x_3) = -\frac{1}{8\pi}r.\qquad(\text{A.3.40'})$$

In case of the equation

$$\Delta u(x_1, x_2, x_3) + k^2 u(x_1, x_2, x_3) = f(x_1, x_2, x_3)\qquad(\text{A.3.41})$$

we may use the basic solution

$$E(x_1, x_2, x_3) = -\frac{1}{4\pi r}\cos kr;\qquad(\text{A.3.41'})$$

we also notice that an integral of *the homogeneous Helmholtz equation*

$$\Delta u(x_1, x_2, x_3) + k^2 u(x_1, x_2, x_3) = 0\qquad(\text{A.3.42})$$

is given by

Appendix

$$u(x_1, x_2, x_3) = \frac{1}{r} \sin kr. \tag{A.3.42'}$$

Replacing k by $\pm ik$, $i = \sqrt{-1}$, we find the basic solution

$$E(x_1, x_2, x_3) = -\frac{1}{4\pi r} \cosh kr \tag{A.3.43}$$

for the equation

$$\Delta u(x_1, x_2, x_3) - k^2 u(x_1, x_2, x_3) = f(x_1, x_2, x_3); \tag{A.3.43'}$$

analogously, we notice that

$$u(x_1, x_2, x_3) = \frac{1}{r} \sinh kr \tag{A.3.44}$$

is an integral of the homogeneous equation

$$\Delta u(x_1, x_2, x_3) - k^2 u(x_1, x_2, x_3) = 0. \tag{A.3.44'}$$

In case of *Poisson's equation in two variables*

$$\Delta u(x_1, x_2) = f(x_1, x_2), \tag{A.3.45}$$

where $f(x_1, x_2)$ is a given distribution, a basic solution is

$$E(x_1, x_2) = -\frac{1}{2\pi} \ln \frac{1}{r}, \quad r = \sqrt{x_1^2 + x_2^2}. \tag{A.3.45'}$$

As well, we get the basic solution

$$E(x_1, x_2) = -\frac{1}{8} r^2 \ln \frac{1}{r} \tag{A.3.46}$$

for the equation

$$\Delta \Delta u(x_1, x_2) = f(x_1, x_2). \tag{A.3.46'}$$

3.3.4 Equations of hyperbolic type

We consider *the wave equations*

$$\Box_i u(x_1, x_2, x_3; t) = f_i(x_1, x_2, x_3; t), \quad i = 1, 2, \tag{A.3.47}$$

where $f_i(x_1, x_2, x_3; t)$ are given distributions; a basic solution of these equations is of the form

$$E_i(x_1,x_2,x_3;t) = -\frac{1}{4\pi r}\delta\left(t-\frac{r}{c_i}\right), \quad i=1,2. \tag{A.3.47'}$$

Analogously, for the equations

$$\Box_i u(x_1,x_2,x_3;t) + 4\pi\varkappa(t)\times\delta(x_1,x_2,x_3) = 0, \quad i=1,2, \tag{A.3.48}$$

where $\varkappa(t)$ is a distribution, one obtains

$$u(x_1,x_2,x_3;t) = \frac{1}{r}\varkappa\left(t-\frac{r}{c_i}\right), \quad i=1,2. \tag{A.3.48'}$$

The solution of the equation

$$\Box_1\Box_2 u(x_1,x_2,x_3;t) + 4\pi\varkappa(t)\times\delta(x_1,x_2,x_3) = 0 \tag{A.3.49}$$

reads

$$u(x_1,x_2,x_3;t) = -\frac{c_1^2 c_2^2}{c_1^2-c_2^2}\left[\left(t-\frac{r}{c_1}\right)_+ - \left(t-\frac{r}{c_2}\right)_+\right]\underset{(t)}{*}\varkappa(t), \tag{A.3.49'}$$

where the convolution product concerns only the time variable. The basic solution of the equation

$$\Box_1\Box_2 u(x_1,x_2,x_3;t) = f(x_1,x_2,x_3) \tag{A.3.50}$$

is of the form

$$E(x_1,x_2,x_3;t) = \frac{c_1^2 c_2^2}{4\pi(c_1^2-c_2^2)}\left[\left(t-\frac{r}{c_1}\right)_+ - \left(t-\frac{r}{c_2}\right)_+\right]. \tag{A.3.50'}$$

In the case of only two space variables, we consider the equations

$$\Box_i u(x_1,x_2;t) = f_i(x_1,x_2;t), \quad i=1,2, \tag{A.3.51}$$
$$\Box_1\Box_2 u(x_1,x_2;t) = f(x_1,x_2;t), \tag{A.3.52}$$

where $f_i(x_1,x_2;t)$ and $f(x_1,x_2;t)$ are given distributions. Introducing the distribution defined by the function

$$f_0\left(t;\frac{r}{c_i}\right) = \mathrm{L}^{-1}\left[K_0\left(p\frac{r}{c_i}\right)\right] = \frac{\theta\left(t-\frac{r}{c_i}\right)}{\sqrt{t^2-\frac{r^2}{c_i^2}}}, \quad t>0, \quad i=1,2, \tag{A.3.53}$$

where K_0 is *the modified Bessel function of order zero*, we obtain the basic solutions

$$E(x_1,x_2;t) = -\frac{1}{2\pi}f_0\left(t;\frac{r}{c_i}\right), \quad i=1,2, \qquad (A.3.51')$$

for the equations (A.3.51). As well, if we introduce the distribution defined by the function

$$f_{-2}\left(t;\frac{r}{c_i}\right) = L^{-1}\left[\frac{1}{p^2}K_0\left(p\frac{r}{c_i}\right)\right] = \theta\left(t-\frac{r}{c_i}\right)\left[t\ln\left(t+\sqrt{t^2-\frac{r^2}{c_i^2}}\right)\right.$$

$$\left.-\sqrt{t^2-\frac{r^2}{c_i^2}}\right], \quad t>0, \quad i=1,2, \qquad (A.3.54)$$

then we get the basic solution

$$E(x_1,x_2;t) = \frac{c_1^2 c_2^2}{2\pi(c_1^2-c_2^2)}\left[f_{-2}\left(t;\frac{r}{c_1}\right) - f_{-2}\left(t;\frac{r}{c_2}\right)\right], \qquad (A.3.52')$$

corresponding to the equation (A.3.52).

3.3.5 Equations of parabolic type

Let be *the caloric equation*

$$\square u(x_1,x_2,x_3;t) = f(x_1,x_2,x_3;t), \qquad (A.3.55)$$

where $f(x_1,x_2,x_3;t)$ is a given distribution; a basic solution is of the form

$$E(x_1,x_2,x_3;t) = \frac{1}{8\pi at\sqrt{\pi at}}\theta(t)e^{-\frac{r^2}{4at}}, \quad t \in (-\infty,\infty), \qquad (A.3.55')$$

or of the form

$$E(x_1,x_2,x_3;t) = \frac{1}{8\pi at\sqrt{\pi at}}e^{-\frac{r^2}{4at}}, \quad t \geq 0. \qquad (A.3.55'')$$

In case of two space variables, let be the equation

$$\square u(x_1,x_2;t) = f(x_1,x_2;t), \qquad (A.3.56)$$

where $f(x_1,x_2;t)$ is a given distribution; a basic solution is of the form

$$E(x_1,x_2;t) = \frac{1}{4\pi at}\theta(t)e^{-\frac{r^2}{4at}}, \quad t \in (-\infty,\infty), \qquad (A.3.56')$$

which can be expressed also in the form

$$E(x_1, x_2; t) = \frac{1}{4\pi at} e^{-\frac{r^2}{4at}}, \quad t \geq 0. \tag{A.3.56"}$$

Analogously, we can consider also other equations which occur in the study of mechanical phenomena.

REFERENCES

Abraham, R. and Marsden, J. (1967). *Foundations of mechanics*, W.A. Benjamin, Inc., New York, Amsterdam.
Abraham, R.H. and Shaw, C.D. (1982). *Dynamics: The geometry of behaviour*. Part one: *Periodic behaviour*. Aerial Press, Santa Cruz.
Abraham, R.H. and Shaw, C.D. (1983). *Dynamics: The geometry of behaviour*. Part two: *Chaotic behaviour*. Aerial Press, Santa Cruz.
Abraham, R.H. and Shaw, C.D. (1985). *Dynamics: The geometry of behaviour*. Part three: *Global behaviour*. Aerial Press, Santa Cruz.
Adhémar, R.d'. (1934). *La balistique extérieure*. Mém. des Sci. Math., fasc. LXV, Paris.
Agostinelli, C. and Pignedoli, A. (1963). *Meccanica razionale*. I, II. Zanichelli, Bologna.
Alembert, J. Le Rond d'. (1743). *Traité de dynamique*. Paris. (1921). Gauthier-Villars, Paris.
Ames, J.S. and Murnaghan, F.D. (1929). *Theoretical mechanics*. Ginn, Boston.
Anđelić, T. and Stojanović, R. (1965). *Racionalna mehanika* (*Rational mechanics*). Zavod. izd. udzbenika, Beograd.
Andonie, G.Şt. (1971). *Istoria matematicilor aplicate clasice din România. Mecanică şi astronomie* (*History of classical applied mathematics in Romania. Mechanics and astronomy*). Ed. Academiei, Bucureşti.
Andronov, A. and Chaikin, C. (1949). *Theory of oscillations*. Princeton Univ. Press, Princeton.
Andronov, A., Leontovich, E., Gordon, I. and Maier, A. (1973). *Theory of bifurcations of dynamical systems on a plane*. J. Wiley and Sons, New York.
Andronov, A., Vitt, A.A. and Haikin, S.E. (1959). *Teoriya kolebaniĭ* (*Theory of vibrations*). IInd ed. (1959). Fizmatgiz, Moskva.
Anosov, D.V. and Arnold, V.I. (ed.). (1988). *Dynamical systems*. I. *Ordinary differential equations and smooth dynamical systems*. Springer-Verlag, Berlin, Heidelberg, New York, Paris, Tokyo.
Appel, P. (1899). *Les mouvements de roulement en dynamique*. " Scientia" no. 4, Gauthier-Villars, Paris.
Appel, P. (1941-1953). *Traité de mécanique rationnelle*. I, II, 6th ed., Gauthier-Villars, Paris.
Appel, P. and Dautheville, S. (1928). *Précis de mécanique rationnelle*. Gauthier-Villars, Paris.
Argyris, J., Faust, G. and Haase, M. (1994). *Die Erforschung des Chaos. Eine Einführung für Naturwissenschaftler und Ingenieure*. (1994). Vieweg, Braunschweig.
Arnold, V.I. (1968). *Lektsii po klasicheskoĭ mehaniki* (*Lessons on classical mechanics*). (1968). Mosk. Gos. Univ., Moskva.
Arnold, V.I. (1976). *Méthodes mathématiques de la mécanique classique*. Mir, Moscou.
Arnold, V.I. (1983). *Geometrical methods in the theory of ordinary differential equations*. Springer-Verlag, New York, Heidelberg, Berlin.

Arnold, V.I. (1984). *Catastrophe theory*. Springer-Verlag, Berlin, Heidelberg, New York, Tokyo.
Arnold V.I. (ed.). (1988). *Dynamical systems*. III. *Mathematical aspects of classical and celestial mechanics*. Springer-Verlag, Berlin, Heidelberg, New York, London, Paris, Tokyo.
Arnold, V.I. and Avez, A. (1967). *Problèmes ergodiques de la mécanique classique*. Gauthier-Villars, Paris.
Arnold, V.I. and Maunder, L. (1961). *Gyrodynamics and its engineering applications*. Academic Press, New York.
Arnold, V.I. and Novikov, S.P. (ed.). (1990). *Dynamical systems*. IV. *Symplectic geometry and its applications*. Springer-Verlag, Berlin, Heidelberg, New York, London, Paris, Tokyo, Hong Kong.
Atanasiu, M. (1969). *Mecanică tehnică (Technical mechanics)*. Ed. Tehnică, București.
Atkin, R.H. (1959). *Classical dynamics*. J. Wiley and Sons, New York.
Babakov, I.M. (1968). *Teoriya kolebaniĭ (Theory of vibrations)*. Izd. Nauka, Moskva.
Babitsky, V.I. (1998). *Theory of vibro-impact systems and applications*. Springer, Berlin, Heidelberg, New York, Barcelona, Budapest, Hong Kong, London, Milan, Paris, Singapore, Tokyo.
Babuška, I., Prager, M. and Vitašek, E. (1966). *Numerical processes in differential equations*. SNTL, Publ. of Tech. Lit., Praha.
Baker, G.L. and Gollub, J.P. (1990). *Chaotic dynamics. An introduction*. Cambridge Univ. Press, Cambridge.
Baker, Jr., R.M.L. (1967). *Astrodynamics*. Academic Press, New York.
Banach, St. (1951). *Mechanics*. Warszawa-Wroclaw.
Barbu, V. (1985). *Ecuații diferențiale (Differential equations)*. Ed. Junimea, Iași.
Barenblatt, G.I., Iooss, G. and Joseph, D.D. (ed.). (1983). *Non-linear dynamics and turbulence*. Pitman, London.
Barnsley, M.F. (1988). *Fractals everywhere*. Academic Press, San Diego.
Barnsley, M.F. and Demko, S.G. (ed.). (1989). *Chaotic dynamics and fractals*. Academic Press, Boston, Orlando, San Diego, New York, Austin, London, Sydney, Tokyo, Toronto.
Bat, M.I., Djanelidze, G.Yu. and Kelzon, A.S. (1963-1973). *Teoreticheskaya mekhanika v primerakh i zadachakh (Theoretical mechanics in examples and problems)*. I-III. "Nauka", Moskva.
Bălan, Șt. (1969). *Lecții complementare de mecanică (Complementary lessons on mechanics)*. Ed. Did. Ped., București.
Bălan, Șt. (1972). *Culegere de probleme de mecanică (Collection of problems on mechanics)*. Ed. Did. Ped., București.
Bălan, Șt. and Ivanov, I. (1966). *Din istoria mecanicii (From the history of mechanics)*. Ed. Științ., București.
Beju, I., Soós, E. and Teodorescu, P.P. (1983). *Euclidean tensor calculus with applications*. Ed. Tehnică, București. Abacus Press, Tunbridge Wells.
Beju, I., Soós, E. and Teodorescu, P.P. (1983). *Spinor and non-Euclidean tensor calculus with applications*. Ed. Tehnică, București. Abacus Press, Tunbridge Wells.
Belenkiĭ, I.M. (1964). *Vvedenie v analiticheskuyu mekhaniku (Introduction to analytical mechanics)*. "Vysshaya shkola", Moskva.
Bellet, D. (1988). *Cours de mécanique générale*. Capadues, Paris.
Bellman, R. (1953). *Stability theory of differential equations*. McGraw-Hill, New York.
Beltrami, E. (1987). *Mathematics for dynamic modelling*. Academic Press, Boston, Orlando, San Diego, New York, Austin, London, Sydney, Tokyo, Toronto.
Berezkin, E.N (1968). *Lektsii po teoreticheskoĭ mekhanike. (Lessons on theoretical mechanics)*. II. Izd. Mosk. Univ., Moskva.

Berge, P., Pomeau, Y. and Vidal, Ch. (1984). *L'ordre dans le chaos*. Hermann, Paris.
Bergmann, P.G. (1949). *Basic theories of physics: Mechanics and electrodynamics*. Prentice-Hall, New York.
Bernousson, J. (1977). *Point mapping stability*. Pergamon Press, Oxford.
Béghin, H. (1921). *Statique et dynamique*. I, II. Armand Colin, Paris.
Béghin, H. (1948). *Cours de mécanique*. Gauthier-Villars, Paris.
Biezeno, C.B. and Grammel, R. (1953). *Technische Dynamik*. I, II. J. Springer, Berlin.
Birkhoff, G.D. (1927). *Dynamical systems*. J.Springer, Berlin.
Bishop, A.R., Gruner, G. and Nicolaenko, B. (1986). *Spatio-temporal coherence and chaos in physical systems*. North Holland, Amsterdam, Oxford, NewYork, Tokyo.
Boerner, H. (1963). *Representation of groups*. North Holland, Amsterdam.
Bogoliubov, N. and Mitropolsky, Y. (1961). *Asymptotic methods in the theory of nonlinear oscillations*. Gordon and Breach, New York.
Bolotin, V.V. (1978). *Vibratsii v tekhnike* (*Vibrations in technics*). Mashinostroienie, Moskva.
Boltzmann, L. (1904). *Vorlesungen über die Prinzipien der Mechanik*. II. J. Barth, Leipzig.
Bone, J., Morel, J. and Boucher, M. (1986). *Mécanique générale*. Dunod, Paris.
Born, M. (1925). *Vorlesungen über Atommechanik*. J. Springer, Berlin.
Borş, C.I. (1983, 1987). *Lecţii de mecanică* (*Lessons on mechanics*).I, II. Univ. "Al. I. Cuza", Iaşi.
Bouligand, G. (1924). *Leçons de géométrie vectorielle*. Vuibert, Paris.
Bourlet, C.E.E. (1898). *Traité des bicycles et biciclettes*. I. Gauthier-Villars, Paris.
Bradbury, T. (1968). *Theoretical mechanics*. J. Wiley and Sons, New York, London, Sydney.
Bratu, P. (2000). *Vibraţiile sistemelor elastice* (*Vibrations of elastic systems*). Ed. Tehnică, Bucureşti.
Bratu, P. (2006). *Mecanica teoretică* (*Theoretical mechanics*). Impuls, Bucureşti.
Brădeanu, P., Pop, I. and Brădeanu, D. (1979). *Probleme şi exerciţii de mecanică teoretică* (*Problems and exercises on theoretical mechanics*). Ed. Tehnică, Bucureşti.
Brelot, M. (1945). *Les principes mathématiques de la mécanique classique*. Arthaud, Paris.
Bricard, R. (1926, 1927). *Leçons de cinématique*. I, II. Gauthier-Villars, Paris.
Brillouin, L. (1938). *Les tenseurs en mécanique et en élasticité*. Masson, Paris.
Brousse, P. (1968). *Mécanique*. Armand Colin, Paris.
Brousse, P. (1973). *Cours de mécanique*. Armand Colin, Paris.
Budó, A. (1965). *Theoretische Mechanik*. VEB Deutscher Verlag der Wiss., Berlin.
Bukhgolts, N.N. (1966). *Osnovnoi kurs teoreticheskoĭ mekhaniki* (*Basic lessons on theoretical mechanics*). II. "Nauka", Moskva.
Bulgakov, B.V. (1954). *Kolebaniya* (*Vibrations*). Gostekhizdat, Moskva.
Bulgakov, B.V. (1955). *Prikladnaya teoriya giroskopa* (*Applied theory of the gyroscope*). IInd ed. Gostekhizdat, Moskva.
Burileanu, Şt. (1942, 1944). *Curs de mecanică raţională* (*Lessons on rational mechanics*). I, II. Socec, Bucureşti.
Butenin, N.V. (1971). *Vvedenie v analiticheskuyu mekhaniku* (*Introduction to analytical mechanics*). "Nauka", Moskva.
Butterfield, H. (1965). *The origins of modern science*, 1300-1800. Macmillan, New York.
Buzdugan, Gh., Fetcu, L. and Radeş, M. (1975). *Vibraţiile sistemelor mecanice* (*Vibrations of mechanical systems*). Ed. Academiei, Bucureşti.
Buzdugan, Gh., Fetcu, L. and Radeş, M. (1979). *Vibraţii mecanice* (*Mechanical vibrations*). Ed. Did. Ped., Bucureşti.
Buzdugan, Gh., Mihăilescu, E. and Radeş, M. (1979). *Măsurarea vibraţiilor* (*Vibrations measurement*). Ed. Academiei, Bucureşti.

Cabannes, H. (1966). *Cours de mécanique générale*. Dunod, Paris.
Caldonazzo, B. (1953). *Meccanica razionale*. Ed. Universitaria, Firenze.
Cartan, E. (1966). *The theory of spinors*. Hermann, Paris.
Cayley, A. (1890). *Collected papers*. II. Cambridge Univ. Press, Cambridge.
Chaplygin, S.A. (1948-1950). *Sobranie sochineniya* (*Collected papers*). I-III. Gostekhizdat, Moskva-Leningrad.
Chaplygin, S.A. (1954). *Izbrannye trudy po mekhanike I matematike* (*Selected works on mechanics and mathematics*). Gostekhizdat, Moskva.
Charlier, C.L. (1902). *Die Mechanik des Himmels*. I, II. V. Veit, Leipzig.
Chazy, J. (1947-1948). *Cours de mécanique rationelle*. IIIrd ed. Gauthier-Villars, Paris.
Chetaev, N.G. (1963). *The stability of motion*. Pergamon Press, Oxford.
Chiroiu, V. and Chiroiu, C. (2003). *Probleme inverse în mecanică* (*Inverse problems in mechanics*). Ed. Academiei, București.
Chiroiu, V., Munteanu, L., Știucă, P. and Donescu, Șt. (2005). *Introducere în nanomecanică* (*Introduction to nanomechanics*). Ed. Academiei, București.
Chow, S.N. and Hale, J.K. (1982). *Methods of bifurcation theory*. Springer-Verlag, New York, Heidelberg, Berlin.
Coe, C.J. (1938). *Theoretical mechanics*. Macmillan, New York.
Collatz, L. (1966). *The numerical treatment of differential equations*. Springer-Verlag, Berlin.
Coppel, W.A. (1978). *Dichotomies in stability theory*. In: Springer Lecture Notes in Math. vol. 629. Springer-Verlag, New York, Heidelberg, Berlin.
Corben, H.C. and Stehle, P. (1950). *Classical mechanics*. J. Wiley and Sons, New York, Chapman and Hall, London.
Courant, R. and Hilbert, D. (1931-1937). *Methoden der mathematischen Physik*. I, II. IInd ed., J. Springer, Berlin.
Crandall, S.H. (1958, 1963). *Random vibrations*. I, II. Prentice Hall, Engelwood Cliffs.
Crăciun, E.M. (1999). *Astrodinamică* (*Astrodynamics*). Ovidius Univ. Press, Constanța.
Curtu, R. (2000). *Introducere în teoria sistemelor dinamice* (*Introduction to the theory of dynamical systems*). Infomarket, Brașov.
Cushing, J.T. (1998). *Philosophical concepts in physics*. The Press Syndicate of the Univ. of Cambridge, Cambridge.
Darboux, G. (1884, 1886). *Cours de mécanique par Despeyrous, avec des notes*. I, II. Hermann, Paris.
Darboux, G. (1887-1896). *Leçons sur la théorie générale des surfaces*. I-IV. Gauthier-Villars, Paris.
Davidson, M. (ed.). (1946). *The gyroscope and its applications*. Hutchinson's Sci. and Techn. Publ., London.
Davis Jr., L., Follin Jr., J.M. and Blitzer, L. (1958). *The exterior ballistics of rockets*. Van Nostrand, Princeton, Toronto, New York, London.
Delassus, E. (1913). *Leçons sur la dynamique des systèmes matériels*. Hermann, Paris.
Destouches, J.L. (1948). *Principes de la mécanique classique*. Ed. C.N.R.S., Paris.
Devaney, R. (1986). *An introduction to chaotic dynamical systems*. Benjamin, Cummings, Menlo Park.
Dincă, F. and Teodosiu, C. (1973). *Nonlinear random vibrations*. Ed. Academiei, București, Academic Press, New York, London.
Doblaré, M., Correas, J.M., Alarcón, E., Gavete, L. and Pastor, M. (ed.). (1996). *Métodos numéricos en ingeniería*. 1, 2. Artes Gráficas Torres, Barcelona.
Dobronravov, V.V. (1976). *Osnovy analiticheskoĭ mekhaniki* (*Fundamentals of analytical mechanics*). "Vysshaya shkola", Moskva.
Dobronravov, V.V., Nikitin, N.N. and Dvor'nkov, A.L. (1974). *Kurs teoreticheskoĭ mekhaniki* (*Lessons on theoretical mechanics*). IIIrd ed. "Vysshaya shkola", Moskva.

Dolapčev, Bl. (1966). *Analitična mehanika (Analytical mechanics)*. Nauka i iskustvo,Sofia.
Dragnea, O. (1956). *Momente de inerţie (Moments of inertia)*. Ed. Tehnică, Bucureşti.
Dragoş, L. (1976). *Principiile mecanicii analitice (Principles of analytical mechanics)*. Ed. Tehnică, Bucureşti.
Dragoş, L. (1983). *Principiile mecanicii mediilor continue (Principles of mechanics of continuous media)*. Ed. Tehnică, Bucureşti.
Draper, C.S. and Horovka, J. (1962). *Inertial guidance*. Pergamon Press, New York.
Drăganu, M. (1957). *Introducere matematică în fizica teoretică modernă (Mathematical introduction to modern theoretical physics)*. I. Ed. Tehnică, Bucureşti.
Drâmbă, C. (1958). *Elemente de mecanică cerească (Elements of celestial mechanics)*. Ed. Tehnică, Bucureşti.
Duffing, G. (1918). *Erzwungene Schwingungen bei veränderlicher Eigenfrequenz*. Vieweg, Braunschweig.
Dugas, R. (1950). *Histoire de la mécanique*. Dunod, Paris.
Eckhaus, W. (1965). *Studies in non-linear stability theory*. In: Springer Tracts in Nat. Phil. vol. 6. Springer-Verlag, New York.
Eddington, A.S. (1921). *Espace, temps et gravitation*. Hermann, Paris.
Edgar, G.A. (1990). *Measure, topology and fractal geometry*. Springer-Verlag, New York.
El Naschie, M.S. (1990). *Stress stability in Chaos*. McGraw-Hill, London.
Emmerson, J.Mc L. (1972). *Symmetry principles in particle physics*. Clarendon Press, Oxford.
Eringen, A.C. (1967). *Mechanics of continua*. J. Wiley and Sons, New York.
Euler, L. (1736). *Mechanica sive motus scientia analytice exposity*. St. Petersburg.
Euler, L. (1765). *Teoria motus corporum solidorum sen rigidorum*. Greifswald.
Euler, L. (1955-1974). *Opera omnia*. Ser. secunda. *Opera mechanica ed astronomica*. Orell Füsli Turici, Lausanne.
Falconer, K.J. (1990). *Fractal geometry – Foundations and applications*. J. Wiley and Sons, Chichester.
Falk, G. and Ruppel, W. (1973). *Mechanik, Relativität, Gravitation*. Springer-Verlag, Berlin, Göttingen, Heidelberg.
Fernandez, M. and Macomber, G.R. (1962). *Inertial guidance*. Engineering. Prentice-Hall, Englewood-Cliffs.
Fertis, D.G. (1998). *Nonlinear mechanics*. IInd ed. C R C Press, Boca Raton, London, New York, Washington.
Feynman, R.P. (1963). *Lectures on physics*. I. *Mechanics, radiation and heat*. Addison-Wesley, Reading.
F.G.-M. (1919). *Cours de mécanique*. A. Mame et fils, Tours, Paris.
Filippov, A.P. (1965). *Kolebaniya mekhanicheskih sistem (Vibrations of mechanical systems)*. "Nauka Dumka", Kiev.
Finkelshtein, G.M. (1959). *Kurs teoreticheskoĭ mekhaniki (Lectures on theoretical mechanics)*. Gostekhizdat, Moskva.
Finzi, B. (1966). *Meccanica razionale*. I, II. Zanichelli, Bologna.
Finzi, B. and Pastori, M. (1961). *Calcolo tensoriale e applicazioni*. Zanichelli, Bologna.
Fischer, U. and Stephan, W. (1972). *Prinzipien und Methoden der Dynamik*. VEB Fachbuchverlag, Leipzig.
Föppl, A. (1925). *Vorlesungen über technische Mechanik*. I. *Einführung in die Mechanik*. 8th ed. B.G. Teubner, Leipzig-Berlin.
Frank, Ph. (1929). *Analytische Mechanik. Die Differential - und Integralgleichungen der Physik*. 2. Vieweg, Braunschweig.
Frank, Ph. and Mises, R. von. (1930, 1935). *Die Differential - und Integralgleichungen der Mechanik und Physik*. I, II. IInd ed. Vieweg, Braunschweig.

French, A. (1971). *Newtonian mechanics*. W.W. Norton, New York.
Gabos, Z., Mangeron, D. and Stan, I. (1962). *Fundamentele mecanicii* (*Fundamentals of mechanics*). Ed. Academiei, București.
Galileo, Galilei (1638). *Discorsi e dimostrazioni matematiche intorno a due nuove scienze attenenti alla meccanica ed ai movimenti locali*. Leiden.
Gallavotti, G. and Zweifel, P.F. (ed.). (1988). *Non-linear evolution and chaotic phenomena*. Plenum Press, New York, London.
Gantmacher, F. (1970). *Lectures on analytical mechanics*. "Mir", Moscow.
Gelfand, I.M. and Fomin, S.V. (1961). *Variatsionnoie ischislenie* (*Variational calculus*). Gos. Izd. Fiz.. - mat. lit. Moskva.
Georgescu, A. (1987). *Sinergetica, o nouă sinteză a științei* (*Synergetics, a new synthesis of science*). Ed. Tehnică, București.
Germain, P. (1962). *Mécanique des milieux continus*. Masson, Paris.
Gibbs, J.W. (1928). *Collected works*. New York.
Gilmore, R. (1981). *Catastrophe theory for scientists and engineers*. J. Wiley and Sons, New York.
Gioncu, V. (2005). *Instabilități și catastrofe în ingineria structurală* (*Instabilities and catastrophes in structural engineering*). Ed. Orizonturi Univ., Timișoara.
Girard, A. and Roy, N. (2003). *Dynamique des structures industrielles*. Hermes-Lavoisier, Paris.
Gleick, J. (1987). *Chaos, making a new science*. Viking Penguin, New York.
Godbillon, C. (1969). *Géométrie différentielle et mécanique analytique*. Hermann, Paris.
Goldstein, H. (1956). *Classical mechanics*. Adison Wesley, Cambridge.
Golubev, V.V. (1953). *Lektsii po integrirovaniyu uravneniĭ dvizheniya tiazhelovo tela okolo napodvijnoĭ tochki* (*Lessons on the integration of the equations of motion of the rigid body around a fixed point*). Gostekhizdat, Moskva.
Golubitsky, M. (1974). *Contact equivalence for Lagrangian submanifolds. Dynamical systems*. Lect. notes in math. 468. Springer-Verlag, Berlin, New York.
Golubitsky, M. (1976). *An introduction to catastrophe theory and its applications*. Lect. notes. Queens College, New York.
Golubitsky, M. and Guillemin, V. (1973). *Stable mappings and their singularities*. Springer-Verlag, New York, Heidelberg, Berlin.
Golubitsky, M. and Schaeffer, D. (1985). *Singularity and groups in bifurcation theory*. Springer-Verlag, New York.
Graffi, D. (1967). *Elementi di meccanica razionale*. R. Patron, Bologna.
Grammel, R. (1927). *Grundlagen der Mechanik*. Geiger-Scheel Handbuch der Physik. V. J. Springer, Berlin.
Grammel, R. (1927). *Kinetik der Massenpunkte*. Geiger-Scheel Handbuch der Physik. V. J. Springer, Berlin.
Gray, A. (1918). *A treatise on gyrostatics and rotational motion. Theory and applications*. Macmillan, London. (1959). Dover Publ., New York.
Grebenikov, E.A. and Ryabov, Yu.A. (1983). *Constructive methods in the analysis of nonlinear systems*. "Mir", Moscow.
Greenhill, A.G. (1982). *The applications of elliptic functions*. Macmillan, London, New York. (1959). Dover Publ., New York.
Greenwood, D.T. (1965). *Principles of dynamics*. Prentice-Hall, Englewood Cliffs.
Guckenheimer, J. and Holmes, P. (1983). *Nonlinear oscillations, dynamical systems and bifurcations of vector fields*. Springer-Verlag, New York, Berlin, Heidelberg.
Gumowski, I. and Mira, C. (1980). *Dynamique chaotique*. Capadues, Toulouse.
Gurel, O. and Rössler, E.E. (ed.). (1979). *Bifurcation theory and application in scientific disciplines*. New York Acad. of Sci., New York.

Gutzwiller, M.C. (1990). *Chaos in classical and quantum mechanics.* Springer-Verlag, New York.
Haikin, S.E. (1962). *Fizicheskie osnovy mekhaniki (Physical fundamentals of mechanics).* Gos. izd. fiz. - mat., Moskva.
Hakan, H. (ed.). (1982). *Evolution of order and chaos in physics, chemistry and biology.* In: Springer Series in Synergetics, 17. Springer-Verlag, Berlin, Heidelberg, New York.
Halanay, A. (1963). *Teoria calitativă a ecuaţiilor diferenţiale (Qualitative theory of differential equations).* Ed. Academiei, Bucureşti.
Hale, J.K. and Koçak, H. (1991). *Dynamics of bifurcations.* Springer-Verlag, New York.
Halliday, D. and Resnick, R. (1960). *Physics.* I. J. Wiley and Sons, New York, London, Sydney.
Halphen, G. (1888). *Traité des fonctions elliptiques.* 2. Gauthier-Villars, Paris.
Hamburger, L. and Buzdugan, Gh. (1958). *Teoria vibraţiilor şi aplicaţiile ei în construcţia maşinilor (Theory of vibrations and its applications to engineering mechanics).* Ed. Tehnică, Bucureşti.
Hamel, G. (1922). *Elementare Mechanik.* B.G. Teubner, Leipzig, Berlin.
Hamel, G (1927). *Die Axiome der Mechanik.* Geiger-Scheel Handbuch der Physik. J. Springer, Berlin.
Hamel, G. (1949). *Theoretische Mechanik.* Springer-Verlag, Berlin.
Hamilton, W.R. (1890). *Collected papers.* III. Cambridge Univ. Press, Cambridge.
Hammermesh, M. (1962). *Group theory and its applications to physical problems.* Pergamon Press, London.
Hangan, S. and Slătineanu, I. (1983). *Mecanică (Mechanics).* Ed. Did. Ped., Bucureşti.
Hao, B.L. (1984). *Chaos.* World. Sci. Publ., Singapore.
Harris, C.M. and Crede, C.E. (ed.). (1961). *Shock and vibration handbook.* I-III. McGraw-Hill, New York, Toronto, London.
Hartog Den, J.P. (1956). *Mechanical vibrations.* McGraw-Hill, New York.
Hassard, B.D., Kazarinoff, N.D. and Wan, Y.-H. (1981). *Theory and application of Hopf bifurcation.* Cambridge Univ. Press, Cambridge.
Hayashi, C. (1964). *Non-linear oscillations in physical systems.* McGraw-Hill, New York, San Francisco, Toronto, London.
Helleman, R.H.G. (ed.). (1980). *Nonlinear dynamics.* Annals of the New York Acad. Sci., vol. 357, New York Acad. Sci., New York.
Hemming, G. (1899). *Billiards mathematically treated.* Macmillan, London.
Henon, M. and Pomeau, Y. (1976). *Two strange attractors with a simple structure.* In: Lectures Notes in Math. Vol. 565. Springer-Verlag, Berlin, Heidelberg, New York.
Herglotz, G. (1941). *Vorlesungen über analitische Mechanik.* Göttingen.
Hertz, H. (1894). *Die Prinzipien der Mechanik in neuem Zusammenhange dargestellt.* Leipzig.
Hirsch, M.W. and Smale, S. (1974). *Differential equations, dynamical systems and linear algebra.* Academic Press, New York.
Hlavaček, V. (ed.). (1985). *Dynamics of nonlinear systems.* Gordon and Breach, New York.
Holden, A.V. (ed.). (1985). *Chaos.* Princeton Univ. Press, Princeton.
Hort, W. and Thoma, A. (1956). *Die differentialgleichungen der Technik und Physik.* J.A. Barth Verlag, Leipzig.
Huseyin, K. (1978). *Vibration and stability of multiple parameter systems.* Noordhoff, Alphen.
Huygens, Cr. (1920). *Oeuvres complètes.* Martinus Nijhoff, La Haye.
Iacob, A. (1973). *Metode topologice în mecanica clasică (Topological methods in classical mechanics).* Ed. Academiei, Bucureşti.
Iacob, C. (1970). *Mecanică teoretică (Theoretical mechanics).* Ed. Did. Ped., Bucureşti.
Ioachimescu, A.G. (1947). *Mecanică raţională (Rational mechanics).* Impr. nat., Bucureşti.

Ioachimescu, A.G. and Țițeica, G. (1945). *Culegere de probleme de mecanică* (*Collection of problems in mechanics*). Ed. Gaz. mat., București.
Ionescu-Pallas, N. (1969). *Introducere în mecanica teoretică modernă* (*Introduction to modern theoretical mechanics*). Ed. Academiei, București.
Iooss, G., Helleman, R.H.G. and Stora, R. (ed.). (1983). *Chaotic behaviour of deterministic systems*. North-Holland, Amsterdam.
Iooss, G. and Joseph, D.D (1980). *Elementary stability and bifurcation theory*. Springer-Verlag, New York.
Irwin, M.C. (1980). *Smooth dynamical systems*. Academic Press, New York.
Isaacson, N. and Keller, H.B. (1966). *Analysis of numerical methods*. J. Wiley and Sons, New York, London.
Ishlinskiĭ, A. Yu. (1963). *Mekhanika giroskopicheskikh sistem* (*Mechanics of gyroscopic systems*). Izd. Akad. Nauk, Moskva.
Jacobi, C.G.J. (1842-1843). *Vorlesungen über Dynamik*. Berlin.
Jacobi, C.G.J. (1866). *Vorlesungen über Dynamik*. Clebsch, Berlin.
Jacobi, C.G.J. (1882). *Gesammelte Werke*. 2. G. Reiner, Berlin.
Jammer, M. (1957). *Concepts of force. A study on the foundations of mechanics*. Harvard Univ. Press, Cambridge, Mass.
Jammer, M. (1961). *Concepts of mass in classical and modern physics*. Harvard Univ. Press, Cambridge, Mass.
Janssens, P. (1968). *Cours de mécanique rationnelle*. I, II. Dunod, Paris.
Jansson, P.-A. and Grahn, R. (1995). *Engineering mechanics. 1. Statics. 2. Dynamics*. Prentice-Hall, New York, London, Toronto, Sydney, Tokyo, Singapore.
Jaunzemis, W. (1967). *Continuum mechanics*. Macmillan, New York.
Jeffreys, W. and Jeffreys, B.S. (1950). *Methods of mathematical physics*. 2nd ed. Cambridge Univ. Press, Cambridge.
Joos, G. (1959). *Lehrbuch der theoretischen Physik*. Akad. Verlags-gesellschaft, Leipzig.
Jukovskiĭ, N.E. (1948). *Sobranie sochinenyia*. (*Collected works*). I. Izd. Tekhn. Teor. Lit., Moskva-Leningrad.
Julia, G. (1928). *Cours de cinématique*. Gauthier-Villars, Paris.
Jung, G. (1901-1908). *Geometrie der Massen*. In: Enc. der math. Wiss. IV/1. B.G. Teubner, Leipzig.
Kahan, Th. (1960). *Précis de physik moderne*. Presses Univ. France, Paris.
Kahan, Th. (1960). *Théorie des groupes en physique classique et quantique*. Dunod, Paris.
Kamke, E. (1967). *Differentialgleichungen. Lösungsmethoden und Lösungen. Gewöhnliche Differentialgleichungen*. Akad. Verlagsgesellschaft Geest und Portig, K.-G., Leipzig.
Kantorovich, L.V. and Krylov, V.I. (1962). *Priblizhenie metody vysshevo analiza* (*Approximate methods of high analysis*). Gos. izd. fiz.-mat. lit., Moskva-Leningrad.
Kapitaniak, T. (1990). *Chaos in systems with noise*. World. Sci., Singapore.
Kapitaniak, T. (1991). *Chaotic oscillations in mechanical systems*. Manchester Univ. Press, Manchester.
Kapitaniak, T. (1998). *Chaos for engineers. Theory, applications and control*. Springer-Verlag, Berlin, Heidelberg, New York.
Kasner, E. (1913). *Differential-geometric aspects of dynamics*. Amer. Math. Soc., New York.
Katok, A. and Hasselblatt, B. (1999). *Introduction to the modern theory of dynamical systems*. Cambridge Univ. Press, Cambridge.
Kauderer, H. (1958). *Nichtlineare Mechanik*. Springer-Verlag, Berlin, Göttingen, Heidelberg.
Kármán, Th. von and Brot, M.A. (1949). *Les méthodes mathématiques de l'ingénieur*. Libr. Polyt. Ch. Bérenger, Paris, Liège.
Kästner, S. (1960). *Vektoren, Tensoren, Spinoren*. Akad.-Verlag, Berlin.

Kecs, W. (1982). *The convolution product and some applications*. Ed. Academiei, București, D. Reidel Publ. Co., Dordrecht, Boston, London.
Kecs, W. and Teodorescu, P.P (1974). *Applications of the theory of distributions in mechanics*. Ed. Academiei, București, Abacus Press, Tunbridge Wells.
Kecs, W. and Teodorescu, P.P (1978). *Vvedenie v teoriyu obobshchennykh funktsiĭ s prilozheniyami v tekhnike* (*Introduction in the theory of distributions with applications to technics*). "Mir", Moskva.
Keller, H.B. (1987). *Lectures on numerical methods in bifurcation problems*. Tate Inst. Fund. Res., Bombay.
Kellog, O.D. (1929). *Potential theory*. J. Springer, Berlin.
Kilchevski, N.A. (1972-1977). *Kurs teoreticheskoĭ mekhaniki* (*Lectures on theoretical mechanics*). I, II. Nauka, Moskva.
Kilcevski, N.A. (1956). *Elemente de calcul tensorial și aplicațiile lui în mecanică* (*Elements of tensor calculus and its applications to mechanics*). Ed. Tehnică, București.
Kirchhoff, G.R. (1876). *Vorlesungen über matematische Physik*. I. *Mechanik*. B.G. Teubner, Leipzig.
Kirchhoff, G.R. (1897). *Vorlesungen über Mechanik*. 4th ed. B.G. Teubner, Leipzig.
Kirpichev, V.L. (1951). *Besedy o mekhanike* (*Conversations on mechanics*). Gostekhizdat, Moskva.
Kittel, Ch., Knight, W.D. and Ruderman, M.A. (1973). *Mechanics*. Berkeley Physics Course. I. McGraw-Hill, New York.
Klein, F. (1897). *The mathematical theory of the top*. Princeton Lect., Charles Scribner's Sons, New York.
Klein, F. and Sommerfeld, A. (1897-1910). *Über die Theorie des Kreisels*. 1-4. B.G. Teubner, Leipzig.
Kocin, N.E. (1954). *Calcul vectorial și introducere în calculul tensorial* (*Vector calculus and introduction to tensor calculus*). Ed. Tehnică, București.
Kooy, J.M.J. and Uytenbogaart, J.W.H. (1946). *Ballistics of the future, with special reference to the dynamical and physical theory of the rocket weapons*. McGraw-Hill, New York, London.
Koshlyakov, V.N. (1972). *Teoriya giroskopicheskikh kompasov* (*Theory of gyroscopic compasses*). "Nauka", Moskva.
Kosmodemyanskiĭ, A.A. (1955). *Kurs teoreticheskoĭ mekhaniki* (*Lectures on theoretical mechanics*). I, II. Gosutsediv, Moskva.
Kovalevskaya, S. (1948). *Nauchnye sochinenya* (*Scientific works*). Izd. Akad. Nauk, Moskva.
Kranz, C. (1923). *Lehrbuch der Ballistik*. I-III. J. Springer, Berlin.
Krbek, F. (1954). *Grundzüge der Mechanik*. Geest und Portig, Leipzig.
Kron, G. (1939). *Tensor analysis of networks*. J. Wiley and Sons, New York.
Kryloff, N. and Bogoliuboff, N. (1949). *Introduction to non-linear mechanics*. Princeton Univ. Press, Princeton.
Krylov, A.N. (1958). *Izbrannye trudy* (*Selected works*). Izd. Akad. Nauk, Moskva.
Kuiper, G.P. and Middlehurst, M.B. (ed.). (1961). *Planets and satellites*. The Univ. Press, Chicago.
Kutterer, R.E. (1959). *Ballistik*. 3rd ed. Braunschweig.
Kuznetsov, Y.A. (1998). *Elements of applied bifurcation theory*. 2nd ed. Springer-Verlag, New York.
Lagrange, J.-L. (1788). *Mécanique analytique*. Paris.
Lakhtin, L.M. (1963). *Svobodnoe dvizhenie v poleznogo sferoida* (*Free motion of the useful spheroid*). Fizmatgiz, Moskva.
Lamb, H. (1929). *Heigher mechanics*. 2nd ed. Cambridge Univ. Press, Cambridge.
Lamb, H. (1929). *Dynamics*. 2nd ed. Cambridge Univ. Press, Cambridge.

Lampariello, G. (1952). *Lezioni di meccanica razionale*. 3rd ed. Ed. Vincenzo Ferrara, Messina.
Landau, L.D. and Lifchitz E. (1960). *Mécanique*. Éd. en langues étrangères, Moscou.
Lantsosh, K. (1965). *Variatsionnye printsipy mekhaniki* (*Variational principles of mechanics*). "Mir", Moskva.
La Salle, J.P. and Lefschetz, S. (1961). *Stability by Lyapunov's direct method with applications*. Academic Press, New York.
La Valée-Poussin, Ch.J. de. (1925). *Leçons de mécanique analytique*. I, II. Paris.
Lecornu, L. (1918). *La mécanique*. Paris.
Leech, J.W. (1965). *Classical mechanics*. Butler and Tanner, London.
Leimanis, E. (1958). *Some recent advances in the dynamics of rigid bodies and celestial mechanics*. Surveys in Appl. Math. Vol. 2. J. Wiley and Sons, New York.
Leimanis, E. (1965). *The general problem of the motion of coupled rigid bodies about a fixed point*. Springer-Verlag, Berlin, Heidelberg, New York.
Leipholz, H. (1970). *Stability theory*. Academic Press, New York.
Levi-Civita, T. (1926). *The absolute differential calculus*. Hafner, New York.
Levi-Civita, T. and Amaldi, U (1950-1952). *Lezioni di meccanica razionale*. I, IIa, IIb. Zanichelli, Bologna.
Lindsay, R.B. (1961). *Physical mechanics*. Van Nostrand, Princeton, New Jersey, Toronto, London, New York.
Lippmann, H. (1970). *Extremum and variational principles in mechanics*. CISM, Udine.
Littlewood, D.E. (1958). *The theory of group characters*. Clarendon Press, Oxford.
Loitsyanski, L.G. and Lurie, A.I. (1955). *Kurs teoreticheskoĭ mekhaniki* (*Lectures on theoretical mechanics*). I, II. Gostekhizdat, Moskva.
Lomont, J.S. (1959). *Applications of finite groups*. Academic Press, New York, London.
Lurie, A.I. (2002). *Analytical mechanics*. Springer, Berlin, Heidelberg, New York, Barcelona, Hong Kong, London, Milan, Paris, Tokyo.
Lyapunov, A.M. (1949). *Problème général de la stabilité du mouvement*. Princeton Univ. Press, Princeton.
Lyubarskiĭ, G.Ya. (1960). *The applications of group theory in physics*. Pergamon Press, London.
Mach, E. (1904). *La mécanique. Exposé historique et critique de son dévelopment*. Hermann, Paris.
Macke, W. (1961). *Wellen. Ein Lehrbuch der theoretischen Physik*. Geest und Portig, K.-G., Leipzig.
Macmillan, W.D. (1927-1936). *Theoretical mechanics*. I, II. McGraw-Hill, New York, London.
Macmillan, W.D. (1936). *Dynamics of rigid bodies*. McGraw-Hill, New York, London.
Madan, R. (1993). *Chua's circuit: paradigm for chaos*. World Sci., Singapore.
Magnus, K. (1971). *Kreisel. Theorie und Anwendungen*. Springer-Verlag, Berlin, Heidelberg, New York.
Malkin, I.G. (1956). *Nekotorye zadachi teoriĭ nelineinykh kolebaniĭ* (*Some problems on the theory of non-linear vibrations*). Gostekhizdat, Moskva.
Malvern, L.E. (1969). *Introduction to the mechanics of continuous media*. Prentice-Hall, New Jersey.
Mandel, J. (1966). *Cours de mécanique des milieux continus*. I, II. Gauthier-Villars, Paris.
Mandelbrot, B.B. (1975). *Les objets fractals: forme, hasard et dimensions*. Flamarion, Paris.
Mandelbrot, B.B. *The fractal geometry of nature*. W.H. Freeman, San Francisco.
Mangeron, D. and Irimiciuc, N. (1978-1981). *Mecanica rigidelor cu aplicaţii în inginerie* (*Mechanics of rigid solids with applications in engineering*). I-III. Ed. Tehnică, Bucureşti.
Marchyuk, G.Z. (1980). *Metody vychislitelnoĭ matematiki* (*Methods of numerical mathematics*). "Nauka", Moskva.
Marcolongo, R. (1922,1923). *Meccanica razionale*. I, II. U. Hoepli, Milano.

References

Marcolongo, R. (1937). *Memorie sulla geometria e la meccanica di Leonardo da Vinci*. S.I.E.M., Napoli.
Marion, J.B. (1965). *Classical dynamics of particles and systems*. Academic Press, New York, London.
Marsden, J.E. and McCracken, M. (1950). *The Hopf bifurcations and its applications*. Springer-Verlag, New York, Heidelberg, Berlin.
McLachlan, N.W. (1950). *Ordinary non-linear differential equations in engineering and physical sciences*. Clarendon Press, London, Oxford.
Meirovitch, I. (1975). *Elements of vibration analysis*. McGraw-Hill, New York.
Mercheş, I. and Burlacu, L. (1983). *Mecanica analitică şi a mediilor deformabile* (*Analytical and deformable media mechanics*). Ed. Did. Ped., Bucureşti.
Mercier, A. (1955). *Principes de mécanique analytique*. Paris.
Merkin, D.N. (1956). *Giroskopicheskie sistemy* (*Gyroscopic systems*). Gostekhizdat, Moskva.
Meshcherskiĭ, I.V. (1949). *Raboty po mekhanike tel peremennoĭ massy* (*Works on mechanics of bodies of variable mass*). Gostekhizdat, Moskva-Leningrad.
Mihăilescu, M. and Chiroiu, V. (2004). *Advanced mechanics on shells and intelligent structures*. Ed. Academiei, Bucharest.
Miller Jr., W. (1972). *Symmetry groups and their applications*. Academic Press, New York, London.
Miller Jr., W. (1977). *Symmetry and separation of variables*. Addison-Wesley, London.
Milne, E.A. (1948). *Vectorial mechanics*. Intersci. Publ., New York.
Minorsky, N. (1947). *Introduction to nonlinear mechanics*. J. W. Edwards, Ann Arbor.
Minorsky, N. (1962). *Nonlinear oscillations*. D. van Nostrand, Princeton.
Miron, R. and Anastasiei, M. (1987). *Fibrate vectoriale. Spaţii Lagrange. Aplicaţii în teoria relativităţii* (*Vector fibre space. Lagrange spaces. Applications in the theory of relativity*). Ed. Academiei, Bucureşti.
Mitropolskiĭ, Yu.A. (1955). *Nestatsionarnye protsesy v nelineinykh kolebatelnykh sistemakh* (*Non-steady processes in non-linear oscillating systems*). Izd. Akad. Nauk Ukr., Kiev.
Moon, F.C. (1988). *Experiments in chaotic dynamics*. CISM, Udine.
Moreau, J.J. (1968,1970). *Mécanique classique*. I, II. Masson, Paris.
Morgenstern, D. and Szabo, I. (1961). *Vorlesungen über theoretische Mechanik*. Springer-Verlag, Berlin.
Morse, P.M. and Feshbach, H. (1953). *Methods of theoretical physics*. I. McGraw-Hill, New York.
Mourre, L. (1953). *Du compass gyroscopique*. Paris.
Munteanu, L. and Donescu, Şt. (2004). *Introduction to soliton theory: applications to mechanics*. Fundamental Theories of Physics. 143. Kluwer Academic Publ., Dordrecht, Boston, London.
Munteanu, M. (1997). *Introducere în dinamica oscilaţiilor solidului rigid şi a sistemelor de solide rigide* (*Introduction to dynamics of oscillations of the rigid solid and of the systems of rigid solids*). Ed. Clusium, Cluj-Napoca.
Murnaghan, F.D. (1962). *The unitary and rotation groups*. Spartan Books, Washington.
Muszyńska, A. (2005). *Rotordynamics*. Taylor&Francis, Boca Raton, London, New York, Singapore.
Naimark, M. and Stern, A. (1979). *Théorie des représentations des groupes*. "Mir", Moscou.
Naumov, A.L. (1958). *Teoreticheskaya mekhanika* (*Theoretical mechanics*). Kiev.
Nayfeh, A.H. and Mook, D.T. (1979). *Non-linear oscillations*. J. Wiley and Sons, New York, Chichester.
Neimark, Ju.I. and Fufaev, N.A. (1972). *Dynamics of nonholonomic systems*. Amer. Math. Soc., Providence.

Nekrasov, A.I. (1961,1962). *Sobranie sochineniĭ (Collected works)*. I, II. Izd. Akad. Nauk, Moskva.
Nelson, W.C. and Loft, E.E. (1962). *Space mechanics*. Prentice-Hall, Englewood Cliffs.
Newbult, H.O. (1946). *Analytical methods in dynamics*. Oxford Univ. Press, Oxford.
Newton, I. (5 July 1686). *Philosophiae naturalis principia mathematica*. S. Pepys, Reg. Soc. Preases, Londini.
Nguyen, Quoc-son. (2000). *Stabilité et mécanique non linéaire*. Hermes Sci. Publ., Paris.
Nguyen, Van Hieu (1967). *Lektsii po teoriĭ unitarnoĭ simetriĭ elementarnykh chastits* (*Lessons on the theory of unitary symmetry of elementary particles*). Atomizdat, Moskva.
Nicolis, G. and Prigogine, I. (1977). *Self-organization in non-equilibrium systems. From dissipative structures to order through fluctuations*. J. Wiley and Sons, New York.
Nielsen, J. (1936). *Vorlesungen über elementare Mechanik*. J. Springer, Berlin.
Nikitin, E.M. and Kapli, D.M. (1957). *Teoreticheskaya mekhanika* (*Theoretical mechanics*). I, II. Gostekhizdat, Moskva.
Nitecki, Z. (1971). *Differentiable dynamics*. M.I.T. Press, Cambridge.
Niţă, M.M. (1972). *Curs de mecanică teoretică. Dinamica* (*Lectures on theoretical mechanics. Dynamics*). Ed. Acad. Mil., Bucureşti.
Niţă, M.M. (1973). *Teoria zborului spaţial* (*Theory of space flight*). Ed. Academiei, Bucureşti.
Niţă, M.M. and Andreescu, D. (1964). *Zborul rachetei nedirijate şi dirijate* (*Flight of the non-directed and directed rocket*). Ed. Acad. Mil., Bucureşti.
Niţă, M.M. and Aron, I.I. (1961). *Pilotul automat* (*The automatic pilot*). Ed. Acad. Mil., Bucureşti.
Niţă, M.M. and Aron, I.I. (1971). *Navigaţia inerţială* (*Inertial navigation*). Ed. Acad. Mil., Bucureşti.
Nordheim, L. (1927). *Die Prinzipien der Mechanik*. Geiger-Scheel Handbuch der Physik. V. J. Springer, Berlin.
Nordheim, L. and Fues, E. (1927). *Die Hamilton-Jacobische Theorie der Dynamik*. Geiger-Scheel Handbuch der Physik. V. J. Springer, Berlin.
Novoselov, V.S. (1969). *Analiticheskaya mekhanika sistem s peremennymi masami* (*Analytical mechanics of the systems with variable masses*).
Nowacki, W. (1961). *Dynamika budowli* (*Dynamics of mechanical systems*). Arkady, Warszawa.
Nyayapathi, V., Swamy, V.J. and Samuel, M.A. (1979). *Group theory made easy for scientists and engineers*. J. Wiley and Sons, New York, Chichester, Brisbane, Toronto.
Obădeanu, V. (1981). *Aplicaţii ale formelor diferenţiale exterioare în mecanică şi electrodinamică* (*Applications of differential exterior forms in mechanics and electrodynamics*). Timişoara Univ., Timişoara.
Obădeanu, V. (1987). *Mecanică. Formalism newtonian, formalism lagrangean* (*Mechanics. Newtonian formalism, Lagrangian formalism*). Timişoara Univ., Timişoara.
Obădeanu, V. (1992). *Introducere în biodinamica analitică* (*Introduction to analytical biodynamics*). Timişoara Univ., Timişoara.
Obădeanu, V. and Marinca, V. (1992). *Problema inversă în mecanica analitică* (*The inverse problem in analytical mechanics*). Timişoara Univ., Timişoara.
Okunev, V.N. (1951). *Svobodnoe dvizhenie giroskopa* (*Free motion of the gyroscope*). Gostekhizdat, Moskva-Leningrad.
Olariu, S. (1987). *Geneza şi evoluţia reprezentărilor mecanicii analitice* (*Genesis and evolution of analytical mechanics representations*). Ed. Ştiinţ. Enc., Bucureşti.
Olkhovskiĭ, I.I. (1970). *Kurs teoreticheskoĭ mekhaniki dlya fizikov* (*Lectures on theoretical mechanics for physicists*). "Nauka", Moskva.

Olkhovskiĭ, I.I., Pavlenko, M.G. and Kuzimenkov, L.S. (1977). *Zadachi po teoreticheskoĭ mekhaniki dlya fizikov* (*Problems of theoretical mechanics for physicists*). Izd. Mosk. Univ., Moskva.
Olson, H.F. (1946). *Dynamical analogies*. Van Nostrand, New York.
Onicescu, O. (1969). *Mecanica* (*Mechanics*). Ed. Tehnică, București.
Onicescu, O. (1974). *Mecanica invariantivă și cosmologia* (*Invariantive mechanics and cosmology*). Ed. Academiei, București.
Osgood, W.F. (1937). *Mechanics*. Macmillan, New York.
Ostrogradskiĭ, M.V. (1946). *Sobranie trudy* (*Collected works*). I. Izd. Akad. Nauk, Moskva.
Ott, E. (1993). *Chaos in dynamical systems*. Cambridge Univ. Press, Cambridge.
Ovsiannikov, V. (1962). *Gruppovye svoistva differentsialnykh uravneniĭ* (*Group properties of differential equations*). Siber. Filial. Akad. Nauk, Novosibirsk.
Painlevé, P. (1895). *Leçons sur l'intégration des équations de la mécanique*. Gauthier-Villars, Paris.
Painlevé, P. (1922). *Les axiomes de la mécanique*. Gauthier-Villars, Paris.
Painlevé, P. (1936). *Cours de mécanique*. I, II. Gauthier-Villars, Paris.
Palis, J. and Melo, W. de. (1982). *Geometric theory of dynamical systems. An introduction*. Springer-Verlag, New York, Heidelberg, Berlin.
Pandrea, N. (2000). *Elemente de mecanica solidelor în coordonate plückeriene* (*Elements of mechanics of solids in Plückerian co-ordinates*). Ed. Academiei, București.
Pandrea, N. and Stănescu, N.-D. (2002). *Mecanica* (*Mechanics*). Ed. Did. Ped., București.
Pandrea, N., Stănescu, N.-D. and Pandrea, M. (2003). *Mecanica. Culegere de probleme* (*Mechanics. Collection of problems*). Ed. Did. Ped., București.
Parker, T. and Chua, L.O. (1989). *Practical numerical algorithms for chaotic systems*. Springer-Verlag, New York.
Pars, L. (1965). *A treatise on analytical dynamics*. Heinemann, London.
Peitgen, H.O., Jurgens, H. and Saupe, D. (1992). *Chaos and fractals. New frontiers of science*. Springer-Verlag, New York, Berlin, Heidelberg.
Peitgen, H.O. and Richter, P.H. (1986). *The beauty of fractals*. Springer-Verlag, New York, Berlin, Heidelberg.
Peitgen, H.O. and Saupe, D. (ed.). (1998). *The science of fractal images*. Springer-Verlag, New York, Berlin, Heidelberg, London, Paris, Tokyo.
Peitgen, H.O. and Walther, H.O. (ed.). (1979). *Functional differential equations and approximation of fixed points*. Springer Lecture Notes in Math. 730. Springer-Verlag, New York, Heidelberg, Berlin.
Peixoto, M.M. (ed.). (1978). *Dynamical systems*. Academic Press, New York.
Perry, J. (1957). *Spinning tops and gyroscopic motion*. Dover Publ., New York.
Petkevich, V.V. (1981). *Teoreticheskaya mekhanika* (*Theoretical mechanics*). "Nauka", Moskva.
Pérès, J. (1953). *Mécanique générale*. Masson, Paris.
Planck, M. (1919). *Einführung in die allgemeine Mechanik*. Leipzig.
Plăcințeanu, I. (1948). *Vectori. Potențiali. Tensori* (*Vectors. Potentials. Tensors*). Ed. Ath. Gheorghiu, Iași.
Plăcințeanu, I. (1958). *Mecanica vectorială și analitică* (*Vector and analytical mechanics*). Ed. Tehnică, București.
Pohl, R.W. (1931). *Einführung in die Mechanik und Akustik*. Berlin.
Poincaré, H. (1892-1899). *Les méthodes nouvelles de la mécanique céleste*. I-III. Gauthier-Villars, Paris.
Poincaré, H. (1902). *La science et l'hypothèse*. Flammarion, Paris.
Poincaré, H. (1908). *La valeur de la science*. Flammarion, Paris.

Poincaré, H. (1952). *Oeuvres*. 7. Gauthier-Villars, Paris.
Poisson, S.-D. (1833). *Traité de mécanique*. I, II. IInd ed. Bachelier, Paris.
Polak, L.S. (ed.). (1959). *Variatsionnye printsipy mekhaniki* (*Variational principles of mechanics*).Gos. izd. fiz. -mat. lit, Moskva.
Pontriaguine, L. (1969). *Equations différentielles ordinaires*. "Mir", Moscou.
Popoff, K. (1954). *Die Hauptprobleme der äusseren Ballistik*. Leipzig.
Popov, V.M. (1966). *Hiperstabilitatea sistemelor automate* (*Hyperstability of automatic systems*). Ed. Academiei, Bucureşti.
Posea, N. (1991). *Calculul dinamic al structurilor* (*Dynamical calculus of structures*). Ed. Tehnică, Bucureşti.
Posea, N., Florian, V., Talle, V. and Tocaci, E. (1984). *Mecanică aplicată pentru ingineri* (*Applied mechanics for engineers*). Ed. Tehnică, Bucureşti.
Poston, T. and Stewart, I. (1978). *Catastrophe theory and its applications*. Pitman, London.
Pöschl, T. (1930). *Lehrbuch der technischen Mechanik*. Berlin.
Prange, G. (1921). *Die allgemeinen Integrationsmethoden der analytischen Mechanik*. Encycl. math. Wiss. IV. B.G. Teubner, Leipzig.
Prigogine, I. (1980). *From being to becoming time and complexity in the physical sciences*. W.H. Freeman, San Francisco.
Pyatnitskiĭ, E.S., Trukhan, N.M., Khanukaev, Yu.I. and Iakovenko, G.N. (1980). *Sbornik zadachi po analiticheskoĭ mekhanike* (*Collection of problems of analytical mechanics*). "Nauka", Moskva.
Racah, G. (1965). *Group theory and spectroscopy*. Springer-Verlag, Berlin, Heidelberg, New York.
Radu, A. (1978). *Probleme de mecanică* (*Problems of mechanics*). Ed. Did. Ped., Bucureşti.
Rausenberger, O. (1888). *Lehrbuch der analytischen Mechanik*. 2. B.G. Teubner, Leipzig.
Rawlings, A.L. (1944). *The theory of the gyroscopic compass and its deviations*. IInd ed. Macmillan, New York.
Rayleigh, lord (Strutt, J.W.). (1893, 1896). *The theory of sound*. Macmillan, London.
Rayleigh, lord (Strutt, J.W.). (1899, 1900). *Scientific papers*. I, II. Cambridge Univ. Press, Cambridge.
Rădoi, M., Deciu, E. and Voiculescu, D. (1973). *Elemente de vibraţii mecanice* (*Elements of mechanical vibrations*). Ed. Tehnică, Bucureşti.
Rădoi, M. and Deciu, E. (1981). *Mecanica* (*Mechanics*). Ed. Did. Ped., Bucureşti.
Richardson, K.I.T. (1954). *The gyroscope applied*. Philosophical Library, New York.
Ripianu, A. (1973). *Mecanica solidului rigid* (*Mechanics of the rigid solid*). Ed. Tehnică, Bucureşti.
Ripianu, A. (1979). *Mecanica tehnică* (*Technical mechanics*). Ed. Did. Ped., Bucureşti.
Rocard, Y. (1943). *Dynamique générale des vibrations*. Masson, Paris.
Rocard, Y. (1954). *L'instabilité en mécanique. Automobiles, avions, ponts suspendus*. Masson, Paris.
Rohrlich, F. (1965). *Classical charged particles*. Addison-Wesley, Reading.
Roitenberg, L.N. (1966). *Teoriya giroskopicheskikh kompasov* (*Theory of gyroscopic compasses*). "Nauka", Moskva.
Roman, P. (1961). *Theory of elementary particles*. North-Holland, Amsterdam.
Rose, N.V. (1938). *Lektsii po analiticheskoĭ mekhanike* (*Lessons on analytical mechanics*). Izd. Univ., Leningrad.
Roseau, M. (1966). *Vibrations nonlinéaires et théorie de la stabilité*. Springer-Verlag, Berlin, Heidelberg, New York.
Rouche, N. and Mawhin, J. (1973). *Equations différentielles ordinaires*. Masson, Paris.

Rouche, N., Habets, P. and Laloy, M. (1977). *Stability theory by Lyapunov's direct method*. Springer-Verlag, New York, Heidelberg, Berlin.
Routh, E.J. (1892). *The advanced part of a treatise on the dynamics of a system of rigid bodies*. Vth ed. Macmillan, New York. (1955). Dover Publ., New York.
Routh, E.J. (1897). *The elementary part of a treatise on the dynamics of a system of rigid bodies*. VIth ed. Macmillan, London, New York. (1955). Dover Publ., New York.
Routh, E.J. (1898). *Die Dynamik der Systeme starrer Körper*. I, II. B.G. Teubner, Leipzig.
Routh, E.J. (1898). *Dynamics of particles*. Cambridge Univ. Press, Cambridge.
Roy, M. (1962). *Dynamique des satélites*. Symposium, Paris.
Roy, M. (1965). *Mécanique. Corps rigides*. Dunod, Paris.
Ruelle, D. (1989). *Chaotic evolution and strange attractors*. Cambridge Univ. Press, Cambridge.
Sabatier, P.C. (1978). *Applied inverse problem*. Springer-Verlag, Berlin, Göttingen, Heidelberg.
Saletan, D.J. and Cromer, A.H. (1971). *Theoretical mechanics*. J. Wiley and Sons, New York.
Sanders, J.A. and Verhulst, V. (1985). *Averaging methods in nonlinear dynamical systems*. Springer-Verlag, New York, Heidelberg, Berlin, Tokyo.
Sansone, G. and Conti, R. (1949). *Non-linear differential equations*. Pergamon Press, Oxford, London, Edinborough, New York, Paris, Frankfurt.
Santilli, R.M. (1984). *Foundations of theoretical mechanics*. I. *The inverse problem in Newtonian mechanics*. Springer-Verlag, New York, Berlin, Heidelberg, Tokyo.
Sauer, R. and Szabo, I. (ed.). (1967-1970). *Mathematische Hilfsmittel des Ingenieurs*. I-IV. Springer-Verlag, Berlin, Heidelberg, New York.
Savarensky, E. (1975). *Seismic waves*. "Mir", Moscow.
Savet, P.H. (1961). *Gyroscops. Theory and design*. McGraw-Hill, New York.
Sburlan, S., Barbu, L. and Mortici, C. (1999). *Ecuaţii diferenţiale, integrale şi sisteme dinamice (Differential and integral equations and dynamical systems)*. Ex Ponto, Constanţa.
Scarborough, J.B. (1958). *The gyroscope. Theory and applications*. Intersci. Publ., New York.
Schaefer, Cl. (1922). *Einführung in die theoretische Physik*. Verein Wiss. Verläger, Berlin.
Scheck, F. (2005). *Mechanics. From Newton's laws to deterministic chaos*. Springer, Berlin, Heidelberg, New York.
Schmutzer, E. (1976). *Osnovnye printsipy klasicheskoĭ mekhaniki i klasicheskoĭ teoriĭ polia (Basic principles of classical mechanics and classical field theory)*. "Mir", Moskva.
Schoenfliess, A. and Grübler, M. (1901,1908). *Kinematik*. In: Encykl. der math., Wiss. IV, B.G. Teubner, Leipzig.
Schuster, H.G. (1984). *Deterministic chaos*. Physik-Verlag, Weinheim.
Sedov, L.I. (1959). *Similarity and dimensional methods in mechanics*. Academic Press, New York.
Sedov, L.I. (1975). *Mécanique des milieux continus*. I, II. "Mir", Moscou.
Seitz, F. (1949). *Théorie moderne des solides*. Masson, Paris.
Sergiescu, V. (1956). *Introducere în fizica solidului (Introduction to physics of solids)*. Ed. Tehnică, Bucureşti.
Sestini, G. (1960). *Lezioni di meccanica razionale*. Ed. Univ., Firenze.
Seydel, R. (1988). *From equilibrium to chaos. Practical bifurcation and stability analysis*. Elsevier, New York, Amsterdam, London.
Shapiro, I.S. (1965). *Vneshnaya balistika (Exterior ballistics)*. Gostekhizdat, Moskva.
Sharp, R.T. and Kolman, B. (1977). *Group theoretical methods in physics*. Academic Press, New York.
Shaw, R. (1984). *The dripping facet as a model of chaotic systems*. Aerial Press, Santa Cruz.
Shigley, J.E. (1967). *Simulation of mechanical systems. An introduction*. McGraw-Hill, New York.

Shub, M. (1987). *Global stability of dynamical systems*. Springer-Verlag, New York, Heidelberg, Berlin.
Siacci, F. (1895-1896). *Lezioni di meccanica*. Tipo.-Lit. R. Cardone, Napoli.
Siegel, C.L. (1956). *Vorlesungen über Himmelsmechanik*. Springer-Verlag, Berlin.
Signorini, A. (1947,1948). *Meccanica razionale con elementi di statica grafica*. I, II. Ed. Perella, Roma.
Silaş, Gh. (1968). *Mecanică. Vibraţii mecanice* (*Mechanics. Mechanical vibrations*). Ed. Did. Ped., Bucureşti.
Slater, J.C. and Frank, N.H. (1947). *Mechanics*. McGraw-Hill, New York.
Smale, S. (1980). *The mathematics of time: essays on dynamical systems, economic processes and related topics*. Springer-Verlag, New York, Heidelberg, Berlin.
Smith, D.E. (1958). *History of mathematics*. I, II. Dover Publ., New York.
Sneddon, I.N. (1951). *Fourier transforms*. McGraw-Hill, New York, Toronto, London.
Sneddon, I.N. (1974). *The use of integral transforms*. Tata, McGraw-Hill, New Delhi.
Soare, M., Teodorescu, P.P. and Toma, I. (1999). *Ecuaţii diferenţiale cu aplicaţii în mecanica construcţiilor* (*Differential equations with applications to mechanics of constructions*). Ed. Tehnică, Bucureşti.
Sofonea, L. (1973). *Principii de invarianţă în teoria mişcării* (*Principles of invariance in the theory of motion*). Ed. Academiei, Bucureşti.
Sofonea, L. (1984). *Geometrii reprezentative şi teorii fizice* (*Representative geometries and physical theories*). Ed. Academiei, Bucureşti.
Sommerfeld, A. (1962). *Mechanik*. 6th ed. Geest und Portig, K.-G., Leipzig.
Somigliana, C. (1936). *Memorie scelte*. S. Lates Ed., Torino.
Soós, E. and Teodosiu, C. (1983). *Calculul tensorial cu aplicaţii în mecanica solidelor* (*Tensor calculus with applications in mechanics of solids*). Ed. Şt. Enc., Bucureşti.
Souriau, J.-M. (1970). *Structure des systèmes dynamiques*. Dunod, Paris.
Souriau, J.-M. (1972). *Mécanique classique et géométrie symplectique*. Dunod, Paris.
Sparrow, C. (1982). *The Lorenz equation*. Springer-Verlag, New York, Heidelberg, Berlin.
Spiegel, M.R. (1967). *Theory and problems of theoretical mechanics*. McGraw-Hill, New York.
Staicu, C.I. (1976). *Analiză dimensională generală* (*General dimensional analysis*).Ed. Tehnică, Bucureşti.
Staicu, C.I. (1986). *Aplicaţii ale calculului matriceal în mecanica solidelor* (*Applications of matric calculus in mechanics of solids*). Ed. Academiei, Bucureşti.
Stan, A. and Grumăzescu, M. (1973). *Probleme de mecanică* (*Problems of mechanics*). Ed. Did. Ped., Bucureşti.
Stäckel, P. (1908). *Elementare Dynamik der Punktsysteme und starren Körper*. Enz. math. Wiss. IV/1, Leipzig.
Sternberg, S. (1965). *Celestial mechanics*. I, II. W. A. Benjamin, New York, Amsterdam.
Stoenescu, Al. (1957). *Elemente de cosmonautică* (*Elements of cosmonautics*). Ed. Tehnică, Bucureşti.
Stoenescu, Al., Buzdugan, Gh., Ripianu, A. and Atanasiu, M. (1958). *Culegere de probleme de mecanică teoretică* (*Collection of problems of theoretical mechanics*). Ed. Tehnică, Bucureşti.
Stoenescu, Al., Ripianu, A. and Atanasiu, M. (1965). *Culegere de probleme de mecanică teoretică* (*Collection of problems of theoretical mechanics*). Ed. Did. Ped., Bucureşti.
Stoenescu, Al. and Silaş, Gh. (1957). *Curs de mecanică teoretică* (*Lectures of theoretical mechanics*). Ed. Tehnică, Bucureşti.
Stoenescu, Al. and Ţiţeica, G. (1961). *Teoria giroscopului şi aplicaţiile sale tehnice* (*The gyroscope theory and its technical applications*). IInd ed. Ed. Tehnică, Bucureşti.

Stoker, J.J. (1950). *Non-linear vibrations in mechanical and electrical systems*. Intersci. Publ., New York, London.
Strelkov, S.P. (1978). *Mechanics*. "Mir", Moskva.
Suslov, G.K. (1950). *Mecanica rațională (Rational mechanics)*. I, II. Ed. Tehnică, București.
Sylvester J.J. (1908). *Collected mathematical papers*. 2. Cambridge Univ. Press, Cambridge.
Synge, J.L. (1936). *Tensorial methods in dynamics*. Toronto Univ. Press, Toronto.
Synge, J.L. (1958). *Classical dynamics*. McGraw-Hill, New York.
Synge, J.L. (1960). *Classical dynamics*. In: Handbuch der Physik. III/1. Springer-Verlag, Berlin, Göttingen, Heidelberg.
Synge, J.L. and Griffith, B.A. (1959). *Principles of mechanics*. IInd ed. McGraw-Hill, New York.
Synge, J.L. and Schild, A. (1978). *Tensor calculus*. Dover Publ., New York.
Szabo, I. (1956). *Höhere technische Mechanik*. J. Springer, Berlin.
Szabo, I. (1966). *Einführung in die technische Mechanik*. 7th ed. Springer-Verlag, Berlin, Heidelberg, New York.
Szava, I., Ciofoaia, V., Luca-Motoc, D. and Curtu, I. (2000). *Metode experimentale în dinamica structurilor mecanice (Experimental methods in dynamics of mechanical structures)*. Ed. Univ., Brașov.
Szemplinska-Stupnika, W. (1988). *Chaotic and regular motion in non-linear vibrating systems*. CISM, Udine.
Tabor, M. (1989). *Chaos and integrability in nonlinear dynamics*. J. Wiley and Sons, New York, Chichester, Brisbane, Toronto, Singapore.
Targ, S. (1975). *Eléments de mécanique rationelle*. "Mir", Moscou.
Tenot, A. (1949). *Cours de mécanique*. Paris.
Teodorescu, N. (1954). *Metode vectoriale în fizica matematică (Vector methods in mathematical physics)*. I, II. Ed. Tehnică, București.
Teodorescu, P.P. and Ille, V. (1976-1980). *Teoria elasticității și introducere în mecanica solidelor deformabile (Theory of elasticity and introduction to mechanics of deformable solids)*. I-III. Dacia, Cluj-Napoca.
Teodorescu, P.P and Nicorovici, N.-A. (2004). *Applications of the theory of groups in mechanics and physics*. Fundamental Theories of physics. 140. Kluwer Academic Publ., Dordrecht, Boston, London.
Ter Haar, D. (1964). *Elements of Hamiltonian mechanics*. North Holland, Amsterdam.
Thirring, W. (1978). *Classical dynamical systems*. Springer-Verlag, New York, Wien.
Thom, R. (1972). *Stabilité structurelle et morphogénèse. Essais d'une théorie générale des modèles*. W. A. Benjamin, Reading.
Thompson, J.M.T. and Hunt, G.W. (1973). *A general theory of elastic stability*. J. Wiley and Sons, London.
Thompson, J.M.T. and Hunt, G.W. (1984). *Elastic instability phenomena*. J. Wiley and Sons, Chichester.
Thompson, J.M.T. and Stewart, H.B. (1986). *Nonlinear dynamics and chaos*. J. Wiley and Sons, Chichester, New York, Brisbane, Toronto, Singapore.
Thomsen, J.J. (1997). *Vibrations and stability*. McGraw-Hill, London.
Thomson, J.J. (1888). *Applications of dynamics to physics and chemistry*. Princeton Univ. Press, Princeton.
Thomson, W. and Tait, P.G. (1912). *Treatise on natural philosophy*. I. Cambridge Univ. Press, Cambridge.
Timoshenko, S.P. (1947). *Théorie de la stabilité élastique*. Libr. Polyt. Ch. Bérenger, Paris, Liège.

Timoshenko, S.P. (1957). *Istoriya nauki o soprotivlenii materialov* (*History of the science on strength of materials*). Gostekhizdat, Moskva.
Timoshenko, S.P. and Young, D.H. (1948). *Advanced dynamics*. McGraw-Hill, New York.
Timoshenko, S.P. and Young, D.H. (1955). *Vibration problems in engineering*. Van Nostrand, Toronto, New York, London.
Timoshenko, S.P. and Young, D.H. (1956). *Engineering mechanics*. McGraw-Hill, New York, Toronto, London.
Tisserand, F. (1891). *Traité de mécanique céleste*. 2. Gauthier-Villars, Paris.
Tocaci, E. (1985). *Mecanica* (*Mechanics*). Ed. Did. Ped., București.
Tokuyama, M. and Oppenheim, I. (ed.). (2004). *Slow dynamics in complex systems*. Melville, New York.
Toma, I. (1995). *Metoda echivalenței lineare și aplicațiile ei* (*Linear equivalence method and its applications*). Flores, București.
Troger, H. (1984). *Application of bifurcation theory to the solution of nonlinear stability problems in mechanical engineering*. Birkhäuser Verlag, Basel.
Truesdell, C. (1968). *Essays in the history of mechanics*. Springer-Verlag, Berlin, Heidelberg, New York.
Truesdell, C. (1972). *A first course in rational continuum mechanics*. The John Hopkins Univ., Baltimore.
Truesdell, C. and Noll, W. (1965). *The non-linear field theories of mechanics*. Handbuch der Physik. III/3. Springer-Verlag, Berlin, Göttingen, Heidelberg.
Tsiolkovskiĭ, K.E. (1962). *Izbrannye trudy* (*Selected works*). Izd. Akad. Nauk, Moskva.
Țăposu, I. (1996). *Teorie și probleme de mecanică newtoniană* (*Theory and problems of Newtonian mechanics*). Ed. Tehnică, București.
Țăposu, I. (1998). *Mecanica analitică și vibrații* (*Analytical mechanics and vibrations*). Ed. Tehnică, București.
Ungureanu, S. (1988). *Sensibilitatea sistemelor mecanice* (*Sensibility of mechanical systems*). Ed. Tehnică, București.
Vâlcovici, V. (1969, 1973). *Opere* (*Works*). I, II. Ed. Academiei, București.
Vâlcovici, V., Bălan, Șt. and Voinea, R. (ed.). (1963). *Mecanică teoretică* (*Theoretical mechanics*). IInd ed. Ed. Tehnică, București.
Verhulst, F. (1990). *Non-linear differential equations and dynamical systems*. Springer-Verlag, Berlin.
Vieru, D., Popescu, D., Poterașu, V.F. and Secu, A. (1999). *Culegere de probleme de mecanică* (*Collection of problems of mechanics*). Ed. "Gh. Asachi", Iași.
Vilenkin, N. Ya. (1969). *Fonctions spéciales et théorie de la représentation des groupes*. Dunod, Paris.
Voinaroski, R. (1967). *Mecanica teoretică* (*Theoretical mechanics*). Ed. Did. Ped., București.
Voinea, R.P. and Atanasiu, M. (1964). *Metode analitice noi în teoria mecanismelor* (*New analytical methods in the theory of mechanisms*). Ed. Tehnică, București.
Voinea, R.P. and Stroe, I.V. (2000). *Introducere în teoria sistemelor dinamice* (*Introduction in the theory of dynamical systems*). Ed. Acad., București.
Voinea, R.P., Voiculescu, D. and Simion, F.P. (1989). *Introducere în mecanica solidului cu aplicații în inginerie* (*Introduction in solid mechanics with applications in engineering*). Ed. Academiei, București.
Volterra, V. (1893, 1899). *Opere matematiche. Memorie e note*. I, II. Accad. Naz. dei Lincei, Roma.
Voronkov, I.M. (1953). *Kurs teoreticheskoĭ mekhaniki* (*Lectures on theoretical mechanics*). Gostekhizdat, Moskva.
Voss, A. (1901). *Die Prinzipien der rationellen Mechanik*. Enz. der math. Wiss. IV/1, Leipzig.

Vrănceanu, Gh. (1936). *Les espaces nonholonomes*. Mém. Sci. Math. 76. Gauthier-Villars, Paris.
Vrănceanu, Gh. (1952-1968). *Lecţii de geometrie diferenţială (Lessons of differential geometry)*. I-IV. Ed. Academiei, Bucureşti.
Vrănceanu, Gh. (1969-1973). *Opera matematică (Mathematical work)*. I-III. Ed. Academiei, Bucureşti.
Waddington, C.H. (ed.). (1968-1972). *Towards a theoretical biology*. I-IV. Edinborough Univ. Press, Edinborough.
Wason, W.R. (1965). *Asymptotic expansions for ordinary differential equations*. Intersci., New York.
Webster, A.G. (1904). *The dynamics of particles and of rigid, elastic and fluid bodies*. B.G. Teubner, Leipzig.
Weeney, R. Mc. (1963). *Symmetry: An introduction to group theory and its applications*. London.
Weierstrass, K. (1915). *Mathematische Werke*. 5. Mayer und Müller, Berlin.
Weizel, W. (1955). *Lehrbuch der theoretischen Physik*. 1. Springer-Verlag, Berlin, Göttingen, Heidelberg.
Weyl, H. (1923). *Raum, Zeit, Materie*. Berlin.
Weyl, H. (1964). *The classical groups*. Princeton Univ. Press, Princeton.
Whittaker, E.T. (1927). *A treatise on the analytical dynamics of particles and rigid bodies*. Cambridge Univ. Press, Cambridge.
Wiggins, S. (1988). *Global bifurcations and chaos. Analytical methods*. Springer-Verlag, New York, Berlin, Heidelberg, London, Paris, Tokyo.
Wiggins, S. (1990). *Introduction to applied nonlinear dynamical systems and chaos*. Springer-Verlag, New York.
Wigner, E. (1959). *Group theory and its applications to the quantum mechanics and atomic spectra*. Academic Press, New York.
Willers, Fr.A. (1943). *Mathematische Instrumente*. Oldenburg, München.
Winkelmann, M. and Grammel, R. (1927). *Kinetik der starren Körper*. Geiger-Scheel Handbuch der Physik. 5. J. Springer, Berlin.
Witner, A. (1941). *The analytical foundations of celestial mechanics*. Princeton Univ. Press, Princeton.
Wittenbauer-Pöschl. (1929). *Aufgaben aus der technischen Mechanik*. J. Springer, Berlin.
Yano Kentaro. (1965). *The theory of Lie derivatives and its applications*. North Holland, Amsterdam, Noordhoff, Groningen.
Yoshizawa, T. (1966). *Stability theory by Lyapunov's second method*. Japan Soc. Mech., Tokyo.
Young, H.D. (1964). *Fundamentals of mechanics and heat*. McGraw-Hill, New York.
Yourgraum, W. and Mandelstam, S. (1955). *Principles in dynamics and quantum mechanics*. Pitman and Sons, London.
Zeeman, E.C. (1977). *Catastrophe theory. Selected papers, 1972-1977*. Addison-Wesley, Reading.
Zeeman, E.C. (1981). *Bibliography on catastrophe theory*. Math. Inst. Univ. of Warwick, Coventry.
Zeveleanu, C. and Bratu, P. (2001). *Vibraţii neliniare (Non-linear vibrations)*. Impuls, Bucureşti.
Ziegler, H. (ed.). (1963). *Gyrodynamics*. IUTAM Symposium Celerina. Springer-Verlag, Berlin, Göttingen, Heidelberg.
Zoretti, L. (1928). *Les principes de la mécanique classique*. Mém. des Sci. Math. XXV, Paris.
*** (1963). *La méthode axiomatique dans les mécaniques classiques et nouvelles*. Colloq. Intern., Paris, 1959. Gauthier-Villars, Paris.

SUBJECT INDEX

Acceleration, 15,293
  areal, 294
  composition, 332,337
  deviation, 296
  generalized, 364
  gravity, 40,46
  in a discontinuous motion, 299
  in curvilinear co-ordinates, 296
    cylindrical, 298
    polar, 298
    spherical, 298
  of a rigid solid, 307
  of higher order, 298
Articulated system, 264
  open, 272
  polygonal line, 272
  spatial, 269
  truss, 265
Axoid, 318
  fixed, 318
  herpolhodic cone, 320
  movable, 318
  polhodic cone, 320

Body, 35
  classification, 35
  incompressible, 32
  rigid solid, 35
    bar, 36
    block, 37
    plate, 36
  state of aggregation, 35

Constraint, 166
  axiom of liberation, 167
  bilateral, 167
  case of a single particle, 176
  catastatic, 183
  critical, 167,183
  degree of freedom, 168
    kinematic, 173
    static, 168,169
  Euler's angles, 170
  Frobenius' theorem, 180
  holonomic, 167,177
  ideal, 185

  non-holonomic, 167,177
  rheonomic, 167,182
  scleronomic, 167,182
  with friction, 194
    of a particle on a curve, 195
    of a particle on a surface, 196
Co-ordinate, 11
  canonical, 78
  Cartesian, 11
  cylindrical, 11
  Euler, 34
  generalized, 364
  Lagrange, 34
  spherical, 11

Displacement, 162
  generalized, 174
    possible, 175
    real, 175
    virtual, 176
  possible, 162
  real, 162
  virtual, 163
Distribution, 18
  basic solution, 736
    ordinary differential equations, 736
    partial differential equations, 738
  characteristic function, 22
  convolution product, 732
  δ Dirac, 19
  δ representative sequence, 23
  direct product, 731
  Fourier transform, 733
  Heaviside function, 21
  Laplace transform, 735
  regular, 19
  singular, 19
  Stieltjes integral, 25

Field, 709
  Almansi's theorem, 722
  Boggio's theorem, 723
  conservative vector, 711
  d'Alembertian, 720
  derivative, 710
    absolute, 719

relative, 718
total, 710
equipotential surface, 712
Gauss-Ostrogradskiĭ formula, 726
Green's formulae, 727
integral depending on a parameter, 728
Laplacian, 719
nabla (del) operator, 710
Newtonian potential, 730
Nicolesco's operator, 721
Stokes' formulae, 725
Force, 37
basis, 77
central, 42,469
Bertrand's theorem, 476
Binet's equation, 470
Binet's formula, 473
capture, 478
diffusion, 479
in a resistent medium, 479
problem of two particles, 476
conservative, 39,355
constraint, 167
Coulombian, 41
damping, 488
decomposition, 75
dissipative, 361
elastic, 42,481
external, 39
field, 38
given, 167
gyroscopic, 361
internal, 39
linear dependence, 76
linear independence, 76
Newtonian, 40
non-stationary, 40
of inertia, 48
perturbing, 513
quasi-conservative, 39
relative, 616
repulsive elastic, 485
stationary, 40
Frame of reference, 4,10
fixed (inertial), 12,615
movable (non-inertial), 615
orthonormed, 11
right-handed, 4

Geometry, 205
Darboux's trihedron, 208
Frenet's trihedron, 206
of a curve, 205
of a surface, 208

Homogeneity, 62
dimensional equation, 63,64
Π theorem, 67

Interaction, 62
electromagnetic, 43

gravitational, 43
strong, 44
weak, 43

Kinematics, 287
kinematic chain, 344
of a particle, 287
Kinetic energy, 358

Mass, 15,18
centre, 115
centre of gravity, 117
continuous medium, 28
gravity, 17
inertial, 16
material line, 28
material point, 28
material surface, 28
Pappus-Guldin theorem, 121, 122
static moment, 118
unit, 29
Mechanical efficiency, 359
Mechanical energy, 358
Mechanical impedance, 519
Mechanism, 190, 345
Membrane analogy, 3
Model, 2
analogic, 3
deterministic, 49
ideal, 2
mathematical, 1,2,3,4,49
of first order, 3
of mechanics, 1
of second order, 3
technical, 2,3
vectorial, 4
Moment, 89
with respect to a pole, 89
with respect to an axis, 91
of a system of vectors, 99
Moment of inertia, 125
ellipse of inertia, 153
ellipsoid of gyration, 150
ellipsoid of inertia, 149
Huygens-Steiner theorems, 158
Land's circle, 157
Mohr's circles, 150,156
plane case, 144
principal directions, 145
tensor of, 142
three-dimensional case, 142
Moment of momentum, 354
Momentum, 18,353
Motion, 2,10
acceleration, 307
bilocal conditions, 361
biological, 1
chemical, 1
circular, 302

Subject index

constraints with friction, 396
cycloidal, 304
elevator problem, 617
first integral, 367
Galileo-Newton group, 372,624
general theorems, 621
helical, 303,452
horary equation, 14
initial conditions, 364
in resistant medium, 410
instantaneous helical, 317
Lagrange's equations, 364
mechanical, 1
meeting problem, 619
of a heavy particle
   in vacuum, 406
   on a surface of rotation, 434
of a particle subjected to constraints, 384
   on a curve, 389
   on a surface, 392
of a rigid solid, 305
   with a fixed point, 319
of Darboux's trihedron, 316
of Frenet's trihedron, 315
of rotation, 311
of rototranslation, 313
of systems of rigid solids, 340
of translation, 310
on a curve, 417
   geodesic, 449
   tautochronous, 443
on a plane, 419
parabolic, 300
physical, 1
plane, 404
plane-parallel, 321
principle of equivalence, 626
rectilinear, 301,401,414
relative, 330,333,615
   equilibrium, 626
transformation, 454
transmission, 346
velocity, 305
with a single degree of freedom, 398
with discontinuity, 595
   basic equations, 599
   collision model, 597
   elastic collision, 611
   general theorems, 606
   generalized force, 598
      constraint, 603
   particle, 595
      free, 595
      heavy in vacuum, 604
      subjected to constraints, 601
with a single degree of freedom, 398

Newtonian universal attraction, 543
   artificial satellite, 578
   Bohr's atom, 591

anomaly
   eccentric, 564
   mean, 569
   true, 564
cosmic velocities, 580
Hertz's oscillator, 588
hyperbolic motion, 571
Kepler's equation, 564
Kepler's laws, 543,566
law of universal gravitation, 547
Newtonian force of attraction
   curvilinear motion, 557
   rectilinear motion, 555
Planck's constant, 590
potential
   Newtonian, 549
   of simple stratum, 552
   of the terrestrial spheroid, 554
   volume, 553
quantic number, 593
Ritz's law, 593
Runge-Lenz vector, 561

Oscillator, 483
   beats, 500
   circular, 485
   composition of orthogonal oscillations, 502
   computation methods, 535
   damped, 504
   electromechanical analogy, 521,522
   elliptic, 483
   forced, 516
   Fresnel's representation, 501
   harmonic analysis, 499
   interference, 499
   linear, 495
   modulated oscillations, 497
   non-linear, 530
   pseudoelliptic, 489
   resonance, 515
   self-sustained, 492,511
   transmissibility, 520
   variable characteristics, 529
   vector representation, 498

Particle of variable mass, 659
   aircraft with jet propulsion, 689
   balloon problem, 681
   central forces, 685
   external ballistic problem, 581
   general theorems, 668
   inverse problem, 665
   Levi-Civita's equation, 661
   mathematical pendulum, 682
   Meshcherskiĭ's generalized equation, 662
   Meshcherskiĭ's model, 660
   motion along a vertical
      in vacuum, 675
      in resistant medium, 679
   propelled ship, 690

Tsiolkovskiĭ's first problem, 666
Tsiolkovskiĭ's second problem, 674
Pendulum, 422
　cycloidal, 432
　elliptic, 432
　in a resistant medium, 430
　in motion of rotation, 453
　spherical, 436
Potential, 39
　generalized, 42
　　quasi-potential, 42
　quasi-potential, 39
Principle
　of d'Alembert, 32
　of Euler, 33
　mass conservation, 31
　of relativity, 370
　of virtual work, 226
Problem
　Abel, 441
　basic, 361
　Puiseux, 442
Power, 359

Quantity
　basic, 59
　derived, 59
　mechanical, 4
　primary, 59
　secondary, 59
　unit, 59

Representative
　point, 169
　space, 169
Rest, 10

Scalar, 8
Science
　interdisciplinary, 2
　of nature, 1
Similitude, 70
　Cauchy's model, 73
　Froude's model, 73
　Newton's model, 72
　Reynold's model, 73
　Weber's model, 74
Space, 9
Stability of equilibrium, 218
　Lagrange-Dirichlet theorem, 219
　of a particle
　　free, 457
　　with constraints, 458
　topological methods, 460
　Torricelli's theorem, 219
Statics, 201
　Lagrange's equations of first kind, 227
　of a particle, 201
　　free, 201
　　with constraints, 204

　　frictionless, 204
　　unilateral, 216
　　with friction, 222
　with ideal constraints, 212
　　on a curve, 214
　　on a surface, 212
　of a rigid solid, 229
　　free, 229
　　with a fixed axis, 236
　　with a fixed point, 236
　　with constraints with friction, 238
　　with ideal constraints, 232
　of discrete mechanical systems, 226
　　free, 226
　　with constraints, 225
　of systems of rigid solids, 241
　simple devices, 246
Support, 190
　built-in, 193
　hinge
　　cylindrical, 192
　　spherical, 191
　of a rigid solid with friction, 197
　　pivoting, 199
　　rolling, 198
　　sliding, 197
　simple, 190
System, 2
　continuous, 1,29
　discrete, 1,29
　dissipative, 400
　dynamics of, 2
　mechanical, 1,4,28
　natural, 43
　of forces, 104
　　coplanar, 104
　　parallel, 105
　of units, 60
　of vectors, 95
　　bound, 97
　　free, 95
　　sliding, 98
　physical, 1
　statically determinate, 190
　statically indeterminate, 190

Tensor, 129
　addition, 133
　antisymmetric (skew-symmetric), 131
　asymmetric, 131
　Cauchy's quadric, 140
　contracted product, 134
　contraction, 134
　deviator, 137
　dyadic product, 133
　Einstein's convention of summation, 5
　equality, 133
　field, 132
　invariants, 139
　Kronecker's symbol, 80

of nth order, 129
of second order, 135
principal directions, 140
product, 133
quotient law, 134
Ricci's symbol, 83,102
spheric, 137
symmetric, 131
trace, 134
Terrestrial mechanics, 629
    acceleration of the centre of the Earth, 636
    Baer's law, 649
    centrifugal force, 639
    Coriolis' force, 647
    cyclone, 651
    deviation towards the east, 645
    Foucault's pendulum, 651
    geocentric frame, 630
    heliocentric frame, 631
    imponderability, 638
    motion at the Earth surface, 631
    plummet's deviation, 640
    relative equilibrium, 634
    terrestrial acceleration, 642
    tide, 637
    trade winds, 650
Theorem
    conservation, 376
    of areas, 383
    of existence and uniqueness, 365
    universal, 372
Thread, 194,276
    on a surface, 283
Time, 9
Torsor of systems of vectors, 99
    central axis, 103
    minimal form, 103
    sliding, 93
Trajectory, 13,287

Vector analysis, 693
    curvilinear co-ordinates, 701
    derivative, 697
        space, 697
        time, 697
        total, 697
    differential form, 708
    differentiation, 694

exterior derivative, 708
external product, 705
integration, 699
limit, 694
sequence, 698
series, 698
    Maclaurin, 698
    Taylor, 698
Vector space, 9
    addition, 6
    axial, 82
    bound, 5
    canonical representation, 81
    Cauchy's inequality, 83
    direction cosines, 79
    equality, 5
    equipollent, 6
    free, 4
    Lagrange's identity, 83
    orthogonal projection, 82
    polar, 82
    product by a scalar, 8
    projection, 7
    scalar product, 79,81
    scalar triple product, 83
    sliding, 5
    vector product, 81
    vector triple product, 86
Velocity, 15,288
    areal, 292
    composition, 331,333
    generalized, 364
    hodograph, 293
    curvilinear co-ordinates, 289
        cylindrical, 291
        polar, 291
        spherical, 291
    of a rigid solid, 305
    possible, 175
    real, 174
    virtual, 175

Weight, 40
    unit, 40
Work, 165,354
    real, 165
    virtual, 185

NAME INDEX

Abel, Niels Henrik, 425,440,441
Abélard, Pierre (Petrus Abaelardus), 53
Abraham, R., 745
Abraham, R.H., 745
Achilles, 51
Adhémar, R.d', 745
Afanasjeva, 67
Agostinelli, Cataldo, 745
Alberti, Leone Battista, 54
Alembert, Jean Baptiste leRond d', 32,55,384, 388,413,720,745
Alexander the Great, 52
Al-Hosein ibn Abdallah ibn al-Hosein ibn Ali Abu Ali al-Sheich al-Rais (Avicenna), 53
Almansi, Emilio, 722
Amaldi, Ugo, 754
Ames, J.S., 745
Ampère, André Marie, 56
Anastasiei, M., 755
Anaxagoras of Clazomene, 51
Anaximender, 51
Anaximenes, 51
Andelić, T., 745
Andonie, George Şt., 745
Andreescu, D., 756
Andronov, A., 745
Anosov, D.V., 745
Apollonius, Pergaeus (Pamphylia), 53
Appel, Paul-Émile, 56,455,456,745
Archimedes, 52,252
Archytas of Tarentum, 52,53
Arcy, Paul d', 55
Argyris, J., 745
Aristotle of Stageira, 52,53,54
Armellini, G., 567
Arnold, V.I., 745
Aron, I.I., 756
Arouet, François Marie (Voltaire), 55
Atanasiu, Mihai, 746,760,762
Atkin, R.H., 746
Atwood, George, 56
Avez, A., 746

Babakov, L.M., 746
Babitsky, V.I., 746
Babuška, Ivo, 746

Bacon, Francis, 54
Bacon, Roger, 53
Baer, Karl Ernst Ritter von, 647,649,650
Baker, G.L., 746
Baker Jr., R.M.L., 746
Balmer, Johann Jakob, 593
Banach, Stephan, 746
Barbu, L., 759
Barbu, Viorel, 746
Barenblatt, G.L., 746
Barnsley, M.F., 746
Bat, M.I., 746
Bălan, Ştefan, 746,762
Beju, Iulian, 746
Belenkiĭ, I.M., 746
Bellet, D., 746
Bellman, R., 746
Beltrami, Enrico, 746
Benedetti, Giovani Battista, 54
Berezkin, E.N., 746
Berge, P., 747
Bergman, P.G., 747
Bernoulli, Daniel, 55
Bernoulli, Jacques (Jacob), 55
Bernoulli, Jean (Johannes), 55
Bernousson Jr., 747
Bertrand, Joseph Louis François, 72,473,476, 543
Bessel, Wilhelm Friedrich, 743
Béghin, H., 747
Biezeno, G.B., 747
Binet, Jacques Phillipe Marie, 56,469,470,473, 480
Birkhoff, Georg David, 747
Bishop, A.R., 747
Blitzer, L., 748
Boerner, H., 747
Boggio, Tomaso, 723
Bogolyubov (Bogoliuboff), Nikolaĭ Nikolaevich, 538,747,753
Bohr, Niels Henrik David, 590
Bolotin, V.V., 747
Boltzmann, Lothar, 747
Boltzmann, Ludwig Eduard, 57
Bone, J., 747
Bonnet, Ossian, 442
Borelli, Giovanni Alfonso, 54

Born, Max, 747
Borş, Constantin, 747
Boucher, M., 747
Bouligand, Georges, 747
Bourlet, C.E.E., 747
Bradbury, T., 747
Bradwardine, Thomas, 53
Brahe, Tycho, 54, 543
Bratu, Polidor, 747,763
Brădeanu, D., 747
Brădeanu, Petre, 747
Brelot, M., 747
Bricard, R., 747
Bridgman, Percy Williams, 67,68
Brillouin, L., 747
Broglie, Louis Victor Pierre Raymond, prince de, 57
Brot, M.A., 752
Brousse, Pierre, 747
Brown, J.L., 400
Bruneleschi, Filippo, 53
Bruno, Giordano, 54
Buckingham, 67
Budó, A., 747
Bukholts, N.N., 747
Bulgakov, B.V., 747
Buridan, Jean, 53
Burileanu, Ştefan, 747
Burlacu, L., 755
Butenin, N.V., 747
Butterfield, H., 747
Buzdugan, Gheorghe, 747,751,760

Cabannes, Henri, 747
Caldonazzo, B., 748
Campanella, Tomaso, 54
Capelli, A., 265
Cardan, Girolamo, 54,325,349
Carnot, Lazare Nicolas Marguerite, 56,613,614
Carnot, Nicolas Léonard Sadi, 56
Cartan, Élie, 708,748
Carus, Lucretius, 52
Cauchy, Augustin Louis, 70,73,83,140,365,461, 738
Cavendish, Henry, 41,546
Cayley, Arthur, 748
Celsius, Anders, 65
Chaikin, C., 745
Chaplygin, Sergei Alekseevich, 56,748
Charcosset, 63
Charlier, C.L., 748
Chasles, Michel, 6,317,318
Chazy, Jean, 47,748
Chebyshev, Pafnutiĭ L'vovich, 345
Chetaev, N.G., 748
Chevilliet, 657
Chiroiu, Călin, 748
Chiroiu, Veturia, 748,755
Chow, S.N., 748
Christoffel, E.B., 364,450,704,705

Chua, L.O., 757
Ciofoaia, Vasile, 701
Clairaut, Alexis Claude, 55,452,643
Coe, C.J., 748
Coleman, Bernard, 57
Collatz, Lothar, 748
Commandin, Frederico, 54
Conti, R., 759
Copernicus, Nicholas, 12,631
Coppel, W.A., 748
Corben, H.C., 748
Coriolis, Gustav Gaspard, 56,332,615,629,647, 649, 650,651
Coulomb, Charles-Augustin, 55,195,396,506
Courant, Richard, 748
Crandall, S.H., 748
Crăciun, Eduard Marius, 748
Crede, C.E., 751
Cremona, Luigi, 269
Ctesibios, 52
Culmann, Carl, 156,274
Curtu, Ioan, 761
Curtu, R., 748
Cushing, J.T., 748

Darboux, Gaston Jean, 208,315,748
Dautheville, S., 745
Davidson, M., 748
Davis Jr., L., 748
Deciu, Eugen, 758
Delassus, E., 748
Demko, S.G., 746
Democritus of Abdera, 52
Descartes, René, 54
Destouches, J.L., 748
Devaney, R., 748
Dicke, 642
Dincă, Florea, 748
Dirac, Paul Adrien Maurice, 19,57
Djanelidze, G.Yu., 746
Doblaré, M., 748
Dobronravov, V.V., 748
Dolapčev, Blagovest, 749
Donescu, Ştefania, 748,755
Dragnea, Ovidiu, 749
Dragoş, Lazăr, 584,749
Draper, C.S., 749
Drăganu, Mircea, 749
Drâmbă, Constantin, 749
Drobov, 467
Duffing, G., 535,748
Dugas, R., 749
Duhamel, Jean-Marie-Constant, 524
Dühring, Eugen, 56
Dvor'nkov, A.L., 748

Eckhaus, W., 748,749
Eddington, A.S., 749
Edgar, G.A., 749
Ehrenfest, 67

Name index

Einstein, Albert, 5,49,57,372,575,626
El Naschie, M.S., 749
Emmerson, J.Mc L.,749
Eötvös, Loránd, 56,626,642
Epicur, 52
Eratostenes of Cyrene, 52
Eringen, A. Cemal, 749
Euclid, 52,703
Euctemon, 51
Eudoxus of Cnidus, 52
Euler, Leonhard, 55,285,356,363,413,442,445, 446,447,448,749

Fahrenheit, Gabriel Daniel, 65
Falconer, K.J., 749
Falk, G., 749
Faust, G., 745
Federman, 67
Fermat, Pierre, 54
Fernandez, M., 749
Fertis, D.G., 749
Feshbach, H., 755
Fetcu, Lucia, 747
Feynman, R.P., 749
Fillipov, A.P., 749
Filon of Byzantium, 52
Finkelshtein, G.M., 749
Finzi, Bruno, 749
Fischer, Ugo, 749
Florian, V., 758
Follin, Jr., J.M., 748
Fomin, S.V., 750
Fontana, Nicolo (Tartaglia), 54
Foucault, Jean Bernard Léon, 56,651,652,655, 658
Fouret, 442,457
Fourier, Jean Baptiste Joseph, 56,501,733
Fowler, 593
Föppl, August, 749
Frank, N.H., 760
Frank, Ph., 749
French, A., 750
Frenet, Frédéric Jean, 205,206,207,315
Fresnel, Augustin Jean, 501
Frère Gabriel-Marie (F.G.-M.), 749
Frobenius, 180
Froude, 70,73,534
Fues, E., 756
Fufaev, N.A., 755

Gabos, Zoltan, 750
Galerkin, Boris Gr'gorevich, 538
Galileo, Galilei, 45,48,54,353,365,370,371,426, 548,624,625,750
Gallavotti, G., 750
Gantmacher, F., 750
Garthe, 659
Gauss, Carl Friedrich, 56,209,551,726
Gelfand, I.M., 750
Georgescu, A., 750

Gerber, 242
Germain, Paul, 750
Gibbs, Josiah Dixon Willard, 57,750
Gilmore, R., 750
Gioncu, Victor, 750
Giorgi, 44
Girard, A., 750
Glashow, Samuel, 44
Gleick, J., 750
Godbillon, C., 750
Goldstein, H., 750
Gollub, J.P., 746
Golubev, V.V., 750
Golubitsky, M., 750
Gordon, I., 745
Graffi, Dario, 750
Grahn, R., 752
Gramm, 85
Grammel, R., 747,750,763
Gray, A., 750
Grebenikov, E.A., 750
Green, George, 523,727
Greenhill, A.G., 439,750
Greenwood, D.T., 750
Griffith, B.A., 761
Gross, Daniel, 44
Grumăzescu, Mircea, 760
Gruner, G., 747
Grübler, M., 759
Guckenheimer, J., 750
Guillemin, V., 750
Guldin, Paul (Habakuk), 53,54,121,122
Gumowski, I., 750
Gurel, O., 750
Gutzwiller, M.C., 751

Haase, M., 745
Habets, P., 759
Haikin, S.E., 745,751
Hakan, H., 751
Halanay, Aristide, 751
Hale, J.K., 748,751
Halley, Edmund, 577
Halliday, D., 751
Halphen, Georges Henri, 439,751
Hamburger, Leon, 751
Hamel, Georg, 56,751
Hamilton, William Rowan, 56,711,751
Hammermesh, M., 751
Hangan, Sanda, 751
Hao, B.L., 751
Harris, C.M., 751
Hartog Den, J.P., 751
Hassard, B.D., 751
Hasselblatt, B., 752
Hayashi, C., 751
Heaviside, 21
Heisenberg, Werner Karl, 57
Helleman, R.H.G., 751,752

Helmholtz, Hermann Ludwig Ferdinand von, 56, 740
Hemming, G., 751
Henneberg, 269
Henon, M., 751
Heraclit of Efes, 51
Herglotz, G., 751
Hermann, Jacob, 55
Hermite, Charles, 439
Herodotus, 51
Heron of Alexandria, 52
Hertz, Heinrich Rudolf, 56,177,588,751
Hilbert, David, 748
Hill, George William, 535
Hirsch, M.W., 751
Hlavaček, V., 751
Holden, A.V., 751
Holmes, P., 750
Hooke, Robert, 55
Horovka, J., 749
Hort, W., 751
Hospital, Guillaume Francis de l', 55
Hunt, G.W., 761
Hurwitz, Adolf, 56
Huseyin, K., 751
Huygens, Christian, 55,158,159,160,434,751
Hypparchus of Rhodes, 52
Hypsicles of Alexandria, 52

Iacob, Andrei, 751
Iacob, Caius, 584,751
Iakovenko, G.N., 758
Ille, Vasile, 761
Ioachimescu, Andrei G., 751,752
Ionescu-Pallas, Nicolae, 752
Iooss, G., 746,752
Irimiciuc, Nicolae, 754
Irwin, M.C., 752
Isaacson, N., 752
Ishlinskiĭ, A.Yu., 752
Ivanov, I., 746

Jacobi, Carl Gustav Jacob, 56,752
Jammer, M., 752
Janssens, P., 752
Jansson, P.-A., 752
Jaunzemis, W., 752
Jeffreys, B.S., 752
Jeffreys, W., 752
Joannes Philiponus (Grammaticus) of Alexandria, 53
Joos, G., 752
Jordan of Namur (Jordanus Nemorarius), 53
Joseph, D.D., 746,752
Jukovskiĭ, Nikolai Egorovich, 752
Julia, Gaston, 752
Jung, G., 752
Jurgen, H., 757

Kahan, Theo, 752
Kamerlingh Onnes, Heike, 658
Kamke, E., 752
Kantorovich, L.V., 752
Kapitaniak, T., 752
Kapli, D.M., 756
Kasner, E., 752
Katok, A., 752
Kauderer, H., 752
Kazarinoff, N.D., 751
Kármán, Theodor von, 752
Kästner, S., 752
Kecs, Wilhelm, 108,281,753
Keller, H.B., 752,753
Kellog, O.D., 753
Kelzon, A.S., 746
Kepler, Johannes, 54,55,72,543,544,562,564, 566,631
Khanukaev, Yu.I., 758
Kilchevski (Kilcevski), N.A., 753
Kirchhoff, Gustav Robert, 522,753
Kirpichev, Viktor L'vovich, 71,753
Kittel, Ch., 753
Klein, Felix, 56,753
Knight, W.D., 753
Kochin, Nikolaĭ Evgrafovich, 753
Koçak, H., 751
Koenig, Samuel, 55
Kolman, B., 759
Kooy, J.M.J., 753
Koppernigk, Niklas (Nicholaus Copernicus), 53
Koshlyakov, V.N., 753
Kosmodemyanskiĭ, A.A., 753
Kranz, C., 753
Krbek, F., 753
Krebs, Nikolaus (Nicolaus Cusanus), 53
Kron, G., 753
Kronecker, Leopold, 80,265,625
Krukowski, Sonya (Kovalevsky, Sophia; Kovalevskaya, Sof'ya Vasil'evna), 56,753
Krylov, Vladimir Ivanovich, 752
Kuiper, G.P., 753
Kutterer, R.E., 753
Kuzimenkov, I.S., 757
Kuznetsov, Y.A., 753

Lagrange, Joseph-Louis, 34,56,83,138,186,219, 227,356,364,384,445,447,457,459,753
Lakhtin, L.M., 753
Laloy, M., 759
Lamb, H., 753
Lamé, Gabriel, 703
Lampariello, G., 754
Land, 157
Landau, Lev Davidovich, 754
Langevin, Paul, 410
Lantsosh, K., 754
Lapin, A.S., 686
Laplace, Pierre Simon de, 55,719,735,736

Name index

La Salle, J.P., 754
La Valée-Poussin, Charles Jean, 754
Lecornu, L., 754
Leech, J.W., 754
Lefschetz, S., 754
Leibnitz, Gottfried Wilhelm, 55
Leimanis, E., 754
Leipholz, H., 754
Lejeune-Dirichlet, Peter Gustav, 56,219,457, 459,501
Lenz, 557,561
Leontovich, E., 745
Le Verrier, Urbain Jean-Joseph, 547
Levi-Civita, Tullio, 56,458,660,661,662,754
Liénard, 400,466
Lifshitz, E., 754
Lindsay, R.B., 754
Liouville, Joseph, 56
Lippmann, Horst, 754
Lipschitz, Rudolf, 365
Lissajous, Jules Antoine, 503
Littlewood, D.E., 754
Loft, E.E., 756
Loitsyanski, L.G., 754
Lomonosov, Mikhail Vasilievich, 55
Lomont, J.S., 754
Lorentz, Hendrick Anton, 57
Luca-Motoc, D., 761
Lurie, Anatoli I., 754
Lyapunov, Aleksandr Mikhailovich, 56,536,754
Lyubarskiĭ, G.Ya., 754

Mach, Ernst, 56,754
Macke, W., 754
Maclaurin, Colin, 55,459,698
Macmillan, W.D., 754
Macomber, G.R., 749
Madan, R., 754
Magnus, K., 754
Maier, A., 745
Ma Kim, 53
Malbranche, Nicole, 55
Malkin, I.G., 754
Malvern, L.E., 754
Mandel, Jean, 754
Mandelbrot, B.B., 754
Mandelstam, S., 763
Mangeron, Dumitru, 750,754
Marchyuk, G.Z., 754
Marcolongo, Roberto, 754,755
Marcus Vitruvius Pollio, 52
Marinca, V., 756
Marion, J.B., 755
Marsden, J., 745
Mathieu, E., 535
Maunder, L., 746
Maupertuis, Pierre Louis Moreau de, 55
Mawhin, J., 758
Maxwell, James Clerk, 56,63,269
McCracken, M., 755
Meirovich, I., 755

Melo, W. de, 757
Mercheş, I., 754
Mercier, A., 755
Merkin, D.N., 755
Mersenne, Marin, 54
Meshcherskiĭ, Ivan Vselovod, 56,660,661,662, 663,665,668,684,689,755
Meton, 51
Meusnier de la Place, Jean Baptiste Marie Charles, 210,285
Michelson, Albert Abraham, 49,57
Mihăilescu, E., 747
Mihăilescu, Mircea, 754
Miller, Dayton Clarence, 49
Miller Jr., W., 755
Milne, E.A., 755
Minkowski, Hermann, 57
Minorsky, N., 755
Mira, C., 750
Miron, Radu, 755
Mises, Richard von, 479
Mitropolskiĭ (Mitropolsky), Yu.A., 538,747,755
Mo Tsy, 51
Mohammed ibn Ahmed ibn Mohammed ibn Roshd Abu Velid (Averroes), 53
Mohammed ibn Mohammed ibn Tarkhan ibn Auzlag Abu Nasr al Farrabi, 53
Mohr, Otto, 150,156,157
Monge, Gaspard, 55
Moon, F.C., 755
Moreau, J.J., 755
Morel, J., 747
Morgenstern, D., 755
Morley, Edward William, 49,57
Morse, P.M., 755
Moses ben Maimum (Maimonides), 53
Mourre, L., 755
Munteanu, Ligia, 748,755
Munteanu, M., 755
Murnaghan, F.D., 745,755
Muszyńska, Agnieszka (Agnes), 756

Naimark, M., 755
Naumov, A.L., 755
Nayfeh, A.H., 755
Neimark, Ju.I., 755
Nekrasov, A.I., 756
Nelson, W.C., 756
Newbull, H.O., 756
Newton, Isaac sir, 45,48,55,71,72,73,353,365, 370,384,388,543,544,546,548,550,553,555, 577,599,624,625,662,730,756
Nguyen, Quoc-son, 756
Nguyen, Van Hieu, 756
Nicolaenko, B., 747
Nicolescu, Miron, 721
Nicolis, G., 756
Nicomachus of Gerasa, 53
Nicorovici, Nicolae-Alexandru, 761
Nielsen, J., 756
Nikitin, E.M., 756

Nikitin, N.N., 748
Nitecki, Z., 756
Niță, Mihai M., 756
Nobel, Alfred Bernhard, 44
Noll, Walter, 57,762
Nordheim, L., 756
Novikov, S.P., 746
Novoselov, V.S., 756
Nowacki, Witold, 756
Nured-din al-Betruji Abu Abdallah (Alpetragius), 53
Nyayapathi, V., 756

Obădeanu, Virgil, 756
Œnopides of Chios, 51
Okunev, V.N., 756
Olariu, S., 756
Olinde, Rodrigues B., 56
Olkhovskiĭ, I.I., 756,757
Olson, H.F., 757
Onicescu, Octav, 57,575,757
Oppenheim, L., 762
Oresme, Nicolas, 53
Osgood, W.F., 757
Ostrogradskiĭ, Mikhail Vasilievich, 56,446,536, 551,726,757
Ott, E., 757
Ovsiannikov, V., 757

Painlevé, Paul, 56,757
Palis, J., 757
Pandrea, Nicolae, 757
Pappus of Alexandria, 53,121,122
Parker, T., 757
Parmenides of Elea, 51
Pars, L., 757
Parseval, 734
Pascal, Blaise, 473
Pastori, Maria, 749
Pavlenko, M.G., 757
Peano, Giuseppe, 366
Pearson, Karl, 56
Peitgen, H.O., 757
Peixoto, M.M., 757
Pelacani, Biagio (Blasius of Parma), 53
Pell, 467
Perry, J., 757
Petkevich, V.V., 757
Pérès, J., 757
Pfaff, Johann Friedrich, 172,374
Phaeinus, 51
Pignedoli, A., 745
Planck, Max Karl Ernst Ludwig, 43,57,588,590, 592,757
Plato, 52,53
Plăcințeanu, Ion, 757
Plücker, Julius, 95
Pohl, R.W., 17,757
Poincaré, Jules Henri, 56,57,463,464,540,757, 758

Poinsot, Louis, 55,149
Poisson, Siméon-Denis, 3,56,310,445,446,719, 740,741
Polak, L.S., 758
Politzer, David, 44
Pomeau, Y., 747,751
Poncelet, Jean Victor, 56
Pontriaguine, L.S., 758
Pop, Ioan, 747
Popescu, D., 762
Popoff, K., 758
Popov, V.M., 758
Poseidonius of Rhodes, 52
Poston, T., 758
Poterașu, V.F., 762
Pöschl, T., 758,763
Prager, Milan, 746
Prandtl, Ludwig, 3
Prange, G., 758
Prigogine, Ilya, 756,758
Proclus of Byzantium (the Successor), 53
Ptolemy (Claudius Ptolemaeus), 11,53,630
Ptolemy's Soter (Ptolemy the Preserver), 52
Puiseux, Pierre Henri, 439,440,441,442
Pyatnitskiĭ, E.S., 758
Pytagoras, 51

Racah, G., 758
Radeș, Mircea, 747
Radu, Anton, 758
Rankine, William-John Macquorn, 56,156
Rausenberger, O., 758
Rawlings, A.L., 758
Rădoi, M., 758
Reich, 646
Resnick, R., 751
Reynolds, Osborne, 70,73
Réaumur, René Antoine Ferchault de, 65
Résal, Amé Henry, 440
Ribaucourt, Albert, 208
Ricci, Gregorio, 83
Richard of Middletown, 53
Richardson, K.I.T., 758
Richter, P.H., 757
Riemann, Georg Friedrich, 27
Ripianu, Andrei, 758,760
Ritter, W., 266
Ritz, Walter, 588,590
Rivals, 308,634
Roberval, Gilles Personne de, 54
Rocard, Y., 758
Rohault, Jacques, 54
Rohrlich, F., 758
Roitenberg, L.N., 758
Roman, P., 758
Rose, N.V., 758
Roseau, Maurice, 758
Rouche, N., 758,759
Routh, E.J., 56,759
Roy, N., 750,759

Rössler, E.E., 750
Ruderman, M.A., 753
Ruelle, D., 759
Runge, 557,561
Ruppel, W., 749
Rutherford, Ernest, 590
Ryabov, Yu.A., 750
Ryabushinsky, 67

Sabatier, P.C., 759
Saint-Germain, 439,457
Saint-Venant, Adhémar Jean Claude Barré de, 3
Saladini, 442
Salam, Abdus, 44
Saletan, D.J., 759
Samuel, M.A., 756
Sanders, J.A., 759
Sansone, Giorgio, 759
Santilli, R.M., 759
Sauer, R., 759
Saupe, D., 757
Savarensky, E., 759
Savet, P.H., 759
Sburlan, Silviu, 759
Scaliger, Jules César, 54
Scarborough, J.B., 759
Schaefer, Cl., 759
Schaeffer, D., 750
Scheck, Florian, 759
Schild, A., 761
Schmutzer, E., 759
Schoenfliess, A., 759
Schrödinger, Erwin, 57
Schuster, H.G., 759
Schwarz, 698
Secu, A., 762
Sedov, Leonid I., 759
Seitz, F., 759
Sergiescu, V., 759
Serret, Joseph Alfred, 206,207
Sestini, G., 759
Seydel, R., 759
Shapiro, I.S., 759
Sharp, R.T., 759
Shaw, C.D., 745
Shaw, R., 759
Shigley, J.E., 759
Shub, M., 760
Siacci, F., 760
Siegel, C.L., 760
Siger of Brabant, 53
Signorini, A., 760
Silaş, Gheorghe, 760
Simion, F.P., 762
Slater, J.C., 760
Slătineanu, Ileana, 751
Smale, S., 751,760
Smith, D.E., 760
Sneddon, Ian N., 760
Soare, Mircea, 760

Sofonea, Liviu, 760
Somigliana, C., 760
Sommerfeld, Arnold, 760
Soós, Eugen, 746
Souriau, J.M., 760
Southern, 626
Sparre, 440
Spiegel, M.R., 760
Staicu, C.I., 760
Stan, Aurelian, 760
Stan, Ion, 750
Stäckel, P., 760
Stănescu, Nicolae-Doru, 757
Stehle, P., 748
Steiner, J., 158,159,160
Stephan, W., 749
Stern, A., 755
Sternberg, S., 760
Stevin, Simon, 54
Stewart, H.B., 761
Stewart, I., 758
Stieltjes, 25
Stoenescu, Alexandru, 760
Stojanović, R., 745
Stoker, J.J., 761
Stokes, George Gabriel sir, 413,725
Stora, R., 752
Strelkov, S.P., 761
Stroe, I.V., 762
Strutt, John William (lord Rayleigh), 534,758
Suslov, G.K., 761
Swamy, V.J., 756
Sylvester, J.J., 761
Synge, J.L., 761
Szabo, I., 755,759,761
Szava, I., 761
Szemplinska-Stupnika, W., 760

Ştiucă, Petru, 748

Tabor, M., 761
Tait, P.G., 761
Talle, V., 758
Targ, S., 761
Taylor, Brook, 541,698
Tenot, A., 761
Teodorescu, Nicolae, 761
Teodorescu, Petre P., 108,281,746,753,760,761
Teodosiu, Cristian, 748,760
Ter Haar, D., 761
Thales at Miletus, 51
Thirring, W., 761
Thom, René, 761
Thoma, A., 751
Thomas d'Aquino, 53
Thompson, J.M.T., 761
Thomsen, J.J., 761
Thomson, William (lord Kelvin), 56,375,376,614, 761

Timoshenko, Stepan (Stephen) Prokof'evich, 536, 761,762
Tisserand, François Félix, 440,762
Tissot, 439
Tocaci, Emil, 758
Todhunter, Isaac, 56
Toma, Ileana, 760,762
Torricelli, Evangelista, 54,219,301,459,677
Troger, H., 762
Truesdell, Clifford Ambrose, 57,762
Trukhan, N.M., 758
Tsaplin, S.A., 269
Tsiolkovskiĭ, Konstantin Eduardovich, 56,660, 666,674,762

Țăposu, I., 762
Țițeica, Gheorghe, 752,760
Țițeica, Șerban, 15

Ulyanov, Vladimir Ilich (Lenin), 56
Ungureanu, S., 762

Valerio, Luca, 54
Valla, Giorgio, 54
Van der Pol, 534,538,539
Varignon, Pierre, 55,91
Vaschy, 67
Vâlcovici, Victor, 47,366,457,762
Verhulst, V., 759,762
Vidal, Ch., 747
Vieru, D., 762
Vilenkin, N.Ya., 762
Vinci, Leonardo da, 54
Vitašek, E., 746
Vitt, A.A., 745
Voiculescu, Dumitru, 758,762
Voinaroski, Rudolf, 762
Voinea, Radu, 762

Volterra, Vito, 762
Voronkov, I.M., 762
Voss, A., 762
Vrănceanu, Gheorghe, 56,180,763

Waddington, C.H., 763
Wallis, John, 54,426
Walther, H.O., 757
Wan, Y.-H., 751
Wason, W.R., 763
Weber, Joseph, 70,73,567
Webster, A.G., 763
Weeney, R.Mc., 763
Weierstrass, Karl, 763
Weil, Hermann, 763
Weinberg, Steven, 44
Weizel, W., 763
Whittaker, E.T., 763
Wiggins, S., 763
Wigner, E., 763
Willers, Fr.A., 763
Winkelmann, M., 763
Witner, A., 763
Wittenbauer, 763

Yano, Kentaro, 763
Yoshizawa, T., 763
Young, D.H., 762,763
Yourgraum, W., 763

Zeeman, 626,642
Zeeman, E.C., 763
Zeno of Elea, 51
Zeveleanu, C., 763
Ziegler, H., 763
Zoretti, L., 763
Zweifel, P.F., 750